T0074440

CRM Series in Mathematical Physics

Springer Science+Business Media, LLC

CRM Series in Mathematical Physics

Jan Felipe van Diejen
Luc Vinet
Editors

Calogero–Moser– Sutherland Models

With 27 Illustrations

Springer

Jan Felipe van Diejen
Departamento de Matematicas
Universidad de Chile
Las Palmeras, 3425, Nunoa
Santiago
Chile
vandiej@abello.dic.uchile.cl
Editorial Board

Luc Vinet
McGill University
James Administration Building, Room 504
Montreal, Quebec H3A 2T5
Canada
vinet@vpa.mcgill.ca

Joel S. Feldman
Department of Mathematics
University of British Columbia
Vancouver, BC V6T 1Z2
Canada
feldman@math.ubc.ca

Duong H. Phong
Department of Mathematics
Columbia University
New York, NY 10027-0029
USA
phong@math.columbia.edu

Yvan Saint-Aubin
Département de Mathématiques
  et Statistique
Université de Montréal
C.P. 6128, Succursale Centre-ville
Montréal, Québec H3C 3J7
Canada
saint@dms.umontreal.ca

Luc Vinet
McGill University
James Administration Building, Room 504
Montreal, Quebec H3A 2T5
Canada
vinet@vpa.mcgill.ca

Library of Congress Cataloging-in-Publication Data
van Diejen, Jan Felipe, 1965–
    Calogero–Moser–Sutherland models / Jan Felipe van Diejan, Luc Vinet.
        p. cm. — (CRM series in mathematical physics)
    Includes bibliographical references.
    ISBN 978-0-387-98968-6   ISBN 978-1-4612-1206-5 (eBook)
    DOI 10.1007/978-1-4612-1206-5
    1. Nuclear reactions.   2. Many–body problem.   I. Vinet, Luc.   II. Title.   III. CRM series
on mathematical physics
QC793.9.V36 2000
530.15—dc21   99-055955

Printed on acid-free paper.

© 2000 Springer Science+Business Media New York
Originally published by Springer-Verlag New York, Inc. in 2000

All rights reserved. This work may not be translated or copied in whole or in part without the
written permission of the publisher Springer Science+Business Media, LLC.
except for brief excerpts in connection with reviews or scholarly analysis. Use
in connection with any form of information storage and retrieval, electronic adaptation, computer
software, or by similar or dissimilar methodology now known or hereafter developed is forbidden.
The use of general descriptive names, trade names, trademarks, etc., in this publication, even if the
former are not especially identified, is not to be taken as a sign that such names, as understood by
the Trade Marks and Merchandise Marks Act, may accordingly be used freely by anyone.

Production managed by MaryAnn Brickner; manufacturing supervised by Jeffrey Taub.
Photocomposed copy prepared from CRM's LaTeX files.

9 8 7 6 5 4 3 2 1

ISBN 978-0-387-98968-6

Series Preface

The Centre de recherches mathématiques (CRM) was created in 1968 by the Université de Montréal to promote research in the mathematical sciences. It is now a national institute that hosts several groups; holds special theme years, summer schools, workshops; and offers a postdoctoral program. The focus of its scientific activities ranges from pure to applied mathematics, and includes statistics, theoretical computer science, mathematical methods in biology and life sciences, and mathematical and theoretical physics. The CRM also promotes collaboration between mathematicians and industry. It is subsidized by the Natural Sciences and Engineering Research Council of Canada, the Fonds FCAR of the Province of Québec, the Canadian Institute for Advanced Research, and has private endowments. Current activities, fellowships, and annual reports can be found on the CRM web page at `http://www.CRM.UMontreal.CA/`.

The CRM Series in Mathematical Physics includes monographs, lecture notes, and proceedings based on research pursued and events held at the Centre de recherches mathématiques.

Yvan Saint-Aubin
Montréal

Preface

In the early seventies Francesco Calogero made the following remarkable observation: for a quantum system that consists of N identical particles on the line interacting pairwise via an inverse-square potential and confined by a harmonic well, the Schrödinger eigenvalue problem is exactly solvable[1] (Calogero, J. Math. Phys., 1971). Soon thereafter, it was demonstrated by Bill Sutherland that the corresponding quantum problem on a circle is also exactly solvable (Sutherland, Phys. Rev. A, 1971, 1972). The revolutionary nature of both discoveries is well illustrated by the fact that, until these works of Calogero and Sutherland, the only known nontrivial example of an exactly solvable quantum N-particle problem consisted of (apart from the textbook case of coupled harmonic oscillators) the quantized nonlinear Schrödinger equation, i.e., the system of N identical particles on the line with a pairwise interaction via a "delta function potential."

The next big advance was made by Jürgen Moser, who proved among other things that the classical mechanical counterpart of the Calogero–Sutherland N-particle model constitutes a completely integrable Hamiltonian system in the sense of Liouville–Arnold, thereby showing that the study of the classical dynamics of the model is also amenable to an exact analytical approach (Moser, Adv. Math., 1975).

Since the appearance of the seminal papers of Calogero, Moser, and Sutherland, the field opened by their works has developed considerably. Important contributions were made in the late seventies by Olshanetsky and Perelomov, who pointed out an intimate relation between the Calogero–Moser–Sutherland (CMS) systems and the study of geodesic motion (at the classical level) and harmonic analysis (at the quantum level) on symmetric spaces of simple Lie groups. This observation led them to a generalization of the CMS systems associated with integral root systems (Olshanetsky and Perelomov, Invent. Math. 1976; Phys. Rep., 1981; Phys. Rep., 1983). Other highly influential developments include the introduction of a relativistic deformation of the CMS system by Ruijsenaars and Schneider (Ruijsenaars and Schneider, Ann. Phys., 1986; Ruijsenaars, Commun. Math. Phys., 1987) and the discovery of a one-dimensional CMS-type long-

---

[1] That is: the spectrum of the Hamiltonian can be exhibited in closed form and the problem of constructing the eigenfunctions has been reduced to elementary operations involving linear algebra and the use of properties of classical special functions.

range spin model by Haldane and Shastry (Haldane, Phys. Rev. Lett., 1988; Shastry, Phys. Rev. Lett., 1988).

The last decade of the 20th century has witnessed a true explosion of activities involving CMS models. These investigations have revealed the extraordinary ubiquitous character of these systems. By now it is well established that CMS systems play a role in investigations in research areas ranging from theoretical physics (such as, e.g., soliton theory, quantum field theory, string theory, solvable models of statistical mechanics, condensed matter physics, quantum chaos, etc.) to pure mathematics (such as representation theory, harmonic analysis, theory of special functions, combinatorics of symmetric functions, dynamical systems, random matrix theory, complex geometry, etc.).

During the period March 10–15, 1997, the Centre de recherches mathématiques in Montréal organized and hosted a workshop on Calogero–Moser–Sutherland models. One of the principal aims of this workshop was to bring together leading researchers from various disciplines in whose work CMS systems appear, either as the topic of investigation itself or as a tool for further applications. This is an attempt to get some kind of an overview of the many branches into which research on CMS systems has diversified in recent years. Apart from the lectures delivered by two of the three founding fathers of the field, F. Calogero (Italy) and B. Sutherland (USA), the workshop program contained 40 more presentations by J. Avan (France), H. Awata (USA), T. Baker (Australia), Y. Berest (Canada), R. Bhaduri (Canada), O. Bogoyavlenskij (Canada), H. Braden (UK), Ph. Choquard (Switzerland), J. van Diejen (Canada), M. Dijkhuizen (Japan), B. Enriquez (France), R. Floreanini (Italy), Ph. Di Francesco (USA), E. Gutkin (USA), F. D. Haldane (USA), K. Hasegawa (Japan), V. Inozemtsev (Russia), A. Kasman (Canada), A. N. Kirillov (Canada), I. Krichever (USA), F. Lesage (USA), P. Mathieu (Canada), N. Nekrasov (USA), M. Olshanetsky (Russia), A. Polychronakos (Greece), C. Quesne (Belgium), S. Ruijsenaars (Netherlands), D. Sen (India), E. Sklyanin (Japan), T. Shiota (Japan), K. Taniguchi (Japan), C. Tracy (USA), A. Turbiner (Mexico), D. Uglov (Japan), K. Vaninsky (USA), A. Varchenko (USA), A. Veselov (UK), M. Wadati (Japan), G. Wilson (UK), and A. Zhedanov (Ukraine).

We thank all attendants for their active participation and the speakers for their interesting state-of-the-art lectures. Special thanks and appreciation goes out to those speakers who took the time and effort to prepare a contribution for these proceedings.

J. F. van Diejen
Santiago

Contents

13 Ruijsenaars's Commuting Difference System from Belavin's Elliptic R-Matrix   193
Koji Hasegawa

14 Invariants and Eigenvectors for Quantum Heisenberg Chains with Elliptic Exchange   203
V. I. Inozemtsev

15 The Bispectral Involution as a Linearizing Map   221
Alex Kasman

16 On Some Quadratic Algebras: Jucys–Murphy and Dunkl Elements   231
Anatol N. Kirillov

36 Oscillator 9j-Symbols, Multdimensional Factorization Method, and Multivariable Krawtchouk Polynomials   549

Alexei Zhedanov

Contributors

Jean Avan Laboratoire de physique théorique et hautes énergies,
 Université Pierre et Marie Curie, 4 place Jussieu, 75252 Paris cedex
 75005, France, `avan@lpthe.jussieu.fr`

Hidetoshi Awata Enrico Fermi Institute and James Frank Institute,
 University of Chicago, 5640 S. Ellis Avenue, Chicago, IL 60637, USA

> *Present address* Yukawa Institute for Theoretical Physics, Kyoto
> University, Sakyo-Ku, Kyoto 606-8502, Japan,
> `awata@yukawa.kyoto-u.ac.jp`

Tim H. Baker Department of Mathematics, The University of
 Melbourne, Parkville, Victoria 3052, Australia,
 `tbaker@mundoe.maths.mu.oz.au`

Yuri Yu. Berest Centre de recherches mathématiques, Université de
 Montréal C.P. 6128, succ. Centre-Ville, Montréal, Québec H3C 3J7,
 Canada

> *Present address* Department of Mathematics, Malott Hall, Cornell
> University, Ithaca, NY 14853-4201 USA,
> `berest@math.cornell.edu`

Oleg I. Bogoyavlenskij Department of Mathematics and Statistics,
 Queen's University, Kingston, Ontario K7L 3N6, Canada,
 `oleg@mast.queensu.ca`

Harry W. Braden Department of Mathematics and Statistics, The
 University of Edinburgh, Mayfield Road, Edinburgh EH9 3JZ, UK,
 `hwb@ed.ac.uk`

Francesco Calogero Istituto di Fisica, Università di Roma I "La
 Sapienza", 00185 Roma, Italy,
 `francesco.calogero@roma1.infn.it`

Philippe Choquard Institut de physique théorique, École Polytechnique
 Fédérale de Lausanne, PHB Ecublens, 1015 Lausanne, Switzerland,
 `Philippe.Choquard@epfl.ch`

Philippe Di Francesco Department of Mathematics, University of North
 Carolina at Chapel Hill, Chapel Hill, NC 27599-3250, USA,
 `philippe@math.unc.edu`; Service de Physique Théorique,

Commissariat à l'énergie atomique/Saclay, 91191 Gif-sur-Yvette
cedex, France

Jan Felipe van Diejen Centre de recherches mathématiques, Université de
Montréal, C.P. 6128, succ. Centre-ville, Montréal (Québec) H3C
3J7, Canada

   Present address Departemento de Matematicas, Universidad de
   Chile, Las Palmeras 3425, Nunoa, Santiago, Chile,
   `vandiej@abello.dic.uchile.cl`

Charles F. Dunkl Department of Mathematics, Kerchof Hall, University
of Virginia, Charlottesville, VA 22903-3199, USA,
`cfd5z@virginia.edu`

Benjamin Enriquez Centre de mathématiques, École Polytechnique,
91128 Palaiseau, France, `enriquez@orphee.polytechnique.fr`

Peter J. Forrester Department of Mathematics, The University of
Melbourne, Parkville, Victoria 3052, Australia,
`matpjf@maths.mu.oz.au`

Eugene Gutkin Mathematics Department, Universty of Southern
California, Los Angeles, CA 90089-1113, USA,
`egutkin@math.usc.edu`

Koji Hasegawa Mathematical Institute, Tohoku University, Sendai
980-77, Japan, `kojihas@math.tohoku.ac.jp`

Vladimir I. Inozemtsev Bogoliubov Laboratory of Theoretical Physics,
Joint Institute of Nuclear Physics, 141980 Dubna, Moscow region,
Russia, `inozv@thsun1.jinr.ru`

Alex Kasman Department of Mathematics and Statistics, Concordia
University, Montréal, Québec, Canada

   Present address Department of Mathematics, College of Charleston,
   66 George Street, Charleston, SC 29424-0001, USA,
   `kasman@math.cofc.edu`

Anatol N. Kirillov Centre de recherches mathématiques, Université de
Montréal, C.P. 6128, succ. Centre-ville, Montréal (Québec), H3C
3J7, Canada

   Present address Department of Mathematics, School of Science,
   Nagoya University, Chikusa-ku, Nagoya 464-01, Japan,
   `kirillov@math.nagoya-u.ac.jp`

Igor M. Krichever Department of Mathematics, Columbia University, 2990 Broadway, New York, NY 10027, USA, krichev@math.columbia.edu; Landau Institute for Theoretical Physics, Kosygina str. 2, 117940 Moscow, Russia

Luc Lapointe Centre de recherches mathématiques, Université de Montréal, C.P. 6128, succ. Centre-ville, Montréal (Québec), H3C 3J7, Canada

   Present address Department of Mathematics, University of California, San Diego, 9500 Gilman Dr., La Jolla, CA 92093-0112, USA, llapointe@math.ucsd.edu

Frédéric Lesage Department of Physics and Astronomy, University of Southern California, Los Angeles, CA 90089-0484, USA

   Present address Centre de recherches mathématiques, Université de Montréal, C.P. 6128, succ. Centre-ville, Montréal (Québec), H3C 3J7, Canada, lesage@crm.umontreal.ca

Andrei M. Levin Institute of Theoretical and Experimental Physics, 117259, Moscow, Russia, andrl@landau.ac.ru

Ziad Maassarani Département de physique, Université Laval, Québec (Québec), G1K 7P4, Canada

   Present address Jesse Beams Laboratory 382 McCormick Rd. Charlottesville, VA 22903, USA, zm4v@virginia.edu

Pierre Mathieu Département de physique, Université Laval, Québec (Québec), G1K 7P4, Canada, pmathieu@phy.ulaval.ca

Evgeny E. Mukhin Department of Mathematics, University of North Carolina at Chapel Hill, Chapel Hill, NC 27599-3250, USA, mukhin@@math.unc.edu

Nikita Nekrasov Institute of Theoretical and Experimental Physics, 117259, Moscow, Russia; Lyman Laboratory of Physics, Harvard University, Cambridge MA 02138, USA, nikita@string.harvard.edu

Mikhail A. Olshanetsky Institute of Theoretical Physics, B. Cheremushkinskaja, 25, Moscow, 117259, Russia, olshanet@heron.itep.ru

Alexios P. Polychronakos Theoretical Physics Department, University of Ioannina, 45110 Ioannina, Greece; Theoretical Physics Department, Uppsala University, S-751 08 Uppsala, Sweden, poly@calypso.teorfys.uu.se

Christiane Quesne Département de physique, Université Libre de Bruxelles, Belgium, `cquesne@ulb.ac.be`

Simon N. M. Ruijsenaars Centrum voor Wiskunde en Informatica, P.O. Box 4079, 1090 AB Amsterdam, The Netherlands, `siru@wxs.nl`

Hubert Saleur Department of Physics and Astronomy, University of Southern California, Los Angeles, CA 90089-0484, USA, `saleur@usc.edu`

Prospero Simonetti Department of Physics and Astronomy, University of Southern California, Los Angeles, CA 90089-0484, USA

Bill Sutherland Department of Physics, University of Utah, Salt Lake City, UT 84112, USA, `suther@mail.physics.utah.edu`

Kenji Taniguchi Graduate School of Mathematical Sciences, University of Tokyo, 3-8-1 Komaba Meguro-ku Tokyo, Japan, `taniken@ms.u-tokyo.ac.jp`

Craig A. Tracy Department of Mathematics and Institute of Theoretical Dynamics, University of California, Davis, CA 95616, USA, `tracy@itd.ucdavis.edu`

Alexander Turbiner Instituto de Ciencias Nucleares, UNAM, Apartado Postal 70-543, 04510 Mexico D.F., Mexico, `turbiner@xochitl.nuclecu.unam.mx`

Denis Uglov Research Institute for Mathematical Sciences, Kyoto University, Kyoto 606, Japan, `duglov@kurims.kyoto-u.ac.jp`

Hideaki Ujino Department of Physics, Faculty of Science, University of Tokyo, Hongo, Bunkyo, Japan, `ujino@monet.phys.s.u-tokyo.ac.jp`

Kirill L. Vaninsky Department of Mathematics, Kansas State University, Manhattan, KS 66502, USA, `vaninsky@math.ias.edu`

Alexander M. Varchenko Department of Mathematics, University of North Carolina at Chapel Hill, Chapel Hill, NC 27599-3250, USA, `av@@math.unc.edu`

Alexander P. Veselov Department of Mathematical Sciences, Loughborough University, Loughborough, Leicestershire, LE11 3TU, UK, `A.P.Veselov@lboro.ac.uk`; Landau Institute for Theoretical Physics, Kosygina 2, Moscow 117940 Russia

Luc Vinet Centre de recherches mathématiques, Université de Montréal, C.P. 6128, succ Centre-ville, Montréal (Québec), H3C 3J7, Canada

Present address McGill University, James Administration Building,
    Room 504, Montréal (Québec), H3A 2T5, Canada,
    vinet@vpa.mcgill.ca

Miki Wadati Department of Physics, Faculty of Science, University of
    Tokyo, Hongo, Bunkyo, Japan,
    wadati@monet.phys.s.u-tokyo.ac.jp

Harold Widom Department of Mathematics, University of California,
    Santa Cruz, CA 95064, USA, widom@math.ucsc.edu

George Wilson Mathematics Department, Imperial College, London, SW7
    2BZ, UK, g.wilson@ic.ac.uk

Alexei Zhedanov Donetsk Institute for Physics and Technology, Donetsk
    340114, Ukraine, zhedanov@host.dipt.donetsk.ua

1

Classical Dynamical r-Matrices for Calogero–Moser Systems and Their Generalizations

J. Avan

1 Introduction

Construction and study of classical (and quantum) dynamical r-matrices are currently undergoing extensive development. Various examples of such objects were recently discussed, for instance, in Refs. 40, 59, and 60. However, at this time there is no general classifying scheme such as exists in the case of constant classical r-matrices thanks to Belavin and Drinfeld [17]. A partial classification scheme has very recently been proposed [53] for dynamical r-matrices obeying the particular version of the dynamical Yang–Baxter equation [16, 28] corresponding to Calogero–Moser models [9].

Consideration of such structures is thus particularly relevant to the study of Calogero–Moser [23, 24, 31, 47] and relativistic Ruijsenaars–Schneider models [25, 39, 51], where they appear systematically. Their occurrence and the particular form they assume, which we shall give in detail in this lecture, are due to the common nature of these models as Hamiltonian reductions of free or harmonic motions on particular symplectic manifolds: cotangent bundle of Lie algebras or Lie groups for rational or trigonometric Calogero–Moser models [48, 49]; double of Lie groups [8] or cotangent bundle to centrally extended loop group [32–34] for Ruijsenaars–Schneider models. The most general elliptic potentials are in turn associated to loop groups over elliptic curves [8, 32–34] and are crucial in understanding the algebraic resolution of these models [36, 37].

This Hamiltonian reduction procedure is also an important tool in the explicit construction of classical r-matrices for BC_n-type systems where the direct resolution of the intricate r-matrices equations, such as was done in Refs. 13, 22, and 57, becomes intractable. We shall exemplify such a construction in the case of trigonometric Calogero–Moser models [10].

The plan of this lecture runs as follows. We first recall the essential results of classical r-matrix theory and introduce the notations to be used throughout it. We then describe the construction of classical r-matrices for trigonometric Calogero–Moser models using the Hamiltonian reduction procedure. This gives us a general formula valid for all nonexceptional Lie groups.

We finally give a systematic overall picture of the classical r-matrix structure of this type that were obtained by various authors for

a) elliptic, trigonometric, and rational A_n Calogero–Moser models [13, 22, 57];

b) elliptic, trigonometric, and rational A_n spin Calogero–Moser models [19, 20];

c) elliptic, trigonometric, and rational A_n Ruijsenaars–Schneider models [11, 14, 58].

The case of spin Ruijsenaars–Schneider systems [39] was recently investigated in the rational case [6] but the general Hamiltonian structure still escapes understanding at this time.

To be complete we must indicate that two alternative approaches were recently described. One, using the same Lax matrices but leading to dynamical r-matrices of a different type (to be commented upon later), was developed in Ref. 21 (see contribution by Prof. Braden). The other uses conjugated Lax matrices which allow us to eliminate the dynamical dependence in the r-matrix at the cost of introducing a more complicated Lax operator. It was developed directly in the quantum case using the formalism of intertwining vectors [3, 35] (see contribution by Prof. Hasegawa).

2 Preliminaries

First of all we need to recall four essential features of the classical r-matrix formalism (see Ref. 26 for a textbook presentation of the Hamiltonian theory of classical integrable systems).

We consider a generic dynamical system described by a set of coordinates $\{x_i\}$ and momenta $\{p_i\}$, $i = 1, \ldots, n$; a Poisson structure $\{\ \}$ and a Hamiltonian $h(x_i, p_j)$.

2.1 Liouville Theorem

The existence of n algebraically independent, globally defined, Poisson commuting quantities such that the Hamiltonian h belongs to the ring generated by this set, guarantees the existence of a canonical transformation $(x_i, p_j) \rightarrow (I_i, \theta_j)$ linearizing the equations of motion [43]. Further assumptions on the topological structure of the phase space allow more precise statements on the geometrical interpretation of the transformed variables known as action-angle variables [4].

2.2 Lax Pair Formulation

The Lax pair formulation of a dynamical system states two elements of a Lie algebra $\mathcal{G}$, $L(x,p)$ and $M(x,p)$ such that the equations of motion for x_i, p_i are equivalent to the isospectral evolution (Lax equation) [41]

$$\frac{dL}{dt} = [L, M]. \qquad (1)$$

It follows that the adjoint-invariant quantities $\operatorname{Tr} L^n$ are time independent. In order to implement the Liouville theorem onto this set of possible action variables we need them to be Poisson-commuting. This is ensured by the classical r-matrix structure.

2.3 The r-Matrix Structure

Defining the decomposition of the Lax operator on a basis $\{t_a\}$ of the Lie algebra $\mathcal{G}$ as $L \equiv \sum_a l^a t_a$, the Poisson commutation of the ad-invariant $\operatorname{Tr} L^n$ is equivalent to the existence of an object $r_{12}(x,p) \in \mathcal{G}_1 \otimes \mathcal{G}_2$ hereafter known as a classical r-matrix [15, 45, 54, 55], such that

$$\{L_1, L_2\} \equiv \sum_{a,b} \{l^a, l^b\} t_a \otimes t_b = [r_{12}, L_1] - [r_{21}, L_2]. \qquad (2)$$

It must immediately be remarked that such an object is by no means unique. Moreover there is no one-to-one correspondence between a given dynamical system and the Lie algebra in which its Lax representation is defined; a same dynamical system may have several Lax representations and several r-matrix structures.

2.4 The Classical Yang–Baxter Equation

The Poisson bracket structure (2) obeys a Jacobi identity that implies an algebraic constraint for the r-matrix. Since r depends a priori on the dynamical variables this constraint takes a complicated form:

$$[L_1, [r_{12}, r_{23}] + [r_{12}, r_{13}] + [r_{32}, r_{13}] + \{L_2, r_{13}\} - \{L_3, r_{12}\}]$$
$$+ \text{cycl. perm.} = 0. \qquad (3)$$

Relevant particular cases of this very general identity are obtained when

a) r is independent of x, q. One is then lead to the general nondynamical Yang–Baxter equation [12, 15, 45, 50, 54, 55]:

$$[r_{12}, r_{23}] + [r_{12}, r_{13}] + [r_{32}, r_{13}] = 0. \qquad (4)$$

b) If, furthermore, r is antisymmetric under permutation of the two copies
of the algebra $\mathcal{G}$, one obtains the better-known and much studied [17,
54, 55] form:

$$[r_{12}, r_{23}] + [r_{12}, r_{13}] + [r_{13}, r_{23}] = 0. \tag{5}$$

c) If, on the contrary, r is dynamical, the supplementary terms in (3) may
take a completely algebraic form owing to the specific structure of the
Lax operator. For instance, the Calogero–Moser models lead to a self-
contained algebraic equation for a classical r-matrix depending only on
position-type canonical variables x_ν:

$$[r_{12}, r_{13}] + [r_{12}, r_{23}] + [r_{13}, r_{23}]$$
$$- \sum_\nu h_\nu^{(1)} \frac{\partial}{\partial x_\nu} r_{23} + \sum_\nu h_\nu^{(2)} \frac{\partial}{\partial x_\nu} r_{13} - \sum_\nu h_\nu^{(3)} \frac{\partial}{\partial x_\nu} r_{12} = 0, \tag{6}$$

where the set $\{h_\nu^{(i)}\}$ is a choice of basis for the Cartan algebra of $\mathcal{G}$
acting on the representation space i. This equation was first derived in
Refs. 16 and 28 and a classification scheme of its solutions was proposed
in Ref. 53, closely connected to the general algebraic scheme in Ref. 17.

Note by the way that there exist canonical examples of dynamical r-ma-
trices, obtained by using the well-known higher Poisson bracket construction for any integrable dynamical system [29, 42, 44, 56] starting from
a constant r-matrix. In particular the quadratic Sklyanin bracket, where
the Poisson structure of a Lax matrix becomes $\{L_1, L_2\} = [r_{12}, L_1 \otimes L_2]$,
may also be described as a linear structure with a dynamical r matrix
$R_{12} \equiv r_{12} L_2$. We shall see on the example of the Ruijsenaars–Schneider
system that dynamical "linear" r-matrix structures may also give rise to
dynamical "quadratic" structures. The *initial* dynamical r-matrices themselves, however, are themselves *not* of the Sklyanin type.

3   Hamiltonian Reduction and r-Matrices

3.1   *General Hamiltonian Reduction*

We begin by recalling some well-known facts concerning the Hamiltonian
reduction of dynamical systems whose phase space is a cotangent bundle [1].
Let M be a manifold and $N = T^*M$ its cotangent bundle. N is equipped
with the canonical 1-form α whose value at the point $p \in T^*M$ is π^*p,
where π is the projection of N on M. If a Lie group G acts on M, each
element $X \in \mathcal{G}$ (the Lie algebra of G) generates a vector field on M that
we shall denote $X.m$ at the point $m \in M$. It lifts to a vector field on N,
leaving α invariant. We shall also denote $X.p \in T_p(N)$, the value at $p \in N$

of this vector field, so that the Lie derivative $\mathcal{L}_{X.p}\alpha$ of the canonical 1-form vanishes.

N is a symplectic manifold equipped with the canonical 2-form $\omega = -d\alpha$. To any function (or Hamiltonian) H on N we associate a vector field X_H such that $dH = i_{X_H}\omega$ and conversely since ω is nondegenerate.

The Hamiltonian associated to the above vector field $X.p$, $X \in \mathcal{G}$ reads

$$H_X(p) = i_{X.p}\alpha = \alpha(X.p). \tag{7}$$

For any two functions F, G on N one defines the Poisson bracket $\{F,G\}$ as a function on N by

$$\{F,G\} = \omega(X_F, X_G). \tag{8}$$

The Poisson bracket of the Hamiltonians associated to the group action has a simple expression. In fact the group action is Poissonnian, that is,

$$\{H_X, H_Y\} = H_{[X,Y]}. \tag{9}$$

Obviously the application $X \in \mathcal{G} \to H_X(p)$, for any $p \in N$, is a linear map from $\mathcal{G}$ to the scalars and so defines an element $\mathcal{P}(p)$ of $\mathcal{G}^*$ that is called the momentum at $p \in N$.

One then restricts oneself to the submanifold N_μ of N with fixed momentum μ such that $N_\mu = \mathcal{P}^{-1}(\mu)$.

Owing to Eq. (7) and the invariance of α, the action of the group G on N is transformed by $\mathcal{P}$ into the coadjoint action of G on $\mathcal{G}^*$:

$$\mathcal{P}(g.p)(X) = \alpha(g.g^{-1}Xg.p) = \mathrm{Ad}_g^* \mathcal{P}(p)(X), \tag{10}$$

where the coadjoint action on an element ξ of $\mathcal{G}^*$ is defined as

$$\mathrm{Ad}_g^* \xi(X) = \xi(g^{-1}Xg)$$

The stabilizer G_μ of $\mu \in \mathcal{G}^*$ acts on N_μ. The reduced phase space is precisely obtained by taking the quotient (assumed to be well behaved):

$$\mathcal{F}_\mu = N_\mu/G_\mu. \tag{11}$$

It is known that this is a symplectic manifold.

We then need to compute the Poisson bracket of functions on $\mathcal{F}_\mu$. These functions are conveniently described as G_μ invariant functions on N_μ. To compute their Poisson bracket we first extend them arbitrarily in the vicinity of N_μ. Two extensions differ by a function vanishing on N_μ. The difference of the Hamiltonian vector fields of two such extensions is controlled by the following:

Lemma 1. *Let f be a function defined in a vicinity of N_μ and vanishing on N_μ. Then the Hamiltonian vector field X_f associated to f is tangent to the orbit $G.p$ at any point $p \in N_\mu$.*

As a consequence of this lemma we have a method to compute the reduced Poisson bracket. We take two functions defined on N_μ and invariant under G_μ and extend them arbitrarily. Then we compute their Hamiltonian vector fields on N and project them on the tangent space to N_μ by adding a vector tangent to the orbit $G.p$. These projections are independent of the extensions and the reduced Poisson bracket is given by the value of the symplectic form on N acting on them.

Proposition 1. *At each point $p \in N_\mu$ one can choose a vector $V_f.p \in \mathcal{G}.p$ such that $X_f + V_f.p \in T_p(N_\mu)$ and $V_f.p$ is determined up to a vector in $\mathcal{G}_\mu.p$.*

One finally gets the consistent general formula for reduced Poisson brackets:

Proposition 2. *The reduced Poisson bracket of two functions on $\mathcal{F}_\mu$ can be computed using any extensions f, g in the vicinity of N_μ according to*

$$\{f, g\}_{\text{reduced}} = \{f, g\} + \tfrac{1}{2}\big((V_g.p).f - (V_f.p).g\big). \tag{12}$$

This is equivalent to the Dirac bracket.

3.2   The Case $N = T^*G$

Let now $M = G$ be a Lie group; one uses the left translations to identify $N = T^*G$ with $G \times \mathcal{G}^*$.

$$\omega \in T_g^*(G) \to (g, \xi), \quad \text{where } \omega = L_{g^{-1}}^* \xi. \tag{13}$$

The Poisson structure on $N = T^*G$ is easily seen to be

$$\{\xi(X), \xi(Y)\} = -\xi([X, Y]); \quad \{\xi(X), g\} = -gX; \quad \{g, g\} = 0. \tag{14}$$

Geodesics on the group G correspond to left translations of 1-parameter groups (the tangent vector is transported parallel to itself), so $d/dt(g^{-1}\dot{g}) = 0$. This is a Hamiltonian system whose Hamiltonian is $\mathrm{H} = \tfrac{1}{2}(\xi, \xi)$, where we have identified $\mathcal{G}^*$ and $\mathcal{G}$ through the invariant Killing metric.

Here H is bi-invariant, so one can reduce this dynamical system using Lie subgroups H_L and H_R of G of Lie algebras $\mathcal{H}_L$ and $\mathcal{H}_R$, acting, respectively, on the left and on the right on T^*G in order to obtain a nontrivial result.

Using the coordinates (g, ξ) on T^*G, this action reads

$$((h_L, h_R), (g, \xi)) \to (h_L g h_R^{-1}, \mathrm{Ad}_{h_R}^* \xi).$$

We have written this action as a left action on T^*G, in order to apply the formalism developed in Section 3.1.

The moments are

$$\mathcal{P}^L(g, \xi) = P_{\mathcal{H}} \cdot \mathrm{Ad}_g^* \xi; \quad \mathcal{P}^R(g, \xi) = -P_{\mathcal{H}} \cdot \xi; \quad \mathcal{P} = (\mathcal{P}^L, \mathcal{P}^R), \tag{15}$$

where we have introduced the projector on $\mathcal{H}^*$ of forms in $\mathcal{G}^*$ induced by the restriction of these forms to $\mathcal{H}$.

3.3   The Calogero–Moser Models

We follow here the derivation of Refs. 48 and 49. Let us consider an involutive automorphism σ of a simple Lie group G and the subgroup H of its fixed points. Then H acts on the right on G defining a principal fiber bundle of total space G and base G/H, which is a global symmetric space. Moreover, G acts on the left on G/H and in particular so does H itself. We shall consider the situation described in Section 3.2 when $H_L = H_R = H$. The Hamiltonian of the geodesic flow on G/H is invariant under the H action, allowing us to construct the Hamiltonian reduction which under suitable choices of the momentum leads to the Calogero–Moser models. As a matter of fact since the phase space of the Calogero model is noncompact, one has to start from a noncompact Lie group G and quotient it by a maximal compact subgroup H so that the symmetric space G/H is of the noncompact type. The derivative of σ at the unit element of G is an involutive automorphism of $\mathcal{G}$ also denoted by σ. Let us consider its eigenspaces $\mathcal{H}$ and $\mathcal{K}$ associated with the eigenvalues $+1$ and -1, respectively. Thus we have a decomposition:

$$\mathcal{G} = \mathcal{H} \oplus \mathcal{K} \tag{16}$$

in which $\mathcal{H}$ is the Lie algebra of H, which acts by inner automorphisms on the vector space $\mathcal{K}$ ($h\mathcal{K}h^{-1} = \mathcal{K}$).

Let $\mathcal{A}$ be a maximal commuting set of elements of $\mathcal{K}$. It is called a Cartan algebra of the symmetric space G/H. It is known that every element in $\mathcal{K}$ is conjugated to an element in $\mathcal{A}$ by an element of H. Moreover, $\mathcal{A}$ can be extended to a maximal commutative subalgebra of $\mathcal{G}$ by adding to it a suitably chosen Abelian subalgebra $\mathcal{B}$ of $\mathcal{H}$. We shall use the radical decomposition of $\mathcal{G}$ under the Abelian algebra $\mathcal{A}$:

$$\mathcal{G} = \mathcal{A} \oplus \mathcal{B} \bigoplus_{e_\alpha, \alpha \in \Phi} \mathbb{R}e_\alpha. \tag{17}$$

These decompositions of $\mathcal{G}$ exponentiate to similar decompositions of G. First $G = KH$ where $K = \exp(\mathcal{K})$. Then $A = \exp(\mathcal{A})$ is a maximal totally geodesic flat submanifold of G/H and any element of K can be written as $k = hQh^{-1}$ with $Q \in A$ and $h \in H$. It follows that any element of G can be written as $g = h_1 Q h_2$ with $h_1, h_2 \in H$.

Of course this decomposition is nonunique. This nonuniqueness is described in the following:

Proposition 3. *If $g = h_1 Q h_2 = h'_1 Q' h'_2$ we have $h'_1 = h_1 d^{-1} h_0^{-1}$, $h'_2 = h_0 d h_2$ and $Q' = h_0 Q h_0^{-1}$ where $d \in \exp(\mathcal{B}) = B$ and $h_0 \in H$ is a representative of an element of the Weyl group of the symmetric space. So if we fix $Q = \exp(q)$ such that q be in a fundamental Weyl chamber, the only ambiguity resides in the element $d \in B$.*

The reduction to Calogero–Moser models is then obtained by an adequate choice of the momentum $\mu = (\mu^L, \mu^R)$ such that $\mathcal{P} = \mu$. We take $\mu^R = 0$ so that the isotropy group of the right component is H_R itself.

The choice of the moment μ^L is of course of crucial importance. It must be fixed so that

- Its isotropy group H_μ is a maximal proper Lie subgroup of H, so that the phase space of the reduced system be of minimal dimension but nontrivial.

- In order to ensure the unicity of the decomposition introduced in Proposition 3 on N_μ we need

$$\mathcal{H}_\mu \cap \mathcal{B} = \{0\}. \tag{18}$$

We choose a complementary maximal isotropic subspace $\mathcal{C}$ so that

$$\mathcal{H} = \mathcal{H}_\mu \oplus \mathcal{B} \oplus \mathcal{C} \tag{19}$$

and χ is a nondegenerate skew-symmetric bilinear form on $\mathcal{B} \oplus \mathcal{C}$, hence $\dim \mathcal{B} = \dim \mathcal{C}$. Notice that $\mathcal{C}$ is defined up to a symplectic transformation preserving $\mathcal{B}$.

- The reduced phase space $\mathcal{F}_\mu$ has dimension $2\dim \mathcal{A}$.

We now construct a section $\mathcal{S}$ of the bundle N_μ over $\mathcal{F}_\mu$ so that one can write

$$N_\mu = H_\mu \mathcal{S} H. \tag{20}$$

To construct this section we take a point Q in A and an $L \in \mathcal{G}^*$ such that the point (Q, L) is in N_μ. In this subsection we shall for convenience identify $\mathcal{G}$ and $\mathcal{G}^*$ under the Killing form, assuming that G is semisimple. Moreover, since the automorphism σ preserves the Killing form, $\mathcal{H}$ and $\mathcal{K}$ are orthogonal, and $P_{\mathcal{H}^*}$ reduces to the orthogonal projection on $\mathcal{H}$. Since $\mu^R = 0$ we have $L \in \mathcal{K}$ and one can write

$$L = p + \sum_{e_\alpha, \alpha \in \Phi'} l_\alpha(e_\alpha - \sigma(e_\alpha)), \tag{21}$$

where $p \in \mathcal{A}$. From Eq. (15) one gets

$$\mu^L = P_{\mathcal{H}}\left(p + \sum_\alpha l_\alpha(Qe_\alpha Q^{-1} - Q\sigma(e_\alpha)Q^{-1})\right)$$

Since $Q = \exp(q)$, $q \in \mathcal{A}$, we have $Qe_\alpha Q^{-1} = \exp(\alpha(q))e_\alpha$ and similarly $Q\sigma(e_\alpha)Q^{-1} = \exp(-\alpha(q))\sigma(e_\alpha)$.

Then the above equation becomes

$$\mu^L = \sum_\alpha l_\alpha \sinh \alpha(q) \big(e_\alpha + \sigma(e_\alpha)\big). \tag{22}$$

One can choose the momentum of the form $\mu^L = \sum_\alpha g_\alpha \big(e_\alpha + \sigma(e_\alpha)\big)$; namely, μ^L has no component in $\mathcal{B}$ where the g_α are such that H_μ is of maximal dimension (we shall see that it essentially fixes them, and obviously if $g_\alpha \neq 0$ for any α Eq. (18) is automatically satisfied) and we have shown the following:

Proposition 4. *The couples* (Q, L) *with* $Q = \exp(q)$ *and*

$$L = p + \sum_\alpha \frac{g_\alpha}{\sinh \alpha(q)} \big(e_\alpha - \sigma(e_\alpha)\big)$$

with $p, q \in \mathcal{A}$ *form a submanifold in* N_μ *of dimension* $2 \dim \mathcal{A}$.

Notice that L is just the Lax operator of the Calogero model and that the section $\mathcal{S}$ depends of $2 \dim \mathcal{A}$ parameters in an immersive way. Hence one can identify N_μ with the set of orbits of $\mathcal{S}$ under $H_\mu \times H$ that is the set of points $(g = h_1 Q h_2, \xi = h_2^{-1} L h_2)$ with $h_1 \in H_\mu$ and $h_2 \in H$ *uniquely defined* owing to condition (18). The variables p and q appearing in Q and L are the dynamical variables of the Calogero model and form a pair of canonically conjugate variables.

We then compute the Poisson bracket of the functions on $\mathcal{F}_\mu$ whose expressions on the section $\mathcal{S}$ are $L(X)$ and $L(Y)$ for $X, Y \in \mathcal{K}$. These functions have uniquely defined $H_\mu \times H$ invariant extensions to N_μ given, respectively, by

$$F_X(g, \xi) = \langle \xi, h_2^{-1} X h_2 \rangle, \quad F_Y(g, \xi) = \langle \xi, h_2^{-1} Y h_2 \rangle, \quad \text{where } g = h_1 Q h_2.$$

Notice that h_2 is a well-defined function of g in N_μ owing to condition (18). According to the prescription given in the Section 3.1, we choose extensions of these functions in the vicinity of N_μ. We *define* these extensions at the point $p = (g, \xi) \in T^*G$ by the *same* formulas in which h_2 is chosen to be a function depending *only* on g and reducing to the above-defined h_2 when $p \in N_\mu$. Because of the nonuniqueness of the decomposition $g = h_1 Q h_2$ outside N_μ one cannot assert that the functions F_X, F_Y are invariant under the action of $H \times H$ and we must appeal to the general procedure to compute the reduced Poisson brackets.

The complete derivation with all its technical subtleties can be found in Ref. 10. The final result gives the general r-matrix for trigonometric Calogero–Moser models in so-called dual form [54, 55]:

Theorem 1. *There exists a linear mapping* $R \colon \mathcal{K} \to \mathcal{H}$ *such that*

$$\{L(X), L(Y)\}_{\text{reduced}} = L([X, RY] + [RX, Y]) \tag{23}$$

and R is given by

$$R(X) = \nabla_g h_2(X) + \frac{1}{2} D_Q(V_X), \tag{24}$$

where

Proposition 5. *On the section S with $Q = \exp(q) \in A$ we have for $X \in \mathcal{K}$ that is $X = X_0 + \sum X_\alpha(e_\alpha - \sigma e_\alpha)$, $X_0 \in \mathcal{A}$,*

$$\nabla_g h_2(X) = -h_0(X) + \sum_\alpha X_\alpha \coth(\alpha(q))(e_\alpha + \sigma e_\alpha). \tag{25}$$

Here $h_0(X)$ is a linear function from $\mathcal{G}$ to $\mathcal{B}$ which is fixed by the condition

$$X_L \equiv h_0(X) - \sum_\alpha \frac{X_\alpha}{\sinh \alpha(q)}(e_\alpha + \sigma e_\alpha) \in \mathcal{H}_\mu \oplus \mathcal{C}. \tag{26}$$

3.4   Two Examples

To illustrate the power of this method we now give two examples of r-matrices. The A_n case has already been treated in Refs. 13, 22, and 57 and serves as a check on the validity of the derivation. The case of $SU(n,n)$ Calogero–Moser model proved to be too intricate for a direct computation; however, this method immediately gives its r-matrix.

The *standard Calogero–Moser model* ($SL(n)$) is obtained by starting from the noncompact group $G = SL(n, \mathbb{C})$ and its maximal compact subgroup $H = SU(n)$ as first shown in Refs. 48 and 49. We choose the momentum μ_L as described in Section 3.1, so that the isotropy group H_μ would be a maximal proper Lie subgroup of H. Obviously one can take μ_L of the form

$$\mu_L = i(vv^+ - 1), \tag{27}$$

where v is a vector in $\mathbb{C}^n$ such that $v^+ v = 1$; hence μ_L is a traceless anti-Hermitian matrix. Then $g\mu_L g^{-1} = \mu_L$ if and only if $gv = cv$ where c is a complex number of modulus 1. Hence $H_\mu = S\big(U(n-1) \times U(1)\big)$, which has the above-stated property.

In this case the automorphism σ is given by $\sigma(g) = (g^+)^{-1}$ (notice that we consider only the real Lie group structure), B is the group of diagonal matrices of determinant 1 with pure phases on the diagonal, and A is the group of real diagonal matrices with determinant 1. The property (18) is then satisfied as soon as the vector v has no zero component. As a matter of fact, v is further constrained by μ_L's being a value of the moment map. Considering Eq. (22) we see that μ_L has no diagonal element, which implies that all the components of v are pure phases $v_j = \exp(i\theta_j)$. These extra phases that will appear in the Lax matrix can, however, be conjugated out by the adjoint action of a constant matrix $\mathrm{diag}\big(\exp(i\theta_j)\big)$; hence we shall from now on set $v_j = 1$ for all j. This is the solution first considered by Olshanetsky and Perelomov.

The Lax matrix L is then given by Proposition 4 and therefore

$$L = p + \sum_{k<l} \frac{1}{\sinh(q_k - q_l)}(iE_{kl} - iE_{lk})$$

The r-matrix can now be deduced straightforwardly from Proposition 5, after reconverting the dual form, where R is an endomorphism of the Lie algebra, into the more usual direct form, where R lives in the tensor product of the Lie algebra by itself. One ends up with

$$R_{12} = \sum_{k \neq l} \coth(q_k - q_l) E_{kl} \otimes E_{lk}$$
$$+ \frac{1}{2} \sum_{k \neq l} \frac{1}{\sinh(q_k - q_l)} (E_{kk} - \frac{1}{n}\mathbf{1}) \otimes (E_{kl} - E_{lk})$$

This gives back the already known r-matrix of the Calogero model for the potential $1/\sinh(x)$, and the other potentials $1/\sin(x)$ and $1/x$ have similar r-matrices obtained by analytic continuation.

The SU(n, n) *Calogero model* is obtained by starting from the noncompact group $G = $ SU(n, n). This is the subgroup of SL$(2n, \mathbb{C})$, which leaves invariant the sesquilinear quadratic form defined by

$$Q((u_1, v_1), (u_2, v_2)) = \begin{pmatrix} u_1^+ & v_1^+ \end{pmatrix} J \begin{pmatrix} cu_2 \\ v_2 \end{pmatrix} = u_1^+ v_2 + v_1^+ u_2, \qquad (28)$$

where u_i, v_i are vectors in $\mathbb{C}^n$ and J is the matrix

$$J = \begin{pmatrix} 0 & 1 \\ 1 & 0 \end{pmatrix}.$$

The Lie algebra of SU(n, n) therefore consists of block matrices

$$\mathcal{G} = \left\{ \begin{pmatrix} a & b \\ c & d \end{pmatrix} \,\middle|\, a = -d^+, \operatorname{Tr}(a + d) = 0, b^+ = -b, c^+ = -c \right\}, \qquad (29)$$

where a, b, c, d are $n \times n$ complex matrices.

We consider again the automorphism σ: $\sigma(g) = (g^+)^{-1}$, which can be consistently restricted to SU(n, n). Its fixed points at the Lie algebra level consist of block matrices

$$\mathcal{H} = \left\{ \begin{pmatrix} a & c \\ c & a \end{pmatrix} \,\middle|\, a^+ = -a, \operatorname{Tr}(a) = 0, c^+ = -c \right\}. \qquad (30)$$

This Lie algebra is isomorphic to the Lie algebra of $S(\mathrm{U}(n) \times \mathrm{U}(n))$, the two $u(n)$'s being realized, respectively, by $a + c$ and $a - c$.

The subalgebra $\mathcal{B}$ consists of matrices of the form (30) with $c = 0$, and a is a diagonal matrix of zero trace and purely imaginary coefficients. The Abelian subalgebra $\mathcal{A}$ consists of matrices of the form (29) with $b = c = 0$ and $a = -d$ being a real diagonal matrix.

To perform the reduction, we choose as above $\mu^R = 0$ and

$$\mu^L = i(vv^+ - 1) + i\gamma J. \tag{31}$$

The vector v has again $2n$ components all equal to 1.

Notice that in Eq. (31) the parameter γ is an arbitrary real number. This will lead to existence of a second coupling constant in the corresponding Calogero model.

Then, from Proposition 4, the Lax matrix is found to be

$$L = p + \sum_{i<j} \frac{1}{\sinh(q_i - q_j)}(1 - \sigma)(iE_{ij} + iE_{j+n,i+n})$$

$$+ \sum_{i<j} \frac{1}{\sinh(q_i + q_j)}(1 - \sigma)(iE_{i,j+n} + iE_{j,i+n})$$

$$+ (\gamma + 1) \sum_i \frac{1}{\sinh(2q_i)}(1 - \sigma)(iE_{i,i+n}),$$

where p is a generic element of $\mathcal{A}$ of the form $\operatorname{diag} p_i, - \operatorname{diag} p_i$.

The r-matrix is then computed straightforwardly:

$$R_{12} = \frac{1}{2} \sum_{k \neq l} \coth(q_k - q_l)(E_{kl} + E_{k+n,l+n}) \otimes (E_{lk} - E_{l+n,k+n})$$

$$+ \frac{1}{2} \sum_{k,l} \coth(q_k + q_l)(E_{k,l+n} + E_{k+n,l}) \otimes (E_{l+n,k} - E_{l,k+n})$$

$$+ \frac{1}{2} \sum_{k \neq l} \frac{1}{\sinh(q_k - q_l)}(E_{kk} + E_{k+n,k+n} - \frac{1}{n}1) \otimes (E_{kl} - E_{k+n,l+n})$$

$$+ \frac{1}{2} \sum_{k,l} \frac{1}{\sinh(q_k + q_l)}(E_{kk} + E_{k+n,k+n} - \frac{1}{n}1) \otimes (E_{k,l+n} - E_{k+n,l}).$$

These dynamical r-matrices depend only on the dynamical variable q. The different approach advocated in Ref. 21 leads to r-matrices depending on both p and q variables, but on a smaller set of algebra generators.

4 The Dynamical r-Matrices of Calogero and Ruijsenaars Models

Dynamical r-matrices have been derived for the Calogero–Moser models and (relativistic) Ruijsenaars–Schneider models using either the technique described here or a direct method starting from an ansatz of the same form. We will now describe the results achieved in this way for A_n models, and indicate interesting and sometimes deep connections between these various r-matrices.

Let us start with *Calogero–Moser models*. The rational and trigonometric matrices were described in the previous section. The elliptic case was solved by Sklyanin [57] and by Braden et al. [13, 22]. The Lax matrix reads

$$L(\lambda) = \sum_{i=1}^{N} p_i e_{ii} + \sum_{\substack{i,j=1 \\ i \neq j}}^{N} l(q_{ij}, \lambda) e_{ij}. \tag{32}$$

Here one has set

$$l(x, \lambda) = -\frac{\sigma(x + \lambda)}{\sigma(x)\sigma(\lambda)}, \quad V(x) = \wp(x), \tag{33}$$

where σ and $\wp$ are Weierstrass elliptic functions. The classical r-matrix reads

$$r_{12}(\lambda, \mu) = \sum_{\substack{i,j=1 \\ i \neq j}}^{N} l(q_{ij}, \lambda - \mu) e_{ij} \otimes e_{ji} + \frac{1}{2} \sum_{\substack{i,j=1 \\ i \neq j}}^{N} l(q_{ij}, \mu)(e_{ii} + e_{jj}) \otimes e_{ij}$$

$$- [\zeta(\lambda - \mu) + \zeta(\mu)] \sum_{i=1}^{N} e_{ii} \otimes e_{ii}. \tag{34}$$

Note that a spectral parameter is now present in L and r. This particular formulation of the spinless elliptic case is due to Krichever [36]. The other known Lax formulation of Olshanetsky and Perelomov [48, 49] has no spectral parameter but requires a p and q dependence in the r-matrix, which was only recently given in Ref. 21 and has a totally different algebraic form.

The spin Calogero–Moser models were introduced in Ref. 31. The Lax operator for the elliptic case reads

$$L(\lambda) = \sum_{i=1}^{N} p_i e_{ii} + \sum_{\substack{i,j=1 \\ i \neq j}}^{N} l(q_{ij}, \lambda) f_{ij} e_{ij}, \tag{35}$$

where f_{ij} are spinlike variables with the Kirillov–Poisson bracket structure

$$\{f_{ij}, f_{kl}\} = \frac{1}{2}(\delta_{il} f_{jk} + \delta_{ki} f_{lj} + \delta_{jk}\, f_{il} + \delta_{lj} f_{ki}). \tag{36}$$

One then needs to introduce a parameterization of f_{ij} so as to be on a coadjoint orbit of $SU(N)$: Introducing vectors

$$(\xi_i)_{i=1,\dots,N} \quad \text{with } \xi_i = (\xi_i^a)_{a=1,\dots,r}$$
$$(\eta_i)_{i=1,\dots,N} \quad \text{with } \eta_i = (\eta_i^a)_{a=1,\dots,r}$$

with the Poisson brackets

$$\{\xi_i^a, \xi_j^b\} = 0, \quad \{\eta_i^a, \eta_j^b\} = 0, \quad \{\xi_i^a, \eta_j^b\} = -\delta_{ij}\delta_{ab}, \qquad (37)$$

we parameterize f_{ij} as follows:

$$f_{ij} = \langle \xi_i \mid \eta_j \rangle = \sum_{a=1}^{r} \xi_i^a \eta_j^a. \qquad (38)$$

The phase space now becomes a true symplectic manifold.

The Hamiltonian takes the form

$$H = \frac{1}{2}\sum_{i=1}^{N} p_i^2 - \frac{1}{2}\sum_{\substack{i,j=1 \\ i\neq j}}^{N} f_{ij}f_{ji}V(q_{ij}), \quad q_{ij} = q_i - q_j \qquad (39)$$

with the Weierstrass function as elliptic potential, as in the spinless case.

The classical r-matrix then reads:

$$r_{12}(\lambda,\mu) = \frac{1}{2}\sum_{\substack{i,j=1 \\ i\neq j}}^{N} l(q_{ij},\lambda-\mu)e_{ij}\otimes e_{ji} + \frac{1}{2}\sum_{\substack{i,j=1 \\ i\neq j}}^{N} l(q_{ij},\lambda+\mu)e_{ij}\otimes e_{ij}$$

$$- \frac{1}{2}[\zeta(\lambda+\mu) + \zeta(\lambda-\mu)]\sum_{i=1}^{N} e_{ii}\otimes e_{ii}. \qquad (40)$$

Trigonometric and rational cases can be derived from the elliptic case by taking suitable limits [19, 20]. The spinless case can also be derived from the spin case by taking $r = 1$ and introducing a further Hamiltonian reduction by the action of U(1) as a phase on the vectors ξ_i, η_i. The supplementary terms in the spinless r matrix arise from the conjugation of the Lax matrix required to bring it in canonical shape (32) after elimination of the vectorlike degrees of freedom. Let us finally remark that this r-matrix structure has yielded a number of important developments: exact classical Yangian symmetry [19, 20] (a quantum version of it had been found beforehand, using heavy direct algebraic computations [18]) and a quantum version of the dynamical r-matrix using the shifted version of the quantum Yang–Baxter equation described in Refs. 27, 30, and 46.

The *spinless relativistic RS models* are described by the Hamiltonian

$$H = mc^2 \sum_{j=1}^{N} (\cosh\theta_j)\prod_{k\neq j} f(q_k - q_j), \qquad (41)$$

where

$$f(q) = \left(1 + \frac{g^2}{q^2}\right)^{1/2} \qquad \text{(rational)}$$

$$f(q) = \left(1 + \frac{\alpha^2}{\sinh^2(\nu q/2)}\right)^{1/2} \qquad \text{(hyperbolic)} \tag{42}$$

$$f(q) = (\lambda + \nu \mathcal{P}(q)) \quad \mathcal{P} = \text{Weierstrass function.} \quad \text{(elliptic).}$$

Here the canonical variables are a set of rapidities $\{\theta_i, i = 1, \ldots, N\}$ and conjugate positions q_i such that $\{\theta_i, q_j\} = \delta_{ij}$.

The dynamical system admits a Lax representation with the Lax operator:

$$L = \sum_{j,k=1}^{N} L_{jk} e_{jk}$$

$$L_{jk} = \exp \beta \theta_j . C_{jk}(q_j - q_k) . \left(\prod_{m \neq j} f(q_j - q_m) \prod_{l \neq k} f(q_l - q_k)\right)^{1/2}, \tag{43}$$

where $\{e_{jk}\}$ is the usual basis for $N \times N$ matrices; f was given in (42) and

$$C_{jk}(q) = \frac{\gamma}{\gamma + iq} \qquad \text{(rational)}$$

$$C_{jk}(q) = \left(\cosh \frac{\nu}{2} q + ia \sinh \frac{\nu}{2} q\right)^{-1} \qquad \text{(trigonometric)} \tag{44}$$

$$C_{jk}(q) = \frac{\Phi(q + \gamma, \lambda)}{\Phi(\gamma, \lambda)} \qquad \text{(elliptic).} \tag{45}$$

Again the function Φ is defined as

$$\Phi(x, \lambda) \equiv \frac{\sigma(x + \lambda)}{\sigma(x)\sigma(\lambda)}, \tag{46}$$

where σ is the Weierstrass function.

The elliptic r-matrix structure is better written as a quadratic expression in terms of the Lax operator [58]:

$$\{L_1(\lambda), L_2(\mu)\} = \left(L_1(\lambda) \otimes L_2(\mu)\right) a_1(\lambda, \mu) - a_2(\lambda, \mu)\left(L_1(\lambda) \otimes L_2(\mu)\right)$$

$$+ \left(1 \otimes L_2(\mu)\right) s_1(\lambda, \mu)(L_1(\lambda) \otimes 1)$$

$$- (L_1(\lambda) \otimes 1) s_2(\lambda, \mu)\left(1 \otimes L_2(\mu)\right). \tag{47}$$

Here one defines

$$a_1(\lambda, \mu) = a(\lambda, \mu) + w, \qquad\qquad s_1(\lambda, \mu) = s(\lambda, \mu) - w$$

$$a_2(\lambda, \mu) = a(\lambda, \mu) + s(\lambda) - s^*(\mu) - w, \quad s_2(\lambda, \mu) = s^*(\mu) + w. \tag{48}$$

The matrices a and s are obtained from the r-matrix of the elliptic Calogero–Moser model given in (34) as $r(\lambda, \mu) \equiv a(\lambda, \mu) + s(\lambda)$, where a is the skew-symmetric matrix:

$$a(\lambda, \mu) = -\zeta(\lambda - \mu) \sum_{k=1}^{N} E_{kk} \otimes E_{kk}$$
$$- \sum_{k \neq j} \Phi(q_j - q_k, \lambda - \mu) E_{jk} \otimes E_{kj} \quad (49)$$

and s, s^* are non-skew-symmetric matrices independent of the second spectral parameter:

$$s(\lambda, \mu) = \zeta(\lambda) \sum_{k=1}^{N} E_{kk} \otimes E_{kk} + \sum_{k \neq j} \Phi(q_j - q_k, \lambda) E_{jk} \otimes E_{kk}$$

$$ \quad (50)$$

$$s^*(\lambda, \mu) = \zeta(\lambda) \sum_{k=1}^{N} E_{kk} \otimes E_{kk} + \sum_{k \neq j} \Phi(q_j - q_k, \lambda) E_{kk} \otimes E_{jk},$$

and finally w is a supplementary matrix, independent of the spectral parameters:

$$w = \sum_{k \neq j} \zeta(q_k - q_j) E_{kk} \otimes E_{jj}. \quad (51)$$

This r-matrix structure is a Sklyanin-type bracket (although realized in the more generic case of an initially dynamical r-matrix) obtained from the Calogero–Moser r-matrix structure viewed as a linear bracket. This can be interpreted from the fact that RS models are obtained not only as Hamiltonian reductions from current algebras on elliptic curves but also alternatively [8] as Hamiltonian reductions from the Heisenberg double of Lie groups [2]. In this case the initial Poisson structure on the large phase space is itself a quadratic bracket instead of the canonical initial linear (Kirillov) bracket, which is the natural structure on the cotangent bundle of a Lie group. This relation is maintained throughout the Hamiltonian reduction procedure and the final r-matrix structures are essentially connected in the same way.

The previously obtained r-matrices [11] can be obtained from this one by sending one period of the elliptic functions to infinity and suitably conjugating the Lax pair in such a way as to get a completely symmetric expression in terms of the momenta θ_i. On the other hand, the r-matrix found in Ref. 14 cannot be easily inserted in this scheme. In fact it corresponds to a very specific value of the parameters where the Lax matrix becomes completely symmetric and the r-matrix may only then take this very special form.

The classical r-matrices admit a quantization scheme on the same lines as the Calogero–Moser case [9]. It was developed in Refs. 5 and 7.

Finally a word about *the spin RS dynamical system*. It was introduced in Ref. 39. It is not clear at this time how to define a consistent Hamiltonian structure in the most general case, although the rational case was solved recently [6]. The task is indeed easier here since there exists a duality symmetry [52] connecting the rational RS model to the trigonometric CM model for which the spin model is well known. Let us finally mention that there exists a general scheme to obtain Hamiltonian structures from Lax representations using the tools of algebraic geometry [38] (see also Prof. Krichever's contribution to this colloquium) and this scheme now appears to be the most promising way to get these elusive Hamiltonian structures.

Acknowledgments: The works presented here were done in collaboration with O. Babelon, E. Billey, G. Rollet, and M. Talon at the LPTHE, Paris VI (URA CNRS 280). I have also drawn upon works by H. W. Braden, Yu. B. Suris, and E. K. Sklyanin. I wish to thank A. Antonov, H. W. Braden, L. Féher, and J. M. Maillet for clarifications of particular points in this presentation; S. N. M. Ruijsenaars for suggesting that I write a summary of the current situation as Section 4; and J. F. van Diejen and L. Vinet for their kind invitation to attend this workshop.

5   REFERENCES

1. R. Abraham and J. E. Marsden, *Foundations of Mechanics*, Benjamin–Cummings, Reading, MA, 1978.

2. A. Yu. Alekseev and A. Z. Malkin, *Symplectic structures associated to Lie–Poisson groups*, Commun. Math. Phys. **162** (1994), No. 1, 147–173.

3. A. Antonov, K. Hasegawa, and A. Zabrodin, *On trigonometric intertwining vectors and nondynamical R-matrix for the Ruijsenaars model*, Nucl. Phys. B **503** (1997), No. 3, 747–770, hep-th/9704074.

4. V. I. Arnold, *Mathematical Methods of Classical Mechanics*, Grad. Texts in Math., Vol. 60, Springer, New York, 1978.

5. G. E. Arutyunov, L. O. Chekhov, and S. A. Frolov, *R-matrix quantization of the elliptic Ruijsenaars–Schneider model*, Commun. Math. Phys. **192** (1998), No. 2, 405–432, q-alg/9612032.

6. G. E. Arutyunov and S. A. Frolov, *On the Hamiltonian structure of the spin Ruijsenaars–Schneider system*, J. Phys. A **31** (1998), No. 18, 4203–4216, hep-th/9703119.

18   J. Avan

7. _____, *Quantum dynamical R-matrices and quantum Frobenius group*, Commun. Math. Phys. **191** (1998), No. 1, 15–29, q-alg/9610009.

8. G. E. Arutyunov, S. A. Frolov, and P. B. Medvedev, *Elliptic Ruijsenaars–Schneider model via the Poisson reduction of the affine Heisenberg double*, J. Phys. A **30** (1997), No. 14, 5051–5063, hep-th/9607170.

9. J. Avan, O. Babelon, and E. Billey, *The Gervais–Neveu–Felder equation and the quantum Calogero–Moser systems*, Commun. Math. Phys. **178** (1996), No. 2, 281–299, hep-th/9505091.

10. J. Avan, O. Babelon, and M. Talon, *Construction of the classical R-matrices for the Toda and Calogero models*, Algebra i Analiz **6** (1994), No. 2, 67–89.

11. J. Avan and G. Rollet, *The classical r-matrix for the relativistic Ruijsenaars–Schneider system*, Phys. Lett. A **212** (1996), No. 1-2, 50–54, hep-th/9510166.

12. J. Avan and M. Talon, *Graded R-matrices for integrable systems*, Nucl. Phys. B **352** (1991), No. 1, 215–249.

13. _____, *Classical R-matrix structure for the Calogero model*, Phys. Lett. B **303** (1993), No. 1-2, 33–37.

14. O. Babelon and D. Bernard, *The sine-Gordon solitons as an N-body problem*, Phys. Lett. B **317** (1993), No. 3, 363–368.

15. O. Babelon and C. M. Viallet, *Hamiltonian structures and Lax equations*, Phys. Lett. B **237** (1990), No. 3-4, 411–416.

16. J. Balog, L. Dąbrowski, and L. Fehér, *Classical r-matrix and exchange algebra in WZNW and Toda theories*, Phys. Lett. B **244** (1990), No. 2, 227–234.

17. A. A. Belavin and V. G. Drinfeld, *Triangle equations and simple Lie algebras*, Mathematical Physics Reviews (S. P. Novikov, ed.), Soviet Sci. Rev. Sect. C Math. Phys. Rev., Vol. 4, Harwood Academic Publ., Chur, 1984, pp. 93–165.

18. D. Bernard, M. Gaudin, F. D. M. Haldane, and V. Pasquier, *Yang–Baxter equation in long-range interacting systems*, J. Phys. A **26** (1993), No. 20, 5219–5236, hep-th/9301084.

19. E. Billey, J. Avan, and O. Babelon, *Exact Yangian symmetry in the classical Euler–Calogero–Moser model*, Phys. Lett. A **188** (1994), No. 3, 263–271, hep-th/9401117.

20. _____, *The r-matrix structure of the Euler–Calogero–Moser model*, Phys. Lett. A **186** (1994), No. 1-2, 114–118, hep-th/9312042.

21. H. W. Braden, *R-matrices and generalized inverses*, J. Phys. A **30** (1997), No. 15, L485–L493, q-alg/9706001, solv-int/9706001.

22. H. W. Braden and T. Suzuki, *R-matrices for elliptic Calogero–Moser models*, Lett. Math. Phys. **30** (1994), No. 2, 147–158, hep-th/9309033.

23. F. Calogero, *Exactly solvable one-dimensional many-body problems*, Lett. Nuovo Cimento **13** (1975), No. 11, 411–416.

24. _____, *On a functional equation connected with integrable many-body problems*, Lett. Nuovo Cimento **16** (1976), No. 3, 77–80.

25. J. F. van Diejen, *Families of commuting difference operators*, Ph.D. thesis, University of Amsterdam, 1994.

26. L. D. Faddeev and L. A. Takhtajan, *Hamiltonian Methods in the Theory of Solitons*, Springer Ser. Soviet Math., Springer, New York, 1987.

27. G. Felder, *Conformal field theory and integrable systems associated to elliptic curves*, Proc. International Congress of Mathematicians (Zürich, 1994) (S. D. Chatterji, ed.), Birkhäuser, Basel, 1994, pp. 1247–1255, hep-th/9407154.

28. G. Felder and C. Wieczerkowski, *Conformal blocks on elliptic curves and the Knizhnik–Zamolodchikov–Bernard equations*, Commun. Math. Phys. **176** (1996), No. 1, 133–161.

29. I. M. Gelfand and I. Ya. Dorfman, *Hamiltonian operators and algebraic structures associated with them*, Funct. Anal. Appl. **13** (1979), No. 4, 248–262.

30. J.-L. Gervais and A. Neveu, *Novel triangle relations and the absence of tachyons in Liouville string field theory*, Nucl. Phys. B **238** (1984), No. 1, 125–141.

31. J. Gibbons and T. Hermsen, *A generalisation of the Calogero–Moser system*, Physica **11D** (1984), No. 3, 337–348.

32. A. Gorsky, *Integrable many body systems in the field theories*, Theoret. and Math. Phys. **103** (1995), No. 3, 681–700, hep-th/9410228.

33. A. Gorsky and N. Nekrasov, *Elliptic Calogero–Moser system from two-dimensional current algebra*, hep-th/9401021.

34. _____, *Relativistic Calogero–Moser model as gauged WZW theory*, Nucl. Phys. B **436** (1995), No. 3, 582–608, hep-th/9401017.

35. K. Hasegawa, *Ruijsenaars' commuting difference operators as commuting transfer matrices*, Commun. Math. Phys. **187** (1997), No. 2, 289–325, q-alg/9512029.

36. I. M. Krichever, *Elliptic solutions of the Kadomtsev–Petviašhvili equation and integrable systems of particles*, Funct. Anal. Appl. **14** (1980), No. 4, 282–290.

37. I. M. Krichever, O. Babelon, E. Billey, and M. Talon, *Spin generalization of the Caloger–Moser system and the matrix KP equation*, Topics in Topology and Mathematical Physics (A. B. Sossinsky, ed.), Amer. Math. Soc. Transl. Ser. 2, Vol. 170, Amer. Math. Soc., Providence, RI, 1995, pp. 83–119, hep-th/9411160.

38. I. M. Krichever and D. H. Phong, *On the integrable geometry of soliton equations and N = 2 supersymmetric gauge theories*, J. Differential Geom. **45** (1997), No. 2, 349–389.

39. I. M. Krichever and A. Zabrodin, *Spin generalization of the Ruijsenaars–Schneider model, non-Abelian 2D Toda chain and representations of Sklyanin algebra*, Russian Math. Surveys **50** (1995), No. 6, 1101–1150, hep-th/9505039.

40. P. P. Kulish, S. Rauch-Wojciechowski, and A. V. Tsiganov, *Stationary problems for equation of the KdV type and dynamical r-matrices*, Modern Phys. Lett. **A9** (1994), No. 7, 2063–2073.

41. P. D. Lax, *Integrals of nonlinear equations of evolution and solitary waves*, Commun. Pure Appl. Math. **21** (1968), 467–490.

42. L. C. Li and S. Parmentier, *Nonlinear Poisson structures and r-matrices*, Commun. Math. Phys. **125** (1989), No. 4, 545–563.

43. J. Liouville, *Note sur l'integration des équations différentielles de la dynamique*, J. Math. Pures Appl. **20** (1855), 137–138.

44. F. Magri, *A simple model of the integrable Hamiltonian equation*, J. Math. Phys. **19** (1978), No. 5, 1156–1162.

45. J.-M. Maillet, *Kac–Moody algebra and extended Yang–Baxter relations in the O(n) nonlinear σ-model*, Phys. Lett. B **162** (1985), No. 1-3, 137–142.

46. J.-M. Maillet and F. W. Nijhoff, *Integrability for multidimensional lattice models*, Phys. Lett. B **224** (1989), No. 4, 389–396.

47. J. Moser, *Three integrable Hamiltonian systems connected to isospectral deformations*, Adv. Math. **16** (1975), 197–220.

48. M. A. Olshanetsky and A. M. Perelomov, *Classical integrable finite-dimensional systems related to Lie algebras*, Phys. Rep. **71** (1981), No. 5, 313–400.

49. _____, *Quantum integrable systems related to Lie algebras*, Phys. Rep. **94** (1983), No. 6, 313–404.

50. A. G. Reyman and M. A. Semenov-Tian-Shansky, *Group-theoretical methods in the theory of finite-dimensional integrable systems*, Dynamical Systems. VII. Integrable Systems, Nonholonomic Dynamical Systems (V. I. Arnold and S. P. Novikov, eds.), Encyclopaedia Math. Sci., Vol. 16, Springer, Berlin, 1994, pp. 190–377.

51. S. N. M. Ruijsenaars, *Complete integrability of relativistic Calogero–Moser systems and elliptic function identities*, Commun. Math. Phys. **110** (1987), No. 2, 191–213.

52. _____, *Action-angle maps and scattering theory for some finite-dimensional integrable systems. I. The pure soliton case*, Commun. Math. Phys. **115** (1988), No. 1, 127–165.

53. O. Schiffmann, *On classification of dynamical r-matrices*, Math. Res. Lett. **5** (1998), No. 1-2, 13–30, q-alg/9706017.

54. M. A. Semenov-Tian-Shansky, *What a classical r-matrix is*, Funct. Anal. Appl. **17** (1983), No. 4, 259–272.

55. E. K. Sklyanin, *On complete integrability of the Landau–Lifschitz equation*, unpublished, 1981.

56. _____, *Quantum variant of the method of the inverse scattering problem*, J. Soviet Math. **19** (1982), No. 5, 1546–1596.

57. _____, *Dynamic r-matrices for the elliptic Calogero–Moser model*, St. Petersburg Math. J. **6** (1995), No. 2, 397–406.

58. Yu. B. Suris, *Why is the Ruijsenaars–Schneider hierarchy governed by the same R-operator as the Calogero–Moser one?*, Phys. Lett. A **225** (1997), No. 4-6, 253–262.

59. A. V. Tsiganov, *Automorphisms of* sl(2) *and dynamical r-matrices*, J. Math. Phys. **39** (1998), No. 1, 650–664, solv-int/9610003.

60. Y. B. Zeng and J. Hietarinta, *Classical Poisson structures and r-matrices from constrained flows*, J. Phys. A **29** (1996), No. 16, 5241–5252, solv-int/9509005.

2

Hidden Algebraic Structure of the Calogero–Sutherland Model, Integral Formula for Jack Polynomial and Their Relativistic Analog

Hidetoshi Awata[1]

ABSTRACT We review some recent results on the Calogero–Sutherland and Ruijsenaars models with emphasis on their algebraic aspects. We give integral formulas for excited states (Jack and Macdonald polynomials) of these models, their relations with the Virasoro singular vectors and its q-analog and obtain free boson realization for level one elliptic affine Lie algebra.

1 Introduction

The Calogero–Sutherland model [19, 20] is a quantum mechanical system with a long-range interaction, and its excited states are known to be Jack polynomials [18]. So far there exist two kinds of explicit formula for them: integral [3–5] and differential formulas [11]. They reveal some hidden algebraic structures of this model; the former relates Jack polynomials to Virasoro algebra; and latter, to Hecke algebra through a Dunkl operator.

The relativistic or q-analog of this model and polynomials are the Ruijsenaars model [16] and Macdonald polynomials [14], respectively. Through the integral formula for these polynomials [6] we found a totally new algebra, called "the" q-deformed Virasoro algebra [17].

The Virasoro algebra is the most important symmetry of string theory and statistical critical phenomena [7] and has played an essential role in their remarkable progress. For the noncritical case, our q-Virasoro algebra is expected to take the place of the Virasoro algebra, and; indeed, it has been found to relate to the sine-Gordon model [12] and RSOS model [13].

On the other hand, when there is an internal symmetry, for example, a

---

[1] JSPS fellow.

XXX spin chain, the affine Lie (Kac-Moody) algebra, takes the place of the Virasoro algebra; and in the noncritical case, for example, the XXZ chain, the quantum affine Lie algebra does [8]. However, because of our scant knowledge of the elliptic affine Lie algebra, the most general (integrable) noncritical case, for example, the XYZ chain, is still underdeveloped. Recently, we found that the q-Virasoro algebra relates with level one elliptic affine Lie algebra and we obtain a free boson realization for it [2], which could open further avenues to the exploration of elliptic models.

In this talk, we start from defining the Jack polynomials at Section 2 and give an integral formula for them in Section 3. In Sections 4 and 5, their one-to-one correspondence with Virasoro singular vectors is shown. Next we turn to the q-analog, and define the q-Virasoro algebra in Section 5. Finally in Section 6, we mention that the screening currents of this new algebra generate an elliptic affine Lie algebra and we give its free boson realization.

For more details, please see our review paper [2].

2   Calogero–Sutherland Model and Jack Polynomial

We start with recapitulating some important properties of the Jack polynomials.

Jack polynomials $J(z) \equiv J(z_1, z_2, \ldots, z_N)$ are defined as symmetric polynomials in z_i's that are eigenfunctions of [14, 18–20]

$$H = \sum_{i=1}^{N} D_i^2 + \beta \sum_{i<j} \frac{z_i + z_j}{z_i - z_j}(D_i - D_j), \qquad (1)$$

with $D_i = z_i \partial/\partial z_i$. Here β is a coupling constant.

This Hamiltonian H relates to the Calogero–Sutherland model, which is an N-body problem on a unit circle with the following Hamiltonian:

$$H_{CS} = \sum_{j=1}^{N} \left(\frac{1}{i}\frac{\partial}{\partial q_j}\right)^2 + \frac{1}{2}\sum_{i<j}\frac{\beta(\beta - 1)}{\sin^2(q_i - q_j)/2}, \qquad (2)$$

by changing the variables of coordinates q_i to $z_j \equiv \exp(iq_j)$ on a complex plane. The vacuum wave-function Δ of H_{CS} is

$$\Delta = \prod_{i<j} \sin^\beta\left(\frac{q_i - q_j}{2}\right) = \prod_{i \neq j}\left(1 - \frac{z_i}{z_j}\right)^{\beta/2}, \qquad (3)$$

and the excited states are written as $J(z)\Delta(z)$ with the Jack polynomial $J(z)$. Thus the relation between H and H_{CS} is $H = \Delta^{-1} H_{CS} \Delta - \varepsilon_0$ with the vacuum energy ε_0.

The reason that the Jack polynomial $J(z)$ should be "symmetric" and "polynomial" is as follows:

i) Because of Eq. (3), the statistic of the particle is governed by the coupling β: if β is even (odd), then the particles become bosonic (fermionic). To possess the same statistic as the vacuum, excited states $J(z)$ have to be symmetric functions.

ii) Because of the periodicity, the power of z_i's should be integers. By multiplying the factor $\prod_{i=1}^{N} z_i^{M}$ with a sufficiently large number M, all powers can become positive.

The Jack polynomial $J(z)$ is known to be labeled by a decreasing set of nonnegative integers, $\lambda = (\lambda_1 \geq \lambda_2 \geq \cdots \geq \lambda_N \geq 0)$, which is identified with a Young diagram:

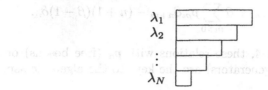

with λ_i (≥ 1) squares in ith row. When $\beta = 1$, the Jack polynomial reduces to the Schur polynomial.

The following orthogonality and the Cauchy formula for the Jack polynomial are very important [14, 18]. First, for an inner product,

$$\langle f(z), g(z) \rangle \equiv \oint \prod_{j=1}^{N} \frac{dz_j}{z_j} \cdot f\left(\frac{1}{z}\right) g(z) \Delta^2(z), \qquad (4)$$

Hamiltonian H is self-adjoint. Here the integration path is along the unit circle around the origin in the complex plane, and $\Delta^2(z) = \prod_{i \neq j} (1 - z_i/z_j)^{\beta}$. From this self-adjointness, orthogonality of the Jack polynomial follows:

$$\langle J_\lambda(z), J_\mu(z) \rangle \propto \delta_{\lambda,\mu}. \qquad (5)$$

When eigenvalues degenerate, the Jack polynomial should be defined with this property.

There exists another inner product for which Hamiltonian is also self-adjoint. From the orthogonality for this the Cauchy formula follows:

$$\Pi(x, y) \equiv \prod_{i=1}^{N} \prod_{j=1}^{M} (1 - x_i y_j)^{-\beta} = \sum_{\lambda} J_\lambda(x) J_\lambda(y) j_\lambda. \qquad (6)$$

Here j_λ is a number (function in β) that depends on the normalization of $J_\lambda(x)$. It is very important that, in Eq. (6), the number of the variables x, N, and that of y, M, be independent.

Let us consider the infinite number of particles case, that is, $N \to \infty$. Since H is symmetric in z_i's, it can be expressed by the power sums $p_n \equiv \sum_{i=1}^{N} z_i^n$ and their derivatives $\partial_n \equiv (n/\beta)\partial/\partial p_n$ as follows [18]:

$$H = \beta^2 \sum_{n,m>0} (p_{n+m}\partial_n\partial_m + p_n p_m \partial_{n+m})$$
$$+ \beta \sum_{n>0}(n - n\beta + N\beta)p_n\partial_n, \quad (7)$$

Here we must treat p_n's as formally independent variables, that is, $\partial_n p_m = (n/\beta)\delta_{n,m}$ for all n, $m > 0$. We can rewrite this as

$$H = \beta \sum_{n>0} p_n \mathcal{L}_n + (\beta - 1 + \beta N)\beta \sum_{n>0} p_n \partial_n,$$
$$(8)$$
$$\mathcal{L}_n = \beta \sum_{m=1}^{n-1} \partial_m \partial_{n-m} + \beta \sum_{m>0} p_m \partial_{n+m} - (n+1)(\beta - 1)\partial_n.$$

As we will see in Section 4, these relations with p_n (free bosons) or $\mathcal{L}_n$ (nonrelativistic Virasoro generators) are the keys to the algebraic aspects of the model.

Finally, we present some examples of Jack polynomials in power sum p_n.

$$J_{(1)} = p_1, \quad (9)$$
$$J_{(2)} = p_2 + \beta p_1^2, \quad (10)$$
$$J_{(22)} = p_4 + \frac{4\beta}{1-\beta}p_3 p_1 - \frac{1+\beta+\beta^2}{1-\beta}p_2^2 - 2\beta p_2 p_1^2 - \frac{\beta^2}{1-\beta}p_1^4. \quad (11)$$

3 Integral Formula for Jack Polynomial

We next try to derive the explicit expression of all excited states. Our strategy is as follows: we introduce two types of (integral) transformations that map the eigenstate into another while changing its energy and the number of particles. We can construct arbitrary state by applying them successively to the vacuum.

First, we introduce the Galilean boost $\mathcal{G}_N$, which uniformly shifts the pseudomomentum of the pseudoparticles from $\lambda = (\lambda_1, \ldots, \lambda_N)$ to $\lambda + (1^N) = (\lambda_1 + 1, \ldots, \lambda_N + 1)$. It can be realized by multiplying the wave function by $\prod_j e^{iq_j} = \prod_j z_j$. When it is applied to the eigenstate, the Young diagram is changed by adding a rectangle (1^N) from the left:

$$\mathcal{G}_N \cdot J_\lambda(z_1, \ldots, z_N) = J_{\lambda+(1^N)}(z_1, \ldots, z_N)$$
$$= J_\lambda(z_1, \ldots, z_N) \cdot \prod_{i=1}^{N} z_i. \quad (12)$$

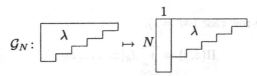

Note that $J_{s^r}(z_1, \ldots, z_r) = \prod_{i=1}^{r} z_i^s$.

The second integral transformation $\mathcal{N}_{NM}$ changes the number of particles from M to N:

$$\mathcal{N}_{NM} \cdot J_\lambda(t_1, \ldots, t_M) = J_\lambda(z_1, \ldots, z_N) \propto \langle \Pi(z, t), J_\lambda(t) \rangle. \qquad (13)$$

Here $\langle \, , \, \rangle$ is the inner product in Eq. (4).

Proof. Just use Eqs. (6) and (5).   □

Therefore, starting from the vacuum with 1-particles, that is 1, and combining these two transformations $\mathcal{G}_n$ and $\mathcal{N}_{nm}$, we obtain all excited states of N particles. For the Young diagram $\lambda = (\lambda_1, \ldots, \lambda_N)$

$$\lambda = \boxed{\begin{array}{c} \lambda_N \quad \lambda_{N-1} - \lambda_N \\ N \qquad \boxed{N-1} \end{array}} \cdots \boxed{\begin{array}{c} \lambda_2 - \lambda_3 \quad \lambda_1 - \lambda_2 \\ 2 \qquad \boxed{1} \end{array}}$$

construct from the right of the diagram as follows:

1) First, acting Galilean transformation $(\mathcal{G}_1)^{\lambda_1 - \lambda_2}$ on the vacuum 1 produces the Jack polynomial with a one-row Young diagram $(\lambda_1 - \lambda_2)$.

2) Next, the particle number changing transformation $\mathcal{N}_{2,1}$ increases the number of particles from 1 to 2.

3) The action by $(\mathcal{G}_2)^{\lambda_2 - \lambda_3}$ adds a new rectangular $(\lambda_2 - \lambda_3, \lambda_2 - \lambda_3)$.

4) Using these transformations iteratively, one gets an integral formula for the Jack polynomials with general Young diagrams.

Theorem 1 (Integral formula [3–5, 15]).

$$J_\lambda(z) = \mathcal{G}_N^{\lambda_N} \mathcal{N}_{N,N-1} \mathcal{G}_{N-1}^{\lambda_{N-1} - \lambda_N} \cdots \mathcal{G}_2^{\lambda_2 - \lambda_3} \mathcal{N}_{2,1} \mathcal{G}_1^{\lambda_1 - \lambda_2} \cdot 1 \qquad (14)$$

Especially, for the rectangular Young diagram $\lambda = (k^M)$, it is

$$J_{(k^M)}(z_1, \ldots, z_N) = \left\langle \Pi(z, t), \prod_{i=1}^{M} t_i^k \right\rangle, \qquad (15)$$

and for the one-row case ($M = 1$),

$$\Pi(z, t) = \sum_{k \geq 0} J_k(z_1, \ldots, z_N) t^k. \tag{16}$$

These formulas reveal new algebraic aspects of the model in the following section.

4   Relation with Virasoro Singular Vectors

Virasoro algebra $\{L_n\}_{n \in \mathbb{Z}}$ is an infinite-dimensional Lie algebra defined as

$$[L_n, L_m] = (n - m)L_{n+m} + \frac{c}{12}n(n^2 - 1)\delta_{n+m,0}. \tag{17}$$

Here the central charge $c \in \mathbb{C}$ is an important parameter.

This algebra is realized by simpler algebra called free boson algebra $\{a_n\}_{n \in \mathbb{Z}}$ defined with

$$[a_n, a_m] = n\delta_{n+m,0}, \tag{18}$$

as follows:

$$L_n = \frac{1}{2} \sum_{m \in \mathbb{Z}} :a_{n-m}a_m: - (n + 1)\alpha_0 a_n, \quad \alpha_0 = \frac{1}{\sqrt{2}}\left(\sqrt{\beta} - \frac{1}{\sqrt{\beta}}\right). \tag{19}$$

Here : : stands for the normal ordering such that positive mode generator should be right, for example, $:a_3 a_{-3}: = a_{-3}a_3$.

Let $\mathcal{V}_h$ be the Verma module over the Virasoro algebra generated by the highest weight state $|h\rangle$, such that

$$L_n|h\rangle = 0, \ (n > 0), \quad L_0|h\rangle = h|h\rangle, \tag{20}$$

that is,

$$\mathcal{V}_h \equiv \mathbb{C}[L_{-1}, L_{-2}, \ldots]|h\rangle. \tag{21}$$

In terms of bosons, $|h\rangle$ is

$$a_n|h\rangle = 0, \ (n > 0), \quad a_0|h\rangle = \alpha|h\rangle, \quad h = \frac{1}{2}\alpha(\alpha - 2\alpha_0). \tag{22}$$

This module $\mathcal{V}_h$ is irreducible if and only if there are no singular vectors $|\chi\rangle \in \mathcal{V}_h$ of grade $N \in \mathbb{Z}_{>0}$ such that

$$L_n|\chi\rangle = 0, \ (n > 0), \quad L_0|\chi\rangle = (h + N)|\chi\rangle, \ (N \in \mathbb{Z}_{>0}). \tag{23}$$

In case of such singular vectors exists, to get an irreducible representation $\mathcal{M}_h$, we should factor out the unnecessary modules generated by them.

$$\mathcal{M}_h = \mathcal{V}_h/(\text{modules generated by } |\chi\rangle\text{'s}). \tag{24}$$

Thus it is important to know when and where such singular vectors appear, and this is described by the following theorem:

Theorem 2. *Singular vector of grade $N = rs$ exists if and only if central charge c and highest weight h are*

$$c = 1 - 6(\beta - 1)\left(1 - \frac{1}{\beta}\right), \quad h_{rs} = \frac{(r\beta - s)^2 - (\beta - 1)^2}{4\beta}, \tag{25}$$

with $\beta \in \mathbb{C}_{\neq 0}$ and $r, s \in \mathbb{Z}_{>0}$, and if β takes generic value, there is only one.

In terms of bosons, the highest weight α_{rs} in Eq. (22) is

$$\alpha_{rs} = \frac{1}{\sqrt{2}}\left((r+1)\sqrt{\beta} - (s+1)\frac{1}{\sqrt{\beta}}\right). \tag{26}$$

Here we present some examples of singular vectors in bosons:

$$|\chi_{11}\rangle = a_{-1}|\alpha_{11}\rangle, \tag{27}$$

$$|\chi_{12}\rangle = (a_{-2} + \sqrt{2\beta}a_{-1}^2)|\alpha_{12}\rangle, \tag{28}$$

$$|\chi_{22}\rangle = \left(a_{-4} + \frac{4\sqrt{2\beta}}{1-\beta}a_{-3}a_{-1} - 2\frac{1+\beta+\beta^2}{\sqrt{2\beta}(1-\beta)}a_{-2}^2 \right.$$
$$\left. - 4a_{-2}a_{-1}^2 - \frac{2\sqrt{2\beta}}{1-\beta}a_{-1}^4\right)|\alpha_{22}\rangle. \tag{29}$$

It is easy to see that these singular vectors correspond to Jack polynomials in power sum in Eq. (11) by the following rule:

$$a_{-n} \to \sqrt{\frac{\beta}{2}}p_n, \quad |\alpha_{rs}\rangle \to 1. \tag{30}$$

This rule holds for the Jack polynomials with rectangular Young diagrams, and we have—

Proposition 1 ([15]). *The Virasoro singular vector $|\chi_{rs}\rangle$ has one-to-one correspondence with the Jack symmetric polynomial with the rectangular Young diagram $\{s^r\}$*

$$|\chi_{rs}\rangle \sim J_{\{s^r\}}(x; \beta), \tag{31}$$

by the above rule (Eq. (30)).

Proof. Compare the integral formula for the Jack polynomial in power sum and that for the Virasoro singular vector in bosons. It is known that the singular vectors in bosons are given as following integral formula:

$$|\chi_{rs}\rangle = \left\langle \prod_{i=1}^{r} \exp\left\{ \sqrt{2\beta} \sum_{n>0} \frac{1}{n} a_{-n} t_i^n \right\}, \prod_{i=1}^{r} t_i^{-s} \right\rangle |\alpha_{rs}\rangle. \qquad (32)$$

When β is nonnegative integer, the integration path is along the circle around the origin such that $|t_i| > |t_{i+1}|$. On the other hand, since

$$\Pi\left(z, \frac{1}{t}\right) = \prod_{i=1}^{r} \exp\left\{ \beta \sum_{n>0} \frac{1}{n} \sum_{j=1}^{N} z_j^n t_i^n \right\}, \qquad (33)$$

from Eq. (15), one can show that they are equivalent.   □

Notice that we can also prove this proposition by using Eq. (8). With Eq. (30), $\mathcal{L}_n$ in Eq. (8) is nothing but the positive mode generator L_n of Virasoro algebra in Eq. (19). Since any singular vectors are annihilated by the $\mathcal{L}_n$ part of H and are trivially eigenfunctions for the remaining part, we complete the proof.

It is very important that the Hamiltonian H in power sum be constructed by positive-mode Virasoro generators.

It can be shown that the Jack polynomials for arbitrary Young diagrams are realized as the singular vector of the $\mathcal{W}_N$ algebra.

Theorem 3 ([3–5]). *The $\mathcal{W}_n$ singular vector has one-to-one correspondence with the Jack symmetric polynomial with arbitrary Young diagram.*

5 Macdonald Polynomial and q-Virasoro Algebra

Let us turn to an analysis of q-analog and define our new algebra. The excited states of trigonometric Ruijsenaars model are called Macdonald symmetric functions $P_\lambda(z)$, which are defined as eigenfunctions of

$$H = \sum_{i=1}^{N} \prod_{j \neq i} \frac{tz_i - z_j}{z_i - z_j} \cdot q^{D_i}, \qquad (34)$$

where $D_i \equiv z_i \partial/\partial z_i$ and $t = q^\beta$. Note that q^{D_i} is the q-shift operator such that $q^{D_i} f(z_1, \dots, z_N) = f(z_1, \dots, qz_i, \dots, z_N)$. In the limit of $q \to 1$, this Macdonald polynomial reduces to the Jack polynomial.

The integral formula for the Macdonald polynomial is the same as the Jack case with the replacement

$$\Pi(x, y) \equiv \prod_{i=1}^{N} \prod_{j=1}^{M} \prod_{k\geq 0} \frac{1 - tq^k x_i y_j}{1 - q^k x_i y_j}, \quad \Delta^2(x) \equiv \prod_{i \neq j} \prod_{k\geq 0} \frac{1 - q^k x_i/x_j}{1 - tq^k x_i/x_j}. \qquad (35)$$

Then the problem below naturally follows;

Problem 1. Find algebras whose singular vectors realize the Macdonald polynomials.

A key property is also the same as in the Jack case; the Hamiltonian should be constructed by positive-mode generators. The solution looks unique and is the following q-deformed Virasoro and $\mathcal{W}_n$ algebra.

Definition 1 ([17]). Let q, t, and $p = q/t$ be complex parameters. The q-Virasoro algebra $\{T_n\}_{n\in\mathbb{Z}}$ is an associative algebra such that

$$[T_n, T_m] = -\sum_{\ell>0} f_\ell(T_{n-\ell}T_{m+\ell} - T_{m-\ell}T_{n+\ell})$$
$$- \frac{(1-q)(1-t^{-1})}{1-p}(p^n - p^{-n})\delta_{m+n,0}, \quad (36)$$

with

$$f(x) \equiv \sum_{\ell\geq0} f_\ell z^\ell = \exp\left\{\sum_{n=1}^{\infty} \frac{(1-q^n)(1-t^{-n})}{1+p^n}\frac{x^n}{n}\right\}. \quad (37)$$

And δ is a delta function $\delta(x) \equiv \sum_{n\in\mathbb{Z}} x^n$ and has a property

$$\delta(x)f(x) = \delta(x)f(1). \quad (38)$$

In the limit of $q \to 1$, this algebra reduces to the Virasoro algebra with center Eq. (25).

In terms of the current $T(z) = \sum_{n\in\mathbb{Z}} T_n z^{-n}$, the above relation is equivalent to

$$f\left(\frac{w}{z}\right)T(z)T(w) - T(w)T(z)f\left(\frac{z}{w}\right) =$$
$$- \frac{(1-q)(1-t^{-1})}{1-p}\left[\delta\left(\frac{pw}{z}\right) - \delta\left(\frac{w}{pz}\right)\right], \quad (39)$$

The structure of the highest-weight module is very similar to the $q = 1$ case. And the q-Virasoro generator $T(z)$ and singular vectors are also realized by free bosons.

Proposition 2 ([17]). *The q-Virasoro singular vector $|\chi_{rs}\rangle$ has one-to-one correspondence with the Macdonald symmetric polynomial with the rectangular Young diagram $\{s^r\}$.*

The q-$\mathcal{W}_n$ algebra can be defined by using the q-deformed Miura transformation [1, 9] and we can show the relation with Macdonald polynomials.

Theorem 4 ([3–5]). *The q-$\mathcal{W}_n$ singular vector has one-to-one correspondence with the Macdonald symmetric polynomial with an arbitrary Young diagram.*

6   Boson Realization for q-Virasoro Algebra and Level One Elliptic Affine Lie Algebra

Let us introduce the fundamental Heisenberg algebra h_n $(n \in \mathbb{Z})$, Q_h, having the commutation relations

$$[h_n, h_m] = \frac{1}{n} \frac{(q^{n/2} - q^{-n/2})(t^{n/2} - t^{-n/2})}{p^{n/2} + p^{-n/2}} \delta_{n+m,0},$$

$$[h_n, Q_h] = \frac{1}{2} \delta_{n,0}. \tag{40}$$

By these, the q-Virasoro current $T(z)$ and the screening current $S_\pm(z)$, which commute with q-Virasoro $[T(z), \oint dt S_\pm(t)] = 0$ and generate the singular vectors, are written as

$$T(z) = \Lambda^+(z) + \Lambda^-(z), \tag{41}$$

$$\Lambda^\pm(z) = :\exp\left\{ \pm \sum_{n \neq 0} h_n p^{\pm n/2} z^{-n} \right\}: q^{\pm \sqrt{\beta} h_0} p^{\pm 1/2}, \tag{42}$$

$$S_+(z) = :\exp\left\{ -\sum_{n \neq 0} \frac{p^{n/2} + p^{-n/2}}{q^{n/2} - q^{-n/2}} h_n z^{-n} \right\}: e^{2\sqrt{\beta} Q_h} z^{2\sqrt{\beta} h_0}, \tag{43}$$

$$S_-(z) = :\exp\left\{ \sum_{n \neq 0} \frac{p^{n/2} + p^{-n/2}}{t^{n/2} - t^{-n/2}} h_n z^{-n} \right\}: e^{-2/\sqrt{\beta} Q_h} z^{-2/\sqrt{\beta} h_0}. \tag{44}$$

Note that under the isomorphism σ such that

$$\sigma: \quad q \leftrightarrow \frac{1}{t}, \quad \sqrt{\beta} \leftrightarrow -\sqrt{\frac{1}{\beta}}, \tag{45}$$

$\sigma \cdot \Lambda^\pm(z) = \Lambda^\pm(z)$ and $\sigma \cdot S_\pm(z) = S_\mp(z)$.

The free boson realization for $T(z)$ is expressed as the following deformed Miura transformation [10]:

$$:(p^D - \Lambda^+(z))(p^D - \Lambda^-(z)): = p^{2D} - T(z)p^D + 1, \tag{46}$$

which has been generalized to define the q-deformed $\mathcal{W}$ algebra [1, 9]. By using this transformation, Frenkel–Reshetikhin [10] proposed a generalization of their quasi-classical q-Virasoro algebra to ABCD-type cases. An analogy to the Baxter's dressed vacuum form Q defined by $:(p^D - \Lambda^-(z))Q(z): = 0$, so $\Lambda^\pm(z) = :Q(zp^{\mp 1})Q(z)^{-1}:$ seems to be of some interest.

The properties of screening currents are quite important in the representation theory of the infinite-dimensional algebra; they govern the irreducibility and the physical states. Moreover, they relate to hidden quantum symmetries.

Here, we show that the screening currents generate an elliptic hidden symmetry, which reduces to the (quantum) affine Lie algebra with a special center. Let us introduce a new current $\Psi(z) \equiv :S_+(q^{\pm 1/2}z)S_-(t^{\pm 1/2}z):$, that is,

$$\Psi(z) = \exp\left\{\sum_{n \neq 0} \frac{p^n - p^{-n}}{(q^{n/2} - q^{-n/2})(t^{n/2} - t^{-n/2})} h_n z^{-n}\right\} e^{2\alpha Q} z^{2\alpha h_0}, \quad (47)$$

with $\alpha = \sqrt{\beta} - 1/\sqrt{\beta}$, then we have—

Proposition 3 ([2]). *Screening currents $S_\pm(z)$ and $\Psi(z)$ generate the following elliptic two-parameter algebra:*

$$f_{00}\left(\frac{w}{z}\right)\Psi(z)\Psi(w) = \Psi(w)\Psi(z)f_{00}\left(\frac{z}{w}\right), \quad (48)$$

$$f_{0\pm}\left(\frac{w}{z}\right)\Psi(z)S_\pm(w) = S_\pm(w)\Psi(z)f_{\pm 0}\left(\frac{z}{w}\right), \quad (49)$$

$$f_{\pm\pm}\left(\frac{w}{z}\right)S_\pm(z)S_\pm(w) = S_\pm(w)S_\pm(z)f_{\pm\pm}\left(\frac{z}{w}\right), \quad (50)$$

$$[S_+(z), S_-(w)] = \frac{1}{(p-1)w}\left[\delta\left(p^{1/2}\frac{w}{z}\right)\Psi(t^{-1/2}w)\right.$$
$$\left. \times \delta\left(p^{-1/2}\frac{w}{z}\right)\Psi(q^{-1/2}w)\right], \quad (51)$$

where $f_{00}(x) = f_{++}(x)f_{+-}^2(xp^{1/2})f_{--}(x)$, $f_{0\pm}(x) = f_{++}(xq^{1/2})f_{-\pm}(xt^{1/2})$ $= f_{\pm 0}(x)$ with

$$f_{+-}(x) = f_{-+}(x) = \exp\left\{-\sum_{n>0}\frac{1}{n}(p^{n/2} + p^{-n/2})x^n\right\}x^{-1}, \quad (52)$$

$$f_{++}(x) = \exp\left\{-\sum_{n>0}\frac{1}{n}\frac{t^{n/2} - t^{-n/2}}{q^{n/2} - q^{-n/2}}(p^{n/2} + p^{-n/2})x^n\right\}x^\beta, \quad (53)$$

and $f_{--}(x) = \omega \cdot f_{++}(x)$.

In the limit q and t tend to 0 with p and $t^{-|n|/2}h_n$ fixed, the relations (48)–(51) reduce to those of level one $U_q(\widehat{\mathfrak{sl}}_2)$. Therefore, the algebra generated by $S^\pm(z)$ and $\Psi(z)$ can be regarded as an elliptic generalization of $U_q(\widehat{\mathfrak{sl}}_2)$ with level one. Note that, in this limit, the algebra with $\Psi_\pm(z) \equiv :S_+(p^{\pm 1/4}z)S_-(p^{\mp 1/4}z):$ is more natural than that with $\Psi(z)$.

In the sense of analytic continuation, these relations are also rewritten by using elliptic theta functions [9],

$$S_\pm(z)S_\pm(w) = U_\pm\left(\frac{w}{z}\right)S_\pm(w)S_\pm(z), \quad (54)$$

with

$$U_\pm(x) = -x^{1-2\beta} \exp\left\{ \sum_{n\neq 0} \frac{1}{n} \frac{q^{n/2}t^{-n} - q^{-n/2}t^n}{q^{n/2} - q^{-n/2}} x^n \right\}$$

$$= -x^{2(1-\beta)} \frac{\vartheta_1(px; q)}{\vartheta_1(px^{-1}; q)}, \tag{55}$$

and $U_-(x) = \omega \cdot U_+(x)$. Note that $U_\pm(x)$ are quasiperiodic functions; namely, for $U_+(x)$, we have

$$U_+(qx) = U_+(x), \quad U_+(e^{2\pi i}x) = e^{-4\pi i\beta} U_+(x). \tag{56}$$

It should be noted that the screening currents of q-$\mathcal{W}$ algebra [1, 9] and $U_q(\widehat{\mathrm{sl}}_N)$ [6] also obey similar elliptic relations.

7 REFERENCES

1. H. Awata, H. Kubo, S. Odake, and J. Shiraishi, *Quantum $\mathcal{W}_N$ algebras and Macdonald polynomials*, Commun. Math. Phys. **179** (1996), No. 2, 401–416, q-alg/950801.

2. _____, *Virasoro-type symmetries in solvable models*, Sūrikaiseki-kenkyūsho Kōkyūroku **1005** (1997), 37–71, hep-th/9612233.

3. H. Awata, Y. Matsuo, S. Odake, and J. Shiraishi, *Collective field theory, Calogero–Sutherland model and generalized matrix models*, Phys. Lett. B **347** (1995), No. 1-2, 49–55, hep-th/9411053.

4. _____, *A note on Calogero–Sutherland model, W_n singular vectors and generalized matrix models*, Soryushiron Kenkyu **91** (1995), A69–A75, hep-th/9503028.

5. _____, *Excited states of the Calogero–Sutherland model and singular vectors of the W_N algebra*, Nucl. Phys. B **449** (1995), No. 1-2, 347–374, hep-th/9503043.

6. H. Awata, S. Odake, and J. Shiraishi, *Integral representations of the Macdonald symmetric polynomials*, Commun. Math. Phys. **179** (1996), No. 3, 647–666, q-alg/9506006.

7. A. A. Belavin, A. M. Polyakov, and A. B. Zamolodchikov, *Infinite conformal symmetry in two-dimensional quantum field theory*, Nucl. Phys. B **241** (1984), No. 2, 333–380.

8. B. Davies, O. Foda, M. Jimbo, T. Miwa, and A. Nakayashiki, *Diagonalization of the XXZ Hamiltonian by vertex operators*, Commun. Math. Phys. **151** (1993), No. 1, 98–153.

9. B. Fegin and E. Frenkel, *Quantum W-algebras and elliptic algebras*, Commun. Math. Phys. **178** (1996), No. 3, 653–678, q-alg/9508009.

10. E. Frenkel and N. Yu. Reshetikhin, *Quantum affine algebras and deformations of the virasoro and W-algebras*, Commun. Math. Phys. **178** (1996), No. 1, 237–264, q-alg/9505025.

11. L. Lapointe and L. Vinet, *Exact operator solution of the Calogero–Sutherland model*, Commun. Math. Phys. **178** (1996), No. 2, 425–452, hep-th/9507073.

12. S. Lukyanov, *A note on the deformed Virasoro algebra*, Phys. Lett. B **367** (1996), No. 1-4, 121–125, hep-th/9509037.

13. S. Lukyanov and Ya. Pugai, *Multipoint local height probabilities in the integrable RSOS model*, Nucl. Phys. B **473** (1996), No. 3, 631–658, hep-th/9602074.

14. I. G. Macdonald, *Symmetric Functions and Hall Polynomials*, 2nd ed., Oxford Math. Monogr., Oxford Univ. Press, New York, 1995.

15. K. Mimachi and Y. Yamada, *Singular vectors of Virasoro algebra in terms of Jack symmetric polynomials*, Sūrikaisekikenkyūsho Kōkyūroku **919** (1995), 68–78 (Japanese).

16. S. N. M. Ruijsenaars, *Complete integrability of relativistic Calogero–Moser systems and elliptic function identities*, Commun. Math. Phys. **110** (1987), No. 2, 191–213.

17. J. Shiraishi, H. Kubo, H. Awata, and S. Odake, *A quantum deformation of the Virasoro algebra and the Macdonald symmetric functions*, Lett. Math. Phys. **38** (1996), No. 1, 33–51, q-alg/9507034.

18. R. P. Stanley, *Some combinatorial properties of Jack symmetric functions*, Adv. Math. **77** (1989), No. 1, 76–115.

19. B. Sutherland, *Exact results for a quantum many-body problem in one dimension*, Phys. Rev. A **4** (1971), 2019–2021.

20. _____, *Exact results for a quantum many-body problem in one dimension. II*, Phys. Rev. A **5** (1972), 1372–1376.

3

Polynomial Eigenfunctions of the Calogero–Sutherland–Moser Models with Exchange Terms

T. H. Baker, C. F. Dunkl, and P. J. Forrester

ABSTRACT We examine eigenfunctions of the periodic and rational Calogero–Sutherland–Moser system with exchange terms. In particular explicit formulae for the normalization of Jack polynomials with prescribed symmetry (i.e., eigenfunctions that can be chosen to be symmetric or antisymmetric in certain variables) are given. In addition Macdonald polynomials of prescribed symmetry are considered, and it is shown that factorization can be achieved in certain cases.

1 Introduction

1.1 The Periodic Model

The type A periodic Calogero–Sutherland–Moser (CSM) model describes a quantum-mechanical system of N particles located on a circle of circumference L interacting via an inverse square potential, and is described by the Hamiltonian

$$H^{(C)} = -\sum_{j=1}^{N} \frac{\partial^2}{\partial x_j^2} + \beta\left(\frac{\beta}{2} - 1\right)\left(\frac{\pi}{L}\right)^2 \sum_{\substack{j,k=1 \\ j<k}}^{N} \frac{1}{\sin^2 \pi(x_j - x_k)/L}, \quad (1)$$

where $0 \le x_j \le L$. The terminology "type A" refers to the relationship between (1) and the classical reflection group of the same name. This system possesses a symmetric ground state that takes the form

$$\psi_0 = \prod_{\substack{j,k=1 \\ j<k}}^{N} \left|e^{2\pi i x_j/L} - e^{2\pi i x_k/L}\right|^{\beta/2} \quad (2)$$

After conjugating the Hamiltonian (1) by the ground state ψ_0, the resulting differential operator can be identified as one that possesses symmetric

Jack polynomials as eigenfunctions [10]. We remark here that one can construct CSM models associated with any (reduced) root system [14], and many articles have appeared dealing with the corresponding eigenfunction problem [3, 7, 8, 11, 17, 19].

The space of states of the model (1), factored by the ground state ψ_0, can be endowed with a natural inner product. Let $z_j = \exp(2\pi i x_j/L)$ introduce the multivariable notation $z := (z_1, \ldots, z_N)$ and suppose $\psi_i(z) = f_i(z)\psi_0(z)$, $i = 1, 2$. Then with * denoting complex conjugate, the inner product is defined as

$$\langle f_1, f_2 \rangle = \int_{[0,L]^N} \psi_1(z)\psi_2^*(z)\,dx_1 \ldots dx_N$$

$$= \int_{[0,L]^N} \prod_{\substack{i,j=1 \\ i<j}}^{N} |z_i - z_j|^{2/\alpha} f_1(z) f_2(z^*)\,dx_1 \ldots dx_N \qquad (3)$$

In the case where $1/\alpha = k \in Z^+$, and the f_i are (Laurent) polynomials, the above trigonometric integral can be replaced by the operator, which extracts the constant term, so that

$$\langle f_1, f_2 \rangle = \mathrm{C.\,T.}\big(|\psi_0(z)|^2 f_1(z) f_2(z^{-1})\big)$$

Returning to the model (1), it turns out that it is more convenient to consider a generalization that includes *exchange* terms, and the corresponding Hamiltonian has the form [16]

$$H^{(C,Ex)} = -\sum_{j=1}^{N} \frac{\partial^2}{\partial x_j^2} + \beta\left(\frac{\pi}{L}\right)^2 \sum_{\substack{j,k=1 \\ j<k}}^{N} \frac{(\beta/2 - s_{jk})}{\sin^2 \pi(x_j - x_k)/L}. \qquad (4)$$

Here s_{jk} is the operator that acts on functions by exchanging the jth and kth coordinates. When acting on symmetric functions, this Hamiltonian reduces to the previous one (1). After conjugation by the symmetric ground state (2) we obtain the conjugated Hamiltonian

$$\widetilde{H}^{(C,Ex)} = \sum_{j=1}^{N} z_j^2 \frac{\partial^2}{\partial z_j^2} + \frac{2}{\alpha} \sum_{\substack{j,k=1 \\ j<k}}^{N} \frac{z_j z_k}{z_j - z_k}\left[\left(\frac{\partial}{\partial z_j} - \frac{\partial}{\partial z_k}\right) - \frac{1 - s_{jk}}{z_j - z_k}\right], \qquad (5)$$

where we have set $\alpha = 2/\beta$. This differential operator also possesses symmetric Jack polynomials as eigenfunctions, but also has eigenfunctions with no fixed symmetry. A convenient basis for these eigenfunctions is the set of *nonsymmetric* Jack polynomials [15]. These polynomials $E_\eta(z)$ are labeled by compositions $\eta := (\eta_1, \ldots, \eta_N)$ and can be defined as the unique polynomials having an expansion of the form

$$E_\eta(z) = z^\eta + \sum_{\nu < \eta} b_{\eta\nu} z^\nu, \qquad (6)$$

which are also simultaneous eigenfunctions of the Cherednik operators

$$\xi_j = \alpha z_j \frac{\partial}{\partial z_j} + \sum_{l<j} \frac{z_j}{z_j - z_l}(1 - s_{lj}) + \sum_{l>j} \frac{z_l}{z_j - z_l}(1 - s_{lj}) - (j - 1). \quad (7)$$

In (6), $<$ is the partial order on compositions defined as follows: For a composition η, let η^+ denote the (unique) partition obtained from η by rearranging the entries in weakly decreasing order. Then let $\nu < \eta$ iff $\nu^+ < \eta^+$ (with regard to the dominance ordering) or, in the case $\nu^+ = \eta^+$, $\sum_{i=1}^p \nu_i \le \sum_{i=1}^p \eta_i$ for all $1 \le p \le N$. The corresponding eigenvalue of E_η under ξ_j is given by

$$\bar{\eta}_i = \alpha \eta_i - \sharp\{k < i \mid \eta_k \ge \eta_i\} - \sharp\{k > i \mid \eta_k > \eta_i\} \quad (8)$$

A direct calculation shows that the Cherednik operators (7) are self-adjoint with respect to the inner product (3), and since the corresponding eigenvalues are distinct for distinct compositions η, a simple argument shows that the nonsymmetric Jack polynomials $\{E_\eta(z)\}$ are orthogonal with regard to the inner product (3). The norm of $E_\eta(z)$ itself can be computed [15] with the result that in the case where $k = 1/\alpha$ is a positive integer

$$\langle E_\eta, E_\eta \rangle = \prod_{\substack{i,j=1 \\ i<j}}^{N} \prod_{p=0}^{k-1} \left(\frac{k(\bar{\eta}_j - \bar{\eta}_i) + p}{k(\bar{\eta}_j - \bar{\eta}_i) - p - 1} \right)^{\epsilon(\bar{\eta}_j - \bar{\eta}_i)}, \quad (9)$$

where $\epsilon(x) = 1$ for $x > 0$, and $\epsilon(x) = -1$ for $x \le 0$.

The Cherednik operators (7) can be used to rewrite (5). We have [4]

$$\widetilde{H}^{(C,Ex)} = \frac{1}{\alpha^2} \sum_{j=1}^{N} \left(\xi_j + \frac{N-1}{2} \right)^2 - \left(\frac{L}{2\pi} \right)^2 E_0^{(C)},$$

where $E_0^{(C)}$ denotes the eigenvalue of (2) for the eigenoperator (4). Hence each $E_\eta(z)$ is an eigenfunction of (5). It can easily be shown that the set of eigenfunctions $\{E_\eta \mid \eta$ a derangement of $\lambda\}$ possess the same eigenvalue, namely,

$$\epsilon_\eta = \sum_{j=1}^{N} \left(\lambda_j^2 + \frac{1}{\alpha}(N + 1 - 2j)\lambda_j \right). \quad (10)$$

This means that an arbitrary linear combination $\sum_\eta a_\eta E_\eta(z)$, where the sum is over all the derangements of a partition λ, will be an eigenfunction of the transformed Hamiltonian (5). One of these linear combinations will be the the symmetric Jack polynomial, denoted $P_\lambda^{(\alpha)}(z)$ [12]. Others will lead to eigenfunctions of prescribed symmetry, and we shall discuss these further in the next section.

1.2   The Linear Model

Another (closely related) CSM model describes a system of N particles on an infinite line with mutual inverse square repulsion in an external harmonic potential. The relevant Hamiltonian is given by

$$H^{(H,Ex)} := -\sum_{j=1}^{N} \frac{\partial^2}{\partial x_j^2} + \frac{\beta^2}{4}\sum_{j=1}^{N} x_j^2 + \beta \sum_{\substack{j,k=1 \\ j<k}}^{N} \frac{(\beta/2 - s_{jk})}{(x_j - x_k)^2}. \tag{11}$$

After conjugation by the ground state $\psi_0^{(H)} = \prod_i e^{-\beta x_i^2/4} \prod_{i<j} |x_i - x_j|^{\beta/2}$ and the change of variables $x_i \rightarrow \sqrt{2/\beta} x_i$, this Hamiltonian is transformed into the operator

$$H^{(H)} = \sum_{j=1}^{N} \left(\frac{\partial^2}{\partial x_j^2} - 2x_j \frac{\partial}{\partial x_j} \right)$$
$$+ \frac{2}{\alpha} \sum_{j<k} \frac{1}{x_j - x_k} \left[\left(\frac{\partial}{\partial x_j} - \frac{\partial}{\partial x_k} \right) - \frac{1 - s_{jk}}{x_j - x_k} \right], \tag{12}$$

where again, $\alpha = 2/\beta$. This operator can be conveniently expressed in terms of the (type A) Dunkl operators [9]:

$$T_i = \frac{\partial}{\partial x_i} + \frac{1}{\alpha} \sum_{p \neq i} \frac{1}{x_i - x_p}(1 - s_{ip}). \tag{13}$$

Indeed, we have $H^{(H,Ex)} = \Delta - 2E_1$, where

$$\Delta = \sum_{i=1}^{N} T_i^2, \quad E_1 = \sum_{i=1}^{N} x_i \frac{\partial}{\partial x_i}. \tag{14}$$

It is convenient to define operators [1]

$$h_i = \xi_i - \frac{\alpha}{2}T_i, \tag{15}$$

which allows us to reexpress $H^{(H,Ex)}$ as

$$H^{(H,Ex)} = -\frac{2}{\alpha}\sum_{i=1}^{N} h_i + \text{const.} \tag{16}$$

It is shown in Ref. 1 that the polynomials

$$E_\eta^{(H)}(x) = \exp(-\Delta/4)E_\eta(x) \tag{17}$$

are eigenfunctions of the h_i (again with eigenvalue $\bar{\eta}_i$) and hence of $H^{(H,Ex)}$ (with eigenvalue $-2|\eta|$, where $|\eta| := \eta_1 + \cdots + \eta_N$). These eigenfunctions are called nonsymmetric *Hermite* polynomials, since in the case of one variable, they reduce to ordinary Hermite polynomials [2]. As in the periodic (i.e., Jack) case, this eigenvalue degeneracy of $H^{(H,Ex)}$ leads one to consider linear combinations of the $E_\eta^{(H)}$ among the derangements of η, again leading to eigenfunctions of prescribed symmetry.

A natural inner product for the polynomial parts of eigenfunctions of the model (11) is given by

$$\langle f,g\rangle_H = \int_{(-\infty,\infty)^N} \prod_{i=1}^N e^{-x_i^2} \prod_{\substack{i,j=1\\i<j}}^N |x_i - x_j|^{2/\alpha} f(x)g(x)\,dx_1\ldots dx_N \quad (18)$$

As in the Jack case, it can be shown that the fundamental operators (15) are self-adjoint with respect to this inner product; and so, because their respective eigenvalues $\bar\eta_i$ are distinct, the nonsymmetric Hermite polynomials $E_\eta^{(H)}(z)$ are orthogonal.

2 Eigenfunctions of Prescribed Symmetry

2.1 Expansions in Terms of E_η

In this section, we derive some results for the normalization of eigenfunctions of (5) with prescribed symmetry. For given compositions $\mu := (\mu_1,\ldots,\mu_{N_0})$, $\rho := (\rho_1,\ldots,\rho_{N_1})$, let (μ,ρ) denote the concatenation of the compositions μ and ρ. Also, for any composition μ, let μ^+ denote the (unique) partition associated to μ, and μ^R denote the composition formed by reversing the entries of the partition μ^+. Note that μ^R is minimal among all the permutations of μ with respect to the partial order on compositions. Finally, let $f_i \equiv f_i(\mu^+)$ denote the number of parts of μ^+ equal to i.

Define the following polynomials of prescribed symmetry,

$$X_{(\mu^+,\rho)}(x,y) = \frac{1}{c_{(\mu^+,\rho)}}\,\mathrm{Sym}^{(x)}\,E_{(\mu^R,\rho)}(x,y)$$

$$Y_{(\mu,\rho^+)}(x,y) = \frac{1}{c_{(\mu,\rho^+)}}\,\mathrm{Asym}^{(y)}\,E_{(\mu,\rho^R)}(x,y)$$

where $c_{(\mu^+,\rho)}$ (respectively $c_{(\mu,\rho^+)}$) is chosen so that the coefficient of $x^{\mu^+} y^\rho$ (respectively $x^\mu y^{\rho^+}$) in $X_{(\mu^+,\rho)}$ (respectively $Y_{(\mu,\rho^+)}$) is unity. Similarly define

$$S_{(\mu^+,\rho^+)}(x,y) = \frac{1}{c_{(\mu^+,\rho^+)}}\,\mathrm{Sym}^{(x)}\,\mathrm{Asym}^{(y)}\,E_{(\mu^R,\rho^R)}(x,y). \quad (19)$$

Note that $Y_{(\mu,\rho^+)}(x,y)$ and $S_{(\mu^+,\rho^+)}(x,y)$ will be zero, unless the parts of ρ are distinct.

Lemma 1. *We have*

$$c_{(\mu^+,\rho)} = \prod_{j=1}^{N_0} f_j(\mu^+)!, \quad c_{(\mu,\rho^+)} = (-1)^{N_1(N_1-1)/2},$$

$$c_{(\mu^+,\rho^+)} = (-1)^{N_1(N_1-1)/2} \prod_{j=1}^{N_0} f_j(\mu^+)!.$$

Proof. We will consider only the first equation; the derivation of the other equations proceeds similarly. We know from (6) that

$$E_{(\mu^R,\rho)}(x,y) = x^{\mu^R} y^\rho + \sum_{(\alpha,\beta)<(\mu^R,\rho)} b_{(\mu^R,\rho),(\alpha,\beta)} x^\alpha y^\beta \qquad (20)$$

for some coefficients $b_{(\mu^R,\rho),(\alpha,\beta)}$. Owing to the definition of the partial order $<$, we see that the only term in (20) of the form $x^{\sigma(\mu)} y^\rho$ where $\sigma(\mu) = (\mu_{\sigma(1)},\ldots,\mu_{\sigma(N_0)})$ is the leading term $x^{\mu^R} y^\rho$. Hence the term $x^{\mu^+} y^\rho$ results from applying Sym to this leading term. The factor $\prod_i f_i(\mu^+)!$ is the coefficient of $x^{\mu^+} y^\rho$ so obtained. $\qquad\square$

Lemma 2.

$$X_{(\mu^+,\rho)}(x,y) = \sum_{\sigma \in S_{N_0}} a_{(\sigma(\mu^+),\rho)} E_{(\sigma(\mu^+),\rho)}(x,y) \qquad (21)$$

for some coefficients $a_{(\sigma(\mu^+),\rho)}$. *In fact* $X_{(\mu^+,\rho)}$ *is uniquely determined as the series of the form* (21) *with* $a_{(\mu^+,\rho)} = 1$ *that is symmetric in* x. *Similarly,*

$$Y_{(\mu,\rho^+)}(x,y) = \sum_{\sigma' \in S_{N_1}} \alpha_{(\mu,\sigma'(\rho^+))} E_{(\mu,\sigma'(\rho^+))}(x,y) \qquad (22)$$

for some coefficients $\alpha_{(\mu,\sigma'(\rho^+))}$ *and* $Y_{(\mu,\rho^+)}$ *is uniquely determined as the series of this form with* $\alpha_{(\mu,\rho^+)} = 1$, *which is antisymmetric in* y.

Proof. We will consider only the statements regarding $X_{(\mu^+,\rho)}$; the statements regarding $Y_{(\mu,\rho^+)}$ proceed similarly.

The existence of an expansion of the form (21) has been noted in Ref. 1; it follows from the action of the elementary transposition operator $s_i := s_{i,i+1}$ on the E_η:

$$s_i E_\eta = \begin{cases} \frac{1}{\delta_{i,\eta}} E_\eta + \left(1 - \frac{1}{\delta_{i,\eta}^2}\right) E_{s_i\eta} & \eta_i > \eta_{i+1} \\ E_\eta & \eta_i = \eta_{i+1} \\ \frac{1}{\delta_{i,\eta}} E_\eta + E_{s_i\eta} & \eta_i < \eta_{i+1}, \end{cases} \qquad (23)$$

where $\delta_{i,\eta} := \bar{\eta}_i - \bar{\eta}_{i+1}$. To see that $a_{(\mu^+,\rho)} = 1$, compare coefficients of $x^{\mu^+} y^\rho$ on both sides. By the definition of $X_{(\mu^+,\rho)}$, we see that the coefficient is unity on the left-hand side. On the right-hand side the only term containing the monomial $x^{\mu^+} y^\rho$ is $E_{(\mu^+,\rho)}$, and the coefficient is also unity.

For the uniqueness property, we use the fact that $\{X_{(\mu^+,\rho)}\}$ is a complete set for analytic functions symmetric in $\{x\}$. Also, $\{X_{(\mu^+,\rho)}\}$ form an orthogonal set with regard to the inner product (3). Since $\{E_\eta\}$ also form an orthogonal set with regard to (3), we see that the right-hand side of (21) is orthogonal to all $\{X_{(\kappa^+,\rho)}\}$ for $\kappa^+ \neq \mu^+$. Hence the right-hand side of (21) must indeed equal $X_{(\mu^+,\rho)}$.

Next, we shall show that the coefficients $a_{(\sigma(\mu^+),\rho)}$ and $a_{(\mu,\sigma'(\rho^+))}$ in (21), and (22) can be written in terms of the constants

$$d_\eta := \prod_{(i,j)\in\eta} (\alpha(a(i,j)+1) + l(i,j) + 1)$$

$$d'_\eta := \prod_{(i,j)\in\eta} (\alpha(a(i,j)+1) + l(i,j)), \tag{24}$$

where $a(i,j) = \eta_i - j$ and $l(i,j) = \#\{k > i | j \leq \eta_k \leq \eta_i\} + \#\{k < i | j \leq \eta_k + 1 \leq \eta_i\}$. These constants have the property [18] that for $\eta_i < \eta_{i+1}$

$$\frac{d'_{s_i\eta}}{d'_\eta} = \frac{\delta_{i,\eta}}{\delta_{i,\eta} - 1}. \tag{25}$$

$\square$

Proposition 1. *We have*

$$X_{(\mu^+,\rho)}(x,y) = d'_{(\mu^+,\rho)} \sum_{\sigma \in S_{N_0}} \frac{1}{d'_{(\sigma(\mu^+),\rho)}} E_{(\sigma(\mu^+),\rho)}(x,y) \tag{26}$$

$$Y_{(\mu,\rho^+)}(x,y) = \frac{1}{d_{(\mu,\rho^+)}} \sum_{\sigma' \in S_{N_1}} (-1)^{\ell(\sigma')} d_{(\mu,\sigma'(\rho^+))} E_{(\mu,\sigma'(\rho^+))}(x,y), \tag{27}$$

where $\ell(\sigma')$ denotes the length of the permutation σ'.

Proof. The right-hand side of (26) and (27) is of the form (21) and (22), respectively, with the coefficients of $E_{(\mu^+,\rho)}$ and $E_{(\mu,\rho^+)}$ equal to unity. It therefore suffices to verify that (26) is symmetric in x and that (27) is antisymmetric in y. We consider only (26), as the verification of (27) proceeds similarly.

Now, $X_{(\mu^+,\rho)}(x,y)$ is symmetric in x if and only if it is invariant under the action of s_i for any $i = 1, \ldots, N_0$. For a given i we can write the right-hand side of (26) as

$$d'_{(\mu^+,\rho)} \sum_{\substack{\sigma \in S_{N_0} \\ \mu_{\sigma(i)} \geq \mu_{\sigma(i+1)}}} \chi_{\sigma(i),\sigma(i+1)} \left(\frac{1}{d'_{(\sigma(\mu^+),\rho)}} E_{(\sigma(\mu^+),\rho)} \right.$$
$$\left. + \frac{1}{d'_{(s_i\sigma(\mu^+),\rho)}} E_{(s_i\sigma(\mu^+),\rho)} \right), \quad (28)$$

where

$$\chi_{\sigma(i),\sigma(i+1)} = \begin{cases} \frac{1}{2} & \mu_{\sigma(i)} = \mu_{\sigma(i+1)}; \\ 1 & \text{otherwise.} \end{cases}$$

Now apply s_i to (28) and rewrite the action of s_i using (23). Straightforward manipulation using (25) then shows that (26) is indeed invariant under s_i.
□

Corollary 1.

$$S_{(\mu^+,\rho^+)}(x,y) = d'_{(\mu^+,\rho^R)} \sum_{\sigma} \frac{1}{d'_{(\sigma(\mu^+),\rho^R)} d_{(\sigma(\mu^+),\rho^+)}}$$
$$\times \sum_{\sigma'} (-1)^{\ell(\sigma')} d_{(\sigma(\mu^+),\sigma'(\rho^+))} E_{(\sigma(\mu^+),\sigma'(\rho^+))}(x,y) \quad (29)$$

$$= \frac{1}{d_{(\mu^R,\rho^+)}} \sum_{\sigma'} (-1)^{\ell(\sigma')} d_{(\mu^R,\sigma'(\rho^+))} d'_{(\mu^+,\sigma'(\rho^+))}$$
$$\times \sum_{\sigma} \frac{1}{d'_{(\sigma(\mu^+),\sigma'(\rho^+))}} E_{(\sigma(\mu^+),\sigma'(\rho^+))}(x,y) \quad (30)$$

Proof. To derive (29), apply $c^{-1}_{(\mu,\rho^+)} \text{Asym}^{(y)}$ to (26) with $\rho = \rho^R$ and use (27). Similarly, to derive (30), apply $c^{-1}_{(\mu^+,\rho)} \text{Sym}^{(x)}$ to (27) with $\mu = \mu^R$ and use (26).
□

Another Jack polynomial with prescribed symmetry consisting of two components is

$$S_{(\rho^+,\tau^+)}(y,z) = (-1)^{N_1(N_1-1)/2+N_2(N_2-1)/2} \text{Asym}^{(y)} \text{Asym}^{(z)} E_{(\rho^R,\tau^R)}(y,z),$$

where $y := (y_1,\ldots,y_{N_1})$, $z := (z_1,\ldots,z_{N_2})$. Proceeding as in the derivation of (27), we see that

$$S_{(\rho^+,\tau^+)}(y,z)$$
$$= \frac{1}{d_{(\rho^+,\tau^+)}} \sum_{\sigma,\sigma'} (-1)^{\ell(\sigma)+\ell(\sigma')} d_{(\sigma(\rho^+),\sigma'(\tau^+))} E_{(\sigma(\rho^+),\sigma'(\tau^+))}(y,z) \quad (31)$$

2.2   Normalization

Let

$$S_\eta(z) := \frac{1}{c_\eta} \mathcal{O}(E_{\eta^*}(z)),$$

where $\mathcal{O}$ denotes some symmetry operation (Sym/Asym in certain variables), c_η is the normalization so that the coefficient of the leading term is unity, and η^* denotes a certain permutation of the composition η. Suppose the coefficient of E_{η^*} in the expansion

$$S_\eta(z) = \sum_{\nu : \nu^+ = \eta^+} u_\nu E_\nu(z)$$

is known. Then, owing to the orthogonality of $\{E_\eta\}$ with regard to (3),

$$\langle \mathcal{O}(E_{\eta^*}(z)), E_{\eta^*}(z) \rangle = c_\eta u_{\eta^*} \langle E_{\eta^*}(z), E_{\eta^*}(z) \rangle.$$

On the other hand, since the weight function in $\langle \cdot, \cdot \rangle$ as defined by (3) is symmetric,

$$\langle \mathcal{O}(E_{\eta^*}), E_{\eta^*} \rangle = \frac{1}{|\mathcal{O}|} \langle \mathcal{O}(E_{\eta^*}), \mathcal{O}(E_{\eta^*}) \rangle$$

where $|\mathcal{O}|$ denotes the order of the symmetry group associated with $\mathcal{O}$. Hence,

$$\langle S_\eta, S_\eta \rangle = |\mathcal{O}| \frac{u_{\eta^*}}{c_\eta} \langle E_{\eta^*}, E_{\eta^*} \rangle. \tag{32}$$

Applying (32) to (29) or (30) gives

$$\langle S_{(\mu^+,\rho^+)}, S_{(\mu^+,\rho^+)} \rangle = \frac{N_0! N_1!}{\prod_i f_i(\mu^+)!} \frac{d'_{(\mu^+,\rho^R)} d_{(\mu^R,\rho^R)}}{d'_{(\mu^R,\rho^R)} d_{(\mu^R,\rho^+)}}$$

$$\times \langle E_{(\mu^R,\rho^R)}, E_{(\mu^R,\rho^R)} \rangle, \tag{33}$$

while applying (32) to (31) shows that

$$\langle S_{(\rho^+,\tau^+)}, S_{(\rho^+,\tau^+)} \rangle = N_1! N_2! \frac{d_{(\rho^R,\tau^R)}}{d_{(\rho^+,\tau^+)}} \langle E_{(\rho^+,\tau^+)}, E_{(\rho^+,\tau^+)} \rangle. \tag{34}$$

The quantities d_η, d'_η and $\langle E_\eta, E_\eta \rangle$ appearing in these expressions can be evaluated using (24) and (9).

For example, in the case $\mu^+ = 0$, $\rho^+ = (N_1 - 1, \dots, 1, 0)$ we have

$$S_{(\mu^+,\rho^+)}(x,y) = \prod_{1 \le i < j \le N_1} (y_i - y_j),$$

and the formula (33) can be simplified to give

$$\langle S_{(\mu^+,\rho^+)}, S_{(\mu^+,\rho^+)} \rangle = \frac{\Gamma(kN_0 + (k+1)N_1 + 1)\Gamma(N_1 + 1)}{\Gamma(k+1)^{N_0}\Gamma(k+2)^{N_1}(1 + kN_0/(k+1))_{N_1}}, \quad (35)$$

where $(a)_n$ denotes the Pochhammer symbol. This reproduces a result of Bressoud and Goulden [5], obtained via combinatorial methods.

We conclude this section with a remark concerning the rational case. We know [9] that the operator Δ as defined in (14) is S_N invariant and hence commutes with the operators Sym and Asym. The simple relationship between the nonsymmetric eigenfunctions of the periodic and rational models, as expressed in (17), tells us that the preceding results concerning the expansion of the eigenfunctions with prescribed symmetry in terms of nonsymmetric Jack polynomials (Proposition 1 and Corollary 1 have rational counterparts with the polynomials $E_\eta(z)$ replaced by the nonsymmetric Hermite polynomials $E_\eta^{(H)}(z)$. In addition, the normalization argument relied solely on the symmetry of the integrand in the integral which defines the inner product (3). The Hermite inner product (18) enjoys the same symmetry property and hence the argument carries over to the rational case immediately.

3 q-Analogs

In this section we shall derive the q-analog of a result contained in Ref. 1, which basically says that the eigenfunction of prescribed symmetry (19), where $\rho^+ = (N_1 - 1, \ldots, 1, 0)$ and $\mu_1^+ \le N_1 - 1$ can be expressed as the product of a Vandermonde product of differences in the variables y and the symmetric Jack polynomial $P_{\mu^+}^{(\alpha+1)}(x)$ (note the shift in parameter). A similar phenomenon occurs when considering the eigenfunctions of prescribed symmetry of the CSM models associated with other root systems.

Let us introduce the necessary ingredients in the q-case. Given two polynomials $f(z; q, t)$, $g(z; q, t)$, let

$$\langle f, g \rangle_{q,t} = \mathrm{C.T.}\big(f(z; q, t)g(z^{-1}; q^{-1}, t^{-1})\Delta(z; q, t)\big), \quad (36)$$

where

$$\Delta(z; q, t) = \prod_{\substack{i,j=1 \\ i<j}}^{N} \left(\frac{z_i}{z_j}; q\right)_k \left(\frac{qz_j}{z_i}; q\right)_k.$$

Here, as usual, $(a; q)_n = (1 - a)(1 - aq) \cdots (1 - aq^{n-1})$, and we consider only the case where $t = q^k$, with k a positive integer.

Furthermore, let τ_i be the q-shift operator in the variable x_i, so that

$$(\tau_i f)(x_1, \ldots, x_i, \ldots, x_N) = f(x_1, \ldots, qx_i, \ldots, x_N),$$

and define the operators

$$T_i = t + \frac{tx_i - x_{i+1}}{x_i - x_{i+1}}(s_i - 1), \quad i = 1, \ldots, N-1 \tag{37}$$

$$\omega := s_{n-1} \cdots s_2 s_1 T_1 = s_{n-1} \cdots s_i T_i s_{i-1} \cdots s_1. \tag{38}$$

These operators satisfy the relations

$$(T_i - t)(T_i + 1) = 0 \tag{39}$$

$$T_i T_{i+1} T_i = T_{i+1} T_i T_{i+1} \tag{40}$$

$$T_i T_j = T_j T_i \quad |i - j| \geq 2 \tag{41}$$

$$\omega T_i = T_{i-1} \omega. \tag{42}$$

For future reference, we note the action of $T_i^{-1} = t^{-1}(1 - t + T_i)$, $1 \leq i \leq N - 1$ on the monomial $x_i^a x_{i+1}^b$:

$$T_i x_i^a x_{i+1}^b = \begin{cases} (t^{-1} - 1)x_i^a x_{i+1}^b + (t^{-1} - 1)x_i^{a-1} x_{i+1}^{b+1} \\ \quad + \cdots + (t^{-1} - 1)x_i^{b+1} x_{i+1}^{a-1} + t^{-1} x_i^b x_{i+1}^a & a > b \\ t^{-1} x_i^a x_{i+1}^a & a = b \\ (1 - t^{-1})x_i^{a+1} x_{i+1}^{b-1} + \cdots + (1 - t^{-1})x_i^{b-1} x_{i+1}^{a+1} \\ \quad + x_i^b x_{i+1}^a & a < b. \end{cases} \tag{43}$$

With these ingredients, we can define [6] the following degree-preserving operators (q-analogs of the operators (7)):

$$Y_i = Y_i^{q,t} = t^{-n+i} T_i \cdots T_{n-1} \omega T_1^{-1} \cdots T_{i-1}^{-1}, \quad 1 \leq i \leq N, \tag{44}$$

which commute amongst themselves: $[Y_i, Y_j] = 0$, $1 \leq i, j \leq N$. Nonsymmetric Macdonald polynomials $E_\eta(z; q, t)$ are defined as being the unique eigenfunctions of the (lower-triangular) operators Y_i, having an expansion of the form (6). Indeed, we have that $Y_i E_\eta = t^{\bar{\eta}_i} E_\eta$, where $\bar{\eta}_i$ is given as in (8). In addition, it follows by the usual argument (utilizing $Y_i^\dagger = Y_i^{-1}$) that the polynomials $E_\eta(z; q, t)$ are orthogonal with respect to the inner product (36).

As in the Jack case, we can consider a "transformed exchange Hamiltonian"

$$\mathcal{H} = \sum_{i=1}^N Y_i^{q,t}, \tag{45}$$

which reduces to the Macdonald operator D_N^1 (see Ref. 12, p. 315, Eq. (3.4)) when acting on symmetric functions The nonsymmetric Macdonald polynomials $\{E_\eta(z; q, t) \mid \eta \text{ a derangement of } \lambda\}$ are a set of degenerate eigenfunctions of this operator, each with eigenvalue $\sum_i t^{N-i} q^{\lambda_i}$. We can therefore consider eigenfunctions that are linear combinations of the $E_\eta(z; q, t)$

having a prescribed symmetry. The following operators are (q, t)-analogs of the symmetrization/antisymmetrization operators that appeared in the previous section [13, 20]:

$$U^+ = \sum_{\sigma} T_{\sigma}, \quad U^- = \sum_{\sigma} (-t)^{-\ell(\sigma)} T_{\sigma},$$

where $T_{\sigma} = T_{i_1} \cdots T_{i_p}$ for a permutation σ with reduced word decomposition $\sigma = s_{i_1} \cdots s_{i_p}$. They have the properties (see Ref. 20 for proofs)

$$T_i^{\pm 1} U^+ = t^{\pm 1} U^+, \quad T_i^{\pm 1} U^- = -U^-$$
$$U^+ T_i^{\pm 1} = t^{\pm 1} U^+, \quad U^- T_i^{\pm 1} = -U^-, \quad U^- s_i = -U^-.$$

In particular for $\delta = (N - 1, \ldots, 2, 1, 0)$, we have

$$t^{N(N-1)/2} U^- z^{\delta} = \prod_{\substack{i,j=1 \\ i<j}}^{N} (t z_i - z_j) =: \Delta_t(z) \tag{46}$$

and

$$U^- z^{\gamma} = 0 \quad \text{if there exists } i, j \text{ such that } \gamma_i = \gamma_j. \tag{47}$$

Let us now consider the function

$$S_{(\delta,\lambda)}(z, w; q, t) = \frac{1}{c_{(\delta,\lambda)}(q, t)} U_z^+ U_w^- E_{(\delta^R, \lambda^R)}(z, w), \tag{48}$$

where $c_{(\delta,\lambda)}(q, t)$ is defined so that the coefficient of $z^{\delta} w^{\lambda}$ is $t^{N_z(N_z - 1)/2}$, which is (q, t)-antisymmetric in the variable $\{z_1, \ldots, z_{N_z}\}$, and symmetric in the variables $\{w_1, \ldots, w_{N_w}\}$. The main result of this section is—

Proposition 2. *If* $\lambda_1 \leq N_z - 1$, *then*

$$S_{(\delta,\lambda)}(z, w; q, t) = \Delta_t(z) P_\lambda(w; qt, t).$$

By its very definition, the polynomial (48) is an eigenfunction of the operators

$$\mathcal{H}_1 = \sum_{i=1}^{N_z} Y_i^{q,t}, \quad \mathcal{H}_2 = \sum_{i=N_z+1}^{N} Y_i^{q,t} \tag{49}$$

separately. Here we consider the operators Y_i as defined on the variables $\{z_i\}_{i=1}^{N}$ where $z_{N_z+i} := w_i$ for $1 \leq i \leq N_w$, and $N = N_z + N_w$. By the definition of the partial order $<$, and the nature of the composition δ^R, the nonsymmetric Macdonald polynomials $E_{(\delta^R, \lambda^R)}(z, w; q, t)$ have an expansion of the form

$$E_{(\delta^R, \lambda^R)}(z, w; q, t) = z^{\delta^R} w^{\lambda^R} + \sum_{\eta < \lambda^R} a_{\lambda^R, \eta} z^{\delta^R} w^{\eta} + \sum_{\gamma, \xi} b_{\gamma, \xi} z^{\gamma} w^{\xi}, \tag{50}$$

where the last sum is over all $\gamma < \delta^R$ (which means in particular that there is a pair i, j such that $\gamma_i = \gamma_j$). It follows from (47) that after application of U_z^- to (50) the terms in the latter sum vanish, and hence

$$U_z^- E_{(\delta^R, \lambda^R)}(z, w; q, t) = (U^- z^{\delta^R})\left(w^{\lambda^R} + \sum_{\eta < \lambda^R} a_{\lambda^R, \eta} w^\eta\right),$$

and so

$$U_z^- U_w^+ E_{(\delta^R, \lambda^R)}(z, w; q, t) = t^{-N_z(N_z-1)/2} \Delta_t(z) g(w) \qquad (51)$$

where $g(w)$ is a symmetric polynomial in w with leading order term $m_\lambda(w)$. To complete the proof of Proposition 2 it suffices to show that $g(w)$ is an eigenfunction of

$$\mathcal{H}_2' = \sum_{i=1}^{N_w} Y_i^{qt, t}, \qquad (52)$$

where $Y_i^{qt, t}$ act on the w variables only. This is aided by the following:

Lemma 3. *Provided $\eta_1 \le N_z - 1$, we have*

$$\omega_q^{(z, w)} T_1^{-1} T_2^{-1} \cdots T_{N_z}^{-1} \Delta_t(z) w^\eta = t^{-N_z} \Delta_t(z) \omega_{qt}^{(w)} w^\eta.$$

Proof. For $1 \le i \le N_z - 1$ we have that

$$T_1^{-1} T_2^{-1} \cdots T_{N_z}^{-1} T_i = T_{i+1} T_1^{-1} T_2^{-1} \cdots T_{N_z}^{-1},$$

so that

$$T_1^{-1} T_2^{-1} \cdots T_{N_z}^{-1} U_{z_1, \ldots, z_{N_z}}^- = U_{z_2, \ldots, z_{N_z}, w_1}^- T_1^{-1} T_2^{-1} \cdots T_{N_z}^{-1}.$$

Thus,

$$T_1^{-1} T_2^{-1} \cdots T_{N_z}^{-1} \Delta_t(z_1, \ldots, z_{N_z}) w_1^{\eta_1} = U_{z_2, \ldots, z_{N_z}, w_1}^- T_1^{-1} T_2^{-1} \cdots T_{N_z}^{-1} z^\delta w_1^{\eta_1}.$$

Now, from the action of T_i^{-1} in (43) we have that

$$T_1^{-1} T_2^{-1} \cdots T_{N_z}^{-1} z^\delta w_1^{\eta_1} = t^{-N_z + \eta_1} (z_2 \cdots z_{N_z} w_1)^\delta (z_1)^{\eta_1} + \cdots,$$

where the additional terms $\cdots$ will have terms $(z_2 \ldots z_{N_z} w_1)^\gamma$ with two parts of γ equal *provided* $\eta_1 \le N_z - 1$ (which will hence vanish under $U_{z_2, \ldots, z_{N_z}, w_1}^-$). Thus

$$T_1^{-1} T_2^{-1} \ldots T_{N_z}^{-1} \Delta_t(z_1, \ldots, z_{N_z}) w_1^{\eta_1} = t^{-N_z} \Delta_t(z_2, \ldots, z_{N_z}, w_1)(t z_1)^{\eta_1}$$

and the result now follows. □

Using the above lemma it follows that when acting on functions of the form $\Delta_t(z)g(w)$, where $g(w)$ is symmetric,

$$(Y_{N_z+i}^{q,t})^{(z)} = t^{-N_z}(Y_i^{qt,t})^{(w)} \quad 1 \le i \le N_w, \tag{53}$$

where the operators $(Y_i)^{(z)}$ act on all variables $\{z_i\}_{i=1}^N$ and the operators $(Y_i)^{(w)}$ act on the variables $\{w_i\}_{i=1}^{N_w}$. Since the function (51) is an eigenfunction of $\mathcal{H}_2$ (recall (49)), it follows that the function $g(w)$ is an eigenfunction of (52). It follows from the leading order behavior of g that $g(w) = P_\lambda(w; qt, t)$, thus concluding the proof of Proposition 2.

Although we won't pursue the matter any further at this time, we remark that the theory of the previous section can be generalized to the Macdonald case, and in particular the normalization of $S_{(\delta,\lambda)}(z, w; q, t)$ with respect to the inner product (36) can be calculated explicitly.

Acknowledgments: T.H.B. and P.J.F. would like to thank Trevor Welsh for useful discussions and making available Ref. 20. The work of T.H.B. and P.J.F. is supported by the Australian Research Council. C.F.D. was partially supported by the National Science Foundation, grant DMS-9401429.

4   References

1. T. H. Baker and P. J. Forrester, *The Calogero–Sutherland model and generalized classical polynomials*, Commun. Math. Phys. **188** (1997), No. 1, 175–216, solv-int/9609010.

2. _____, *Nonsymmetric Jack polynomials and integral kernels*, Duke Math. J. **95** (1998), No. 1, 1–50, q-alg/9612003.

3. R. J. Beerends and E. M. Opdam, *Certain hypergeometric series related to the root system BC*, Trans. Amer. Math. Soc. **339** (1993), No. 2, 581–609.

4. D. Bernard, M. Gaudin, F. D. M. Haldane, and V. Pasquier, *Yang–Baxter equation in long-range interacting systems*, J. Phys. A **26** (1993), No. 20, 5219–5236, hep-th/9301084.

5. D. M. Bressoud and I. P. Goulden, *The generalized plasma in one dimension: evaluation of a partition function*, Commun. Math. Phys. **110** (1987), No. 2, 287–291.

6. I. Cherednik, *A unification of the Knizhnik–Zamolodchikov and Dunkl operators via affine Hecke algebras*, Invent. Math. **106** (1991), No. 2, 411–432.

7. J. F. van Diejen, *Commuting difference operators with polynomial eigenfunctions*, Compositio Math. **95** (1995), No. 2, 183–233.

8. _____, *Properties of some families of hypergeometric orthogonal polynomials in several variables*, Trans. Amer. Math. Soc. **351** (1999), No. 1, 233–270, q-alg/9604004.

9. C. F. Dunkl, *Differential-difference operators associated to reflection groups*, Trans. Amer. Math. Soc. **311** (1989), No. 1, 167–183.

10. P. J. Forrester, *Selberg correlation integrals and the $1/r^2$ quantum many-body system*, Nucl. Phys. B **388** (1992), No. 3, 671–699.

11. M. Lassalle, *Polynômes de Hermite généralisés*, C. R. Acad. Sci. A **313** (1991), No. 9, 579–582.

12. I. G. Macdonald, *Symmetric Functions and Hall Polynomials*, 2nd ed., Oxford Math. Monogr., Oxford Univ. Press, New York, 1995.

13. _____, *Affine Hecke algebras and orthogonal polynomials*, Séminaire Bourbaki. Vol. 1994/95, Astérisque, Vol. 237, Soc. Math. France, Paris, 1996, pp. 189–207.

14. M. A. Olshanetsky and A. M. Perelomov, *Quantum integrable systems related to Lie algebras*, Phys. Rep. **94** (1983), No. 6, 313–404.

15. E. M. Opdam, *Harmonic analysis for certain representations of graded Hecke algebra*, Acta Math. **175** (1995), No. 1, 75–121.

16. A. P. Polychronakos, *Exchange operator formalism for integrable systems of particles*, Phys. Rev. Lett. **69** (1992), No. 5, 703–705.

17. M. Rösler, *Generalized Hermite polynomials and the heat equation for Dunkl operators*, Commun. Math. Phys. **192** (1998), No. 3, 519–542, q-alg/9703006.

18. S. Sahi, *A new scalar product for nonsymmetric Jack polynomials*, Internat. Math. Res. Notices (1996), No. 20, 997–1004.

19. H. Ujino and M. Wadati, *Algebraic construction of the eigenstates for the second conserved operator of the quantum Calogero model*, J. Phys. Soc. Japan **65** (1996), No. 3, 653–656.

20. T. A. Welsh, *Representations of Hecke algebras and Young tableaux*, Bayreuth. Math. Schr. (to appear).

4

The Theory of Lacunas and Quantum Integrable Systems

Yuri Yu. Berest

ABSTRACT We show how recent developments in the theory of (quantum) integrable systems can be applied to the study of lacunas of hyperbolic equations, one of the classical problems in analysis of linear differential operators. This report is based mostly on results of our recent work [3].

1 Introduction

Linear wave propagation in continuous media with finite velocities in all directions is governed by hyperbolic differential operators. The wave field induced by an instantaneous point source may happen to vanish in certain regions inside the propagation domain. These regions, called (interior) *lacunas*, are transitory in the sense that both before and after them the perturbation of the medium is nonzero. Lacunas are traveling zones of absolute quiet.[1]

Mathematically, the wave movement is described by a distribution $\Phi = \Phi(\mathcal{L}, x, x_0)$, the Green function (fundamental solution) of a given hyperbolic operator $\mathcal{L}: \mathcal{L}[\Phi] = \delta(x - x_0)$. It turns out that Φ is smooth (or even holomorphic) outside a system of possibly criss-crossing wave fronts $W = W(P, x_0)$ and vanishes outside the cone of propagation $K = K(P, x_0)$, a conical region bounded by the fastest fronts. The existence of interior lacunas amounts to the fact that Φ is identically zero in some open components of $K \setminus W$. In general, this is a rare analytic property related inherently to (the parity of) the space dimension and to the structure of $\mathcal{L}$'s symmetry group. In fact, the "lacunary" operators (i.e., those ones whose fundamental solutions admit nontrivial lacunas) constitute an exceptional subclass among all hyperbolic operators. The question of their characterization is nowadays a classical problem in analysis of linear partial differential operators, usually referred to as the *problem of lacunas*.

---

[1]The simplest and the most important example of such a phenomenon is related to the famous *Huygens' principle*, according to which the nature of observed "clean cut" light signals can be explained by perfect cancellation of secondary waves, namely, those that occur off the leading wave front.

For scalar operators with constant coefficients, this problem has been investigated in detail, basically owing to the fundamental work of Petrovsky [14] and Atiyah et al. [1, 2]. Underlying this work is a deep and beautiful link between the theory of lacunas and topology of complex projective algebraic surfaces. This link allows one to get a lot of interesting information on lacunas, essentially by the use of topological and algebro-geometrical tools.

In spite of considerable efforts, the study of lacunas remains much less complete in the case of *variable* coefficients, particularly for higher-order hyperbolic operators $(p > 2)$. The purpose of this report is to discuss some new results in this direction [3]. The key idea of our work is to extend the classical Petrovsky–Atiyah–Bott–Gårding theory of lacunas to certain *commutative* rings of partial differential operators *other than* the algebra of differential operators with constant coefficients. The main class of examples developed systematically in Ref. 3 is provided by algebras of "quantum integrals" of generalized Calogero–Moser models associated with finite reflection groups [13]. Our attempt here is to illuminate this connection and to indicate some further developments. For more details and proofs the reader is referred to Ref. 3.

2   Hyperbolic Operators on Root Systems

2.1   Hyperbolic Polynomials and Associated Riesz Integrals

We recall some basic facts from the theory of hyperbolic equations with constant coefficients [6] (see also Refs. 1 and 10).

Let $P(\zeta) = P(\zeta_1, \ldots, \zeta_n)$ be a complex homogeneous polynomial of degree p, and let $P(D)$, $D = -i\partial/\partial x$, be the corresponding constant coefficient differential operator. The polynomial $P(\zeta)$ and the operator $P(D)$ are said to be *hyperbolic* [6] if $P(D)$ has a fundamental solution $\Phi(P, x) \in \mathcal{D}'(\mathbb{R}^n)$ (necessarily unique) whose support $\operatorname{supp} \Phi(P, \cdot)$ is contained in a proper cone K with its vertex at the origin. If $\mathring{K} := K \setminus \{0\}$ lies in a half-space, say, $(x, \vartheta) > 0$, the hyperbolicity is equivalent to the following algebraic condition: $P(\xi + t\vartheta) = 0$ has only *real* roots $t = t_j(\xi, \vartheta)$, $j = 1, \ldots, p$, for every $\xi \in \mathbb{R}^n$. We write $\operatorname{Hyp}(\vartheta, p)$ for the set of all such P's, and denote by $\operatorname{Hyp}^\circ(\vartheta, p) \subset \operatorname{Hyp}(\vartheta, p)$ its proper subset consisting of *strongly hyperbolic* polynomials, i.e., those whose characteristic roots $t_j(\xi, \vartheta)$ are all pairwise distinct (unless ξ is proportional to ϑ). Put $\operatorname{Hyp}(\vartheta) := \bigcup_{p \geq 0} \operatorname{Hyp}(\vartheta, p)$. Clearly, $\operatorname{Hyp}(\vartheta) = \operatorname{Hyp}(-\vartheta)$, and every $P(\zeta) \in \operatorname{Hyp}(\vartheta)$ is actually real (maybe, up to a complex factor). Algebraically, $\operatorname{Hyp}(\vartheta)$ is a multiplicative semigroup in the polynomial ring $\mathbb{R}[\zeta_1, \ldots, \zeta_n]$, while, topologically, $\operatorname{Hyp}(\vartheta, p) \subset \overline{\operatorname{Hyp}^\circ(\vartheta, p)}$, $\operatorname{Hyp}^\circ(\vartheta, p)$ being an open set in the space of all real homogeneous polynomials of degree p.

Let $\Xi \subset \mathbb{R}^n$ be an algebraic hypersurface $P(\xi) = 0$ of real zeros of P.

It consists of p possibly intersecting conical sheets, and $\vartheta \notin \Xi$. We write $\Gamma(P, \vartheta)$ for the component of $\mathbb{R}^n \setminus \Xi$ that contains ϑ, and call it the (global) *hyperbolicity cone* of P. It is a basic property of a hyperbolic polynomial $P \in \text{Hyp}(\vartheta)$ that $\Gamma(P, \vartheta)$ is a convex open cone, and $P \in \text{Hyp}(\eta)$ for any $\eta \in \Gamma(P, \vartheta)$. The cone dual to $\Gamma(P, \vartheta)$ with the vertex at x_0: $K(P, \vartheta, x_0) := \{x \in \mathbb{R}^n \mid (\eta, x - x_0) \geq 0, \forall \eta \in \Gamma(P, \vartheta)\}$ is a closed convex affine set in $\mathbb{R}^n$ referred to as the *propagation cone*. If $P(\zeta) \in \text{Hyp}(\vartheta)$, then $P_\xi(\zeta) \in \text{Hyp}(\vartheta)$, where $P_\xi(\zeta)$ is a localization of P at the point $\xi \in \mathbb{R}^n$ (i.e., P_ξ is the lowest-order term of the polynomial $t \mapsto P(\xi + t\zeta)$). We define the *local hyperbolicity* and the *local propagation* cones of P at ξ by setting $\Gamma_\xi(P, \vartheta) := \Gamma(P_\xi, \vartheta)$ and $K_\xi(P, \vartheta, x_0) := K(P_\xi, \vartheta, x_0)$, respectively. The *wave front surface* W of a hyperbolic polynomial $P \in \text{Hyp}(\vartheta)$ is generated by the union of all local propagation cones $W(P, \vartheta, x_0) := \bigcup K_\xi(P, \vartheta, x_0)$ with $\xi \notin \Lambda(P)$. Here, we denote by $\Lambda(P)$ a real linearity of P, that is, the maximal linear subspace of $\mathbb{R}^n$ such that P can be restricted to a polynomial on the quotient $\mathbb{R}^n / \Lambda(P)$. When P is complete, that is, $\Lambda(P) = \{0\}$, $K(P, \vartheta, x_0)$ has a nonempty interior $K^\circ(P, \vartheta, x_0)$, and $W(P, \vartheta, x_0)$ is a closed semialgebraic part of K containing its boundary, $\partial K \subseteq W$. As we will see, the cones K and W have an important analytical meaning; namely, $K(P, \vartheta, x_0)$ coincides with the convex hull of the support of the fundamental solution $\Phi(P, x, x_0)$ of $P(D)$, while $W(P, \vartheta, x_0)$ is a carrier of its singularities.

Given $P \in \text{Hyp}(\vartheta)$, we construct an analytic family of homogeneous distributions (called *Riesz integrals*) $\Phi^\pm(P) : \mathbb{C} \to \mathcal{D}'(\mathbb{R}^n)$, letting

$$\lambda \mapsto \Phi_\lambda^\pm(P, x) := (2\pi)^{-n} \int_{\mathbb{R}^n} P(\zeta)^{-\lambda} e^{i(x, \zeta)} \, d\xi, \qquad (1)$$

where $\zeta := \xi + i\eta \in T_\pm(P, \vartheta)$ and $T_\pm(P, \vartheta) := \mathbb{R}^n \mp i\Gamma(P, \vartheta)$ is a tube domain in $\mathbb{C}^n$ spanned over the (global) hyperbolicity cone of P. Since $|P(\zeta)|$ has no zeroes and $\arg P(\zeta)$ is continuous and single-valued on $T_\pm(P, \vartheta)$ (once $\arg P(\pm\vartheta)$ is fixed), the definition of the complex power $P(\zeta)^{-\lambda}$ offers no difficulties, and the inverse Fourier–Laplace integral in (1) is absolutely convergent (in the distribution sense) for all $\lambda \in \mathbb{C}$. By Cauchy's theorem, this integral is independent of $\text{Im} \, \zeta \in \pm\Gamma(P, \vartheta)$.

Lemma 1 ([1, 2, 6]). *Let* $P \in \text{Hyp}(\vartheta, p)$. *Then,* $\Phi_\lambda^\pm(P)$ *is entire analytic in* λ *and the following properties hold:*

$$\Phi_\lambda^\pm(P, \kappa x) = \kappa^{\lambda p - n} \Phi_\lambda^\pm(P, x), \quad \kappa \in \mathbb{R}_+, \qquad (2)$$

$$\text{sing supp} \, \Phi_\lambda^\pm(P, x) \subseteq \pm W(P, \vartheta), \qquad (3)$$

$$\text{supp} \, \Phi_\lambda^\pm(P, x) \subseteq \pm K(P, \vartheta), \qquad (4)$$

$$P(D)\Phi_\lambda^\pm(P, x) = \Phi_{\lambda-1}^\pm(P, x), \qquad (5)$$

$$\Phi_0^\pm(P, x) = \delta(x). \qquad (6)$$

It follows from (4)–(6) that $\Phi^\pm(P, x, x_0) := \Phi_1^\pm(P, x - x_0)$ is a *fundamental solution* of $P(D)$ with support in $\pm K(P, \vartheta, x_0)$. Such a solution is clearly unique, and we call it the *principal fundamental solution* of $P(D)$.

2.2 Invariant Hyperbolic Operators

Let $(G, \mathfrak{R}, m)$ be a triple consisting of a finite group $G \subset O(n)$ generated by reflections in $\mathbb{R}^n$, a related root system $\mathfrak{R} \subset \mathbb{R}^n$, and a G-invariant real-valued multiplicity function $m \colon \mathfrak{R} \to \mathbb{R}$. Associated to $(G, \mathfrak{R}, m)$ there is a commutative ring $\mathfrak{D}^{G,m}$ of G-invariant partial differential operators, the algebra of "quantum integrals" of a generalized Calogero–Moser model [13]. It is known that $\mathfrak{D}^{G,m}$ is isomorphic to the ring $\mathbb{R}[\zeta]^G$ of all invariant polynomials of the group G, and the corresponding isomorphism (Harish-Chandra) $\mathbf{D} \colon \mathbb{R}[\zeta]^G \to \mathfrak{D}^{G,m}$ can be given explicitly, for example, in terms of Dunkl operators [5, 8].

Let $\vartheta \in \mathbb{R}^n$ be a real nonzero vector. Put

$$\mathrm{Hyp}(\vartheta)^G := \mathrm{Hyp}(\vartheta) \cap \mathbb{R}[\zeta]^G. \tag{7}$$

Here, we assume ϑ to be chosen in such a way that $\mathrm{Hyp}(\vartheta)^G$ contains some polynomials of positive degrees (not only constants). A few concrete examples of such polynomials are given below.

1. The set $\mathrm{Hyp}(\vartheta, 1)^G$ consists of all linear forms $v_*(\zeta) := (v, \zeta)$ such that $(v, \vartheta) \neq 0$ and $(v, \alpha) = 0$ for all $\alpha \in \mathfrak{R}(G)$. If $\mathrm{rk}\, G = n$, $\mathrm{Hyp}(\vartheta, 1)^G$ is empty.

2. A quadratic polynomial $P \in \mathbb{R}[\zeta]$ belongs to $\mathrm{Hyp}(\vartheta, 2)$ if and only if the quadric P has a Lorentzian signature and ϑ is a timelike vector relative to it. If P is complete, then $P \in \mathrm{Hyp}^\circ(\vartheta, 2)$, and it can be reduced to the canonical form $P(\zeta) := -(\zeta_1)^2 + (\zeta_2)^2 + \cdots + (\zeta_n)^2$ with $\vartheta := (\pm 1, 0, \ldots, 0)$ as a hyperbolicity direction. Clearly, $P \in \mathrm{Hyp}(\vartheta, 2)$ if and only if $G \subset O(n) \cap O(n-1, 1)$. For example, if $\mathrm{rank}\, G \leq n - 1$, we may choose ϑ to be G-invariant vector.

3. Let $G \cong S_n$ be a symmetric group with a root system $\mathfrak{R}$ canonically realized in $\mathbb{R}^n$. Fix a vector $\vartheta := (1, 1, \ldots, 1)$. Then, a semigroup $\mathrm{Hyp}(\vartheta)^{S_n}$ is generated by the *elementary* symmetric functions $s_k(\zeta) := \sum \zeta_1 \zeta_2 \cdots \zeta_k$, which all are hyperbolic polynomials in direction ϑ.

4. Let G be any finite reflection group, and let $d(\zeta) := \prod_{\alpha \in \mathfrak{R}} (\alpha, \zeta)$ be its *discriminant* polynomial. Then, d is both G-invariant and hyperbolic, its hyperbolicity cone being a Weyl chamber $\mathrm{Ch}(G)$ of the group G, that is, $d \in \mathrm{Hyp}(\vartheta, r)^G$ with $\vartheta \in \mathrm{Ch}(G)$ and $r = |\mathfrak{R}|$.

The object of our study is a class of differential operators defined by

$$\mathfrak{D}^{G,m}_{\mathrm{Hyp}} := \{ \mathcal{L} = \mathbf{D}(P) \mid P \in \mathrm{Hyp}(\vartheta)^G \}. \tag{8}$$

Unless $P \in \mathrm{Hyp}^\circ(\vartheta)$, the fact that $\mathbf{D}(P)$ is hyperbolic *is not* immediate. In case of multiple characteristics, the algebraic hyperbolicity of the symbol of $\mathcal{L}$ does not guarantee the existence of its fundamental solution with a conic support. Some additional constraints, either to singular points of the hyperbolic hypersurface $\Xi(P)$ or to the lower terms added, may have to be imposed (cf., e.g., Refs. 9–12). For our purposes the following result will be sufficient.

Lemma 2 ([3]). *Let* $m_\alpha \in \mathbb{Z}$ *for all* $\alpha \in \mathfrak{R}$. *Then,* $\mathfrak{D}_{\mathrm{Hyp}}^{G,m}$ *is a multiplicative semigroup of* hyperbolic *differential operators isomorphic to* $\mathrm{Hyp}(\vartheta)^G$.

2.3 The Representation Theorem

Under the assumption of Lemma 2, we construct an explicit representation for the principal fundamental solution $\Phi(\mathcal{L}, x, x_0)$ of a hyperbolic operator $\mathcal{L} \in \mathfrak{D}_{\mathrm{Hyp}}^{G,m}$ in terms of Riesz integrals associated to its symbol.

Let $G \in \mathrm{O}(n)$ be a finite reflection group in $\mathbb{R}^n$ with a root system $\mathfrak{R}$, and let $\pi(x) := \prod_{\alpha \in \mathfrak{R}_+} (\alpha, x) \in \mathbb{R}[x]$ be its fundamental alternating polynomial. We write $V_{\mathrm{reg}} := \mathbb{R}^n \setminus \pi^{-1}(0)$ for the n-dimensional real affine variety obtained from $\mathbb{R}^n$ by removing the reflection hyperplanes of G. The ring of differential operators $\mathfrak{D} := \mathfrak{D}(V_{\mathrm{reg}})$ with regular coefficients on V_{reg} can be identified with the localization of the Weyl algebra $W_n(\mathbb{R}) := \mathbb{R}\langle x, \partial/\partial x\rangle$ at $\pi = \pi(x)$, that is $\mathfrak{D} \cong \mathbb{R}[x, \pi^{-1}] \otimes_{\mathbb{R}[x]} W_n(\mathbb{R})$. It turns out that $\mathfrak{D}^{G,m} \subset \mathfrak{D}$, and each $\mathcal{L} \in \mathfrak{D}^{G,m}$ has a constant G-invariant principal symbol.

Notation. Let $\mathrm{ad}\colon \mathfrak{D} \times \mathfrak{D} \to \mathrm{End}_{\mathbb{R}}(\mathfrak{D})$ denote a bilinear map that associates to a pair of operators $\mathcal{L}$, $\mathcal{L}_0 \in \mathfrak{D}$ a $\mathbb{R}$-linear endomorphism $\mathrm{ad}(\mathcal{L}, \mathcal{L}_0)$ on the space $\mathfrak{D}$, such that

$$\mathrm{ad}(\mathcal{L}, \mathcal{L}_0)[Q] := \mathcal{L}Q - Q\mathcal{L}_0, \quad Q \in \mathfrak{D}. \tag{9}$$

Given $k \in \mathbb{Z}_+$, we write $\mathrm{ad}_k(\mathcal{L}, \mathcal{L}_0)\colon \mathfrak{D} \to \mathfrak{D}$ for the kth iteration of (9). Namely, by definition we set

$$\mathrm{ad}_k(\mathcal{L}, \mathcal{L}_0) := \mathrm{ad}(\mathcal{L}, \mathcal{L}_0) \circ \mathrm{ad}_{k-1}(\mathcal{L}, \mathcal{L}_0)$$

with the additional convention $\mathrm{ad}_0(\mathcal{L}, \mathcal{L}_0) \equiv \mathrm{id}$.

Theorem 1 ([3]). *Let* $\mathfrak{m}$ *be a lattice of* $\mathbb{Z}$-*valued* G-*invariant multiplicity functions on* $\mathfrak{R}$, *and let* $\mathfrak{m}_+$ *be its positive part. Then, each differential operator* $\mathcal{L}_m \in \mathfrak{D}_{\mathrm{Hyp}}^{G,m}$ *is hyperbolic on* V_{reg} *for* $m \in \mathfrak{m}_+$, *and its principal fundamental solution is given by the formula*

$$\Phi^{\pm}(\mathcal{L}_m, x, x_0) \\ = \theta_m(x_0)^{-1} \sum_{k=0}^{M} (-1)^k \, \mathrm{ad}_k(\mathcal{L}_m, \mathcal{L}_0)[\theta_m(x)]\Phi_{k+1}^{\pm}(P, x - x_0). \tag{10}$$

Here, $\Phi_\lambda^\pm(P,\cdot)$ are the Riesz distributions (1) associated to the principal part $\mathcal{L}_0 := P(D)$ of $\mathcal{L}_m$, and $\theta_m(x) := \prod_{\alpha \in \mathfrak{R}_+}(\alpha, x)^{m_\alpha}$ is a polynomial of degree $M := \sum_{\alpha \in \mathfrak{R}_+} m_\alpha$ regarded as a multiplication operator in $\mathfrak{D}$.

The proof of this theorem is based on the study of certain algebraic identities which occur in the representation theory of Dunkl operator algebras.

3   Lacunas and Topology of Algebraic Surfaces

Let $\mathcal{L} \in \mathfrak{D}$ be a hyperbolic differential operator with a constant principal symbol $P \in \mathrm{Hyp}(\vartheta)$, and let $x_0 \in V_{\mathrm{reg}}$ be fixed. Consider a fundamental solution $\Phi^\pm(\mathcal{L}, \cdot, x_0) \in \mathcal{D}'(V_{\mathrm{reg}})$ of $\mathcal{L}$ with support in $\pm K(P, \vartheta, x_0)$.

Definition 1. An open connected component $\Omega^\pm(x_0)$ of $\pm K(P, \vartheta, x_0) \setminus \pm W(P, \vartheta, x_0)$ is called a *regular lacuna* of the operator $\mathcal{L}$ at the point x_0 if $\Omega^\pm(x_0) \cap \mathrm{supp}\,\Phi^\pm(\mathcal{L}, \cdot, x_0)$ is empty.

The main concern in the classical theory of lacunas [1, 2, 14] is to study the support structure of the principal fundamental solution $\Phi^\pm(P, \cdot)$ of a hyperbolic operator $\mathcal{L}_0 = P(D)$ with constant coefficients. According to Petrovsky, the basic idea in the theory is to represent the function $\Phi^\pm(P, \cdot)$ locally (*outside* the wave front surface $\pm W(P, \vartheta)$) as an Abelian integral of a rational form (with poles at complex zeroes of P) over properly constructed cycles in the complex projective space $\mathbb{C}P^{n-1}$. The vanishing of the cycle relative to a point $x \in \pm K^\circ(P, \vartheta)$ will then imply the *vanishing* of $\Phi^\pm(P, x)$ in the vicinity of x and, by analyticity, also inside the whole component $\Omega^\pm \subset \pm K(P, \vartheta) \setminus \pm W(P, \vartheta)$ containing this point. This provides an effective link between the theory of lacunas and topology of projective algebraic surfaces in $\mathbb{C}P^{n-1}$.

Theorem 1 allows us to develop a similar approach to study regular lacunas of invariant hyperbolic operators in $\mathfrak{D}_{\mathrm{Hyp}}^{G,m}$. In what follows we briefly outline the key points of this construction.

3.1   Vector Fields and Cycles

Given $P \in \mathrm{Hyp}(\vartheta, p)$ and $x \in \mathbb{R}^n \setminus \{0\}$ we define a family $\mathfrak{V} = \mathfrak{V}(x, P, \vartheta)$ of $\mathcal{C}^\infty$-smooth real vector fields $v \colon \mathbb{R}^n \setminus \{0\} \to \mathbb{R}^n$, $\xi \mapsto v(\xi)$, such that

(i) $v(\xi) \in \Gamma_\xi(P, \vartheta) \cap X$, where X is a real hyperplane dual to x;

(ii) $v(\kappa\xi) = |\kappa| v(\xi)$, $\kappa \in \mathring{\mathbb{R}}$; and

(iii) $P\big(\xi \pm i\epsilon v(\xi)\big) \neq 0$, when $0 < \epsilon \leq 1$.

Unless $x \in \pm W(P, \vartheta)$, this family turns out to be *nonempty*, any two vector fields in it being *homotopic* to each other. Hence, when $x \notin \pm W(P, \vartheta)$, we can replace the constant vector field $\xi \mapsto \mp \operatorname{Im} \zeta$ in the Riesz integral (1) by a smooth field $\xi \mapsto \mp v(\xi) \in \mathfrak{V}(x, P, \vartheta)$ homotopic to it, and then perform a radial integration with the use of homogeneity (ii) of $v(\xi)$. By Cauchy's theorem, this does not alter the integral, since the properties (i) and (iii) guarantee for the exponential in (1) to stay bounded and for $\xi \mp i v(\xi)$ to stay away from the complex zeroes of P. If $\lambda = k \in \mathbb{Z}_+ \setminus \{0\}$ and $x \notin \mp K(P, \vartheta) \cup \pm W(P, \vartheta)$, the result of this procedure reads

$$
\begin{aligned}
&\Phi_k^\pm(P, x) \\
&= \frac{i^{pk-n+1}}{2(2\pi)^{n-1}(pk-n)!} \int_{\gamma(\xi)=1} P(\zeta)^{-k}(x, \xi)^{pk-n} \operatorname{sgn}(x, \xi)\omega(\zeta),
\end{aligned} \quad (11)
$$

for $n \leq pk$, and

$$
\Phi_k^\pm(P, x) = -\frac{(-i)^{n-pk+1}}{(2\pi)^{n-1}} \int_{\gamma(\xi)=1} P(\zeta)^{-k}\delta^{(n-pk-1)}((x, \xi))\omega(\zeta), \quad (12)
$$

for $n > pk$. Here, integration is understood in the distribution sense with a measure given by Kronecker's $(n-1)$-form

$$
\omega(\zeta) := \sum_{k=1}^{n}(-1)^{k-1}\zeta^k d\zeta^1 \wedge \cdots \wedge d\zeta^{k-1} \wedge d\zeta^{k+1} \wedge \cdots \wedge d\zeta^n, \quad (13)
$$

and $\xi \mapsto \gamma(\xi)$ is any smooth real positive function on $\mathbb{R}^n \setminus \{0\}$ absolutely homogeneous of degree 1, for example, $\gamma(\xi) = |\xi|$.

The next step is to associate to $\mathfrak{V}(x, P, \vartheta)$ certain cycles and relative cycles in $\mathbb{C}P^{n-1}$ and to rewrite (11), (12) as rational integrals over them. Let $P \in \operatorname{Hyp}(\vartheta)$, $x \in \mathbb{R}^n \setminus \pm W(P, \vartheta)$ and $v(\xi) \in \mathfrak{V}(x, P, \vartheta)$ be fixed as above. Assume the integration chain $\gamma^\sim := \{\xi \in \mathbb{R}^n \setminus \{0\} \mid \gamma(\xi) = 1\}$ in (11), (12) to be oriented in such a way that the $(n-1)$-form $(x, \xi)\omega(\xi)$ stays positive on $\gamma^\sim$, while the space $\mathbb{R}^n$ being equipped with the standard orientation $d\xi > 0$. Following Refs. 1 and 2, we define a smooth "shift" map $\sigma_\pm(x, v) \colon \mathbb{R}^n \setminus \{0\} \to \mathbb{C}^n$, $\xi \mapsto \xi \mp i v(\xi)$, and consider the image of $\gamma^\sim$ under $\sigma_\pm(x, v)$:

$$
\sigma_\pm(x, v; \gamma^\sim) := \operatorname{Im}_{\sigma_\pm}[\gamma^\sim]. \quad (14)
$$

Let $\Xi_C := \{\zeta \in \mathbb{C}^n \mid P(\zeta) = 0\}$, $X_C := \{\zeta \in \mathbb{C}^n \mid (x, \zeta) = 0\}$ be complexifications of the real surfaces $\Xi(P)$ and X, respectively. Then, $\sigma_\pm(x, v; \gamma^\sim)$ is a relative cycle of the pair $(\mathbb{C}^n \setminus \Xi_C, X_C \setminus (X_C \cap \Xi_C))$ oriented by $\omega(\zeta)(x, \operatorname{Re}\zeta) > 0$. By construction, $\sigma_\pm(x, v; \gamma^\sim)$ is homologous to $\gamma^\sim$ in $\mathbb{C}^n \setminus \Xi_C$ for any $v \in \mathfrak{V}(x, P, \vartheta)$. Since the mapping $\xi \mapsto v(\xi)$ is absolutely homogeneous, it is convenient to project the cycle (14) onto $\mathbb{C}P^{n-1}$:

$$
\sigma_\pm^*(x, v; \gamma^\sim) := \operatorname{Im}_\pi[\sigma_\pm(x, v; \gamma^\sim)] \quad (15)
$$

via canonical identification $\pi: \mathbb{C}^n \setminus \{0\} \to \mathbb{CP}^{n-1}$. Denote by Ξ_C^* and X_C^* the projective images of Ξ_C and X_C in $\mathbb{CP}^{n-1}$. A representative of the relative homology class

$$[\sigma_\pm^*(x)] \in H_{n-1}(\mathbb{CP}^{n-1} \setminus \Xi_C^*, X_C^* \setminus (X_C^* \cap \Xi_C^*); \mathbb{C}) \qquad (16)$$

generated by (15) is called the *relative Petrovsky cycle*. Since $\mathfrak{V}(x, P, \vartheta)$ is a single homotopy class, $[\sigma_\pm^*(x)]$ does not depend on the choice of $v \in \mathfrak{V}$. Moreover, the homology class $[\sigma_\pm^*(x)]$ is also independent of $\gamma^\sim$ and *locally independent* of $x \in V \setminus \pm W(P, \vartheta)$. This justifies the notation (16). Taking the boundary of $[\sigma_\pm^*(x)]$

$$\partial[\sigma_\pm^*(x)] \in H_{n-2}(X_C^* \setminus (X_C^* \cap \Xi_C^*); \mathbb{C}) \qquad (17)$$

gives an (absolute) homology in $X_C^* \setminus (X_C^* \cap \Xi_C^*)$. The *absolute Petrovsky cycle* $\beta_\pm^*(x)$ is then defined as a representative of the class

$$[\beta_\pm^*(x)] := \tfrac{1}{2} \mathfrak{t}_x^* \partial[\sigma_\pm^*(x)] \in H_{n-1}(\mathbb{CP}^{n-1} \setminus (X_C^* \cup \Xi_C^*); \mathbb{C}), \qquad (18)$$

where $\mathfrak{t}_x^*$ stands for the map induced on homologies by the Leray tube operation $\mathfrak{t}_x: X_C^* \setminus (X_C^* \cap \Xi_C^*) \to \mathbb{CP}^{n-1} \setminus (X_C^* \cup \Xi_C^*)$. The latter associates to each point $\zeta \in X_C^* \setminus (X_C^* \cap \Xi_C^*)$ the boundary of a small neighborhood of this point in the real 2-plane in $\mathbb{CP}^{n-1}$ orthogonal to X_C^* at ζ. More precisely, given a compact chain $\sigma \subset X_C \setminus (X_C \cap \Xi_C)$, $\mathfrak{t}_x \sigma$ can be regarded as a product $\{|(x, \zeta)| = \rho\} \times \sigma$ with ρ so small that $\mathfrak{t}_x \sigma \subset \mathbb{C}^n \setminus \Xi_C$, the orientation of $\mathfrak{t}_x \sigma$ being a product of orientation of the $\mathbb{C}$-plane with (x, ζ) as a coordinate and the orientation of σ.

3.2   The Herglotz–Petrovsky–Leray Formulas

With definitions (16) and (18), the integrals (11), (12) can be rewritten in the form

$$\Phi_k^\pm(P, x) = \frac{\pi i^{pk-n+1}}{(2\pi)^n (pk-n)!} \int_{\sigma_\pm^*(x)} P(\zeta)^{-k} (x, \zeta)^{pk-n} \omega(\zeta), \qquad (19)$$

and

$$\Phi_k^\pm(P, x) = -\frac{i^{n-pk}(n-pk-1)!}{(2\pi)^n} \int_{\beta_\pm^*(x)} P(\zeta)^{-k} (x, \zeta)^{pk-n} \omega(\zeta), \qquad (20)$$

according to whether $n \le pk$ or $n > pk$, respectively. Both in (19) and (20), the result of integration depends *only* on homology classes $[\sigma_\pm^*(x)]$ and $[\beta_\pm^*(x)]$ rather than on the choice of their individual representatives. Equations (19), (20) are essentially the classical *Herglotz–Petrovsky–Leray formulas* [1, 2]). The following theorem generalizes these formulas to the case of hyperbolic operators in $\mathfrak{D}_{\mathrm{Hyp}}^{G,m}$.

Theorem 2 ([3]). *Let $\mathcal{L}_m \in \mathfrak{D}_{\text{Hyp}}^{G,m}$, $m \in \mathfrak{m}_+$, and let $P \in \text{Hyp}(\vartheta, p)^G$ be the principal symbol of $\mathcal{L}_m$. Fix $x_0 \in V_{\text{reg}}$, and assume that $x \in V_{\text{reg}}$ and $x \notin \mp K(P, \vartheta, x_0) \cup \pm W(P, \vartheta, x_0)$. Then, the principal fundamental solution $\Phi^{\pm}(\mathcal{L}_m) = \Phi^{\pm}(\mathcal{L}_m, x, x_0)$ is holomorphic in the vicinity of x and*

$$\Phi^{\pm}(\mathcal{L}_m) = \sum_{k=1}^{M+1} C_k \, \text{ad}_{k-1}(\mathcal{L}_m, \mathcal{L}_0)[\Theta_m] \int_{\beta_{\pm}^*} P(\zeta)^{-k}(\bar{x}, \zeta)^{pk-n} \omega(\zeta), \quad (21)$$

when $M < n/p - 1$,

$$\Phi^{\pm}(\mathcal{L}_m) = \sum_{k=1}^{[(n-1)/p]} C_k \, \text{ad}_{k-1}(\mathcal{L}_m, \mathcal{L}_0)[\Theta_m] \int_{\beta_{\pm}^*} P(\zeta)^{-k}(\bar{x}, \zeta)^{pk-n} \omega(\zeta)$$

$$+ \sum_{k=[(n-1)/p]+1}^{M+1} \tilde{C}_k \, \text{ad}_{k-1}(\mathcal{L}_m, \mathcal{L}_0)[\Theta_m] \int_{\sigma_{\pm}^*} P(\zeta)^{-k}(\bar{x}, \zeta)^{pk-n} \omega(\zeta) \quad (22)$$

when $0 < n/p - 1 \leq M$, and

$$\Phi^{\pm}(\mathcal{L}_m) = \sum_{k=1}^{M+1} \tilde{C}_k \, \text{ad}_{k-1}(\mathcal{L}_m, \mathcal{L}_0)[\Theta_m] \int_{\sigma_{\pm}^*} P(\zeta)^{-k}(\bar{x}, \zeta)^{pk-n} \omega(\zeta), \quad (23)$$

when $n \leq p$. Here, the integration is taken over an absolute $\beta_{\pm}^* = \beta_{\pm}^*(\bar{x})$ and a relative $\sigma_{\pm}^* = \sigma_{\pm}^*(\bar{x})$ Petrovsky cycles associated to the point $\bar{x} = x - x_0$, the constants C_k and $\tilde{C}_k$ are given explicitly by

$$C_k := \frac{(-1)^k i^{n-pk}(n - pk - 1)!}{(2\pi)^n}, \quad \tilde{C}_k := \frac{(-1)^{k-1} i^{pk-n+1} \pi}{(2\pi)^n (pk - n)!},$$

and $\Theta_m \in \mathfrak{D}$ is a multiplication operator by $\theta_m(x)/\theta_m(x_0)$ with the same $\theta_m(x)$ and M as in Theorem 1.

As in the classical case, the importance of Herglotz–Petrovsky-type formulas (21)–(23) is that they provide some sufficient conditions for existence of regular lacunas.

Corollary 1. *Let $\mathcal{L}_m \in \mathfrak{D}_{\text{Hyp}}^{G,m}, m \in \mathfrak{m}_+$, and let $P \in \text{Hyp}(\vartheta, p)^G$ be a complete polynomial. Suppose that*

$$\sum_{\alpha \in \mathfrak{R}_+} m_\alpha < \frac{n}{p} - 1, \quad (24)$$

and

$$\partial[\sigma_{\pm}^*(\bar{x})] = 0 \quad in \quad H_{n-2}(\bar{X}_C^* \setminus (\bar{X}_C^* \cap \Xi_C^*); \mathbb{C}) \quad (25)$$

relative to some point $x \in \pm K(P, \vartheta, x_0) \setminus \pm W(P, \vartheta, x_0)$. Then, the open component $\Omega^{\pm}(x_0) \subset \pm K(P, \vartheta, x_0) \setminus \pm W(P, \vartheta, x_0)$ containing this point is a regular lacuna of the distribution $\Phi^{\pm}(\mathcal{L}_m, x, x_0)$. Equation (25) is referred to as Petrovsky's condition *in the form of Refs. 1 and 2.*

In general, formulas (21)–(23) make it possible to carry out a close analysis of the fundamental distribution $\Phi^{\pm}(\mathcal{L}_m)$ and to extend many results of the classical lacuna theory to hyperbolic operators under consideration. It is not our intention here to enter these details.[2] Instead, we will conclude this report with one simple illustrative example.

3.3   Products of Wave Operators

The topology of Petrovsky cycles depends drastically on the parity of space dimension. Indeed, let $\iota \colon \xi \mapsto -\xi$ be the antipodal involution, and let $\sigma_{\pm} = \sigma_{\pm}(x, v; \gamma^{\sim})$ be a relative cycle (14) associated to $P \in \mathrm{Hyp}(\vartheta)$ and $x \in \mathbb{R}^n \setminus \pm W(P, \vartheta)$. In view of orientation of $\sigma_{\pm}$ and absolute homogeneity of $v(\xi)$, we have $\iota(\sigma_{\pm}) = (-1)^{n-1}\sigma_{\mp}$ and, hence, $\iota(\partial\sigma_{\pm}) = (-1)^{n-1}\partial\sigma_{\mp}$. It follows that $\partial[\sigma_{\pm}^*(x)] = (-1)^{n-1}\partial[\sigma_{\mp}^*(x)]$, and the inclusion

$$\partial\sigma_{\pm}^*(x, v; \gamma^{\sim}) + (-1)^{n-1}\partial\sigma_{\mp}^*(x, v; \gamma^{\sim}) \in 2\partial[\sigma_{\pm}^*(x)] \qquad (26)$$

holds for all $x \in \mathbb{R}^n \setminus \pm W(P, \vartheta)$. The latter leads to a useful sufficient condition for (25) to be satisfied.

Suppose that a hyperplane $X \subset \mathbb{R}^n$ meets the real characteristic surface $\Xi = \Xi(P)$ at the origin *only*, that is, $X \cap \Xi = \{0\}$. Then, we can choose a vector field $v(\xi) \in \mathfrak{V}(x, P, \vartheta)$ in the definition of $\sigma_{\pm}$ in such a way that $v \equiv 0$ in some small real conical neighborhood of $X \setminus \{0\}$, and, hence, $\partial\sigma_{\pm} = \partial\sigma_{\mp}$ in that case. In view of (26) this implies the vanishing of the class $\partial[\sigma_{\pm}^*(x)]$ when n is *even*.

As an application, consider a product of wave polynomials with different light velocities $P(\zeta) := \prod_{k=1}^{l}(-\zeta_1^2/c_k^2 + \zeta_2^2 + \cdots + \zeta_n^2)$, $c_k \in \mathbb{R}_+$. Clearly, P is both hyperbolic in $\vartheta = (\pm 1, 0, \ldots, 0)$ and invariant under any reflection group G preserving this direction. For a fixed $m \in \mathfrak{m}_+$, the operator $\mathcal{L}_m \in \mathfrak{D}_{\mathrm{Hyp}}^{G,m}$ takes the form

$$\mathcal{L}_m = \prod_{k=1}^{l}\left(\frac{1}{c_k^2}\frac{\partial^2}{\partial x_1^2} - \frac{\partial^2}{\partial x_2^2} - \cdots - \frac{\partial^2}{\partial x_n^2} + \sum_{\alpha \in \mathfrak{R}_+}\frac{m_\alpha(m_\alpha + 1)(\alpha, \alpha)}{(\alpha, x)^2}\right). \qquad (27)$$

The surface $\pm W(P, \vartheta)$ consists of l telescoping conics and splits the space on $2l+1$ connected components, one of them being the exterior of $\pm K(P, \vartheta)$ and one being "the most inner" convex cone $\pm K_{\mathrm{in}}$ inside $\pm K(P, \vartheta)$. It is easy to see that $X \cap \Xi = \{0\}$ for *every* $x \in \pm K_{\mathrm{in}}$. Hence, we conclude by (24) and (25) that any open subset $\Omega^{\pm}(x_0)$ in $\pm K_{\mathrm{in}}(x_0) \cap V_{\mathrm{reg}}$ is a regular lacuna of operator (27) provided n is even and

$$\sum_{\alpha \in \mathfrak{R}_+} m_\alpha < \frac{n}{2l} - 1. \qquad (28)$$

---

[2]The interested reader is referred to Ref. 3, Section III.

The existence of lacuna in "the most inner" component of the propagation cone implies the *absence of diffusion of waves*, the phenomenon which, according to Petrovsky, may be viewed as an analog of Huygens' principle for higher-order hyperbolic operators. For the second-order wave equation (27) (with $l = 1$) the existence of Huygens' principle can be proved [4] by local methods based on the classical Hadamard's construction [7]. However, a similar explicit approach seems hardly applicable for operators of order $p \geq 3$ since fundamental solutions in that case have much more complicated singularities.

4 REFERENCES

1. M. F. Atiyah, R. Bott, and L. Gårding, *Lacunas for hyperbolic differential operators with constant coefficients.* I, Acta Math. **124** (1970), 109–189.

2. _____, *Lacunas for hyperbolic differential operators with constant coefficients.* II, Acta Math. **131** (1973), 145–206.

3. Yu. Yu. Berest, *The problem of lacunae and analysis on root systems*, Tech. Report CRM-2468, Centre de recherches mathématiques, 1997.

4. Yu. Yu. Berest and A. P. Veselov, *The Hadamard problem and Coxeter groups: new examples of the Huygens equations*, Funct. Anal. Appl. **28** (1994), No. 1, 3–12.

5. C. F. Dunkl, *Differential-difference operators associated to reflection groups*, Trans. Amer. Math. Soc. **311** (1989), No. 1, 167–183.

6. L. Gårding, *Linear hyperbolic partial differential equations with constant coefficients*, Acta Math. **85** (1951), 1–62.

7. J. Hadamard, *Lectures on Cauchy's Problem in Linear Partial Differential Equations*, Yale Univ. Press, New Haven, CT, 1923.

8. G. J. Heckman, *A remark on the Dunkl differential-difference operators*, Harmonic Analysis on Reductive Groups (Bowdoin, 1989) (W. Barker and P. Sally, eds.), Progr. Math., Vol. 101, Birkhäuser, Boston, MA, 1991, pp. 181–191.

9. L. Hörmander, *The Analysis of Linear Partial Differential Operators. I. Distribution Theory and Fourier Analysis*, Grundlehren Math. Wiss., Vol. 256, Springer, New York, 1983.

10. _____, *The Analysis of Linear Partial Differential Operators. II. Differential Operators with Constant Coefficients*, Grundlehren Math. Wiss., Vol. 257, Springer, New York, 1983.

11. _____, *The Analysis of Linear Partial Differential Operators*. III. *Pseudodifferential Operators*, Grundlehren Math. Wiss., Vol. 274, Springer, New York, 1985.

12. _____, *The Analysis of Linear Partial Differential Operators*. IV. *Fourier Integral Operators*, Grundlehren Math. Wiss., Vol. 275, Springer, New York, 1985.

13. M. A. Olshanetsky and A. M. Perelomov, *Quantum integrable systems related to Lie algebras*, Phys. Rep. **94** (1983), No. 6, 313–404.

14. I. G. Petrowsky, *On the diffusion of waves and the lacunas for hyperbolic equations*, Mat. Sbornik **17** (1945), 289–370.

5

Canonical Forms for the C-Invariant Tensors

Oleg I. Bogoyavlenskij

ABSTRACT The canonical forms are derived for the C-invariant differential 1-forms, for $(0,2)$ tensors, $(1,1)$ tensors, and $(1,2)$ tensors on a symplectic manifold M^{2k}. It is proved that the characteristic polynomial of any C-invariant $(1,1)$ tensor A_β^α is a perfect square; therefore, its eigenvalues have even multiplicities. Any C-invariant metric $g_{\alpha\beta}$ is indefinite and has signature $\sigma \leq k$.

1 Introduction

This paper continues the study initiated in Ref. 4 where a complete classification was derived for the invariant Poisson structures for the nondegenerate Liouville-integrable Hamiltonian systems. The concept of the nondegeneracy of integrable Hamiltonian systems was introduced by Poincaré in Ref. 13, Vol. I, p. 206, along with the concept of the isoenergetic nondegeneracy. In Ref. 5, the theorem on symmetries was proved that states that the Lie algebra of symmetries S of a C-integrable Hamiltonian system V is Abelian if and only if the system is nondegenerate in the Poincaré sense. We investigate the Hamiltonian systems

$$\dot{x}^\alpha = V^\alpha(x) = \sum_{\beta=1}^{2k} P^{\alpha\beta} \frac{\partial H(x)}{\partial x_\beta}, \qquad (1)$$

on a Poisson manifold M^{2k} with a nondegenerate Poisson structure $P^{\alpha\beta}$.

Recall that the Hamiltonian system (1) on a Poisson manifold M^{2k} is called integrable in Liouville's sense [1, 3, 9] if it possesses k involutive functionally independent first integrals $F_1(x), \ldots, F_k(x)$:

$$\{F_j, F_\ell\} = P^{\alpha\beta} \frac{\partial F_j(x)}{\partial x_\alpha} \frac{\partial F_\ell(x)}{\partial x_\beta} = 0.$$

In this paper all Greek indices α, β, γ, δ range between 1 and $n = 2k$. All Latin indices i, j, ℓ, m range between 1 and k. The summation with respect to repeated indices is understood everywhere.

Liouville's classical theorem [9] implies that almost all points of the manifold M^{2k} (excluding an invariant subset $S \subset M^{2k}$, $\dim S \leq 2k - 1$) are covered by a system of open toroidal domains $\mathcal{O}_m \subset M^{2k}$ with the action-angle coordinates $I_1, \ldots, I_k, \varphi_1, \ldots, \varphi_k$. In these coordinates, the Liouville-integrable system (1) has the form

$$\dot{I}_1 = 0, \ldots, \dot{I}_k = 0, \quad \dot{\varphi}_1 = \frac{\partial H(I)}{\partial I_1}, \ldots, \dot{\varphi}_k = \frac{\partial H(I)}{\partial I_k}, \tag{2}$$

and the symplectic structure $\omega = P^{-1}$ has the canonical form

$$\omega = \sum_{j=1}^{k} dI_j \wedge d\varphi_j.$$

The action coordinates $I_1, \ldots, I_k$ are defined in a ball

$$B_a : \sum_{j=1}^{k} (I_j - I_{j0})^2 < a^2. \tag{3}$$

The angle coordinates $\varphi_1, \ldots, \varphi_k$ run over a torus $\mathbb{T}^k$, $0 \leq \varphi_j \leq 2\pi$, in the compact case and over a toroidal cylinder $\mathbb{T}^m \times \mathbb{R}^{k-m}$, $0 \leq m < k$ if the manifold $I_j(x) = I_{j0}$ is noncompact.

The completely integrable Hamiltonian system (1), (2) is called nondegenerate in the Poincaré sense [13] if the condition for the Hessian

$$\det \left\| \frac{\partial^2 H(I)}{\partial I_\alpha \partial I_\beta} \right\| \neq 0 \tag{4}$$

is met in a dense open domain in the action-angle coordinates

$$I_1, \ldots, I_k, \quad \varphi_1, \ldots, \varphi_k, \quad \varphi_i = \varphi_i \pmod{2\pi}.$$

The Hamiltonian system (1) is called C-integrable if it is completely integrable in Liouville's sense and all invariant submanifolds of constant level of k involutive first integrals are compact [4].

The invariant submanifolds of a C-integrable Hamiltonian system (1) are tori $\mathbb{T}^k$:

$$\mathbb{T}^k : I_1 = c_1, \ldots, I_k = c_k, \quad 0 \leq \varphi_i \leq 2\pi. \tag{5}$$

The corresponding tangent subspaces define the k-dimensional distribution

$$\mathcal{L}_x = T_x(\mathbb{T}^k) \subset T_x(M^{2k}). \tag{6}$$

Exactly the class of the C-integrable and Poincaré-nondegenerate Hamiltonian systems was the starting point for the Kolmogorov–Arnold–Moser (KAM) theory [2, 7, 8, 10] that studies small Hamiltonian perturbations of these systems. Our main objective in the series of papers 4 and 5 is to develop a theory of tensor invariants for the systems of this class. This theory implies several necessary conditions for the nondegenerate C-integrability of dynamical systems and has applications [4, 5] connected with the KAM theory.

A (ℓ, m) tensor T_m^ℓ on the manifold M^{2k} is called C-invariant if it is invariant with respect to a C-integrable Hamiltonian system (1) that is nondegenerate in the Poincaré sense. Results of Refs. 4 and 5 lead to the following general problem.

> *What are the canonical forms for the C-invariant (ℓ, m) tensors T_m^ℓ?*

In this paper, we present the canonical forms for the differential 1-forms θ_α, for $(0, 2)$ tensors $h_{\alpha\beta}$, $(1, 1)$ tensors A_β^α, and $(1, 2)$ tensors $N_{\beta, \gamma}^\alpha$ that are invariant with respect to the C-integrable and Poincaré-nondegenerate Hamiltonian system (2).

The Poincaré-nondegeneracy condition (4) implies that the k functions

$$J_\ell(I) = \frac{\partial H(I)}{\partial I_\ell}$$

form a system of local coordinates in a ball B_a (3). The Hamiltonian system (2) takes the form

$$\dot{J}_1 = 0, \dots, \dot{J}_k = 0, \quad \dot{\varphi}_1 = J_1, \dots, \dot{\varphi}_k = J_k \tag{7}$$

in the toroidal coordinates

$$J_1, \dots, J_k, \quad \varphi_1, \dots, \varphi_k, \quad \varphi_i = \varphi_i \pmod{2\pi}. \tag{8}$$

Remark 1. The advantage of using the noncanonical coordinates J_i, φ_i is that the Hamiltonian system (1), (2) takes the most elegant form (7) in coordinates (8). Our results concerning the algebraic properties of the invariant tensors are proven in the toroidal domains $\mathcal{O}_m$ that cover the manifold M^{2k} excluding the invariant subset S, $\dim S \leq 2k - 1$. The smoothness of the tensors under consideration implies that the established algebraic properties of the C-invariant tensors are true everywhere on the manifold M^{2k} and are independent of any choice of local coordinates.

The trajectories of the dynamical system (7) are everywhere dense on a torus $\mathbb{T}^k$ (5) if and only if the k coordinates $J_1, \dots, J_k$ are rationally independent. It means that for arbitrary integers $m_1, \dots, m_k$ we have

$$m_1 J_1 + \cdots + m_k J_k \neq 0.$$

This condition is true almost everywhere in the ball B_a (3). Hence trajectories of the C-integrable nondegenerate Hamiltonian system (2), (7) are everywhere dense on almost all tori (5). Therefore any smooth first integral $F(J, \varphi)$ of system (7) is constant on all tori $\mathbb{T}^k$ and hence any first integral F is a function of the action variables only:

$$\frac{\mathrm{d}F}{\mathrm{d}t} = 0 \implies F = F(J_1, \ldots, J_k). \tag{9}$$

The Calogero–Moser system [6, 11] has Hamiltonian

$$H(p, q) = \frac{1}{2} \sum_{i=1}^{2k} p_i^2 + \alpha \sum_{i \neq j}^{2k} (q_i - q_j)^{-2}$$

and is Liouville-integrable. This system possesses the Hamiltonian conformal symmetry [5]

$$F(p, q) = p_1 q_1 + \cdots + p_n q_n.$$

For the two functions $H(p, q)$ and $F(p, q)$, the equations

$$\{H, \{H, F\}\} = 0, \quad \{H, F\} = 2H \neq 0$$

hold. The latter equation implies that the Calogero–Moser system has non-compact invariant submanifolds.

2 C-Invariant Differential 1-Forms

Any differential 1-form θ has a form

$$\theta = \theta_i(J, \varphi) \, \mathrm{d}J_i + \theta_{i+k}(J, \varphi) \, \mathrm{d}\varphi_i \tag{10}$$

in the toroidal coordinates (8).

Proposition 1. *A differential 1-form θ is invariant with respect to the C-integrable and Poincaré-nondegenerate Hamiltonian system* (1) *if and only if it has the form*

$$\theta = \theta_i(J) \, \mathrm{d}J_i \tag{11}$$

in coordinates J_i, φ_i (8).

Proof. The vector field V that is defined by the dynamical system (7) has the following components:

$$V^i = 0, \quad V^{i+k} = J_i.$$

These formulas imply

$$V^\alpha_{,\beta} = \frac{\partial V^\alpha(x)}{\partial x_\beta} = \delta^\alpha_{i+k} \delta^i_\beta, \tag{12}$$

where summation with respect to the index $i = 1, \ldots, k$ is understood. For any 1-form θ, the invariance equation $L_V \theta = 0$ has the form

$$(L_V \theta)_\beta = \dot{\theta}_\beta + V^\alpha_{,\beta} \theta_\alpha = 0. \tag{13}$$

Here L_V is the Lie derivation operator. After substituting formulas (10) and (12), Eq. (13) implies

$$\dot{\theta}_i = -\theta_{i+k}, \quad \dot{\theta}_{i+k} = 0. \tag{14}$$

In view of Eq. (9), solutions to Eq. (14) have the form

$$\theta_i(t) = -\tilde{\theta}_{i+k}(J)t + \tilde{\theta}_i(J), \quad \theta_{i+k}(t) = \tilde{\theta}_{i+k}(J), \tag{15}$$

where $\tilde{\theta}_i(J)$ and $\tilde{\theta}_{i+k}(J)$ are arbitrary smooth functions. Components $\theta_\alpha(J, \varphi)$ of any smooth invariant 1-form (10) are bounded on any torus $\mathbb{T}^k$ (5). The solutions (15) are bounded for all t if and only if $\tilde{\theta}_{i+k}(J) = 0$ for $i = 1, \ldots, k$. Therefore using (15) and the fact that generic trajectories of the C-integrable nondegenerate Hamiltonian system (1), (7) are everywhere dense on the tori $\mathbb{T}^k$, we obtain that the 1-form θ is invariant if and only if it has the form (11). $\square$

Let a dynamical system

$$\dot{x}^\alpha = V^\alpha(x^1, \ldots, x^{2k}) \tag{16}$$

on a smooth manifold M^{2k} be a C-integrable and Poincaré-nondegenerate Hamiltonian system.

Theorem 1. *Any C-invariant differential 1-form θ satisfies the equations*

$$\theta(V) = 0, \quad i_V d\theta = 0, \quad \text{rank} \, d\theta \leq k, \tag{17}$$
$$\theta \wedge d\theta \wedge \cdots \wedge d\theta = 0, \tag{18}$$

where the number of factors in Eq. (18) is $\ell + 1$ if $k = 2\ell$, and equation

$$d\theta \wedge \cdots \wedge d\theta = 0 \tag{19}$$

if $k = 2\ell + 1$, where the number of factors is $\ell + 1$.

Proof. Applying Proposition 1, we obtain that if dynamical system (16) is C-integrable and Poincaré-nondegenerate, then any invariant 1-form θ takes the form (11) in the toroidal coordinates (8). Hence Eq. (17) follows.

The wedge products (18) and (19) have degrees $k + 1$. Therefore, for the 1-form θ (11) we have

$$\theta \wedge d\theta \wedge \cdots \wedge d\theta = 0, \quad d\theta \wedge \cdots \wedge d\theta = 0$$

for $k = 2\ell$ and $k = 2\ell + 1$, respectively. $\square$

3   General C-Invariant $(0,2)$ Tensors

In the toroidal coordinates J_i, φ_i (8), any $(0,2)$ tensor $h_{\alpha\beta}(J,\varphi)$ has the following block form:

$$h_{\alpha\beta}(J,\varphi) = \begin{pmatrix} b_1 & b_3 \\ b_2 & b_4 \end{pmatrix}, \qquad (20)$$

where b_1, b_2, b_3, $b_4(J_i,\varphi_i)$ are $k \times k$ matrices.

Proposition 2. *The (0,2) tensor $h_{\alpha\beta}$ is invariant with respect to the C-integrable and Poincaré-nondegenerate Hamiltonian system* (1) *if and only if it has the block form*

$$h_{\alpha\beta} = \begin{pmatrix} a(J) & b(J) \\ -b(J) & 0 \end{pmatrix} \qquad (21)$$

in coordinates (8). *Here $a(J)$ and $b(J)$ are arbitrary $k \times k$ matrices.*

Proof. For the $(0,2)$ tensor h the invariance equation has the form

$$(L_V h)_{\alpha\beta} = \dot{h}_{\alpha\beta} + V^\gamma_{,\alpha} h_{\gamma\beta} + V^\gamma_{,\beta} h_{\alpha\gamma} = 0. \qquad (22)$$

In view of Eqs. (12) and (20), system (22) is equivalent to the linear system of four matrix equations

$$\dot{b}_1 = -b_2 - b_3, \quad \dot{b}_2 = -b_4, \quad \dot{b}_3 = -b_4, \quad \dot{b}_4 = 0. \qquad (23)$$

It is evident that this system has an upper-triangular structure. Applying Eq. (9), we obtain that all solutions to Eq. (23) have the form

$$b_1(t) = \tilde{b}_4(J)t^2 - (\tilde{b}_2(J) + \tilde{b}_3(J))t + \tilde{b}_1(J),$$
$$b_2(t) = -\tilde{b}_4(J)t + \tilde{b}_2(J), \quad b_3(t) = -\tilde{b}_4(J)t + \tilde{b}_3(J), \qquad (24)$$
$$b_4(t) = \tilde{b}_4(J).$$

The components $h_{\alpha\beta}(J,\varphi)$ (20) of the smooth C-invariant $(0,2)$ tensor h are bounded on any torus $\mathbb{T}^k$ (5). The solutions (24) are bounded for all t if and only if

$$\tilde{b}_2(J) = -\tilde{b}_3(J), \quad \tilde{b}_4(J) = 0. \qquad (25)$$

Hence the block form (21) follows.

The generic trajectories of the C-integrable nondegenerate Hamiltonian system (1) are everywhere dense on the tori (5). Therefore, formulas (24) and (25) imply that components $h_{\alpha\beta}$ (20) depend on coordinates J_i only. $\qquad \square$

The C-invariant $(0,2)$ tensor $h_{\alpha\beta}$ (21) is symmetric if the equations

$$a^t = a, \quad b^t = -b \tag{26}$$

are satisfied. Denote $g_{\alpha\beta}$ the arising metric on the manifold M^{2k}.

Theorem 2. *Any C-invariant metric $g_{\alpha\beta} = g_{\beta\alpha}$ is indefinite with signature $\sigma = |n_+ - n_-| \le k$. The numbers n_+ and n_- of its positive and negative squares satisfy the inequalities*

$$n_+ \le k, \quad n_- \le k. \tag{27}$$

For $n = 4\ell + 2$, any C-invariants metric $g_{\alpha\beta}$ is degenerate everywhere on the manifold M^n.

Proof. Proposition 2 implies that any C-invariant metric $g_{\alpha\beta}$ has the block structure (21) in the toroidal coordinates (8). In view of Eq. (26), the matrix $a(J)$ is symmetric and the matrix $b(J)$ is skew-symmetric. Hence metric $g_{\alpha\beta}$ (21) on $T_x(M^{2k})$ vanishes on the k-dimensional linear subspace $\mathcal{L}_x \subset T_x(M^{2k})$ and the inequalities (27) hold along with $\sigma = |n_+ - n_-| \le k$.

For any odd $k = 2\ell + 1$, the skew-symmetric $k \times k$ matrix $b(J)$ is degenerate. The formula (21) implies

$$\det \|g\| = (\det \|b\|)^2.$$

Therefore, for $n = 2k = 4\ell + 2$ any C-invariant metric g is degenerate. $\square$

In Ref. 4, we presented a complete classification of all C-invariant closed differential 2-forms and Poisson structures.

4   Characteristic Polynomial of Any C-Invariant $(1,1)$ Tensor Is a Perfect Square

In the toroidal coordinates (8), any $(1,1)$ tensor $A_\beta^\alpha(J,\varphi)$ has the following block structure:

$$A_\beta^\alpha(J,\varphi) = \begin{pmatrix} B_2 & B_4 \\ B_1 & B_3 \end{pmatrix} \tag{28}$$

where $B_1, B_2, B_3, B_4(J,\varphi)$ are $k \times k$ matrices.

Proposition 3. *The $(1,1)$ tensor $A_\beta^\alpha(J,\varphi)$ is invariant with respect to the C-integrable and Poincaré-nondegenerate Hamiltonian system (1) if and only if it has the block form*

$$A_\beta^\alpha = \begin{pmatrix} B(J) & 0 \\ \sigma(J) & B(J) \end{pmatrix} \tag{29}$$

in coordinates (8). Here $B(J)$ and $\sigma(J)$ are arbitrary $k \times k$ matrices.

Proof. For the $(1,1)$ tensor $A_\beta^\alpha(J, \varphi)$ (28), the invariance equation $L_V A = 0$ has the form

$$(L_V A)_\beta^\alpha = \dot{A}_\beta^\alpha - V_{,\gamma}^\alpha A_\beta^\gamma + V_{,\beta}^\gamma A_\gamma^\alpha = 0. \qquad (30)$$

Using the formulas (12) and (28), we find that the invariance equation (30) is equivalent to the following linear dynamical system in space of matrices

$$\dot{B}_1 = B_2 - B_3, \quad \dot{B}_2 = -B_4, \quad \dot{B}_3 = B_4, \quad \dot{B}_4 = 0. \qquad (31)$$

It is evident that this system has an upper-triangular structure. In view of Eq. (9), solutions to Eq. (31) have the form

$$B_1(t) = -\tilde{B}_4(J)t^2 + \left(\tilde{B}_2(J) - \tilde{B}_3(J)\right)t + \tilde{B}_1(J),$$
$$B_2(t) = -\tilde{B}_4(J)t + \tilde{B}_2(J), \quad B_3(t) = \tilde{B}_4(J)t + \tilde{B}_3(J), \qquad (32)$$
$$B_4(t) = \tilde{B}_4(J).$$

All components of the smooth invariant $(1,1)$ tensor $A_\beta^\alpha(J, \varphi)$ are bounded on any torus $\mathbb{T}^k$ (5). The solutions (32) are bounded for all t if and only if

$$\tilde{B}_2(J) = \tilde{B}_3(J), \quad \tilde{B}_4(J) = 0.$$

Therefore, using the fact that generic trajectories of the integrable nondegenerate Hamiltonian system (1) are everywhere dense on the tori $\mathbb{T}^k$, we obtain that any C-invariant $(1,1)$ tensor A_β^α has the block form (29).   □

Theorem 3. *If a $(1,1)$ tensor $A_\beta^\alpha(x)$ is invariant with respect to the C-integrable and Poincaré-nondegenerate Hamiltonian system (1), then the characteristic polynomial*

$$P(\lambda, x) = \det \|A(x) - \lambda\| \qquad (33)$$

is a perfect square

$$P(\lambda, x) = Q^2(\lambda, x). \qquad (34)$$

All eigenvalues of any C-invariant $(1,1)$ tensor have even multiplicities.

Proof. The characteristic polynomial (33) is an invariant of the $(1,1)$ tensor A_β^α; it does not depend on a choice of local coordinates. Using the block form (29) of the C-invariant $(1,1)$ tensor A_β^α in the toroidal coordinates (8), we obtain the formula (34) where

$$Q(\lambda, x) = \det \|B(J) - \lambda\|.$$

Formula (34) implies that each root of the characteristic polynomial $P(\lambda, x)$ or each eigenvalue of the $(1,1)$ tensor A_β^α has an even multiplicity.   □

5 Nijenhuis Tensor and C-Invariant $(1,2)$ Tensors

The $(1,2)$ Nijenhuis tensor $N_{A\beta\gamma}^{\alpha}$ is defined by the formula [12]

$$N_{A\beta\gamma}^{\alpha} = A_{\gamma,\tau}^{\alpha}A_{\beta}^{\tau} - A_{\beta,\tau}^{\alpha}A_{\gamma}^{\tau} + (A_{\beta,\gamma}^{\tau} - A_{\gamma,\beta}^{\tau})A_{\tau}^{\alpha}. \qquad (35)$$

The Nijenhuis tensor N_A is C-invariant if the $(1,1)$ tensor A is. The $(1,2)$ Nijenhuis tensor N_A possesses remarkable algebraic properties, which can be derived by a direct calculation using the definition (35) and formula (29). This calculation shows that the main algebraic properties of the Nijenhuis tensor N_A are manifestations of the generic algebraic properties of all C-invariant $(1,2)$ tensors $N_{\beta\gamma}^{\alpha}$.

Any $(1,2)$ tensor $N_{\beta\gamma}^{\alpha}$ defines a bilinear product of tangent vectors $u, v \in T_x(M^{2k})$

$$\big(N(u,v)\big)^{\alpha} = N_{\beta\gamma}^{\alpha} u^{\beta} v^{\gamma} \in T_x(M^{2k}). \qquad (36)$$

This formula defines an algebraic structure in the tangent space $T_x(M^{2k})$.

Let us collect the tensor components $N_{\beta\gamma}^{\alpha}(J,\varphi)$ in the toroidal coordinates J_i, φ_i (8) into the following eight groups:

$$N_{j\ell}^{i+k} = N_{1j\ell}^{i}, \quad N_{j+k.\ell}^{i+k} = N_{2j\ell}^{i}, \quad N_{j.\ell+k}^{i+k} = N_{3j\ell}^{i},$$

$$N_{j\ell}^{i} = N_{4j\ell}^{i}, \quad N_{j+k.\ell}^{i} = N_{5j\ell}^{i}, \quad N_{j.\ell+k}^{i} = N_{6j\ell}^{i}, \qquad (37)$$

$$N_{j+k.\ell+k}^{i+k} = N_{7j\ell}^{i}, \quad N_{j+k.\ell+k}^{i} = N_{8j\ell}^{i},$$

where indices i, j, ℓ range between 1 and k.

Proposition 4. *The $(1,2)$ tensor $N_{\beta\gamma}^{\alpha}(J,\varphi)$ is invariant with respect to the C-integrable and Poincaré-nondegenerate Hamiltonian system (1) if and only if all its components (37) depend on the variables J_i only and the equations*

$$N_4(J) = N_2(J) + N_3(J), \quad N_5 = N_6 = N_7 = N_8 = 0 \qquad (38)$$

hold. The components of $N_1(J)$, $N_2(J)$ and $N_3(J)$ are arbitrary smooth functions of $J_1, \ldots, J_k$.

Proof. For the $(1,2)$ tensor $N_{\beta\gamma}^{\alpha}(J,\varphi)$, the invariance equation $L_V N = 0$ has the form

$$(L_V N)_{\beta\gamma}^{\alpha} = \dot{N}_{\beta\gamma}^{\alpha} - V_{,\tau}^{\alpha}N_{\beta\gamma}^{\tau} + V_{,\beta}^{\tau}N_{\tau\gamma}^{\alpha} + V_{,\gamma}^{\tau}N_{\beta\tau}^{\alpha} = 0. \qquad (39)$$

Substituting formulas (12) and (37), we find that Eq. (39) is equivalent to the following linear dynamical system in the space of tensors N_a, $a = 1$, $\ldots$, 8:

$$\dot{N}_1 = N_4 - N_2 - N_3, \quad \dot{N}_2 = N_5 - N_7, \quad \dot{N}_3 = N_6 - N_7,$$

$$\dot{N}_4 = -N_5 - N_6, \quad \dot{N}_5 = -N_8, \quad \dot{N}_6 = -N_8, \quad \dot{N}_7 = N_8, \quad \dot{N}_8 = 0. \qquad (40)$$

Using the evident upper-triangular structure of system (40) and the basic property of first integrals (9), we present all solutions to Eq. (40) in the explicit form

$$N_1(t) = \tilde{N}_8(J)t^3 + (\tilde{N}_7(J) - \tilde{N}_5(J) - \tilde{N}_6(J))t^2$$
$$+ (\tilde{N}_4(J) - \tilde{N}_2(J) - \tilde{N}_3(J))t + \tilde{N}_1(J),$$
$$N_2(t) = -\tilde{N}_8(J)t^2 + (\tilde{N}_5(J) - \tilde{N}_7(J))t + \tilde{N}_2(J),$$
$$N_3(t) = -\tilde{N}_8(J)t^2 + (\tilde{N}_6(J) - \tilde{N}_7(J))t + \tilde{N}_3(J), \qquad (41)$$
$$N_4(t) = \tilde{N}_8(J)t^2 - (\tilde{N}_5(J) + \tilde{N}_6(J))t + \tilde{N}_4(J),$$
$$N_5(t) = -\tilde{N}_8(J)t + \tilde{N}_5(J), \qquad N_6(t) = -\tilde{N}_8(J)t + \tilde{N}_6(J),$$
$$N_7(t) = -\tilde{N}_8(J)t + \tilde{N}_7(J), \qquad N_8(t) = \tilde{N}_8(J),$$

where $\tilde{N}_a(J)$ are arbitrary smooth functions of $J_1, \ldots, J_k$.

The components $N_{\beta\gamma}^\alpha(J, \varphi)$ of any smooth C-invariant $(1,2)$ tensor are bounded on any torus $\mathbb{T}^k$ (5). The solutions (41) are bounded for all t if and only if Eq. (38) holds. Hence, using the fact that generic trajectories of the integrable nondegenerate Hamiltonian system (1) are everywhere dense on the tori $\mathbb{T}^k$, we obtain that all components $N_{\beta\gamma}^\alpha$ depend on the variables J_i only and satisfy Eq. (38).   □

For any tangent vector $u \in T_x(M^{2k})$, we define the linear operator

$$N_u : T_x(M^{2k}) \to T_x(M^{2k}), \quad N_u w = N(u, w). \qquad (42)$$

In Ref. 4, we introduced the following polynomial-valued function on the tangent bundle $T(M^{2k})$:

$$P_N(u, \lambda) = \det(N_u - \lambda). \qquad (43)$$

Theorem 4. *Any C-invariant $(1,2)$ tensor N possesses the following properties:*

1. *The k-dimensional linear subspaces $\mathcal{L}_x$ (6) are the commutative ideals with respect to the algebraic structure (36):*

$$N(\mathcal{L}_x, \mathcal{L}_x) = 0, \quad N(T_x(M^{2k}), \mathcal{L}_x) \subset \mathcal{L}_x,$$
$$N(\mathcal{L}_x, T_x(M^{2k})) \subset \mathcal{L}_x. \qquad (44)$$

2. *The polynomial $P_N(u, \lambda)$ is reducible and is a product of two polynomials of degree k. The identities*

$$P_N(u + v, \lambda) = P_N(u, \lambda), \quad P_N(v, \lambda) = \lambda^{2k} \qquad (45)$$

hold for any tangent vectors $u \in T_x(M^{2k})$ and $v \in \mathcal{L}_x = T_x(\mathbb{T}^k) \subset T_x(M^{2k})$.

Proof. 1. The formulas (44) follow from the last four components of Eq. (38) because the tangent vectors $u \in T_x(M^{2k})$ and $v \in \mathcal{L}_x$ have the form

$$u = u^i e_i + u^{i+k} e_{i+k}, \quad v = v^{i+k} e_{i+k} \tag{46}$$

in the basis

$$e_i = \frac{\partial}{\partial J_i}, \quad e_{i+k} = \frac{\partial}{\partial \varphi_i}. \tag{47}$$

2. The formulas (44) imply that the linear operators N_{e_α} (42) have the following $k \times k$ block structures:

$$N_{e_i} = \begin{pmatrix} V_i & 0 \\ U_i & W_i \end{pmatrix}, \quad N_{e_{i+k}} = \begin{pmatrix} 0 & 0 \\ Q_i & 0 \end{pmatrix}$$

in the basis (47). Using formula $u = u^i e_i + u^{i+k} e_{i+k}$, we obtain

$$P_N(u, \lambda) = \det \left\| \sum_{i=1}^k u^i V_i - \lambda \right\| \cdot \det \left\| \sum_{i=1}^k u^i W_i - \lambda \right\|. \tag{48}$$

This formula provides the factorization of the polynomial $P_N(u, \lambda)$. The identities (45) follow from Eqs. (46) and (48). $\square$

Acknowledgments: This research is supported by the Natural Sciences and Engineering Research Council of Canada.

6   REFERENCES

1. R. Abraham and J. E. Marsden, *Foundations of Mechanics*, Benjamin–Cummings, Reading, MA, 1978.

2. V. I. Arnold, *Proof of a theorem of A. N. Kolmogorov on the preservation of conditionally periodic motions under small perturbation of the Hamiltonian*, Russian Math. Surveys **18** (1963), No. 5, 9–36.

3. _____, *Mathematical Methods of Classical Mechanics*, Grad. Texts in Math., Vol. 60, Springer, New York, 1978.

4. O. I. Bogoyavlenskij, *Theory of tensor invariants of integrable Hamiltonian systems. I. Incompatible Poisson structures*, Commun. Math. Phys. **180** (1996), No. 3, 529–586.

5. _____, *Theory of tensor invariants of integrable Hamiltonian systems. II. Theorem on symmetries and its application*, Commun. Math. Phys. **184** (1997), No. 2, 301–365.

6. F. Calogero, *Solution of the one-dimensional N-body problems with quadratic and/or inversely quadratic pair potentials*, J. Math. Phys. **12** (1971), 419–436.

7. A. N. Kolmogorov, *On conservation of conditionally periodic motions for small changes in Hamilton functions*, Dokl. Akad. Nauk SSSR **98** (1954), 527–530 (Russian).

8. _____, *Théorie générale des systèmes dynamiques et mécanique classique*, Proc. International Congress of Mathematicians (Amsterdam, 1954), North-Holland, Amsterdam, 1957, pp. 315–333.

9. J. Liouville, *Note sur l'integration des équations différentielles de la dynamique*, J. Math. Pures Appl. **20** (1855), 137–138.

10. J. Moser, *On invariant curves of area-preserving mappings of an annulus*, Nachr. Akad. Wiss. Göttingen Math.-Phys. Kl. II (1962), 1–20.

11. _____, *Three integrable Hamiltonian systems connected to isospectral deformations*, Adv. Math. **16** (1975), 197–220.

12. A. Nijenhuis, X_{n-1}-*forming sets of eigenvectors*, Nederl. Akad. Wetensch. Proc. Ser. A **54** (1951), 200–212.

13. H. Poincaré, *Les méthodes nouvelles de la mécanique céleste*, Gauthier–Villars, Paris, 1892.

6

R-Matrices, Generalized Inverses, and Calogero–Moser–Sutherland Models

H. W. Braden

ABSTRACT Four results are given that address the existence, ambiguities, and construction of a classical R-matrix given a Lax pair. They enable the uniform construction of R-matrices in terms of any generalized inverse of $\mathrm{ad}(L)$. For generic L a generalized inverse (and indeed the Moore–Penrose inverse) is explicitly constructed. The R-matrices are in general momentum dependent and dynamical. We apply our construction to the elliptic Calogero–Moser–Sutherland models associated with gl_n.

1 Introduction

I'd like to thank the organizers of the meeting for bringing together so many workers interested in the Calogero–Moser–Sutherland models and related areas. As we have heard, the area is rich in connections to many areas of mathematics and a variety of problems remain outstanding. Indeed, within my talk I will highlight several problems that have yet to be answered but for which (widely believed) "folk" answers exist. These, if you wish, constitute my class exam for those interested in the subject.

My interests in the Calogero–Moser–Sutherland models may broadly be grouped under

a) connections with functional equations,

b) connections with affine Toda theory,

c) connections with geometry, and

d) R-matrix constructions.

Although I will focus on the latter here I'd like to begin with some general comments.

Connections with Functional Equations

Functional equations arise in several contexts with the mechanics and quantum mechanics of Calogero–Moser–Sutherland models. By making various ansatz for a Lax pair L, M one may reduce the constraints implicit in $\dot{L} = [L, M]$ to that of a functional equation [16]. Similarly, assuming the ground state of a quantum-mechanical problem factorizes leads to functional equations [15, 35–37]. In the paper *Integrable systems with pairwise interactions and functional equations* with V. Buchstaber [9] we enlarge the class of ansatz considered to arrive at the functional equation

$$\phi_1(x+y) = \frac{\begin{vmatrix} \phi_2(x) & \phi_2(y) \\ \phi_3(x) & \phi_3(y) \end{vmatrix}}{\begin{vmatrix} \phi_4(x) & \phi_4(y) \\ \phi_5(x) & \phi_5(y) \end{vmatrix}}, \tag{1}$$

and this is then solved in Ref. 8 *The general analytic solution of a functional equation of addition type.* Interestingly, the solutions found are those described by Krichever from his rather different starting point. Several unsolved functional equations may be found in this work. Related to the question of ground-state factorization we confront—

Problem 2. Determine the most general meromorphic solution to

$$\begin{vmatrix} h'(a) & h'(b) & h'(c) \\ h(a) & h(b) & h(c) \\ 1 & 1 & 1 \end{vmatrix} = 0, \quad a+b+c = 0. \tag{2}$$

Observe

a) this equation is invariant under $h(x) \to \alpha h(\delta x) + \beta$,

b) if h has a pole, then $h(x) = \alpha \wp(x+d) + \beta$ where $3d$ lies in the lattice, and

c) that x and e^x satisfy (2).

The difficulty lies with the entire solutions. The problem has been addressed in Ref. 14; I would like to see a direct answer. Further, within a suitable class of distributions the delta function satisfies (2): can this be obtained as a limit of some suitable class of functions that satisfy (2)?

Connections with Affine Toda Theory

The solitons of the sine-Gordon theory are described by the generalization of Ruijsenaars and Schneider [31]of the Calogero–Moser–Sutherland models. The affine Toda systems are an important relativistically invariant, integrable system of PDE's [10] possessing soliton solutions [25] that generalize the sine-Gordon equation. The fundamental solitons of these theories are characterized by a label or type. (Different types often have different masses.) An outstanding problem is to identify completely the finite dimensional system associated with the multisolitons of the affine Toda equations that are composed of several fundamental solitons of different types. By introducing the spin-generalization of the Ruijsenaars–Schneider models Braden and Hone [11] in *affine Toda solitons and systems of Calogero–Moser type* have shown how to incorporate the various fundamental solitons of the affine Toda theories. This work, however, is not the complete answer, for the affine Toda solitons can only form a subspace of the spin-generalization Ruijsenaars–Schneider models (the latter having too many degrees of freedom). Krichever and Zabrodin [22] had earlier introduced these spin-generalizations and discussed certain solutions but the connections with affine Toda solitons are still unclear. Ruijsenaars [30] has suggested applying the "Fusing"-type arguments of Hollowood [20] to the nonspin Ruijsenaars–Schneider models to generate solitons of different types from a given basic soliton, but this has yet to be done for the multisolitons constructed from differing fundamental solitons.

It is perhaps worth mentioning that there is disagreement as to the *physical* interpretation of the Ruijsenaars–Schneider models, the issue being whether one should describe these as *relativistically invariant* classical mechanical theories. These issues are taken up in *The Ruijsenaars–Schneider model* with Ryu Sasaki [12].

Connections with Geometry

The Calogero–Moser–Sutherland models arise naturally as Hamiltonian reductions [21] and with J. Robbins I have been considering the geometry of the spin-generalized Calogero–Moser–Sutherland models. There are connections between the differential equations associated with the Calogero–Moser–Sutherland models and equations of Bogomolny type. Wilson here has argued that the phase space of the (complexified, rational, spin-generalized) Calogero–Moser model is to be identified with the moduli space of instantons over S^4. Similar (partial) results exist for monopoles.

R-Matrix Constructions

Some years ago T. Suzuki and I constructed [13] *R-matrices for elliptic Calogero–Moser models* with spectral parameter; an equivalent result was obtained simultaneously by Sklyanin [33], though by different means. Following Ref. 3 we initially sought momentum-independent R-matrices for the spectral-parameter-independent elliptic Calogero–Moser models but found obstructions to this. Here I take up this theme and present four general results that address the existence, ambiguities, and construction of a classical R-matrix given a Lax pair. These enable the uniform construction of

R-matrices, though these are in general momentum dependent and dynamical. We apply our construction to the elliptic Calogero–Moser–Sutherland models associated with $\mathfrak{gl}_n$.

2 R-Matrices

The modern approach to completely integrable systems is in terms of Lax pairs L, M and R-matrices. Here the consistency of the matrix equation $\dot{L} = [L, M]$ expresses the equations of motion of the system under consideration. The great merit of this approach is that it provides a unified framework for treating the many disparate completely integrable systems known. Given a $2n$-dimensional phase space, Liouville's theorem [1, 24], which ensures the existence of action-angle variables, requires that we have n independent conserved quantities in involution; that is, they mutually Poisson commute. As a consequence of the Lax equation the traces $\operatorname{Tr} L^k$ are conserved and these are natural candidates for the action variables of Liouville's theorem. (In practice the action variables are typically transcendental functions of these traces.) It remains, however, to verify that these traces provide enough independent quantities in involution. Verifying the number of independent quantities is usually straightforward and the remaining step is then to show they mutually Poisson commute. The final ingredient of the modern approach, the R-matrix [32], guarantees their involution. If L is in a representation E of a Lie algebra $\mathfrak{g}$ (here taken to be semisimple), the classical R-matrix is a $E \otimes E$ valued matrix such that

$$[R, L \otimes 1] - [R^T, 1 \otimes L] = \{L \underset{,}{\otimes} L\}. \tag{3}$$

(The notation is amplified below.) Then

$$\{\operatorname{Tr}_E L^k, \operatorname{Tr}_E L^m\} = \operatorname{Tr}_{E \otimes E}\{L^k \underset{,}{\otimes} L^m\}$$

$$= km \operatorname{Tr}_{E \otimes E} L^{k-1} \otimes L^{m-1}\{L \underset{,}{\otimes} L\} = 0,$$

which vanishes owing to the cyclicity of the trace. By a result of Babelon and Viallet [4], such an R-matrix is guaranteed to exist if the eigenvalues of L are in involution. The Liouville integrability of a system represented by a Lax pair has been reduced then to finding any solution to (3) and counting the number of independent traces. Further, the R-matrix is an essential ingredient when examining the separation of variables of such integrable systems [23, 34].

Unfortunately the construction of R-matrices has hitherto been somewhat of an arcane art and many have been obtained in a case-by-case manner [3]. Here I will present four new results that address the existence and construction of solutions of (3) and hence the Liouville integrability

of the system under consideration. They yield a uniform construction of
R-matrices. In fact the *R*-matrices satisfying (3) are by no means unique
and our construction characterizes this ambiguity. Finally we will apply
these results to the elliptic Calogero–Moser models without spectral parameters whose *R*-matrices presented here are new. At the outset let me
remark that the *R*-matrix solutions to (3) are generically momentum dependent. Within this family of solutions some may be particularly simple:
they may, for example, be constant (as in the Toda system [18]) or momentum independent. We are content here with providing the construction
of an *R*-matrix given a Lax matrix *L* and so answering the question of
Liouville integrability: we do not seek to specify further the momentum
or position dependence of the solution. For the chosen example of the elliptic Calogero–Moser model there are in fact [13] no *R*-matrices that are
independent of both momentum and spectral parameter (for more than
four particles) and this illustrates the fact that simple assumptions on the
parameter dependence of an *R*-matrix need not be natural.

Our approach is as follows. First we rewrite (3) in the form of the matrix
equation

$$A^T X - X^T A = B. \tag{4}$$

Here *A* is built out of *L* and the Lie algebra, the unknown matrix *X* being
solved for is essentially the *R*-matrix in a given basis and *B* represents the
right-hand side of (3). Our first result is to give necessary and sufficient
conditions for (4) to admit solutions together with its general solution. This
general solution encodes the possible ambiguities of the *R*-matrix. Because
A is (in general) singular our solution is in terms of a generalized inverse
G satisfying

$$AGA = A \quad \text{and} \quad GAG = G. \tag{5}$$

Such a generalized inverse always exists. (Accounts of generalized inverses
may be found in Ref. 5, 17, 28, and 29.) Indeed the Moore–Penrose inverse—
which is unique and always exists—further satisfies $(AG)^\dagger = AG$, $(GA)^\dagger =$
GA. Observe that given a *G* satisfying (5) we have at hand projection
operators $P_1 = GA$ and $P_2 = AG$ which satisfy

$$AP_1 = P_2 A = A, \quad P_1 G = GP_2 = G. \tag{6}$$

Our second result shows that the choice of generalized inverse *G* only al-
ters the *R*-matrix within the ambiguities specified by the general solution,
and so any generalized inverse suffices to solve (4) and hence construct
an *R*-matrix. At this stage we have reduced the problem of constructing
an *R*-matrix to that of constructing a generalized inverse *G* and our third
result constructs such for a generic element *L* of $\mathfrak{g}$. Because the Moore–Penrose inverse is unique, our fourth result is to present this inverse for generic
L, though we shall not need to use this in our application.

In the next section we present the four results. The proofs of the first two are somewhat lengthy and algebraic and will be presented elsewhere [7]; the proofs of the remaining two are easier to outline. In Section 4 we consider the Calogero–Moser models. We conclude with a brief discussion.

3 Four Results

Our first task is to identify (3) with (4). Let T_μ denote a basis for the (finite dimensional) Lie algebra $\mathfrak{g}$ with $[T_\mu, T_\nu] = c^\lambda_{\mu\nu} T_\lambda$ defining the structure constants of $\mathfrak{g}$. Set $\phi(T_\mu) = X_\mu$, where ϕ yields the representation E of the Lie algebra $\mathfrak{g}$; we may take this to be a faithful representation. With $L = \sum_\mu L^\mu X_\mu$ the left-hand side of (3) becomes

$$\{L \overset{\otimes}{,} L\} = \sum_{\mu,\nu} \{L^\mu, L^\nu\} X_\mu \otimes X_\nu,$$

while upon setting $R = R^{\mu\nu} X_\mu \otimes X_\nu$ and $R^T = R^{\nu\mu} X_\mu \otimes X_\nu$ the right-hand side yields

$$
\begin{aligned}
[R, L \otimes 1] - [R^T, 1 \otimes L] &= R^{\mu\nu}([X_\mu, L] \otimes X_\nu - X_\nu \otimes [X_\mu, L]) \\
&= R^{\mu\nu} L^\lambda ([X_\mu, X_\lambda] \otimes X_\nu - X_\nu \otimes [X_\mu, X_\lambda]) \\
&= R^{\tau\nu} c^\mu_{\tau\lambda} L^\lambda - R^{\tau\mu} c^\nu_{\tau\lambda} L^\lambda) X_\mu \otimes X_\nu.
\end{aligned}
$$

By identifying $A^{\mu\nu} = c^\nu_{\mu\lambda} L^\lambda \equiv -ad(L)^\nu_\mu$, $B^{\mu\nu} = \{L^\mu, L^\nu\}$ and $X^{\mu\nu} = R^{\mu\nu}$ we see that (3) is an example of (4).

Having shown how to identify (3) with the matrix equation (4), we may now state our first result.

Result 1. *The matrix equation (4) has solutions if and only if*

(C1) $B^T = -B$,

(C2) $(1 - P_1^T) B (1 - P_1) = 0$,

in which case the general solution is

$$X = \tfrac{1}{2} G^T B P_1 + G^T B (1 - P_1) + (1 - P_2^T) Y + (P_2^T Z P_2) A, \qquad (7)$$

where Y is arbitrary and Z is only constrained by the requirement that $P_2^T Z P_2$ be symmetric.

Although the general solution appears to depend on the generalized inverse G, we in fact find—

Result 2. *If $\overline{G}$ is any other solution of (5) with attendant projection operators $\overline{P}_{1,2}$ then (7) may also be written*

$$X = \tfrac{1}{2} \overline{G}^T B \overline{P}_1 + \overline{G}^T B (1 - \overline{P}_1) + (1 - \overline{P}_2^T) \overline{Y} + \overline{P}_2^T \overline{Z} \overline{P}_2 A,$$

where

$$\overline{Y} = (1 - P_2^T)Y + P_2^T Z P_2 A + G^T B(1 - \tfrac{1}{2}P_1)$$
$$\overline{Z} = Z + \tfrac{1}{2}(G^T B\overline{G} - \overline{G}^T BG).$$

Thus $\overline{Z}$ is again symmetric and we have a solution of the form (7).

In the *R*-matrix context the matrix B is manifestly antisymmetric because of the antisymmetry of the Poisson bracket and so (C1) is clearly satisfied. We have thus reduced the existence of an *R*-matrix to the single consistency equation (C2) and the construction of a generalized inverse to ad(L). We turn now to the construction of the generalized inverse of $-\,\text{ad}(L)_\mu^\nu \equiv c_{\mu\lambda}^\nu L^\lambda = A^{\mu\nu}$.

Let X_μ denote a Cartan–Weyl basis for the Lie algebra $\mathfrak{g}$. That is $\{X_\mu\} = \{H_i, E_\alpha\}$, where $\{H_i\}$ is a basis for the Cartan subalgebra $\mathfrak{h}$ and $\{E_\alpha\}$ is the set of step operators (labeled by the root system Φ of $\mathfrak{g}$). The structure c constants are found from $[H_i, E_\alpha] = \alpha_i E_\alpha$, $[E_\alpha, E_{-\alpha}] = \alpha^\vee \cdot H$ and $[E_\alpha, E_\beta] = N_{\alpha,\beta}E_{\alpha+\beta}$ if $\alpha + \beta \in \Phi$. Here $N_{\alpha,\beta} = c_{\alpha\beta}^{\alpha+\beta}$. With these definitions we then have that

$$\text{ad}(L) = \begin{matrix} & \overset{j}{\downarrow} & \overset{\beta}{\downarrow} \\ i \to & \\ \alpha \to & \end{matrix} \begin{pmatrix} 0 & -\beta_i^\vee L^{-\beta} \\ -\alpha_j L^\alpha & \Lambda_\beta^\alpha \end{pmatrix} = \begin{pmatrix} 0 & u^T \\ v & \Lambda \end{pmatrix}, \tag{8}$$

where we index the rows and columns first by the Cartan subalgebra basis $\{i, j : 1, \ldots, \text{rank}\,\mathfrak{g}\}$ then the root system $\{\alpha, \beta \in \Phi\}$. We will use this block decomposition of matrices throughout. Here u and v are $|\Phi| \times \text{rank}\,\mathfrak{g}$ matrices and we have introduced the $|\Phi| \times |\Phi|$ matrix

$$\Lambda_\beta^\alpha = \alpha \cdot L\delta_\beta^\alpha + c_{\alpha-\beta\beta}^\alpha L^{\alpha-\beta}, \tag{9}$$

where $\alpha \cdot L = \sum_{i=1}^{\text{rank}\,\mathfrak{g}} \alpha_i L^i$. With these definitions we have—

Result 3. *For generic L the matrix Λ is invertible and a generalized inverse of* ad(L) *is given by*

$$\begin{pmatrix} 1 & 0 \\ -\Lambda^{-1}v & 1 \end{pmatrix} \begin{pmatrix} 0 & 0 \\ 0 & \Lambda^{-1} \end{pmatrix} \begin{pmatrix} 1 & -u^T\Lambda^{-1} \\ 0 & 1 \end{pmatrix} = \begin{pmatrix} 0 & 0 \\ 0 & \Lambda^{-1} \end{pmatrix}. \tag{10}$$

We establish the result by first showing that for generic L

$$\text{ad}(L) = \begin{pmatrix} 1 & u^T\Lambda^{-1} \\ 0 & 1 \end{pmatrix} \begin{pmatrix} 0 & 0 \\ 0 & \Lambda \end{pmatrix} \begin{pmatrix} 1 & 0 \\ \Lambda^{-1}v & 1 \end{pmatrix}; \tag{11}$$

it then follows that (10) is a generalized inverse for ad(L) by direct multiplication.

Now, for any matrices m and Λ we have the general factorization [5, 19]

$$\begin{pmatrix} m & u^T \\ v & \Lambda \end{pmatrix} = \begin{pmatrix} 1 & u^T \Xi \\ 0 & 1 \end{pmatrix} \begin{pmatrix} m - u^T \Xi v & u^T(1 - \Xi \Lambda) \\ (1 - \Lambda \Xi) v & \Lambda \end{pmatrix} \begin{pmatrix} 1 & 0 \\ \Xi v & 1 \end{pmatrix},$$

where Ξ is a generalized inverse of Λ. In particular, when $m = 0$ and Λ is invertible (and so $\Xi = \Lambda^{-1}$) this shows that

$$\begin{pmatrix} 0 & u^T \\ v & \Lambda \end{pmatrix} = \begin{pmatrix} 1 & u^T \Lambda^{-1} \\ 0 & 1 \end{pmatrix} \begin{pmatrix} -u^T \Lambda^{-1} v & 0 \\ 0 & \Lambda \end{pmatrix} \begin{pmatrix} 1 & 0 \\ \Lambda^{-1} v & 1 \end{pmatrix}. \tag{12}$$

Thus (11) and hence the results follow by establishing that Λ is generically invertible and that

$$u^T \Lambda^{-1} v = 0. \tag{13}$$

From (9) we see that Λ is the perturbation of a diagonal matrix and so is generically invertible: the zero locus $\det \Lambda = 0$ is a polynomial in the coefficients of $ad(L)$ and so the complement of this set is dense and open. For such an invertible Λ we thus have

$$\operatorname{rank} \Lambda = \dim \Lambda = \dim \mathfrak{g} - \operatorname{rank} \mathfrak{g}. \tag{14}$$

Now the maximum rank[1] of the matrix $ad(L)$ is $\dim \mathfrak{g} - \operatorname{rank} \mathfrak{g}$ [38]. From (12) we see that $\operatorname{rank} \Lambda + \operatorname{rank}(u^T \Lambda^{-1} v) = \operatorname{rank}\big(ad(L)\big) \le \dim \mathfrak{g} - \operatorname{rank} \mathfrak{g}$ and so from (14) we deduce that $\operatorname{rank} \Lambda = \operatorname{rank}\big(ad(L)\big)$. Therefore $0 = \operatorname{rank}(u^T \Lambda^{-1} v)$ and consequently (13) must hold. The result then follows.

An alternate factorization of $ad(L)$ is possible for the generic L under consideration. Utilizing (13) we find that

$$ad(L) = \begin{pmatrix} u^T \Lambda^{-1} \\ 1 \end{pmatrix} \Lambda \begin{pmatrix} \Lambda^{-1} v & 1 \end{pmatrix} = E \Lambda F.$$

Employing a result of MacDuffee [5] this full-rank factorization then yields—

Result 4. *For generic L the Moore–Penrose inverse of* $ad(L)$ *is given by*

$$F^\dagger (F F^\dagger)^{-1} \Lambda^{-1} (E^\dagger E)^{-1} E^\dagger,$$

where $E = \begin{pmatrix} u^T \Lambda^{-1} \\ 1 \end{pmatrix}$ *and* $F = \begin{pmatrix} \Lambda^{-1} v & 1 \end{pmatrix}$.

---

[1]If $\det\big(t - ad(L)\big) = \sum_{j=0}^{\dim \mathfrak{g}} p_j(L) t^j$ is the characteristic polynomial of $ad(L)$, the regular semisimple elements of a semisimple Lie algebra $\mathfrak{g}$ are those elements for which $p_{\operatorname{rank} \mathfrak{g}}(L) \ne 0$. These elements are also of rank $\dim \mathfrak{g} - \operatorname{rank} \mathfrak{g}$ and form an open dense set in $\mathfrak{g}$, but this condition is different from $\det \Lambda \ne 0$.

4   Application

We now apply our theory to the gl_n Calogero–Moser models. For any root system [26, 27] the Calogero–Moser models are the natural Hamiltonian systems

$$H = \frac{1}{2} \sum_i p_i^2 + \sum_{\alpha \in \Phi} U(\alpha \cdot x); \tag{15}$$

where (up to a constant) the potential $U(z)$ is the Weierstrass $\wp$-function or a degeneration that will be specified below. For the root systems of classical algebras a Lax pair may be associated with the models; in the exceptional setting the existence of a Lax pair is still an open problem.

Problem 3. Construct a Lax pair that yields the Calogero–Moser model associated with the E_8 root system (or more generally any exceptional simple Lie algebra).

In fact we don't have a direct proof of—

Problem 4. Show the complete integrability of the Calogero–Moser model associated with the E_8 root system (or more generally any exceptional simple Lie algebra).

Let us consider Lax pairs of the following form [16]:

$$L = p \cdot H + \sum_{\alpha \in \Phi} f^\alpha E_\alpha, \quad M = b \cdot H + \sum_{\alpha \in \Phi} w^\alpha E_\alpha, \tag{16}$$

and where the functions f^α, w^α ($\alpha \in \Phi$) are such that

$$f^\alpha = f^\alpha(\alpha \cdot x), \quad w^\alpha = w^\alpha(\alpha \cdot x). \tag{17}$$

Then

$$\dot{L} = \dot{p} \cdot H + \sum_{\alpha \in \Phi} \alpha \cdot \dot{x} f^{\alpha\prime} E_\alpha \tag{18}$$

and

$$[L, M] = \sum_{\alpha \in \Phi} ((\alpha \cdot p w^\alpha - \alpha \cdot b f^\alpha) E_\alpha - f^{-\alpha} w^\alpha \alpha^\vee \cdot H)$$
$$+ \sum_{\substack{\beta,\gamma \in \Phi \\ \beta+\gamma=\alpha}} c_{\beta\gamma}^\alpha f^\beta w^\gamma E_\alpha. \tag{19}$$

We further assume that b is momentum independent. Upon utilizing $\dot{x}_i = p_i$ and comparing (18) and (19), we find that the Lax equation $\dot{L} = [L, M]$

yields the equations of motion for (15) provided the following consistency conditions (for each $\alpha \in \Phi$) are satisfied:

a) $\qquad\qquad w^\alpha = f^{\alpha\prime},$

b) $\qquad\qquad \dot{p} = -\sum_{\alpha \in \Phi} f^{-\alpha} w^\alpha \alpha^\vee = -\sum_{\alpha \in \Phi} f^{-\alpha} f^{\alpha\prime} \alpha^\vee$

$$= -\frac{d}{dx} \frac{1}{2} \sum_{\alpha \in \Phi} \frac{2}{\alpha \cdot \alpha} f^{-\alpha} f^\alpha = -\frac{d}{dx} \sum_{\alpha \in \Phi} U(\alpha \cdot x),$$

c) $\qquad\qquad \alpha \cdot b = \sum_{\substack{\beta,\gamma \in \Phi \\ \beta+\gamma=\alpha}} c_{\beta\gamma}^\alpha \frac{f^\beta w^\gamma}{f^\alpha} = \sum_{\substack{\beta,\gamma \in \Phi \\ \beta+\gamma=\alpha}} c_{\beta\gamma}^\alpha \frac{f^\beta f^{\gamma\prime}}{f^\alpha},$

The second of the equations determines the potential in terms of the unknown functions f^α. It is the final constraint that is the most difficult to satisfy.

Let us now focus on the Lie algebra gl_n. Here $\Phi = \{e_i - e_j : 1 \leq i \neq j \leq n\}$, where the e_i form an orthonormal basis of $\mathbb{R}^n$. If e_{rs} denotes the elementary matrix with (r,s)th entry one and zero elsewhere, then the $n \times n$ matrix representation $H_i = e_{ii}$ and $E_\alpha = e_{ij}$ when $\alpha = e_i - e_j$ gives the usual representation of L. Working with the simple algebra a_n corresponds to the center of mass frame. Here the Calogero–Moser models are built from the functions

$$f^\alpha = \lambda \frac{\sigma(u - \alpha \cdot x)}{\sigma(u)\sigma(\alpha \cdot x)} e^{\zeta(u)\alpha \cdot x}. \tag{20}$$

These functions satisfy the addition formula

$$f^\alpha f^{\beta\prime} - f^\beta f^{\alpha\prime} = (z_\alpha - z_\beta) f^{\alpha+\beta}, \tag{21}$$

where

$$z_\alpha = \frac{f^{\alpha\prime\prime}}{2f^\alpha} = \lambda \wp(\alpha \cdot x) + \frac{\lambda}{2} \wp(u). \tag{22}$$

Here $\sigma(x)$ and $\zeta(x) = \sigma'(x)/\sigma(x)$ are the Weierstrass sigma and zeta functions [39]. The quantity u in (20) is known as the spectral parameter. We find

$$U(\alpha \cdot x) = -\frac{\lambda^2}{2} \left(\wp(\alpha \cdot x) - \wp(u) \right) \tag{23}$$

and that

$$b = \frac{1}{2(n+1)} \sum_{\substack{\beta,\gamma \in \Phi \\ \beta+\gamma=\alpha}} c_{\beta\gamma}^\alpha z_\beta \alpha,$$

which in components becomes

$$b_i = \lambda \sum_{j \neq i} \wp(x_i - x_j). \tag{24}$$

The Weierstrass $\wp$-function includes as degenerations the potentials

$$U(z) = \frac{\lambda^2}{z^2}, \quad \frac{\lambda^2}{\sin z^2}, \quad \frac{\lambda^2}{\sinh z^2}. \tag{25}$$

We now begin unraveling (3) in the present setting. Using $\{p_j, f^\alpha\} = \{p_j, \alpha \cdot q\} f^{\alpha\prime} = \alpha_j f^{\alpha\prime}$, the right-hand side becomes

$$\{L \otimes L\} = \sum_{\mu,\nu} \{L^\mu, L^\nu\} X_\mu \otimes X_\nu = \sum_{j,\alpha} \alpha_j f^{\alpha\prime}(H_j \otimes E_\alpha - E_\alpha \otimes H_j).$$

This means that we have

$$B = \begin{pmatrix} 0 & \beta_i f^{\beta\prime} \\ -\alpha_j f^{\alpha\prime} & 0 \end{pmatrix} = -B^T$$

and the constraint (C1) is manifestly satisfied.

Let us temporarily defer discussion of the constraint (C2). From the fact that $A = -\operatorname{ad}(L)^T$, a generalized inverse of A is given by minus the transpose of the generalized inverse (10). Using our earlier notation, this means that we have projectors

$$P_1 = \begin{pmatrix} 0 & 0 \\ \Lambda^{-1T}u & 1 \end{pmatrix}, \quad P_2 = \begin{pmatrix} 0 & v^T\Lambda^{-1T} \\ 0 & 1 \end{pmatrix}$$

and from (7) the general *R*-matrix

$$R = \begin{pmatrix} 0 & 0 \\ (\Lambda^{-1})^\alpha_\beta f^{\beta\prime}\beta_j & 0 \end{pmatrix} + \begin{pmatrix} p & q \\ -\Lambda^{-1}vp - Fu & -\Lambda^{-1}vq - F\Lambda^T \end{pmatrix}. \tag{26}$$

The second term characterizes the ambiguity in R. Here p, q are arbitrary matrices of the appropriate sizes coming from $Y = \left(\begin{smallmatrix} p & q \\ r & s \end{smallmatrix}\right)$, while the entries of $Z = \left(\begin{smallmatrix} a & b \\ c & d \end{smallmatrix}\right)$ are such that

$$F = \Lambda^{-1}vav^T\Lambda^{-1T} + d + \Lambda^{-1}vb + cv^T\Lambda^{-1T}$$

is symmetric.

It is instructive to consider how the minimal solution given by the first term of (26) satisfies (3). We have

$$R^{\alpha j} = (\Lambda^{-1})^\alpha_\beta f^{\beta\prime}\beta_j, \quad R^{ij} = 0, \quad R^{i\alpha} = 0, \quad R^{\alpha\beta} = 0. \tag{27}$$

This is to be compared with the previously known *R*-matrix [3]

$$R^{\alpha j} = 0, \quad R^{ij} = 0, \quad R^{i\alpha} = -\frac{|\alpha_i|}{2}f^\alpha, \quad R^{\alpha\beta} = \delta_{\alpha+\beta,0}\frac{f^{\alpha\prime}}{f^\alpha},$$

which exists only [13] for the potentials $U(z) = \lambda^2/z^2$, $\lambda^2/\sin z^2$, $\lambda^2/\sinh z^2$. Now Eq. (3) yields three different equations, depending on the range of $\{\mu, \nu\}$. For $(\mu, \nu) = (i, j)$, (i, α), and (α, β) respectively, these are

$$0 = \sum_\alpha (R^{\alpha j} \alpha_i - R^{\alpha i} \alpha_j) f^{-\alpha}, \tag{28}$$

$$\alpha_i f^{\alpha \prime} = \alpha \cdot p \, R^{\alpha i} - \sum_j \alpha_j R^{ji} f^\alpha$$
$$+ \sum_\beta (\beta_i f^\beta R^{-\beta \alpha} + f^{\alpha - \beta} R^{\beta i} c^\alpha_{\alpha - \beta \, \beta}), \tag{29}$$

and

$$0 = \sum_i (\alpha_i R^{i\beta} f^\alpha - \beta_i R^{i\alpha} f^\beta) - (\alpha \cdot p \, R^{\alpha \beta} - \beta \cdot p \, R^{\beta \alpha})$$
$$+ \sum_\gamma (R^{\gamma \beta} c^\alpha_{\gamma \, \alpha - \gamma} f^{\alpha - \gamma} - R^{\gamma \alpha} c^\beta_{\gamma \, \beta - \gamma} f^{\beta - \gamma}). \tag{30}$$

Employing (27), we see that the final two equations are automatically satisfied. The first is less obvious until we realize it just expresses the remaining constraint (C2) necessary for a solution to exist. We require

$$0 = (1 - P_1^T) B (1 - P_1) = \begin{pmatrix} u^T \Lambda^{-1} f^{\beta \prime} \beta - \alpha f^{\alpha \prime} \Lambda^{-1T} u & 0 \\ 0 & 0 \end{pmatrix},$$

and so (taking the (i, j)th entry), this becomes

$$\text{(C2)} \qquad 0 = \sum_{\alpha, \beta} (\alpha_i f^{-\alpha} (\Lambda^{-1})^\alpha_\beta f^{\beta \prime} \beta_j - \alpha_i f^{\alpha \prime} (\Lambda^{-1})^\beta_\alpha f^{-\beta} \beta_j). \tag{31}$$

Upon using $R^{\alpha j} = (\Lambda^{-1})^\alpha_\beta f^{\beta \prime} \beta_j$ we may identify this with (28).

At this stage we have reduced the existence of an R-matrix for Calogero–Moser systems to the consistency equation (31). Although the inverse matrices here look somewhat daunting we may use the cofactor expansion of an inverse to give

$$\begin{vmatrix} 0 & k \\ l & \Lambda \end{vmatrix} = -|\Lambda| \, k^T \Lambda^{-1} l.$$

Thus (31) is equivalent to showing that (for each i, j) the $(|\Phi| + 1) \times (|\Phi| + 1)$ determinants satisfy

$$\begin{vmatrix} 0 & \alpha_i f^{-\alpha} \\ \beta_j f^{\beta \prime} & \Lambda \end{vmatrix} = \begin{vmatrix} 0 & \alpha_j f^{-\alpha} \\ \beta_i f^{\beta \prime} & \Lambda \end{vmatrix}, \tag{32}$$

where[2] $\Lambda = (\Lambda_{\alpha \beta})$. Note that (13) yields in the present setting

$$0 = (u^T \Lambda^{-1} v)_{ij} = \sum_{\alpha, \beta} \alpha_i^\vee L^{-\alpha} (\Lambda^{-1})^\alpha_\beta L^\beta \beta_j = \begin{vmatrix} 0 & \alpha_i f^{-\alpha} \\ \beta_j f^\beta & \Lambda \end{vmatrix}. \tag{33}$$

---

[2]Note the adjugate matrix of Λ involves the transpose of the cofactors and hence the perhaps puzzling interchange of rows and columns here.

Whereas (33) is true for any functions f^α, Eq. (32) will only hold for a more restricted class of functions.

Problem 5. Determine which functions of the form (20) satisfy (32).

I expect that (32) is satisfied when the spectral parameter corresponds to one of the half-periods but have yet to show this. This would mean that a spectral-parameter-independent *R*-matrix exists for these values. I am not convinced (32) is true for arbitrary values of the spectral parameter u but we know that for such values an *R*-matrix with spectral parameter exists [13, 33].

5 Discussion

We have presented a uniform construction for a classical *R*-matrix given a Lax pair, thus answering the question of the Liouville integrability of the system in terms of the invariants of the matrix L. The method not only gives necessary and sufficient conditions for the *R*-matrix to exist and describes its ambiguities, but is algorithmic as well. Given L, first construct $ad(L)$. Next construct any generalized inverse to $ad(L)$ and verify (C2); this is the necessary and sufficient condition for an *R*-matrix to exist: it is given explicitly by (7). Further we have given a generalized inverse for generic L in (10); genericity is easily checked by evaluating $\det \Lambda \neq 0$, where Λ is the restriction of $ad(L)$ to the root space (given by (9)).

Having presented our uniform construction we then applied this in the context of the Calogero–Moser–Sutherland models, the focus of this meeting. Here $ad(L)$ is generic and our construction of the generalized inverse applies. The necessary and sufficient condition for the existence of the *R*-matrix (C2) is then (31) and with some rearrangement this becomes the question of the equality of two determinants (32). I have still to prove this holds, though I believe it to be true for the spectral parameter equaling one of the half-periods.

For simplicity I have presented the construction for Lax pairs with no spectral parameter: it is straightforward to incorporate the spectral parameter. This extension has been considered in Ref. 6, where our approach is also shown to be relevant to the construction of quadratic *R*-matrices.

Finally let me mention that for systems obtained by Hamiltonian reduction an alternative geometric construction of classical *R*-matrices exists [2] in terms of Dirac brackets. This suggests there is a correspondence between Dirac brackets and generalized inverses. This is indeed the case, and I will present this elsewhere.

Acknowledgments: I would like to thank the workshop organizers and participants for such a stimulating meeting and the opportunity to present this material. I have benefitted from comments by J. Avan, J. Harnad, A. N. W. Hone, I. Krichever, V. Kuznetsov, M. Olshanetsky, and E. K. Sklyanin.

6 REFERENCES

1. V. I. Arnold, *Mathematical Methods of Classical Mechanics*, Grad. Texts in Math., Vol. 60, Springer, New York, 1978.

2. J. Avan, O. Babelon, and M. Talon, *Construction of the classical R-matrices for the Toda and Calogero models*, Algebra i Analiz **6** (1994), No. 2, 67–89.

3. J. Avan and M. Talon, *Classical R-matrix structure for the Calogero model*, Phys. Lett. B **303** (1993), No. 1-2, 33–37.

4. O. Babelon and C. M. Viallet, *Hamiltonian structures and Lax equations*, Phys. Lett. B **237** (1990), No. 3-4, 411–416.

5. A. Ben-Israel and T. N. E. Greville, *Generalized Inverses: Theory and Applications*, Krieger, Huntington, NY, 1980.

6. H. W. Braden, *R-matrices and generalized inverses*, J. Phys. A **30** (1997), No. 15, L485–L493, q-alg/9706001, solv-int/9706001.

7. _____, *The equation $A^T X - X^T A = B$*, SIAM J. Matrix Anal. Appl **20** (1999), No. 2, 295–302, Edinburgh preprint MS-97-007.

8. H. W. Braden and V. M. Buchstaber, *The general analytic solution of a functional equation of addition type*, SIAM J. Math. Anal. **28** (1997), No. 4, 903–923.

9. _____, *Integrable systems with pairwise interactions and functional equations*, Reviews in Mathematics and Mathematical Physics (to appear).

10. H. W. Braden, E. Corrigan, P. E. Dorey, and R. Sasaki, *Affine Toda field theory and exact S-matrices*, Nucl. Phys. B **338** (1990), No. 3, 689–746.

11. H. W. Braden and A. N. W. Hone, *Affine Toda solitons and systems of Calogero-Moser type*, Phys. Lett. B **380** (1996), No. 3-4, 296–302.

12. H. W. Braden and R. Sasaki, *The Ruijsenaars–Schneider model*, Progr. Theor. Phys. **97** (1997), No. 6, 1003–1017, hep-th/9702182.

13. H. W. Braden and T. Suzuki, *R-matrices for elliptic Calogero–Moser models*, Lett. Math. Phys. **30** (1994), No. 2, 147–158, hep-th/9309033.

14. V. M. Buchstaber and A. M. Perelomov, *On the functional equation related to the quantum three-body problem,* Contemporary Mathematical Physics (A. B. Sossinsky, ed.), Amer. Math. Soc. Transl. Ser. 2, Vol. 175, Amer. Math. Soc., Providence RI, 1996, pp. 15–34.

15. F. Calogero, *One-dimensional many-body problems with pair interactions whose ground-state wavefunction is of product type,* Lett. Nuovo Cimento **13** (1975), 507–511.

16. ———, *On a functional equation connected with integrable many-body problems,* Lett. Nuovo Cimento **16** (1976), No. 3, 77–80.

17. S. R. Caradus, *Generalized Inverses and Operator Theory,* Queen's Papers in Pure and Appl. Math., Vol. 50, Queen's Univ., Kingston, ON, 1978.

18. L. A. Ferreira and D. I. Olive, *Noncompact symmetric spaces and the Toda molecule equations,* Commun. Math. Phys. **99** (1985), No. 3, 365–384.

19. T. N. E. Greville, *Some applications of the pseudoinverse of a matrix,* SIAM Rev. **2** (1960), 15–22.

20. T. Hollowood, *Quantizing* sl(n) *solitons and the Hecke algebra,* Internat. J. Modern Phys. A **8** (1993), No. 5, 947–981.

21. D. Kazhdan, B. Kostant, and S. Sternberg, *Hamiltonian group actions and dynamical systems of Calogero type,* Commun. Pure Appl. Math. **31** (1978), No. 4, 481–507.

22. I. M. Krichever and A. Zabrodin, *Spin generalization of the Ruijsenaars–Schneider model, non-Abelian 2D Toda chain and representations of Sklyanin algebra,* Russian Math. Surveys **50** (1995), No. 6, 1101–1150, hep-th/9505039.

23. V. B. Kuznetsov, F. W. Nijhoff, and E. K. Sklyanin, *Separation of variables for the Ruijsenaars system,* Commun. Math. Phys. **189** (1997), No. 3, 855–877, solv-int/9701004.

24. J. Liouville, *Note sur l'integration des équations différentielles de la dynamique,* J. Math. Pures Appl. **20** (1855), 137–138.

25. D. I. Olive, N. Turok, and J. W. R. Underwood, *Affine Toda solitons and vertex operators,* Nucl. Phys. B **409** (1993), No. 3, 509–546.

26. M. A. Olshanetsky and A. M. Perelomov, *Completely integrable Hamiltonian systems connected with semisimple Lie algebras,* Invent. Math. **37** (1976), No. 2, 93–108.

27. _____, *Classical integrable finite-dimensional systems related to Lie algebras*, Phys. Rep. **71** (1981), No. 5, 313–400.

28. R. M. Pringle and A. A. Rayner, *Generalized Inverse Matrices with Applications to Statistics*, Griffin's Statist. Monogr. Courses, Hafner Publishing Co., New York, 1971.

29. C. R. Rao and S. K. Mitra, *Generalized Inverse of Matrices and Its Applications*, Wiley, New York–London–Sidney, 1971.

30. S. N. M. Ruijsenaars, *Systems of Calogero–Moser type*, Particles and Fields (Banff, 1994) (G. Semenoff and L. Vinet, eds.), CRM Series in Mathematical Physics, Springer, New York, 1999, pp. 251–352.

31. S. N. M. Ruijsenaars and H. Schneider, *A new class of integrable systems and its relation to solitons*, Ann. Phys. (NY) **170** (1986), No. 2, 370–405.

32. M. A. Semenov-Tian-Shansky, *What a classical r-matrix is*, Funct. Anal. Appl. **17** (1983), No. 4, 259–272.

33. E. K. Sklyanin, *Dynamic r-matrices for the elliptic Calogero–Moser model*, St. Petersburg Math. J. **6** (1995), No. 2, 397–406.

34. _____, *Separation of variables—new trends*, Quantum Field Theory, Integrable Models and Beyond (Kyoto, 1994) (T. Inami and R. Sasaki, eds.), Progr. Theoret. Phys. Suppl., Vol. 118, Progr. Theoret. Phys., Kyoto, 1995, pp. 35–60.

35. B. Sutherland, *Exact results for a quantum many-body problem in one dimension*, Phys. Rev. A **4** (1971), 2019–2021.

36. _____, *Exact results for a quantum many-body problem in one dimension*. II, Phys. Rev. A **5** (1972), 1372–1376.

37. _____, *Exact ground-state wave function for a one-dimensional plasma*, Phys. Rev. Lett. **34** (1975), 1083–1085.

38. V. S. Varadarajan, *Lie Groups, Lie Algebras, and their Representations*, Prentice-Hall Series in Modern Analysis, Prentice-Hall, Englewood Cliffs, NJ, 1974.

39. E. T. Whittaker and G. N. Watson, *A Course of Modern Analysis*, Cambridge Univ. Press, Cambridge, 1927.

7

Tricks of the Trade: Relating and Deriving Solvable and Integrable Dynamical Systems

Francesco Calogero[1]

ABSTRACT A heuristic/pedagogical presentation of various tricks that can be used to relate different solvable and integrable dynamical systems to each other, hence also to derive new ones from known ones. Several examples are exhibited.

1 Introduction

This paper is characterized by a heuristic approach and a pedagogical purpose. Our aim is to survey the various tricks that allow us to relate different dynamical systems to each other. None of these tricks is new; but perhaps there is no other place where an attempt has been made to survey several of them, as we try to do here (of course, with no pretense at completeness).

Our attention focuses on *solvable* and *integrable* models: in the heuristic spirit that, in the interest of easy readability, dominates our presentation, we generally qualify as *solvable* the dynamical systems whose equations of motion can be "solved," generally by some appropriate change of variables (typically, from the n coefficients of a monic polynomial of degree n to its n zeros, and vice versa); and we qualify instead as "integrable" the dynamical systems for which some kind of "Lax type" [17] reformulation of the equations of motion exist, entailing the possibility of evincing immediately the existence of n constants of the motion (typically related to the n eigenvalues of a square matrix of rank n whose entries are explicitly given in terms of the dynamical variables; eventually, of course, this approach also allows us to "solve" the equations of motion). Generally the technique of solution is "easier" for *solvable* systems than for *integrable* systems; hence *solvable* systems are generally also *integrable*, although there is not much of an incentive to display this fact (if you know how to solve the equations of motion, what more do you want?). Perhaps some read-

---

[1]On leave while serving as Secretary General, Pugwash Conferences on Science and World Affairs, Geneva, London, Rome.

ers will recognize in this distinction between *solvable* and *integrable* dynamical systems an analogy with the distinction between *C-integrable* and *S-integrable* nonlinear partial differential equations [7, 12, 13]; although in the case of finite-dimensional systems the distinction is perhaps even less clear-cut than for infinite-dimensional systems [7, 12, 13]. In any case, in the heuristic/pedagogical spirit in which this paper is presented, we advise any reader who feels confused by this distinction to simply ignore it.

We also tend to qualify as *integrable* only *Hamiltonian* systems (whose *equations of motion* can be cast in *Hamiltonian* form); although in the following we focus our attention mainly on the *equations of motion*, without paying much attention to their origin, as long as they can be interpreted as describing the motion of a (nonquantal) *many-body problem of Newtonian type* ("acceleration = force"), or some variation on this theme. And we only consider models in which the number n of "particles" is *arbitrary*.

In the following Section 2 we survey several such known *solvable* and *integrable* dynamical systems; our treatment is generally limited to a display, with minimal comments, of the relevant *equations of motion*, which are then used as raw material in the subsections of the subsequent section (i.e., Section 3, the core part of this paper), to illustrate how the various tricks described there work. In the interest of the hasty reader, we also identify at the end of Section 2, without any elaboration, those *solvable* and *integrable* models, obtained from the known ones in Section 3, that are perhaps *new*. The hasty reader might also like to skim through the headings of the subsections of Section 3 in order to get a quick idea of the material covered in this paper.

2 Survey of Solvable and Integrable Many-Body Models

In this section we mention some known *solvable* or *integrable* systems (without any pretense to provide a complete list), and at the end we identify those systems, obtained via the tricks of the following Section 3 and displayed there, which are, to the best of our knowledge, *new*.

Given the context of this contribution, the first models we list are the *integrable* equations of motion:

$$\ddot{x}_j + \Omega^2 x_j = 2g^2 \sum_{\substack{k=1 \\ k \neq j}}^{n} (x_j - x_k)^{-3}, \tag{1a}$$

$$\ddot{x}_j = 2g^2 a^3 \sum_{\substack{k=1 \\ k \neq j}}^{n} \frac{\cos[a(x_j - x_k)]}{\sin^3[a(x_j - x_k)]}, \tag{1b}$$

$$\ddot{x}_j = -g^2 a^3 \sum_{\substack{k=1 \\ k \neq j}}^{n} \wp'[a(x_j - x_k)].$$ (1c)

Here, and always below unless otherwise indicated, j takes all integer values from 1 to n, and the "number of particles" n is an *arbitrary* positive integer $(n \geq 2)$.

These equations of motion are *Hamiltonian*. The Hamiltonian function that yields (1a) was first shown to be "solvable" in the *quantal* context (its complete spectrum was obtained, in the $\Omega^2 > 0$ case, and the scattering behavior completely characterized, in the $\Omega = 0$ case) [3]; and on this basis the conjecture was formulated, that in the *classical* case all its motions would be completely periodic (if Ω is *real* and nonvanishing) [3]. The complete integrability of (1a) with $\Omega = 0$ was then proved by J. Moser [19], by recasting these equations of motion in Lax form. Hence, models of type (1a) are often referred to as "Calogero–Moser systems." The extension of the treatment of Moser to include the $\Omega \neq 0$ case is due to M. Adler [1], who thereby validated the conjecture made in Ref. 3; but in fact there is a simple trick, due to A. M. Perelomov [20], that allows transforming (1a) with $\Omega = 0$ into (1a) with $\Omega \neq 0$ (and vice versa; see below, Section 3.1. The Hamiltonian which, in the *classical* context, yields the equations of motion (1b), was introduced, and shown to be "solvable," by B. Sutherland, also in the *quantal* context (its complete spectrum was obtained) [23, 24]. Hence the model (1b) is often referred to as "Sutherland system." The equations of motion (1c) were introduced in the *classical, Hamiltonian*, context, and shown to be *integrable* in Ref. 4 by casting them in Lax form, using an *ansatz* based on the Lax matrix introduced by Moser for the system (1a) [19], and then solving a functional equation (the first appearance of functional equations, and of elliptic functions, in the context of classical integrable dynamical systems of this type). Note that in (1c) $\wp(z) \equiv \wp(z \mid \omega, \omega')$ is the Weierstrass function, of *arbitrary* semiperiods ω, ω', and of course $\wp'(z) \equiv d\wp(z)/dz$. It is easily seen that (1b) and (1a) (with $\Omega = 0$) are special cases of (1c), since $\wp(z \mid \pi/(2a), i\infty) = -a^2/3 + a^2/\sin^2(az)$ and $\wp(z \mid \infty, i\infty) = z^{-2}$ (and of course (1a) is the special case of (1b) corresponding to $a = 0$). But it is also possible to obtain (1b) and (1c) from (1a), by using one of the tricks described below (see Section 3.3.4).

The next system we exhibit is *solvable*, and it features several *arbitrary* "coupling constants":

$$\ddot{x}_j = a\dot{x}_j + \beta_0 + \beta_1 x_j$$
$$+ \sum_{\substack{k=1 \\ k \neq j}}^{n} (x_j - x_k)^{-1}[2\dot{x}_j \dot{x}_k + (\dot{x}_j + \dot{x}_k)(\lambda_0 + \lambda_1 x_j)$$
$$+ \lambda_2(\dot{x}_j x_k + \dot{x}_k x_j)x_j$$
$$+ \mu_{-1} + \mu_0 x_j + \mu_1 x_j^2 + \mu_2 x_j^2 x_k].$$ (2)

Its solvability is connected with the *change of variables* among the n coefficients of a (monic) polynomial of degree n and its n zeros [6]. Let us also mention that, while one generally assumes the 10 "coupling constants" that appear in the right-hand side of (2) to be time independent, the method of solution works even if these quantities are (arbitrarily!) time dependent [6].

The most natural environment to investigate the zeros of polynomials is of course the *complex plane* rather than the *real line*. By considering systems of type (2) in such a context, and by suitably restricting their generality (namely, setting to zero all the "coupling constants" with subscript different from 1, $\beta_0 = \lambda_0 = \lambda_2 = \mu_{-1} = \mu_0 = \mu_2 = 0$), a system has been recently obtained which is naturally interpretable as a *solvable* many-body problem *in the plane* [9]:

$$\ddot{\vec{r}}_j = (\alpha + \alpha'\hat{z}\wedge)\dot{\vec{r}}_j + (\beta + \beta'\hat{z}\wedge)\vec{r}_j$$
$$+ \sum_{\substack{k=1 \\ k\neq j}}^{n} r_{jk}^{-2}\Big\langle 2[\dot{\vec{r}}_j(\vec{r}_k \cdot \vec{r}_{jk}) + \dot{\vec{r}}_k(\vec{r}_j \cdot \vec{r}_{jk}) - \vec{r}_{jk}(\dot{\vec{r}}_j \cdot \vec{r}_k)]$$
$$+ (\lambda + \lambda'\hat{z}\wedge)\Big\{(\dot{\vec{r}}_j + \dot{\vec{r}}_k)[r_j^2 - (\vec{r}_j \cdot \vec{r}_k)]$$
$$- \vec{r}_j[\vec{r}_k \cdot (\dot{\vec{r}}_j + \dot{\vec{r}}_k)] + \vec{r}_k[\vec{r}_j \cdot (\dot{\vec{r}}_j + \dot{\vec{r}}_k)]\Big\}$$
$$+ (\mu + \mu'\hat{z}\wedge)\{\vec{r}_j[r_j^2 - 2(\vec{r}_j \cdot \vec{r}_k)] + \vec{r}_k r_j^2\}\Big\rangle,$$
$$\vec{r}_{jk} \equiv \vec{r}_j - \vec{r}_k. \quad (3)$$

Here $\vec{r}_j$ is a vector *in the plane* for which we use the three-dimensional notation $\vec{r}_j \equiv (x_j, y_j, 0)$, with $\hat{z} \equiv (0, 0, 1)$ the unit vector orthogonal to the plane, so that $\hat{z} \wedge \vec{r}_j \equiv (-y_j, x_j, 0)$. This notation has the advantage of exhibiting the rotation-invariant character of these equations of motion, which we deem essential to interpret them as describing a many-body problem *in the plane* (see Section 3.7). Note the *arbitrariness* of the eight "coupling constants" that appear in the right-hand side of (3); it entails the existence of a large family of models, whose motions display a rich phenomenology [9].

The next system we display is also *solvable* by a variation of the technique based on the nonlinear mapping between the n coefficients and the n zeros of a (monic) polynomial of degree n [6]:

$$\ddot{x}_j = \alpha\dot{x}_j + \beta + \omega \sum_{\substack{k=1 \\ k\neq j}}^{n} \coth[\omega(x_j - x_k)]\{2\dot{x}_j\dot{x}_k + \lambda(\dot{x}_j + \dot{x}_k) + \mu\}. \quad (4)$$

Here we consider again the *real* line as natural environment for such a system.

Next we display an *integrable* dynamical system that extends [10] those originally introduced by S. N. M. Ruijsenaars and H. Schneider [22],[2] as reported in Ref. 2:

$$\ddot{x}_j + i\Omega \dot{x}_j = \sum_{\substack{k=1 \\ k \neq j}}^{n} \dot{x}_j \dot{x}_k f(x_j - x_k),$$ (5a)

$$f(x) = \frac{2}{x},$$ (5b)

$$f(x) = \frac{2}{x(1 + r^2 x^2)},$$ (5c)

$$f(x) = 2a \coth(ax),$$ (5d)

$$f(x) = \frac{2a}{\sinh(ax)},$$ (5e)

$$f(x) = \frac{2a \coth(ax)}{1 + r^2 \sinh^2(ax)},$$ (5f)

$$f(x) = -a\wp'(ax)[\wp(ax) - \wp(ab)].$$ (5g)

As explained in Ref. 10, these systems are *Hamiltonian* and *integrable*, and presumably they only feature *completely periodic* motions (with period $T = 2\pi/\Omega$, or some finite multiple of it; of course iff the constant Ω is *real* and it does not vanish). It is clear that the functions $f(x)$ given in (5b)–(5f) are all special cases of (5g) (see above, the end part of the second paragraph after (1c)), and accordingly, that in all cases, (5b)–(5g), the function $f(x)$ has a simple pole at $x = 0$ with residue equal to 2. Also note that (5) with (5a) and (5c) are special cases of, respectively, (2) and (4); hence in these two special cases (the first of which is of course itself a special case of the second) the system (5) is actually not only *integrable* but also *solvable*.

The last instance we report is the well-known *integrable* system featuring only "nearest neighbor interactions":

$$\ddot{x}_j = 2ag^2 \left[\exp\{-a(x_j - x_{j-1})\} - \exp\{a(x_j - x_{j+1})\} \right].$$ (6)

This system was originally introduced in the $n = \infty$ case by M. Toda [25–27]; hence, it is generally referred to as "Toda lattice" or "Toda system." Note that, for finite n, the equations of motion (6) are incomplete, since one must additionally specify the behavior at the end points (namely, how to interpret x_{j+1} for $j = n$ and x_{j-1} for $j = 1$). In fact, the system (6) is generally *integrable* for any ("reasonable") such assignment; a key step in demonstrating this has been the reformulation of the equations of motion (6) in Lax form, first done (independently and practically simultaneously) by H. Flaschka [15, 16] and S. V. Manakov [18].

We end this section by identifying, without any additional comment, those among the systems obtained and displayed below that are, to the

---

[2]See also Ref. 21 and the other papers by S. N. M. Ruijsenaars referred to there.

best of our knowledge, *new*. Since they are obtained starting from *solvable* or *integrable* systems, these systems are of course themselves *solvable* or *integrable*, as the case may be. The list of the equations that display these systems reads as follows: (29a)–(29c) or equivalently (30a)–(30c); (32); (49a)–(49c); (51a)–(51c); (72)–(74); and perhaps (59) with (60), as well as (61) with (62) (these two latter cases are examples of new solutions rather than new systems). This list is given here merely for the convenience of the hasty reader; if any interest is evoked by it, it should be pursued by following up the appropriate treatment below.

3 Tricks

In the following subsections various tricks are introduced that allow us to obtain a different dynamical system from a known one. It is generally the case that, if such a trick is applied to a *solvable* or to an *integrable* dynamical system, as we do below, it yields a system that is as well *solvable* or *integrable*. We take this statement for granted here, without further ado; in fact, it is generally easy, in each case, to demonstrate its validity by applying the trick not only to the equations of motion (to which our attention is mainly limited below), but also to the mechanism that underpins the property of *solvability* or *integrability*.

3.1 *Time-Dependent Rescalings of Dependent Variables and of Time*

Consider the transformation

$$x_j(t) = \varphi(t)\xi_j(\tau), \quad \tau = \tau(t), \tag{7a}$$

which of course entails

$$\dot{x}_j(t) = \varphi(t)\dot{\tau}(t)\xi_j'(\tau) + \dot{\varphi}(t)\xi_j(\tau), \tag{7b}$$

$$\ddot{x}_j(t) = \varphi(t)[\dot{\tau}(t)]^2\xi_j''(\tau) + [2\dot{\varphi}(t)\dot{\tau}(t) + \varphi(t)\ddot{\tau}(t)]\xi_j'(\tau) + \ddot{\varphi}(t)\xi_j(\tau). \tag{7c}$$

Here (and throughout this paper) a superimposed dot denotes differentiation with respect to the time t, while in this section an appended prime denotes differentiation with respect to τ.

Clearly, by using these formulas with appropriate choices for the two *a priori arbitrary* functions $\varphi(t)$ and $\tau(t)$, a dynamical system gets transformed into a different one.

A convenient choice sets

$$\dot{\tau}(t) = [\varphi(\tau)]^{-2}, \tag{8a}$$

so that the previous equations take the neater form

$$\dot{x}_j(t) = [\varphi(t)]^{-1}\xi_j'(\tau) + \dot{\varphi}(t)\xi_j(\tau), \tag{8b}$$

$$\ddot{x}_j(t) = [\varphi(t)]^{-3}\xi_j''(\tau) + \ddot{\varphi}(t)\xi_j(\tau). \tag{8c}$$

As already mentioned above, this trick has been used by A. M. Perelomov [20] to transform (1a) with $\Omega \neq 0$ into the same equation with $\Omega = 0$ (and, more generally, to solve (1a) with a time-dependent Ω). Indeed the insertion of (7a) and (8c) in (1a) yields

$$\xi_j'' + [(\ddot{\varphi} + \Omega^2\varphi)\varphi^3]\xi_j = 2g^2 \sum_{\substack{k=1 \\ k\neq j}}^{n} (\xi_j - \xi_k)^{-3}; \tag{9}$$

hence the choice $\varphi(t) = \cos(\Omega t)$ accomplishes the task if Ω is time independent; and choosing for $\varphi(t)$ any solution of the linear second-order ODE $\ddot{\varphi} + \Omega^2\varphi = 0$ achieves the same goal in the more general case of a time-dependent Ω.

Another application of this trick allows us to transform the systems (1b) and (1c) with a constant into analogous systems with $a = a(t) = 1/(A + Bt)$, with A and B constant (and, of course, vice versa) [8].

Another convenient choice sets

$$\varphi(t) = 1, \tag{10a}$$

so that

$$\dot{x}_j(t) = \dot{\tau}(t)\xi_j'(\tau), \tag{10b}$$

$$\ddot{x}_j(t) = [\dot{\tau}(\tau)]^2\xi_j''(\tau) + \ddot{\tau}(t)\xi_j'(\tau). \tag{10c}$$

By applying this transformation to the system

$$\ddot{x}_j(t) + \alpha(t)\dot{x}_j(t) = \sum_{k=1}^{n} \dot{x}_j(t)\dot{x}_k(t)f_{jk}[x_j(t), x_k(t)], \tag{11}$$

one gets

$$\xi_j''(\tau) + [\dot{\tau}(t)]^{-2}[\ddot{\tau}(t) + \alpha(t)\dot{\tau}(t)]\xi_j'(\tau)$$

$$= \sum_{k=1}^{n} \xi_j'(\tau)\xi_k'(\tau)f_{jk}[\xi_j(\tau), \xi_k(\tau)]. \tag{12}$$

Hence a choice of $\tau(t)$ such that

$$\ddot{\tau}(t) + \alpha(t)\dot{\tau}(t) = 0 \tag{13}$$

transforms (11) into

$$\xi_j''(\tau) = \sum_{k=1}^{n} \xi_j'(\tau)\xi_k'(\tau) f_{jk}[\xi_j(\tau), \xi_k(\tau)]. \tag{14}$$

In particular the transformation

$$x_j(t) = \xi_j(\tau), \quad \tau(t) = \frac{i}{\Omega}[\exp(-i\Omega t) - 1], \tag{15}$$

transforms the system (5a) into

$$x_j''(\tau) = \sum_{\substack{k=1 \\ k \neq j}}^{n} \xi_j'(\tau)\xi_k'(\tau) f[\xi_j(\tau) - \xi_k(\tau)]. \tag{16}$$

Hence every nonsingular and univalent (complex) solution of (16) yields a periodic solution, with period $T = 2\pi/\Omega$, of (5a).

3.2   From One to More Kinds of Particles by Shifting the Dependent Variables

A convenient context to illustrate this trick [4] is the one-dimensional many-body problem:

$$\ddot{x}_j = 2g^2 a^3 \sum_{\substack{k=1 \\ k \neq j}}^{n} \frac{\cosh[a(x_j - x_k)]}{\sinh^3[a(x_j - x_k)]}, \tag{17}$$

which obtains from (1b) via the replacement $a \to ia$. The corresponding Hamiltonian reads of course

$$H = \frac{1}{2}\sum_{j=1}^{n} p_j^2 + g^2 a^2 \sum_{j=2}^{n}\sum_{k=1}^{j-1} \{\sinh[a(q_j - q_k)]\}^{-2}, \tag{18}$$

with p_j the canonical momentum conjugated to the canonical variable $q_j = x_j$. It is convenient to think of this model as describing n equal particles interacting pairwise via the short-range repulsive potential $V_e(x)$,

$$V_e(x) = \frac{g^2 a^2}{\sinh^2(ax)}. \tag{19}$$

Now set

$$x_{m+k} = y_k + \frac{i\pi}{2a}, \quad k = 1, \ldots, n - m, \tag{20}$$

with $0 \leq m \leq n$. Then (18) takes the form

$$
H = \frac{1}{2}\sum_{j=1}^{m}(p_j^{(1)})^2 + \frac{1}{2}\sum_{j=1}^{n-m}(p_j^{(2)})^2 + \sum_{j=2}^{m}\sum_{k=1}^{j-1}V_e(q_j^{(1)} - q_k^{(1)})
$$

$$
+ \sum_{j=2}^{n-m}\sum_{k=1}^{j-1}V_e(q_j^{(2)} - q_k^{(2)}) + \sum_{j=1}^{m}\sum_{k=1}^{n-m}V_d(q_j^{(1)} - q_k^{(2)}), \quad (21)
$$

with

$$
V_d(x) = -\frac{g^2 a^2}{\cosh^2(ax)}. \tag{22}
$$

Here of course $x_j = q_j^{(1)}$, $y_j = q_j^{(2)}$, and $p_j^{(1)}$ resp. $p_j^{(2)}$ are the canonical momenta conjugate to the canonical variables $q_j^{(1)}$ resp. $q_j^{(2)}$.

The new Hamiltonian (21) clearly describes an n-body problem on the line, with m particles of one type and $n - m$ of another, the (singular repulsive) potential $V_e(x)$, see (19), characterizing the interaction acting among *equal* particles, and the (nonsingular attractive) potential $V_d(x)$, see (22), characterizing the interaction among *different* particles.

3.3 "Duplications": Configurations that are Preserved Throughout the Motion

The general idea of these tricks is to identify special solutions of a dynamical system that describe configurations which are preserved throughout the motion, over time, and which may therefore be interpreted as providing the solutions of new dynamical systems characterized by less degrees of freedom than the original one. What we mean will perhaps be clearer from the various examples we now describe.

3.3.1 Finite Duplications that Reproduce the Same System Again

A good example to illustrate this phenomenon is the many-body system characterized by the *integrable* Hamiltonian (21). Assume $n = 2m$, and consider the special configuration characterized by the restrictions

$$
q_j^{(1)} = q_j^{(2)} = q_j, \ p_j^{(1)} = p_j^{(2)} = p_j, \quad j = 1, \ldots m. \tag{23}
$$

It is easily seen that this configuration is preserved over time; hence it gives rise to an m-body system characterized by the Hamiltonian

$$
H = \frac{1}{2}\sum_{j=1}^{m}p_j^2 + \sum_{j=2}^{m}\sum_{k=1}^{j-1}V(q_j - q_k), \tag{24}
$$

with

$$V(x) = V_e(x) + V_d(x), \tag{25}$$

which obtains from (21) (up to an irrelevant *additive* constant) by noting that one needs now only consider the motion of the particles of one kind, since those of the other simply follow, as implied by the constraints (23). It might appear at first sight that in this manner a *new* system has been obtained, characterized by the *new* interaction potential $V(x)$, see (25); but in fact the insertion of (19), (22) in this formula yields

$$V(x) = \frac{4g^2 a^2}{\sinh^2(2ax)}; \tag{26}$$

hence $V(x)$ coincides with $V_e(x)$ with a replaced by $2a$. One has in this manner obtained again essentially the same system (17), (18) one started from.

An amusing consequence of this fact is the possibility to construct, for the *integrable* system (17) (and of course also for the system (1b); indeed, for the more general system (1c) as well), Lax pairs given by square matrices of order $2^p n$ rather than n, with p an arbitrary positive integer [5].

3.3.2 Finite Duplications that Yield New Systems

There are many variations on this theme. Here we treat a new example rather than reproducing known cases [14]; the diligent reader will have no difficulty in practicing with other new examples based on the material reported in this paper.

Consider the simple system

$$\ddot{\xi}_j = \alpha \dot{\xi}_j + \beta \xi_j + 2 \sum_{\substack{k=1 \\ k \neq j}}^{n+2m} \frac{\dot{\xi}_j \dot{\xi}_k}{\xi_j - \xi_k}, \quad j = 1, \ldots, n+2m, \tag{27}$$

which is obtained from (2) by setting to zero all coupling constants in the right-hand side except α and β_1 (and performing the trivial notational changes $x_j \to \xi_j$, $n \to n + 2m$, $\beta_1 \to \beta$).

Now consider for this system a special configuration characterized as follows:

$$\xi_j = z_j, \qquad\qquad j = 1, \ldots, n, \tag{28a}$$

$$\xi_{j+n} = x_j + i y_j, \qquad j = 1, \ldots, m, \tag{28b}$$

$$\xi_{j+n+m} = x_j - i y_j, \qquad j = 1, \ldots, m. \tag{28c}$$

Note that in this subsection we moreover assume that the quantities z_j, $j = 1, \ldots, n$, and x_k, y_k, $k = 1, \ldots, m$, are *real* (of course this requires that the two constant α and β be also *real*).

By inserting this *ansatz* in (27) one easily verifies its compatibility with the equations of motion, and obtains the following (*new, real*) equations of motion:

$$\ddot{z}_j = \alpha \dot{z}_j + \beta z_j + 2\sum_{\substack{k=1 \\ k \neq j}}^{n} \frac{\dot{z}_j \dot{z}_k}{z_j - z_k} + 4\sum_{k=1}^{m} \frac{\dot{z}_j [\dot{x}_k(z_j - x_k) - \dot{y}_k y_k]}{(z_j - x_k)^2 + y_k^2},$$

$$j = 1, \ldots, n, \quad (29a)$$

$$\ddot{x}_j = \alpha \dot{x}_j + \beta x_j + 4\sum_{\substack{k=1 \\ k \neq j}}^{m} \frac{A_{jk}}{D_{jk}} + 4\sum_{k=1}^{n} \frac{\dot{z}_k [\dot{x}_j(x_j - z_k) + \dot{y}_j y_j]}{(x_j - z_k)^2 + y_j^2},$$

$$j = 1, \ldots, m, \quad (29b)$$

$$\ddot{y}_j = \alpha \dot{y}_j + \beta y_j - \frac{\dot{x}_j^2 + \dot{y}_j^2}{y_j} + 4\sum_{\substack{k=1 \\ k \neq j}}^{m} \frac{B_{jk}}{D_{jk}} + 4\sum_{k=1}^{n} \frac{\dot{z}_k [-\dot{x}_j y_j + \dot{y}_j(x_j - z_k)]}{(x_j - z_k)^2 + y_j^2},$$

$$j = 1, \ldots, m, \quad (29c)$$

$$A_{jk} = \dot{x}_j \dot{x}_k (x_j - x_k)[(x_j - x_k)^2 + y_j^2 + y_k^2] - 2\dot{y}_j \dot{y}_k (x_j - x_k) y_j y_k$$
$$- \dot{x}_j \dot{y}_k y_k [(x_j - x_k)^2 - y_j^2 + y_k^2] + \dot{y}_j \dot{x}_k y_j [(x_j - x_k)^2 + y_j^2 - y_k^2], \quad (29d)$$

$$B_{jk} = -\dot{x}_j \dot{x}_k y_j [(x_j - x_k)^2 + y_j^2 - y_k^2] - \dot{y}_j \dot{y}_k y_k [(x_j - x_k)^2 - y_j^2 + y_k^2]$$
$$+ 2\dot{x}_j \dot{y}_k (x_j - x_k) y_j y_k + \dot{y}_j \dot{x}_k (x_j - x_k)[(x_j - x_k)^2 + y_j^2 + y_k^2], \quad (29e)$$

$$D_{jk} = (x_j - x_k)^2 [(x_j - x_k)^2 + 2y_j^2 + y_k^2] + (y_j^2 - y_k^2)^2. \quad (29f)$$

These may be interpreted as the equations of motion of a many-body problem on the line featuring $n + 2m$ particles of three different kinds, namely, n of one kind (represented by the coordinates z_j, $j = 1, \ldots, n$), and m each of two other kinds (represented by the coordinates x_k and y_k, $k = 1, \ldots, m$). Note, however, that while, for $\beta = 0$, the system (27) is translation-invariant, the system (29) is generally not invariant under translations (this loss of translation invariance is a general feature of this type of duplications [14]). Also note that, contrary to the other models treated in this paper, this many-body system involves "3-body forces": there are terms in the right-hand sides of (29a)–(29c) that involve the coordinates of three different particles.

The system (29) may be reformulated via the position $Y_j = y_j^2$:

$$\ddot{z}_j = \alpha \dot{z}_j + \beta z_j + 2\sum_{\substack{k=1 \\ k \neq j}}^{n} \frac{\dot{z}_j \dot{z}_k}{z_j - z_k} + 2\sum_{k=1}^{m} \frac{\dot{z}_j [2\dot{x}_k(z_j - x_k) - \dot{Y}_k]}{(z_j - x_k)^2 + Y_k},$$

$$j = 1, \ldots, n, \quad (30a)$$

$$\ddot{x}_j = \alpha \dot{x}_j + \beta x_j + 4\sum_{\substack{k=1 \\ k \neq j}}^{m} \frac{\tilde{A}_{jk}}{\tilde{D}_{jk}} + 2\sum_{k=1}^{n} \frac{\dot{z}_k [2\dot{x}_j(x_j - z_k) + \dot{Y}_j]}{(x_j - z_k)^2 + Y_j},$$

$$j = 1, \ldots, m, \quad (30b)$$

$$\ddot{Y}_j = \alpha \dot{Y}_j + 2\beta Y_j - 2\dot{x}_j^2 + 4\sum_{\substack{k=1 \\ k \neq j}}^{m} \frac{\widetilde{B}_{jk}}{\widetilde{D}_{jk}} + 4\sum_{k=1}^{n} \frac{\dot{z}_k[-2\dot{x}_j Y_j + \dot{Y}_j(x_j - z_k)]}{(x_j - z_k)^2 + Y_j},$$
$$j = 1, \ldots, m, \quad (30c)$$

$$\tilde{A}_{jk} = \dot{x}_j \dot{x}_k (x_j - x_k)[(x_j - x_k)^2 + Y_j + Y_k] - \frac{1}{2}\dot{Y}_j \dot{Y}_k (x_j - x_k)$$
$$- \frac{1}{2}\dot{x}_j \dot{Y}_k [(x_j - x_k)^2 - Y_j + Y_k] + \frac{1}{2}\dot{Y}_j \dot{x}_k [(x_j - x_k)^2 + Y_j - Y_k], \quad (30d)$$

$$\widetilde{B}_{jk} = -2\dot{x}_j \dot{x}_k Y_j [(x_j - x_k)^2 + Y_j - Y_k] - \frac{1}{2}\dot{Y}_j \dot{Y}_k [(x_j - x_k)^2 - Y_j + Y_k]$$
$$+ 2\dot{x}_j \dot{Y}_k (x_j - x_k) Y_j + \dot{Y}_j \dot{x}_k (x_j - x_k)[(x_j - x_k)^2 + Y_j + Y_k], \quad (30e)$$

$$\widetilde{D}_{jk} = (x_j - x_k)^2[(x_j - x_k)^2 + 2Y_j + 2Y_k] + (Y_j - Y_k)^2. \quad (30f)$$

The diligent reader may look at the simpler (yet not quite trivial) case with $m = n = 1$, whose general solution can of course be obtained in completely explicit form (by solving a cubic equation).

3.3.3 Finite Duplications that Yield Other "Root Systems"

To provide a simple illustration of this well-known trick, we rewrite the system (4) with x_j replaced by z_j (and of course as well x_k by z_k), and then set $z_j = x_j$ for $j = 1, \ldots, m$ and $z_{j+m} = y_j$ for $j = m+1, \ldots, n$; namely, we rewrite (4) as follows:

$$\ddot{x}_j = \alpha \dot{x}_j + \beta + \omega \sum_{\substack{k=1 \\ k \neq j}}^{m} \coth[\omega(x_j - x_k)]\{2\dot{x}_j \dot{x}_k + \lambda(\dot{x}_j + \dot{x}_k) + \mu\}$$
$$+ \omega \sum_{k=1}^{n-m} \coth[\omega(x_j - y_k)]\{2\dot{x}_j \dot{y}_k + \lambda(\dot{x}_j + \dot{y}_k) + \mu\},$$
$$j = 1, \ldots, m, \quad (31a)$$

$$\ddot{y}_j = \alpha \dot{y}_j + \beta + \omega \sum_{\substack{k=1 \\ k \neq j}}^{m} \coth[\omega(y_j - y_k)]\{2\dot{y}_j \dot{y}_k + \lambda(\dot{y}_j + \dot{y}_k) + \mu\}$$
$$+ \omega \sum_{k=1}^{m} \coth[\omega(y_j - x_k)]\{2\dot{y}_j \dot{x}_k + \lambda(\dot{y}_j + \dot{x}_k) + \mu\},$$
$$j = 1, \ldots, n - m. \quad (31b)$$

It is now easily seen that, for $n = 2m$, and provided $\beta = \lambda = 0$, the special configuration characterized by the conditions $y_j = -x_j$, $\dot{y}_j = -\dot{x}_j$, $j = 1, \ldots, m$ is compatible with this system. Thus one obtains the new model

$$\ddot{x}_j = \alpha \dot{x}_j + \omega \coth(2\omega x_j)[-2\dot{x}_j^2 + \mu]$$

$$+ \omega \sum_{\substack{k=1 \\ k \neq j}}^{m} \langle \coth[\omega(x_j - x_k)]\{2\dot{x}_j\dot{x}_k + \mu\} + \coth[\omega(x_j + x_k)]\{-2\dot{x}_j\dot{x}_k + \mu\}\rangle,$$

$$j = 1,\ldots,m, \quad (32)$$

which features the combination $x_j + x_k$ in addition to $x_j - x_k$.

It is amusing to apply this same trick to the system (2). One finds first of all that it can be done iff $\beta_0 = \lambda_0 = \lambda_2 = \mu_0 = \mu_2 = 0$. But then one also finds that the new system obtained in this manner can be recast, via the position $x_j^2 = \tilde{x}_j$, back again exactly in the same form (2) (of course for the new variables $\tilde{x}_j$), with the following new coupling constants:

$$\tilde{\alpha} = \alpha, \quad \tilde{\beta}_0 = \mu_{-1}, \quad \tilde{\beta}_1 = 2\beta_1 + \mu_1, \quad \tilde{\lambda}_0 = 0, \quad \tilde{\lambda}_1 = 2\lambda_1,$$

$$\tilde{\lambda}_2 = \tilde{\mu}_{-1} = 0, \quad \tilde{\mu}_0 = 4\mu_{-1}, \quad \tilde{\mu}_1 = 4\mu_1, \quad \tilde{\mu}_2 = 0.$$

$$(33)$$

In particular, if one starts from the special case of the model (2) with all coupling constants except (possibly) α vanishing, one finds that the end result of this trick is to reproduce *exactly the same model*, for the new variable $\tilde{x}_j = x_j^2$.

3.3.4 Infinite Duplications: From Rational Functions to Hyperbolic (or Trigonometric) Functions to Elliptic Functions

To illustrate this trick we consider first the standard system (1a) with $\Omega = 0$, which we rewrite in the form (see [8])

$$\ddot{z}_J(t) = 2g^2 \sum_{\substack{K=1 \\ K \neq J}}^{N} [z_J(t) - z_K(t)]^{-3}. \quad (34)$$

Note that we are using here capital letters for the subscripts and for the "number of bodies" N.

We now consider a configuration of this "N-body system" characterized by the following *ansatz*:

$$z_J(t) = x_j(t) - \frac{i\pi s}{a}, \quad J = \{j,s\}, \quad j = 1,\ldots,n, \quad s = 0, \pm 1, \pm 2, \ldots; \quad (35)$$

namely, to every "particle coordinate" x_j we associate an infinite retinue of other bodies located in the complex plane of the z_J variable equidistantly on a vertical line with abscissa x_j. It is clear that such a configuration, if it makes sense, is compatible with the time evolution: the translation invariance of the system (34), the *odd* character of the forces in its right-hand side and the overall symmetry of the configuration entail that no force acts on any body in the vertical direction, and that the forces acting on all the bodies sitting on a vertical line are the same. The qualification "if it makes sense" originates from the fact that we are now implying that $N = \infty$; hence everything makes sense provided all sums converge (in particular the infinite sum over the index s, see below). It is moreover clear that we only need to follow the movement of the "particles" x_j, $j = 1$, ..., n. Hence we obtain the new system

$$\ddot{x}_j = 2g^2 x_j^{-3} Z + 2g^2 \sum_{\substack{k=1 \\ k \neq j}}^{n} S(x_j - x_k), \tag{36}$$

with

$$Z = \sum_{s=-\infty, s \neq 0}^{\infty} \left(\frac{i\pi s}{a}\right)^{-3}, \tag{37a}$$

$$S(z) = \sum_{s=-\infty}^{\infty} \left(z + \frac{i\pi s}{a}\right)^{-3}. \tag{37b}$$

It is now clear that the sums in the right-hand sides of (37a) and (37b) converge, and moreover that they entail

$$Z = 0, \tag{38a}$$

$$S(z) = a^3 \frac{\cosh(az)}{\sinh^3(az)}. \tag{38b}$$

This last formula is implied by the observation that the right-hand sides of (37b) and (38b) must coincide, because they both represent a meromorphic function with poles at $z = -i\pi s/a$, $s = 0, \pm 1, \pm 2, \ldots$, with unit residues there, which remains bounded as z tends to infinity in the complex plane in all directions (away from the poles).

But the insertion of (37b), (38b) in (36) yields precisely (1b) (after the additional substitution $a \to ia$, to go from "hyperbolic forces" to "trigonometric forces"); hence we have obtained the system (1b) with "hyperbolic or trigonometric forces," starting from the system (1a) (with $\Omega = 0$) featuring "rational forces."

It is easily seen that a repetition of this argument, performed taking as starting point (1b) rather than (1a), yields the system (1c) featuring "elliptic forces." An alternative route to obtain this system goes from (1a) (with $\Omega = 0$) directly to (1c), by replacing the *ansatz* (35) with

$$z_J(t) = x_j(t) - \frac{\omega s + \omega' s'}{a}, \quad J = \{j, s, s'\},$$

$$j = 1, \ldots, n, \quad s = 0, \pm 1, \pm 2, \ldots, \quad s' = 0, \pm 1, \pm 2, \ldots, \tag{39}$$

using the formula

$$\wp'(z \mid \omega, \omega') = -2 \sum_{s=-\infty}^{\infty} \sum_{s'=-\infty}^{\infty} [z - (\omega s + \omega' s')]^{-3}, \tag{40}$$

which can be justified (if need be!) by an argument analogous to that reported above after (38b).

It is clear that one cannot continue this game any more: in the complex plane there exist singly-periodic ("trigonometric" or "hyperbolic") and doubly-periodic ("elliptic") functions, but there is no room for triple-periodic functions!

It is easily seen that this same trick, applied to the *translation-invariant* version of the *solvable* system (2) (characterized by the vanishing of some of the coupling constants in the right-hand side of this equation, namely, $\beta_1 = \lambda_1 = \lambda_2 = \mu_0 = \mu_1 = \mu_2 = 0$), yields the *solvable* system (4). The question of applying once more this trick to (4), in order to obtain a *solvable* system featuring *elliptic functions*, is more tricky; work on this issue is now in progress in collaboration with J.-P. Françoise; its outcome, if successful, will be reported elsewhere.

3.4   How to Get Rid of Velocity-Dependent Forces

To illustrate this trick, we start from the *integrable* system (5a) with (5c) and $\Omega = 0$ (which features "velocity-dependent forces"), namely (up to a trivial notational change),

$$\ddot{z}_j = 2 \sum_{\substack{k=1 \\ k \neq j}}^{n} \frac{\dot{z}_j \dot{z}_k}{(z_j - z_k)[1 + r^2 (z_j - z_k)^2]}. \tag{41}$$

We now set

$$z_j(t) = x_j(t) + Vt, \tag{42}$$

as well as

$$r^2 = \left(\frac{V}{g}\right)^2. \tag{43}$$

In this manner we get, from (41),

$$\ddot{x}_j = 2 \sum_{\substack{k=1 \\ k \neq j}}^{n} \frac{(\dot{x}_j + V)(\dot{x}_k + V)}{(z_j - z_k)[1 + (V/g)^2 (z_j - z_k)^2]}. \tag{44}$$

It is then clear that, in the limit $V \to \infty$, we get (1a), which of course contains no velocity-dependent forces.

By a completely analogous procedure one can obtain (1b) (with a replaced by ia), starting from (5a) with (5e) and $\Omega = 0$. Indeed one can obtain as well (1c), starting from (5a) with (5g) and $\Omega = 0$ and following the same route, except for the replacement of (43) by

$$b = \frac{iV}{ag}. \tag{45}$$

3.5 From Two-Body to Nearest-Neighbor Forces

I learned this trick many years ago from Simon Ruijsenaars; perhaps it was discovered simultaneously by him and by Bill Sutherland (see the parenthetical remark after Eq. (2.1.17) in Ref. 21). To illustrate it, let us take as starting point the many-body system (5a) with (5f), which we rewrite here as follows:

$$\ddot{z}_j + i\Omega \dot{z}_j = 2a \sum_{\substack{k=1 \\ k \neq j}}^{n} \frac{\dot{z}_j \dot{z}_k \coth[a(z_j - z_k)]}{1 + r^2 \sinh^2[a(z_j - z_k)]}. \tag{46}$$

We now set

$$z_j(t) = x_j(t) + Lj, \tag{47}$$

and

$$r^2 = 4g^2 \exp(-|2aL|). \tag{48}$$

It is then clear that, in the limit $L \to \infty$, we get from (46) the following system:

$$\ddot{x}_j + i\Omega \dot{x}_j = 2a \left\langle \frac{\dot{x}_j \dot{x}_{j-1}}{1 + g^2 \exp[2a(x_j - x_{j-1})]} - \frac{\dot{x}_j \dot{x}_{j+1}}{1 + g^2 \exp[-2a(x_j - x_{j+1})]} \right\rangle,$$
$$j = 2, \ldots, n-1, \tag{49a}$$

which features "nearest-neighbor" forces. Note that this derivation also implies the following equations for the "end points":

$$\ddot{x}_1 + i\Omega \dot{x}_1 = -2a \frac{\dot{x}_1 \dot{x}_2}{1 + g^2 \exp[-2a(x_1 - x_2)]}, \tag{49b}$$

$$\ddot{x}_n + i\Omega \dot{x}_n = 2a \frac{\dot{x}_n \dot{x}_{n-1}}{1 + g^2 \exp[2a(x_n - x_{n-1})]}. \tag{49c}$$

It is easily seen that, by applying essentially this same trick starting from (1b) (with a replaced by ia and g by h), with (48) replaced by

$$h^2 = 2 \left(\frac{g}{a}\right)^2 \exp|aL|, \tag{50}$$

one obtains precisely the Toda system (6); which can also be obtained by applying the trick of the previous Section 3.4 to (49a) (with $\Omega = 0$, a replaced by $a/2$ and g^2 replaced by $r^2/2$).

Note that setting $g^2 = -1$, and then letting a tend to zero, (49) become

$$\ddot{x}_j + i\Omega \dot{x}_j = -\dot{x}_j \left[\frac{\dot{x}_{j-1}}{x_j - x_{j-1}} + \frac{\dot{x}_{j+1}}{x_j - x_{j+1}} \right], \quad j = 2, \ldots, n-1, \quad (51a)$$

$$\ddot{x}_1 + i\Omega \dot{x}_1 = -\frac{\dot{x}_1 \dot{x}_2}{x_1 - x_2}, \tag{51b}$$

$$\ddot{x}_n + i\Omega \dot{x}_n = -\frac{\dot{x}_n \dot{x}_{n-1}}{x_n - x_{n-1}}. \tag{51c}$$

3.6 Infinite Rescalings, and the Use of Special Solutions, to Identify Solvable Nonautonomous Models

In this subsection we take as starting point the *solvable* model

$$\ddot{z}_j = \alpha \dot{z}_j + \beta z_j + \sum_{\substack{k=1 \\ k \neq j}}^{n} (z_j - z_k)^{-1} [2\dot{z}_j \dot{z}_k + \lambda(\dot{z}_j + \dot{z}_k)z_j + \mu z_j^2], \tag{52}$$

which is clearly a special case of (2); and we illustrate one or two (possibly interesting) tricks, whose final outcome is, however, merely to produce a version of this same model with the constants α and β replaced by specific time-dependent functions. In this manner one has obtained a *solvable nonautonomous* system; hardly a remarkable result, if one remembers that, as mentioned in Section 2, the system (2) is *solvable* even if all the "coupling constants" it features are (arbitrarily!) time dependent. However, the techniques we illustrate below might be useful to *identify* solvable nonautonomous systems whose solutions are more easily displayed in explicit form (although we do not pursue this matter below, except by the display of a simple example); moreover these techniques, as well as some of the intermediate findings (such as special explicit solutions of (52), see below) are perhaps of interest in their own right.

Indeed let us immediately exhibit a special ("similarity") solution of (52). It reads

$$z_j(t) = F(t) \exp\left(\frac{2\pi i j}{n} \right), \tag{53a}$$

$$F(t) = [f(t)]^{1/n}, \tag{53b}$$

$$f(t) = c_+ \exp(\gamma_+ t) + c_- \exp(\gamma_- t), \tag{53c}$$

$$\gamma_\pm = \frac{\alpha}{2n} \left\langle 1 \pm \left\{ 1 + \left(\frac{2n}{\alpha} \right)^2 \left[\beta + \mu \frac{n-1}{2} \right] \right\}^{1/2} \right\rangle. \tag{53d}$$

To check that these formulas do provide a (complex) solution of (52) one must use the identities

$$\sum_{\substack{k=1 \\ k\neq j}}^{n} \frac{\exp(2\pi ij/n)}{\exp(2\pi ij/n)-\exp(2\pi ik/n)} = \frac{n-1}{2}, \tag{54a}$$

$$\sum_{\substack{k=1 \\ k\neq j}}^{n} \frac{\exp(2\pi ik/n)}{\exp(2\pi ij/n)-\exp(2\pi ik/n)} = -\frac{n-1}{2}, \tag{54b}$$

and some elementary algebra (as suggested by the notation (53b), (53c)).
Now let us set

$$z_k = x_k, \qquad\qquad k=1,\dots,n, \tag{55a}$$
$$z_{k+m} = y_k, \qquad\quad k=m+1,\dots,n-m, \tag{55b}$$

where we have of course chosen a value of m inside the interval from 1 to n, $1 < m < n$. With this *ansatz* (52) gets rewritten in the form

$$\ddot{x}_j = \alpha\dot{x}_j + \beta x_j + \sum_{\substack{k=1 \\ k\neq j}}^{m}(x_j-x_k)^{-1}[2\dot{x}_j\dot{x}_k + \lambda(\dot{x}_j+\dot{x}_k)x_j + \mu x_j^2]$$
$$+ \sum_{k=1}^{n-m}(x_j-y_k)^{-1}[2\dot{x}_j\dot{y}_k + \lambda(\dot{x}_j+\dot{y}_k)x_j + \mu x_j^2],$$
$$j=1,\dots,m, \tag{56a}$$

$$\ddot{y}_j = \alpha\dot{y}_j + \beta y_j + \sum_{\substack{k=1 \\ k\neq j}}^{n-m}(y_j-y_k)^{-1}[2\dot{y}_j\dot{y}_k + \lambda(\dot{y}_j+\dot{y}_k)y_j + \mu y_j^2]$$
$$+ \sum_{k=1}^{m}(y_j-x_k)^{-1}[2\dot{y}_j\dot{x}_k + \lambda(\dot{y}_j+\dot{x}_k)y_j + \mu y_j^2].$$
$$j=1,\dots,n-m. \tag{56b}$$

We now perform the formal replacement $y_k \to Ly_k$, $k=1,\dots,n-m$, and then let $L \to \infty$. We thus get

$$\ddot{x}_j = [\alpha - 2\varphi(t)]\dot{x}_j + [\beta - \varphi(t)]x_j$$
$$+ \sum_{\substack{k=1 \\ k\neq j}}^{m}(x_j-x_k)^{-1}[2\dot{x}_j\dot{x}_k + \lambda(\dot{x}_j+\dot{x}_k)x_j + \mu x_j^2],$$
$$j=1,\dots,m, \tag{57a}$$

$$\ddot{y}_j = \alpha\dot{y}_j + \beta y_j + \sum_{\substack{k=1 \\ k\neq j}}^{n-m}(y_j-y_k)^{-1}[2\dot{y}_j\dot{y}_k + \lambda(\dot{y}_j+\dot{y}_k)y_j + \mu y_j^2],$$
$$j=1,\dots,n-m, \tag{57b}$$

$$\varphi(t) = \sum_{k=1}^{n-m}\frac{\dot{y}_k(t)}{y_k(t)}. \tag{57c}$$

Now one can use for $y_j(t)$ the special solution (53) (of course with $z_j(t)$ replaced by $y_j(t)$ and n by $n - m$), obtaining thereby the following explicit expression for $\varphi(t)$:

$$\varphi(t) = \gamma(1 + \delta) \cot[\delta(t + t_0)], \tag{58a}$$

$$\gamma = \frac{\alpha}{2(n - m)}, \tag{58b}$$

$$\delta^2 = 1 + \left[\frac{2(n - m)}{\alpha}\right]^2 \left[\beta + \mu\frac{n - m - 1}{2}\right], \tag{58c}$$

with t_0 an arbitrary constant (possibly complex).

In this manner we have indeed reobtained the system (52) (up to trivial notational changes), but now in a nonautonomous version: see(57a) with (58).

It may be of interest to point out the particularly simple form some of these results take in the special case $\alpha = \beta = \mu = 0$, namely, if (52) is replaced by

$$\ddot{z}_j = \sum_{\substack{k=1 \\ k \neq j}}^{n} (z_j - z_k)^{-1}[2\dot{z}_j\dot{z}_k + \lambda(\dot{z}_j + \dot{z}_k)z_j]. \tag{59}$$

Then the special solution (53) gets replaced by

$$z_j(t) = c(t + t_0)^{1/n} \exp\left(\frac{2\pi i j}{n}\right), \tag{60}$$

and the "new system" (57a) gets replaced by

$$\ddot{x}_j = \frac{1}{t + t_0)}\dot{x}_j + \sum_{\substack{k=1 \\ k \neq j}}^{m} (x_j - x_k)^{-1}[2\dot{x}_j\dot{x}_k + \lambda(\dot{x}_j + \dot{x}_k)x_j],$$
$$j = 1,\ldots,m, \tag{61}$$

which itself admits the simple "similarity" solution

$$x_j(t) = c(t + t_0)^{-1/n} \exp\left(\frac{2\pi i j}{n}\right). \tag{62}$$

3.7 Two-Dimensional Models via Complexification

Consider a dynamical system characterized, as indeed all those considered in this paper, by equations of motion of Newtonian type ("acceleration equal force"),

$$\ddot{z}_j = F_j(z_k, \dot{z}_k; \{k = 1, \ldots, n\}), \tag{63}$$

with "forces" F_j being analytic functions of the variables z_k, $\dot{z}_k$, $k = 1$, ..., n (as indeed it is the case for all the systems considered in this paper). It is then clear that by making the formal replacement ("analytic continuation")

$$z_j = x_j + iy_j, \tag{64}$$

one can obtain a "two-dimensional" system describing the motion *in the plane* of particles described by *two-vectors* $\vec{r}_j$ of Cartesian components x_j, y_j:

$$\vec{r}_j \equiv (x_j, y_j). \tag{65}$$

It is actually convenient to use for two-vectors the "three-dimensional notation"

$$\vec{r} \equiv (x, y, 0), \tag{66a}$$

which entails

$$\hat{z} \wedge \vec{r} \equiv (-y, x, 0), \tag{66b}$$

having introduced the unit-vector $\hat{z} \equiv (0, 0, 1)$ orthogonal to the plane. The advantage of this notation, which is used hereafter (and also in Section 2 above), is to identify the two quantities (66) that behave as *vectors* under rotations (actually (66a) is a vector and (66b) a pseudovector), as well as the (only) two (rotation-invariant) scalars that can be constructed with two-vectors,

$$\vec{r} \cdot \vec{r}' \equiv xx' + yy', \tag{67a}$$

$$\hat{z} \cdot \vec{r} \wedge \vec{r}' \equiv xy' - yx' \tag{67b}$$

(actually (67a) is a scalar and (67b) a pseudoscalar: the first is invariant, and the second changes sign, under the inversions $x \to -x$, $y \to y$ or $x \to x$, $y \to -y$).

However, the models obtained by this trivial trick generally lack the important property of *rotation invariance* while we opine that the two-dimensional models qualify as *many-body systems in the plane* iff they are *rotation invariant*, namely, iff their equations of motion have the form

$$\ddot{\vec{r}}_j = \vec{F}_j(\vec{r}_k, \dot{\vec{r}}_k, \{k = 1, \ldots, n\}), \tag{68a}$$

with $\vec{F}_j$ a *two-dimensional vector* constructed out of the $2n$ two-vectors $\vec{r}_k$, $\dot{\vec{r}}_k$, $k = 1, \ldots, n$, namely,

$$\vec{F}_j = \sum_{k=1}^{n} [f_{jk}^{(1)} \vec{r}_k + f_{jk}^{(2)} \dot{\vec{r}}_k + f_{jk}^{(3)} \hat{z} \wedge \vec{r}_k + f_{jk}^{(4)} \hat{z} \wedge \dot{\vec{r}}_k], \tag{68b}$$

with $f_{jk}^{(s)}$, $s = 1, 2, 3, 4$; j, $k = 1, \ldots, n$, *rotation-invariant* ("scalar") functions. If we, moreover, require the many-body system (68a) to feature only "one-body" and "two-body" forces, each of the four functions $f_{jk}^{(s)}$, $s = 1, 2, 3, 4$, can only depend on the eight *scalar* quantities $(\vec{r}_j \cdot \vec{r}_k)$, $(\dot{\vec{r}}_j \cdot \vec{r}_k)$, $(\vec{r}_j \cdot \dot{\vec{r}}_k)$, $(\dot{\vec{r}}_j \cdot \dot{\vec{r}}_k)$, $(\hat{z} \cdot \vec{r}_j \wedge \vec{r}_k)$, $(\hat{z} \cdot \dot{\vec{r}}_j \wedge \vec{r}_k)$, $(\hat{z} \cdot \vec{r}_j \wedge \dot{\vec{r}}_k)$, $(\hat{z} \cdot \dot{\vec{r}}_j \wedge \dot{\vec{r}}_k)$.

But while, as we just wrote, the "complexification" trick, when applied to systems of type (63), yields models that generally do not qualify as *bona fide* many-body problems *in the plane* because they lack rotation invariance, there exists (as recently noticed [9]) an easily identifiable class of systems of type (63) that yield, via complexification, two-dimensional problems that are *rotation invariant*. Assume indeed to start from a system (63) that is *invariant* under the (complex) transformation

$$z_j \rightarrow z_j \exp(i\theta) \tag{69}$$

with θ an *arbitrary real* constant. It is then clear that the corresponding model *in the plane*, obtained via the transformation (64) and (65), is *rotation invariant*, since to the phase transformation (69) of the complex variable z_j there corresponds, via (64) and (65), the rotation in the plane

$$x_j \rightarrow x_j \cos(\theta) - y_j \sin(\theta), \quad y_j \rightarrow x_j \sin(\theta) + y_j \cos(\theta). \tag{70}$$

In fact, as already indicated in Section 2, in this manner one obtains from the *solvable* system (2) (rewritten with x_j replaced by z_j, and with the restrictions $\beta_0 = \lambda_0 = \lambda_2 = \mu_{-1} = \mu_0 = \mu_2 = 0$, which are necessary and sufficient to guarantee invariance under the transformation (69)), precisely the system (3) (the following notational changes are also used: $\alpha \rightarrow \alpha + i\alpha'$, $\beta_1 \rightarrow \beta + i\beta'$, $\lambda_1 \rightarrow \lambda + i\lambda'$, $\mu_1 \rightarrow \mu + i\mu'$).

The diligent reader may find it amusing to derive by analogous techniques other *solvable many-body problems in the plane*, starting, for instance, from the system (29) (say, in the case $n = m = 1$; merely for the sake of simplicity!). The following hints outline the calculations to be performed: $\alpha \rightarrow \alpha + i\alpha'$, $\beta \rightarrow \beta + i\beta'$, $z_1 \rightarrow z^{(1)}$, $x_1 \rightarrow z^{(2)}$, $y_1 \rightarrow z^{(3)}$; notice then the invariance under the transformation $z^{(s)} \rightarrow z^{(s)} \exp(i\theta)$, $s = 1$, 2, 3; set $z^{(s)} \equiv x^{(s)} + iy^{(s)}$, $\vec{r}^{(s)} \equiv (x^{(s)}, y^{(s)}, 0)$, $s = 1, 2, 3$; multiply the numerators and denominators in the right-hand sides of (29a)–(29c) by the complex conjugate of the expression $[(z^{(1)} - z^{(2)})^2 + (z^{(3)})^2]$ that appears in the denominators; then evaluate everything in terms of the (real) two-vectors $\vec{r}^{(s)}$, $s = 1, 2, 3$, for instance,

$$|(z^{(1)} - z^{(2)})^2 + z^{(3)2}|^2 = |\vec{r}^{(1)} - \vec{r}^{(2)}|^4 + |\vec{r}^{(3)}|^4$$
$$+ 2[\vec{r}^{(3)} \cdot (\vec{r}^{(1)} - \vec{r}^{(2)})]^2 + 2[\hat{z} \cdot (\vec{r}^{(1)} - \vec{r}^{(2)}) \wedge \vec{r}^{(3)}]^2 \tag{71}$$

(incidentally notice that this scalar quantity, which will appear at the denominator in the right-hand side of the final equations of motion, ranges in value between $[(\vec{r}^{(1)} - \vec{r}^{(2)})^2 - (\vec{r}^{(3)})^2]^2$, attained when the two-vectors $(\vec{r}^{(1)} - \vec{r}^{(2)})$ and $\vec{r}^{(3)}$ are orthogonal, and $[(\vec{r}^{(1)} - \vec{r}^{(2)})^2 + (\vec{r}^{(3)})^2]^2$, attained when the two-vectors $(\vec{r}^{(1)} - \vec{r}^{(2)})$ and $\vec{r}^{(3)}$ are parallel).

Three other solvable or integrable *rotation-invariant* many-body problems *in the plane* are exhibited in Ref. 11.

Finally let us mention three other (real) integrable *rotation-invariant* many-body problems *in the plane* that can be obtained by the complexification technique illustrated in this section, and that only feature "nearest neighbor" interactions (for simplicity we ignore here the "end-points" issue):

$$\ddot{\vec{r}}_j = (\alpha + \alpha'\hat{z}\wedge)\dot{\vec{r}}_j - \sum_{k=j\pm1} \frac{\dot{\vec{r}}_j(\dot{\vec{r}}_k \cdot \vec{r}_{jk}) + \dot{\vec{r}}_k(\dot{\vec{r}}_j \cdot \vec{r}_{jk} - \vec{r}_{jk}(\dot{\vec{r}}_j \cdot \dot{\vec{r}}_k))}{r_{jk}^2}, \quad (72)$$

$$\ddot{\vec{r}}_j = (\alpha + \alpha'\hat{z}\wedge)\dot{\vec{r}}_j + \sum_{k=j\pm1} \frac{2\dot{\vec{r}}_j(\dot{\vec{r}}_j \cdot \vec{r}_{jk}) - \vec{r}_{jk}\dot{r}_j^2}{r_{jk}^2}, \quad (73)$$

$$\ddot{\vec{r}}_j = (\beta + \beta'\hat{z}\wedge)\vec{r}_j$$
$$+ \sum_{k=j\pm1} \frac{2\vec{r}_j(\vec{r}_j\cdot\vec{r}_{jk}) - \vec{r}_{jk}r_j^2 - \frac{1}{2}(\beta+|1\beta'\hat{z}\wedge)[\vec{r}_j(r_j^2-2\vec{r}_j\cdot\vec{r}_k)+\vec{r}_k r_j^2]}{r_{jk}^2}, \quad (74)$$

where of course $\vec{r}_{jk} \equiv \vec{r}_j - \vec{r}_k$, $r_{jk} = |\vec{r}_{jk}|$.

None of these three models is merely the restriction of (some appropriate special case of) (3) to nearest-neighbor forces only (no such integrable case is known).

The model (72) is obtained by complexification from (51a), using moreover the trick described at the end of Section 3.1. It is natural to conjecture that, for $\alpha = 0$ (and of course α' real and nonvanishing), *all its trajectories are completely periodic.*

The model (73) is obtained by complexifying an appropriate special case of Eqs. (6) or (7) of Ref. 28, and then again using the trick described at the end of Section 3.1. It is again natural to conjecture that, for $\alpha = 0$ (and of course α' real and nonvanishing) *all its trajectories are completely periodic.*

The model (74) is obtained by complexifying a special case of Eq. (7) of Ref. 28.

4   Envoi

In this paper we have presented several tricks and we have illustrated their effectiveness by exhibiting how they work in several examples, but we have by no means tried to review *all* the cases in which such techniques have been employed. The diligent reader may profit by practicing further applications of these tricks after studying those exhibited herein; but without any guarantee that such endeavors necessarily break *new* ground.

Acknowledgments: This paper was largely drafted while I served as visiting professor with the UFR 920 at the Université Paris VI. It is a pleasure to thank my colleagues there for their kind hospitality, and in particular J.-P. Françoise, who also provided much useful advice.

5 REFERENCES

1. M. Adler, *Some finite-dimensional integrable systems and their scattering behavior*, Commun. Math. Phys. **55** (1977), No. 3, 195–230.

2. M. Bruschi and F. Calogero, *The Lax representation for an integrable class of relativistic dynamical systems*, Commun. Math. Phys. **109** (1987), No. 3, 481–492.

3. F. Calogero, *Solution of the one-dimensional N-body problems with quadratic and/or inversely quadratic pair potentials*, J. Math. Phys. **12** (1971), 419–436.

4. _____ , *Exactly solvable one-dimensional many-body problems*, Lett. Nuovo Cimento **13** (1975), No. 11, 411–416.

5. _____ , *A sequence of Lax matrices or certain integrable Hamiltonian systems*, Lett. Nuovo Cimento **16** (1976), No. 1, 22–24.

6. _____ , *Motion of poles and zeros of special solutions of nonlinear and linear partial differential equations and related "solvable" many-body problems*, Nuovo. Cimento **43B** (1978), No. 2, 177–241.

7. _____ , *Why are certain nonlinear PDEs both widely applicable and integrable?*, What Is Integrability? (V. E. Zakharov, ed.), Springer Ser. Nonlinear Dynam., Springer, Berlin, 1991, pp. 1–62.

8. _____ , *Remarks on certain integrable one-dimensional many-body problems*, Phys. Lett. A **183** (1993), No. 1, 85–88.

9. _____ , *A solvable N-body problem in the plane. I*, J. Math. Phys. **37** (1996), No. 4, 1735–1759.

10. _____ , *A class of integrable Hamiltonian systems whose solutions are (perhaps) all completely periodic*, J. Math. Phys. **38** (1997), No. 11, 5711–5719.

11. _____ , *Three solvable many-body problems in the plane*, Acta Appl. Math. **51** (1998), No. 1, 93–111.

12. F. Calogero and W. Eckhaus, *Nonlinear evolution equations, rescalings, model PDEs and their integrability. I*, Inverse Problems **3** (1987), No. 1, 229–262.

13. _____, *Nonlinear evolution equations, rescalings, model PDEs and their integrability*. II, Inverse Problems **4** (1988), No. 2, 11–33.

14. F. Calogero and J.-P. Françoise, *Integrable dynamical systems obtained by duplication*, Ann. Inst. Henri Poincaré A **57** (1992), No. 2, 167–181.

15. H. Flaschka, *The Toda lattice*. I. *Existence of integrals*, Phys. Rev. B **9** (1974), 1924–1925.

16. _____, *The Toda lattice*. II. *Inverse-scattering solution*, Progr. Theor. Phys. **51** (1974), 703–716.

17. P. D. Lax, *Integrals of nonlinear equations of evolution and solitary waves*, Commun. Pure Appl. Math. **21** (1968), 467–490.

18. S. V. Manakov, *Complete integrability and stochastization of discrete dynamical systems*, Sov. Phys.–JETP **40** (1974), No. 2, 269–274.

19. J. Moser, *Three integrable Hamiltonian systems connected to isospectral deformations*, Adv. Math. **16** (1975), 197–220.

20. A. M. Perelomov, *The simple relation between certain dynamical systems*, Commun. Math. Phys. **63** (1978), No. 1, 9–11.

21. S. N. M. Ruijsenaars, *Systems of Calogero-Moser type*, Particles and Fields (Banff, 1994) (G. Semenoff and L. Vinet, eds.), CRM Series in Mathematical Physics, Springer, New York, 1999, pp. 251–352.

22. S. N. M. Ruijsenaars and H. Schneider, *A new class of integrable systems and its relation to solitons*, Ann. Phys. (NY) **170** (1986), No. 2, 370–405.

23. B. Sutherland, *Exact results for a quantum many-body problem in one dimension*, Phys. Rev. A **4** (1971), 2019–2021.

24. _____, *Exact results for a quantum many-body problem in one dimension*. II, Phys. Rev. A **5** (1972), 1372–1376.

25. M. Toda, *One-dimensional dual transformation*, J. Phys. Soc. Japan **20** (1967), 2095.

26. _____, *Vibration of a chain with nonlinear interaction*, J. Phys. Soc. Japan **20** (1967), 431–435.

27. _____, *Theory of Nonlinear Lattices*, Springer Ser. Solid-State Sci., Vol. 20, Springer, Berlin, 1981.

28. R. I. Yamilov, *Classification of Toda-type scalar lattices*, Nonlinear Evolution Equations and Dynamical Systems NEEDS '92 (Dubna, 1992) (V. Makhankov, I. Puzynin, and O. Pashaev, eds.), World Scientific, River Edge, NJ, 1993, pp. 423–431.

8

Classical and Quantum Partition Functions of the Calogero–Moser–Sutherland Model

Ph. Choquard

ABSTRACT A new and elementary method is proposed for obtaining the equations of state of the classical and quantum CSM models as solutions of first-order nonlinear differential equations satisfied by auxiliary functions constructed from the corresponding canonical partition functions. For the classical case, the differential equation (18) is new. For the quantum case the differential equation (34) is shown to be equivalent with Yang–Yang's type of solution obtained by Sutherland. This equivalence provides an explanation for the local character, in momentum space, of our differential equation.

1 Preamble and Summary

The present work originated in discussions held during the Montreal workshop on the Calogero–Moser–Sutherland models with S. Ruijsenaars and B. Sutherland concerning the partition functions and equations of state of the CMS, which we call the CSM model in the classical and quantum cases, notably the fact that the thermodynamics of the classical case was not known.

A first step toward the solution of this problem is made in this paper. Starting from the well-known fact that in soluble one-dimensional models of statistical mechanics the pressure equation of state in the grand-ensemble formalism is expressed in terms of a momentum integral over R_1 of some given integrand, our approach is based on the idea, inspired by some early work on the statistical mechanics of one-dimensional classical Coulomb systems [1, 2], to introduce an auxiliary function that is to cut off the above integration at some finite momentum and to look for a first-order differential equation satisfied by this auxiliary function. Now a remarkable property of the CSM model that we have discovered and that will be explained later, is that the integrand is a function of precisely the first derivative of the same auxiliary function. It follows that the equation state of the model is related to a well-defined solution of a nonlinear first-order differential equation, the transcendental equation (18) in the classical case.

In developing a similar approach for the quantum case we have found that for integer values of the coupling constant, the equation of state is also related to the solution of a first-order nonlinear equation, the algebraic equation (36). Furthermore, the link has been found between our auxiliary function and Yang–Yang's implicit solution of the CSM model given by Sutherland [6, 8].

The following elementary steps are implied in the derivation of these differential equations:

- Introduce the notion of an *Incomplete Canonical Partition Function* (ICPF) and write the recurrence equation (11), (12) and (24), (25), (26);

- Go over to the *Incomplete Grand Canonical Partition Functions* (IGCPF) and write the fundamental recurrence equations (14) and (27), (28), (29);

- Proceed to the thermodynamic limit and obtain the first-order nonlinear equations (18) and (34), (35), (36).

This program is carried out in Section 3 for the classical case and in Section 4 for the quantum case. In the next section we introduce briefly what we need to know about the model.

2   The CSM Model

This is a system of N particles of mass m of coordinates x_j and momenta p_j ($= -i\hbar\partial/\partial x_j$) moving on a circle of circumference L and interacting pairwise via the periodized inverse quadratic potential. Setting the strength of the interaction equal to g^2/m for g to have the dimension of an action, the Hamiltonian of the system reads

$$H = \sum_{j=1}^{N} \frac{p_j^2}{2m} + \sum_{\substack{i,j=1 \\ i \leq j}}^{N} \frac{g^2\pi^2}{mL^2 \sin^2(\pi(x_i - x_j)/L)}. \tag{1}$$

In the classical case and although Moser [4] had proved the integrability of the model and Hénon [3] had expressed the constants of motion in a compact form in the $(\mathbf{p}, \mathbf{x})$ variables more than 20 years ago, it is only recently that Ruijsenaars [5, Sections 1.2, 2.1, and 5.1] succeeded in finding the action-angle map of the model and, in diagonalizing an appropriate Lax matrix, established its separable energy spectrum, namely,

$$E_c(\mathbf{p}^0) = \sum_{j=1}^{N} \frac{1}{2m} \left(p_j^0 - b\left(j - \frac{N+1}{2}\right)\right)^2, \tag{2}$$

where $\infty \geq p_1^0 \geq p_2^0 \geq \cdots \geq p_N^0 \geq -\infty$ and where the momentum shift

$$b = \frac{2\pi g}{L}. \tag{3}$$

In the quantum case, Sutherland [7] found many years ago that the energy spectrum of the Schrödinger equation associated to the Hamiltonian (1) was given by

$$E_q(\mathbf{k^0}) = \sum_{j=1}^{N} \frac{\hbar^2}{2m} \left(k_j^0 - \frac{2\pi\lambda}{L} \left(j - \frac{N+1}{2} \right) \right)^2, \tag{4}$$

where $k_1^0 \geq k_2^0 \geq k_2^0 \geq \cdots \geq k_N^0$, with $k_j^0 = 2\pi n_j / L$, $n_j \in \mathbb{Z}$, and where

$$g^2 = \hbar^2 \lambda(\lambda - 1), \tag{5}$$

the acceptable root being

$$\hbar\lambda = \frac{1}{2}\hbar + \frac{1}{2}\sqrt{\hbar^2 + 4g^2}. \tag{6}$$

Notice here that $E_c(\mathbf{p^0})$ can be obtained from $E_q(\mathbf{k^0})$ and from (6) in going over to the classical limit defined by $h \to 0$, $n_j \to \infty$, $\lambda \to \infty$, $\hbar k_j^0 \to p_j^0$ and $\hbar\lambda \to g$.

3 Thermodynamics: Classical Case

By analogy with the so-called *Incomplete Gamma Function* we define below an Incomplete Canonical Partition Function (ICPF). Expressed in terms of the momenta

$$p_j = p_j^0 - b\left(j - \frac{N+1}{2} \right), \quad j = 1, \ldots, N, \tag{7}$$

where the p_j's vary from $-\infty$ to $p_{j-1} - b$, with $\beta \equiv 1/(k_B T)$, with

$$a(p) = e^{-\beta p^2/(2m)}, \tag{8}$$

and with the factor L^N resulting from the configuration integrals, the ICPF of the model reads

$$Q_c(p, \beta, N, L)$$
$$= \left(\frac{L}{h} \right)^N \int_{-\infty}^{p} dp_1\, a(p_1) \int_{-\infty}^{p_1 - b} dp_2\, a(p_2) \cdots \int_{-\infty}^{p_{N-1} - b} dp_N\, a(p_N). \tag{9}$$

Clearly the CPF is

$$Q_c(\beta, N, L) = Q_c(\infty, \beta, N, L). \tag{10}$$

Equation (9) tells us that we have the relation

$$Q_c(p, \beta, N, L) = \frac{L}{h} \int_{-\infty}^{p} dp_1\, a(p_1) Q_c(p_1 - b, \beta, N - 1, L), \tag{11}$$

which is the formal solution of the differential and recurrence equation with displaced argument,

$$\frac{d}{dp} Q_c(p, \beta, N, L) \equiv \dot{Q}(p, \beta, N, L)_c = \frac{L}{h} a(p) Q_c(\beta, N - 1, p - b, L), \tag{12}$$

and with initial condition $Q_c(-\infty, \beta, N, L) = 0$.

At this point it is convenient to introduce the Incomplete Grand Canonical Partition Function (IGCPF),

$$G_c(p, \beta, z, L) = \sum_{N=0}^{\infty} z^N Q_c(p, \beta, N, L), \tag{13}$$

where z is the fugacity. This function satisfies the fundamental differential equation with displacement argument

$$\dot{G}_c(p, \beta, z, L) = \frac{Lz}{h} a(p) G_c(p - b, \beta, z, L), \tag{14}$$

the initial condition being that $G_c(-\infty, \beta, z, L) = 0$.

This equation in momentum space is reminiscent of Baxter's equation in configuration space [1, Eq. 26], which had been established for the one-dimensional one-component plasma, a collection of ordered and equidistanced harmonic oscillators. The differences are that $p^2/(2m)$ replaces $\frac{1}{2} m\omega_p^2 x^2$ (where ω_p is the plasma frequency), that L/h replaces the inverse Debye length $\sqrt{2\pi\hbar^2\beta/m}$, that $p - b$ replaces $x + a$, where a is the lattice constant whereas the momentum shift $b \to 0$ in the thermodynamic limit.

Clearly again the GCPF is defined by

$$\mathcal{G}_c(\beta, z, L) = G_c(\infty, \beta, N, L). \tag{15}$$

We proceed with the investigation of the thermodynamic limit in setting

$$G_c(p, \beta, z, L) = \exp(\beta \mathcal{P}_c(p, \beta, z, L)L). \tag{16}$$

Then Eq. (14) becomes, in dividing by G_c,

$$\beta h \dot{\mathcal{P}}_c(p, \beta, z, L)$$
$$= za(p) \exp\left\{ \beta\left[\mathcal{P}_c\left(p - \frac{2\pi g}{L}, \beta, z, L\right) - \mathcal{P}_c(p, \beta, z, L)\right]L \right\}. \tag{17}$$

The limit $L \to \infty$ can now be taken assuming the differentiability of the auxiliary function and its size independence.

The result is

$$\beta h \dot{\mathcal{P}}_c(p, \beta, z) = z a(p) \exp\{-2\pi\beta g \dot{\mathcal{P}}_c(p, \beta, z)\}. \tag{18}$$

The initial condition is $\mathcal{P}_c(-\infty, \beta, z) = 0$ and the value of $\mathcal{P}_c(\infty, \beta, z)$ yields the pressure equation of state, $P(\beta, z)$. We have the relation

$$\beta P_c(\beta, z) = \frac{z}{h} \int_{-\infty}^{+\infty} dp\, a(p) \exp\{-2\pi\beta g \dot{\mathcal{P}}_c(p, \beta, z)\}. \tag{19}$$

Equation (18) is the first-order transcendental equation announced in the first section. Clearly it constitutes the basis for the forthcoming investigation concerning the equation of state of the CSM model in the classical case.

An obvious extension of the above approach will be to define also an auxiliary function for the density through

$$\rho_c(p, \beta, z) = z \frac{d}{dz} \beta \mathcal{P}_c(p, \beta, z). \tag{20}$$

4   Thermodynamics: Quantum Case

In the quantum case, we consider also new quantum numbers

$$n_j = n_j^0 - \lambda\left(j - \frac{N+1}{2}\right), \quad j = 1, \dots, N, \tag{21}$$

which vary from $-\infty$ to $n_{j-1} - \lambda$; and since the n_j's are integers we consider integer λ only, the effect of enhanced exclusion statistics being particularly transparent in these cases. We define next the quantum ICPF of the model. With

$$a(n) = e^{-\beta h^2 n^2/(mL^2)}, \tag{22}$$

we have

$$Q_q(n, \beta, N, L) = \sum_{n_1=-\infty}^{n} a(n_1) \sum_{n_2=-\infty}^{n_1-\lambda} a(n_2) \cdots \sum_{n_N=-\infty}^{n_{N-1}-\lambda} a(n_N). \tag{23}$$

In order to establish the recurrence relation for any integer λ it is instructive to consider first the special cases $\lambda = 0$ and $\lambda = 1$.

For $\lambda = 0$, Eq. (23) describes the ICPF of the perfect Bose gas in which the N particles can occupy all the states up to the nth, which can be occupied by $0, 1, \dots, N$ particles. In this case the recurrence relation reads

$$Q_q^{(0)}(n, \beta, N, L) = \sum_{k=0}^{N} a(n)^k Q_q^{(0)}(n-1, \beta, N-k, L). \tag{24}$$

For $\lambda = 1$, which corresponds to the spinless fermion case, the nth state can be empty or occupied by one particle and the recurrence relation becomes simply

$$Q_q^{(1)}(n, \beta, N, L) = Q_q^{(1)}(n - 1, \beta, N, L)$$
$$+ a(n)Q_q^{(1)}(n - 1, \beta, N - 1, L). \quad (25)$$

It is clear now that for $\lambda > 1$ corresponding to an enhanced exclusion principle, we have,

$$Q_q(n, \beta, N, L) = Q_q(n - 1, \beta, N, L) + a(n)Q_q(n - \lambda, \beta, N - 1, L). \quad (26)$$

This is the fundamental recurrence relation in the canonical ensemble for the CSM model in the quantum case and for integer λ.

We proceed with the construction of the quantum IGCPF. In multiplying Eqs. (24), (25), and (26) by z^N, in summing over N from zero to infinity and in permuting the summation over k and N in (23), we find

$$G_q^{(0)}(n, \beta, z, L) = \frac{1}{1 - za(n)} G_q^{(0)}(n - 1, \beta, z, L), \quad (27)$$

$$G_q^{(1)}(n, \beta, z, L) = 1 + za(n)G_q^{(1)}(n - 1, \beta, z, L), \quad (28)$$

and

$$G_q(n, \beta, z, L) = G_q(n - 1, \beta, z, L) + za(k)G_q(n - \lambda, \beta, z, L), \quad (29)$$

for the Bose, Fermi, and Sutherland case, respectively. Here we can check that the Bose (27) and Fermi (28) cases are indeed recovered from (29) with $\lambda = 0$ and $\lambda = 1$, respectively. It remains to define the familiar GPF through

$$\mathcal{G}_q(\beta, z, L) = G_q(\infty, \beta, z, L). \quad (30)$$

We have now to proceed to the thermodynamic limit. To this end the quantum number n is replaced by the quantized momentum $p = hn/L$ and, as in the classical case, we set

$$G_q(p, \beta, z, L) = \exp(\beta \mathcal{P}_q(p, \beta, z, L)L). \quad (31)$$

Then, in dividing (29) by $G_q(p, \beta, z, L)$ and introducing (31), we get

$$1 = \exp\left(\beta \mathcal{P}_q\left(p - \frac{h}{L}, \beta, z, L\right)L - \beta \mathcal{P}_q(p, \beta, z, L)L\right)$$
$$+ za(p) \exp\left(\beta \mathcal{P}_q\left(p - \frac{h\lambda}{L}, \beta, z, L\right)L - \beta \mathcal{P}_q(p, \beta, z, L)L\right). \quad (32)$$

It transpires from (32) that, as $L \to \infty$, and with the same assumptions as in the classical case we get

$$1 = \exp(-\beta h \dot{P}_q(p, \beta, z)) + za(p) \exp(-\lambda h \beta \dot{P}_q(p, \beta, z)) \qquad (33)$$

or

$$\exp(\beta h \dot{P}_q) = 1 + za(p) \exp((1 - \lambda)\beta h \dot{P}_q). \qquad (34)$$

If we set

$$X = \exp(\beta h \dot{P}_q), \qquad (35)$$

then equation (34)can be written equivalently as

$$X^\lambda - X^{\lambda - 1} - za(p) = 0. \qquad (36)$$

This is the algebraic equation announced in the first section and it is its real and positive root that interests us. Explicit solutions of (35) can be given for a few values of λ, both fractional and integer ones [6, Section F]. The classical limit (18) is obtained from (33) in replacing $h\lambda$ by $2\pi g$ in its second term and in retaining the contribution linear in $\dot{P}_q$ in the power series expansion of its first term. From (34) we can write the implicit solution for the pressure equation of state, namely,

$$\beta P_q(\beta, z) = \frac{1}{h} \int_{-\infty}^{+\infty} dp \ln\left(1 + z \exp\left[-\frac{\beta p^2}{2m} + (1 - \lambda)\beta h \dot{P}_q(p, \beta, z)\right]\right). \qquad (37)$$

At this point we can make contact with Yang–Yang's self-consistent solution of the CSM model as given by Sutherland. In our notation and in introducing the energy spectrum $\mathcal{E}(p, \beta, z)$, Sutherland's equations are

$$\beta P_q(\beta, z) = \frac{1}{h} \int_{-\infty}^{+\infty} dp \ln(1 + z \exp[-\beta \mathcal{E}(p, \beta, z)]) \qquad (38)$$

and

$$\beta \mathcal{E}(p, \beta, z) = \beta \frac{p^2}{2m} + (\lambda - 1) \ln(1 + z \exp[-\beta \mathcal{E}(p, \beta, z)]). \qquad (39)$$

Then it is clear from (38) and (39) that

$$\beta h \dot{P}_q(p, \beta, z) = \ln(1 + z \exp[-\beta \mathcal{E}(p, \beta, z)]) \qquad (40)$$

and that

$$\beta \mathcal{E}(p, \beta, z) = \beta \frac{p^2}{2m} + (\lambda - 1)\beta h \dot{P}_q(p, \beta, z). \qquad (41)$$

It remains to explain why the Eqs. (34) and (39) are local in $\dot{\mathcal{P}}_q(p, \beta, z)$, or $\mathcal{E}(p, \beta, z)$. The reason is that the quantity $\mathcal{E}(p, \beta, z)$ that maximizes Sutherland's pressure functional satisfies an integral equation [8, Eq. 2.22] with a kernel proportional to the derivative of the two-body phase-shift $\theta(p' - p)$. Now, Sutherland has shown that $\theta(p' - p) = \pi(\lambda - 1)\,\mathrm{sgn}(p' - p)$, its derivative is therefore $\pi(\lambda-1)\delta(p'-p)$ and the integral equation collapses to the local nonlinear relation (39).

For the one-dimensional system of bosons with delta function interaction, solved by Yang and Yang [9], the nonlinear equation (34) for $\dot{\mathcal{P}}_q$ would become nonlocal, the term $(1 - \lambda)\beta h\dot{\mathcal{P}}_q$ being replaced by a convolution of $\dot{\mathcal{P}}_q$ with a Lorentzian kernel.

Acknowledgments: We are very much indebted to S. Ruijsenaars and B. Sutherland for their continuing and active interest in the work presented here. We are also indebted to M. Mobilia for his careful reading and typing of the manuscript.

5 REFERENCES

1. R. J. Baxter, *Statistical mechanics of a one-dimensional Coulomb system with a uniform charge background*, Proc. Cambridge Philos. Soc. **59** (1963), 779–787.

2. P. Choquard, *On the statistical mechanics of one-dimensional Coulomb systems*, Helv. Phys. Acta **48** (1975), No. 4, 585–598.

3. M. Hénon, *Integrals of the Toda lattice*, Phys. Rev. B **9** (1974), 1921–1923.

4. J. Moser, *Three integrable Hamiltonian systems connected to isospectral deformations*, Adv. Math. **16** (1975), 197–220.

5. S. N. M. Ruijsenaars, *Action-angle maps and scattering theory for some finite-dimensional integrable systems. III. Sutherland type systems and their duals*, Publ. Res. Inst. Math. Sci. **31** (1995), No. 2, 247–353.

6. B. Sutherland, *Quantum many-body problem in one dimension: thermodynamics*, J. Math. Phys. **12** (1971), 251–256.

7. _____, *Exact results for a quantum many-body problem in one dimension. II*, Phys. Rev. A **5** (1972), 1372–1376.

8. _____, *A microscopic theory of exclusion statistics*, Phys. Rev. B **56** (1997), No. 8, 4422–4431.

9. C. M. Yang and C. P. Yang, *Thermodynamics of a one-dimensional system of bosons with repulsive delta-function interaction*, J. Math. Phys. **10** (1969), 1115–1122.

9
The Meander Determinant and Its Generalizations

P. Di Francesco

ABSTRACT We investigate various generalizations of meanders, that is, configurations of non-self-intersecting loops crossing a line through a given number of points. In all cases, we derive explicit formulas for corresponding meander determinants.

1 Introduction

In this chapter we review various generalizations of the concept of meander [1, 7, 9–11, 16]. The original meander problem consists in counting the number M_n of meanders of order n, that is, of topologically inequivalent configurations of a closed non-self-intersecting loop crossing an infinite line through $2n$ points. One can also define the corresponding multicomponent meander problem, by demanding that the road have a given number of connected components. The meander problem probably first arose in the work of Poincaré on differential geometry, then reemerged in various contexts, such as the classification of 3-manifolds [8], or the physics of compact polymer folding [4]. The latter problems exhibit highly nonlocal interactions, which is probably responsible for the lack of integrability and, so far, solvability of the meander problem.

In the present work, we extend the purely algebraic approach advocated in Ref. 5, which relates multicomponent meanders to *pairs* of reduced elements of the Temperley–Lieb algebra [15] (see also P. Martin's book [12] for an elementary introduction), or ideals thereof. The idea is to define generalized multicomponent meanders as *pairs* of reduced elements of the SU(N) quotients of the Hecke algebra [13, 14] that generalize the Temperley–Lieb SU(2) quotient, or of ideals thereof. The notion of "component" here still awaits a good combinatorial interpretation. We trade it for a piece of information on any given generalized meander, provided by the Markov trace of the corresponding product of reduced elements. Given a reduced basis of these Hecke algebra quotients or ideals, this information is summarized by the Gram matrix of the basis.In this note, we present results for the "meander determinants," namely, the determinants of these Gram matrices.

The paper is organized as follows. In Section 2, we recall basic definitions of meanders, arch configurations and walk diagrams, and present the meander determinant formula. Section 3 is devoted to generalizations to semimeanders and crossing meanders, both expressed as problems of counting roads crossing a river. For each generalization, we obtain an explicit meander determinant formula. In Section 4, the concept of meander is generalized to SU(N), through a connection with the Hecke algebra. We express the corresponding determinants, as well as a duality relation connecting them.

2   Meanders: Definitions and Reformulations

2.1 Definitions

Meanders

The *meander* problem of order $2n$ is that of enumerating the topologically inequivalent configurations of a planar nonintersecting closed loop (road) crossing a line (river) through $2n$ distinct points (bridges). We denote by M_{2n} the number of meanders of order $2n$.

Arch Configurations

The line cuts the meander into an upper and a lower part, which are both made of n nonintersecting arches (pieces of the loop) connecting the $2n$ bridges by pairs. Such an upper (or lower) configuration of a meander is called an *arch configuration of order* $2n$. The set of arch configurations of order $2n$, A_{2n}, has cardinal equal to the Catalan number

$$c_n = \frac{(2n)!}{(n+1)!n!}. \tag{1}$$

Multicomponent Meanders

A *multicomponent meander* of order $2n$ is the superposition of two arbitrary upper and lower arch configurations $a, b \in A_{2n}$ (note that the lower configuration b is reflected with regard to the river). This results a priori in a configuration of k different nonintersecting roads crossing the river through a total of $2n$ bridges: k is called the number of connected components of the meander, also denoted by $k = \kappa(a, b)$. We define the meander polynomial as

$$m_{2n}(\beta) = \sum_{a,b \in A_{2n}} \beta^{\kappa(a,b)} = \sum_{k=1}^{n} M_{2n}^{(k)} \beta^k \tag{2}$$

where $M_{2n}^{(k)}$ stands for the number of meanders of order $2n$ with k connected components.

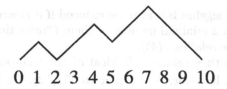

$$0\ 1\ 2\ 3\ 4\ 5\ 6\ 7\ 8\ 9\ 10$$

FIGURE 9.1. A sample walk diagram of order 10.

Walk Diagrams

We adopt an alternative description of meanders in terms of SU(2) *walk diagrams of order* $2n$, namely, closed paths of length $2n$ on the semi-infinite line $\{1, 2, 3, \ldots\}$, identified with the Weyl chamber of the sl(2) Lie algebra. More precisely, a walk diagram is a sequence $\{h(i) \mid i = 0, 1, 2, \ldots, 2n\}$ of positive integer "heights," such that

$$h(i+1) - h(i) \in \{1, -1\}, \quad h(0) = h(2n) = 1. \tag{3}$$

A pictorial representation for a walk diagram is presented in Figure 9.1: it consists of the graph of the corresponding function $i \to h(i)$, whose points are joined by consecutive segments. We denote by $W_{2n}^{(2)}$ the set of walk diagrams of order $2n$.

The walk diagrams of order $2n$ are in one-to-one correspondence with the arch configurations of order $2n$. Starting from an arch configuration of order $2n$, let us label by $0, 1, 2, \ldots, 2n$, respectively, the portions of river left of the leftmost bridge, between the first and second, $\ldots$, right of the rightmost bridge along the river. We define the map $i \to h(i)$ by assigning to the portion of river labelled i the number $h(i)$ of arches passing above it, plus one. The constraints (3) are satisfied by h; hence we have constructed a walk diagram for each arch configuration. The process is clearly bijective, as an arch configuration is entirely determined by the numbers $h(i+1) - h(i)$, with the value $+1$ if an arch originates from the left bridge of the portion i of river and passes above it, and the value -1 if an arch terminates at the left bridge of the portion i of river.

A multicomponent meander of order $2n$ is therefore equivalently given by a couple (a, b) of walk diagrams of order $2n$, and we still denote by $\kappa(a, b)$ its number of connected components of road.

2.2   Temperley–Lieb Algebra

The Temperley–Lieb algebra $\mathrm{TL}_n(\beta)$ is defined by generators $1, e_1, e_2, \ldots, e_{n-1}$ and relations

$$e_i^2 = \beta\, e_i \quad \text{for } i = 1, 2, \ldots, n-1$$
$$e_i e_j = e_j e_i \quad \text{for } |i - j| > 1 \tag{4}$$
$$e_i e_{i\pm 1} e_i = e_i \quad \text{for } i = 1, 2, \ldots, n-1.$$

An element of this algebra is said to be reduced if it is written as a product of generators, with a minimal number of them ("reduction" is achieved by repeated use of the relations (4)).

We will work with a certain left ideal of the Temperley–Lieb algebra $TL_{2n}(\beta)$, which is, however, iisomorphic to $TL_n(\beta)$. We denote by $\mathcal{I}_{2n}^{(2)}(\beta)$ the left ideal generated by the element $e_1 e_3 e_5 \cdots e_{2n-1}$ of $TL_{2n}(\beta)$.

There is a one-to-one correspondence between reduced elements of the ideal $\mathcal{I}_{2n}^{(2)}(\beta)$ and the walk diagrams of order $2n$. To best see this, let us first reconsider the walk diagrams of order $2n$. We start from the "fundamental" walk diagram $a_0^{(2)} \in W_{2n}^{(2)}$, such that

$$h(1) = h(3) = \cdots = h(2n-1) = 2 \text{ and } h(0) = h(2) = \cdots = h(2n) = 1. \quad (5)$$

This is the walk with the smallest values of the height $h(i)$. Now any other walk diagram of order $2n$ may be constructed by successive "box additions" on $a_0^{(2)}$. By *box addition on a walk diagram a at position i*, which we denote by $a + \diamond_i$, we mean the following transformation. For the box addition to be possible, a must have a minimum at the position i, namely,k $h(i+1) = h(i-1) = h(i) + 1$. The box addition then simply amounts to transform this minimum into a maximum, namely, change $h(i) \to h(i) + 2$, without altering the other values of h. By successive box additions on $a_0^{(2)}$, it is easy to describe all the set of walk diagrams of order $2n$. We are now in position to construct a map φ from $W_{2n}^{(2)}$ to a basis of reduced elements of $\mathcal{I}_{2n}^{(2)}(\beta)$. We start with

$$\varphi(a_0^{(2)}) = e_1 e_3 \cdots e_{2n-1} \quad (6)$$

and proceed recursively, using box additions, by setting

$$\varphi(a + \diamond_i) = e_i \varphi(a). \quad (7)$$

The map is well defined, as two distinct sequences of box additions leading to the same walk diagram correspond to different products of the same commuting e_i's (at each step, if two distinct box additions are possible, they take place at positions i and j with $|j-i| > 1$; hence the corresponding e_i and e_j commute, owing to (4)). It exhausts all the reduced elements of $\mathcal{I}_{2n}^{(2)}(\beta)$, which has the dimension c_n (1) as a vector space. A meander is therefore equivalently given by a pair of reduced elements of $\mathcal{I}_{2n}^{(2)}(\beta)$.

The Temperley–Lieb algebra $TL_n(\beta)$ is endowed with a natural scalar product attached to the Markov trace, denoted by Tr. The latter is defined by the normalization $\mathrm{Tr}(1) = \beta^n$, and the Markov property that for any element $E(e_1, e_2, \ldots, e_{j-1})$ involving only e_i's with $i < j$, we have

$$\mathrm{Tr}(E(e_1, e_2, \ldots, e_{j-1})e_j) = \eta \, \mathrm{Tr}\big(E(e_1, e_2, \ldots, e_{j-1})\big). \quad (8)$$

The standard choice for the constant η, compatible with (4), is

$$\eta = \frac{1}{\beta}. \tag{9}$$

The trace extends linearly to any element of $\mathrm{TL}_n(\beta)$. We also need to define the transposed e^t of an element $e \in \mathrm{TL}_n(\beta)$, as $1^t = 1$, $e_i^t = e_i$ for $i = 1$, $2, \ldots, n$, and $(ef)^t = f^t e^t$ for any two elements e, $f \in \mathrm{TL}_n(\beta)$; again, the definition extends to any element by linearity. This leads to the scalar product

$$(e, f) = \mathrm{Tr}(ef^t). \tag{10}$$

Remarkably, when restricted to the ideal $\mathcal{I}_{2n}^{(2)}(\beta)$, and when expressed between two reduced elements say $\varphi(a)$ and $\varphi(b)$, a, b two walk diagrams of order $2n$, this scalar product reads

$$(\varphi(a), \varphi(b)) = \beta^{\kappa(a,b)+n} \tag{11}$$

thus making the contact with our initial road/river picture of meanders. Defining the normalized reduced basis elements $(a)_1 = \beta^{-n/2}\varphi(a)$ (this basis is referred to as basis 1 in the following), we have $((a)_1, (b)_1) = \beta^{\kappa(a,b)}$. The $c_n \times c_n$ matrix $\mathcal{G}_{2n}^{(2)}(\beta)$ with entries

$$\left[\mathcal{G}_{2n}^{(2)}(\beta) \right]_{a,b} = ((a)_1, (b)_1) = \beta^{\kappa(a,b)} \tag{12}$$

is the Gram matrix of the basis 1 for the scalar product $(\cdot, \cdot)$. It carries all the relevant information about the multicomponent meanders. In particular, the meander polynomials (2) read

$$m_{2n}(\beta) = \mathrm{Tr} \left[\mathcal{G}_{2n}^{(2)}(\sqrt{\beta}) \right]^2. \tag{13}$$

Unfortunately, the spectrum of the matrix $\mathcal{G}_{2n}^{(2)}(\beta)$ does not display any simple property that would give access, say, to large n asymptotics of the meander polynomial. The only quantity we have been able to express so far is the determinant of this matrix.

2.3 Meander Determinant

The meander determinant $\Delta_{2n}^{(2)}(\beta)$ is defined as the determinant of the Gram matrix (12) of the basis 1 above. In Refs. 3 and 5 we have derived an exact formula for $\Delta_{2n}^{(2)}(\beta)$ based on the explicit Gram–Schmidt orthogonalization of the matrix $\mathcal{G}_{2n}^{(2)}(\beta)$. The formula reads

$$\Delta_{2n}^{(2)}(\beta) = \prod_{m=1}^{n} [U_m(\beta)]^{a_{m,n}^{(2)}}, \tag{14}$$

TABLE 9.1. The powers $a_{m,n}^{(2)}$ of U_m in the meander determinant of order $2n$, $\Delta_{2n}^{(2)}(\beta)$, for $n = 1, 2, \ldots, 10$.

m \ n	1	2	3	4	5	6	7	8	9	10
1	1	2	4	8	15	22	0	−208	−1326	−6460
2		1	4	13	40	121	364	1092	3264	9690
3			1	6	26	100	364	1288	4488	15504
4				1	8	43	196	820	3264	12597
5					1	10	64	336	1581	6954
6						1	12	89	528	2755
7							1	14	118	780
8								1	16	151
9									1	18
10										1

where $U_m(\beta)$ are Chebyshev polynomials, with

$$U_m(2\cos\theta) = \frac{\sin(m+1)\theta}{\sin\theta} \tag{15}$$

and

$$a_{m,n}^{(2)} = C_{2m+1}^{(2n)} - C_{2m+3}^{(2n)}, \tag{16}$$

where $C_{2m+1}^{(2n)}$ counts the number of paths of length $2n$ on the half-line, starting from the origin $(h(0) = 1)$ and terminating at height $2m + 1$ $(h(2n) = 2m + 1)$, easily computed as

$$C_{2m+1}^{(2n)} = \binom{2n}{n-m} - \binom{2n}{n-m-1} \tag{17}$$

with in particular $C_1^{(2n)} = c_n$ of (1). The first few numbers $a_{m,n}^{(2)}$ are displayed in Table 9.1.

The derivation of (14) involves the orthogonalization of the basis 1 with regard to the scalar product $(\cdot, \cdot)$, namely, the construction of an orthonormal basis 2, whose elements are denoted by $(a)_2$, $a \in W_{2n}^{(2)}$. In the same spirit as for the basis 1, it is defined by induction on box additions:

$$(a_0)_2 = \beta^{-n/2}(a_0)_1$$

$$(a + \diamond_{i,m})_2 = \sqrt{\frac{\mu_{m+1}}{\mu_m}}(e_i - \mu_m)(a)_2, \tag{18}$$

where the two indices of the box refer to its position i and the height m of its center, and

$$\mu_m = \frac{U_{m-1}(\beta)}{U_m(\beta)}. \tag{19}$$

Comparing the definition (18) with that of the basis 1, we see that the change of basis $1 \to 2$ is triangular with regard to to the inclusion of walk diagrams

$$(a)_2 = \sum_{b \subset a} P_{a,b}(b)_1, \tag{20}$$

where $b \subset a$ iff a is obtained from b by some box additions. This basis has also the remarkable property that $(a)_2(b)_2^t = 0$ unless $a = b$, and the prefactors in (18) ensure that $((a)_2, (a)_2) = 1$ for all a. The formula (14) follows from the triangularity of the change of basis, namely,

$$\Delta_{2n}^{(2)}(\beta) = \prod_{a \in W_{2n}^{(2)}} P_{a,a}^{-2}, \tag{21}$$

where the diagonal normalization factors $P_{a,a}$ of (20) satisfy the recursion relation $P_{a+\diamond_{i,m}, a+\diamond_{i,m}} = \sqrt{\mu_{m+1}/\mu_m} P_{a,a}$, a direct consequence of (18). This is easily solved as

$$P_{a,a}^2 = \prod_{\text{edges } e \text{ of } a} w(e), \tag{22}$$

where the product extends over the consecutive edges $e = (i, h(i))$, $(i + 1, h(i+1))$ of the walk diagram a, and

$$w(e) = \sqrt{\mu_{\max(h(i), h(i+1))}}. \tag{23}$$

The factor μ_m will appear in the product (21) $\alpha_{m,n}$ times, where $\alpha_{n,m}$ is the number of walk diagrams with an edge whose highest end is at height m. The result is $\alpha_{n,m} = \binom{2n}{n-m} - \binom{2n}{n-m-1}$ and (14) follows from the definition of $\mu_m = U_{m-1}/U_m$ in terms of Chebyshev polynomials, with $a_{n,m} = \alpha_{n,m} - \alpha_{n,m+1}$.

Note that the zeros of the determinant $\Delta_{2n}^{(2)}(\beta)$ are exactly the numbers $2\cos[\pi k/(p+1)]$, $1 \leq k \leq p \leq n$. When n is large these numbers cover the interval $]-2, 2[$.

3 Road/River Generalizations

3.1 Semimeanders

The meander problem is equivalent to that of counting compact folding configurations of a *closed* polymer of $2n$ identical constituents [4]. The corresponding *open* polymer folding problem leads to the definition of semime-anders of order n as topologically inequivalent configurations of a loop (road) crossing a half-line (semi-infinite river with a source) through n points (bridges). The novelty is that the road may freely wind around the source of the river. The winding number w of such a semimeander is defined as the minimal number of bridges created by continuing the half-line into a line (note that $w = n \bmod 2$).

As in the meander case, the river separates the semimeander into an upper and lower part, connected through the n bridges and the w pieces of loop that go around the source of the river. This leads to the notion of *open arch configuration of order n with w open arches*, made of $(n - w)/2$ arbirtary nonintersecting arches connecting $n - w$ bridges by pairs, and w open arches attached to only one bridge. Let $A_n^{(w)}$ denote the set of open arch configurations of order n with w open arches. Its cardinal is the generalized Catalan number

$$|A_n^{(w)}| = c_{n,w} = \binom{n}{(n - w)/2} - \binom{n}{(n - w)/2 - 1} \qquad (24)$$

with $c_n = c_{2n,0}$.

Superposing any two open arch configurations a, $b \in A_n^{(w)}$, and connecting the w upper and lower open arches around the source of the river, one after the other from left to right, we obtain a multicomponent semimeander. We still denote by $\kappa(a, b)$ its number of connected components (loops). We may form the matrix $\mathcal{G}_{n,w}^{(2)}(\beta)$ with entries

$$\left[\mathcal{G}_{n,w}^{(2)}(\beta)\right]_{a,b} = \beta^{\kappa(a,b)} \quad \text{for } a, b \in A_n^{(w)}. \qquad (25)$$

This matrix contains again all the relevant information on multicomponent semimeanders.

In Ref. 3, we have derived the following formula for the determinant $\Delta_{n,w}(\beta)$ of the matrix (25):

$$\Delta_{n,w}(\beta) = \prod_{m=1}^{(n-w)/2+1} [U_m(\beta)]^{a_{m,n,w}^{(2)}}$$

$$a_{m,n,w}^{(2)} = c_{n,w+2m} - c_{n,w+2m+2} + w(c_{n,w+2m-2} - c_{n,w+2m}).$$

The proof imitates strongly that of (14), and may be found in Ref. 3.

3.2   Crossing Meanders and Brauer Algebra

Another generalization of meanders (called *crossing meanders*) consists in allowing for the roads to cross each other or themselves through additional bridges. The upper and lower halves of such a crossing meander are *crossing arch configurations* of order $2n$, made of n possibly crossing arches connecting the $2n$ bridges by pairs. We denote by CA_{2n} the set of crossing arches of order $2n$. The cardinal of this set is the number of pairings of the $2n$ bridges, namely,

$$|CA_{2n}| = (2n-1)!! = (2n-1)(2n-3)\cdots 5\cdot 3\cdot 1. \tag{26}$$

For $a \in CA_{2n}$, we denote by $\gamma(a)$ the number of road crossings in a. Superposing any two $a, b \in CA_{2n}$ leads to a multicomponent crossing meander, with a number of connected components of road denoted by $\kappa(a, b)$, and a total number $\gamma(a) + \gamma(b)$ of upper and lower bridges for road crossings. We define the crossing meander polynomial as

$$m_{2n}(\beta, t) = \sum_{a,b \in CA_{2n}} \beta^{\kappa(a,b)} t^{\gamma(a)+\gamma(b)}. \tag{27}$$

It reduces to the meander polynomial (2) when $t = 0$, in which case only nonintersecting roads are considered.

We now consider the matrix $\mathcal{CG}_{2n}(\beta, t)$ with entries

$$[\mathcal{CG}_{2n}(\beta, t)]_{a,b} = \beta^{\kappa(a,b)} t^{\gamma(a)+\gamma(b)}. \tag{28}$$

Remarkably, this matrix (at $t = 1$) may again be interpreted as the Gram matrix of a reduced basis of the Brauer algebra $B_n(\beta)$, defined as follows. It is obtained by supplementing the Temperley–Lieb algebra (4) above with a new set of generators f_i, $i = 1, \ldots, n-1$, with the relations

$$f_i^2 = 1 \quad e_i f_i = f_i e_i = e_i \quad f_i f_{i+1} f_i = f_{i+1} f_i f_{i+1}$$
$$e_i f_{i\pm 1} e_i = e_i \quad f_i e_{i+1} f_i = f_{i+1} e_i f_{i+1} \tag{29}$$
$$[f_i, f_j] = [e_i, e_j] = 0 \text{ for } |i - j| > 1.$$

Pictorially, f_i generates the crossing between two consecutive pieces of loop at position i. As in the meander case, we consider a certain left ideal $CI_{2n}(\beta)$ of $B_{2n}(\beta)$, generated by $e_1 e_3 \cdots e_{2n-1}$. This ideal is isomorphic to $B_n(\beta)$. Moreover, the reduced elements of this ideal are in one-to-one correspondence with crossing arch configurations of order $2n$. The standard Markov trace of $TL_n(\beta)$ extends naturally to $B_n(\beta)$ and the matrix $\mathcal{CG}_{2n}(\beta, 1) \equiv \mathcal{CG}_{2n}(\beta)$ is nothing but the Gram matrix for a reduced basis of $CI_{2n}(\beta)$.

The determinant of the matrix (28) follows from the study of Ref. 6, in which the spectrum of the matrix is explicitly derived. The crucial property of $\mathcal{CG}_{2n}(\beta)$ allowing for this diagonalization is a simple action of the permutation group S_n on the reduced elements of $CI_{2n}(\beta)$, which basically commutes with $\mathcal{CG}_{2n}(\beta)$. The results read as follows. The eigenvalues and eigenspaces of $\mathcal{CG}_{2n}(\beta)$ are indexed by even Young tableaux

$Y = \{2f_1, \ldots, 2f_s\}$ with even numbers of boxes $2f_i$ in the ith row, and $|Y| = 2\sum_i f_i = 2n$. The corresponding eigenvalue reads

$$\lambda_Y(\beta) = \prod_{i=1}^{s} \prod_{j=0}^{f_i-i} (\beta + 2j - i + 1) \tag{30}$$

and it has the degeneracy

$$d_Y = \frac{(2n)!}{\prod_{i=1}^{s} \ell_i!} \prod_{\substack{i,j=1 \\ i<j}}^{s} (\ell_j - \ell_i) \tag{31}$$

where $\ell_i = f_i + s - i$, equal to the dimension of the corresponding irreducible representation of the symmetric group S_n. The resulting determinant

$$C\Delta_{2n}(\beta) = \prod_{\substack{\text{even } Y \\ |Y|=2n}} \lambda_Y(\beta)^{d_Y} \tag{32}$$

is again a product of simple factors. Its zeros are relative integers in the interval $[-2n + 2, n - 1]$, hence very different from those of $\Delta_{2n}^{(2)}(\beta)$.

It is also interesting to note that

$$m_{2n}(\beta, 1) = (2n - 1)!!\beta(\beta + 2) \cdots (\beta + 2n - 2) \tag{33}$$

is identified as the sum over all the entries of $C\mathcal{G}_{2n}(\beta)$, and each of the $(2n-1)!!$ lines contributes the same $\beta(\beta+2) \cdots (\beta+2n-2)$, to be contrasted with the meander polynomial (2).

4   SU(N) Generalizations

4.1   Hecke Algebra and Its Quotients

Meanders are strongly related to the sl(2) Lie algebra. Apart from the fact that we considered paths on the Weyl chamber of sl(2) (the half-line), the Temperley–Lieb algebra is known to be a certain quotient of the Hecke algebra $H_n(\beta)$. The latter is defined by generators $1, e_1, e_2, \ldots, e_{n-1}$ and relations

$$\begin{aligned} e_i^2 &= \beta e_i & \text{for } i = 1, 2, \ldots, n-1 \\ e_i e_j &= e_j e_i & \text{for } |i - j| > 1 \\ e_i e_{i+1} e_i - e_i &= e_{i+1} e_i e_{i+1} - e_{i+1} & \text{for } i = 1, 2, \ldots, n-2. \end{aligned} \tag{34}$$

This algebra is usually defined through the generators $T_i = q^{1/2}(q^{1/2} - e_i)$, where $\beta = q^{1/2} + q^{-1/2}$, as a deformation of the symmetric group algebra (in particular, the three-term relation reads simply $T_iT_{i+1}T_i = T_{i+1}T_iT_{i+1}$). In terms of these latter generators, the quantites $e_ie_{i+1}e_i - e_i$, by which we have to quotient the algebra to recover $\mathrm{TL}_n(\beta)$ (see (4)), are simply the generalized Young antisymmetrizers of order 3, namely,

$$A(T_i, T_{i+1}) = 1 - q^{-1}T_i - q^{-1}T_{i+1} + q^{-2}T_iT_{i+1} + q^{-2}T_{i+1}T_i$$
$$- q^{-3}T_iT_{i+1}T_i \quad (35)$$

easily reexpressed in terms of the e_i's as

$$Y(e_i, e_{i+1}) = q^{3/2}A(T_i, T_{i+1}) = e_ie_{i+1}e_i - e_i. \quad (36)$$

Requiring the vanishing of (35) bears a strong analogy with the SU(2) representations.

The natural SU(N) generalizations of the Temperley–Lieb algebra are obtained by performing quotients of the Hecke algebra by the generalized Young antisymmetrizer of order $N + 1$, expressed in terms of the e_i's, through the following recursion (see Ref. 12). Starting with $Y(e_i) = e_i$, we set

$$Y(e_i, e_{i+1}, \ldots, e_{i+p}) = Y(e_i, \ldots, e_{i+p-1})(e_{i+p} - \mu_p)Y(e_i, \ldots, e_{i+p-1})$$
$$(37)$$

for all i, $p \geq 1$, with μ_p as in (19). The SU(N) quotient of the Hecke algebra, $H_n^{(N)}(\beta)$ is defined by the relations (34), supplemented by

$$Y(e_i, e_{i+1}, \ldots, e_{i+N-1}) = 0 \quad (38)$$

for $i = 1, 2, \ldots, n - N$.

4.2 SU(N) Meanders

The notions of walk diagram, meander, and meander determinant extend to SU(N) as follows.

SU(N) Walk Diagrams

Let us denote by $\Lambda = \sum \lambda_i\omega_i = (\lambda_1, \ldots, \lambda_{N-1})$, $\lambda_i \in \mathbb{Z}$, the elements of the weight lattice of the $sl(N)$ algebra, generated by the fundamental weights $\omega_1, \omega_2, \ldots, \omega_{N-1}$ in $\mathbb{R}^{N-1}$, with the scalar products

$$\omega_i \cdot \omega_j = \frac{i(N-j)}{N} \quad (39)$$

for $1 \leq i \leq j \leq N - 1$. The Weyl chamber $P_+ \subset P$ is defined as the set of weights

$$P_+ = \{(\lambda_1, \ldots, \lambda_{N-1}) \mid \lambda_i \geq 1 \text{ for all } i\}. \quad (40)$$

The Weyl group of $\mathrm{sl}(N)$ is the group generated by the reflections s_i with regard to the walls of the Weyl chamber, that is, the hyperplanes $\lambda_i = 0$. It is isomorphic to the permutation group S_N of N objects. The Weyl chamber is nothing but the quotient of the weight lattice by the action of this group.

The weight lattice and Weyl chamber are made into simplices respectively denoted by Π and Π_+, by the adjunction of oriented links between the weights, along the vectors

$$\epsilon_i = \omega_i - \omega_{i-1}, \quad i = 2, 3, \ldots, N - 1 \tag{41}$$

and $\epsilon_1 = \omega_1$, $\epsilon_N = -\omega_{N-1}$, with the property that $\sum \epsilon_i = 0$. Let us denote by $\rho = (1, 1, \ldots, 1)$ the origin (apex) of the Weyl chamber P_+.

A $\mathrm{SU}(N)$ *walk diagram of order* Nn is a closed path of Nn steps on Π_+ starting and ending at ρ. It is uniquely determined by a sequence $\Lambda_0 = \rho, \Lambda_1, \ldots, \Lambda_{Nn-1}, \Lambda_{Nn} = \rho$ of weights in P_+, satisfying $\Lambda_i - \Lambda_{i-1} \in \{\epsilon_1, \epsilon_2, \ldots, \epsilon_N\}$ for all $i = 1, 2, \ldots, Nn$. As before, the index i in Λ_i is referred to as the position of the weight Λ_i in the walk diagram. The set of $\mathrm{SU}(N)$ walk diagrams of order Nn is denoted by $W_{Nn}^{(N)}$. Imitating the $\mathrm{SU}(2)$ case, we can represent the $\mathrm{SU}(N)$ walk diagrams pictorially in the plane, by replacing each step ϵ_i by an edge of unit length, making an angle of $\pi(N - 2i + 1)/(2N)$ with the horizontal axis, and connecting the successive edges of each walk diagram. We also define the *fundamental* $\mathrm{SU}(N)$ walk diagram $a_0^{(N)}$ with the successive weights $\Lambda_{i+Nj} - \Lambda_{i+Nj-1} = \epsilon_i$, for $i = 1$, $2, \ldots, N$ and $j = 0, 1, \ldots, n - 1$.

The number $c_{Nn}^{(N)}$ of $\mathrm{SU}(N)$ walk diagrams of order Nn reads [2]

$$c_{Nn}^{(N)} = (Nn)! \prod_{i=0}^{N-1} \frac{i!}{(n+i)!}. \tag{42}$$

It is a natural N-dimensional generalization of the Catalan numbers (1). Similarly, the number $C_\Lambda^{(M)}$ of paths of M steps on Π_+ from the origin ρ to a given weight Λ reads

$$C_\Lambda^{(M)} = M! \frac{\prod_{1 \leq i < j \leq N}((\epsilon_i - \epsilon_j) \cdot \Lambda)!}{\prod_{i=0}^{N-1}(M/N + \epsilon_i \cdot (\Lambda - \rho) + i)!}. \tag{43}$$

$\mathrm{SU}(N)$ Ideal of $H_{Nn}^{(N)}(\beta)$

We now concentrate on the $\mathrm{SU}(N)$ quotient $H_{Nn}^{(N)}(\beta)$ of the Hecke algebra $H_{Nn}(\beta)$. We now consider the left ideal $\mathcal{I}_{Nn}^{(N)}(\beta)$, generated by the element

$$Y_{Nn}^{(N)} = \prod_{i=0}^{n-1} Y(e_{iN+1}, e_{iN+2}, \ldots, e_{iN+N-1}). \tag{44}$$

There is a one-to-one correspondence between the $SU(N)$ walk diagrams of order Nn and the reduced elements of $\mathcal{I}_{Nn}^{(N)}(\beta)$. Let us first express the $SU(N)$ walk diagrams as the results of successive box additions on the fundamental diagram $a_0^{(N)}$. Given a walk diagram $a \in W_{Nn}^{(N)}$, the process of box addition at position i on a, producing a diagram $b = a + \diamond_i$, is allowed only if a has a minimum at i, namely, $N \geq r > s \geq 1$, if $\Lambda_{i+1} - \Lambda_i = \epsilon_s$ and $\Lambda_i - \Lambda_{i-1} = \epsilon_r$. The box addition amounts to replacing $\Lambda_i \to \Lambda_i + \epsilon_s - \epsilon_r$, that is, exchanging the two steps ϵ_r and ϵ_s in the corresponding path on Π_+. In the above pictorial representation, a box addition amounts to adding to a a parallelogram (the "box"), with edges corresponding to the vectors ϵ_r and ϵ_s. This gives rise to $N(N-1)/2$ different types of boxes. It is clear that any walk $a \in W_{Nn}^{(N)}$ can be obtained from the fundamental one $a_0^{(N)}$ by successive box additions. The box decomposition of a given walk a is, however, not unique, owing to all possible "hexagon" ambiguities: namely, for any three integers $N \geq i > j > k \geq 1$, there are two possibilities to change the succession of steps $(\epsilon_i, \epsilon_j, \epsilon_k)$ into $(\epsilon_k, \epsilon_j, \epsilon_i)$ by three successive box additions.

$$\tag{45}$$

To resolve these ambiguities, we forbid all the box additions of the form

$$\tag{46}$$

for all $N \geq i > j > k \geq 1$. With this last rule, each walk $a \in W_{Nn}^{(N)}$ has a unique box decomposition, represented as the set of boxes in between $a_0^{(N)}$ and a.

We can now construct the map φ from $W_{Nn}^{(N)}$ to the set of reduced elements of $\mathcal{I}_{Nn}^{(N)}(\beta)$, through the recursion

$$\varphi(a_0^{(N)}) = Y_{Nn}^{(N)} \qquad \varphi(a + \diamond_i) = e_i \varphi(a) \tag{47}$$

with $Y_{Nn}^{(N)}$ as in (44), and for any $a \in W_{Nn}^{(N)}$. As before, we introduce the basis 1, with elements $(a)_1 = \gamma_{N,n} \varphi(a)$ where $\gamma_{N,n}$ is a specific normalization factor (see Ref. 2).

$SU(N)$ Meanders

The $SU(N)$ meanders are defined as pairs of walks $(a, b) \in W_{Nn}^{(N)}$, or equivalently of elements of the basis 1 above. To the latter, we attach the quantity

$((a)_1, (b)_1)$, where the scalar product is attached to the Markov trace Tr on $H_{Nn}^{(N)}(\beta)$, defined by the normalization $\mathrm{Tr}(1) = (U_{N-1})^{Nn}$ and the recursion (8), with $\eta = \mu_{N-1} = U_{N-2}(\beta)/U_{N-1}(\beta)$. This leads to the $c_{Nn}^{(N)} \times c_{Nn}^{(N)}$ Gram matrix $\mathcal{G}_{Nn}^{(N)}(\beta)$, with entries

$$\left[\mathcal{G}_{Nn}^{(N)}(\beta)\right]_{a,b} = ((a)_1, (b)_1) \quad \text{for } a, b \in W_{Nn}^{(N)}. \tag{48}$$

4.3 SU(N) Meander Determinants

SU(N) Meander Determinants

The SU(N) meander determinant $\Delta_{Nn}^{(N)}(\beta)$ of the Gram matrix (48) reads

$$\Delta_{Nn}^{(N)} = \prod_{m=1}^{n+N-2} [U_m(\beta)]^{a_{m,n}^{(N)}}, \tag{49}$$

where $a_{m,n}^{(N)}$ are some positive integers, defined as follows. We first introduce the vectors $u_0 = v_0 = \bar{u}_0 = \bar{v}_0 = 0$, and

$$\begin{aligned}
u_j &= j\epsilon_1 - (\epsilon_j + \epsilon_{j+1} + \cdots + \epsilon_{N-1}) \\
v_j &= -j\epsilon_N + (\epsilon_2 + \epsilon_3 + \cdots + \epsilon_{j+1}) \\
\bar{u}_j &= j(\epsilon_1 - \epsilon_N) - u_{N-1-j} \\
\bar{v}_j &= j(\epsilon_1 - \epsilon_n) - v_{N-1-j}
\end{aligned} \tag{50}$$

for $j = 1, 2, \ldots, N - 2$. We define the following difference operators Δ_N, and $\bar{\Delta}_N$ acting on any function $f(\alpha)$, of $\alpha \in P$ by the alternate sums

$$\begin{aligned}
(\alpha) &= \sum_{\substack{i,j \geq 0 \\ i+j \leq N-2}} (-1)^{i+j} f(\alpha + (u_i + v_j)) \\
[\bar{\Delta}_N f](\alpha) &= \sum_{\substack{i,j \geq 0 \\ i+j \leq N-2}} (-1)^{i+j} f(\alpha - (\bar{u}_i + \bar{v}_j)).
\end{aligned} \tag{51}$$

We also define Δ_N^* as the same expression as for Δ_N, except that the point $i = j = 0$ is excluded from the sum (51). Now, we use the function $f(\Lambda) = C_\Lambda^{(Nn)}$, $\Lambda \in P_+$, and $C_\Lambda^{(Nn)}$ as in (43), to write the integers $a_{m,n}^{(N)}$ as

$$a_{m,n}^{(N)} = \Delta_N f(\rho + (m - N + 2)(\epsilon_1 - \epsilon_N))$$
$$- \bar{\Delta}_N f(\rho + (m + 1)(\epsilon_1 - \epsilon_N)) \tag{52}$$

for any integer $m \neq N - 2$ and

$$a_{N-2,n}^{(N)} = \Delta_N^* f(\rho) - \bar{\Delta}_N f(\rho + (N - 1)(\epsilon_1 - \epsilon_N)). \tag{53}$$

TABLE 9.2. The powers $a_{m,n}^{(3)}$ of U_m in the SU(3) meander determinant of order $3n$, $\Delta_{3n}^{(3)}(\beta)$, for $n = 1, 2, \ldots, 8$.

m \ n	1	2	3	4	5	6	7	8
1	1	6	42	297	1430	−14586	−764218	−21246940
2	1	6	63	814	11583	175032	2762942	45108888
3		4	42	506	7306	119340	2098208	38571368
4			21	374	5707	89352	1495490	26803832
5				121	3276	65790	1218356	22309287
6					728	27336	701879	15622750
7						4488	218994	6931694
8							28101	1701678
9								177859

TABLE 9.3. The powers $a_{m,n}^{(4)}$ of U_m in the SU(4) meander determinant $\Delta_{4n}^{(4)}(\beta)$, for $n = 1, 2, \ldots, 6$.

m \ n	1	2	3	4	5	6
1	1	20	627	24024	831402	−8498776
2	1	15	572	36036	2922504	274085526
3	1	22	880	48048	3375996	291900268
4		13	550	36036	2910876	265913626
5			341	24024	1951566	192203088
6			12012	1372104	139085738
7				492252	85314636
8					22064130

TABLE 9.4. The powers $a_{m,n}^{(5)}$ of U_m in the SU(5) meander determinant $\Delta_{5n}^{(5)}(\beta)$, for $n = 1, 2, \ldots, 5$.

m \ n	1	2	3	4	5
1	1	69	10582	2494206	701149020
2	1	44	6435	2065908	1051723530
3	1	58	10712	3275220	1402298040
4	1	76	12311	3740340	1752872550
5		41	8736	3036846	1402298040
6			5278	1953504	1051723530
7				1170552	701149020
8					350574510

We see that the generalization (52) of the SU(2) case is an alternate sum over a set of $N(N-1)$ shifted weights. The first few values for the integers $a_{m,n}^{(N)}$ are given in Tables 9.2–9.4 for the cases $N = 3, 4, 5$ respectively.

As in the SU(2) case, the formula (49) is proved by direct orthogonalization of the basis 1. The basis 2 is given by the usual inductive definition on box additions, $(a_0^{(N)})_2 = \gamma'_{N,n}(a_0^{(N)})_1$ with some proportionality coefficient $\gamma'_{N,n}$, and

$$(a + \diamond_{i,m})_2 = \sqrt{\frac{\mu_{m+1}}{\mu_m}}(e_i - \mu_m)\,(a)_2, \tag{54}$$

where m stands for the "height" of the box addition, defined as

$$m = \Lambda_i \cdot (\Lambda_{i+1} + \Lambda_{i-1} - 2\Lambda_i) \tag{55}$$

where Λ_{i-1}, Λ_i, Λ_{i+1} are the weights of a with respective positions $i - 1$, i, $i + 1$. The change of basis $1 \to 2$ is again triangular, with $(a)_2 = \sum_{b \subset a} P_{a,b}(b)_1$. The determinant $\Delta_{Nn}^{(N)}(\beta) = \prod_a P_{a,a}^{-2}$ follows from a careful reexpression of $P_{a,a}$, using (54) (we refer the reader to Ref. 2 for a detailed proof).

SU(N) ↔ SU(k) Duality

The compact definitions (54) for the basis 2 elements of $\mathcal{I}_{Nn}^{(N)}(\beta)$ lead us to a simple formula, relating the SU(N) meander determinant of order Nk to the SU(k) meander determinant of same order, namely,

$$\Delta_{Nk}^{(N)}(\beta)\Delta_{kN}^{(k)}(\beta) = (\Phi_{N,k})^{c_{Nk}^{(N)}} \tag{56}$$

where, $\Phi_{N,k}$ is symmetric in $k \leftrightarrow N$, and for $k \leq N$

$$\Phi_{N,k} = \prod_{m=1}^{k-1} (U_m)^{m+1} \prod_{m=k}^{N-1} (U_m)^k \prod_{m=N}^{N+k-2} (U_m)^{N+k-1-m} \tag{57}$$

in terms of the Chebyshev polynomials (15). Note also that from the definition (42),

$$c_{Nk}^{(N)} = (Nk)!\frac{\prod_{i=1}^{k-1} i! \prod_{i=1}^{N-1} i!}{\prod_{i=1}^{N+k-1} i!} = c_{Nk}^{(k)} \tag{58}$$

which makes the right-hand side of (56) symmetric under $k \leftrightarrow N$. The formula (56) is due to the existence of a duality map $\delta: W_{Nk}^{(N)} \to W_{Nk}^{(k)}$, defined by $\delta(a + \diamond_i) = \delta(a) - \diamond_i$; that is, it maps the fundamental $a_0^{(N)}$ to the diagram of $W_{Nk}^{(k)}$ with maximal number of boxes, and removes boxes from the latter whenever a box of same position is added to $a_0^{(N)}$. Equation (56) follows from the recursion $P_{\delta(a+\diamond),\delta(a+\diamond)}P_{a+\diamond,a+\diamond} = P_{\delta(a),\delta(a)}P_{a,a}$, implied by (54).

A simple consequence of (56) is that the "self-dual" determinants, with $k = N$, read

$$\Delta_{N^2}^{(N)}(\beta) = (\Phi_{N,N})^{c_{N^2}^{(N)}/2}$$
$$= (U_1^2 U_2^3 \cdots U_{N-2}^{N-1} U_{N-1}^N U_N^{N-1} \cdots U_{2N-3}^2 U_{2N-2})^{c_{N^2}^{(N)}/2}, \qquad (59)$$

which is considerably simpler than (49)–(52). This is readily checked for $N = 2, 3, 4, 5$ in Tables 9.1–9.4.

5 REFERENCES

1. V. I. Arnold, *The branched covering of* $\mathrm{CP}_2 \to S_4$, *hyperbolicity and projective topology*, Siberian Math. J. **29** (1988), No. 5, 717–726.

2. P. Di Francesco, SU(n) *meander determinants*, J. Math. Phys. **38** (1997), No. 11, 5905–5943, hep-th/9702181.

3. ———, *Meander determinants*, Commun. Math. Phys. **191** (1998), No. 3, 543–583, hep-th/9612026.

4. P. Di Francesco, O. Golinelli, and E. Guitter, *Meander, folding and arch statistics*, Math. Comput. Modelling **26** (1997), No. 8-10, 97–147, hep-th/9506030.

5. ———, *Meanders and the Temperley–Lieb algebra*, Commun. Math. Phys. **86** (1997), No. 1, 1–59, hep-th/9602025.

6. P. Hanlon and D. Wales, *On the decomposition of Brauer's centralizer algebras*, J. Algebra **121** (1989), No. 2, 409–445.

7. K. Hoffmann, K. Mehlhorn, P. Rosenstiehl, and R. E. Tarjan, *Sorting Jordan sequences in linear time using level-linked search trees*, Inform. and Control **68** (1986), No. 1-3, 170–184.

8. K. H. Ko and L. Smolinsky, *A combinatorial matrix in 3-manifold theory*, Pacific Math. J. **149** (1991), No. 2, 319–336.

9. S. K. Lando and A. K. Zvonkin, *Meanders*, Selecta Math. Soviet. **11** (1992), No. 2, 117–144.

10. ———, *Plane and projective meanders*, Theoret. Comput. Sci. **117** (1993), No. 1-2, 227–241.

11. W. Lunnon, *A map-folding problem*, Math. Comp. **22** (1968), 193–199.

12. P. Martin, *Potts Models and Related Problems in Statistical Mechanics*, Ser. Adv. Statist. Mech., Vol. 5, World Scientific, River Edge, NJ, 1991.

13. N. Yu. Reshetikhin, *Quantized universal enveloping algebras, the Yang–Baxter equation and invariants of links 1 and 2*, Tech. Report E-4-87, Steklov Mathematical Institute, St. Petersburg, 1988.

14. _____, *Quantized universal enveloping algebras, the Yang–Baxter equation and invariants of links 1 and 2*, Tech. Report E-17-87, Steklov Mathematical Institute, St. Petersburg, 1988.

15. H. N. V. Temperley and E. H. Lieb, *Relations between the "percolation" and "colouring" problem and other graph-theoretical problems associated with regular planar lattices: some exact results for the "percolation" problem*, Proc. Roy. Soc. London, Ser. A **322** (1971), No. 1549, 251–280.

16. J. Touchard, *Contributions à l'étude du problème des timbres poste*, Canad. J. Math. **2** (1950), 385–398.

10
Differential Equations for Multivariable Hermite and Laguerre Polynomials

J. F. van Diejen

ABSTRACT Systems of differential equations for multivariable Hermite and Laguerre polynomials are discussed associated to the integrals of the rational quantum Calogero models with harmonic confinement invariant under the action of a classical Weyl group.

1 Introduction

It is well known that the eigenfunctions for the rational quantum Calogero model with harmonic confinement can be written as the product of a factorized ground-state wave function and certain multivariable polynomials [2]. In general the eigenfunctions are not orthogonal because of a large degeneracy in the spectrum. However, if one exploits the quantum integrability and requires that the wave functions be joint eigenfunctions for a complete set of commuting quantum integrals for the model, then the degeneracy is removed and one ends up with an orthogonal eigenbasis. The polynomial parts of the orthogonalized eigenfunctions turn out to be given by multivariable generalizations of the Hermite polynomials that were introduced by Macdonald [13] and Lassalle [11]. The eigenvalue equations associated to the commuting quantum integrals give rise in this connection to a system of differential equations for the multivariable Hermite polynomials. One convenient way to describe these differential equations is to employ Dunkl's differential-reflection operators [9, 19]. A different approach was followed by Baker and Forrester [1], who presented expressions for the differential equations at issue involving nested commutators of Sekiguchi's commuting differential operators for (i.e., diagonalized by) the Jack symmetric functions [14].

Below we will consider still another method to characterize the differential equations for the multivariable Hermite polynomials. The starting point is a system of difference equations arising as eigenvalue equations for the commuting quantum integrals of a Ruijsenaars-type difference version of the quantum Calogero model [5, 17, 18]. By sending the difference step size to zero the difference version reduces to the ordinary quantum Calogero model and the corresponding difference equations pass over to differential equations for the multivariable Hermite polynomials. Further generalization to quantum Calogero models with classical Weyl-group symmetry [16] and their Ruijsenaars-type difference versions [5] leads us in a similar way to a system of differential equations for the Macdonald–Lassalle multivariable Laguerre polynomials [12, 13]. A corresponding treatment of the Laguerre case in terms of Dunkl's differential-reflection operators or in terms of nested commutators of Sekiguchi's differential operators may be found in Refs. 10 and 1, respectively.

The structure of the chapter is as follows. Section 2 recalls the definition of the multivariable Hermite and Laguerre polynomials. In Section 3 the corresponding systems of differential equations are introduced. The proof that the polynomials indeed satisfy the differential equations in question is relegated to Section 4. The chapter is closed in Section 5 by briefly detailing the connection between the differential equations for the Hermite/Laguerre families and the quantum integrals for the harmonically confined rational quantum Calogero models with classical Weyl-group symmetry.

2   Multivariable Hermite and Laguerre Polynomials

Let

$$m_\lambda^A(x) = \sum_{\mu \in S_n(\lambda)} x_1^{\mu_1} \cdots x_n^{\mu_n}, \quad \lambda \in \Lambda \tag{1a}$$

and

$$m_\lambda^B(x) = \sum_{\mu \in S_n(\lambda)} x_1^{2\mu_1} \cdots x_n^{2\mu_n}, \quad \lambda \in \Lambda, \tag{1b}$$

where Λ denotes the cone of partitions

$$\Lambda = \{\lambda \in \mathbb{Z}^n \mid \lambda_1 \geq \lambda_2 \geq \cdots \geq \lambda_n \geq 0\} \tag{2}$$

and the summation in (1a), (1b) is over the orbit of λ with respect to the action of the permutation group S_n (which permutes the parts $\lambda_1, \ldots, \lambda_n$). The monomial basis $\{m_\lambda^C(x)\}_{\lambda \in \Lambda}$ (here and below C stands for A or B) inherits a partial ordering from the dominance-type partial ordering of the cone Λ (2) defined for $\lambda, \mu \in \Lambda$ by

$$\lambda \leq \mu \quad \text{iff} \quad \sum_{j=1}^k \lambda_j \leq \sum_{j=1}^k \mu_j \text{ for } k = 1, \ldots, n \tag{3}$$

($\lambda < \mu$ iff $\lambda \leq \mu$ and $\lambda \neq \mu$).

Let us next consider the L^2-inner product

$$\langle f, g \rangle_C = \int_{-\infty}^{\infty} \cdots \int_{-\infty}^{\infty} f(x)\overline{g(x)}\Delta^C(x)\, dx_1 \cdots dx_n, \quad C = \mathrm{A}, \mathrm{B} \quad (4)$$

(for f, g in $L^2(\mathbb{R}^n, \Delta^C\, dx_1 \cdots dx_n)$) with

$$\Delta^{\mathrm{A}}(x) = \prod_{\substack{j,k=1 \\ j<k}}^{n} |x_j - x_k|^{2g_0} \prod_{j=1}^{n} e^{-\omega x_j^2}, \qquad (5\mathrm{a})$$

$$\Delta^{\mathrm{B}}(x) = \prod_{\substack{j,k=1 \\ j<k}}^{n} |(x_j - x_k)(x_j + x_k)|^{2g_0} \prod_{j=1}^{n} |x_j|^{2g_1} e^{-\omega x_j^2} \qquad (5\mathrm{b})$$

and define polynomials $p_\lambda^C(x)$, $\lambda \in \Lambda$ determined by the conditions

i) $p_\lambda^C(x) = m_\lambda^C(x) + \sum_{\mu \in \Lambda, \mu < \lambda} c_{\lambda,\mu}^C m_\mu(x)$, $c_{\lambda,\mu}^C \in \mathbb{C}$;

ii) $\langle p_\lambda^C, m_\mu \rangle_C = 0$ for $\mu < \lambda$.

(For technical reasons it is assumed in this definition that g_0, $g_1 \geq 0$ and that $\omega > 0$.) In the case $C = \mathrm{A}$ the polynomials $p_\lambda^C(x)$, $\lambda \in \Lambda$ constitute a basis for the space of permutation-invariant polynomials in the variables $x_1, \ldots, x_n$ whereas for $C = \mathrm{B}$ one has a basis for the even subsector of this space (i.e., the subspace of permutation-invariant polynomials in $x_1^2, \ldots, x_n^2$). The following theorem states that the basis $\{p_\lambda^C(x)\}_{\lambda \in \Lambda}$ is orthogonal with respect to the inner product $\langle \cdot, \cdot \rangle_C$.

Theorem 1 (Orthogonality). *Let λ, $\mu \in \Lambda$ (2). One has that*

$$\langle p_\lambda^C, p_\mu^C \rangle_C = \int_{-\infty}^{\infty} \cdots \int_{-\infty}^{\infty} p_\lambda^C(x)\overline{p_\mu^C(x)}\Delta^C(x)dx_1 \ldots dx_n = 0 \quad \text{if } \lambda \neq \mu$$

(where $C = A$ or B).

Various alternative proofs of the orthogonality theorem may be found in Refs. 1, 7, 9, and 10. The polynomials $p_\lambda^{\mathrm{A}}(x)$ and $p_\lambda^{\mathrm{B}}(x)$ are referred to as multivariable Hermite and Laguerre polynomials, respectively. The distinction between the type A (Hermite) and type B (Laguerre) case corresponds to symmetry with respect to the action of a Weyl group of type A_{n-1} ($\cong S_n$) or type B_n ($\cong S_n \ltimes (\mathbb{Z}_2)^n$), respectively.

3 Differential Equations

In this section a system of differential equations for the multivariable Hermite and Laguerre polynomials is discussed. To this end it is necessary to first introduce a certain family of (commuting) difference operators $D_{1,\beta}^C$, $\ldots$, $D_{n,\beta}^C$. For $C = A$ operators are given by

$$D_{r,\beta}^A = \sum_{\substack{J_+,J_- \subset \{1,\ldots,n\} \\ J_+ \cap J_- = \varnothing \\ |J_+|+|J_-| \le r}} U_{J_+^c \cap J_-^c, r-|J_+|-|J_-|}^A V_{J_+,J_-;J_+^c \cap J_-^c}^A \exp\left(\frac{\beta}{i}(\partial_{J_+} - \partial_{J_-})\right) \quad (6a)$$

$r = 1, \ldots, n$, with

$$\exp\left(\frac{\beta}{i}(\partial_{J_+} - \partial_{J_-})\right) = \prod_{j \in J_+} \exp\left(\frac{\beta}{i}\frac{\partial}{\partial x_j}\right) \prod_{j \in J_-} \exp\left(-\frac{\beta}{i}\frac{\partial}{\partial x_j}\right),$$

$$V_{J_+,J_-;K}^A = \prod_{j \in J_+} w^A(x_j) \prod_{j \in J_-} w^A(-x_j) \prod_{\substack{j \in J_+ \\ j' \in J_-}} v^A(x_j - x_{j'})v^A(x_j - x_{j'} - i\beta)$$

$$\times \prod_{\substack{j \in J_+ \\ k \in K}} v^A(x_j - x_k) \prod_{\substack{j \in J_- \\ k \in K}} v^A(x_k - x_j),$$

$$U_{K,p}^A = (-1)^p \sum_{\substack{L_+,L_- \subset K \\ L_+ \cap L_- = \varnothing \\ |L_+|+|L_-|=p}} \left(\prod_{l \in L_+} w^A(x_l) \prod_{l \in L_-} w^A(-x_l) \right.$$

$$\times \prod_{\substack{l \in L_+ \\ l' \in L_-}} v^A(x_l - x_{l'})v^A(x_{l'} - x_l + i\beta)$$

$$\left. \times \prod_{\substack{l \in L_+ \\ k \in K \setminus L_+ \cup L_-}} v^A(x_l - x_k) \prod_{\substack{l \in L_- \\ k \in K \setminus L_+ \cup L_-}} v^A(x_k - x_l) \right)$$

and

$$v^A(z) = \left(1 + \frac{\beta g_0}{iz}\right), \quad w^A(z) = (1 + i\beta\omega z).$$

For $C = B$ the operators in question take the form

$$D_{r,\beta}^B = \sum_{\substack{J \subset \{1,\ldots,n\} \\ 0 \le |J| \le r \\ \varepsilon_j = \pm 1, j \in J}} U_{J^c, r-|J|}^B V_{\varepsilon J, J^c}^B \exp\left(\frac{\beta}{i}\partial_{\varepsilon J}\right) \quad (6b)$$

$r = 1, \ldots, n$, with

$$\exp\left(\frac{\beta}{i}\partial_{\varepsilon J}\right) = \prod_{j \in J} \exp\left(\varepsilon_j \frac{\beta}{i}\frac{\partial}{\partial x_j}\right) \quad (\varepsilon_j \in \{-1,1\}),$$

$$V_{\varepsilon J,K}^B = \prod_{j \in J} w^B(\varepsilon_j x_j) \prod_{\substack{j,j' \in J \\ j < j'}} v^B(\varepsilon_j x_j + \varepsilon_{j'} x_{j'})v^B(\varepsilon_j x_j + \varepsilon_{j'} x_{j'} - i\beta)$$

$$\times \prod_{\substack{j \in J \\ k \in K}} v^{\mathrm{B}}(\varepsilon_j x_j + x_k) v^{\mathrm{B}}(\varepsilon_j x_j - x_k),$$

$$U^{\mathrm{B}}_{K,p} = (-1)^p \sum_{\substack{L \subset K \\ |L|=p \\ \varepsilon_l = \pm 1, l \in L}} \left(\prod_{l \in L} w^{\mathrm{B}}(\varepsilon_l x_l) \prod_{\substack{l,l' \in L \\ l < l'}} v^{\mathrm{B}}(\varepsilon_l x_l + \varepsilon_{l'} x_{l'}) v^{\mathrm{B}}(-\varepsilon_l x_l - \varepsilon_{l'} x_{l'} + i\beta) \right. \\ \left. \times \prod_{\substack{l \in L \\ k \in K \setminus L}} v^{\mathrm{B}}(\varepsilon_l x_l + x_k) v^{\mathrm{B}}(\varepsilon_l x_l - x_k) \right)$$

and

$$v^{\mathrm{B}}(z) = \left(1 + \frac{\beta g_0}{iz}\right), \quad w^{\mathrm{B}}(z) = \left(1 + \frac{\beta g_1}{iz}\right)(1 + i\beta \omega z).$$

In (6a) the sum is meant over all disjoint pairs of index sets $J_+, J_- \subset \{1, \ldots, n\}$ with the sum of the cardinalities being $\leq r$, and in (6b) the sum is over all index sets $J \subset \{1, \ldots, n\}$ of cardinality $\leq r$ and all configurations of signs $\varepsilon_j \in \{+1, -1\}$, $j \in J$. We have furthermore assumed the conventions that empty products are equal to one and that $U^C_{K,p}$ is equal to one for $p = 0$. The exponentials $\exp(\pm \frac{\beta}{i} \frac{\partial}{\partial x_j})$ act on analytic functions of $x_1, \ldots, x_n$ by means of a shift of the jth argument in the (for β real) imaginary direction

$$\left(\exp\left(\pm \frac{\beta}{i} \frac{\partial}{\partial x_j}\right) f \right)(x_1, \ldots, x_n) = f(x_1, \ldots, x_{j-1}, x_j \mp i\beta, x_{j+1}, \ldots, x_n).$$

Thus, the operators $D^C_{r,\beta}$ (6a), (6b) are analytic difference operators of order $2r$ in the elementary shifters $\exp(\frac{\beta}{i} \frac{\partial}{\partial x_1}), \ldots, \exp(\frac{\beta}{i} \frac{\partial}{\partial x_n})$.

If we act with $D^C_{r,\beta}$ ($C = \mathrm{A}, \mathrm{B}$) on an (arbitrary) analytic function f of the variables $x_1, \ldots, x_n$, then we end up with an expression that is holomorphic in the step size parameter β around $\beta = 0$. The coefficients of the Taylor expansion in β around zero can be written in terms of partial differential operators applied to the function f. We will call the first nonzero differential operator in this expansion the leading differential part of the difference operator. One obtains the leading differential part $D^C_{r,0}$ of $D^C_{r,\beta}$ ($C = \mathrm{A}, \mathrm{B}$) by expanding the analytic difference operator in powers of β using the formal identity $\exp(\pm \frac{\beta}{i} \frac{\partial}{\partial x_j}) = \sum_{m=0}^{\infty} (\pm \frac{\beta}{i} \frac{\partial}{\partial x_j})^m / m!$. The result is a formal expansion of the difference operator in terms of differential operators that gets its precise meaning (as a Taylor expansion) upon acting with both sides on an (arbitrary) analytic function.

The first part of the next theorem describes the structure of the highest-order symbol of the leading differential parts $D^C_{1,0}, \ldots, D^C_{n,0}$ of the difference operators $D^C_{1,\beta}, \ldots, D^C_{n,\beta}$ (6a), (6b); the second part of the theorem states that the leading differential parts at issue commute and are diagonal with respect to the basis $\{p^C_\lambda\}_{\lambda \in \Lambda}$ (i.e., the polynomials p^C_λ, $\lambda \in \Lambda$ introduced in the previous section are joint eigenfunctions for the commuting differential operators $D^C_{1,0}, \ldots, D^C_{n,0}$).

150   J. F. van Diejen

Theorem 2 (Differential equations). i. *The formal Taylor expansion of the difference operator $D_{r,\beta}^C$ ($r = 1, \ldots, n$; $C = A, B$) in the step size parameter β has the form*

$$D_{r,\beta}^C = D_{r,0}^C \beta^{2r} + O(\beta^{2r+1})$$

with

$$D_{r,0}^C = (-1)^r \sum_{\substack{J \subset \{1,\ldots,n\} \\ |J|=r}} \prod_{j \in J} \frac{\partial^2}{\partial x_j^2} + \text{l.o.},$$

(where l.o. stands for the terms of lower order in the partial derivatives $\partial/\partial x_1, \ldots, \partial/\partial x_n$).

ii. *The leading differential parts $D_{1,0}^C, \ldots, D_{n,0}^C$ commute and are simultaneously diagonalized by the multivariable Hermite (type A) and Laguerre (type B) polynomials*

$$D_{r,0}^C p_\lambda^C = E_r^C(\lambda) p_\lambda^C, \quad \lambda \in \Lambda,$$

where the corresponding eigenvalues are given explicitly by

$$E_r^C(\lambda) = (\omega M)^r \sum_{\substack{J \subset \{1,\ldots,n\} \\ |J|=r}} \prod_{j \in J} \lambda_j$$

with $M = 2$ if $C = A$ and $M = 4$ if $C = B$.

Corollary 1. *We have that*

$$\lim_{\beta \to 0} \beta^{-2r} (D_{r,\beta}^C p_\lambda^C)(x) = E_r^C(\lambda) p_\lambda^C(x), \quad \lambda \in \Lambda.$$

Corollary 2 (Symmetry). *The leading differential operators $D_{1,0}^C, \ldots, D_{n,0}^C$ map the space of permutation-invariant polynomials in $x_1, \ldots, x_n$ (type A) or $x_1^2, \ldots, x_n^2$ (type B) into itself and are symmetric with respect to the inner product $\langle \cdot, \cdot \rangle_C$ (4)*

$$\langle D_{r,0}^C m_\lambda^C, m_\mu^C \rangle_C = \langle m_\lambda^C, D_{r,0}^C m_\mu^C \rangle_C, \quad \lambda, \mu \in \Lambda$$

($C = A, B$).

The eigenvalue equations,

$$(D_{r,0}^C p_\lambda^C)(x) = \lim_{\beta \to 0} \beta^{-2r} (D_{r,\beta}^C p_\lambda^C)(x) = E_r^C(\lambda) p_\lambda^C(x),$$

$r = 1, \ldots, n$, constitute a system of differential equations for the multivariable Hermite/Laguerre polynomials $p_\lambda^C(x)$, $\lambda \in \Lambda$. It is instructive to exhibit these differential equations in a more explicit form for the simplest case $r = 1$ (which amounts to the order of the equation being equal to two). For $r = 1$, the difference operators $D^{\mathrm{A}}_{r,\beta}$ (6a) and $D^{\mathrm{B}}_{r,\beta}$ (6b) reduce to

$$
D^{\mathrm{A}}_{1,\beta} = \sum_{j=1}^{n}\left(w^{\mathrm{A}}(x_j)\prod_{\substack{k=1\\k\neq j}}^{n} v^{\mathrm{A}}(x_j - x_k)\left(\exp\left(\frac{\beta}{i}\frac{\partial}{\partial_j}\right) - 1\right)\right.
$$
$$
\left. + w^{\mathrm{A}}(-x_j)\prod_{\substack{k=1\\k\neq j}}^{n} v^{\mathrm{A}}(-x_j + x_k)\left(\exp\left(-\frac{\beta}{i}\frac{\partial}{\partial_j}\right) - 1\right)\right)
$$

and

$$
D^{\mathrm{B}}_{1,\beta} = \sum_{j=1}^{n}\left(w^{\mathrm{B}}(x_j)\prod_{\substack{k=1\\k\neq j}}^{n} v^{\mathrm{B}}(x_j - x_k)v^{\mathrm{B}}(x_j + x_k)\left(\exp\left(\frac{\beta}{i}\frac{\partial}{\partial_j}\right) - 1\right)\right.
$$
$$
\left. + w^{\mathrm{B}}(-x_j)\prod_{\substack{k=1\\k\neq j}}^{n} v^{\mathrm{B}}(-x_j + x_k)v^{\mathrm{B}}(-x_j - x_k)\left(\exp\left(-\frac{\beta}{i}\frac{\partial}{\partial_j}\right) - 1\right)\right)
$$

(with $v^{\mathrm{A}}(\cdot)$, $w^{\mathrm{A}}(\cdot)$ and $v^{\mathrm{B}}(\cdot)$, $w^{\mathrm{B}}(\cdot)$ taken the same as in (6a) and (6b), respectively). A formal Taylor expansion in β reveals that $D^{C}_{1,\beta} = D^{C}_{1,0}\beta^2 + O(\beta^3)$ with

$$
D^{\mathrm{A}}_{1,0} = \sum_{j=1}^{n}\left(-\frac{\partial^2}{\partial x_j^2} + 2\omega x_j\frac{\partial}{\partial x_j}\right) - 2g_0\sum_{\substack{j,k=1\\j<k}}^{n}\frac{1}{x_j - x_k}\left(\frac{\partial}{\partial x_j} - \frac{\partial}{\partial x_k}\right), \qquad (7a)
$$

$$
D^{\mathrm{B}}_{1,0} = \sum_{j=1}^{n}\left(-\frac{\partial^2}{\partial x_j^2} - \frac{2g_1}{x_j}\frac{\partial}{\partial x_j} + 2\omega x_j\frac{\partial}{\partial x_j}\right)
$$
$$
- 2g_0\sum_{\substack{j,k=1\\j<k}}^{n}\left(\frac{1}{x_j - x_k}\left(\frac{\partial}{\partial x_j} - \frac{\partial}{\partial x_k}\right) + \frac{1}{x_j + x_k}\left(\frac{\partial}{\partial x_j} + \frac{\partial}{\partial x_k}\right)\right). \qquad (7b)
$$

The explicit formulas for $D^{\mathrm{A}}_{1,0}$ and $D^{\mathrm{B}}_{1,0}$ combined with the expressions for the corresponding eigenvalues taken from Theorem 2

$$
E^{\mathrm{A}}_1(\lambda) = 2\omega(\lambda_1 + \cdots + \lambda_n), \qquad E^{\mathrm{B}}_1(\lambda) = 4\omega(\lambda_1 + \cdots + \lambda_n), \qquad (8)
$$

render the second-order differential equations $D^{\mathrm{A}}_{1,0}p^{\mathrm{A}}_\lambda = E^{\mathrm{A}}_1(\lambda)p^{\mathrm{A}}_\lambda$ and $D^{\mathrm{B}}_{1,0}p^{\mathrm{B}}_\lambda = E^{\mathrm{B}}_1(\lambda)p^{\mathrm{B}}_\lambda$ for the Hermite and Laguerre family in a fully explicit form.

In principle it is possible to compute for given r the differential operator $D^{C}_{r,0}$ algorithmically as the leading part of the (explicit) difference operator $D^{C}_{r,\beta}$ (6a), (6b) (via a formal expansion in the step size parameter β). It seems quite a nontrivial combinatorial exercise, however, to derive along these lines a closed expression for the differential operator $D^{C}_{r,0}$ for arbitrary

$r \in \{1, \ldots, n\}$. In Refs. 1, 9, 10, and 19, expressions for the differential equations satisfied by the multivariable Hermite/Laguerre families were given in terms of Dunkl's differential-reflection operators and nested commutators of Sekiguchi's differential operators for the Jack symmetric functions, respectively. Those approaches have in common with the above method that is difficult to extract detailed information regarding the explicit form of the differential equations. It would be interesting to try to arrive at more explicit formulas similar to those available for the differential equations for the Jack symmetric functions [14] by comparing with the results of Refs. 3 and 15.

4   Proof of the Differential Equations

Let

$$
\Delta_\beta^{\mathrm{A}}(x) = \prod_{\substack{j,k=1 \\ j<k}}^n \left| \frac{\Gamma(g_0 + i\beta^{-1}(x_j - x_k))}{\Gamma(i\beta^{-1}(x_j - x_k))} \right|^2
$$
$$
\times \prod_{j=1}^n \left| \Gamma\left(\frac{1}{\tilde{\omega}\beta^2} + i\beta^{-1}x_j\right) \Gamma\left(\frac{1}{\omega'\beta^2} + i\beta^{-1}x_j\right) \right|^2, \qquad (9a)
$$

$$
\Delta_\beta^{\mathrm{B}}(x) = \prod_{\substack{j,k=1 \\ j<k}}^n \left| \frac{\Gamma(g_0 + i\beta^{-1}(x_j - x_k))\Gamma(g_0 + i\beta^{-1}(x_j + x_k))}{\Gamma(i\beta^{-1}(x_j - x_k))\Gamma(i\beta^{-1}(x_j + x_k))} \right|^2
$$
$$
\times \prod_{j=1}^n \left| \frac{\Gamma(\tilde{g}_1 + i\beta^{-1}x_j)\Gamma(g_1' + \frac{1}{2} + i\beta^{-1}x_j)}{\Gamma(i\beta^{-1}x_j)\,\Gamma(\frac{1}{2} + i\beta^{-1}x_j)} \right.
$$
$$
\left. \times \Gamma\left(\frac{1}{\tilde{\omega}\beta^2} + i\beta^{-1}x_j\right)\Gamma\left(\frac{1}{\omega'\beta^2} + i\beta^{-1}x_j\right) \right|^2, \qquad (9b)
$$

and let $p_{\lambda,\beta}^C(x)$ (with $C = \mathrm{A}$ or B and $\lambda \in \Lambda$) denote the polynomials determined by the conditions i and ii of Section 2 with the weight functions $\Delta^C(x)$ (5a), (5b) replaced by $\Delta_\beta^C(x)$ (9a), (9b). The polynomials in question coincide up to a reparametrization and a rescaling of the variables with the multivariable continuous Hahn and Wilson polynomials associated to the weight functions [6, 8]

$$
\Delta^{cH}(x) = \prod_{\substack{j,k=1 \\ j<k}}^n \left| \frac{\Gamma(g_0 + i(x_j - x_k))}{\Gamma(i(x_j - x_k))} \right|^2 \prod_{j=1}^n \left| \Gamma(a + ix_j)\Gamma(b + ix_j) \right|^2 \qquad (10)
$$

and

$$\Delta^W(x) = \prod_{\substack{j,k=1 \\ j<k}}^{n} \left| \frac{\Gamma(g_0 + i(x_j - x_k))\Gamma(g_0 + i(x_j + x_k))}{\Gamma(i(x_j - x_k))\Gamma(i(x_j + x_k))} \right|^2$$

$$\times \prod_{j=1}^{n} \left| \frac{\Gamma(a + ix_j)\Gamma(b + ix_j)\Gamma(c + ix_j)\Gamma(d + ix_j)}{\Gamma(2ix_j)} \right|^2. \quad (11)$$

Indeed, if we rescale the variables by substituting

$$x_j \to \frac{x_j}{\beta}, \quad j = 1, \ldots, n \quad (12)$$

and reparametrize the parameters in the following way,

$$a = (\beta^2 \tilde{\omega})^{-1}, \quad b = (\beta^2 \omega')^{-1}, \quad c = \tilde{g}_1, \quad d = g'_1 + \frac{1}{2}, \quad (13)$$

then the weight functions $\Delta^{cH}(x)$ and $\Delta^W(x)$ go over in $\Delta_\beta^A(x)$ and $\Delta_\beta^B(x)$, respectively.

In Ref. 8 it was shown that the multivariable continuous Hahn and Wilson polynomials satisfy a system of difference equations. For the rescaled and reparametrized polynomials $p_{\lambda,\beta}^C(x)$ the difference equations at issue become of the form

$$D_{r,\beta}^C p_{\lambda,\beta}^C = E_{r,\beta}^C(\lambda) p_{\lambda,\beta}^C \quad (r = 1, \ldots, n; C = A, B), \quad (14)$$

where $D_{r,\beta}^C$ is the difference operator of Section 3 with $w^A(z)$, $w^B(z)$ replaced by

$$w^A(z) = (1 + i\beta\tilde{\omega}z)(1 + i\beta\omega'z), \quad (15a)$$

$$w^B(z) = \left(1 + \frac{\beta\tilde{g}_1}{iz}\right)\left(1 + \frac{\beta g'_1}{iz + \beta/2}\right)(1 + i\beta\tilde{\omega}z)(1 + i\beta\omega'z); \quad (15b)$$

and the eigenvalues on the right-hand side are given by

$$E_{r,\beta}^A(\lambda) = (\beta^4 \tilde{\omega}\omega')^r$$
$$\times E_r((\rho_1^A + \lambda_1)^2, \ldots, (\rho_n^A + \lambda_n)^2; (\rho_r^A)^2, \ldots, (\rho_n^A)^2), \quad (16a)$$
$$E_{r,\beta}^B(\lambda) = (4\beta^4 \tilde{\omega}\omega')^r$$
$$\times E_r((\rho_1^B + \lambda_1)^2, \ldots, (\rho_n^B + \lambda_n)^2; (\rho_r^B)^2, \ldots, (\rho_n^B)^2) \quad (16b)$$

with

$$E_r(\zeta_1, \ldots, \zeta_n; \eta_r, \ldots, \eta_n)$$

$$= \sum_{\substack{J \subset \{1,\ldots,n\} \\ 0 \le |J| \le r}} (-1)^{r-|J|} \prod_{j \in J} \zeta_j \sum_{\substack{l_1,\ldots,l_{r-|J|}=r \\ l_1 \le \cdots \le l_{r-|J|}}}^{n} \eta_{l_1} \cdots \eta_{l_{r-|J|}} \quad (17)$$

and

$$\rho_j^A = (n-j)g_0 + \beta^{-2}\left(\frac{1}{\tilde{\omega}} + \frac{1}{\omega'}\right) - \frac{1}{2}, \tag{18a}$$

$$\rho_j^B = (n-j)g_0 + \frac{\tilde{g}_1 + g_1'}{2} + \beta^{-2}\left(\frac{1}{2\tilde{\omega}} + \frac{1}{2\omega'}\right) - \frac{1}{4}. \tag{18b}$$

It was demonstrated in Ref. 7 that for $\beta \to 0$ the rescaled and reparametrized continuous Hahn and Wilson polynomials $p_{\lambda,\beta}^A(x)$ and $p_{\lambda,\beta}^B(x)$ converge to the multivariable Hermite and Laguerre polynomials $p_\lambda^A(x)$ and $p_\lambda^B(x)$ with parameters

$$\omega = \tilde{\omega} + \omega', \quad g_1 = \tilde{g}_1 + g_1'. \tag{19}$$

In order to derive the differential equations for the multivariable Hermite and Laguerre families, we shall now analyze the behavior of the difference equations for the polynomials $p_{\lambda,\beta}^C$ in the limit when the step size parameter β tends to zero. Essential is the limiting behavior of the eigenvalues $E_{r,\beta}^C(\lambda)$, which is governed by the following lemma.

Lemma 1. *One has that*

$$\lim_{\beta \to 0} \beta^{-2r} E_{r,\beta}^C(\lambda) = E_r^C(\lambda) \qquad (r = 1, \ldots, n; C = A, B),$$

where $E_r^C(\lambda)$ is given by the expression in Theorem 2 (with $\omega = \tilde{\omega} + \omega'$).

Proof. The proof hinges on the following decomposition of $E_r(\ldots;\ldots)$ (17) (cf. Ref. 4, Lemma B.2):

$$E_r(\zeta_1, \ldots, \zeta_n; \eta_r, \ldots, \eta_n)$$

$$= (\zeta_n - \eta_n)\left(\sum_{\substack{J \subset \{1,\ldots,n-1\} \\ 0 \le |J| \le r-1}} (-1)^{r-1-|J|} \prod_{j \in J} \zeta_j \sum_{\substack{l_1,\ldots,l_{r-1-|J|}=r \\ l_1 \le \cdots \le l_{r-1-|J|}}}^{n} \eta_{l_1} \cdots \eta_{l_{r-1-|J|}} \right)$$

$$+ \sum_{\substack{J \subset \{1,\ldots,n-1\} \\ 0 \le |J| \le r}} (-1)^{r-|J|} \prod_{j \in J} \zeta_j \sum_{\substack{l_1,\ldots,l_{r-|J|}=r \\ l_1 \le \cdots \le l_{r-|J|}}}^{n-1} \eta_{l_1} \cdots \eta_{l_{r-|J|}},$$

with the convention that the sum on the third line equals one if $r = 1$ and that the sum on the fourth line equals zero if $r = n$. Notice that both the sum between brackets and the sum on the fourth line are of the same type as on the right-hand side of (17) but with one ζ-variable less (there is no longer dependence on ζ_n). With the aid of this formula and induction on the number of variables it is not difficult to infer that

$$\lim_{\beta \to 0} \frac{E_{r,\beta}^C(\lambda)}{\beta^{2r}(\tilde{\omega} + \omega')^r M^r} = \lambda_n \left(\sum_{\substack{J \subset \{1,\dots,n-1\} \\ |J|=r-1}} \prod_{j \in J} \lambda_j \right) + \sum_{\substack{J \subset \{1,\dots,n-1\} \\ |J|=r}} \prod_{j \in J} \lambda_j$$

$$= \sum_{\substack{J \subset \{1,\dots,n\} \\ |J|=r}} \prod_{j \in J} \lambda_j,$$

where $M = 2$ for $C = \mathrm{A}$ and $M = 4$ for $C = \mathrm{B}$. □

It is immediate from the above lemma and the limit formula $\lim_{\beta \to 0} p_{\lambda,\beta}^C = p_\lambda^C$ [7] that $E_{r,\beta}^C(\lambda)p_{\lambda,\beta}^C = \beta^{2r} E_r^C(\lambda)p_\lambda^C + o(\beta^{2r})$ (assuming the identification of parameters in (19)). If we plug this asymptotics into the eigenvalue equation (14), then we arrive at the following asymptotic behavior of the corresponding left-hand side for $\beta \to 0$

$$(D_{r,\beta}^C p_{\lambda,\beta}^C)(x) = \beta^{2r}(D_r^C p_\lambda^C)(x) + o(\beta^{2r}), \tag{20}$$

with D_r^C a certain partial differential operator. In other words, for $\beta \to 0$ the difference equation for $p_{\lambda,\beta}^C$ goes over (cf. Corollary 1) into a differential equation for the polynomial p_λ^C of the form

$$D_r^C p_\lambda^C = E_r^C(\lambda)p_\lambda^C, \tag{21}$$

where D_r^C is a certain differential operator and $E_r^C(\lambda)$ is of the form given in Theorem 2 (with $\omega = \tilde{\omega} + \omega'$).

To complete the proof of Theorem 2 we will now show that the differential operators $D_1^C, \dots, D_n^C$ commute and are indeed of the form described in the first part of the theorem (i.e., we show that the operator D_r^C is equal to the leading part $D_{r,0}^C$ of the formal Taylor expansion of the difference operator $D_{r,\beta}^C$ in the step size parameter β and its leading symbol is given by the rth elementary symmetric function in the partials $-\partial^2/\partial x_1^2, \dots, -\partial^2/\partial x_n^2$).

Let us to this end first observe that one may replace $p_{\lambda,\beta}^C$ in the left-hand side of (20) by p_λ^C (using once more that $\lim_{\beta \to 0} p_{\lambda,\beta}^C = p_\lambda^C$). Moreover, since the polynomials $\{p_\lambda^C\}_{\lambda \in \Lambda}$ constitute a basis for the space of (even) symmetric polynomials, we have in fact that $(D_{r,\beta}^C p)(x) = \beta^{2r}(D_r^C p)(x) + O(\beta^{2r+1})$ for arbitrary (even) symmetric polynomial $p(x)$ in the variables $x_1, \dots, x_n$. But then the same holds true for arbitrary analytic (not necessarily symmetric or even) function of $x_1, \dots, x_n$ and we have (formally)

$$D_{r,\beta}^C = \beta^{2r} D_r^C + O(\beta^{2r+1}). \tag{22}$$

Here we have used the fundamental property that the vanishing of a (linear) partial differential operator on the space of (even) symmetric polynomials implies that it is zero on an arbitrary (analytic) function (i.e., all its coefficients must be zero). In Appendix C of Ref. 4 a proof of this general property was given in a trigonometric context. (Specifically, we assumed there that the differential operator vanishes on the space of symmetric polynomials in $\sin^2(x_1), \ldots, \sin^2(x_n)$.) The present case follows after an appropriate substitution of the variables turning the relevant space of trigonometric polynomials into the space of (even) symmetric polynomials in $x_1, \ldots, x_n$. (Specifically, the type A case is recovered by the substitution $\sin^2(x_j) \to x_j$ and the type B case by $\sin^2(x_j) \to x_j^2$.) A consequence of this general property is that the leading part of the formal expansion of the difference operator $D^C_{r,\beta}$ in β is completely determined by its action on the (even) symmetric polynomials. (This excludes the (a priori) possibility of a lower-order leading part in the formal expansion (22) corresponding to a term determined by a nontrivial differential operator that vanishes on the space of (even) symmetric polynomials.) Notice also that this property may be used to arrive at a proof for the commutativity of $D^C_1, \ldots, D^C_n$. Indeed, the commutators $[D^C_r, D^C_s]$ obviously vanish on the simultaneous eigenbasis $\{p^C_\lambda\}_{\lambda \in \Lambda}$ (and hence on all (even) symmetric polynomials), from which it then follows that they must be zero identically.

Equation (22) expresses the fact that the differential operator D^C_r on the left-hand side of the differential equation (21) equals the leading differential part $D^C_{r,0}$ of $D^C_{r,\beta}$. To determine the highest-order symbol of the leading differential part $D^C_{r,0}$, we use that the functions v^C, w^C governing the coefficients of the difference operator $D^C_{r,\beta}$ are of the form $1 + O(\beta)$ and that for v^C, $w^C = 1$

$$D^C_{r,\beta} \overset{(v^C,w^C=1)}{=} \sum_{\substack{J \subset \{1,\ldots,n\} \\ |J|=r}} \prod_{j \in J} \left(\exp\left(\frac{\beta}{i} \frac{\partial}{\partial x_j} \right) + \exp\left(-\frac{\beta}{i} \frac{\partial}{\partial x_j} \right) - 2 \right)$$

$$= \beta^{2r} \left((-1)^r \sum_{\substack{J \subset \{1,\ldots,n\} \\ |J|=r}} \prod_{j \in J} \frac{\partial^2}{\partial x_j^2} \right) + O(\beta^{2r+1}). \tag{23}$$

(Notice in this connection that in the situation where v^C and w^C are taken to be equal to one, the operators $D^C_{r,\beta}$ (6a), (6b) reduce to

$$D^A_{r,\beta} = \sum_{\substack{J_+,J_- \subset \{1,\ldots,n\} \\ J_+ \cap J_- = \varnothing \\ |J_+|+|J_-| \leq r}} (-2)^{r-|J_+|-|J_-|} \binom{n-|J_+|-|J_-|}{r-|J_+|-|J_-|} \exp\left(\frac{\beta}{i}(\partial_{J_+} - \partial_{J_-}) \right)$$

and

$$D^B_{r,\beta} = \sum_{\substack{J \subset \{1,\ldots,n\} \\ 0 \leq |J| \leq r \\ \varepsilon_j = \pm 1, j \in J}} (-2)^{r-|J|} \binom{n-|J|}{r-|J|} \exp\left(\frac{\beta}{i} \partial_{\varepsilon J} \right),$$

which both can be rewritten in the form given by the first line of (23).) It then follows (i.e., from the asymptotics v^C, $w^C = 1 + O(\beta)$ and (23)) that the leading part D_r^C in (22) is of the form

$$D_{r,0}^C = (-1)^r \sum_{\substack{J \subset \{1,\ldots,n\} \\ |J|=r}} \prod_{j \in J} \frac{\partial^2}{\partial x_j^2} + \text{l.o.},$$

as advertised.

Thus far we have shown that the statements of Theorem 2 hold when taking difference operators $D_{r,\beta}^C$ (6a), (6b) with $w^C(z)$ given by (15a), (15b) and ω, g_1 given by (19). The formulation in the theorem corresponds to choosing the specialization $\tilde{\omega} = \omega$, $\omega' = 0$ and $\tilde{g}_1 = g_1$, $g_1' = 0$.

The symmetry of the differential operator $D_{r,0}^C$ with respect to the inner product $\langle \cdot, \cdot \rangle_C$ in the space of permutation-invariant polynomials (type A) or permutation-invariant and even polynomials (type B) (Corollary 2) stems from the fact that the operator in question is diagonal on an orthogonal basis (viz. $\{p_\lambda^C\}_{\lambda \in \Lambda}$) with eigenvalues that are real.

5   Relation to the Quantum Calogero Model

The second-order differential operator $D_{1,0}^C$ (7a), (7b) coincides—up to conjugation with the square root of the weight function $\Delta^C(x)$ (5a), (5b)—with the Hamiltonian of a harmonically confined rational quantum Calogero model with classical Weyl-group symmetry [16]:

$$H^A = (\Delta^A)^{1/2} D_{1,0}^A (\Delta^A)^{-1/2}$$

$$= \sum_{j=1}^n \left(-\frac{\partial^2}{\partial x_j^2} + \omega^2 x_j^2 \right) + 2g_0(g_0 - 1) \sum_{\substack{j,k=1 \\ j<k}}^n (x_j - x_k)^{-2} - E_0^A, \quad (24)$$

$$H^B = (\Delta^B)^{1/2} D_{1,0}^B (\Delta^B)^{-1/2}$$

$$= \sum_{j=1}^n \left(-\frac{\partial^2}{\partial x_j^2} + g_1(g_1 - 1)x_j^{-2} + \omega^2 x_j^2 \right)$$

$$+ 2g_0(g_0 - 1) \sum_{\substack{j,k=1 \\ j<k}}^n \left((x_j - x_k)^{-2} + (x_j + x_k)^{-2} \right) - E_0^B, \quad (25)$$

where $E_0^A = \omega n(1 + g_0(n - 1))$ and $E_0^B = \omega n(1 + 2g_0(n - 1) + 2g_1)$. More generally, the commuting differential operators $D_{r,0}^C$, $r = 1, \ldots, n$ of Theorem 2 conjugate to a complete set of commuting quantum integrals $H_r^C = (\Delta^C)^{1/2} D_{r,0}^C (\Delta^C)^{-1/2}$, $r = 1, \ldots, n$ for the Calogero models in question. It would be interesting to compare these quantum integrals for the B-type with the explicit formulas obtained by Oshima et al. [15].

A consequence of the results of the present paper is that the quantum integrals for the rational quantum Calogero models with harmonic term may be recovered as the first nontrivial coefficients in the formal Taylor expansion with respect to the step size parameter of the quantum integrals for the Ruijsenaars-type difference version of the models presented in Ref. [5]. A similar result for the models without confining external field was obtained in Refs. 17, 18 (type A: root system A_{n-1}) and 5 (type B: root systems B_n and D_n).

6 REFERENCES

1. T. H. Baker and P. J. Forrester, *The Calogero–Sutherland model and generalized classical polynomials*, Commun. Math. Phys. **188** (1997), No. 1, 175–216, solv-int/9609010.

2. F. Calogero, *Solution of the one-dimensional N-body problems with quadratic and/or inversely quadratic pair potentials*, J. Math. Phys. **12** (1971), 419–436.

3. A. Debiard, *Système différentiel hypergéométrique et parties radiales des opérateurs invariants des espaces symétriques de type BC_p*, Séminaire d'algèbre Paul Dubreil et Marie-Paule Malliavin (M.-P. Malliavin, ed.), Lecture Notes in Math., Vol. 1296, Springer, New York, 1988, pp. 42–124.

4. J. F. van Diejen, *Commuting difference operators with polynomial eigenfunctions*, Compositio Math. **95** (1995), No. 2, 183–233.

5. ———, *Difference Calogero–Moser systems and finite Toda chains*, J. Math. Phys. **36** (1995), No. 3, 1299–1323.

6. ———, *Multivariable continuous Hahn and Wilson polynomials related to integrable difference systems*, J. Phys. A **28** (1995), No. 13, L369–L374.

7. ———, *Confluent hypergeometric orthogonal polynomials related to the rational quantum Calogero system with harmonic confinement*, Commun. Math. Phys. **188** (1997), No. 2, 467–497.

8. ———, *Properties of some families of hypergeometric orthogonal polynomials in several variables*, Trans. Amer. Math. Soc. **351** (1999), No. 1, 233–270, q-alg/9604004.

9. S. Kakei, *Common algebraic structure for the Calogero–Sutherland models*, J. Phys. A **29** (1996), No. 24, L619–L624.

10. ———, *An orthogonal basis for the B_n-type Calogero model*, J. Phys. A **30** (1997), No. 15, L535–L541.

11. M. Lassalle, *Polynômes de Hermite généralisés*, C. R. Acad. Sci. A **313** (1991), No. 9, 579–582.

12. _____, *Polynômes de Laguerre généralisés*, C. R. Acad. Sci. A **312** (1991), No. 10, 725–728.

13. I. G. Macdonald, *Hypergeometric functions*, manuscript.

14. _____, *Symmetric Functions and Hall Polynomials*, 2nd ed., Oxford Math. Monogr., Oxford Univ. Press, New York, 1995.

15. H. Ochiai, T. Oshima, and H. Sekiguchi, *Commuting families of symmetric differential operators*, Proc. Japan Acad. Ser. A Math. Sci. **70** (1994), No. 2, 62–66.

16. M. A. Olshanetsky and A. M. Perelomov, *Quantum integrable systems related to Lie algebras*, Phys. Rep. **94** (1983), No. 6, 313–404.

17. S. N. M. Ruijsenaars, *Finite-dimensional soliton systems*, Integrable and Superintegrable Systems (B. A. Kupershmidt, ed.), World Scientific, Singapore, 1990, pp. 165–206.

18. _____, *Systems of Calogero-Moser type*, Particles and Fields (Banff, 1994) (G. Semenoff and L. Vinet, eds.), CRM Series in Mathematical Physics, Springer, New York, 1999, pp. 251–352.

19. H. Ujino and M. Wadati, *Rodrigues formula for hi-Jack symmetric polynomials associated with the quantum Calogero model*, J. Phys. Soc. Japan **65** (1996), No. 8, 2423–2439.

11

Quantum Currents Realization of the Elliptic Quantum Groups $E_{\tau,\eta}(\mathfrak{sl}_2)$

B. Enriquez

ABSTRACT We review the construction by G. Felder and the author of the realization of the elliptic quantum groups by quantum currents.

The elliptic quantum groups were introduced by G. Felder in Refs. 13 and 14. These are algebraic objects based on a solution $R(z, \lambda)$ of the dynamical Yang–Baxter equation. Here dynamical means that in addition to the spectral parameter z, the R-matrix depends on a parameter λ, which belongs to a product of elliptic curves, and that these parameters undergo shifts in the various terms of the equation.

The aim of this paper is to review the construction by G. Felder and the author [9] of the realization of elliptic quantum groups $E_{\tau,\eta}(\mathfrak{sl}_2)$ by quantum current algebras. This construction relies on quasi-Hopf algebra techniques. We introduce (Section 2) a quantum loop algebra $U_\hbar\mathfrak{g}(\tau)$ (τ is the elliptic parameter, $\mathfrak{g} = \mathfrak{sl}_2$) that presents analogies with $E_{\tau,\eta}(\mathfrak{sl}_2)$. Namely, it has the property that the image in representations of its classical r-matrix coincides with the classical limit of $R(z, \lambda)$. $U_\hbar\mathfrak{g}(\tau)$ is endowed with "Drinfeld-type" coproducts Δ and $\bar{\Delta}$ (see Ref. 6), conjugated by a twist F (Section 2.3). Then our goal is to construct, in this algebra, a solution of the DYBE yielding $R(z, \lambda)$ in finite-dimensional representations. For that, we make use of a result of O. Babelon, D. Bernard, and E. Billey (BBB). This result extends to the dynamical situation the theory of Drinfeld twists (see Ref. 7): it states that a solution of the so-called twisted Hopf cocycle equation, in some quasi-triangular Hopf algebra, yields a solution of the DYBE at the algebra level. To construct such a solution, we solve a factorization problem for the twist F (Section 4.3). This factorization in turn relies on some results on Hopf algebra pairings within the quantum loop algebra. After we have obtained a DYBE solution in $U_\hbar\mathfrak{g}(\tau)^{\otimes 2}$, we study its representations and construct from it L-operators satisfying RLL relations that are exactly the elliptic quantum groups relations (Section 4.6).

To explain the strategy we are following, we first review (Section 4.1) the treatment of Ref. 8 of the rational analog of our elliptic situation. In that case, the solution of a factorization problem allows to construct a twist conjugating the Drinfeld coproduct for Yangians with the usual coproduct. Another derivation of this twist was earlier obtained by S. Khoroshkin and V. Tolstoy [21].

The interest of this construction lies in that it permits to embed the elliptic quantum group in an algebra "with central extension." This allows to apply quantum Kac–Moody algebra techniques to the study of this algebra. For example, one may expect that the elliptic quantum KZB equations of Ref. 15 are obtained in terms of intertwiners or coinvariants from the algebras presented here (such a study should be related to the apporach of Refs. 1 and 2 to the Ruijsenaars–Schneiders models). Another interesting application would be the study of the center of the algebra $U_\hbar \mathfrak{g}(\tau)$ at the critical level, as a Poisson algebra, and its possible deformation as some kind of $\mathcal{W}$-algebra, in the spirit of Ref. 19. Another subject of possible interest would be to study the connection with other types of elliptic quantum groups—those arising from the Belavin–Baxter solution or those studied in [18]. Finally, it would be interesting to find relations satisfied by a finite set of generators of the quantum loop algebra $U_\hbar \mathfrak{g}(\tau)$, or its subalgebra $U_\hbar \mathfrak{g}_\mathcal{O}$ (see 4.1) analogous to the Drinfeld presentation for Yangians in terms of generators $I(a)$ and $J(a)$.

1   Elliptic Quantum Groups

1.1   Function Spaces Associated with Elliptic Curves

Let τ be a complex number of positive imaginary part, and E be the elliptic curve $\mathbb{C}/\Gamma$, where $\Gamma = \mathbb{Z} + \tau \mathbb{Z}$. The basic theta function θ associated to E is defined by the conditions that it is holomorphic on $\mathbb{C}$, $\theta'(0) = 1$, the only zeros of θ are the points of Γ, $\theta(z + 1) = -\theta(z)$, and $\theta(z + \tau) = -e^{-i\pi\tau} e^{-2i\pi z} \theta(z)$. θ is then odd.

For $\lambda \in \mathbb{C}$, define L_λ as follows. If λ does not belong to Γ, define L_λ to be the set of holomorphic functions on $\mathbb{C} \setminus \Gamma$, 1-periodic and such that $f(z + \tau) = e^{-2i\pi\lambda} f(z)$. For $\lambda = 0$, define L_λ as the maximal isotropic subspace of $\mathbb{C}((z))$ (for the pairing $\langle f, g \rangle = \mathrm{res}_0(fg\,dz)$) containing all holomorphic functions f on $\mathbb{C} \setminus \Gamma$, Γ-periodic, such that $\oint_a f(z)\,dz = 0$, where a is the cycle $(i\epsilon, i\epsilon + 1)$ (with ϵ small and > 0). Finally, define $L_\lambda = e^{-2i\pi mz} L_0$ for $\lambda = n + m\tau$.

We then have

$$L_\lambda = \bigoplus_{j \geq 0} \mathbb{C} \left(\frac{\theta'}{\theta} \right)^{(j)} e^{-2i\pi mz}, \quad \text{if } \lambda = n + m\tau, \tag{1}$$

$$L_\lambda = \bigoplus_{i \geq 0} \mathbb{C} \left(\frac{\theta(\lambda + z)}{\theta(z)} \right)^{(i)}, \quad \text{if } \lambda \in \mathbb{C} \setminus \Gamma, \tag{2}$$

where we let $g' = \partial_z g$ and $g^{(i)} = \partial_z^i g$.

We will set

$$e_{i,\lambda}(z) = \left(\frac{\theta(\lambda + z)}{\theta(z)} \right)^{(i)}. \tag{3}$$

1.2 Elliptic R–Matrix

Let (v_1, v_{-1}) be the standard basis of $\mathbb{C}^2$ and E_{ij} the endomorphism of $\mathbb{C}^2$ defined by $E_{ij} v_\alpha = \delta_{j\alpha} v_i$. Let $\hbar$ be a formal variable and set

$$\begin{aligned}
R(z, \lambda) = {}& E_{11} \otimes E_{11} + E_{-1,-1} \otimes E_{-1,-1} \\
&+ \frac{\theta(z)}{\theta(z + \hbar)} \frac{\theta(\lambda + \hbar)\theta(\lambda - \hbar)}{\theta(\lambda)^2} E_{1,1} \otimes E_{-1,-1} \\
&+ \frac{\theta(z)}{\theta(z + \hbar)} E_{-1,-1} \otimes E_{11} + \frac{\theta(z + \lambda)\theta(\hbar)}{\theta(z + \hbar)\theta(\lambda)} E_{1,-1} \otimes E_{-1,1} \\
&- \frac{\theta(z - \lambda)\theta(\hbar)}{\theta(z + \hbar)\theta(\lambda)} E_{-1,1} \otimes E_{1,-1}. \tag{4}
\end{aligned}$$

1.3 The Dynamical Yang–Baxter Equation

This matrix satisfies the equation

$$\begin{aligned}
R^{(12)}&(z_{12}, \lambda) R^{(13)}(z_{13}, \lambda + \hbar \bar{h}^{(2)}) R^{(23)}(z_{23}, \lambda) \\
&= R^{(23)}(z_{23}, \lambda + \hbar \bar{h}^{(1)}) R^{(13)}(z_{13}, \lambda) R^{(12)}(z_{12}, \lambda + \hbar \bar{h}^{(3)}), \tag{5}
\end{aligned}$$

where $z_{ij} = z_i - z_j$ and z_i, $i = 1, 2, 3$ are generic complex numbers. We set $\bar{h} = E_{11} - E_{22}$, and $\bar{h}^{(k)}$ is the image of $\bar{h}$ in the kth factor of $\text{End}(\mathbb{C}^2)^{\otimes 3}$. For i, j, k a permutation of 1, 2, 3, and any $v \in (\mathbb{C}^2)^{\otimes 3}$ such that $\bar{h}^{(k)} v = \mu v$, we set $R^{(ij)}(z, \lambda + \hbar h^{(k)}) v = R^{(ij)}(z, \lambda + \hbar \mu) v$.

Equation (5) is called the dynamical Yang–Baxter equation.

1.4 Elliptic Quantum Groups

Set $\eta = \hbar/2$. The elliptic quantum group $E_{\tau,\eta}(\mathfrak{sl}_2)$ is defined as the algebra generated by h and the $a_i(\lambda)$, $b_i(\lambda)$, $c_i(\lambda)$, $d_i(\lambda)$, $i \geq 0$, $\lambda \in \mathbb{C} \setminus \Gamma$, subject to the relations

$$[h, a_i(\lambda)] = [h, d_i(\lambda)] = 0, \quad [h, b_i(\lambda)] = -2b_i(\lambda), \quad [h, c_i(\lambda)] = 2c_i(\lambda),$$

and if we set

$$a(z, \lambda) = \sum_{i \geq 0} a_i(\lambda) e_{i, +\hbar h/2}(z), \qquad b(z, \lambda) = \sum_{i \geq 0} b_i(\lambda) e_{i, \lambda + \hbar(h-2)/2}(z),$$

$$c(z, \lambda) = \sum_{i \geq 0} c_i(\lambda) e_{i, -\lambda - \hbar(h+2)/2}(z), \quad d(z, \lambda) = \sum_{i \geq 0} d_i(\lambda) e_{i, -\hbar h/2}(z),$$

and

$$L(z, \lambda) = \begin{pmatrix} a(z, \lambda) & b(z, \lambda) \\ c(z, \lambda) & d(z, \lambda) \end{pmatrix}, \tag{6}$$

the relations

$$R^{(12)}(z_1 - z_2, \lambda + \hbar h) L^{(1)}(z_1, \lambda) L^{(2)}(z_2, \lambda + \hbar h^{(1)})$$
$$= L^{(2)}(z_2, \lambda) L^{(1)}(z_1, \lambda + \hbar h^{(2)}) R^{(12)}(\lambda, z_1 - z_2) \tag{7}$$

and

$$\mathrm{Det}(z, \lambda) = d(z - \hbar, \lambda) a(z, \lambda - \hbar)$$
$$- b(z - \hbar, \lambda) c(z, \lambda - \hbar) \frac{\theta(\lambda + \hbar h + \hbar)}{\theta(\lambda + \hbar h)} = 1. \tag{8}$$

(These relations are made explicit in Ref. 16.)

We use the convention that $x_{\lambda + \hbar h}(z) = \sum_{i \geq 0} \partial_\lambda^i x_\lambda(z)(\hbar h)^i / i!$.

1.5   Another Presentation

The formulas defining the quantum groups of Refs. 13 and 14 are based on an R-matrix $\overline{R}$ slightly different from R. Let φ be a solution to the equation

$$\frac{\varphi(\lambda - \hbar)}{\varphi(\lambda + \hbar)} = \frac{\theta(\lambda)}{\theta(\lambda + \hbar)};$$

then we have

$$\overline{R}(z, \lambda) = \varphi(\lambda + \hbar h^{(2)}) R(z, \lambda) \varphi(\lambda + \hbar h^{(1)})^{-1}.$$

The L-matrix of the elliptic quantum group based on $\overline{R}$ can be connected with that of $E_{\tau, \eta}(\mathfrak{sl}_2)$ using the transformation

$$\overline{L}(z, \lambda) = \varphi(\lambda + \hbar h) L(z, \lambda) \varphi(\lambda + \hbar h^{(1)})^{-1}.$$

2   Quantum Currents Algebra

The quantum currents algebra used in Ref. 9 is an example of a family of algebras that were introduced in Ref. 12. These algebras are associated to the data of an algebraic curve and a rational differential. In Ref. 11, it was shown that these algebras can be endowed with quasi-Hopf structures, quantizing certain Manin pairs.

2.1   Classical Structures

2.1.1   Manin Pairs

Let $\mathcal{K} = \mathbb{C}((z))$ be the completed local field of E at its origin 0, and $\mathcal{O} = \mathbb{C}[[z]]$ the completed local ring at the same point. Endow $\mathcal{K}$ with the scalar product $\langle\ ,\ \rangle_{\mathcal{K}}$ defined by

$$\langle f, g \rangle_{\mathcal{K}} = \mathrm{res}_0(fg\,dz).$$

Define on $\mathcal{K}$ the derivation ∂ to be equal to d/dz. Then ∂ is invariant with regard to $\langle\ ,\ \rangle_{\mathcal{K}}$, and $\mathcal{O}$ is a maximal isotropic subring of $\mathcal{K}$.

Let us set $\mathfrak{a} = \mathfrak{sl}_2(\mathbb{C})$, and denote by $\langle , \rangle_{\mathfrak{a}}$ an invariant scalar product on $\mathfrak{a}$. Let us set $\mathfrak{g} = (\mathfrak{a} \otimes \mathcal{K}) \oplus \mathbb{C}D \oplus \mathbb{C}K$; let us define on $\mathfrak{g}$ the Lie algebra stucture defined by the central extension of $\mathfrak{a} \otimes \mathcal{K}$

$$c(x \otimes f, y \otimes g) = \langle x, y \rangle_{\mathfrak{a}} \langle f, \partial g \rangle_{\mathcal{K}} K$$

and by the derivation $[D, x \otimes f] = x \otimes \partial f$.

Let $\mathfrak{g}_{\mathcal{O}}$ be the Lie subalgebra of $\mathfrak{g}$ equal to $(\mathfrak{a} \otimes \mathcal{O}) \oplus \mathbb{C}D$. Define $\langle\ ,\ \rangle_{\mathfrak{a} \otimes \mathcal{K}}$ as the tensor product of $\langle\ ,\ \rangle_{\mathfrak{a}}$ and $\langle\ ,\ \rangle_{\mathcal{K}}$, and $\langle\ ,\ \rangle_{\mathfrak{g}}$ as the scalar product on $\mathfrak{g}$ defined by $\langle\ ,\ \rangle_{\mathfrak{g}}|_{\mathfrak{a} \otimes \mathcal{K}} = \langle\ ,\ \rangle_{\mathfrak{a} \otimes \mathcal{K}}$, $\langle D, \mathfrak{a} \otimes \mathcal{K} \rangle_{\mathfrak{g}} = \langle K, \mathfrak{a} \otimes \mathcal{K} \rangle_{\mathfrak{g}} = 0$, and $\langle D, K \rangle_{\mathfrak{g}} = 1$. Then $\mathfrak{g}_{\mathcal{O}}$ is a maximal isotropic Lie subalgebra of $\mathfrak{g}$.

A maximal isotropic supplementary to $\mathfrak{g}_{\mathcal{O}}$ on $\mathfrak{g}$ is defined by

$$\mathfrak{g}_\lambda = (\mathfrak{h} \otimes L_0) \oplus (\mathfrak{n}_+ \otimes L_\lambda) \oplus (\mathfrak{n}_- \otimes L_{-\lambda}) \oplus \mathbb{C}K. \tag{9}$$

Therefore,

$$\mathfrak{g} = \mathfrak{g}_{\mathcal{O}} \oplus \mathfrak{g}_\lambda \tag{10}$$

define a Lie quasi-bialgebra structure on $\mathfrak{g}_{\mathcal{O}}$, and (as in Ref. 11), of double Lie quasi-bialgebra on $\mathfrak{g}$. Its classical r-matrix is given by the formula

$$r_\lambda = D \otimes K + \sum_i \frac{1}{2} h \left[\frac{z^i}{i!} \right] \otimes h[e_{i;0}] + e \left[\frac{z^i}{i!} \right] \otimes f[e_{i;\lambda}] + f \left[\frac{z^i}{i!} \right] \otimes e[e_{i;-\lambda}],$$

because $(z^i/i!)_{i \geq 0}$, $(e_{i;\lambda})_{i \geq 0}$ are dual bases of $\mathcal{O}$ and L_λ; we denote $x \otimes f$ by $x[f]$; in other terms,

$$r_\lambda(z, w) = \frac{1}{2}(h \otimes h)\frac{\theta'}{\theta}(z - w) + (e \otimes f)\frac{\theta(z - w + \lambda)}{\theta(z - w)\theta(\lambda)}$$

$$+ (f \otimes e)\frac{\theta(z - w - \lambda)}{\theta(z - w)\theta(-\lambda)} + D \otimes K. \tag{11}$$

In what follows, we will set

$$e^i = \frac{z^i}{i!}. \tag{12}$$

Remark 1. The expansion of $R(z-w, \lambda)$ in powers of $\hbar$ is $(\pi_z \otimes \pi_w)(1 + \hbar r_\lambda + \cdots)$, where π_z and π_w are the two-dimensional evaluation representations of $\mathfrak{g}$ at the points z and w. This may be viewed as an indication that the quantization of the Manin pair (10) yields a realization of the elliptic quantum group relations.

Remark 2. The role played by $r_\lambda(z, w)$ in the elliptic KZB equations and the Hitchin systems is explained in Refs. 10 and 17.

2.1.2   Manin Triples

It is useful to consider the following twists of the Lie quasi-bialgebra structures provided by (10).

Let us set

$$
\begin{aligned}
\mathfrak{g}_+ &= (\mathfrak{n}_+ \otimes \mathcal{K}) \oplus (\mathfrak{h} \otimes L_0) \oplus \mathbb{C}K, \\
\mathfrak{g}_- &= (\mathfrak{n}_- \otimes \mathcal{K}) \oplus (\mathfrak{h} \otimes L_0) \oplus \mathbb{C}D,
\end{aligned}
\tag{13}
$$

and

$$
\begin{aligned}
\bar{\mathfrak{g}}_+ &= (\mathfrak{n}_- \otimes \mathcal{K}) \oplus (\mathfrak{h} \otimes L_0) \oplus \mathbb{C}K, \\
\bar{\mathfrak{g}}_- &= (\mathfrak{n}_+ \otimes \mathcal{K}) \oplus (\mathfrak{h} \otimes L_0) \oplus \mathbb{C}D.
\end{aligned}
\tag{14}
$$

Then the decompositions

$$
\mathfrak{g} = \mathfrak{g}_+ \oplus \mathfrak{g}_-, \quad \mathfrak{g} = \bar{\mathfrak{g}}_+ \oplus \bar{\mathfrak{g}}_-
\tag{15}
$$

are decompositions of $\mathfrak{g}$ as direct sums of isotropic subalgebras. This defines two Lie bialgebra structures on $\mathfrak{g}$ Manin triples; both are connected by a twist. Also, there is some twist connecting them with the Lie quasi-bialgebra structure (10). The problem that we are going to solve is to quantize these structures.

2.2   Quantum Algebra

We now present the algebra $U_\hbar \mathfrak{g}(\tau)$ deforming the enveloping algebra of $\mathfrak{g}$. We will also denote this algebra by $A(\tau)$. Generators of $U_\hbar \mathfrak{g}(\tau)$ are D, K and the $x[\epsilon]$, $x = e, f, h, \epsilon \in \mathcal{K}$; they are subject to the relations

$$
x[\alpha \epsilon] = \alpha x[\epsilon], \quad x[\epsilon + \epsilon'] = x[\epsilon] + x[\epsilon'], \quad \alpha \in \mathbb{C}, \epsilon, \epsilon' \in \mathcal{K}.
$$

They serve to define the generating series

$$
x(z) = \sum_{i \in \mathbb{Z}} x[\epsilon^i] \epsilon_i(z), \quad x = e, f, h,
$$

$(\epsilon^i)_{i \in \mathbb{Z}}$, $(\epsilon_i)_{i \in \mathbb{Z}}$ dual bases of $\mathcal{K}$; recall that $(e^i)_{i \in \mathbb{N}}$, $(e_{i;0})_{i \in \mathbb{N}}$ are dual bases of $\mathcal{O}$ and L_0 and set

$$
h^+(z) = \sum_{i \in \mathbb{N}} h[e^i] e_{i;0}(z), \quad h^-(z) = \sum_{i \in \mathbb{N}} h[e_{i;0}] e^i(z).
$$

We will also use the series

$$K^+(z) = e^{((q^\partial - q^{-\partial})h^+/(2\partial))(z)}, \quad K^-(z) = q^{h^-(z)},$$

where $q = e^\hbar$. The relations presenting $U_\hbar \mathfrak{g}(\tau)$ are then

K is central, $\quad [K^+(z), K^+(w)] = [K^-(z), K^-(w)] = 0,$ $\qquad$ (16)

$$\theta(z - w - \hbar)\theta(z - w + \hbar + \hbar K)K^+(z)K^-(w)$$
$$= \theta(z - w + \hbar)\theta(z - w - \hbar + \hbar K)K^-(w)K^+(z), \qquad (17)$$

$$K^+(z)e(w)K^+(z)^{-1} = \frac{\theta(z - w + \hbar)}{\theta(z - w - \hbar)}e(w) \qquad (18)$$

$$K^-(z)e(w)K^-(z)^{-1} = \frac{\theta(w - z + \hbar K + \hbar)}{\theta(w - z + \hbar K - \hbar)}e(w), \qquad (19)$$

$$K^+(z)f(w)K^+(z)^{-1} = \frac{\theta(w - z + \hbar)}{\theta(w - z - \hbar)}f(w),$$
$$\qquad\qquad\qquad\qquad\qquad\qquad\qquad\qquad (20)$$
$$K^-(z)f(w)K^-(z)^{-1} = \frac{\theta(z - w + \hbar)}{\theta(z - w - \hbar)}f(w),$$

$$\theta(z - w - \hbar)e(z)e(w) = \theta(z - w + \hbar)e(w)e(z), \qquad (21)$$
$$\theta(w - z - \hbar)f(z)f(w) = \theta(w - z + \hbar)f(w)f(z), \qquad (22)$$

$$[e(z), f(w)] = \frac{1}{\hbar}\left(\delta(z, w)K^+(z) - \delta(z, w - \hbar K)K^-(w)^{-1}\right). \qquad (23)$$

Here δ denotes, as usual, the formal series $\sum_{i \in \mathbb{Z}} z^i w^{-i-1}$.
Similar relations were presented in Ref. 5.

2.3   Coproducts

The algebra $U_\hbar \mathfrak{g}(\tau)$ is endowed with a Hopf structure given by the coproduct Δ defined by

$$\Delta\big(K^+(z)\big) = K^+(z) \otimes K^+(z),$$
$$\Delta\big(K^-(z)\big) = K^-(z) \otimes K^-(z + \hbar K_1), \qquad (24)$$
$$\Delta\big(e(z)\big) = e(z) \otimes K^+(z) + 1 \otimes e(z), \qquad (25)$$
$$\Delta\big(f(z)\big) = f(z) \otimes 1 + K^-(z)^{-1} \otimes f(z + \hbar K_1), \qquad (26)$$
$$\Delta(D) = D \otimes 1 + 1 \otimes D, \qquad \Delta(K) = K \otimes 1 + 1 \otimes K, \qquad (27)$$

the counit ε, and the antipode S defined by them; we set $K_1 = K \otimes 1$, $K_2 = 1 \otimes K$.

$U_\hbar \mathfrak{g}(\tau)$ is also endowed with another Hopf structure given by the coproduct $\bar{\Delta}$ defined by

$$\bar{\Delta}\big(K^+(z)\big) = K^+(z) \otimes K^+(z),$$
$$\bar{\Delta}\big(K^-(z)\big) = K^-(z) \otimes K^-(z + \hbar K_1), \qquad (28)$$

$$\bar{\Delta}\big(e(z)\big) = e(z - \hbar K_2) \otimes K^-(z - \hbar K_2)^{-1} + 1 \otimes e(z), \qquad (29)$$

$$\bar{\Delta}\big(f(z)\big) = f(z) \otimes 1 + K^+(z) \otimes f(z), \qquad (30)$$

$$\bar{\Delta}(D) = D \otimes 1 + 1 \otimes D, \qquad \bar{\Delta}(K) = K \otimes 1 + 1 \otimes K, \qquad (31)$$

the counit ε, and the antipode $\bar{S}$ defined by them.

The Hopf structures associated with Δ and $\bar{\Delta}$ are connected by a twist,

$$F = \exp\Big(\hbar \sum_{i \in \mathbf{Z}} e[\epsilon_i] \otimes f[\epsilon^i] \Big), \qquad (32)$$

where $(\epsilon^i)_{\in \mathbf{Z}}$ is the basis of $\mathcal{K}$ dual to $(\epsilon_i)_{i \in \mathbf{Z}}$ with regard to $\langle \, , \, \rangle_{\mathcal{K}}$; that is, we have $\bar{\Delta} = \mathrm{Ad}(F) \circ \Delta$.

Then F satisfies the cocycle equation

$$(F \otimes 1)(\Delta \otimes 1)(F) = (1 \otimes F)(1 \otimes \Delta)(F) \qquad (33)$$

(see Ref. 11, Proposition 3.1).

Proposition 1. $(U_\hbar \mathfrak{g}(\tau), \Delta)$ *and* $(U_\hbar \mathfrak{g}(\tau), \bar{\Delta})$ *are quantizations of the Manin triple structures defined by* (15). *The universal R-matrix of* $(U_\hbar \mathfrak{g}(\tau), \Delta)$ *is*

$$\mathcal{R}_\infty = q^{D \otimes K} q^{(\sum_{i \geq 0} h[e^i] \otimes h[e_{i;0}])/2} q^{\sum_{i \in \mathbf{Z}} e[\epsilon^i] \otimes f[\epsilon_i]},$$

and for $(U_\hbar \mathfrak{g}(\tau), \bar{\Delta})$ *it is* $F^{(21)} \mathcal{R}_\infty F^{-1}$.

3   The Realization

3.1   Half-Currents

Fix a complex number λ and set for $x = e, f, K^+$,

$$x_\lambda^+(z) = \sum_i x[e^i] e_{i;\lambda}(z), \qquad (34)$$

and for $x = e, f, K^-$,

$$x_\lambda^-(z) = \sum_i x[e_{i;-\lambda}] e^i(z); \qquad (35)$$

recall that (e^i), $(e_{i;\lambda})$ are dual bases of $\mathcal{O}$ and L_λ.

The fields $e(z)$ and $f(z)$ are then split according to

$$e(z) = e_\lambda^+(z) + e_\lambda^-(z), \quad f(z) = f_{-\lambda}^+(z) + f_{-\lambda}^-(z); \qquad (36)$$

we call the expression $x_\lambda^\pm(z)$ "half-currents." Let us introduce the generating series $k^+(z)$ and $k^-(z)$, defined by

$$k^+(z) = e^{((q^\partial - 1)h^+/2\partial)(z)}, \quad k^-(z) = q^{(h^-/(1+q^{-\partial}))(z)}; \qquad (37)$$

they satisfy the relations

$$K^+(z) = k^+(z)k^+(z - \hbar), \quad K^-(z) = k^-(z)k^-(z - \hbar). \qquad (38)$$

3.2 Realization

Introduce the L-operators

$$L_\lambda^+(\zeta) = \begin{pmatrix} 1 & \theta(\hbar)f_{\lambda+\hbar h-\hbar}^+(\zeta) \\ 0 & 1 \end{pmatrix} \begin{pmatrix} k^+(z - \hbar) & 0 \\ 0 & k^+(z)^{-1} \end{pmatrix} \begin{pmatrix} 1 & 0 \\ \hbar e_{-\lambda}^+(\zeta) & 1 \end{pmatrix} \qquad (39)$$

$$L_\lambda^-(\zeta) = \begin{pmatrix} 1 & 0 \\ \hbar e_{-\lambda}^-(\zeta - K\hbar) & 1 \end{pmatrix} \begin{pmatrix} k^-(\zeta - \hbar) & 0 \\ 0 & k^-(\zeta)^{-1} \end{pmatrix} \begin{pmatrix} 1 & \theta(\hbar)f_{\lambda+\hbar h-\hbar}^-(\zeta) \\ 0 & 1 \end{pmatrix} \qquad (40)$$

The main result of this paper is—

Theorem 1. *Set*

$$R^-(z, \lambda) = E_{11} \otimes E_{11} + E_{-1,-1} \otimes E_{-1,-1} + \frac{\theta(z)}{\theta(z - \hbar)} E_{11} \otimes E_{-1,-1}$$

$$+ \frac{\theta(\lambda + \hbar)\theta(\lambda - \hbar)}{\theta(\lambda)^2} \frac{\theta(z)}{\theta(z - \hbar)} E_{-1,-1} \otimes E_{11}$$

$$- \frac{\theta(z + \lambda)\theta(\hbar)}{\theta(z - \hbar)\theta(\lambda)} E_{1,-1} \otimes E_{-1,1} + \frac{\theta(z - \lambda)\theta(\hbar)}{\theta(z - \hbar)\theta(\lambda)} E_{-1,1} \otimes E_{1,-1}, \qquad (41)$$

and $R^+(z, \lambda) = R(z, \lambda)$. The L-operators $L_\lambda^\pm(\zeta)$ satisfy the relations

$$R^\pm(\zeta - \zeta', \lambda) L_{\lambda+\hbar h^{(2)}}^{\pm(1)}(\zeta) L_\lambda^{\pm(2)}(\zeta')$$

$$= L_{\lambda+\hbar h^{(1)}}^{\pm(2)}(\zeta') L_\lambda^{\pm(1)}(\zeta) R^\pm(\zeta - \zeta', \lambda + \hbar h) \qquad (42)$$

$$L_\lambda^{-(1)}(\zeta) R^-(\zeta - \zeta', \lambda + \hbar h) L_\lambda^{+(2)}(\zeta')$$

$$= L_{\lambda+\hbar h^{(1)}}^{+(2)}(\zeta') R^-(\zeta - \zeta' - K\hbar, \lambda) L_{\lambda+\hbar h^{(2)}}^{-(1)}(\zeta) \frac{A(\zeta, \zeta' + K\hbar)}{A(\zeta, \zeta')}, \qquad (43)$$

where

$$A(\zeta, \zeta') = \exp\left(\sum_{i \geq 0} \left(\frac{1}{\partial} \frac{q^\partial - 1}{q^\partial + 1} e^i \right) (\zeta) e_{i;0}(\zeta') \right). \qquad (44)$$

This result can be viewed as an elliptic analog of the result of Ref. 4.

4 About the Proof

4.1 A Rational Analog

In Ref. 8, we constructed a Hopf algebra cocycle in the Yangian double $A = DY(\mathfrak{sl}_2)$, conjugating Drinfeld's coproduct to the usual one. $DY(\mathfrak{sl}_2)$ is a rational version of $U_\hbar \mathfrak{g}(\tau)$ (it was first introduced in Ref. 20). This means that it has the same presentation as $U_\hbar \mathfrak{g}(\tau)$, replacing the theta functions by their arguments.

For $x = e, f, h$, let x_n be the analog of $x[z^n]$. $DY(\mathfrak{sl}_2)$ contains two "negative modes" and "nonnegative modes" subalgebras $A^{<0}$ and $A^{\geq 0}$, respectively generated by K and the $x_n, n < 0$ and D and the $x_n, n \geq 0$, $x = e, f, h$.

The Yangian coproduct Δ_{Yg} on $DY(\mathfrak{sl}_2)$ defines a Hopf algebra structure on $DY(\mathfrak{sl}_2)$, for which $A^{\geq 0}$ and $A^{<0}$ are both Hopf subalgebras. On the other hand, the rational analogs Δ_{rat} and $\bar{\Delta}_{\mathrm{rat}}$ of Δ and $\bar{\Delta}$ have the following properties:

$$\Delta_{\mathrm{rat}}(A^{\geq 0}) \subset A \otimes A^{\geq 0}, \quad \Delta_{\mathrm{rat}}(A^{<0}) \subset A^{<0} \otimes A,$$
$$\bar{\Delta}_{\mathrm{rat}}(A^{\geq 0}) \subset A^{\geq 0} \otimes A, \quad \bar{\Delta}_{\mathrm{rat}}(A^{<0}) \subset A \otimes A^{<0}.$$

Based on the study of Hopf algebra duality within $DY(\mathfrak{sl}_2)$, we show that we have a decomposition of the rational analog of F, F_{rat}, as a product $F_{\mathrm{rat}} = F_2 F_1$, with $F_1 \in A^{<0} \otimes A^{\geq 0}$ and $F_2 \in A^{\geq 0} \otimes A^{<0}$.

Then the twist $\Delta = F_1 \Delta_{\mathrm{rat}} F_1^{-1}$ coincides with the conjugation $F_2^{-1} \bar{\Delta}_{\mathrm{rat}} F_2$. It follows that Δ satisfies both

$$\Delta(A^{\geq 0}) \subset A^{\geq 0} \otimes A^{\geq 0} \quad \text{and} \quad \Delta(A^{<0}) \subset A^{<0} \otimes A^{<0}.$$

On the other hand, Δ defines a Hopf algebra structure on $DY(\mathfrak{sl}_2)$. Indeed, the associator corresponding to F_1 is expressed as

$$\Phi = F_1^{(12)}(\Delta \otimes 1)(F_1)\big(F_1^{(23)}(1 \otimes \Delta)(F_1)\big)^{-1}.$$

We have clearly $\Phi \in A^{<0} \otimes A \otimes A^{\geq 0}$. Since we also have

$$\Phi = \big((\bar{\Delta} \otimes 1)(F_2)F_2^{(12)}\big)^{-1}(1 \otimes \bar{\Delta})(F_2)F_2^{(23)},$$

we also see that $\Phi \in A^{\geq 0} \otimes A \otimes A^{<0}$. Therefore, $\Phi = 1 \otimes a \otimes 1$, for a certain $a \in A$. On the other hand, as Φ is obtained by twisting a quasi-Hopf structure, it should satisfy the compatibility condition (see Ref. 7)

$$(\Delta \otimes \mathrm{id} \otimes \mathrm{id})(\Phi)(\mathrm{id} \otimes \mathrm{id} \otimes \Delta)(\Phi) = (\Phi \otimes 1)(\mathrm{id} \otimes \Delta \otimes \mathrm{id})(\Phi)(1 \otimes \Phi),$$

so that $a = 1$. Therefore, $\Phi = 1$ and F_1 satisfies the Hopf cocycle equation. It follows that Δ defines Hopf algebra structure on $DY(\mathfrak{sl}_2)$, which we can show to coincide with Δ_{Yg} (see Ref. 8).

The strategy followed in Ref. 9 can be described as follows. $U_\hbar \mathfrak{g}(\tau)$ contains an analog of $A^{\geq 0}$, its subalgebra $U_\hbar \mathfrak{g}_\mathcal{O}$ generated by the $x[\epsilon]$, $\epsilon \in \mathcal{O}$, $x = e, f, h$. It contains no analog of $A^{<0}$. However, there are certain subalgebras of $\mathrm{Hol}(\mathbb{C} \setminus \Gamma, U_\hbar \mathfrak{g}(\tau)^{\otimes 2})$ and $\mathrm{Hol}(\mathbb{C} \setminus \Gamma, U_\hbar \mathfrak{g}(\tau)^{\otimes 3})$, playing the roles of $A^{\geq 0} \otimes A^{<0}$ and $A^{<0} \otimes A^{\geq 0}$, respectively, $A^{\geq 0, <0} \otimes A^{\otimes 2}$ and $A^{\otimes 2} \otimes A^{\geq 0, <0}$. We will decompose F as a product of elements of these algebras. This will give rise to a solution of the twisted cocycle equation, and by BBB, to a solution of the DYBE in $\mathrm{Hol}(\mathbb{C} - \Gamma, U_\hbar \mathfrak{g}(\tau)^{\otimes 2})$. This solution will happen to coincide with $R(z, \lambda)$ in suitable representations. Therefore, the corresponding L-operators will satisfy the elliptic quantum group relations. Let us see now in more detail how this program is realized.

4.2 Subalgebras of $\mathrm{Hol}(\mathbb{C} \setminus \Gamma, U_\hbar \mathfrak{g}(\tau)^{\otimes 2})$ and $\mathrm{Hol}(\mathbb{C} \setminus \Gamma, U_\hbar \mathfrak{g}(\tau)^{\otimes 3})$

For X a vector space, call $\mathrm{Hol}(\mathbb{C} \setminus \Gamma, X)$ the space of holomorphic functions from $\mathbb{C} \setminus \Gamma$ to X and set $\mathrm{Hol}(\mathbb{C} - \Gamma) = \mathrm{Hol}(\mathbb{C} \setminus \Gamma, \mathbb{C})$. The parameter in $\mathbb{C} \setminus \Gamma$ will be identified with the spectral parameter λ.

Definition 1. Let A^{-+} be the subalgebra of $\mathrm{Hol}(\mathbb{C} \setminus \Gamma, A(\tau)^{\otimes 2})$ generated (over $\mathrm{Hol}(\mathbb{C} \setminus \Gamma)$) by $h^{(2)}$ and the $e_{-\lambda - \hbar h^{(2)}}^{-(1)}[\epsilon] f^{(2)}[r]$, with $\epsilon \in \mathcal{K}$ and $r \in \mathcal{O}$, and A^{+-} as the subalgebra of $\mathrm{Hol}(\mathbb{C} \setminus \Gamma, A(\tau)^{\otimes 2})$ generated (over $\mathrm{Hol}(\mathbb{C} \setminus \Gamma)$) by $h^{(2)}$ and the $e^{(1)}[r] f_{\lambda + \hbar h^{(2)} - 2\hbar}^{-(2)}[\epsilon]$, with $r \in \mathcal{O}$, $\epsilon \in \mathcal{K}$.

Definition 2. Let $A^{-,\cdot,\cdot}$ be the subspace of the algebra $\mathrm{Hol}(\mathbb{C} \setminus \Gamma, A(\tau)^{\otimes 3})$, linearly spanned (over $\mathrm{Hol}(\mathbb{C} \setminus \Gamma)$) by the elements of the form

$$\xi' = e_{-\lambda - \hbar(h^{(2)} + h^{(3)})}^{-(1)}[\eta_1] \cdots e_{-\lambda - \hbar(h^{(2)} + h^{(3)}) - 2(n-1)\hbar}^{-(1)}[\eta_n](1 \otimes a \otimes b), \quad (45)$$

$n \geq 0$ (recall that the empty product is equal to 1), where $\eta_i \in \mathcal{K}$, and $a, b \in A(\tau)$ are such that $[h^{(1)} + h^{(2)} + h^{(3)}, \xi'] = 0$; and $A^{\cdot,\cdot,+}$ as the subspace of $\mathrm{Hol}(\mathbb{C} \setminus \Gamma, A(\tau)^{\otimes 3})$ spanned (over $\mathrm{Hol}(\mathbb{C} \setminus \Gamma)$) by the elements of the form

$$\eta' = (a' \otimes b' \otimes 1) f^{(3)}[r_1] \cdots f^{(3)}[r_n](h^{(3)})^s, \quad n, s \geq 0,$$

where $a', b' \in A(\tau)$, $r_i \in \mathcal{O}$, and such that $[h^{(1)} + h^{(2)} + h^{(3)}, \eta'] = 0$.

Definition 3. $A^{+,\cdot,\cdot}$ is the subspace of the algebra $\mathrm{Hol}(\mathbb{C} \setminus \Gamma, A(\tau)^{\otimes 3})$ linearly spanned (over $\mathrm{Hol}(\mathbb{C} \setminus \Gamma)$) by the elements of the form

$$\xi' = e^{(1)}[r_1] \ldots e^{(1)}[r_n](1 \otimes a \otimes b), \quad n \geq 0, \quad (46)$$

where $r_i \in \mathcal{O}$, and $a, b \in A(\tau)$ are such that $[h^{(1)} + h^{(2)} + h^{(3)}, \xi'] = 0$.

$A^{\cdot,\cdot,-}$ is the subspace of the algebra $\mathrm{Hol}(\mathbb{C} \setminus \Gamma, A(\tau)^{\otimes 3})$ linearly spanned (over $\mathrm{Hol}(\mathbb{C} \setminus \Gamma)$) by the elements of the form

$$\eta' = (a' \otimes b' \otimes 1) f_{\lambda + \hbar h^{(3)} - 2\hbar}^{-(3)}[\eta_1] \cdots f_{\lambda + \hbar h^{(3)} - 2\hbar}^{-(3)}[\eta_n](h^{(3)})^s, \quad n, s \geq 0, \quad (47)$$

where $\eta_i \in \mathcal{K}$, and a', $b' \in A(\tau)$ are such that $[h^{(1)} + h^{(2)} + h^{(3)}, \eta'] = 0$.

Proposition 2. $A^{-\cdots}$, $A^{\cdots+}$, $A^{+\cdots}$, and $A^{\cdots-}$ are all four subalgebras of $\mathrm{Hol}(\mathbb{C} \setminus \Gamma, A(\tau)^{\otimes 3})$. We have

$$(\Delta \otimes 1)(A^{-+}) \subset A^{-\cdots} \cap A^{\cdots+}, \quad (1 \otimes \Delta)(A^{-+}) \subset A^{-\cdots} \cap A^{\cdots+}, \quad (48)$$

$$(\bar{\Delta} \otimes 1)(A^{+-}) \subset A^{+\cdots} \cap A^{\cdots-}, \quad (1 \otimes \bar{\Delta})(A^{+-}) \subset A^{+\cdots} \cap A^{\cdots-}. \quad (49)$$

The first statement is a consequence of the following relations between half-currents:

Lemma 1. The generating series $e_\lambda^\pm(z)$, $f_\lambda^\pm(z)$ satisfy the following relations:

$$\frac{\theta(z-w-\hbar)}{\theta(z-w)} e_{\lambda+\hbar}^\epsilon(z) e_{\lambda-\hbar}^{\epsilon'}(w) + \epsilon\epsilon' \frac{\theta(w-z-\lambda)\theta(-\hbar)}{\theta(w-z)\theta(-\lambda)} e_{\lambda+\hbar}^{\epsilon'}(w) e_{\lambda-\hbar}^{\epsilon'}(w)$$

$$= \frac{\theta(z-w+\hbar)}{\theta(z-w)} e_{\lambda+\hbar}^\epsilon(w) e_{\lambda-\hbar}^\epsilon(z)$$

$$+ \epsilon\epsilon' \frac{\theta(z-w-\lambda)\theta(-\hbar)}{\theta(z-w)\theta(-\lambda)} e_{\lambda+\hbar}^\epsilon(z) e_{\lambda-\hbar}^\epsilon(z), \quad (50)$$

$$\frac{\theta(z-w+\hbar)}{\theta(z-w)} f_{\lambda-\hbar}^\epsilon(z) f_{\lambda+\hbar}^{\epsilon'}(w) + \epsilon\epsilon' \frac{\theta(w-z-\lambda)\theta(\hbar)}{\theta(w-z)\theta(-\lambda)} f_{\lambda-\hbar}^{\epsilon'}(w) f_{\lambda+\hbar}^{\epsilon'}(w)$$

$$= \frac{\theta(z-w-\hbar)}{\theta(z-w)} f_{\lambda-\hbar}^{\epsilon'}(w) f_{\lambda+\hbar}^\epsilon(z)$$

$$+ \epsilon\epsilon' \frac{\theta(z-w-\lambda)\theta(\hbar)}{\theta(z-w)\theta(-\lambda)} f_{\lambda-\hbar}^\epsilon(z) f_{\lambda+\hbar}^\epsilon(z), \quad (51)$$

where ϵ, ϵ' take the values $+$, $-$.

To prove the statement about coproducts in Proposition 2, it is then sufficient to prove it from generators of A^{+-} and A^{-+} (see Refs. 9, Lemmas 1.2–4).

A useful argument in Section 4.1 was that the intersection of $A^{\geq 0} \otimes A \otimes A^{<0}$ and $A^{<0} \otimes A \otimes A^{\geq 0}$ is reduced to $1 \otimes A \otimes 1$. Similarly, we have—

Proposition 3. We have

$$A^{+\cdots} \cap A^{-\cdots} = \mathrm{Hol}(\mathbb{C} \setminus \Gamma, 1 \otimes (A(\tau)^{\otimes 2})^\flat),$$

$$A^{\cdots+} \cap A^{\cdots-} = \mathrm{Hol}(\mathbb{C} \setminus \Gamma, (A(\tau)^{\otimes 2})^\flat \otimes \mathbb{C}[h]),$$

where $(A(\tau)^{\otimes 2})^\flat$ are the elements of $A(\tau)^{\otimes 2}$ commuting with $h^{(1)} + h^{(2)}$.

4.3 Decomposition of F

We then have—

Proposition 4. *There is a unique a decomposition of F as*

$$F = F_\lambda^2 F_\lambda^1, \quad \text{with } F_\lambda^1 \in A^{-+} \text{ and } F_\lambda^2 \in A^{+-}, \tag{52}$$

with $(\varepsilon \otimes 1)(F_i) = (1 \otimes \varepsilon)(F_i) = 1$, $i = 1, 2$.

As in the rational case, the proof of this fact relies on a study of the Hopf duality between the opposite quantum nilpotent current subalgebras of $U_\hbar\mathfrak{g}(\tau)$.

Lemma 2. *We have an expansion*

$$F_\lambda^1 \in 1 + \hbar \sum_{i \geq 0} e_{-\lambda-\hbar h^{(2)}}^{-(1)} [e_{i;-\lambda}] f^{(2)}[e^i] + U_\hbar\mathfrak{n}_+(\tau)^{\geq 2} \otimes U_\hbar\mathfrak{n}_-(\tau)^{\geq 2}\mathbb{C}[\hbar],$$

where $U_\hbar\mathfrak{n}_\pm(\tau)^{\geq 2} = \bigoplus_{i \geq 2} U_\hbar\mathfrak{n}_\pm(\tau)^{[i]}$, *and* $U_\hbar\mathfrak{n}_\pm(\tau)^{[i]}$ *is the linear span of the products of i terms of the form $e[\epsilon]$ in $\pm = +$, and $f[\epsilon]$ if $\pm = -$.*

4.4 Twisted Cocycle Equation

Proposition 5. *The family* $(F_\lambda^1)_{\lambda \in \mathbb{C}\backslash\Gamma}$ *satisfies the twisted cocycle equation*

$$F_{\lambda+\hbar h^{(3)}}^{1(12)}(\Delta \otimes 1)(F_\lambda^1) = F_\lambda^{1(23)}(1 \otimes \Delta)(F_\lambda^1). \tag{53}$$

To show this result, we proceed as follows. Define for $\lambda \in \mathbb{C}\backslash\Gamma$,

$$\Phi_\lambda = F_{\lambda+\hbar h^{(3)}}^{1(12)}(\Delta \otimes 1)(F_\lambda^1)\left(F_\lambda^{1(23)}(1 \otimes \Delta)(F_\lambda^1)\right)^{-1}.$$

Then (48) implies that $(\Phi_\lambda)_{\lambda \in \mathbb{C}\backslash\Gamma}$ belongs to $A^{-,\cdot,\cdot} \cap A^{\cdot,\cdot,+}$. On the other hand, using (33), we may rewrite Φ_λ as

$$\Phi_\lambda = \left((\bar\Delta \otimes 1)(F_\lambda^2)(F_{\lambda+\hbar h^{(3)}}^{2(12)})\right)^{-1}(1 \otimes \bar\Delta)(F_\lambda^2)(1 \otimes F_\lambda^{2(23)}).$$

Then (49) then implies that $(\Phi_\lambda)_{\lambda \in \mathbb{C}\backslash\Gamma}$ belongs to $A^{+,\cdot,\cdot} \cap A^{\cdot,\cdot,-}$.
Proposition 3 then implies that

$$\Phi_\lambda = \sum_{i \geq 0} 1 \otimes a_\lambda^{(i)} \otimes h^i. \tag{54}$$

Define now for $x \in A(\tau)$,

$$\Delta_\lambda(x) = F_\lambda^1 \Delta(x)(F_\lambda^1)^{-1};$$

$(\Delta_\lambda)_{\lambda \in \mathbb{C}\backslash\Gamma}$ and $(\Phi_\lambda)_{\lambda \in \mathbb{C}\backslash\Gamma}$ satisfy the compatibility condition

$$(\Delta_{\lambda+\hbar(h^{(3)}+h^{(4)})} \otimes 1 \otimes 1)(\Phi_\lambda)(1 \otimes 1 \otimes \Delta_\lambda)(\Phi_\lambda)$$

$$= \Phi_{\lambda+\hbar h^{(4)}}^{(123)}(1 \otimes \Delta_{\lambda+\hbar h^{(4)}} \otimes 1)(\Phi_\lambda)\Phi_\lambda^{(234)}. \tag{55}$$

It follows that $\Phi_\lambda = 1$.

4.5   The BBB Result

Reference 3, Section 2, contains the following result:

Proposition 6 (see Ref. 3). *Let $(\mathcal{A}, \Delta_\infty^\mathcal{A}, \mathcal{R}_\infty^\mathcal{A})$ be a quasi-triangular Hopf algebra, with a fixed element $\tilde{h}$. Let $F(\lambda)$ be a family of invertible elements of $\mathcal{A} \otimes \mathcal{A}$, parametrized by some subset $U \subset \mathbb{C}$. Set $\Delta(\lambda) = \mathrm{Ad}(F(\lambda)) \circ \Delta_\infty^\mathcal{A}$. Suppose that the identity*

$$F^{(12)}(\lambda + \hbar\tilde{h}^{(3)})(\Delta_\infty^\mathcal{A} \otimes 1)(F(\lambda)) = F^{(23)}(\lambda)(1 \otimes \Delta_\infty^\mathcal{A})(F(\lambda)) \qquad (56)$$

is satisfied. Then we have

$$(\Delta(\lambda + \hbar\tilde{h}^{(3)}) \otimes 1) \circ \Delta(\lambda) = (1 \otimes \Delta(\lambda)) \circ \Delta(\lambda), \qquad (57)$$

and if we set $\mathcal{R}(\lambda) = F^{(21)}(\lambda)\mathcal{R}_\infty^\mathcal{A} F(\lambda)^{-1}$, we have the identity

$$\mathcal{R}^{(12)}(\lambda)\mathcal{R}^{(13)}(\lambda + \hbar\tilde{h}^{(2)})\mathcal{R}^{(23)}(\lambda)$$
$$= \mathcal{R}^{(23)}(\lambda + \hbar\tilde{h}^{(1)})\mathcal{R}^{(13)}(\lambda)\mathcal{R}^{(12)}(\lambda + \hbar\tilde{h}^{(3)}). \qquad (58)$$

Equation (58) is the algebraic version of the DYBE.

4.6   End of the Proof

Now apply Proposition 6 to $\mathcal{A} = U_\hbar\mathfrak{g}(\tau)$, $\Delta_\infty^\mathcal{A} = \Delta$, $\mathcal{R}_\infty^\mathcal{A} = \mathcal{R}_\infty$ and $F(\lambda) = F_\lambda^1$. Set $\mathcal{R}_\lambda = (F_\lambda^1)^{(21)}\mathcal{R}_\infty(F_\lambda^1)^{-1}$. We have—

Corollary 1. *The family $(\mathcal{R}_\lambda)_{\lambda \in \mathbb{C}\backslash\Gamma}$ satisfies the dynamical Yang–Baxter relation*

$$\mathcal{R}_\lambda^{(12)}\mathcal{R}_{\lambda+\hbar h^{(2)}}^{(13)}\mathcal{R}_\lambda^{(23)} = \mathcal{R}_{\lambda+\hbar h^{(1)}}^{(23)}\mathcal{R}_\lambda^{(13)}\mathcal{R}_{\lambda+\hbar h^{(3)}}^{(12)}. \qquad (59)$$

The procedure to derive RLL relations from a Yang–Baxter-like equation is to first study finite-dimensional representations of the algebra, and then take the image of the YBE in these representations (see Ref. 22).

Proposition 7 (see Ref. 12, Proposition 9). *There is a morphism of algebras $\pi_\zeta \colon A(\tau) \to \mathrm{End}(\mathbb{C}^2) \otimes \mathcal{K}_\zeta[\partial_\zeta][[\hbar]]$, defined by the formulas*

$$\pi_\zeta(K) = 0, \quad \pi_\zeta(D) = \mathrm{Id}_{\mathbb{C}^2} \otimes \partial_\zeta,$$

$$\pi_\zeta(h[r]) = E_{11} \otimes \left(\frac{2}{1+q^\partial}r\right)(\zeta) - E_{-1-1} \otimes \left(\frac{2}{1+q^{-\partial}}r\right)(\zeta), \quad r \in \mathcal{O},$$

$$\pi_\zeta(h[\lambda]) = E_{11} \otimes \left(\frac{1-q^{-\partial}}{\hbar\partial}\lambda\right)(\zeta) - E_{-1-1} \otimes \left(\frac{q^\partial - 1}{\hbar\partial}\lambda\right)(\zeta), \quad \lambda \in L_0,$$

$$\pi_\zeta(e[\epsilon]) = \frac{\theta(\hbar)}{\hbar}E_{1,-1} \otimes \epsilon(\zeta), \quad \pi_\zeta(f[\epsilon]) = E_{-1,1} \otimes \epsilon(\zeta), \quad \epsilon \in \mathcal{K}.$$

The image of $\mathcal{R}_\lambda$ by these representations is computed as follows:

Lemma 3. *The image of* $\mathcal{R}_\lambda$ *by* $\pi_\zeta \otimes \pi_{\zeta'}$ *is*

$$(\pi_\zeta \otimes \pi_{\zeta'})(\mathcal{R}_{\lambda+\hbar}) = A(\zeta, \zeta') R^-(\zeta - \zeta', \lambda), \tag{60}$$

where $R^-(z, \lambda)$ *has been defined in* (41).

The computation relies on Lemma 2.

Appling $\mathrm{id} \otimes \pi_\zeta \otimes \pi_{\zeta'}$, $\pi_\zeta \otimes \pi_{\zeta'} \otimes \mathrm{id}$ and $\pi_\zeta \otimes \mathrm{id} \otimes \pi_{\zeta'}$ to (59), after the change of λ into $\lambda + \hbar$, we find (42) and (43).

5   REFERENCES

1. G. E. Arutyunov, L. O. Chekhov, and S. A. Frolov, *R-matrix quantization of the elliptic Ruijsenaars–Schneider model*, Commun. Math. Phys. **192** (1998), No. 2, 405–432, q-alg/9612032.

2. J. Avan, O. Babelon, and E. Billey, *The Gervais–Neveu–Felder equation and the quantum Calogero–Moser systems*, Commun. Math. Phys. **178** (1996), No. 2, 281–299, hep-th/9505091.

3. O. Babelon, D. Bernard, and E. Billey, *A quasi-Hopf algebra interpretation of quantum 3-j and 6-j symbols and difference equations*, Phys. Lett. B **375** (1996), No. 1-4, 89–97, q-alg/9511019.

4. J. Ding and I. B. Frenkel, *Isomorphism of two realizations of quantum affine algebras* $U_q(\widehat{\mathfrak{gl}}_n)$, Commun. Math. Phys. **156** (1993), No. 2, 277–300.

5. J. Ding and K. Iohara, *Generalization and deformation of Drinfeld quantum affine algebras*, q-alg/9608002.

6. V. G. Drinfeld, *A new realization of Yangians and quantum affine algebras*, Soviet Math. Dokl. **36** (1988), No. 2, 212–216.

7. _____, *Quasi-Hopf algebras*, Leningrad Math. J. **1** (1990), 1419–1457.

8. B. Enriquez and G. Felder, *A construction of Hopf algebra cocycles for Yangian double* $DY(\mathfrak{sl}_2)$, J. Phys. A **31** (1998), No. 10, 2401–2413, q-alg/9703012.

9. _____, *Elliptic quantum groups* $E_{\tau,\eta}(\mathfrak{sl}_2)$ *and quasi-Hopf algebras*, Commun. Math. Phys. **195** (1998), No. 3, 651–689, q-alg/9703018.

10. B. Enriquez and V. N. Rubtsov, *Hitchin systems, higher Gaudin operators and R-matrices*, Math. Res. Lett. **3** (1996), No. 3, 343–357.

11. _____, *Quasi-Hopf algebras associated with* $\mathfrak{sl}_2$ *and complex curves*, Tech. Report 1145, École Polytechnique, 1996, q-alg/9608005.

12. _____, *Quantum groups in higher genus and Drinfeld's new realizations method (sl₂ case)*, Ann. Sci. École Norm. Sup. **30** (1997), No. 6, 821–846, q-alg/9601022.

13. G. Felder, *Conformal field theory and integrable systems associated to elliptic curves*, Proc. International Congress of Mathematicians (Zürich, 1994) (S. D. Chatterji, ed.), Birkhäuser, Basel, 1994, pp. 1247–1255, hep-th/9407154.

14. _____, *Elliptic quantum groups*, XIth International Congress of Mathematical Physics (Paris, 1994) (D. Iagolnitzer, ed.), Internat. Press, Cambridge, MA, 1995, pp. 211–218.

15. G. Felder, V. O. Tarasov, and A. N. Varchenko, *Solutions of the elliptic qKZB equations and Bethe ansatz*. I, Topics in Singularity Theory: V. I. Arnold's 60th Anniversary Collection (A. Khovanskii, A. Varchenko, and V. Vassiliev, eds.), Amer. Math. Soc. Transl. Ser. 2, Vol. 180, Amer. Math. Soc., Providence, RI, 1997, pp. 45–76, q-alg/9606005.

16. G. Felder and A. N. Varchenko, *On representations of the elliptic quantum group $E_{\tau,\eta}(sl_2)$*, Commun. Math. Phys. **181** (1996), No. 3, 741–761, q-alg/9601003.

17. G. Felder and C. Wieczerkowski, *Conformal blocks on elliptic curves and the Knizhnik–Zamolodchikov–Bernard equations*, Commun. Math. Phys. **176** (1996), No. 1, 133–161.

18. O. Foda, K. Iohara, M. Jimbo, T. Miwa, and H. Ya, *An elliptic algebra for sl₂*, Lett. Math. Phys. **32** (1994), No. 3, 259–268.

19. E. Frenkel and N. Yu. Reshetikhin, *Quantum affine algebras and deformations of the virasoro and W-algebras*, Commun. Math. Phys. **178** (1996), No. 1, 237–264, q-alg/9505025.

20. S. Khoroshkin, *Central extension of the Yangian double*, Algèbre non commutative, groupes quantiques et invariants (Reims, 1995) (J. Alev and G. Cauchon, eds.), Sém. Congr., Vol. 2, Soc. Math. France, Paris, 1997, pp. 119–135.

21. S. M. Khoroshkin and V. N. Tolstoy, *Yangian double*, Lett. Math. Phys. **36** (1996), No. 4, 373–402.

22. N. Yu. Reshetikhin and M. A. Semenov-Tian-Shansky, *Central extensions of quantum current groups*, Lett. Math. Phys. **19** (1990), No. 2, 133–142.

12
Heisenberg–Ising Spin Chain: Plancherel Decomposition and Chebyshev Polynomials

Eugene Gutkin

ABSTRACT We present an exact solution of the infinite quantum XXZ spin chain (at zero temperature), whose Hamiltonian, $H_{XXZ} = H_N(\Delta)$, depends on the magnon number, $N \geq 1$, and the anisotropy parameter $\Delta \in \mathbb{R}$. The Heisenberg spin chain corresponds to $\Delta = 1$. We show the completeness of the Bethe ansatz eigenstates, and give an explicit Plancherel decomposition of $H_N(\Delta)$ for all N and Δ. We observe a critical spectral phenomenon in the anisotropy regime $|\Delta| < 1$. The critical values of the anisotropy parameter are zeros of Chebyshev polynomials.

Introduction

The Heisenberg model of a magnet has a long and rich history [3, 8]. The first exact results are due to H. Bethe [5], who originated an approach that became widely known under the name of *Bethe ansatz* [8]. Using this approach, Yang and Yang found the eigenstates of the anisotropic version of the Heisenberg magnet, the Heisenberg–Ising model, also known as the XXZ spin chain [19]. Exploiting a far-reaching generalization of the Bethe ansatz, R. Baxter solved the XYZ model [3], which reduces to the XXZ spin chain in a special case.

The phrase "exactly solved model" has many different meanings. For most applications, an exact expression for the eigenstates of a Hamiltonian does not suffice. Let $|z\rangle$, $z \in Z$, be a family of eigenstates of a Hamiltonian H: $H|z\rangle = E(z)|z\rangle$. In analysis, one needs a *Plancherel decomposition* of H, that is, a *Plancherel measure*, $d\mu(z)$ on Z, so that for any state f in the Hilbert space

$$\|f\|^2 = \int_Z |\langle f \mid z \rangle|^2 \, d\mu(z).$$

The equation above implies, in particular, the completeness of the eigenstates $|z\rangle$, $z \in Z$. The subject of this paper is an explicit Plancherel decomposition of the infinite Heisenberg–Ising spin chain (at zero temperature, in the ground state representation). The XXZ Hamiltonian decomposes as a direct sum of the *N-magnon Hamiltonians*, $H_N(\Delta)$, $N \geq 1$, $\Delta \in \mathbb{R}$. We distinguish two regions of the *anisotropy parameter* Δ: the *hyperbolic regime* $1 < |\Delta|$, and the *elliptic regime* $|\Delta| < 1$, with the "parabolic boundary" $|\Delta| = 1$ in between. The Plancherel decompositions of $H_N(\Delta)$ in the two anisotropy regimes are qualitatively different. In the hyperbolic region, $|\Delta| \geq 1$, the Plancherel measure $d\mu_\Delta(z)$ analytically depends on Δ. In a way, the regime $|\Delta| \geq 1$ is similar to the Heisenberg case: $\Delta = 1$ [1, 2].

The Plancherel decomposition in the elliptic region $|\Delta| < 1$ presents a new phenomenon: it is discontinuous at the critical values of Δ (depending on the magnon number N). As a result, the Plancherel decomposition in the elliptic regime is quite involved.

In this paper we obtain an explicit Plancherel decomposition of the XXZ Hamiltonian $H(\Delta)$ for all values of Δ. The support of the Plancherel measure in the N-magnon sector is the disjoint union of the "Plancherel sets" $\Gamma_\beta(\Delta)$ where β runs through the family, B_N, of *N-bindings* (i.e., partitions: $N = n_1 + \cdots + n_\ell$). The wave functions supported on $\Gamma_\beta(\Delta)$, $\beta \in B_N$, correspond to the clustering of N magnons into ℓ *clusters* of the sizes $n_1, \ldots, n_\ell$.

In the hyperbolic region, $|\Delta| \geq 1$, the sets $\Gamma_\beta(\Delta)$ analytically depend on Δ. In the elliptic region, $|\Delta| < 1$, each binding β has a finite number of critical values of the anisotropy parameter: $\Delta_1^{(\beta)}, \ldots, \Delta_{k(\beta)}^{(\beta)}$. The points $\Delta_i^{(\beta)}$ divide the interval $(-1, 1)$ into *bands* and *gaps*. The Plancherel set $\Gamma_\beta(\Delta)$ is nontrivial when Δ runs through a band, and becomes empty in a gap. Thus $\Gamma_\beta(\Delta)$ appear and disappear as the anisotropy parameter crosses $\Delta_i^{(\beta)}$, which is a *critical spectral phenomenon*.

A complete description of $\Delta_i^{(\beta)}$ is complicated, and we don't present it here. Theorem 2 gives the basic information about bands and gaps. In particular, the critical values of the anisotropy parameter in the N-magnon sector are contained in the set of zeros of the Chebyshev polynomials, P_k, of the second kind, $1 \leq k \leq N-1$. These numbers have the form $\cos \pi m/n$, $1 \leq m < n \leq N$, that is, $\Delta = \cos \theta$ where $e^{i\theta}$ is a root of unity.

This observation suggests possible connections between the spectral decomposition of the XXZ Hamiltonian, and the seemingly unrelated questions of mathematical physics. Here are some examples. In the Jones' theory of subfactors [14] the possible values of the index (less than four) are $4 \cos^2 \pi/n$. The same numbers appear in the Beraha conjecture on the zeros of chromatic polynomials of planar graphs [4]. They are essential in the classification of unitary conformal field theories [7]. These numbers play an important role in quantum groups and their representations [6]. Some connections between quantum groups and the XXZ model have been discussed in the literature [15–17].

Takahashi and Suzuki [18] have also encountered a special phenomenon in the thermodynamic limit of the XXZ spin chain. It occurs for the anisotropy parameters $\Delta = \cos \pi k / n$, which happen to be the zeros of Chebyshev polynomials. It would be interesting to uncover a direct connection between the material of Ref. 18 and the spectral decomposition of the XXZ Hamiltonian.

The work is organized as follows. In Sections 1 and 2 we define the Hamiltonians $H_N(\Delta)$, and introduce the *Bethe ansatz eigenstates*. The main tool is the *intertwining operators* [11, 12]. In Section 3 we establish a connection between the Plancherel sets, $\Gamma_N(\Delta) \subset \mathbb{C}^N$, parameterizing the Bethe ansatz eigenstates in the N-magnon sector, and *Chebyshev polynomials*, $P_k(\Delta), k \leq N$, of the second kind. We define *Chebyshev circles* and *Chebyshev arcs* that are the building blocks for $\Gamma_N(\Delta)$. We pinpoint the critical values of the anisotropy parameter and describe the *gaps and bands* for arbitrary bindings. In Section 4 we present the *Plancherel measure* and the *Plancherel decomposition* for the Hamiltonian $H_N(\Delta)$. The proof is outlined in Section 5. In Section 6 we illustrate it with simple special cases. Some of the material presented here was announced in Ref. 1.

1 The XXZ Hamiltonian and the Intertwining Operators

The infinite Heisenberg–Ising spin chain (in the ground state representation, at zero temperature) corresponds to the XXZ Hamiltonian [8],

$$H_{\text{XXZ}}(\Delta) = \sum_{-\infty < i < \infty} S_i^x S_{i+1}^x + S_i^y S_{i+1}^y + \Delta S_i^z S_{i+1}^z, \tag{1}$$

where S^x, S^y, S^z are the spin $\frac{1}{2}$ operators, and Δ is the *anisotropy parameter*. The Hamiltonian (1) commutes with the magnon number operator $\widehat{N}$, and decomposes into the direct sum of N-magnon Hamiltonians, $H_N(\Delta)$, acting on the N-magnon sectors, $\mathcal{H}_N = l_2(\mathbb{Z}_+^N)$, $\mathbb{Z}_+^N = \{(q_1 < \cdots < q_N) = q\}$. The renormalized N-magnon Hamiltonian is given by

$$[H_N(\Delta)f(q) = \sum_{i=1}^{N} f(\ldots, q_i - 1, \ldots) + f(\ldots, q_i + 1, \ldots)$$

$$+ 2\Delta |i : q_i + 1 = q_{i+1} | f(q), \tag{2}$$

where the terms with $q_i + 1 = q_{i+1}$ are omitted in the sum. The permutation group W_N naturally acts on $\mathbb{Z}^N$, and let $\mathcal{H}_N^- \subset l_2(\mathbb{Z}^N)$ be the subspace of W_N-antisymmetric functions. We identify $\mathcal{H}_N$ with $\mathcal{H}_N^-$ via the antisymmetrization operator, $E_N : \mathcal{H}_N \to \mathcal{H}_N^- \subset \ell_2(\mathbb{Z}^N)$, and the restriction operator, $E_N^* : \ell_2(\mathbb{Z}^N) \to \mathcal{H}_N$. The Hamiltonian $H_N(0)$ is the $\mathbb{Z}^N$

Laplacian, and $H_N(\Delta)$ is the Laplacian plus potential, $2\Delta V(q)$, where

$$V(q_1, \ldots, q_N) = \#\{i : q_i + 1 = q_{i+1}\}. \tag{3}$$

Definition 1. An operator $P = P_N(\Delta)$ on $\mathcal{H}_N$ is *intertwining* if it satisfies

$$H_N(\Delta)P = PH_N(0). \tag{4}$$

We will use notation $\det(w) = (-1)^w$. With any N-tuple, $z = (z_1, \ldots, z_N)$ $\in \mathbb{C}^N$, $z_i \neq 0$, we associate the functions z^q and $|z\rangle_0$ on $\mathbb{Z}^N$:

$$z^q = z_1^{q_1} \cdots z_N^{q_N}, \quad \langle q \mid z \rangle_0 = \sum_{w \in W_N} (-1)^w z^{w^{-1}(q)}. \tag{5}$$

Let T_i, $1 \le i \le N$, be the translation operators: $(T_i f)(q) = f(\ldots, q_i + 1, \ldots)$. The functions z^q are the *algebraic eigenfunctions* of the operators

$$p(T_1, T_1^{-1}, \ldots, T_N, T_N^{-1}),$$

where p is an arbitrary polynomial. We have

$$p(T_1, T_1^{-1}, \ldots, T_N, T_N^{-1})z^q = p(z_1, z_1^{-1}, \ldots, z_N, z_N^{-1})z^q. \tag{6}$$

The operators $p(T_1, T_1^{-1}, \ldots, T_N, T_N^{-1})$ naturally act from $\mathcal{H}_N$ to $\ell_2(\mathbb{Z}^N)$.

Theorem 1. *Set*

$$p_\Delta(T) = \prod_{\substack{i,j=1 \\ i<j}}^{N} (1 + T_i T_j - 2\Delta T_i). \tag{7}$$

Then $P_N(\Delta) = E_N^* p_\Delta(T)$ is an intertwining operator.

Theorem 1 is proved in Refs. 12 and 9. The method of intertwining operators is an analytic version of the Bethe ansatz (see below). It is useful for the quantum nonlinear Schrödinger equation and the delta-potential Bose gas [10, 11, 13]. Connections between the intertwining operators for the Heisenberg magnet and for the delta-gas are discussed in Ref. 9.

Set $|z\rangle_\Delta = P_N(\Delta)|z\rangle_0$. Then $|z\rangle_\Delta$ is a function on $\mathbb{Z}_+^N$, and we denote by $(q \mid z)_\Delta$ its value at $q \in \mathbb{Z}_+^N$. For $w \in W_N$ set $i_1 = w(1), \ldots, i_N = w(N)$. The corollary below is immediate from Theorem 1.

Corollary 1. *For any $z \in \mathbb{C}^N$ we have*

1) $$(q \mid z)_\Delta = \sum_{w \in W_N} (-1)^w \prod_{r<s} (1 + z_{i_r} z_{i_s} - 2\Delta z_{i_r}) z_{i_1}^{q_1} \cdots z_{i_N}^{q_N}. \tag{8}$$

2) *The function $|z\rangle_\Delta$ is an eigenfunction of $H_N(\Delta)$ in the algebraic sense:*

$$H_N(\Delta)|z\rangle_\Delta = \left[\sum_{i=1}^{N} (z_i + z_i^{-1}) \right] |z\rangle_\Delta. \tag{9}$$

Note that Eq. (8) is a version of the *Bethe ansatz* [5, 8].

2   Bethe Ansatz Eigenstates

In this section we fix the magnon number N and the anisotropy parameter Δ, and suppress them from notation, where it does not lead to confusion. Note that $|z\rangle = \sum_w a(w,z) z^{w^{-1}(q)}$, where the coefficients $a(w,z)$ are given by Eq. (8). We normalize: $|z\rangle = a(1,z)^{-1}|z\rangle$. Equation (8) implies

$$\langle q \mid z \rangle = \sum_{\substack{w \in W}} \prod_{\substack{i<j \\ w(i)>w(j)}} \left(-\frac{1 + z_i z_j - 2\Delta z_j}{1 + z_i z_j - 2\Delta z_i} \right) z^{w(q)}. \tag{10}$$

Definition 2. The functions $|z\rangle$ (for any $z \in \mathbb{C}^N$) are the Bethe ansatz (BA) eigenfunctions in the algebraic sense. The functions $|z\rangle$ that are bounded on $\mathbb{Z}_+^N$ are the *Bethe ansatz eigenstates* (BAEs).

For a typical $z \in \mathbb{C}^N$ the function $|z\rangle$ grows exponentially in certain directions on $\mathbb{Z}_+^N$. If $|z\rangle$ contributes to the Plancherel decomposition of the Hamiltonian, then it is bounded. In what follows we define a subset, $\Gamma = \Gamma_N(\Delta) \subset \mathbb{C}^N$, such that $|z\rangle$ is bounded for $z \in \Gamma$, and the Bethe ansatz eigenstates $|z\rangle$, $z \in \Gamma$, suffice to obtain the Plancherel decomposition. Note that $|z\rangle \notin \mathcal{H}$, that is, they are generalized eigenfunctions.

Denote by $t = t_\Delta$ the fractional-linear transformation:

$$t(z) = \frac{2\Delta z - 1}{z}. \tag{11}$$

For $n \geq 1$ we denote by $\Gamma_{(n)}(\Delta) \subset \mathbb{C}^n$ the set of $z = (z_1, \ldots, z_n)$ such that

$$z_1 = t(z_2), \ldots, z_{n-1} = t(z_n); \tag{12}$$

$$|z_n| < 1, |z_{n-1} z_n| < 1, \ldots, |z_2 \ldots z_n| < 1; \tag{13}$$

$$|z_1 \ldots z_n| = 1; \tag{14}$$

$$\prod_{i<j} (1 + z_i z_j - 2\Delta z_i) \neq 0. \tag{15}$$

Note that $\gamma_{(n)}(\Delta)$ may be empty (see Theorem 3 below and Section 6), and that $\Gamma_1(\Delta) = S^1$, the unit circle in $\mathbb{C}$.

Let $\beta = (n_1, \ldots n_\ell)$ be a *partition* of N, that is $N = n_1 + \cdots + n_\ell$, $0 < n_1 \leq \cdots \leq n_\ell$. Then β determines a N-*binding*, that is, a partition of the set $\{1, \ldots, N\}$ into ℓ parts: $\{1, \ldots, N\} = \{1, \ldots, n_1\} \cup \{n_1 + 1, \ldots, n_1 + n_2\} \cup \cdots \cup \{n_1 + \cdots + n_{\ell-1} + 1, \ldots, N\}$. We denote this binding also by β. We set $N_j = n_1 + \cdots + n_j$, $1 \leq j \leq \ell$, $N_0 = 0$ and $I_j = \{N_{j-1} + 1, \ldots, N_j\}$. The sets $\{I_j\}$ are called the *clusters* of β. We introduce the *Plancherel sets*

$$\Gamma_\beta(\Delta) = \Gamma_{(n_1)}(\Delta) \times \cdots \times \Gamma_{(n_\ell)}(\Delta) \subset \mathbb{C}^N$$

where $\Gamma_{(n_i)}(\Delta) \subset \mathbb{C}^{n_i}$, $i = 1, \ldots, \ell$, are defined by Eqs. (12)–(15) with the obvious relabeling. The sets $\Gamma_\beta(\Delta)$ are disjoint for distinct β, and we define $\Gamma_N(\Delta) \subset \mathbb{C}^N$ by $\Gamma_N(\Delta) = \bigcup_{N-\text{bindings } \beta} \Gamma_\beta(\Delta)$. Set

$$b(u,v) = -\frac{1+uv-2\Delta v}{1+uv-2\Delta u}, \quad b_w(z) = \prod_{\substack{i<j \\ w(i)>w(j)}} b(z_i, z_j). \tag{16}$$

With this notation, Eq. (10) becomes

$$\langle q \mid z \rangle = \sum_w b_w(z) z^{w(q)}. \tag{10'}$$

Let $W_\beta = \{w \in W \mid w \text{ preserves the order of each cluster of } \beta\}$. For any N-binding β, and $q \in \mathbb{Z}_+^N$, we define a function $\langle q \mid z \rangle_\beta$, of $z_\beta = (z_{N_1}, \ldots, z_{N_\ell}) \in \mathbb{C}^\ell$, as follows:

$$\langle q \mid z \rangle_\beta = \sum_{w \in W_\beta} b_w(z) z^{w(q)} \tag{17}$$

where $z = (z_1, \ldots, z_N)$ is the unique vector in $\mathbb{C}^N$ such that $(z_{N_1}, \ldots, z_{N_\ell})$ $= z_\beta$, and satisfying Eq. (12) with the obvious relabelings.

Proposition 1. (i) *If $z \in \Gamma_\beta(\Delta)$, then $\langle q \mid z \rangle = \langle q \mid z \rangle_\beta$ as functions on $\mathbb{Z}_+^N$;*

(ii) *$\langle q \mid z \rangle_\beta$, as a function of $q \in \mathbb{Z}_+^N$, is bounded for any $z \in \Gamma_\beta(\Delta)$.*

Set $W^\beta = \{w \in W \mid w \text{ reverses the order of each cluster of } \beta\}$. We then define

$$^\beta\langle z \mid q \rangle = \sum_{w \in W^\beta} (b_w(z))^{-1} z^{-w(q)}. \tag{18}$$

We will refer to the functions above as the *dual Bethe ansatz* (DBA) eigenfunctions.

Proposition 2. *Let β be an N-binding, let $\langle q \mid z \rangle_\beta$ and $^\beta\langle z \mid q \rangle$ be as above. Then for any $z \in \Gamma_\beta(\Delta)$ and $q \in \mathbb{Z}_+^N$:*

$$\overline{\langle q \mid z \rangle_\beta} = {}^\beta\langle z \mid q \rangle.$$

3   Chebyshev Circles and Plancherel Sets

We will geometrically describe the Plancherel sets $\Gamma_\beta(\Delta)$. Let us write t_Δ of Eq. (11) in matrix form:

$$t_\Delta = \begin{pmatrix} 2\Delta & -1 \\ 1 & 0 \end{pmatrix}. \tag{19}$$

Then, for $n \geq 1$, we have

$$(t_\Delta)^n = \begin{pmatrix} P_n(\Delta) & -P_{n-1}(\Delta) \\ P_{n-1}(\Delta) & -P_{n-2}(\Delta) \end{pmatrix}, \tag{20}$$

where $P_\ell(\Delta)$ is a polynomial of degree ℓ in Δ. By definition,

$$P_1(\Delta) = 2\Delta, \quad P_0(\Delta) = 1, \quad P_{-1}(\Delta) = 0, \tag{21}$$

and the polynomials P_n satisfy the recurrence relation:

$$P_{n+1}(\Delta) + P_{n-1}(\Delta) = 2\Delta P_n(\Delta). \tag{22}$$

Equations (21) and (22) say that $\{P_n(\Delta) : n = 0, 1, \dots\}$ are the *Chebyshev polynomials of the second kind*. We define the following sets for $n = 1, 2, \dots$:

$$D_n(\Delta) = \{z \in \mathbb{C} : |P_{n-1}(\Delta)z - P_{n-2}(\Delta)| < 1\},$$
$$S_n(\Delta) = \{z \in \mathbb{C} : |P_{n-1}(\Delta)z - P_{n-2}(\Delta)| = 1\}.$$

Note that $S_1(\Delta) = S^1$, the unit circle and $D_1(\Delta) = D^1$, the unit disc. If $P_{n-1}(\Delta) \neq 0$, these are (open) discs and circles. We call them the *Chebyshev discs and circles*. If $P_{n-1}(\Delta) = 0$, then $|P_{n-2}(\Delta)| = 1$, and therefore $S_n(\Delta) = \mathbb{C}, D_n(\Delta) = \varnothing$.

The following proposition establishes a connection between the Chebyshev discs and circles, on one hand, and the sets $\Gamma_{(n)}(\Delta)$, on the other hand.

Proposition 3. *Let $\Gamma_{(n)}(\Delta) \subset \mathbb{C}^n$ be defined by Eqs. (12)–(15). Then*

1) *If $P_j(\Delta) \neq 0$ for $0 < j < n$, then*

$$\Gamma_{(n)}(\Delta) = \{z = (z_1, \dots, z_n) \mid z_j = t_\Delta^{n-j}(z_n), 1 \leq j \leq n-1\},$$

where $z_n \in (\bigcap_{j=1}^{n-1} D_j(\Delta)) \cap S_n(\Delta)$, the (possibly empty) intersection of $n-1$ Chebyshev discs and a Chebyshev circle;

2) *If $P_j(\Delta) = 0$ for some j, $0 < j < n-1$, then $\Gamma_{(n)}(\Delta) = \varnothing$;*

3) *If $P_j(\Delta) \neq 0$ for $0 < j < n-1$, and $P_{n-1}(\Delta) = 0$, then $\Gamma_{(n)}(\Delta) = \bigcap_{j=1}^{n-1} D_j(\Delta)$, a nonempty intersection of open discs.*

From now on we identify $\Gamma_{(n)}(\Delta)$ with the circular arc $(\bigcap_{j=1}^{n-1} D_j(\Delta)) \cap S_n(\Delta) \subset \mathbb{C}$ and call it *the Chebyshev arc*. In order to determine when $\Gamma_{(n)}(\Delta) \neq \varnothing$, we further analyze Chebyshev discs and circles. As we will see, the results are qualitatively different, depending on whether $|\Delta| < 1$ or $|\Delta| \geq 1$. Let $\lambda_\pm = \lambda_\pm(\Delta) = \Delta \pm \sqrt{\Delta^2 - 1}$ be the fixed points of t_Δ.

Proposition 4.   i) *If* $|\Delta| < 1$, *then* $|\lambda_\pm| = 1$, *and all circles* $S_n(\Delta)$, $n = 1, 2, \ldots$, *pass through* $\lambda_\pm$.

ii) *If* $|\Delta| = 1$, *then* $\lambda_\pm = \Delta$ *is the unique fixed point of* t_Δ *and* $\Delta \in S_n(\Delta)$, $D_{n+1}(\Delta) \subset D_n(\Delta)$, *for* $n = 1, 2, \ldots$.

iii) *If* $\Delta > 1$, *the fixed points of* t_Δ *are real,* $0 < \lambda_- < 1 < \lambda_+$, *with* $D_{n+1}(\Delta) \subset D_n(\Delta)$, $n = 1, 2, \ldots$, *and the discs* $D_n(\Delta)$, $n = 1, 2$, $\ldots$, *converge to the point* λ_-, *as* $n \to \infty$. *If* $\Delta < -1$ *the analogous statements hold with* $\lambda_- < -1 < \lambda_+ < 0$, *and the discs* $D_n(\Delta)$ *converge to* λ_+.

Proposition 4 immediately implies the corollary below.

Corollary 2. *Let* $|\Delta| \geq 1$. *Then*

i) *We have* $\Gamma_{(n)}(\Delta) = S_n(\Delta)$;

ii) *Fix a positive integer* N, *and let* $\beta = (n_1, \ldots, n_\ell)$ *be a* N-*binding. Then*

$$\Gamma_\beta(\Delta) = S_{n_1}(\Delta) \times \cdots \times S_{n_\ell}(\Delta).$$

The description of $\Gamma_{(n)}(\Delta)$ when $|\Delta| < 1$ is much more complicated. For $n \geq 1$ we set

$$R_n = \{r = p/q : p, q \in \mathbb{N}, 0 < p < q \leq n\},$$
$$R_n^* = \{r \in R_n : r = p/n, (p, n) = 1\}.$$

Let $0 = r_0 < r_1 < \cdots < r_{k(n)} = 1$ be the elements of R_n.

Definition 3. We say that the interval (r_j, r_{j+1}), where $r_j, r_{j+1} \in R_{n-1}$ is an n-*gap* if $(r_j, r_{j+1}) \cap R_n^* = \varnothing$. Otherwise (r_j, r_{j+1}) is an n-*band*.

Note that $\Delta = \cos(\pi r_j)$, $r_j \in R_n$ if and only if Δ is a zero of some P_k: $1 \leq k \leq n-1$. If $r_j \in R_n^*$, then $P_{n-1}(\Delta) = 0$ and $P_j(\Delta) \neq 0$, for $j < n-1$.

Theorem 2. *Let* $|\Delta| < 1$ *and set* $\Delta = \cos(\pi r)$. *If* $r \notin R_{n-1}$, *then*

$$\Gamma_{(n)}(\Delta) = \begin{cases} \varnothing & \text{if } r \text{ belongs to a } n\text{-gap} \\ D_1 \cap S_n(\Delta) & \text{if } r \text{ belongs to a } n\text{-band} \end{cases}$$

Besides, $\Gamma_{(n)}(\Delta) = \varnothing$ *if* $P_j(\Delta) = 0$ *for* $1 < j < n-1$.

Definition 4. Let $|\Delta| < 1$, and set $\Delta = \cos(\pi r)$, $0 < r < 1$. We say that an integer $n \geq 1$ is Δ-*admissible* if r is in a n-band and Δ does not satisfy condition 3 of Proposition 3. We say that an N-*binding* $\beta = (n_1, \ldots, n_\ell)$ is Δ-*admissible* if all n_j are Δ-admissible.

Corollary 3. *Fix a positive integer N, and let $\Delta \neq 0$. Let $\beta = (n_1, \dots, n_\ell)$ be an N-binding. Then*

$$\Gamma_\beta(\Delta) = \begin{cases} \varnothing & \text{if } \beta \text{ is not } \Delta\text{-admissible} \\ \Gamma_{(n_1)}(\Delta) \times \cdots \times \Gamma_{(n_\ell)}(\Delta) & \text{if } \beta \text{ is } \Delta\text{-admissible} \end{cases}$$

In the latter case we have $\Gamma_{(n_j)}(\Delta) = D_1 \cap S_{n_j}(\Delta)$.

Thus if a N-binding β is not Δ-admissible, there are no BAE's corresponding to β.

4   The Plancherel Formula

Fix $\Delta \neq 0$. For a Δ-admissible $n > 1$ we define a meromorphic differential form on C by

$$d\mu_{(n)}^{(\Delta)}(z) = \frac{P_1(\Delta)^2 \cdots P_{n-2}(\Delta)^2}{2\pi i} \prod_{j=1}^{n-1} (z_j z_{j+1}^{-1} - 1)$$

$$\times \frac{d[P_{n-1}(\Delta)z - P_{n-2}(\Delta)]}{[P_{n-1}(\Delta)z - P_{n-2}(\Delta)]}, \quad (23)$$

where $z_j = t_\Delta^{n-j}(z)$, $j = 1, \dots, n$. Set $d\mu_{(1)} = (2\pi i)^{-1} dz/z$.

Proposition 5. *Let Δ and n be as above. Then*

$$d\mu_{(n)}^{(\Delta)}\big|_{\Gamma_{(n)}(\Delta)} > 0$$

that is, $d\mu_{(n)}^{(\Delta)}$ defines a measure on $\Gamma_{(n)}(\Delta)$.

Fix $N > 1$. For a Δ-admissible binding $\beta = (n_1, \dots, n_\ell)$ we define a meromorphic differential form on $\mathbb{C}^\ell$ by

$$d\mu_\beta^{(\Delta)}(z_\beta) = \prod_{j=1}^{\ell} d\mu_{(n_j)}^{(\Delta)}(z_{N_j}) \quad (24)$$

where $z_\beta = (z_{N_1}, \dots, z_{N_\ell})$ with $N_j = n_1 + \cdots + n_j$, $1 \leq j \leq \ell$.

Theorem 3. *Let Δ and β be as above. Then*

$$d\mu_\beta^{(\Delta)}\big|_{\Gamma_\beta(\Delta)} > 0.$$

that is, $d\mu_\beta^{(\Delta)}$ defines a measure on $\Gamma_\beta(\Delta)$.

Definition 5. *The measure $d\mu_\beta^{(\Delta)}\big|_{\Gamma_\beta(\Delta)}$ is called the Plancherel measure for the binding β.*

Now let $\{\langle \cdot \mid z \rangle_\beta, {}^\beta\langle z \mid \cdot \rangle, z \in \Gamma_\beta(\Delta)\}$ be the BA eigenfunctions and the DBA eigenfunctions defined in Section 2. For $\beta = (n_1, \ldots, n_\ell)$ we let $m_j \geq 0$ be the number of clusters of size j, $1 \leq j \leq N$. Note that $m_1 + 2m_2 + \cdots + Nm_N = N$. We set $\beta! = m_1! \cdots m_\ell!$.

Theorem 4 (Plancherel formula; first version). *Fix $N \geq 1$ and $\Delta \neq 0$. For each Δ-admissible N-binding β the integral*

$$\int_{\Gamma_\beta(\Delta)} \langle p \mid z \rangle_\beta {}^\beta\langle z \mid q \rangle d\mu_\beta^{(\Delta)}(z_\beta)$$

exists and

$$\sum_{\Delta\text{-admissible } \beta} \frac{1}{\beta!} \int_{\Gamma_\beta(\Delta)} \langle p \mid z \rangle_\beta {}^\beta\langle z \mid q \rangle \, d\mu_\beta^{(\Delta)}(z_\beta) = \delta(p - q). \quad (25)$$

From this and Proposition 2 we immediately obtain

Theorem 5 (Plancherel formula; second version). *Let the notation be as above. Then*

$$\sum_{\Delta\text{-admissible } \beta} \frac{1}{\beta!} \int_{\Gamma_\beta(\Delta)} \langle p \mid z \rangle_\beta \overline{\langle z \mid q \rangle_\beta} \, d\mu_\beta^{(\Delta)}(z_\beta) = \delta(p - q). \quad (26)$$

Remark 1. Note that when Δ crosses a zero of a Chebyshev polynomial, the family of Δ-admissible bindings changes discontinuously. By Theorem 5, so does the spectral decomposition of $H_N(\Delta)$, which is a *critical phenomenon*.

5   Sketch of the Proof of Plancherel Decomposition

5.1   Induction on the Magnon Number

We prove Theorem 4 by induction on the magnon number N. The goal is to reduce the Plancherel formula in the N-magnon sector to the corresponding formula in the $(N-1)$-magnon sector. Introducing the notation $L_\beta(\Delta)$ for the "Plancherel integrals" in Eq. (25), we rewrite it as

$$\sum_{N\text{-bindings } \beta} L_\beta(\Delta) = \delta(p - q) \quad (27)$$

where $p = (p_1, \ldots, p_N)$, $q = (q_1, \ldots, q_N)$. The idea is to integrate z_1 out in those $L_\beta(\Delta)$ where $m_1 \geq 1$. Since the "Plancherel integrands" are rational in all variables, we can use the residue formula to evaluate the integrals. To obtain the desired reduction from N to $N - 1$ we need the identity

$$\sum_{N\text{-bindings } \beta} L_\beta(\Delta) = \delta(p_1 - q_1) \sum_{(N-1)\text{-bindings } \gamma} L_\gamma(\Delta). \quad (28)$$

Since for $N = 1$, Eq. (25) becomes (p, q are integers)

$$\int_{|z|=1} z^p z^{-q} (2\pi i z)^{-1} \, dz = \delta(p - q), \tag{29}$$

Theorem 4 would follow by induction on N.

5.2   Hyperbolic Region of the Anisotropy Parameter

For $\Delta = 1$ the program above has been carried out in Ref. 2. In this case the contours $\Gamma_{(n)}$ are nested circle. Since the latter holds for $|\Delta| \geq 1$ (see Proposition 4 and Corollary 2), the argument extends to the whole hyperbolic region, $|\Delta| \geq 1$. To complete the argument, we need to know the location of poles of the integrands with respect to the contours of integration in Eq. (25).

The location of poles of the Plancherel integrands is a difficult question even in the region $|\Delta| \geq 1$. The reason is that most of the singularities of the summands in Plancherel integrands cancel out in summation (the so-called "spurious poles"). The exposition in Ref. 2 relies on a complicated combinatorial technique (unpublished) to detect spurious poles. Our approach is to reduce the analysis of singularities of the Plancherel integrands to that of the rational function

$$G(z_1, \ldots, z_N) = \prod_{i<j} \frac{1 + z_i z_j - 2\Delta z_i}{z_j - z_i}. \tag{30}$$

We use Eq. (30) to analyze the singularities of Plancherel integrals for all values of the anisotropy parameter.

5.3   Elliptic Region of the Anisotropy Parameter

The structure of the Plancherel sets $\Gamma_\beta(\Delta)$ is much more complicated when $|\Delta| < 1$. As Δ varies, the sets $\Gamma_\beta(\Delta)$ vary discontinuously, which makes keeping track of the Plancherel integrals $L_\beta(\Delta)$ very difficult (see examples below). In order to establish Eq. (28) in the elliptic regime, we introduce a "linearizing change of variables" setting

$$z = \frac{\lambda_+(\Delta)\zeta - \lambda_-(\Delta)}{\zeta - 1}. \tag{31}$$

In the ζ-plane the family of circles through $\lambda_+(\Delta), \lambda_-(\Delta)$ becomes the set of lines through the origin. In particular, the Chebyshev circles $S_n(\Delta)$ become the lines $\{\zeta = te^{-in\theta} : t \in \mathbb{R}\}$, where $\Delta = \cos\theta$, and t_Δ becomes the rotation by 2θ. In the ζ-plane it is simpler to keep track of the Chebyshev arcs $\Gamma_{(n)}(\Delta)$ (they become rays emanating from the origin) and of the Plancherel sets $\Gamma_\beta(\Delta)$. The function $G(z_1, \ldots, z_N)$ of Eq. (30) also simplifies:

$$G(\zeta_1, \ldots, \zeta_N) = \prod_{i<j} \frac{\lambda_+ \zeta_i - \lambda_- \zeta_j}{\zeta_i - \zeta_j}. \tag{32}$$

To prove Eq. (28), we split the integrals $L_\beta(\Delta)$ into the "partial Plancherel integrals" $L(\beta, I, \Delta)$, where I is a cluster of the binding β. Denote by A' the complement of a set A, and set

$$X_n(\Delta) = D_1 \cap D_2(\Delta) \cap \cdots \cap D_{n-1}(\Delta) \cap D'_n(\Delta).$$

We have $\partial X_n(\Delta) = \bigcup_{k=1}^n Z_n^k(\Delta)$ where $Z_n^k(\Delta)$ are circular arcs (only two of them are nonempty), and $\Gamma_{(n)}(\Delta) = Z_n^n(\Delta)$. Let $|I| = n$. Replacing formally in the integral $L(\beta, I, \Delta)$ the Chebyshev arc $\Gamma_{(n)}(\Delta)$ by $Z_n^k(\Delta)$, $1 \leq k \leq n$, we define the "virtual integrals" $L(\beta, I, k, \Delta)$, $1 \leq k \leq |I|$. For notational convenience, we set $L(\beta, I, k, \Delta) = 0$ if $k > |I|$. A crucial step in the proof of the reduction from N to $N - 1$ is the following identity $(1 \leq n \leq N)$:

$$\sum_{\substack{N\text{--bindings } \beta \\ |I|>n,\ 0\leq k}} L(\beta, I, k, \Delta) = 0. \tag{33}$$

Eq. (33) holds for all Δ, but in the hyperbolic regime, $|\Delta| \geq 1$, it is not needed for the proof of Theorem 4.

6   Examples

6.1   2-Magnon Sector

There are two bindings β: $(1,1)$ and (2), and Eq. (25) becomes

$$L_{(1,1)}(\Delta) + L_{(2)}(\Delta) = \delta(p_1 - q_1)\delta(p_2 - q_2).$$

We fix z_2 in $L_{(1,1)}(\Delta)$ and evaluate the integral over the unit circle $\{|z_1|=1\}$ via the residues in $\{|z_1| \geq 1\}$. We obtain

$$L_{(1,1)}(\Delta) + L_{(2)}(\Delta) = \delta(p_1 - q_1)L_{(1)}(\Delta)$$
$$+ \int_{\gamma(\Delta)} d\mu_{(2)}(z)\langle p \mid z \rangle_{(2)}\, {}^{(2)}\langle z \mid q \rangle, \tag{34}$$

where $L_{(1)}(\Delta)$ is the Plancherel integral for $N = 1$ (see Eq. (29)) and $\gamma(\Delta)$ is a closed contour in the z_2-plane. The factor $\delta(p_1 - q_1)$ in Eq. (34) is due to the residue at $z_1 = \infty$. For $0 < |\Delta| < 1$ the contour $\gamma(\Delta)$ is simple and consists of two circular arcs with ends at $\lambda_{\pm}(\Delta)$. For $|\Delta| \geq 1$, $\gamma(\Delta)$ consists of two Chebyshev circles (with opposite orientations). In either case, $\gamma(\Delta)$ bounds a domain where the integrand has no singularities, hence the integral in Eq. (34) is zero (this is a particular case of Eq. (33)). Thus Eq. (34) becomes

$$L_{(1,1)}(\Delta) + L_{(2)}(\Delta) = \delta(p_1 - q_1)L_{(1)}(\Delta),$$

which completes the reduction.

6.2   3-Magnon Sector

There are three bindings β: $(1,1,1)$, $(1,2)$ and (3). In our notation, the left hand side of Eq. (28) is $L_{(1,1,1)}(\Delta) + L_{(1,2)}(\Delta) + L_{(3)}(\Delta)$. The first two of the three Plancherel integrals contain integration over the the unit circle $\{|z_1| = 1\}$. We fix (z_2, z_3) and evaluate the z_1-integral using the residues in $\{|z_1| > 1\}$. The residue at infinity produces the factor $\delta(p_1 - q_1)$, and we get

$$L_{(1,1,1)}(\Delta) + L_{(1,2)}(\Delta) + L_{(3)}(\Delta) = \delta(p_1 - q_1)[L_{(1,1)}(\Delta) + L_{(2)}(\Delta)]$$

$$+ \int_{\gamma(\Delta)} d\mu_{(3)}(z)\langle p \mid z \rangle_{(3)} \, {}^{(3)}\langle z \mid q \rangle. \quad (35)$$

The closed contour $\gamma(\Delta)$ depends on the position of Δ with respect to the points 0, $\pm\frac{1}{2}$, ± 1. For $|\Delta| < 1$, and $|\Delta| \neq 0$, $\frac{1}{2}$, $\gamma(\Delta)$ is a simple closed contour formed by two circular arcs with ends at $\lambda_{\pm}(\Delta)$. Each arc belongs to one of the Chebyshev circles $S_1(\Delta)$, $S_2(\Delta)$, $S_3(\Delta)$, and there is a "bifurcation" of $\gamma(\Delta)$ as Δ crosses the points $0, \pm\frac{1}{2}$ (the zeros of P_1 and P_2). Another bifurcation of $\gamma(\Delta)$ occurs when Δ crosses the values ± 1 into the hyperbolic region $|\Delta| \geq 1$. Then $\gamma(\Delta)$ becomes the difference of the Chebyshev circles $S_2(\Delta)$ and $S_3(\Delta)$. In every case, $\gamma(\Delta)$ is the boundary of a domain where the integrand has no singularities, hence the integral vanishes (a particular case of Eq. (33)), and Eq. (35) becomes

$$L_{(1,1,1)}(\Delta) + L_{(1,2)}(\Delta) + L_{(3)}(\Delta) = \delta(p_1 - q_1)[L_{(1,1)}(\Delta) + L_{(2)}(\Delta)]. \quad (36)$$

The symbols $L_{(1,1)}(\Delta)$ and $L_{(2)}(\Delta)$ in the right-hand side of Eq. (36) are the Plancherel integrals for $N = 2$ corresponding to the 2-tuples (p_2, p_3), (q_2, q_3). Hence Eq. (36) provides the reduction to $N = 2$.

6.3   4-Magnon Sector

There are 5 bindings: $(1,1,1,1)$, $(1,1,2)$, $(1,3)$, $(2,2)$ and (4). The critical values of Δ are 0, $\pm 1/\sqrt{2}$, $\pm\frac{1}{2}$, ± 1 ($\pm 1/\sqrt{2}$ are the roots of P_3). The corresponding regions of the anisotropy parameter are $0 < |\Delta| < \frac{1}{2}$, $\frac{1}{2} <$

$|\Delta| < 1/\sqrt{2}$, $1/\sqrt{2} < |\Delta| < 1$ and $|\Delta| \geq 1$. For the first time we observe that $\Gamma_\beta(\Delta)$ can be empty for an open Δ-interval: $\Gamma_{(4)}(\Delta) = \varnothing$ for $|\Delta| < \frac{1}{2}$. Thus $-\frac{1}{2} < \Delta < 0$ and $0 < \Delta < \frac{1}{2}$ are the gaps.

As for $N < 4$, we integrate z_1 out in $L_{(1,1,1,1)}(\Delta)$, $L_{(1,1,2)}(\Delta)$ and $L_{(1,3)}(\Delta)$ evaluating the integral via the residues in $\{|z_1| > 1\}$. We obtain:

$$L_{(1,1,1,1)}(\Delta) + L_{(1,1,2)}(\Delta) + L_{(1,3)}(\Delta) + L_{(2,2)}(\Delta) + L_{(4)}(\Delta)$$
$$= \delta(p_1 - q_1)[L_{(1,1,1)}(\Delta) + L_{(1,2)}(\Delta) + L_{(3)}(\Delta)]$$
$$+ \int_{\gamma(\Delta)} d\mu_{(4)}(z) \langle p \mid z \rangle_{(4)}{}^{(4)}\langle z \mid q \rangle. \quad (37)$$

The Plancherel integrals inside the square brackets are the 3-magnon integrals corresponding to (p_2, p_3, p_4), (q_2, q_3, q_4). The closed contour $\gamma(\Delta)$ analytically depends on Δ inside each region $0 < |\Delta| < \frac{1}{2}$, $\frac{1}{2} < |\Delta| < 1/\sqrt{2}$, $1/\sqrt{2} < |\Delta| < 1$, $|\Delta| \geq 1$, and bifurcates at the boundaries between the regions. In all cases, $\gamma(\Delta)$ bounds a region where the integrand has no singularities, hence $\int_{\gamma(\Delta)} d\mu_{(4)}(z) \langle p \mid z \rangle_{(4)}{}^{(4)}\langle z \mid q \rangle = 0$. This identity is a special case of Eq. (33), and can be verified directly. Hence

$$L_{(1,1,1,1)}(\Delta) + L_{(1,1,2)}(\Delta) + L_{(1,3)}(\Delta) + L_{(2,2)}(\Delta) + L_{(4)}(\Delta)$$
$$= \delta(p_1 - q_1)[L_{(1,1,1)}(\Delta) + L_{(1,2)}(\Delta) + L_{(3)}(\Delta)],$$

which completes the reduction to $N = 3$.

Acknowledgments: This work was partially supported by NSF grants DMS-9013220, DMS-9107314, DMS-9400295.

7 REFERENCES

1. D. Babbitt and E. Gutkin, *The Plancherel formula for the infinite XXZ Heisenberg spin chain*, Lett. Math. Phys. **20** (1990), No. 2, 91–99.

2. D. Babbitt and L. Thomas, *Ground state representation of the infinite one-dimensional Heisenberg ferromagnet. II. An explicit Plancherel formula*, Commun. Math. Phys. **54** (1977), No. 3, 255–278.

3. R. J. Baxter, *Exactly Solved Models in Statistical Mechanics*, Academic Press, London, 1982.

4. S. Beraha, *Infinite nontrivial families of maps and chromials*, Ph.D. thesis, Johns Hopkins University, 1977.

5. H. Bethe, *Theorie der Metalle. I. Eigenwerte und Eigenfunktionen der lineare Atomkette*, Z. Phys. **71** (1931), 205–226.

6. V. G. Drinfeld, *Quantum groups*, Proc. International Congress of Mathematicians (Berkeley, CA, 1986) (A. M. Gleason, ed.), Amer. Math. Soc., Providence, RI, 1987, pp. 798–820.

7. D. Friedan, Z. Qiu, and S. Shenker, *Conformal invariants, unitarity and critical exponents in two dimensions*, Phys. Rev. Lett. **52** (1984), No. 18, 1575–1578.

8. M. Gaudin, *La fonction d'onde de Bethe*, Masson, Paris, 1983.

9. E. Gutkin, *Heisenberg–Ising spin chain and the nonlinear Schrödinger equation*, Rep. Math. Phys. **24** (1986), No. 1, 121–127.

10. ———, *Bethe ansatz and generalized Yang–Baxter equations*, Ann. Phys. (NY) **176** (1987), No. 1, 22–48.

11. ———, *Harmonic analysis of the quantum nonlinear Schrödinger equation*, Harmonic Analysis and Operator Algebras (Canberra, 1987) (M. Cowling, C. Meaney, and W. Moran, eds.), Proc. Centre Math. Anal. Austral. Nat. Univ., Vol. 16, Austr. Nation. Univ., Canberra, 1988, pp. 113–132.

12. ———, *Method of interwining operators for integrable models of quantum mechanics*, Finite-Dimensional Integrable Nonlinear Dynamical Systems (Johannesburg, 1988) (P. G. L. Leach and W.-H. Steeb, eds.), World Scientific, Singapore, 1988, pp. 215–228.

13. ———, *Quantum nonlinear Schrödinger equation: two solutions*, Phys. Rep. **167** (1988), No. 1-2, 1–131.

14. V. Jones, *Index for the subfactors*, Invent. Math. **72** (1983), No. 1, 1–25.

15. V. Pasquier, *Continuum limits of lattice models built on quantum groups*, Nucl. Phys. B **295** (1988), No. 4, FS 21, 491–510.

16. V. Pasquier and H. Saleur, *Symmetries of the XXZ chain and quantum SU(2)*, Champs, cordes et phénomènes critiques (Les Houches, 1988) (. Brézin and J. Zinn-Justin, eds.), North-Holland, Amsterdam, 1990, pp. 281–289.

17. H. Saleur, *Zeroes of chromatic polynomials: a new approach to Beraha conjecture using quantum groups*, Commun. Math. Phys. **132** (1990), No. 3, 657–678.

18. M. Takahashi and M. Suzuki, *One-dimensional anisotropic Heisenberg model at finite temperatures*, Progr. Theor. Phys. **48** (1972), 2187–2209.

19. C. N. Yang and C. P. Yang, *One-dimensional chain of anisotropic spin-spin interactions. I. Proof of Bethe's hypothesis for ground state in a finite system*, Phys. Rev. **150** (1966), 321–337.

13
Ruijsenaars's Commuting Difference System from Belavin's Elliptic R-Matrix

Koji Hasegawa

ABSTRACT For Belavin's elliptic R-matrix, we constract an "L-operator" as a set of difference operators acting on the functions on a type A weight space. According to Baxter's argument for commuting transfer matrices, the trace of the L-operator gives a commutative difference system. We show that for our L-operator this approach gives the elliptic Macdonald-type operators, actually equivalent to Ruijsenaars's operators. The calculation of the trace uses a one parameter generalization of the Frobenius determinant formula for Jacobi theta function. We also investigate the invariant subspace for the operators spanned by the symmetric theta functions, showing that it is isomorphic to the symmetric tensor representation of the sl_n-generalization of the Sklyanin algebra. Unlike the trigonometric case, the action of the Ruijsenaars's operators on this space is not triangular with respect to the "affine orbitsum" basis.

1 Introduction

As the master family among the integrable n-body systems on a line, or as the generalized spherical function equation on a simple Lie group, much attention has been paid for the Macdonald system [23, 24] and the Ruijsenaars elliptic system [27] these days. In both cases it is remarkable that the commutativity of the operators of the system is nontrivial. As is observed by Ruijsenaars, a direct computation of the commutator will lead to the Cauchy-type identity of the function appeared in the coefficient of the operators. On the other hand, Macdonald showed in the trigonometric case that the system is explicitly diagonalizable, which implies the commutativity of the system as a corollary. Many authors have made efforts to elucidate the nature of the commutativity, and at least two ways of understanding for Macdonald's commuting operators have been established. One is due to Cherednik [8], who uses the double affine Hecke algebra, its representation via difference operators (Dunkl operators), and the center of the algebra. Restricting ourselves to the type A case, the other is the work by Etingof and Kirillov [11], who obtained these operators as the image of central ele-

ments of the quantum universal enveloping algebra. These two works have
related the system to the two natural, universal, commuting, but different
objects. (There should be a question that these two approaches can be
related or not by any means, which seems to be not yet clarified so far.)
From the conformal field theory point of view, the former approach can
be regarded as the reformulation of the theory of the Knizhnik–Zamolodchikov equation for the n-pointed conformal block, while the latter can be
considered as the analogue of the Bernard equation for the one-pointed
conformal block on the torus at the critical level [12, 14].

In both approaches above, the Yang–Baxter equation can be regarded
as the important background of the theory. Interesting enough, in Baxter's
study of solvable lattice statistical models, the Yang–Baxter equation arise
as the condition to provide sufficiently many commuting operators. This is
done by taking the traces of the so-called L-operators, the operators which
satisfy the "$RLL = LLR$ relation" (1), which is nothing but a variant of
the Yang–Baxter equation. In Ref. 17 we considered a realization of the
L-operator for Belavin's elliptic R-matrix [4] in terms of difference operators, and studied the commuting operators arising from the traces thereof.
It turns out that one can recover Ruijsenaars's elliptic commuting difference operators in this way. In other words, the argument of "commuting
transfer matrices" in statistical lattice model theory provides the way to
make the commutativity apparent in the elliptic case. In this brief chapter
we will summarize the results of Ref. 17 and will report some further facts.

Our difference L-operator can be identified with the representation of
Felder's elliptic algebra [12] after some suitable similarity transformation
and reformulation. This suggests that our L-operator would allow a natural formulation in a q-deformed conformal field theory in a way that our
approach could be regarded as an elliptic generalization of the above mentioned work of Etingof and Kirillov.

2   The Difference Operators

For $n > 1$ let $V = \bigoplus_{k \in \mathbb{Z}/n\mathbb{Z}} \mathbb{C}e^k$ ($e^k = e^{k+n}$) and let $g, h \in \mathrm{GL}(V)$ to be

$$ge^k := e^k \exp\left(\frac{2\pi i k}{n}\right), \qquad he^k := e^{k+1}.$$

We have $gh = hg \exp(2\pi i/n)$. Let $\hbar, \tau \in \mathbb{C}$, $\hbar \neq 0$, $\mathrm{Im}\,\tau > 0$.

Belavin's R-matrix $R(u) = R_\hbar(u)$ is characterized as the unique solution
of the following five conditions:

- $R_\hbar(u)$ is a holomorphic $\mathrm{End}(\mathbb{C}^n \otimes \mathbb{C}^n)$-valued function in u,
- $R_\hbar(u) = (x \otimes x)R_\hbar(u)(x \otimes x)^{-1}$ for $x = g, h$,

- $R_\hbar(u+1) = (g \otimes 1)^{-1} R_\hbar(u)(g \otimes 1) \times (-1)$,

- $R_\hbar(u+\tau) = (h \otimes 1) R_\hbar(u)(h \otimes 1)^{-1} \times (-\exp 2\pi i[u + \hbar/n + \tau/2])^{-1}$,

- $R_\hbar(0) = P : x \otimes y \mapsto y \otimes x$.

It holds that 1) there is a unique solution to the above conditions and 2) the solution satisfies the Yang–Baxter equation [4].

In the trigonometric limit, up to a certain simple "gauge transformation" [26], this R-matrix degenerates to the image of the universal R-matrix for the quantum affine enveloping algebra $U_q(A_{n-1}^{(1)})$ [10, 19] in the vector representation.

By an L-operator we mean the matrix $L(u) = [L(u)_j^i]_{i,j=1}^n$ of operators (noncommutative letters) that satisfy

$$\check{R}(u-v)L(u) \otimes L(v) = L(v) \otimes L(u)\check{R}(u-v), \tag{1}$$

where $\check{R}(u) := PR(u)$.

For Belavin's R-matrix we shall construct such an L-operator in the following way. Let $\mathbf{h}^*$ be the weight space for $\mathrm{sl}_n(\mathbb{C})$ and realize $\mathbf{h}^*$ in $\mathbb{C}^n = \bigoplus_{i=1}^n \mathbb{C}\epsilon_i$, $\langle \epsilon_i, \epsilon_j \rangle = \delta_{i,j}$, as the orthogonal complement to $\sum_{i=1}^n \epsilon_i$. We denote the orthogonal projection of ϵ_i by $\bar{\epsilon}_i$. Then the root lattice is $Q = \sum_{i=1}^{n-1} \mathbb{Z}(\bar{\epsilon}_i - \bar{\epsilon}_{i+1})$ and the weight lattice is $P = \sum_{i=1}^n \mathbb{Z}\bar{\epsilon}_i$.

For each $\lambda, \mu \in \mathbf{h}^*$ and $j = 1, \ldots, n$ we put

$$\phi(u)_{\lambda}^{\mu}{}_j = \begin{cases} \theta_j(u/n - \langle \lambda, \bar{\epsilon}_k \rangle), & \mu - \lambda = \hbar\bar{\epsilon}_k \text{ for some } k = 1, \ldots, n, \\ 0, & \text{otherwise} \end{cases} \tag{2}$$

where

$$\theta_j(u) := \sum_{\mu \in n/2 - j + n\mathbb{Z}} \exp 2\pi i \left[\mu \left(u + \frac{1}{2} \right) + \frac{\mu^2}{2n}\tau \right].$$

Also we define $\bar{\phi}(u)_{\mu}^{\mu + \hbar\bar{\epsilon}_k j}$ to be the (k, j)-entry in the inverse matrix of $[\phi(u)_{\mu}^{\mu + \hbar\bar{\epsilon}_k}{}_j]_{j,k=1}^n$, namely,

$$\sum_{j=1}^n \bar{\phi}(u)_{\mu}^{\mu + \hbar\bar{\epsilon}_k j} \phi(u)_{\mu}^{\mu + \hbar\bar{\epsilon}_{k'}}{}_j = \delta_{k,k'},$$

$$\sum_{k=1}^n \phi(u)_{\mu}^{\mu + \hbar\bar{\epsilon}_k}{}_j \bar{\phi}(u)_{\mu}^{\mu + \hbar\bar{\epsilon}_k j'} = \delta_{j,j'}. \tag{3}$$

Then, generalizing a result in the celebrated paper [28, 29], we have

Theorem 1 ([16, 18]). *For a function f on $\mathbf{h}^*$, put*

$$(L(c \mid u)_j^i f)(\mu) := \sum_{k=1}^n \phi(u + c\hbar)_{\mu}^{\mu + \hbar\bar{\epsilon}_k}{}_j \bar{\phi}(u)_{\mu}^{\mu + \hbar\bar{\epsilon}_k i} f(\mu + \hbar\bar{\epsilon}_k). \tag{4}$$

Then for any $c \in \mathbb{C}$, the collection of difference operaors

$$L(c \mid u) = [L(c \mid u)_j^i]_{i,j=1}^n$$

satisfies the desired relation (1). *That is, $L(c \mid u)$ gives a one parameter family (in c) of L-operators.*

Remark 1. The objects ϕ and $\bar{\phi}$ are called the "intertwining vectors" [20]. The trigonometric version of this construction of L-operator for the standard (nondynamical) R-matrix is rather obscure and not trivial from the above [1].

Recall $V = \bigoplus_{j=1}^n \mathbb{C}e^j = \mathbb{C}^n$ and let $\mathcal{M}(\mathbf{h}^*)$ be the field of meromorphic functions on $\mathbf{h}^*$. Then the above L-operator is an endomorphism on the space $V \otimes \mathcal{M}(\mathbf{h}^*)$,

$$L(c \mid u) \in \mathrm{End}(V \otimes \mathcal{M}(\mathbf{h}^*)).$$

Here the first space $V = \mathbb{C}^n$ can be regarded as the space of defining comodule (vector "co"representation) for the bialgebra $A(R)$ defined by the relation (1). We can consider more complicated comodules for this bialgebra as well: for each Young diagram Y we can construct an $A(R)$-comodule $V(Y)$ whose dimension is just the same as for the GL_n-module that corresponds to Y. This is an early result known as the fusion technique [5–7, 22] and done by taking the appropriate sub/quotient of the tensor comodule of $V(\square) = V = \mathbb{C}^n$. It follows that we get a collection (a matrix) of difference operators

$$L^Y(c \mid u) \in \mathrm{End}(V(Y) \otimes \mathcal{M}(\mathbf{h}^*))$$

for each Y and they satisfy the relation

$$\check{R}^{Y,Y'}(u-v)L^Y(c \mid u) \otimes L^{Y'}(c \mid v)$$
$$= L^{Y'}(c \mid v) \otimes L^Y(c \mid u)\check{R}^{Y,Y'}(u-v), \quad (5)$$

where $\check{R}^{Y,Y'}$ stands for the "fused R-matrices" which gives the isomorphism between the $A(R)$-comodules $V(Y) \otimes V(Y') \to V(Y') \otimes V(Y)$.

These structures are of course now well understood for the trigonometric R-matrix case, where we have the quantised enveloping algebra and its universal R-matrix as the origin of these fused R-matrix.

Now we will state the main result, concerning the trace of the operator $L^Y(c \mid u)$. We shall consider the case $Y = 1^k$, the vertical k boxes case. Then L^{1^k} is a matrix of size $\dim \bigwedge^k \mathbb{C}^n$ whose matrix element is a difference operator. We denote the Jacobi theta function by

$$\theta(u) = \sqrt{-1}p^{1/8}(z^{1/2} - z^{-1/2}) \prod_{m=1}^{\infty} (1 - zp^m)(1 - z^{-1}p^m)(1 - p^m)$$

with $p^{1/8} = \exp(2\pi i\tau/8)$ and $z^{1/2} = \exp(2\pi iu/2)$.

Theorem 2.   i) *Let* $M^{(k)}(c \mid u) := \mathrm{Trace}_{V(1^k)} L^{1^k}(c \mid u)$, $k = 1, \dots, n$. *Then we have*

$$M^{(k)}(c \mid u) = \frac{\theta(u + kc\hbar/n)}{\theta(u)} \sum_{\substack{I \subset \{1,\dots,n\} \\ |I|=k}} \left(\prod_{\substack{s \notin I \\ t \in I}} \frac{\theta(\lambda_{s,t} + c\hbar/n)}{\theta(\lambda_{s,t})} \right) T_I^\hbar, \quad (6)$$

where $\lambda_{s,t} := \langle \lambda, \bar{\epsilon}_s - \bar{\epsilon}_t \rangle$ *and* $T_I^\hbar$ *stands for the $\hbar$-shift operator:*

$$(T_i^\hbar f)(\lambda) := f(\lambda + \hbar\bar{\epsilon}_i), \quad T_I^\hbar := \prod_{i \in I} T_i^\hbar.$$

ii) *For any* $j, k = 1, \dots, n$, *we have*

$$[M^{(j)}(c \mid u), M^{(k)}(c \mid u)] = 0.$$

iii) *(cf. [9]) Suppose* $q := \exp 2\pi i\hbar$ *satisfies* $|q| < 1$ *and put* $g := c/n$. *Let*

$$d^+(z) := \prod_{k=0}^\infty \prod_{m=0}^\infty \frac{1 - zq^{m+1}p^k}{1 - zq^{m+g+1}p^k} \frac{1 - z^{-1}q^{m-g}p^{k+1}}{1 - z^{-1}q^m p^{k+1}}, \quad (7)$$

and define a function on $\mathfrak{h}^*$ *by* $\Phi(\lambda) := \prod_{k \neq k'} d^+(z_k/z_{k'})$, *where* $p = \exp 2\pi i\tau$ *and* $z_j := \exp 2\pi i\langle \lambda, \bar{\epsilon}_j \rangle$. *Then the conjugation by the square root* $\Phi^{1/2}$ *yields the Ruijenaars' [27] commuting operators:*

$$\left(\frac{\theta(u + kc\hbar/n)}{\theta(u)} \right)^{-1} \Phi^{-1/2} M^{(k)}(c \mid u) \Phi^{1/2}$$

$$= \sum_{|I|=k} \left(\prod_{\substack{s \notin I \\ t \in I}} \sqrt{\frac{\theta(c\hbar + \lambda_{s,t})}{\theta(\lambda_{s,t})}} \right) T_I^\hbar \left(\prod_{\substack{s \notin I \\ t \in I}} \sqrt{\frac{\theta(c\hbar + \lambda_{t,s})}{\theta(\lambda_{t,s})}} \right).$$

In the operator $M^{(k)}(c \mid u)$ (6), the spectral parameter u appears only in the overall factor. It is easy to see that, in the trigonometric limit $p \to 0$ the commuting system (6) falls into the Macdonald one with the parameters $q = \exp 2\pi i\hbar$ and $t = q^{-c/n}$.

An important remark is that these operators obviously commute, as is mentioned in the introduction. This is because the extended "$RLL = LLR$" relation (5) can be rewritten as

$$\check{R}^{Y,Y'}(u - v)L^Y(c \mid u) \otimes L^{Y'}(c \mid v)\check{R}^{Y,Y'}(u - v)^{-1}$$

$$= L^{Y'}(c \mid v) \otimes L^Y(c \mid u),$$

and then taking the trace simply gives

$$M^Y(c \mid u)M^{Y'}(c \mid v) = M^{Y'}(c \mid v)M^Y(c \mid u).$$

198   Koji Hasegawa

where $M^Y(c \mid u) := \mathrm{Trace}_{V(Y)} L^Y(c \mid u)$. This simple argument and the
resulting operators ("commuting transfer matrices") were effectively used
in Baxter's analysis of the spin chain models [2, 3]. From the viewpoint of
spin chain models, a solution of the Yang–Baxter equation is believed to
define a solvable lattice model. Since our L-operator solves the Yang–Baxter
equation, we may regard the function space on which the operators $\{L(c \mid
u)^i_j\}$ will act as the space of freedom for the site variables. Actually we can
consider the tensor products ("chain") of the function spaces consisting of
arbitrary number (say N) of components, on which the Nth tensor product
of the L-operator will acts (using the coproduct of the algebra $A(R)$). That
is, the operator $L(c \mid u)$ gives a function space realizaion (a Schrödinger
picture) of the generalized Heisenberg finite spin chain (with the periodic
boundary condition). Thus we may state the ideology:

commuting transfer matrices = commuting difference system.

Functions in $\lambda \in \mathbf{h}^*$ of course form an infinite-dimensional vector space, but
as we will see in the next section there exists a certain finite-dimensional
subspace that can be regarded as a function space realization of the sym-
metrically fused generalized XYZ spin chain.

Among the statements of the theorem, the computation of the trace
(Theorem 2 i) is rather nontrivial and uses the formula below, which is
interesting itself.

Lemma 1. *Let $Y_{r<s} := 1$ if $r < s$ holds, $Y_{r<s} := 0$ otherwise. The following formula holds:*

$$\det\left[\prod_{r=1}^{d} \theta\big(\mu_r - \lambda_{s'} + \hbar Y_{r<s} + \delta_{r,s}(u-(s-1)\hbar)\big)\right]_{s,s'=1}^{d}$$
$$= \prod_{s=1}^{d-1} \theta(u - s\hbar) \prod_{\substack{s,s'=1 \\ s<s'}}^{d} \theta(\lambda_{s'} - \lambda_s)\theta(\hbar + \mu_s - \mu_{s'}). \quad (8)$$

Proof of this formula can be done by using the standard complex analysis
and the induction on k. The $\hbar = 0$ case of (8) is easily transformed into
the Cauchy-type determinant formula of Frobenius [15]:

$$\det\left[\frac{\theta(\mu_s - \lambda_{s'} + u)}{\theta(\mu_s - \lambda_{s'})\theta(u)}\right]_{s,s'=1}^{d} = \frac{\prod_{\substack{s,s'=1 \\ s<s'}}^{d} \theta(\mu_s - \mu_{s'})\theta(\lambda_{s'} - \lambda_s)}{\prod_{s,s'=1}^{d} \theta(\mu_s - \lambda_{s'})}.$$

The formula (8) also turns out to be useful to derive the complete generators
of commuting differential operators in the limit $\hbar \to 0$.

Rescently I learned from V. Tarasov that (8) can be obtained as a special
case of the series of determinant formulas shown in Ref. 31, Appendix B.

3   An Invariant Subspace Spanned by Symmetric Theta Functions

Fix a nonnegative integer $l \in \mathbb{Z}_{>0}$. Let $Th_l^{S_n}$ be the vector space consisting of level l integrable $A_{n-1}^{(1)}$-characters [21]. This space is of dimension $\binom{l+n-1}{l}$ and spanned by the theta series on $\mathbf{h}^*$ that are invariant under the action of the symmetric group S_n. Then we have—

Theorem 3 ([16]). *For any i, $j = 1, \ldots, n$ we have $L(l \mid u)_j^i Th_l^{S_n} \subset Th_l^{S_n}$, hence $M^{(k)}(l \mid u) Th_l^{S_n} \subset Th_l^{S_n}$. As an $A(R)$-module, $Th_l^{S_n}$ is isomorphic to the representation $V(\square \overset{l}{\cdots} \square)$, the module corresponding to the Young diagram of l horizontal boxes.*

As an elliptic analogue of Macdonald's theory, we may define a family of orthogonal polynomials as the simultaneous eigenfunctions for our operators $M^{(k)}(c \mid u)$. In the trigonometric case, this diagonalization problem can be solved as one observes the triangularity of the action of the Macdonald operators with respect to the so-called orbitsum symmetric polynomials. In the elliptic case, we have the analogue of the orbitsum polynomials, namely, the symmetric theta series for the affine Weyl group of type $A_{n-1}^{(1)}$. One can calculate the action of the Ruijsenaars's operators for this basis as follows, but it turns out that the triangularity property fails unfortunately.

Recall that Q and P stand for the root lattice and the weight lattice, respectively. For $\mu \in \mathbf{h}^*$, $l \in \mathbb{Z}_{>0}$, put

$$\Theta_{\mu,l}(\lambda) := \sum_{\nu \in \mu + lQ} \exp 2\pi \sqrt{-1} \left[\langle \nu, \lambda \rangle + \frac{\langle \nu, \nu \rangle}{2l} \tau \right]$$

and define the affine orbitsum function, the affine Schur function, respectively, by

$$m_{\mu,l}(\lambda) := \sum_{w \in S_n} \Theta_{w(\mu),l}(\lambda), \quad S_{\mu,l}(\lambda) := \frac{A_{\mu+\rho,l+n}(\lambda)}{A_{\rho,n}(\lambda)},$$

where

$$A_{\mu,l}(\lambda) := \sum_{w \in S_n} \mathrm{sgn}(w) \Theta_{w(\mu),l}(\lambda),$$

and $\rho := \sum_{i=1}^{n-1}(n-i)\bar{\epsilon}_i$ stands for the half-sum of positive roots for A_{n-1}. It is easy to see that the space $Th_l^{S_n}$ has the basis $\{m_{\mu,l} \mid \mu \in P^+(l)\}$, where $P^+(l)$ denotes the set of dominant integral weights $\mu \in P^+$ that satisfy $\langle \mu, \rho \rangle \le l$.

Proposition 1. *For $\mu \in P^+(l)$, $k = 1, \ldots, n$, we have*

$$
M^{(k)}(l \mid u)m_{\mu,l}
$$
$$
= \frac{\theta(u + lk\hbar/n)}{\theta(u)} \sum_{w \in S_n, \alpha} \left(\sum_{|K|=k} T_K^{\hbar/n} \Theta_{l\rho - nw(\mu+l\alpha), nl(n+l)}(0) \right) S_{w(\mu+l\alpha),l},
$$

where α runs over $Q/(l+n)Q$, K runs over the subsets of $\{1, \ldots, n\}$ consisting of k elements.

One can show this in the same manner as the calculation performed in Ref. 25, pp. 315, 316. However, the transition matrix between the orbit-sums $m_{\nu,l}$ and the Schur functions $S_{\nu,l}$ is not triangular any more. (Since the entries of the transition matrix are nothing but the so-called string functions [21], the above formula provides some information on the string functions.) The summation over α also reflects the affine nature of the problem; it comes from the quotient of the Cartesian product of two affine Weyl groups by a certain subgroup. According to these two facts the naive triangularity fails for the elliptic Ruijsenaars operators. This difficulty is not an accident but can be regarded as another form of the "absense of the vacuum" in the XYZ spin chain. This model was solved by Takhtajan and Faddeev [30] in the face formulation, which is applied to the two-body ($n = 2$) case of Ruijsenaars's system by Felder and Varchenko [13].

Acknowledgments: It is a great pleasure of the author to express the gratitude for the hospitality at the Centre de recherches mathématiques, Université de Montréal, especially for Dr. J. F. van Diejen. He also thanks Professor B. Enriquez, Dr. B. Jurčo, Professor S. Ruijsenaars, Professor V. Tarasov, and Professor A. Varchenko for discussion.

4   References

1. A. Antonov, K. Hasegawa, and A. Zabrodin, *On trigonometric intertwining vectors and nondynamical R-matrix for the Ruijsenaars model*, Nucl. Phys. B **503** (1997), No. 3, 747–770, hep-th/9704074.

2. R. J. Baxter, *Partition function of the eight-vertex lattice model*, Ann. Phys. (NY) **70** (1972), 193–228.

3. _____, *Exactly Solved Models in Statistical Mechanics*, Academic Press, London, 1982.

4. A. A. Belavin, *Dynamical symmetry of integrable quantum systems*, Nucl. Phys. B **180** (1981), No. 2, FS 2, 189–200.

5. I. Cherednik, *On the properties of factorized S-matrices in elliptic functions*, Sov. J. Nucl. Phys. **36** (1982), No. 2, 320–324.

6. _____, *Some finite-dimensional representation of generalized Sklyanin algebras*, Funct. Anal. Appl. **19** (1985), No. 1, 77–79.

7. _____, *"Quantum" deformations of irreducible finite-dimensional representations of* $\mathfrak{gl}_N$, Soviet Math. Dokl. **33** (1986), No. 2, 507–510.

8. _____, *Double affine Hecke algebras, Knizhnik–Zamolodchikov equations, and Macdonald's operators*, Internat. Math. Res. Notices (1992), No. 9, 171–180.

9. J. F. van Diejen, *Commuting difference operators with polynomial eigenfunctions*, Compositio Math. **95** (1995), No. 2, 183–233.

10. V. G. Drinfeld, *Quantum groups*, Proc. International Congress of Mathematicians (Berkeley, CA, 1986) (A. M. Gleason, ed.), Amer. Math. Soc., Providence, RI, 1987, pp. 798–820.

11. P. I. Etingof and A. A. Kirillov, Jr., *Macdonald's polynomials and representations of quantum groups*, Math. Res. Lett. **1** (1994), No. 3, 279–296.

12. G. Felder, *The KZB equations on Riemann surfaces*, Symétries quantiques (Les Houches, 1995) (A. Connes, K. Gawedzki, and J. Zinn-Justin, eds.), North-Holland, Amsterdam, 1998, pp. 687–725.

13. G. Felder and A. N. Varchenko, *Algebraic Bethe ansatz for the elliptic quantum group* $E_{\tau,\eta}(\mathrm{sl}_2)$, Nucl. Phys. B **480** (1996), No. 1-2, 485–503, q-alg/9605024.

14. G. Felder and C. Wieczerkowski, *Conformal blocks on elliptic curves and the Knizhnik–Zamolodchikov–Bernard equations*, Commun. Math. Phys. **176** (1996), No. 1, 133–161.

15. G. Frobenius, *Über die elliptischen functionen zweiter art*, J. Reine Angew. Math. **93** (1882), 53–68.

16. K. Hasegawa, *L-operator for Belavin's R-matrix acting on the space of theta functions*, J. Math. Phys. **35** (1994), No. 11, 6158–6171.

17. _____, *Ruijsenaars' commuting difference operators as commuting transfer matrices*, Commun. Math. Phys. **187** (1997), No. 2, 289–325, q-alg/9512029.

18. T. Hollowood, *Quantizing* sl(n) *solitons and the Hecke algebra*, Internat. J. Modern Phys. A **8** (1993), No. 5, 947–981.

19. M. Jimbo, *A q-difference analogue of U($\mathfrak{g}$) and the Yang–Baxter equation*, Lett. Math. Phys. **10** (1985), No. 1, 63–69.

20. M. Jimbo, T. Miwa, and M. Okado, *Local state probabilities of solvable lattice models: an $A_{n-1}^{(1)}$ family*, Nucl. Phys. B **300** (1988), No. 1, FS 22, 74–108.

21. V. G. Kac, *Infinite-dimensional Lie algebras*, 3rd ed., Cambridge Univ. Press, Cambridge, 1990.

22. P. P. Kulish, N. Yu. Reshetikhin, and E. K. Sklyanin, *Yang–Baxter equation and representation theory.* I, Lett. Math. Phys. **5** (1981), No. 5, 393–403.

23. I. G. Macdonald, *A new class of symmetric functions*, Publ Inst. Res. Math. Av. (1988), 131–171.

24. _____, *Orthogonal polynomials associated with root systems*, Orthogonal Polynomials (Columbus, OH, 1989) (P. Nevai, ed.), NATO Adv. Sci. Inst. Ser. C Math. Phys. Sci., Vol. 294, Kluwer Acad. Publ., Dordrecht, 1990, pp. 311–318.

25. _____, *Symmetric Functions and Hall Polynomials*, 2nd ed., Oxford Math. Monogr., Oxford Univ. Press, New York, 1995.

26. N. Yu. Reshetikkin, *Multiparameter quantum groups and twisted quasitriangular Hopf algebras*, Lett. Math. Phys. **20** (1990), No. 4, 331–335.

27. S. N. M. Ruijsenaars, *Complete integrability of relativistic Calogero–Moser systems and elliptic function identities*, Commun. Math. Phys. **110** (1987), No. 2, 191–213.

28. E. K. Sklyanin, *Some algebraic structure connected with the Yang–Baxter equation*, Funct. Anal. Appl. **16** (1982), No. 4, 263–270.

29. _____, *Some algebraic structures connected with the Yang–Baxter equation. Representations of quantum algebras*, Funct. Anal. Appl. **17** (1983), No. 4, 273–284.

30. L. A. Takhtajan and L. D. Faddeev, *The quantum method of the inverse problem and the Heisenberg XYZ model*, Russian Math. Surveys **34** (1979), No. 5, 11–68.

31. V. O. Tarasov and A. N. Varchenko, *Geometry of q-Hypergeometric Functions, Quantum Affine Algebras and Elliptic Quantum Groups*, Astérisque, Vol. 246, Soc. Math. France, Paris, 1997, q-alg/9703044.

14

Invariants and Eigenvectors for Quantum Heisenberg Chains with Elliptic Exchange

V. I. Inozemtsev

ABSTRACT It is shown that for a one-parameter set $\mathcal{H}_N$ of linear combinations of $N(N-1)/2$ elementary transpositions $\{P_{jk}\}$ $(1 \leq j < k \leq N)$ at arbitrary natural $N \geq 3$, one can construct a variety $\{I_m\}$ $(3 \leq m \leq N)$ of operators that commute with $\mathcal{H}_N$. Applied to SU(2) spin representations of the permutation group, this proves the integrability of 1D periodic spin chains with elliptic short-range interaction. The eigenvectors of the spin Hamiltonian are connected with the double Bloch meromorphic solutions of the quantum continuous elliptic Calogero–Moser problem.

1 Introduction

Several years ago, in Ref. 11 I proposed the Lax representation and found examples of invariants of motion for the model of Heisenberg ferromagnet on a 1D periodic N-site lattice with SU(2) spin exchange given by the elliptic Weierstrass $\wp$ function. The symmetry of two limiting cases of this one-parameter model, that is the Bethe lattice with nearest-neighbor interaction [2] and long-range r^{-2} exchange [9, 17], is now well understood and regular procedures of finding invariants are described in the literature [1, 7, 8, 18]. These procedures, based on either the general Yang–Baxter scheme [1, 7] or simple form of exchange integrals and special properties of one of admissible Lax pairs in the long-range limit [8, 18], still cannot be applied to the general elliptic interaction.

The problem is that, contrary to the situation in classical dynamics, the quantum Lax relation $[\mathcal{H}, L] = [L, M]$ does not produce invariants of the motion in usual form of $\mathrm{tr}(L^l)$ since operator entries of the matrices L and M do not commute.

Here I would like to describe a possible way of constructing invariants in the case of elliptic exchange from the elements of L matrix for any representation of the permutation group π_N. This gives the direct proof of the integrability of the spin model proposed in Ref. 11.

In Section 2 I introduce the notation and formulate some properties of the solutions to the basic functional equation (statements A–D), the main statement (Proposition 1) and one of its consequences (Proposition 2). The proofs are given in Section 3. In Section 4, I describe the eigenvectors of the quantum spin chains with elliptic exchange which are connected with double Bloch solutions to the quantum continuous Calogero–Moser problem.

2   Notation and Statements

Let P_{jk} $(1 \leq j \neq k \leq N)$ be operators of elementary transpositions in ordered sequences of N symbols. Introduce the notation $\psi_{jk} \equiv \psi(j - k)$ for any function of the difference of numbers j and k. Define the Hamiltonian $\mathcal{H}_N$ as a linear combination of P's,

$$\mathcal{H}_N = \frac{1}{2} \sum_{\substack{j,k=1 \\ j \neq k}}^{N} h_{jk} P_{jk}. \tag{1}$$

As has been shown in Ref. 11, the Lax relation holds for 1 if L and M are chosen as

$$L_{jk} = (1 - \delta_{jk}) f_{jk} P_{jk}, \quad M_{jk} = (1 - \delta_{jk}) g_{jk} P_{jk} - \delta_{jk} \sum_{s \neq j}^{N} h_{jk} P_{jk}$$

and f, g, h obey the functional Calogero–Moser equation

$$f_{pq} g_{qr} - g_{pq} f_{qr} = f_{pr}(h_{qr} - h_{pq}) \tag{2}$$

supplemented by the periodicity condition

$$h'_{pq} = h'_{p,q+N}, \tag{3}$$

where $h'(x)$ is an odd function of its argument,

$$h'_{pq} = f_{qp} g_{pq} - f_{pq} g_{qp}. \tag{4}$$

It is worth noting for convenience the properties of f and h that have been described earlier [11, 14]. Namely, it follows directly from (2) that there is the normalization of f and h that allows one to write the relations

$$h(x) = f(x) f(-x), \quad g(x) = \frac{df(x)}{dx}, \quad h'(x) = \frac{dh(x)}{dx}. \tag{5}$$

The most general analytic solution to (2), (3) (up to an unessential factor e^{bx} of f) normalized as in (5) is

$$f(x) = \frac{\sigma(x + \alpha)}{\sigma(x)\sigma(\alpha)} \exp(-x\zeta(\alpha)), \quad h(x) = \wp(\alpha) - \wp(x). \tag{6}$$

The Weierstrass functions σ, ζ and $\wp$ in (6) are defined on the torus $\mathbb{T}_N = \mathbb{C}/N\mathbb{Z} + ia\mathbb{Z}$, $a \in \mathbb{R}_+$ is the free parameter of the model and $\alpha \in \mathbb{T}_N$ is the "spectra" parameter which does not contribute to exchange dynamics given by (1) but, as it will be shown later, is essential for constructing the whole set of invariants.

Let C_l be a group of cyclic permutations of subindices $(1, \ldots, l)$ in the objects of the form $B_{s_1 \cdots s_l}$. The operators $P_{s_1 \cdots s_l} \equiv P_{s_1 s_2} P_{s_2 s_3} \cdots P_{s_{l-1} s_l}$ and functions $F_{s_1 \cdots s_l} = f_{s_1 s_2} f_{s_2 s_3} \cdots f_{s_{l-1} s_l} f_{s_l s_1}$ are invariant under the action of elements of C_l. Denote as $\Phi(s_1, \ldots, s_l)$ the functions that are completely symmetric in their arguments.

The symbol $\sum_{C \in C_l} B_{s_1 \cdots s_l}$ will be used for summation over all cyclic permutations of the subindices of $B_{s_1 \ldots, s_l}$. The sum over all sets $(s_1, \ldots, s_l)$, $1 \leq s_1, \ldots, s_l \leq N$, such that the product $\prod_{\substack{\alpha,\beta=1 \\ \alpha<\beta}}^{l} (s_\alpha - s_\beta)$ does not vanish, will be denoted as $\sum_{s_1 \neq \cdots \neq s_l}$.

Among the properties of combinations of the objects introduced above I mention the following to be used in the subsequent section:

A. *The functions $h(x)$ and $h'(x)$ obey the relation*

$$\sum_{C \in C_3} h'_{s_1 s_2} (h_{s_1 s_3} - h_{s_2 s_3}) = 0. \tag{7}$$

B. *The sum $F^{(C)}_{s_1 \cdots s_{l+1}} = \sum_{C \in C_l} F_{s_1 \cdots s_l s_{l+1}}$ does not depend on s_{l+1}.*

C. *The sum $\sum_{s_1 \neq \cdots \neq s_{l+1}} \Phi(s_1, \ldots, s_{l+1}) F_{s_1 \cdots s_l} (h_{s_{l+1} s_l} - h_{s_{l+1} s_1}) P_{s_1 \cdots s_{l+1}}$ vanishes for any symmetric function Φ.*

D. *The sum*

$$S_l(\Phi) = \sum_{s_1 \neq \cdots \neq s_l} F_{s_1 \cdots s_l} \Phi(s_1, \ldots, s_l)(h_{s_l s_{l-1}} - h_{s_l s_1}) P_{s_1 \cdots s_{l-1}}$$

has a representation in the form $S_{l-1}^{(1)}(\Phi) + S_{l-1}^{(2)}(\Phi)$, where

$$S_{l-1}^{(1)}(\Phi) = \sum_{s_1 \neq \cdots \neq s_{l-1}} F_{s_1 \cdots s_{l-1}}$$
$$\times \left(\frac{1}{l-1} \sum_{p \neq s_1, \ldots, s_{l-1}}^{N} \Phi(s_1, \ldots, s_{l-1}, p) \sum_{\nu=1}^{l-1} h'_{ps_\nu} \right) P_{s_1 \cdots s_{l-1}} \tag{8}$$

and

$$S_{l-1}^{(2)}(\Phi) = \sum_{s_1 \neq \cdots \neq s_{l-1}} F_{s_1 \cdots s_{l-2}} (h_{s_{l-1} s_{l-2}} - h_{s_{l-1} s_1})$$
$$\times \left(\sum_{p \neq s_1, \ldots, s_{l-1}} h_{s_{l-1} p} \Phi(s_1, \ldots, s_{l-1}, p) \right) P_{s_1 \cdots s_{l-1}} \quad \text{if } l > 3, \tag{9}$$

$$S_2^{(2)}(\Phi) = -\frac{1}{2}\sum_{s_1 \neq s_2} h'_{s_1 s_2} \sum_{p \neq s_1, s_2}^{N} (h_{s_1 p} - h_{s_2 p})\Phi(s_1, s_2, p)P_{s_1 s_2}. \tag{10}$$

The following main statement can be proved without using the specific form (6) of the solutions to the Calogero–Moser equation.

Let I_m ($3 \leq m \leq N$) be the following linear combinations of the operators of cyclic permutations in ordered sequences of N symbols,

$$I_m = \sum_{l=0}^{[m/2]-1} \frac{(-1)^l}{m-2l} \sum_{s_1 \neq \cdots \neq s_{m-2l}} \Phi^{(l)}(s_1, \ldots, s_{m-2l})F_{s_1 \cdots s_{m-2l}}P_{s_1 \cdots s_{m-2l}}. \tag{11}$$

Proposition 1. *The operators I_m commute with $\mathcal{H}_N$ given by (1) if the functions $\Phi^{(l)}$ are determined by the recurrence relation*

$$\Phi^{(0)} = 1,$$

$$\Phi^{(l)}(s_1, \ldots, s_{m-2l}) = l^{-1} \sum_{\substack{1 \leq j < k \leq N \\ j,k \neq s_1, \ldots, s_{m-2l}}} h_{jk}\Phi^{(l-1)}(s_1, \ldots, s_{m-2l}, j, k) \tag{12}$$

or, equivalently, are given by sums over $2l$ indices

$$\Phi^{(l)}(s_1, \ldots, s_{m-2l}) = (l!)^{-1} \sum_{\substack{1 \leq j_\alpha < k_\alpha \leq N \\ \{j,k\} \neq s_1, \ldots, s_{m-2l}}} \lambda_{\{jk\}} \prod_{\alpha=1}^{l} h_{j_\alpha k_\alpha}, \tag{13}$$

where $\lambda_{\{jk\}}$ equals 1 if the product $\prod_{\alpha \neq \beta}^{l}(j_\alpha - j_\beta)(k_\alpha - k_\beta)(j_\alpha - k_\beta)$ differs from zero and vanishes otherwise.

The use of (6) in the construction (11)–(13) provides a more detailed information about the structure of the whole set of invariants.

Proposition 2. *The operators (11) can be written in the form*

$$I_m = w_m(\alpha)P_m + \sum_{\mu=1}^{m-2} w_{m-\mu}(\alpha)I_{m,\mu} + I_{m,m}. \tag{14}$$

Here P_m commute with all the operators of transpositions $\{P_{jk}\}$ and $I_{m,\mu}$ are linear combinations of $\{P_{s_1 \cdots s_{m-2l}}\}$. The coefficients in these combinations, treated as functions of their indices, are linearly independent. They do not contain the spectral parameter α entering into formula (6) for $f(x), h(x)$. The elliptic functions $w_l(\alpha)$ have the form

$$w_l(\alpha) = [\wp'(\alpha)]^{\lambda_l}[\wp(\alpha)]^{\nu_l} \tag{15}$$

with $\lambda_l = l - 2[l/2]$, $\nu_l = 3[l/2] - l$.

Remark 1. Some of the operators $I_{m,\mu}$ can be reduced to invariants with lower values of the first index or vanish if one considers only some fixed irreducible representation of π_N. The problem of finding the number of nonvanishing functionally independent terms in (14) for a given irreducible representation remains unsolved. The same concerns the problem of proving the conjecture $[I_{m,\mu}, I_{m',\mu'}] = 0$ supported by explicit calculation of the commutator for $m = m' = 3$.

3   Proofs

3.1   Proofs of Auxiliary Statements

A. By using Eq. (2) and the definition (4) the summand in (7) can be written in the form

$$A_{s_1 s_2 s_3} - A_{s_2 s_3 s_1} + A_{s_2 s_1 s_3} - A_{s_1 s_3 s_2},$$

where $A_{s_1 s_2 s_3} = g_{s_1 s_2} f_{s_2 s_3} g_{s_3 s_1}$. The statement is now easily checked by straightforward summation over $C \in C_3$.   □

B. It follows from Eq. (2) that

$$F_{s_1 \cdots s_l}(h_{s_1 s_{l+1}} - h_{s_l s_{l+1}}) = \varphi_{s_1 s_2 \cdots s_{l+1}} - \varphi_{s_{l+1} s_1 \cdots s_l}, \qquad (16)$$

where

$$\varphi_{s_1 s_2 \cdots s_{l+1}} = f_{s_1 s_2} f_{s_2 s_3} \cdots f_{s_l s_{l+1}} g_{s_{l+1} s_1}. \qquad (17)$$

Consider s_{l+1} as a continuous variable. Then

$$\frac{\partial F^{(C)}_{s_1 \cdots s_{l+1}}}{\partial s_{l+1}} = \sum_{C \in C_l} (\varphi_{s_1 s_2 \cdots s_{l+1}} - \varphi_{s_{l+1} s_1 \cdots s_l})$$

$$= F_{s_1 \cdots s_l} \sum_{C \in C_l} (h_{s_1 s_{l+1}} - h_{s_l s_{l+1}}).$$

It is easy to see that the last sum vanishes, which validates the statement.   □

C. As a result of (16), one can write the statement in the form

$$\sum_{s_1 \neq \cdots \neq s_{l+1}} (\varphi_{s_1 s_2 \cdots s_{l+1}} - \varphi_{s_{l+1} s_1 \cdots s_l}) \Phi(s_1, \ldots, s_{l+1}) P_{s_1 \cdots s_{l+1}} = 0. \qquad (18)$$

The second term in the brackets can be reduced to the first one by cyclic change of summation variables: $s_{l+1} \to s_1, s_1 \to s_2, \ldots, s_l \to s_{l+1}$. Since both the factors behind the brackets are invariant under this operation, the left-hand side of (18) vanishes. The statement is proved.   □

D. As a consequence of Eq. 2, one can write

$$F_{s_1 \cdots s_l}(h_{s_l s_{l-1}} - h_{s_{l-1} s_1} + h_{s_{l-1} s_1} - h_{s_l s_1}) = \psi_l^{(1)} + \psi_l^{(2)}, \tag{19a}$$

where

$$\psi_l^{(1)} = F_{s_1 \cdots s_{l-1}}(f_{s_{l-1} s_l} g_{s_l s_{l-1}} - f_{s_l s_1} g_{s_1 s_l}), \tag{19b}$$

$$\psi_l^{(2)} = \varphi_{s_1 \cdots s_{l-1}}(h_{s_1 s_l} - h_{s_{l-1} s_l}). \tag{19c}$$

Substituting (19) into the definition of $S_l(\Phi)$, one finds

$$S_{l-1}^{(1)}(\Phi) = \sum_{s_1 \neq \cdots \neq s_l} \Phi(s_1, \ldots, s_l) F_{s_1 \cdots s_{l-1}} P_{s_1 \cdots s_{l-1}}$$
$$\times (f_{s_{l-1} s_l} g_{s_l s_{l-1}} - f_{s_l s_1} g_{s_1 s_l}), \tag{20}$$

$$S_{l-1}^{(2)}(\Phi) = \sum_{s_1 \neq \cdots \neq s_l} \Phi(s_1, \ldots, s_l) P_{s_1 \cdots s_{l-1}} \varphi_{s_1 \cdots s_{l-1}}(h_{s_1 s_l} - h_{s_{l-1} s_l}). \tag{21}$$

Let us first transform the right-hand side of (20). Note that the second term in the brackets acquires the form $f_{s_l s_{l-1}} g_{s_{l-1} s_l}$ if one makes the cyclic change σ_{l-1} of summation variables as $s_1 \to s_{l-1}, \ldots, s_{l-1} \to s_{l-2}$, which does not influence the factors in front of the brackets. Then, combining the terms in the brackets into $h'_{s_l s_{l-1}}$, symmetrizing the summand by all cyclic shifts of the summation variables $s_1, \ldots, s_{l-1}$ and changing s_l to p, one arrives at formula (8).

It is easy to see that the right-hand side of (21) at $l = 3$ acquires the form (10) after symmetrizing it in s_1, s_2 and making the change $s_3 \to p$. The case of $l > 3$ needs a bit more work. Namely, note that the operation σ_{l-1} transforms $\varphi_{s_1 s_2 \cdots s_{l-1}} h_{s_1 s_l}$ into $\varphi_{s_{l-1} s_1 \cdots s_{l-2}} h_{s_{l-1} s_l}$ leaving $\Phi(s_1, \ldots, s_l) P_{s_1 \cdots s_{l-1}}$ invariant. Then one arrives at formula (9) with the use of a relation of the type (16) and change of s_l to p. □

3.2   Proof of Proposition 1

Consider the commutator

$$J_n = \sum_{s_1 \neq \cdots \neq s_n} [\Phi(s_1, \ldots, s_n) F_{s_1 \cdots s_n} P_{s_1 \cdots s_n}, \mathcal{H}_N], \tag{22}$$

where Φ is symmetric in its variables. With the use of invariance of $F_{s_1 \cdots s_n}$ and $P_{s_1 \cdots s_n}$ under cyclic changes of summation variables it is easy to show that this commutator can be written as

$$J_n = n \left[J_n^{(1)} + J_n^{(2)} + \sum_{\nu=2}^{[n/2]} \left(1 - \left(\frac{n-1}{2} - \left[\frac{n}{2} \right] \right) \delta_{\nu, [n/2]} \right) J_{n,\nu}^{(3)} \right], \tag{23}$$

where

$$J_n^{(1)} = \sum_{s_1 \neq \cdots \neq s_{n+1}} \Phi(s_1, \ldots, s_n) F_{s_1 \cdots s_n}(h_{s_n s_{n+1}} - h_{s_1 s_{n+1}}) P_{s_1 \cdots s_{n+1}}, \qquad (24)$$

$$J_n^{(2)} = \sum_{s_1 \neq \cdots \neq s_n} \Phi(s_1, \ldots, s_n) F_{s_1 \cdots s_n}(h_{s_{n-1} s_n} - h_{s_1 s_n}) P_{s_1 \cdots s_{n-1}}, \qquad (25)$$

$$J_{n,\nu}^{(3)} = \sum_{s_1 \neq \cdots \neq s_n} \Phi(s_1, \ldots, s_n) F_{s_1 \cdots s_n}(h_{s_\nu s_n} - h_{s_1 s_{\nu+1}}) P_{s_1 \cdots s_\nu} P_{s_{\nu+1} \cdots s_n}. \quad (26)$$

Let us transform $J_{n,\nu}^{(3)}$ as follows. By using Eq. (2) one can write for $\nu \geq 2$

$$F_{s_1 \cdots s_n}(h_{s_\nu s_n} - h_{s_{\nu+1} s_n} + h_{s_{\nu+1} s_n} - h_{s_1 s_{\nu+1}})$$
$$= \varphi_{s_{\nu+1} \cdots s_{n-1} s_n}(F_{s_1 \cdots s_\nu s_{\nu+1}} - F_{s_1 \cdots s_\nu s_n})$$
$$+ F_{s_{\nu+1} \cdots s_n}(\varphi_{s_n s_1 \cdots s_\nu} - \varphi_{s_1 \cdots s_{\nu+1}}). \quad (27)$$

Substituting (27) into (26) and taking into account the invariance of $P_{s_1 \cdots s_\nu}$ and $P_{s_{\nu+1} \cdots s_n}$ under cyclic shifts of indexes, one finds

$$J_{n,\nu}^{(3)} = \sum_{s_1 \neq \cdots \neq s_n} \Phi(s_1, \ldots, s_n) [\nu^{-1} \varphi_{s_{\nu+1} \cdots s_{n-1} s_n}(F_{s_1 \cdots s_\nu s_{\nu+1}}^{(C)} - F_{s_1 \cdots s_\nu s_n}^{(C)})$$
$$+ F_{s_{\nu+1} \cdots s_n}(\varphi_{s_{\nu+1} s_1 \cdots s_\nu} - \varphi_{s_1 \cdots s_{\nu+1}})] P_{s_1 \cdots s_\nu} P_{s_{\nu+1} \cdots s_n}. \quad (28)$$

The term in the first brackets in 28 disappears owing to statement B. The term in the second brackets is reduced with the use of (16) to the form $F_{s_1 \cdots s_\nu}(h_{s_\nu s_{\nu+1}} - h_{s_1 s_{\nu+1}})$ and the corresponding sum vanishes, as it can be seen after symmetrization of the summand in all cyclic changes of $s_1, \ldots, s_\nu$. One concludes that $J_{n,\nu}^{(3)}$ vanishes, that is the commutator J_n is a linear combination of the operators of one-cycle permutations.

Now let us consider the commutator of $\mathcal{H}_N$ and the operator I_m defined by Eq. (11). It is easy to show with the use of (23)–(25) and statement D that it acquires the form

$$[I_m, \mathcal{H}_N] = W^{(1)} + \sum_{l=0}^{[m/2]-2} (-1)^l W_l^{(2)} + W^{(3)}, \qquad (29)$$

where

$$W^{(1)} = \sum_{s_1 \neq \cdots \neq s_m} \Phi^{(0)}(s_1, \ldots, s_m) F_{s_1 \cdots s_m}(h_{s_m s_{m+1}} - h_{s_1 s_{m+1}}) P_{s_1 \cdots s_{m+1}}, \qquad (30)$$

$$W_l^{(2)} = \sum_{s_1 \neq \cdots \neq s_{m-2l-1}} P_{s_1 \cdots s_{m-2l-1}} \{ F_{s_1 \cdots s_{m-2l-2}}(h_{s_{m-2l-2} s_{m-2l-1}} - m h_{s_1 s_{m-2l-1}}) U_l^{(1)}$$
$$+ (m - 2l - 1)^{-1} F_{s_1 \cdots s_{m-2l-1}} U_l^{(2)} \}, \qquad (31)$$

$$U_l^{(1)} = -\Phi^{(l+1)}(s_1, \ldots, s_{m-2l-2})$$
$$+ \sum_{p \neq s_1, \ldots, s_{m-2l-1}} h_{p s_{m-2l-1}} \Phi^{(l)}(s_1, \ldots, s_{m-2l-1}, p), \qquad (32)$$

$$U_l^{(2)} = \sum_{p\neq s_1,\ldots,s_{m-2l-1}} \Phi^{(l)}(s_1,\ldots,s_{m-2l-1},p) \sum_{\nu=1}^{m-2l-1} h'_{ps_\nu}. \tag{33}$$

The last term $W^{(3)}$ is present only for odd $m = 2k + 1$. It is written as

$$W^{(3)} = \frac{(-1)^k}{2} \sum_{s_1\neq s_2} P_{s_1 s_2} \Big\{ h'_{s_1 s_2} \sum_{p\neq s_1,s_2} \Phi^{(k-1)}(s_1,s_2,p)(h_{s_1 p} - h_{s_2 p})$$
$$- h_{s_1 s_2} U_{k-1}^{(2)} \Big\}. \tag{34}$$

Note that $W^{(1)}$ can vanish, in accordance with the statement C, if $\Phi^{(0)}(s_1,\ldots,s_m)$ is symmetric in $(s_1,\ldots,s_{m+1})$. It takes place if and only if $\Phi^{(0)}$ does not depend on summation indices, that is, can be chosen as 1 for convenience.

With the use of the statement C it is easy to see that the first term in braces in (31) also vanishes if $U_l^{(1)}$ is symmetric in its arguments. This results in the recurrence relation

$$\sum_{p\neq s_1,\ldots,s_{m-2l-1}} (h_{ps_{m-2l-1}} - h_{ps_1})\Phi^{(l)}(s_1,\ldots,s_{m-2l-1},p)$$

$$= \Phi^{(l+1)}(s_1,s_2,\ldots,s_{m-2l-2}) - \Phi^{(l+1)}(s_{m-2l-1},s_2,\ldots,s_{m-2l-2}). \tag{35}$$

One concludes that all the terms $W_l^{(2)}$ in (31) disappear if the functions $\Phi^{(l)}$ obey the relation (35) supplemented by the condition

$$U_l^{(2)} = 0. \tag{36}$$

It can be shown by straightforward calculation that (35) is satisfied for $l = 0$ provided that

$$\Phi^{(0)} = 1, \quad \Phi^{(1)} = \sum_{j<k; j, k\neq s_1,\ldots s_{m-2}} h_{jk}.$$

Taking this into account, it is easy to prove by induction that the solution to (35) for all $l \leq [m/2] - 2$ is given by formula (12).

The validity of (36) for $\Phi^{(0)}=1$ is simply a consequence of the periodicity (3) of an odd function $h'(x)$. Now let us suppose that (36) is satisfied for $\Phi^{(l)}(s_1,\ldots,s_{m-2l})$. Then the use of the recurrence relation (12) yields

$$U_{l+1}^{(2)} = [2(l+1)]^{-1}$$

$$\times \sum_{\substack{p_1\neq p_2\neq p_3 \\ p_{1,2,3}\neq s_1,\ldots,s_{m-2l-3}}} \Phi^{(l)}(s_1,\ldots,s_{m-2l-3},p_1,p_2,p_3)h_{p_2 p_3}(h'_{p_2 p_1} + h'_{p_3 p_1}). \tag{37}$$

Symmetrizing the summand in the right-hand side of (37) in indices p_1, p_2, p_3, one obtains the factor $\sum_{C \in C_3} h'_{p_1 p_2} (h_{p_1 p_3} - h_{p_2 p_3})$ in it. This, owing to statement A, implies that $U^{(2)}_{l+1}$ vanishes and (36) is valid for all the elements of the sequence given by (12).

It is enough now to consider the first term in the braces in (34). Let us combine the relations (35) with $l = k - 1$ and (12) with $l = k$ and write

$$\sum_{p \neq s_1, s_2} \Phi^{(k-1)}(s_1, s_2, p)(h_{s_1 p} - h_{s_2 p}) = \Phi^{(k)}(s_2) - \Phi^{(k)}(s_1),$$

where

$$\Phi^{(k)}(s_1) = k^{-1} \sum_{\substack{1 \leq j_1 < j_2 \leq N \\ j_{1,2} \neq s_1}} h_{j_1 j_2} \Phi^{(k-1)}(s_1, j_1, j_2).$$

The function $\Phi^{(k-1)}$ is periodic and depends only on the differences of its arguments as it can be seen from (12) or (13). Therefore, $\Phi^{(k)}(s_1)$ does not vary with s_1 and $W^{(3)}$ vanishes. This completes the proof of Proposition 1. ☐

3.3 Proof of Proposition 2

It follows from the explicit expression (6) for $f(x)$ that $F_{s_1 \cdots s_{m-2l}} = f_{s_1 s_2} \times \cdots f_{s_{m-2l-1} s_{m-2l}} f_{s_{m-2l} s_1}$ is the elliptic function of the spectral parameter α which has the pole of the order $(m - 2l)$ at $\alpha = 0$ as the only singularity on the torus $\mathbb{T}_N$. Therefore, one can write

$$F_{s_1 \cdots s_{m-2l}} = c_{m-2l} w_{m-2l}(\alpha) + \sum_{\nu=1}^{m-2l-2} f_\nu(s_1, \ldots, s_{m-2l}) w_{m-2l-\nu}(\alpha) + f_{m-2l}(s_1, \ldots, s_{m-2l}), \quad (38)$$

where $w_\lambda(\alpha)$ are elementary elliptic functions given by (15), c_{m-2l} is a constant, and $f_1, \ldots, f_{m-2l-2}, f_{m-2l}$ are $(m - 2l - 1)$ functions of s_1, $\ldots, s_{m-2l}$, that are elliptic in each argument and do not depend on α. Being substituted into (10) and (11), the decompositions (38) provide the structure (14) of the operators I_m. As one can see from (38) and the explicit formula (6) for $h(x)$, the factor behind $w_m(\alpha)$ in the right-hand side of (14) is a linear combination of the uniform sums $\sum_{s_1 \neq \cdots \neq s_{m-2l}} P_{s_1 \cdots s_{m-2l}}$, i.e., the Casimir operators that commute with all $\{P_{jk}\}$. This completes the proof. ☐

4 The Eigenvectors

The Hamiltonian of the isotropic $S = \frac{1}{2}$ Heisenberg chain is

$$\mathcal{H}^{(s)} = \frac{J}{4} \sum_{\substack{j,k=1 \\ j \neq k}}^{N} h(j-k)(\vec{\sigma}_j \vec{\sigma}_k - 1) \quad h(j) = h(j+N), \tag{39}$$

where $\vec{\sigma}_j$ are Pauli matrices acting on spin at jth site. The corresponding mathematical problem consists in finding the proper analytic tool for its diagonalization.

At finite N, it has been successfully treated in the integrable cases of nearest-neighbor coupling solved by Bethe [2],

$$h(j) = \delta_{|j \,(\text{mod}\,N)|,1}, \tag{40}$$

and long-range trigonometric exchange proposed independently by Haldane [9] and Shastry [17],

$$h(j) = \left(\frac{N}{\pi} \sin \left[\frac{\pi j}{N} \right] \right)^{-2}. \tag{41}$$

At present, a number of impressive results are known for both these models. In particular, they include the additivity of the spectrum under proper choice of "rapidity" variables [2, 9, 17], the description of underlying symmetry [1, 7], construction of thermodynamics in the limit $N \to \infty$ [10, 19], the connection to the continuum integrable many-body problems [9, 15, 17], and closed-form expressions of correlations in the antiferromagnetic ground state. The rich collection of various generalizations and physical applications of Bethe and Haldane–Shastry models can be found in recent review papers [13, 20].

There is a similarity of Heisenberg equations of motion for continuum and lattice models. In the former case, the most general translationally invariant integrable Hamiltonian with elliptic pairwise particle interaction has been found by Calogero [3] and Moser [16],

$$H_{CM} = \frac{1}{2} \left[- \sum_{\beta=1}^{L} \frac{\partial^2}{\partial x_\beta^2} + \lambda(\lambda+1) \sum_{\beta \neq \gamma}^{L} \wp(x_\beta - x_\gamma) \right]. \tag{42}$$

Recently, the eigenvalue problem for the elliptic Calogero–Moser operator received much attention owing to its relation to the representations of double affine algebras and solutions of Knizhnik–Zamolodchikov–Bernard equations [5, 6].

The lattice analog of (42) is given by (39) with

$$h(j) = \left(\frac{\omega}{\pi} \sin \left[\frac{\pi}{\omega} \right] \right)^2 \left[\wp_N(j) + \frac{2}{\omega} \zeta_N \left(\frac{\omega}{2} \right) \right], \tag{43}$$

where $\wp_N(x)$, $\zeta_N(x)$ are the Weierstrass functions defined on the torus $T_N = \mathbb{C}/\mathbb{Z}N + \mathbb{Z}\omega$, $\omega = i\kappa$, $\kappa \in \mathbb{R}_+$. Remarkably, it turned out that the exchange (43) comprises both (40) and (41) [11]: in fact, the factor in (43) is chosen as to reproduce the nearest-neighbor coupling under periodic boundary conditions (40) in the limit $\kappa \to 0$ and the long-range exchange (41) in the limit $\kappa \to \infty$.

However, till now much less was known about the lattice model with the exchange (5) in comparison with its limiting forms owing to the mathematical complexities caused by the presence of the Weierstrass functions. The simpler case of infinite chain $N \to \infty$, $h(j) \to [\sinh(\pi/\kappa)/\sinh(\pi j/\kappa)]^2$ has been considered in detail in [12]. As for finite N, the description of the spectrum has been performed only for simplest two- and three-magnon excitations over ferromagnetic vacuum.

I would like to demonstrate the remarkable correspondence between the highest-weight eigenstates of the lattice Hamiltonian with the elliptic exchange (43) and double quasiperiodic meromorphic eigenfunctions of the Calogero–Moser operator (42) that allows us to formulate the equations of the Bethe ansatz type for calculating the whole spectrum.

The Hamiltonian (39) commutes with the operator of total spin $\vec{S} = \frac{1}{2}\sum_{j=1}^{N} \vec{\sigma}_j$. Then the eigenproblem for it is decomposed into the problems in the subspaces formed by the common eigenvectors of $\mathbf{S}_3$ and $\vec{\mathbf{S}}^2$ such that $S = S_3 = N/2 - M$, $0 \le M \le [N/2]$,

$$\mathcal{H}^{(s)}|\psi^{(M)}\rangle = E_M|\psi^{(M)}\rangle. \qquad (44)$$

The eigenvectors $|\psi^{(M)}\rangle$ are written in the usual form

$$|\psi^{(M)}\rangle = \sum_{n_1,\ldots,n_M} \psi_M(n_1 \cdots n_M) \prod_{\beta=1}^{M} s_{n_\beta}^- |0\rangle, \qquad (45)$$

where $|0\rangle = |\uparrow\uparrow \cdots \uparrow\rangle$ is the ferromagnetic ground state with all spins up and the summation is taken over all combinations of integers $\{n\} \le N$ such that $\prod_{\mu<\nu}^{M}(n_\mu - n_\nu) \ne 0$. The substitution of (45) into (44) results in the lattice Schrödinger equation for completely symmetric wave function ψ_M

$$\sum_{s \ne n_1,\ldots,n_M} \sum_{\beta=1}^{M} \wp_N(n_\beta - s)\psi_M(n_1,\ldots,n_{\beta-1},s,n_{\beta+1},\ldots,n_M)$$
$$+ \left[\sum_{\beta \ne \gamma}^{M} \wp_N(n_\beta - n_\gamma) - \mathcal{E}_M\right]\psi_M(n_1,\ldots,n_M) = 0. \qquad (46)$$

The eigenvalues $\{E_M\}$ are given by

$$E_M = J\left(\frac{\omega}{\pi}\sin\left[\frac{\pi}{\omega}\right]\right)^2$$
$$\times \left\{\mathcal{E}_M + \frac{2}{\omega}\left[\frac{2M(2M-1) - N}{4}\zeta_N\left(\frac{\omega}{2}\right) - M\zeta_1\left(\frac{\omega}{2}\right)\right]\right\}, \qquad (47)$$

where $\zeta_1(x)$ is the Weierstrass zeta function defined on the torus $T_1 = \mathbb{C}/\mathbb{Z} + \mathbb{Z}\omega$.

To find the solutions to (46), let us consider the following ansatz for ψ_M:

$$\psi_M(n_1,\ldots,n_M) = \sum_{P \in \pi_M} \varphi_M^{(p)}(n_{P1},\ldots,n_{PM}), \tag{48}$$

$$\varphi_M^{(p)}(n_1,\ldots,n_M) = \exp\left(-i\sum_{\nu=1}^{M} p_\nu n_\nu\right)\chi_M^{(p)}(n_1,\ldots,n_M), \tag{49}$$

where π_M is the group of all permutations $\{P\}$ of the numbers from 1 to N and $\chi_M^{(p)}$ is the solution to the *continuum* quantum many-particle problem

$$\left[-\frac{1}{2}\sum_{\beta=1}^{M}\frac{\partial^2}{\partial x_\beta^2} + \sum_{\beta \neq \lambda}^{M}\wp_N(x_\beta - x_\lambda) - \mathsf{E}_M(p)\right]\chi_M^{(p)}(x_1,\ldots,x_M) = 0. \tag{50}$$

It is specified up to a normalization factor by the particle pseudomomenta $(p_1,\ldots,p_M)$. The standard argumentation of the Floquet–Bloch theory shows that owing to perodicity of the potential term in (50) $\chi_M^{(p)}$ obeys the quasiperiodicity conditions [12]

$$\chi_M^{(p)}(x_1,\ldots,x_\beta + N,\ldots,x_M) = \exp(ip_\beta N)\chi_M^{(p)}(x_1,\ldots,x_M), \tag{51}$$

$$\chi_M^{(p)}(x_1,\ldots,x_\beta + \omega,\ldots,x_M) = \exp(q_\beta(p) + ip_\beta\omega)\chi_M^{(p)}(x_1,\ldots,x_M), \tag{52}$$
$$0 \leq \mathrm{Im}(q_\beta) < 2\pi, \ 1 \leq \beta \leq M.$$

The eigenvalue $\mathsf{E}_M(p)$ is some symmetric function of $(p_1,\ldots,p_M)$. The set $\{q_\beta(p)\}$ is also completely determined by $\{p\}$.

The structure of the singularity of $\wp_N(x)$ at $x = 0$ implies that $\chi_M^{(p)}$ can be presented in the form

$$\chi_M^{(p)} = \frac{F^{(p)}(x_1,\ldots,x_M)}{G(x_1,\ldots,x_M)}, \quad G(x_1,\ldots,x_M) = \prod_{\alpha < \beta}^{M} \sigma_N(x_\alpha - x_\beta), \tag{53}$$

where $\sigma_N(x)$ is the Weierstrass sigma function on the torus T_N. The only simple zero of $\sigma_N(x)$ on T_N is located at $x = 0$. Thus $[G(x_1,\ldots,x_M)]^{-1}$ absorbs all the singularities of $\chi_M^{(p)}$ on the hypersurfaces $x_\alpha = x_\beta$. The numerator $F^{(p)}$ in (52) is analytic on $(T_N)^M$ and obeys the equation

$$\sum_{\alpha=1}^{M}\frac{\partial^2 F^{(p)}}{\partial x_\alpha^2} + \left[2\mathsf{E}_M(p) - \frac{M}{2}\sum_{\alpha \neq \beta}^{M}(\wp_N(x_\alpha - x_\beta) - \zeta_N^2(x_\alpha - x_\beta))\right]F^{(p)}$$

$$= \sum_{\alpha \neq \beta}\zeta_N(x_\alpha - x_\beta)\left(\frac{\partial F^{(p)}}{\partial x_\alpha} - \frac{\partial F^{(p)}}{\partial x_\beta}\right). \tag{54}$$

The regularity of the left-hand side of (54) as $x_\mu \to x_\nu$ implies that

$$\left(\frac{\partial}{\partial x_\mu} - \frac{\partial}{\partial x_\nu} \right) F^P(x_1, \dots, x_M)|_{x_\mu} = x_\nu = 0 \qquad (55)$$

for any pair (μ, ν).

The remarkable fact is that the properties (51)–(53), and (55) of $\chi_M^{(p)}$ allow one to validate the ansatz (48) and (49) for the eigenfunctions of the lattice Schrödinger equation (46). Substitution of (48) into (46) yields

$$\sum_{P \in \pi_M} \left\{ \sum_{\beta=1}^{M} \mathcal{S}_\beta(n_{P1}, \dots, n_{PM}) \right.$$

$$\left. + \left[\sum_{\beta \neq \gamma}^{M} \wp_N(n_{P\beta} - n_{P\gamma}) - \mathcal{E}_M \right] \varphi_M^{(p)}(n_{P1}, \dots, n_{PM}) \right\} = 0, \qquad (56)$$

where

$$\mathcal{S}_\beta(n_{P1}, \dots, n_{PM}) = \sum_{s \neq n_{P1}, \dots, n_{PM}}^{N} \wp_N(n_{P\beta} - s) \hat{Q}_\beta^{(s)} \varphi_M^{(p)}(n_{P1}, \dots, n_{PM}).$$

$$(57)$$

The operator $\hat{Q}_\beta^{(s)}$ in (57) replaces βth argument of the function of M variables to s.

To calculate the sum (57), let us introduce, following the consideration of the hyperbolic exchange in Ref. 12, the function of one complex variable x,

$$W_P^{(\beta)}(x) = \sum_{s=1}^{M} \wp_N(n_{P\beta} - s - x) \hat{Q}_\beta^{(s+x)} \varphi_M^{(p)}(n_{P1}, \dots, n_{PM}). \qquad (58)$$

As a consequence of (49), (51), (52), it obeys the relations

$$W_P^{(\beta)}(x+1) = W_P^{(\beta)}(x), \qquad W_P^{(\beta)}(x+\omega) = \exp(q_\beta(p)) W_P^{(\beta)}(x). \qquad (59)$$

The only singularity of $W_P^{(\beta)}$ on the torus $T_1 = \mathbb{C}/\mathbb{Z} + \mathbb{Z}\omega$ is located at the point $x = 0$. It arises from the terms in (58) with $s = n_{P1}, \dots, n_{PM}$. The Laurent decomposition of (58) near $x = 0$ has the form

$$W_P^{(\beta)}(x) = w_{-2} x^{-2} + w_{-1} x^{-1} + w_0 + O(x). \qquad (60)$$

The explicit expressions for w_{-i} can be found from (20),

$$w_{-2} = \varphi_M^{(p)}(n_{P1}, \dots, n_{PM}) \qquad (61a)$$

$$w_{-1} = \frac{\partial}{\partial n_{P\beta}} \varphi_M^{(p)}(n_{P1}, \ldots, n_{PM})$$

$$+ (-1)^P G(n_1, \ldots, n_M) \sum_{\lambda \neq \beta} T_{\beta\lambda}(n_{P1}, \ldots, n_{PM}) \hat{Q}_\beta^{(n_{P\lambda})}$$

$$\times \exp\left(-i \sum_{\nu=1}^M p_\nu n_{P\nu}\right) F^{(p)}(n_{P1}, \ldots, n_{PM}) \qquad (61b)$$

$$w_0 = \mathcal{S}_\beta(n_{P1}, \ldots, n_{PM}) + \frac{1}{2} \frac{\partial^2}{\partial n_{P\beta}} \varphi_M^{(p)}(n_{P1}, \ldots, n_{PM})$$

$$+ (-1)^P G(n_1, \ldots, n_M) \sum_{\lambda \neq \beta} T_{\beta\lambda}(n_{P1}, \ldots, n_{PM})$$

$$\times [U_{\beta\lambda}(n_{P1}, \ldots, n_{PM}) \hat{Q}_\beta^{(n_{P\lambda})} + \wp_N(n_{P\beta} - n_{P\lambda}) \partial \hat{Q}_\beta^{(n_{P\lambda})}]$$

$$\times \exp\left(-i \sum_{\nu=1}^M p_\nu n_{P\nu}\right) F^{(p)}(n_{P1}, \ldots, n_{PM}), \qquad (61c)$$

where

$$T_{\beta\lambda}(n_{P1}, \ldots, n_{PM}) = \sigma_N(n_{P\lambda} - n_{P\beta}) \prod_{\rho \neq \beta, \lambda}^M \frac{\sigma_N(n_{P\rho} - n_{P\beta})}{\sigma_N(n_{P\rho} - n_{P\lambda})},$$

$$U_{\beta\lambda}(n_{P1}, \ldots, n_{PM}) = \wp_N'(n_{P\lambda} - n_{P\beta})$$

$$- \wp_N(n_{P\beta} - n_{P\lambda}) \sum_{\rho \neq \beta, \lambda} \zeta_N(n_{P\rho} - n_{P\lambda}),$$

$(-1)^P$ means the parity of the permutation P and the action of the operator $\partial \hat{Q}_\beta^{(n_{P\lambda})}$ on the function Y of M variables is defined as

$$\partial Q_\beta^{(n_{P\lambda})} Y(z_1, \ldots, z_M) = \frac{\partial}{\partial z_\beta} Y(z_1, \ldots, z_M)\Big|_{z_\beta = n_{P\lambda}}. \qquad (62)$$

The next step consists in writing the explicit expression for the function $W_P^{(\beta)}(x)$ obeying the relations (59) and (60) [12],

$$W_P^{(\beta)}(x) = \exp(a_\beta x) \frac{\sigma_1(r_\beta + x)}{\sigma_1(r_\beta - x)} \{w_{-2}(\wp_1(x) - \wp_1(r_\beta))$$

$$+ (w_{-2}(a_\beta + 2\zeta_1(r_\beta)) - w_{-1})$$

$$\times [\zeta_1(x - r_\beta) - \zeta_1(x) + \zeta_1(r_\beta) - \zeta_1(2r_\beta)]\}. \qquad (63)$$

The Weierstrass functions $\wp_1, \zeta_1$ and σ_1 in (63) are defined on the torus T_1 and the parameters a_β, r_β are chosen as to satisfy the conditions (59),

$$a_\beta = (\pi i)^{-1} q_\beta(p) \zeta_1(\tfrac{1}{2}) \qquad r_\beta = -(4\pi i)^{-1} q_\beta(p).$$

By expanding (63) in powers of x one can find w_0 in terms of w_{-2}, w_{-1}, q_β and obtain the explicit expression for $\mathcal{S}_\beta(n_{P1}, \ldots, n_{PM}$ with the use of (61a). It turns out that the equation (56) can be recast in the form

$$
\sum_{P \in \pi_M} \left[-\frac{1}{2} \sum_{\beta=1}^{M} \left(\frac{\partial}{\partial n_{P\beta}} - f_\beta(p) \right)^2 + \sum_{\beta \neq \gamma}^{M} \wp_N(n_{P\beta} - n_{P\gamma}) - \mathcal{E}_M + \sum_{\beta=1}^{M} \varepsilon_\beta(p) \right]
$$
$$
\times \, \varphi^{(p)}(n_{P1}, \ldots, n_{PM})
$$
$$
= \frac{1}{2} G(n_1, \ldots, n_M) \sum_{P \in \pi_M} (-1)^P \sum_{\beta \neq \lambda} [Z_{\beta\lambda}(n_{P1}, \ldots, n_{PM})
$$
$$
\qquad\qquad\qquad\qquad\qquad + Z_{\lambda\beta}(n_{P1}, \ldots, n_{PM})], \quad (64)
$$

where

$$
f_\beta(p) = (\pi i)^{-1} q_\beta(p) \zeta_1(\tfrac{1}{2}) - \zeta_1((2\pi i)^{-1} q_\beta(p)), \tag{65}
$$
$$
\varepsilon_\beta(p) = \frac{1}{2} \wp_1((2\pi i)^{-1} q_\beta(p)) \tag{66}
$$

and $Z_{\beta\lambda}(n_{P1}, \ldots, n_{PM})$ is defined by the relation

$$
Z_{\beta\lambda}(n_{P1}, \ldots, n_{PM}) = T_{\beta\lambda}(n_{P1}, \ldots, n_{PM})
$$
$$
\times [U_{\beta\lambda}(n_{P1}, \ldots, n_{PM}) \hat{Q}_\beta^{(n_{P\lambda})} \wp_N(n_{P\lambda} - n_{P\beta})(\partial \hat{Q}_\beta^{(n_{P\lambda})} - f_\beta(p) \hat{Q}_\beta^{(n_{P\lambda})})]
$$
$$
\times \exp\left(-i \sum_{\nu=1}^{M} p_\nu n_{P\nu} \right) F^{(p)}(n_{P1}, \ldots, n_{PM}). \tag{67}
$$

Turning to the definition (49) of $\varphi^{(p)}$, one observes that each term of the left-hand side of (64) has the same structure as the left-hand side of the many-particle Schrödinger equation (50) and vanishes if $\mathcal{E}_M$ and $f_\beta(p)$ are chosen as

$$
f_\beta(p) = -ip_\beta, \quad \beta = 1, \ldots, M, \tag{68}
$$
$$
\mathcal{E}_M = \mathsf{E}_M(p) + \sum_{\nu=1}^{M} \varepsilon_\beta(p). \tag{69}
$$

Now let us prove that that the right-hand side of (67) also vanishes. The crucial observation is that the sum over permutations in it can be recast in the form

$$
\sum_{P \in \pi_M} (-1)^P \sum_{\beta \neq \lambda} [Z_{\beta\lambda}(n_{P1}, \ldots, n_{PM}) - Z_{\lambda\beta}(n_{PR1}, \ldots n_{PRM})],
$$

where R is the transposition $(\beta \leftrightarrow \lambda)$ which leaves other numbers from 1 to M unchanged. The term in square brackets is simplified drastically with the use of the identities

$$T_{\lambda\beta}(n_{PR1},\ldots,n_{PRM}) = T_{\beta\lambda}(n_{P1},\ldots,n_{PM}),$$
$$U_{\lambda\beta}(n_{PR1},\ldots,n_{PRM}) = U_{\beta\lambda}(n_{P1},\ldots,n_{PM})$$
$$\hat{Q}_{\lambda}^{(n_{P\beta})}F(n_{PR1},\ldots,n_{PRM}) = \hat{Q}_{\beta}^{(n_{P\lambda})}F(n_{P1},\ldots,n_{PM}).$$

Taking into account the relations (67) and (68), one finds

$$Z_{\beta\lambda}(n_{P1},\ldots,n_{PM}) - Z_{\lambda\beta}(n_{PR1},\ldots,n_{PRM})$$

$$= T_{\beta\lambda}(n_{P1},\ldots,n_{PM})\wp_N(n_{P\lambda}-n_{P\beta})\exp\left[-i\left((p_\beta+p_\lambda)n_{P\lambda}+\sum_{\rho\neq\beta,\lambda}^{M}p_\rho n_{P\rho}\right)\right]$$
$$\times\left(\frac{\partial}{\partial n_{P\beta}}-\frac{\partial}{\partial n_{P\lambda}}\right)F^{(p)}(n_{P1},\ldots,n_{PM})\Bigg|_{n_{P\beta}=n_{P\lambda}}. \qquad (70)$$

The last factor in (70) vanishes owing to the condition (55) imposed by the regularity of the left-hand side of the Schrödinger equation (54).

It remains to show that the states of the spin lattice given by (45) with the functions ψ_M of the form (48) and (49) are highest-weight states with $S = S_3$. This statement is equivalent to the relation $\mathbf{S}_+|\psi^{(M)}\rangle = 0$, which can be rewritten as

$$\sum_{\beta=1}^{M}\sum_{P\in\pi_M^{(\beta)}}\sum_{s\neq n_1,\ldots,n_{M-1}}\hat{Q}_{\beta}^{(s)}\varphi_M^{(p)}(n_{P1},\ldots,n_{PM}) = 0, \qquad (71)$$

where $\{\pi_M^{(\beta)}\}$ are the subsets of π_M: $P \in \pi_M^{(\beta)} \leftrightarrow P\beta = M$. The sums over s in (71) can be reduced and presented in the closed form by using the technique described above. It turns out that the left-hand side of (71) contains the factors similar to the last factor in (70) and vanishes as a result of condition (55).

The descendant states with $S_3 < S$ are obtained by acting with $\mathbf{S}_-$ on the basic states $|\psi^{(M)}\rangle$ (45). The equations (68) for the pseudomomenta $\{p\}$ constitute the analog of the usual Bethe ansatz. The spectrum is given by the relations (47) and (69).

In conclusion, it is demonstrated that the procedure of the exact diagonalization of the lattice Hamiltonian with the non-nearest-neighbor elliptic exchange can be reduced in each sector of the Hilbert space with given magnetization to the construction of the special double quasiperiodic eigenfunctions of the many-particle Calogero–Moser problem on a continuous line. The equations of the Bethe ansatz form appear very naturally as a set of restrictions to the particle pseudomomenta. The proof of this correspodence between lattice and continuum integrable models is based only on analytic properties of the eigenfunctions. One can expect that the set of spin lattice states constructed by this way is complete. This is supported by exact analytic proof in the two-magnon case.

The analysis of explicit form of the equations (68) available for $M = 2, 3$ shows that the spectrum of the lattice Hamiltonian with the exchange (43) is *not* additive, being given in terms of pseudomomenta $\{p\}$ or phases that parametrize the sets $\{p, q\}$ [4]. The problem of finding an appropriate set of parameters that gives the "separation" of the spectrum remains open. It would also be of interest to consider various limits $(N \to \infty, \kappa \to 0, \infty)$ so as to recover the results of the papers [2, 9, 17] and prove the validity of the approximate methods of asymptotic Bethe ansatz.

5 REFERENCES

1. D. Bernard, M. Gaudin, F. D. M. Haldane, and V. Pasquier, *Yang–Baxter equation in long-range interacting systems*, J. Phys. A **26** (1993), No. 20, 5219–5236, hep-th/9301084.

2. H. Bethe, *Theorie der Metalle. I. Eigenwerte und Eigenfunktionen der lineare Atomkette*, Z. Phys. **71** (1931), 205–226.

3. F. Calogero, *Exactly solvable one-dimensional many-body problems*, Lett. Nuovo Cimento **13** (1975), No. 11, 411–416.

4. J. Dittrich and V. I. Inozemtsev, *On the structure of eigenvectors of the multidimensional Lamé operator*, J. Phys. A **26** (1993), No. 16, L753–L756.

5. P. I. Etingof, *Quantum integrable systems and representations of lie algebras*, J. Math. Phys. **36** (1995), No. 6, 2636–2651, hep-th/9311132.

6. P. I. Etingof and A. A. Kirillov, Jr., *Representations of affine Lie algebras, parabolic differential equations, and Lamé functions*, Duke Math. J. **74** (1994), No. 3, 585–614.

7. L. D. Faddeev, *Integrable models of $(1+1)$-dimensional quantum field theory*, Développements récents en théorie des champs et mécanique statistique (Les Houches, 1982) (J.-B. Zuber and R. Stora, eds.), North-Holland, Amsterdam–New York, 1984, pp. 561–608.

8. M. Fowler and J. A. Minahan, *Invariants of the Haldane–Shastry SU(n) chain*, Phys. Rev. Lett. **70** (1993), 2325–2328.

9. F. D. M. Haldane, *Exact Jastrow–Gutzwiller resonating-valence-bond ground state of the spin-$\frac{1}{2}$ antiferromagnetic Heisenberg chain with $1/r^2$ exchange*, Phys. Rev. Lett. **60** (1988), 635–638.

10. _____, *"Spinon gas" description of the $S = \frac{1}{2}$ Heisenberg chain with inverse-square exchange: exact spectrum and thermodynamics*, Phys. Rev. Lett. **66** (1991), No. 11, 1529–1532.

11. V. I. Inozemtsev, *On the connection between the one-dimensional $S = \frac{1}{2}$ Heisenberg chain and Haldane-Shastry model*, J. Stat. Phys. **59** (1990), No. 5-6, 1143–1155.

12. _____, *The extended Bethe ansatz for infinite $S = \frac{1}{2}$ quantum spin chains with non-nearest-neighbor interaction*, Commun. Math. Phys. **148** (1992), No. 2, 359–376.

13. N. Kawakami, *Critical properties of quantum many-body systems with $1/r^2$ interaction*, Progr. Theor. Phys. **91** (1994), 189–218.

14. I. M. Krichever, *Elliptic solutions of the Kadomtsev–Petviašhvili equation and integrable systems of particles*, Funct. Anal. Appl. **14** (1980), No. 4, 282–290.

15. E. H. Lieb and W. Liniger, *Exact analysis of an interacting Bose gas. I. The general solution and the ground state*, Phys. Rev. **130** (1963), 1605–1616.

16. J. Moser, *Three integrable Hamiltonian systems connected to isospectral deformations*, Adv. Math. **16** (1975), 197–220.

17. B. S. Shastry, *Exact solution of an $S = \frac{1}{2}$ Heisenberg antiferromagnetic chain with long-ranged interactions*, Phys. Rev. Lett. **60** (1988), 639–642.

18. B. S. Shastry and B. Sutherland, *Solution of some integrable one-dimensional quantum systems*, Phys. Rev. Lett. **71** (1993), No. 1, 5–8.

19. M. Takahashi, *One-dimensional Heisenberg chain at finite temperature*, Progr. Theor. Phys. **46** (1971), 401–415.

20. _____, *Analytical and numerical investigations of spin chains*, Progr. Theor. Phys. **91** (1994), 1–15.

15

The Bispectral Involution as a Linearizing Map

Alex Kasman

ABSTRACT In 1975, Moser introduced the map σ to linearize the motion of the Calogero–Moser–Sutherland model with rational inverse square potential. This linearizing map was unusual in that it turned out to be an involution on the phase space. In 1990, Wilson introduced an involution on a part of the Sato Grassmannian to demonstrate the bispectrality of certain Kadomtsev–Petviashvili (KP) solutions. Using the correspondence between particle systems and the poles of KP solutions, one may then also view Wilson's bispectral involution as a map on the phase space of the particle system and compare them. As a result, we find that the bispectral involution is the linearizing map of the Calogero–Moser particle system and that it generates nonisospectral master symmetries of the KP flows. This chapter will review this material and then update it with reference to recent papers of Wilson, Shiota, Kasman–Rothstein, and Rothstein that simplify the proof and extend the results to include other integrable Hamiltonian systems of particle systems with linearizing involutions.

1 Moser's Linearizing Involution

For the purposes of this talk, the "Calogero–Moser Particle System" will mean the Hamiltonian system of N particles interacting on the complex plane with the Hamiltonian functions

$$H_n \operatorname{tr} \Lambda^n \quad n = 1, 2, 3, \ldots \tag{1}$$

where the matrix Λ written in terms of the phase space variables is given by

$$\Lambda_{ij} = y_i \delta_{ij} + \frac{1 - \delta_{ij}}{x_i - x_j}.$$

Since the motion under the different Hamiltonians commutes, it is possible to introduce time variables $\mathbf{t} = (t_1, t_2, t_3, \ldots)$ in such a way that the phase space variables are functions of the t_n and so that translation in the nth time variable corresponds to the flow under the nth Hamiltonian. The velocity $y_i(\mathbf{t})$ is, in fact,

$$y_i(\mathbf{t}) = \frac{1}{2}\frac{\partial}{\partial t_2}x_i(\mathbf{t}).$$

(The change of variables between this work and [1] introduced by the factor of $\frac{1}{2}$ here is useful for the correspondence to the KP hierarchy to be discussed below.)

One may determine that the eigenvalues of the matrix Λ are independent constants of motion in involution. In 1977, Airault–McKean–Moser [1] (working in the more restrictive case that all phase space variables were real and the eigenvalues of the matrix Λ were distinct) introduced the map σ to take advantage of these constants and "linearize" the motion of the system. The map σ is defined in terms of asymptotic properties of the particle system. Specifically,

$$\sigma(x_i, y_i) = (\xi_i, \eta_i),$$

where ξ_i is defined as

$$\xi_i = \lim_{t_2 \to +\infty} y_i$$

and η_i is defined by the existence of an asymptotic representation of x_i as a function of t_2 in the form

$$x_i(t_2) = 2\xi_i t_2 + \eta_i + O(t_2^{-1}).$$

It is clear from these formulae that the map σ simplifies the motion of the particle system under H_2, since ξ_i is clearly constant in t_2 and η_i is clearly linear in t_2. Furthermore, ξ_i is also constant and η_i is also linear in all of the other time variables. In this way, the map σ is a linearizing map since it replaces the complicated motion of the system that one gets from solving Hamilton's equations with N constants and N linear functions. This is not a surprise; in fact, it really depends only on the fact that the particles are asymptotically free.

However, there is something unusual about this case, namely, that σ is an involution [1],

$$\sigma = \sigma^{-1},$$

a fact that is not at all obvious from the definition. Note that the eigenvalues of the matrix Λ are in fact the constants ξ_i. (This can be determined merely by taking the limit of the matrix as $t_2 \to \infty$ and noting that the off-diagonal terms tend to zero while the diagonal terms, by definition, are the ξ_i.) Then, by the involutory nature of the map σ, the eigenvalues of the matrix Λ^σ (gotten by replacing x_i with ξ_i and y_i with η_i in the matrix Λ) are the x_i.

Note. Thus, σ is a map that exchanges the positions of the particles (a spatial parameter) with the eigenvalues of the matrix Λ (a spectral parameter). The significance of this statement here is that a similar "philosophical" interpretation can be made of the bispectral involution [19].

2   The Bispectral Involution

2.1   Bispectrality of Ordinary Differential Operators

Given an ordinary differential operator L in the variable x, one is often led to consider an eigenvalue problem of the form

$$Lf(x, z) = p(z)f(x, z)$$

where $p(z)$ is some nonconstant function and we say that f is an eigenfunction depending on the spectral parameter z.

The operator L is said to be *bispectral* [5, 6] if it has such an eigenfunction f that is *also* an eigenfunction for an operator M in the variable z with spectral parameter x. In other words, L is bispectral if it is part of a "bispectral triple" (L, M, f) where f is a simultaneous eigenfunction of L and M with nonconstant eigenvalues depending on z and x, respectively. The bispectral problem is the problem of identifying and classifying all bispectral operators.

A major advance in the (as yet incomplete) classification of bispectral operators was the geometric classification of all "rank one" bispectral operators [19]. In order to describe this achievement, it is necessary briefly to review some basic facts about the Sato Grassmannian [14, 15], which is most frequently used in the study of the KP hierarchy.

2.2   Brief Review of Sato Grassmannian

The Sato Grassmannian Gr is a Grassmannian of the Hilbert space H with basis given by all integer powers of the variable z. There is a decomposition of H into $H_+ \oplus H_-$, where these are the closed subspaces spanned by all nonnegative powers of z and all negative powers of z, respectively. Then Gr is defined to be the Grassmannian of all subspaces $W \subset H$ such that the projection to H_- is compact, and the projection to H_+ is Fredholm of index 0.

Associated to a point $W \in$ Gr is the wave (or Baker–Akhiezer) function $\psi_W(x, z)$, which is the unique function of the form

$$\psi_W(x, z) = \left(1 + \alpha_1(x)z^{-1} + \alpha_2(x)z^{-2} + \cdots\right)e^{xz}$$

that is contained in W for each fixed value x in its domain. This function can also be written as $K_W e^{xz}$ for some monic, zero order, pseudo-differential operator K_W in the variable x. Finally, also associated to W is the ring, $\mathcal{A}_W$, of ordinary differential operators that commute with the pseudo-differential operator $\mathcal{L}_W = K_W \partial K_W^{-1}$.

In the context of the KP hierarchy, one is interested in studying the dependence of the operator $\mathcal{L}_W$ on additional time variables t_i as determined by the nonlinear PDEs

$$\frac{\partial}{\partial t_i}\mathcal{L}_W = [(\mathcal{L}_W^i)_+, \mathcal{L}_W]. \qquad (2)$$

For the application to bispectrality, however, it is sufficient to note that any differential operator $L \in \mathcal{A}_W$ has $\psi_W(x,z)$ as an eigenfunction with spectral parameter z.

2.3   The Bispectral Involution and $\mathrm{Gr}^{\mathrm{ad}}$

Let $\beta\colon \mathrm{Gr} \to \mathrm{Gr}$ be the map that takes W with wave function $\psi_W(x,z)$ to $\beta(W)$ with wave function

$$\psi_{\beta(W)}(x,z) = \psi_W(z,x).$$

In other words, this map exchanges the role of x and z in the wave function.

In fact, I am abusing notation here since β is not well defined on most of the Grassmannian. In general, $\psi_W(z,x)$ is not another wave function. So, let us restrict our attention to a subset on which it is well defined. Following [15] we denote by Gr_1 the subset of points $W \in \mathrm{Gr}$ such that there exists polynomials p, $q \in \mathbb{C}[z]$ satisfying

$$pH_+ \subset W \subset q^{-1}H_+.$$

Let $\mathrm{Gr}^{\mathrm{ad}} \subset \mathrm{Gr}_1$ be the subset of points W such that $\psi_W(z,x)$ is the wave function of some other point $\beta(W) \in \mathrm{Gr}_1$ [19]. By construction β is an involution on $\mathrm{Gr}^{\mathrm{ad}}$, the *bispectral involution*.

It is then immediately obvious that any operator in $\mathcal{A}_W$ is bispectral if $W \in \mathrm{Gr}^{\mathrm{ad}}$. (In particular, its eigenfunction ψ_W is also an eigenfunction for any element of $\mathcal{A}_{\beta(W)}$ after the substitution $x \mapsto z$.) So, there is no question that this subset of the Grassmannian is useful for studying bispectrality. A more powerful statement, whose proof I cannot summarize here, is that all rank one[1] bispectral operators (up to renormalization) are determined in this way by a point of $\mathrm{Gr}^{\mathrm{ad}}$ [19].

2.4   Dynamics: the Bispectral Flow

It is well known that the flows of the KP hierarchy (2) correspond to multiplication by functions of the form $\exp(tz^i)$ on the Grassmannian. The Grassmannian $\mathrm{Gr}^{\mathrm{ad}}$ is closed under the flows of the KP hierarchy as well as being closed under the bispectral involution. Thus, one may consider the image under the involution of the vector fields generating the KP flows.

---

[1]The rank of an ordinary differential operator is the greatest common divisor of the orders of all operators that commute with it.

This alternate set of vector fields generate flows that we may call the *bispectral flows* [8]. Unlike the KP flows, which preserve the spectral curve, the bispectral flows on $\mathrm{Gr}^{\mathrm{ad}}$ are *nonisospectral*. In fact, as will be explained below, the deformation of the spectral curve corresponding to these flows is determined by the Hamiltonian of the Calogero–Moser particle system. As a consequence, we will be able to determine that these flows are in fact master symmetries of the KP flows.

3   Relating the Two Involutions

Thus, we have two involutions to consider: σ and β. However, as was shown by Krichever in 1978 [11], there is a strong relationship between the KP equation and Calogero–Moser particle systems. Thus, there is reason to expect a relationship between these two involutions.

3.1   Tau Functions

Just like finite dimensional Grassmannians [7], the Sato Grassmannian is equipped with projective *Plücker coordinates*. These coordinates π_S are indexed by the set of all Young diagrams. Another object that can be associated to a Young diagram is the Schur polynomial, $p_S(t_i)$ which is a weighted homogeneous polynomial whose order is given by the number of blocks in the diagram. (A more specific description can be found, for instance, in [15].) Thus, we may associate to each point in the Grassmannian a formal sum

$$\tau_W(t_i) = \sum_S \pi_S p_S(t_i). \tag{3}$$

Since the Schur polynomials are independent, this is merely a "faithful" way to represent the Plücker coordinates of Gr as a function[2].

Definition 1. Let $\mathrm{Gr}^b \subset \mathrm{Gr}$ be the set of points $W \in \mathrm{Gr}$ such that $\tau_W(t_i)$ is a monic polynomial in the variable t_1. That Gr^b is nonempty is clear since it contains the sub-Grassmannian Gr_0 [15] of all of the points with only finitely many nonzero Plücker coordinates. However, it also contains many "interesting" points for which the series 3 contains sufficient cancellation to allow τ to be a polynomial.

Shiota [16] showed that for $W \in \mathrm{Gr}^b$ the roots x_j of the polynomial τ_W

---

[2]In fact, this representation of the Plücker coordinates as a Grassmannian is physically significant when viewed as an example of *bosonization* [17].

$$\tau_W(t_i) = \prod_{j=1}^{n} (t_1 - x_j(t_2, t_3, \ldots)), \tag{4}$$

depend on the variable t_i so as to satisfy the ith system in the Calogero–Moser hierarchy 1. Conversely, taking any point in the Calogero–Moser phase space and writing the polynomial 4, there exists a $W \in \mathrm{Gr}^b$ such that this polynomial is the corresponding tau function. This can be seen as an extension of the results of Krichever [11]. Moreover, a detailed analysis of the collision manifolds in this correspondence has only recently been undertaken in [20].

However, this is not the only significance of Gr^b. In fact, it is an object we already know: $\mathrm{Gr}^b = \mathrm{Gr}^{ad}$ [8].

Consequently, there is a bijection between Gr^{ad} (the space upon which the involution σ acts) and the Calogero–Moser phase space[3] (upon which the linearizing map σ acts). Thus, we are able to pull β back to the phase space, view it as another involution on Calogero–Moser particles and compare it to σ.

4   Conclusions

Just as σ was constructed to be a linearizing map and had the surprising property of being an involution, it turns out that β—an involution by construction—is a linearizing map of the Calogero–Moser particle system[4].

There are several ways in which one can interpret this result:

• The bispectral involution is the extension of the original linearizing map to Wilson's entire "Completed Calogero–Moser Phase Space."

• The involution β was a rediscovery of the linearizing map σ in a different context.

• Under the bispectral flows (see Section 2.4) the cusps of the spectral curves move as a Calogero–Moser particle system.

• Combining these results with Theorem VIII in Ref. 12, one concludes that the bispectral flows are master symmetries of the KP flows.

---

[3] Here I am making use of results from [20], without which I would have to exclude collision points from the phase space and the corresponding points from Gr^{ad}.

[4] The original proof of this statement [8] was achieved by an unenlightening analysis of the wave function ψ_W and the differential equations it is known to satisfy. However, recently Wilson has determined another description of Gr^{ad} as a quotient of the set of pairs of square matrices satisfying $\mathrm{rank}([X, Z] + Id) = 1$ [20]. In this context, the equivalence of σ and β is so apparent that no proof is necessary.

• Although not apparent from its definition, the bispectral involution has a symplectic significance for the Hamiltonian system 1 determining the motion of the roots of the τ-function.

Of these, the last is of the greatest interest to me since it applies as well to extensions of Wilson's bispectral involution. Generalizations of the bispectral involution to higher rank have now been discovered [3, 10]. In many ways, these higher-rank results are not as nice as the rank one case. For example, they apparently cannot be used to characterize completely all bispectral operators. Also, unlike $\mathrm{Gr^{ad}}$, which is closed under all flows of the KP hierarchy, the subsets of the Grassmannian on which the higher-rank bispectral involutions are defined are only closed under some of the KP flows. However, it is interesting to note that once again the bispectral involution has a dynamical significance as a linearizing map for the motion of the roots of a polynomial.

In the higher-rank constructions, the polynomial of interest is not a KP τ-function, but rather a quotient of τ-functions [2, 9]. Under a subhierarchy of isospectral flows that preserve bispectrality, the roots of this polynomial move as particles in a Hamiltonian system and, just as the rank one bispectral involution linearizes the motion of the Calogero–Moser system, the higher-rank involution linearizes the motion of these "particles" [10, 13]. The Hamiltonians for these other systems can be written nicely in terms of the usual Lax matrices X and Z for the Calogero–Moser particle system [13]. I have been able to identify special cases of these systems as previously known. In particular, some are apparently recognizable as Calogero–Moser systems in the presence of an external field [4]. Furthermore, some of the Hamiltonians have appeared previously in Ref. 18 as determining the motion of poles of nonvanishing rational KP solutions. Each of these systems shares with the rational Calogero–Moser system the property of being linearized by a "bispectral involution."

5 References

1. H. Airault, H. P. McKean, and J. Moser, *Rational and elliptic solutions of the Korteweg–de Vries equation and a related many-body problem*, Commun. Pure Appl. Math. **30** (1977), No. 1, 95–148.

2. B. Bakalov, E. Horozov, and M. Yakimov, *Bäcklund–Darboux transformations in Sato's Grassmannian*, Serdica Math. J. **22** (1996), No. 4, 571–588.

3. _____, *Highest weight modules of $W_{1+\infty}$, Darboux transformations and the bispectral problem*, Serdica Math. J. **23** (1997), No. 2, 95–112.

4. J. F. van Diejen, *Deformations of Calogero–Moser systems*, Theoret. and Math. Phys. (1994), No. 2, 549–554.

5. J. J. Duistermaat and F. Grünbaum, *Differential equations in the spectral parameter*, Commun. Math. Phys. **103** (1986), No. 2, 177–240.

6. F. A. Grünbaum, *The limited angle problem in tomography and some related mathematical problems*, Bifurcation Theory, Mechanics and Physics (C. P. Bruter, A. Aragnol, and A. Lichnerowicz, eds.), Math. Appl., Reidel, Dordrecht, 1983, pp. 317–329.

7. W. V. D. Hodge and D. Pedoe, *Methods of Algebraic Geometry*. II, Cambridge Univ. Press, Cambridge, 1952.

8. A. Kasman, *Bispectral KP solutions and linearization of Calogero–Moser particle systems*, Commun. Math. Phys. **172** (1995), No. 2, 427–448.

9. _____, *Darboux transformations from n-KdV to KP*, Acta Appl. Math. **49** (1997), No. 2, 179–197.

10. A. Kasman and M. Rothstein, *Bispectral Darboux transformations*, Physica **102D** (1997), No. 3-4, 159–177.

11. I. M. Krichever, *Rational solutions of the Kadomtsev–Petviashvili equation and integrable systems of N particles on a line*, Funct. Anal. Appl. **12** (1978), No. 1, 59–61.

12. W. Oevel and M. Falck, *Master symmetries for finite-dimensional integrable systems: the Calogero–Moser system*, Progr. Theor. Phys. **75** (1986), No. 6, 1328–1341.

13. M. Rothstein, *Explicit formulas for the Airy and Bessel bispectral involutions in terms of Calogero–Moser pairs*, The Bispectral Problem (Montréal, QC, 1997) (J. Harnad and A. Kasman, eds.), CRM Proc. Lecture Notes, Vol. 14, Amer. Math. Soc., Providence, RI, 1998, pp. 105–110, q-alg/9611027.

14. M. Sato and Y. Sato, *Soliton equations as dynamical systems on infinite-dimensional Grassmann manifold*, Nonlinear Partial Differential Equations in Applied Science (Tokyo, 1982) (H. Fujita, P. D. Lax, and G. Strang, eds.), Lecture Notes Numer. Appl. Anal., Vol. 5, Kinokuniya, Tokyo, 1983, pp. 259–271.

15. G. Segal and G. Wilson, *Loop groups and equations of KdV type*, Inst. Hautes Études Sci. Publ. Math. (1985), No. 61, 5–65.

16. T. Shiota, *Calogero–Moser hierarchy and Kp hierarchy*, J. Math. Phys. **35** (1994), No. 11, 5844–5849.

17. M. Stone (ed.), *Bozonization*, World Sci. Publ., Singapore, 1994.

18. A. P. Veselov, *Rational solutions of the Kadomtsev–Petviashvili equation and Hamiltonian systems*, Russian Math. Surveys **35** (1980), No. 1, 239–240.

19. G. Wilson, *Bispectral commutative ordinary differential operators*, J. Reine Angew. Math. **442** (1993), 177–204.

20. ———, *Collisions of Calogero–Moser particles and an adelic Grassmannian*, Invent. Math. **133** (1998), No. 1, 1–41.

16

On Some Quadratic Algebras: Jucys–Murphy and Dunkl Elements

Anatol N. Kirillov[1]

ABSTRACT We study some quadratic algebras that appeared in low-dimensional topology and Schubert calculus. We introduce Jucys–Murphy elements in the braid algebra and in the pure braid group, as well as Dunkl elements in the extended affine braid group. Relationships between the Dunkl elements, Dunkl operators and Jucys–Murphy elements are described.

1 Introduction

Quadratic algebras play a role in various branches of mathematics; see, for example, Refs. 12 and 15. Our motivation for the study of quadratic algebras is based on their connections with the celebrated Dunkl operators and Jucys–Murphy elements, Calogero–Moser–Sutherland models and Toda lattices, and Vassiliev invariants of braids.

The Jucys–Murphy elements [16, 27] play an important role in the representation theory of the symmetric group and that of the Hecke algebras (see, e.g., Ref. 28). The Dunkl operators [8, 9], and their generalizations (see, e.g., Ref. 5), play an important role in the theory of orthogonal polynomials on the sphere [9], in the theory of Macdonald polynomials [4, 5, 26], in the theory of integrable systems (see, e.g., Refs. 5 and 13), and Schubert calculus [10, 20].

The starting point for applications of Dunkl's operators to the Schubert calculus is the following remarkable theorem of C. Dunkl [9, Section 3]:

> For any finite Coxeter group the algebra generated by the "truncated" Dunkl operators [9, Definition 2.1] is canonically isomorphic to the algebra of coinvariants of the Coxeter group.

It is well known (see, e.g., Ref. 14), that for a finite crystallographic Coxeter group the algebra of coinvariants is canonically isomorphic to the

---

[1] On leave from Steklov Mathematical Institute, St. Petersburg, Russia.

cohomology ring of the corresponding flag variety.

Our general program is to construct for any Coxeter group a certain non-commutative algebra (the so-called *bracket algebra*) with following properties:

- It contains a commutative subalgebra generated by "Dunkl's elements" that is canonically isomorphic to the algebra of coinvariants of the Coxeter group,

- It admits a representation by operators acting on the ring of polynomials such that the "Dunkl elements" become the (truncated) Dunkl operators,

- It admits a representation in the group ring of the Coxeter group.

This program has been realized for the symmetric group in Ref. 10, and for finite Coxeter groups in Ref. 20.

In this paper we study some quadratic algebras that are naturally appeared in the low-dimensional topology [2, 7, 23] and Schubert calculus [10], and investigate some of their properties. We define a quadratic algebra $\mathcal{G}_n$, which is a further generalization of the quadratic algebras $\mathcal{E}_n$ and $\mathcal{E}_n^t$ introduced in Ref. 10, Sections 2 and 15. Our main idea is to apply the results obtained for the braid algebra $\mathcal{B}_n$ [7, 23] to the algebra $\mathcal{G}_n$. The main observation is that some results which are well-known for the braid algebra $\mathcal{B}_n$, can be (re)proven for the quadratic algebra $\mathcal{G}_n$. For example, it happens that the quadratic algebra $\mathcal{G}_n$ and the braid algebra $\mathcal{B}_n$ have the same Hilbert series

$$H(\mathcal{B}_n;t) = H(\mathcal{G}_n;t) = \frac{1}{(1-t)(1-2t)\cdots(1-(n-1)t)},$$

and have the commutative subalgebras $\mathcal{K}_n \subset \mathcal{B}_n$ and $\mathcal{H}_n \subset \mathcal{G}_n$ which are both isomorphic to the ring of polynomials. The algebra $\mathcal{K}_n$ is generated by the Jucys–Murphy elements, whereas the algebra $\mathcal{H}_n$ is generated by the Dunkl ones. Both algebras $\mathcal{G}_n$ and $\mathcal{B}_n$ have the additive bases consisting of all monomials in normal form (Theorems 1 and 4; cf. Ref. 25). We expect also that the certain quotients $\mathcal{E}_n^0$ and $\mathcal{B}_n^0$ of the quadratic algebras $\mathcal{G}_n$ and $\mathcal{B}_n$ respectively, have the same Hilbert polynomials as well (see Ref. 19 for more details).

One of the main goals of this paper is to describe the relationships between the Dunkl and Jucys–Murphy elements.

The structure of the paper is the following:

In Section 2 we define the braid algebra $\mathcal{B}_n$ as the infinitesimal deformation of the pure braid group P_n (cf. T. Kohno [23]). The algebra $\mathcal{B}_n$ and its completion $\widehat{\mathcal{B}}_n$ have many important applications to the low-dimensional topology; see, for example, Refs. 2, 7, 11, 24, and 25.

In Section 3 we define the Jucys–Murphy elements d_k in the braid algebra $\mathcal{B}_n$, and the multiplicative Jucys–Murphy elements D_k in the pure braid group P_n. Following A. Ram [28] we prove that the Jucys–Murphy element $d_k \in \mathcal{B}_n$ is the quasiclassical limit of the element $D_k \in P_n$. We prove also, that the infinitesimal deformation of the multiplicative Jucys–Murphy element D_k coincides with the element d_k.

In Section 4 we define the quadratic algebra $\mathcal{G}_n$ and compute the Hilbert series for this algebra. In Section 5 we define the Dunkl elements θ_j in the quadratic algebra $\mathcal{G}_n$ and describe a commutative subalgebra generated by θ_j, $1 \le j \le n$.

In Section 6, which contains one of the main results of this paper, we study a relationship between the Jucys–Murphy elements and the Dunkl elements. We define the dual Dunkl elements $Y_1^*, \ldots, Y_n^*$ as certain elements in the extended affine braid group $\widetilde{B}_n$, and prove that under a natural homomorphism $\pi \colon \widetilde{B}_n \to B_n$, the dual Dunkl element Y_k^* maps to the multiplicative Jucys–Murphy element D_k. We explain also a connection between Dunkl element $Y_k \in \widetilde{B}_n$ and the Dunkl operator $\mathbf{Y}_k \in \mathcal{D}_{q,x}[W]$, where $\mathcal{D}_{q,x}[W]$ is the algebra of q-difference operators with permutations.

2 Quadratic Algebra $\mathcal{B}_n$

We start with consideration of the quadratic algebra $\mathcal{B}_n$ which is the infinitesimal deformation of the pure braid group P_n [23]. This quadratic algebra is well known as the algebra of (finite order) Vassiliev invariants of braids [2, 11, 24]. The completion $\widehat{\mathcal{B}}_n$ of $\mathcal{B}_n$ with respect to the grading was considered in the papers of V. Drinfeld [6, 7], in his study of quasi-Hopf algebras, and in the papers of T. Kohno [22–24], in his study of monodromy representations of braid groups.

Definition 1 ([7, 23]). Define the braid algebra $\mathcal{B}_n$ as the quadratic algebra (say over $\mathbb{Z}$) with generators X_{ij}, $1 \le i < j \le n$, which satisfy the following relations

$$X_{ij} \cdot X_{jk} - X_{jk} \cdot X_{ij} = X_{ik} \cdot X_{ij} - X_{ij} \cdot X_{ik}$$
$$= X_{jk} \cdot X_{ik} - X_{ik} \cdot X_{jk}; \tag{1}$$
$$X_{ij} \cdot X_{kl} = X_{kl} \cdot X_{ij}, \quad \text{if all } i, j, k, l \text{ are distinct.} \tag{2}$$

Remark 1. The algebra $\mathcal{B}_n$ is denoted by $\mathcal{P}_n$ in Ref. 25, and as A^n in Refs. 22–24.

The algebra $\mathcal{B}_n$ has a natural structure of a co-commutative Hopf algebra. Namely, a comultiplication $\Delta \colon \mathcal{B}_n \to \mathcal{B}_n \otimes \mathcal{B}_n$, antipode $S \colon \mathcal{B}_n \to \mathcal{B}_n$, and counit $\epsilon \colon A \to \mathbb{Z}$, can be defined as follows:

$$\Delta(X_{ij}) = 1 \otimes X_{ij} + X_{ij} \otimes 1,$$
$$S(X_{ij}) = -X_{ij},$$
$$\epsilon(X_{ij}) = 0, \quad \epsilon(1) = 1.$$

Let us explain briefly an origin of the relations (1) and (2). For more details and proofs see Refs. 6, 7, 11, 22–24, and 25. Let us denote by

$$\mathcal{M} = \mathbb{C}^n \setminus \bigcup_{\substack{i,j=1 \\ i<j}}^{n} \{z_i = z_j\},$$

the configuration space of n distinct ordered points in $\mathbb{C}$. It is well known that the fundamental group $\pi_1(\mathcal{M})$ of the space $\mathcal{M}$ coincides with the pure braid group P_n:

$$\pi_1(\mathcal{M}) = P_n.$$

Let

$$w_{ij} = w_{ji} = \frac{1}{2\pi\sqrt{-1}} d\log(z_i - z_j)$$

be closed 1-form on $\mathcal{M}$. Then, $\{w_{ij} \mid 1 \le i < j \le n\}$ represents a basis for $H^1(\mathcal{M})$, and the relations among w_{ij}, $1 \le i < j \le n$, are generated by the Arnold relations

$$w_{ij} \wedge w_{jk} + w_{jk} \wedge w_{ik} + w_{ik} \wedge w_{ij} = 0$$

for $i < j < k$ (see, e.g., Ref. 1).

Let $\{X_{ij} = X_{ji} \mid 1 \le i < j \le n\}$ be a set of noncommutative variables, and consider a formal connection

$$\Omega = \sum_{\substack{i,j=1 \\ i,j}}^{n} w_{ij} X_{ij}. \tag{3}$$

Lemma 1 (T. Kohno). *The connection Ω is integrable if and only if the following conditions are satisfied:*

$$[X_{ij}, X_{ik} + X_{jk}] = [X_{ij} + X_{ik}, X_{jk}] = 0, \quad \text{for } i < j < k,$$
$$[X_{ij}, X_{kl}] = 0, \quad\quad\quad\quad \text{for distinct } i, j, k, l.$$

Let us remember that $[a, b] = ab - ba$ is the usual commutator.

It is clear that these conditions are equivalent to the relations (1) and (2). A proof of Lemma 1 follows from the observations that $d\Omega = 0$, and the integrability condition for the connection Ω is equivalent to the following one: $\Omega \wedge \Omega = 0$. Notice that the connection (3) is the formal version of the so-called Knizhnik–Zamolodchikov connection in conformal field theory.

The relations (1) and (2) can be also interpreted [5, 7, 23] as the self-consistency conditions

$$\frac{\partial A_j}{\partial z_i} - \frac{\partial A_i}{\partial z_j} = [A_i, A_j], \quad 1 \le i, j \le n,$$

of the Knizhnik–Zamolodchikov (system of) equation(s)

$$\frac{\partial \Phi}{\partial z_i} = k A_i \Phi, \quad 1 \leq i \leq n,$$

where $A_i = \sum_{j \neq i} X_{ij}/(z_i - z_j)$, and a function $\Phi := \Phi(z)$, $z = (z_1, \ldots, z_n)$, takes the values in a certain algebra (say over $\mathbb{C}((z))$) generated by the elements X_{ij}.

Below we formulate the basic properties of the quadratic algebra $\mathcal{B}_n$.

Theorem 1 ([22, 23]). *The Hilbert series of the algebra $\mathcal{B}_n$ is given by*

$$H(\mathcal{B}_n; t) = \frac{1}{(1-t)(1-2t)\cdots(1-(n-1)t)}. \tag{4}$$

Corollary 1 ([6, 25]). *A linear basis in the algebra $\mathcal{B}_n$ is given by the following set of noncommutative monomials:*

$$Z_2 \cdot \cdots \cdot Z_n, \tag{5}$$

where

$$Z_2 = \{1, X_{12}\},$$

$$\vdots$$

$$Z_k = \{1\} \cup \{X_{i_1 k} \ldots X_{i_l k} \mid 1 \leq i_j < k, 1 \leq j \leq l\}, \quad 2 \leq k \leq n.$$

A monomial in X_{ij}'s is called to be in *normal form* if it belongs to the set (5).

Remark 2. Corollary 1 was stated without a proof in Ref. 6. The formula (4) was obtained in Ref. 22 by constructing a free resolution of $\mathbb{C}$ as a trivial $\mathcal{B}_n$-module. Also, the formula (4) is a consequence of the fact that $\mathcal{B}_n$ is a semidirect product of free associative algebras as shown in Ref. 7; see also Ref. 25. Note, that the relations (1) can be interpreted as the horizontal version of the 4-term relations in the theory of Vassiliev link invariants.

Let us consider a completion $\widehat{\mathcal{B}}_n$ of the algebra $\mathcal{B}_n \otimes_{\mathbb{Z}} \mathbb{C}$ with respect to the powers of the ideal $\mathcal{I} = (X_{ij})$ generated by X_{ij}, $1 \leq i < j \leq n$. More precisely, let us denote by $\mathbb{C}\langle X_{ij} \rangle$ the ring of noncommutative formal power series with indeterminates X_{ij}, $1 \leq i < j \leq n$, and let J be an ideal generated by the relations (1) and (2). Then $\widehat{\mathcal{B}}_n = \mathbb{C}\langle X_{ij} \rangle / J$. It is well known [22] that there exists an isomorphism of complete Hopf algebras

$$\widehat{\mathbb{C}[P_n]} \cong \widehat{\mathcal{B}}_n,$$

where $\widehat{\mathbb{C}[P_n]}$ stands for the completion of the group ring of the pure braid group P_n with respect to the powers of the augmentation ideal.

3 Jucys–Murphy Elements

Definition 2. The Jucys–Murphy elements d_j, for $j = 2, \ldots, n$, in the quadratic algebra $\mathcal{B}_n$ are defined by

$$d_j = \sum_{1 \le i < j} X_{ij}. \tag{6}$$

It is clear that d_i is a primitive element, that is, $\Delta(d_i) = 1 \otimes d_i + d_i \otimes 1$.

Lemma 2. *Relations* $[X_{ik} + X_{jk}, X_{ij}] = 0$ *for* $i < j < k$, *together with commutativity relations* (2), *imply that the Jucys–Murphy elements* d_j *commute pairwise.*

Proof. For $j < l$,

$$[d_j, d_l] = \sum_{\substack{1 \le i < j \\ 1 \le k < l}} [X_{ij}, X_{kl}] = \sum_{1 \le i < j < l} \{[X_{ij}, X_{il}] + [X_{ij}, X_{jl}]\} = 0. \qquad \square$$

Let us assume that additionally to the relations (1) and (2) the following relations are satisfied:

$$X_{ij}^2 = 1, \quad 1 \le i < j \le n. \tag{7}$$

Then a map $(i < j)$

$$X_{ij} \to (i, j),$$

where $(i, j) \in S_n$ is the transposition that interchanges i and j and fixes each $k \ne i, j$, defines a representation of the relations (1), (2), and (7). In other words, there exists a homomorphism p from the braid algebra $\mathcal{B}_n$ to the group ring of the symmetric group, $p \colon \mathcal{B}_n \to \mathbb{Z}[S_n]$, such that $p(X_{ij}) = (i, j)$. Under this homomorphism p the element d_j maps to the Jucys–Murphy element in the group ring of the symmetric group S_n (see, e.g., Refs. 16, 27, and 28). A proof follows from the following relations between the transpositions in the symmetric group S_n: if $1 \le i < j < k \le n$, then

$$(ij)(ik) = (jk)(ij) = (ik)(jk), \quad (ij)(jk) = (ik)(ij) = (jk)(ik).$$

Theorem 2. *Let* $\mathcal{K}_n$ *be a commutative subalgebra of* $\mathcal{B}_n$ *generated by the Jucys–Murphy elements* d_j, $j = 2, \ldots, n$. *Then a map* $d_j \mapsto x_{j-1}$ *defines an isomorphism*

$$\mathcal{K}_n \cong \mathbb{Z}[x_1, \ldots, x_{n-1}].$$

There exists a multiplicative analog, denoted by D_k, of the Jucys–Murphy element d_k. Our construction of the elements D_k follows according to Ref. 28.

Let g_i, $1 \leq i \leq n-1$, be the standard generators of the braid group B_n. Thus, the generators g_i satisfy the following relations:

$$g_i g_j = g_j g_i, \qquad\qquad \text{if } |i-j| \geq 2, \qquad (8)$$

$$g_i g_{i+1} g_i = g_{i+1} g_i g_{i+1}, \qquad 1 \leq i \leq n-2. \qquad (9)$$

Now let us define the multiplicative Jucys–Murphy elements D_k:

Definition 3 (cf. Ref. 28 (3.16)).   The multiplicative Jucys–Murphy elements D_k are defined by the following formulae:

$$D_2 = g_1^2; \quad D_k = g_{k-1} g_{k-2} \cdots g_2 g_1^2 g_2 \cdots g_{k-2} g_{k-1}, \ 3 \leq k \leq n. \qquad (10)$$

Lemma 3. *The elements D_k commute pairwise.*

Proof. First of all,

$$D_2 D_3 = g_1^2 g_2 g_1^2 g_2 = g_1 g_2 g_1 g_2 g_1 g_2 = g_2 g_1^2 g_2 g_1^2 = D_3 D_2.$$

Now, using induction, we have $(3 \leq k \leq n-1)$

$$\begin{aligned}
D_k D_{k+1} &= g_{k-1} D_{k-1} g_{k-1} g_k g_{k-1} D_{k-1} g_{k-1} g_k \\
&= g_{k-1} g_k D_{k-1} g_{k-1} D_{k-1} g_{k-1} g_k g_{k-1} g_k \\
&= g_{k-1} g_k D_{k-1} g_{k-1} D_{k-1} g_{k-1} g_k g_k g_{k-1} \\
&= g_{k-1} g_k D_{k-1} D_k g_k g_{k-1} \\
&= g_{k-1} g_k D_k D_{k-1} g_k g_{k-1} = D_{k+1} D_k.
\end{aligned}$$

Finally, if $l - k \geq 2$, then

$$\begin{aligned}
D_k D_l &= D_k g_{l-1} \cdots g_{k+1} D_{k+1} g_{k+1} \cdots g_{l-1} \\
&= g_{l-1} \cdots g_{k+1} D_k D_{k+1} g_{k+1} \cdots g_{l-1} \\
&= g_{l-1} \cdots g_{k+1} D_{k+1} D_k g_{k+1} \cdots g_{l-1} = D_l D_k. \qquad\square
\end{aligned}$$

The elements D_k, $2 \leq k \leq n$, generate a commutative subgroup K_n in the braid group B_n. Now we are going to consider the following two problems: what is the infinitesimal deformation of the commutative subgroup K_n, and what is the quasiclassical limit

$$\mathbb{Z}[B_n] \to H_n(q) \to \mathbb{Z}[S_n] \qquad (11)$$

of the multiplicative Jucys–Murphy elements D_k? In the quasiclassical limit (11) the group ring $\mathbb{Z}[B_n]$ of the braid group B_n is degenerated at first to the Iwahori–Hecke algebra $H_n(q)$, and then to the group ring $\mathbb{Z}[S_n]$ of the symmetric group S_n.

Proposition 1 (A. Ram). *The quasiclassical limit* (11) *of the multiplicative Jucys–Murphy element* $D_k \in B_n$ *is equal to the Jucys–Murphy element* $d_k \in \mathbb{Z}[S_n]$.

Proof. Let us compute the quasiclassical limit (11) of the element D_k. The first step is to consider D_k as an element of the Iwahori–Hecke algebra $H_n(q)$. In other words, we have to add to (8) and (9) the new relations

$$g_i^2 = (q^{1/2} - q^{-1/2})g_i + 1, \quad 1 \le i \le n - 1. \tag{12}$$

The next step is to consider the limit

$$\lim_{q \to 1} \frac{D_k - 1}{q^{1/2} - q^{-1/2}}.$$

For this goal it is enough to compute $dD_k/dq|_{q=1}$. This can be done using the formula

$$\frac{dg_i}{dq} = \frac{(q^{1/2} - q^{-1/2})g_i + 2}{2q(q^{1/2} + q^{-1/2})}.$$

Thus,

$$\frac{dD_k}{dq} = \sum_{i=2}^{k-1} \left\{ g_{k-1} \cdots g_{i+1} \left(\frac{dg_i}{dq}\right) g_{i-1} g_2 g_1^2 g_2 \cdots g_{k-1} \right.$$

$$\left. + g_{k-1} \cdots g_2 g_1^2 g_2 \cdots g_{i-1} \left(\frac{dg_i}{dq}\right) g_{i+1} \cdots g_{k-1} \right\}$$

$$+ g_{k-1} \cdots g_2 \left(\frac{d}{dq} g_1^2\right) g_2 \cdots g_{k-1}$$

$$\overset{q=1}{=} \sum_{i=2}^{k-1} \left\{ \frac{1}{2} s_{k-1} \cdots s_{i+1} s_i s_{i+1} \cdots s_{k-1} \right.$$

$$\left. + \frac{1}{2} s_{k-1} \cdots s_{i+1} s_i s_{i+1} \cdots s_{k-1} \right\}$$

$$+ s_{k-1} \cdots s_2 s_1 s_2 \cdots s_{k-1}$$

$$= \sum_{i=1}^{k-1} (i, k) = d_k. \qquad \square$$

Finally, let us explain a connection between the pure braid group P_n and the multiplicative Jucys–Murphy elements D_k. More precisely, let us consider the infinitesimal deformations of the pure braid group P_n, and the multiplicative Jucys–Murphy elements D_k. To start, it is convenient to remind a few definitions.

By definition, the pure braid group P_n is a kernel of the natural homomorphism

$$B_n \to S_n, \quad g_i \mapsto s_i = (i, i+1),$$

where $s_i = (i, i+1) \in S_n$, $1 \leq i \leq n-1$, denote the transposition that interchanges i and $i+1$, and fixes all other elements of $[1, n]$. We have an exact sequence:

$$1 \to P_n \to B_n \to S_n \to 1.$$

It is well known (see, e.g., Ref. 3), that P_n is generated by the following elements:

$$g_{ij} = (g_{j-1}g_{j-2}\cdots g_{i+1})g_i^2(g_{j-1}g_{j-2}\cdots g_{i+1})^{-1}, \quad 1 \leq i < j \leq n,$$

subject to the following relations:

$$g_{ij}g_{kl} = g_{kl}g_{ij}, \qquad \text{if all } i, j, k, l \text{ are distinct,} \quad (13)$$
$$g_{ij}g_{ik}g_{jk} = g_{ik}g_{jk}g_{ij} = g_{jk}g_{ij}g_{ik}, \qquad \text{if } 1 \leq i < j < k \leq n, \quad (14)$$
$$g_{ik}g_{jk}g_{jl}g_{ij} = g_{jk}g_{jl}g_{ij}g_{ik}, \qquad \text{if } 1 \leq i < j < k < l \leq n. \quad (15)$$

One can show using only the relations (13)–(15) that the elements

$$D_k = g_{1,k}g_{2,k}\cdots g_{k-1,k} \in P_n, \quad 2 \leq k \leq n,$$

mutually commute.

Now let us consider the infinitesimal deformation $g_{ij} \mapsto 1 + \epsilon X_{ij}$, of the pure braid group P_n. It is easy to see that the coefficients of ϵ^2 on both sides of relations (13)–(15) coincide with the defining relations (1) and (2) for the braid algebra $\mathcal{B}_n$, and the element D_k transforms to the Jucys–Murphy elements d_k (more precisely, $D_k = 1 + \epsilon d_k + o(\epsilon^2)$).

4   Quadratic Algebra $\mathcal{G}_n$

In the recent paper of the author and S. Fomin [10] the quadratic algebra $\mathcal{E}_n^t$ was introduced. This algebra is closely related to the theory of quantum Schubert polynomials [10] and their multiparameter deformation [18]. In this section we introduce the quadratic algebra $\mathcal{G}_n$, which is a further generation of the quadratic algebra $\mathcal{E}_n^t$.

Definition 4. Define the algebra $\mathcal{G}_n$ (of type A_{n-1}) as the quadratic algebra (say, over $\mathbb{Z}$) with generators $[ij]$, $1 \leq i < j \leq n$, which satisfy the following relations:

$$[ij][jk] = [jk][ik] + [ik][ij], \quad [jk][ij] = [ik][jk] + [ij][ik], \quad \text{for } i<j<k; \quad (16)$$
$$[ij][kl] = [kl][ij], \quad \text{whenever } \{i,j\} \cap \{k,l\} = \varnothing, \, i < j, \text{ and } k < l. \quad (17)$$

The quadratic algebra $\mathcal{E}_n^t$ [10] is the quotient of the algebra $\mathcal{G}_n$ by the two-side ideal generated by the relations

$$[ij]^2 - t_{ij} = 0,$$

where the t_{ij}, for $1 \leq i < j \leq n$, are a set of commuting parameters.

From (22) follows that the generators $[ij]$ satisfy the classical Yang–Baxter equation (CYBE)

$$[[ij], [ik] + [jk]] + [[ik], [jk]] = 0, \quad i < j < k, \tag{18}$$

where the external brackets stand for the usual commutator: $[a, b] = ab - ba$.

Theorem 3. *The Hilbert series of the algebra $\mathcal{G}_n$ is given by*

$$H(\mathcal{G}_n; t) = \frac{1}{(1 - t)(1 - 2t) \cdots (1 - (n - 1)t)}. \tag{19}$$

Theorem 3 is a corollary of the following result, which gives a description of an additive basis in the algebra $\mathcal{G}_n$:

Theorem 4. *A linear basis of the algebra $\mathcal{G}_n$ is given by the following set of noncommutative monomials:*

$$Z_2 \cdot Z_3 \cdot \cdots \cdot Z_n, \tag{20}$$

where

$$Z_2 = \{\varnothing, [12]\}$$

$$\vdots$$

$$Z_k = \{\varnothing\} \cup \{[i_1 k] \cdots [i_l k] \mid 1 \leq i_j < k, 1 \leq j \leq l\}, \quad 2 \leq k \leq n.$$

A monomial in $[ij]$'s is called to be in *normal form*, if it belongs to the set (20).

A proof of Theorem 3 is similar to a proof of Corollary 1 given by X.-S. Lin [25, Theorem 2.3]. For this reason we omit a proof.

Remark 3. It is clear that $H(Z_k; t) = \frac{1}{1-(k-1)t}$, and

$$H(\mathcal{G}_n; t) = \prod_{k=2}^{n} H(Z_k; t). \tag{21}$$

Corollary 2. *Let $\mathcal{G}_n'$ be a commutative version of the algebra $\mathcal{G}_n$; that is, let us assume that the relations (17) are valid for all i, j, k, l. Then*

$$H(\mathcal{G}_n'; t) = (1 - t)^{-\binom{n}{2}}.$$

5 Dunkl Elements

The Dunkl elements θ_j, for $j = 1, \ldots, n$, in the quadratic algebra $\mathcal{G}_n$ are defined by

$$\theta_j = -\sum_{1 \leq i < j} [ij] + \sum_{j < k \leq n} [jk]. \tag{22}$$

For the quadratic algebra $\mathcal{E}_n^t$ the Dunkl elements (22) coincide with those introduced in Ref. 10, Section 5. The following result is well known (cf., e.g., Ref. [17], Theorem 1.4, or Ref. 10, Lemma 5.1)

Lemma 4. *The classical Yang–Baxter equation* (18), *together with the commutativity relation* (17), *imply that the Dunkl elements* θ_j *commute pairwise.*

There exists an obvious relation between Dunkl's elements $\theta_1, \ldots, \theta_n$ in the quadratic algebra $\mathcal{G}_n$, namely, $\theta_1 + \cdots + \theta_n = 0$. The result below shows that in the algebra $\mathcal{G}_n$ the Dunkl elements $\theta_1, \ldots, \theta_{n-1}$ are algebraically independent.

Theorem 5. *Let* $\mathcal{H}_n$ *be a commutative subalgebra of* $\mathcal{G}_n$ *generated by the Dunkl elements* θ_j, $j = 1, \ldots, n$. *Then a map* $\theta_i \to x_i$, $1 \le i \le n-1$, *defines an isomorphism*

$$\mathcal{H}_n \cong \mathbb{Z}[x_1, \ldots, x_{n-1}].$$

Problem 6. To describe the two-side ideals $\mathcal{R} \subset \mathcal{G}_n$ such that

$$H\left(\mathcal{H}_n / \mathcal{R} \cap \mathcal{H}_n; t\right) = [n]! := \prod_{j=1}^{n} \frac{1 - t^j}{1 - t}. \tag{23}$$

The examples of two-side ideals in the algebra $\mathcal{G}_n$ with the property (23) will be given in [19, Sects. 7 and 8]. These examples are related to the quantum cohomology ring of the flag variety [10], and the multiparameter deformation of Schubert polynomials [18].

Example 1. Let us take $n = 3$. In the algebra $\mathcal{G}_3$ we have the following relations:

$$\theta_1 + \theta_2 + \theta_3 = 0,$$
$$\theta_1\theta_2 + \theta_1\theta_3 + \theta_2\theta_3 + [12]^2 + [13]^2 + [23]^2 = 0,$$
$$\theta_1\theta_2\theta_3 + [12]^2\theta_3 + \theta_1[23]^2 - [13]^2\theta_1 - \theta_3[13]^2 = 0,$$
$$[\theta_2, [23]^2] = [\theta_1, [13]^2],$$

where $[a, b] = ab - ba$ is the usual commutator.

6   Dunkl and Jucys–Murphy Elements

Let us start with the definition of the extended affine braid group $\widetilde{B}_n$.

Definition 5. The extended affine braid group $\widetilde{B}_n$ is a group with generators

$$g_0, \quad g_1, \quad \ldots, \quad g_{n-1}, \quad w,$$

which satisfy the following relations

$$g_i g_j = g_j g_i, \qquad |i - j| \geq 2, \ 0 \leq i, j \leq n - 1, \qquad (24)$$

$$g_i g_{i+1} g_i = g_{i+1} g_i g_{i+1}, \qquad 0 \leq i \leq n - 1, \qquad (25)$$

$$w g_i = g_{i-1} w, \qquad 0 \leq i \leq n - 1, \qquad (26)$$

with indices understood as elements of $\mathbb{Z}/n\mathbb{Z}$.

It follows from (26) that w^n is a central element.

There exists a canonical homomorphism π from the extended affine braid group $\widetilde{B}_n$ to the classical braid group B_n. On the generators g_i and w the homomorphism π is given by the following rules

$$\pi(g_i) = g_i, \quad 1 \leq i \leq n - 1,$$
$$\pi(w) = g_{n-1} g_{n-2} \cdots g_2 g_1. \qquad (27)$$

It follows from (27) that

$$\pi(g_0) = \pi(w) g_1 \pi(w)^{-1} = g_{n-1} g_{n-2} \cdots g_2 g_1 g_2^{-1} \cdots g_{n-2}^{-1} g_{n-1}^{-1}.$$

Now we are going to define the Dunkl elements $Y_1, \ldots, Y_n$, and the dual Dunkl elements $Y_1^*, \ldots, Y_n^*$ in the extended affine braid group.

Definition 6. For each $i = 1, \ldots, n$, we define (cf. Refs. 21 and 4) the Dunkl and dual Dunkl elements Y_i and Y_i^*, respectively, by the following formulae:

$$Y_i = g_i g_{i+1} \cdots g_{n-1} w g_1^{-1} \cdots g_{i-1}^{-1}, \qquad (28)$$

$$Y_i^* = g_i^{-1} g_{i+1}^{-1} \cdots g_{n-1}^{-1} w g_1 \cdots g_{i-1}. \qquad (29)$$

Note that $Y_1 = g_1 \cdots g_{n-1} w$, $Y_n = w g_1^{-1} \cdots g_{n-1}^{-1}$, $Y_1^* = w g_1 \cdots g_{n-1}$ and $Y_n^* = g_1^{-1} \cdots g_{n-1}^{-1} w$.

The Dunkl elements satisfy the following commutation relations with g_1, $\ldots$, g_{n-1}:

$$g_i Y_{i+1} g_i = Y_i, \ g_i Y_j = Y_j g_i \qquad (j \neq i, i + 1), \qquad (30)$$

$$g_i Y_i^* g_i = Y_{i+1}^*, \quad g_i Y_j^* = Y_j^* g_i \qquad (j \neq i, i + 1). \qquad (31)$$

It is clear from our construction, that the image in the braid group B_n of the dual Dunkl element Y_k^*, $2 \leq k \leq n$, under the canonical homomorphism $\pi \colon \widetilde{B}_n \to B_n$ coincides with the multiplicative Jucys–Murphy element D_k, namely,

$$\pi(Y_k^*) = g_k^{-1} g_{k+1}^{-1} \cdots g_{n-1}^{-1} \pi(w) g_1 \cdots g_{k-1}$$
$$= g_{k-1} \cdots g_2 g_1^2 g_2 \cdots g_{k-1}$$
$$\overset{(10)}{=} D_k, \quad 2 \leq k \leq n,$$

and $\pi(Y_1^*) = 1$.

Lemma 5. *The Dunkl elements Y_k (resp. the dual Dunkl elements Y_k^*) commute pairwise.*

Let us illustrate the main idea of the proof of Lemma 5 on some example. Let us take $n = 5$ and prove that $Y_2^* Y_3^* = Y_3^* Y_2^*$. Indeed,

$$Y_2^* Y_3^* = g_2^{-1} g_3^{-1} g_4^{-1} w g_1 g_3^{-1} g_4^{-1} w g_1 g_2 = g_2^{-1} g_3^{-1} g_4^{-1} g_0 g_2^{-1} g_3^{-1} g_4 g_0 w^2$$
$$= g_2^{-1} g_3^{-1} g_2^{-1} g_4 g_3^{-1} g_0 g_4 g_0 w^2 = g_3^{-1} g_2^{-1} g_3^{-1} g_4 g_3^{-1} g_4 g_0 g_4 w^2.$$

Similarly,

$$Y_3^* Y_2^* = g_3^{-1} g_4^{-1} w g_1 g_3^{-1} g_4^{-1} w g_1 = g_3^{-1} g_2^{-1} g_4^{-1} g_3^{-1} g_0 g_4 w^2.$$

It is easy to see that in $\widetilde{B}_5$ we have

$$g_3^{-1} g_2^{-1} g_3^{-1} g_4 g_3^{-1} g_4 g_0 g_4 = g_3^{-1} g_2^{-1} g_4^{-1} g_3^{-1} g_0 g_4.$$

Now let us study the quasiclassical limit of the Dunkl element Y_k. The first step is to consider Y_k as an element of the extended affine Hecke algebra $H(\widetilde{W})$, [21, Section 2], [4, 5]. In other words, we have to add to the relations (24)–(26) the new ones

$$g_i^2 = (t - 1)g_i + t, \quad 0 \leq i \leq n, \tag{32}$$

where t is a new parameter. The next step is to consider a representation of the extended affine Hecke algebra $H(\widetilde{W})$ in the algebra $\mathcal{D}_{q,x}[W]$ of q-difference operators with permutations, see, for example, Refs 4 and 21. Following Ref. 21, we define the elements T_i, $i = 0, 1, \ldots, n-1$, in $\mathcal{D}_{q,x}[W]$ by

$$T_i = t + \frac{x_{i+1} - tx_i}{x_{i+1} - x_i}(s_i - 1), \quad i = 1, \ldots, n-1,$$

$$T_0 = t + \frac{x_1 - tqx_n}{x_1 - qx_n}(s_0 - 1).$$

Here $s_1, \ldots, s_{n-1}$ are the standard generators of the symmetric group S_n, and $s_0 = w s_1 w^{-1}$, where $w = s_{n-1} s_{n-2} \cdots s_1 \tau_1$, and $\tau_1 = T_{q,x_1}$ is the q-shift operator:

$$(\tau_1 f)(x_1, \ldots, x_n) = f(qx_1, x_2, \ldots, x_n).$$

One can check that the elements T_i, $0 \leq i \leq n - 1$, and w, satisfy the relations (24)–(26) and (32), and define a representation of the extended affine Hecke algebra $H(\widetilde{W})$ in the algebra $\mathcal{D}_{q,x}[W]$ of q-difference operators with permutations. In this representation, the Dunkl element Y_k (respectively Y_k^*), $1 \leq k \leq n$, up to a power of t coincides (see Refs. 4, 5, and 21) with the Dunkl–Cherednik operator $\mathbf{Y}_k$:

$$\mathbf{Y}_k = t^{-n+2k-1} T_k T_{k+1} \cdots T_{n-1} w t_1^{-1} \cdots T_{k-1}^{-1}, \tag{33}$$

$$\mathbf{Y}_k^* = t^{n-2k+1} T_k^{-1} T_{k+1}^{-1} \cdots T_{n-1}^{-1} w T_1 \cdots T_{k-1}. \tag{34}$$

In order to understand better the quasiclassical behavior of the Dunkl-Cherednik operator $\mathbf{Y}_k$, it is convenient to rewrite slightly the formulae (33) and (34). Namely, let us define ($1 \le i < j \le n$) the following operators acting on the ring of polynomials $\mathbb{Z}[t, t^{-1}][x_1, \ldots, x_n]$

$$\overline{T}_{ij} = 1 + \frac{(1 - t^{-1})x_j}{x_i - x_j}(1 - s_{ij}) = t^{-1}T_{ij}s_{ij},$$

where s_{ij} is the exchange operator for variables x_i, x_j. Then

$$\mathbf{Y}_i = \overline{T}_{i,i+1}\overline{T}_{i,i+2} \cdots \overline{T}_{i,n}\tau_i \overline{T}_{1,i}^{-1} \cdots \overline{T}_{i-1,i}^{-1}, \tag{35}$$

where $\tau_i = s_i s_{i+1} \cdots s_{n-1} w s_1 \cdots s_{i-1}$ is the q-shift operator: $\tau_i = \tau_{x_i} = T_{q,x_i}$. Finally, in the quasiclassical limit $q \to 1$ with rescaling $t = q^\beta$, we obtain from (35) that

$$\mathcal{D}_j := \lim_{q \to 1} \frac{1 - \mathbf{Y}_j}{1 - q} = x_j \frac{\partial}{\partial x_j} + \beta \sum_{i<j} x_j \partial_{ij} - \beta \sum_{k>j} \partial_{jk} x_k. \tag{36}$$

Note that these Dunkl operators $\mathcal{D}_k$ commute with each other; that is, $[\mathcal{D}_i, \mathcal{D}_j] = 0$.

More generally, let us define the affine extension of the algebra $\mathcal{G}_n$:

Definition 7. Define the algebra $\widetilde{\mathcal{G}}_n$ as the algebra (say over $\mathbb{Z}[q, q^{-1}]$) with generators

$$[ij], \quad 0 \le i < j \le n; \quad x_i, \; 1 \le i \le n, \quad \text{and} \quad w,$$

which satisfy the relations (16) and (17) for i, j, k, $l \in [1, n]$, and the following additional relations:

$$x_i x_j = x_j x_i, \qquad\qquad\qquad\qquad 1 \le i, j \le n, \tag{37}$$
$$x_i[ab] = [ab]x_i, \qquad\qquad\qquad\qquad \text{if } i \ne a, b, \tag{38}$$
$$x_j[ij] = [ij]x_i + 1, \; x_i[ij] = [ij]x_j - 1 \quad 1 \le i < j \le n, \tag{39}$$
$$w x_i = q^{\delta_{1,i}} x_{i-1} w, \tag{40}$$
$$w[ij] = q^{-\delta_{i,0}}[i-1, j-1]w, \tag{41}$$

with indices understood as elements of $\mathbb{Z}/n\mathbb{Z}$.

It is clear from (39), that

$$[x_i + x_j, [ij]] = 0. \tag{42}$$

We define the Dunkl elements $\tilde{\theta}_j$ in the algebra $\widetilde{\mathcal{G}}_n$ by the following rule:

$$\tilde{\theta}_j = x_j + \theta_j, \quad 1 \le j \le n. \tag{43}$$

The following result is well known; see, for example, Ref. 17:

Lemma 6. *The classical Yang–Baxter equation* (18), *together with commutativity relation* (17) *and relation* (42), *imply that the Dunkl elements* $\tilde{\theta}_j = x_j + \theta_j$ *commute pairwise.*

Proof. If $i < j$, then

$$[\tilde{\theta}_i, \tilde{\theta}_j] = [x_i, x_j] + [x_i, \theta_j] + [\theta_i, x_j] + [\theta_i, \theta_j] = [x_i + x_j, [ij]] = 0. \qquad \square$$

Let us introduce the algebra $\widetilde{\mathcal{G}}_n^0$ which is the quotient of the algebra $\widetilde{\mathcal{G}}_n$ by the two-side ideal generated by the relation

$$[ij]^2 = 0, \quad 1 \le i < j \le n. \tag{44}$$

Definition 8. For $1 \le i < j \le n$ let us define an element $T_{ij} \in \widetilde{\mathcal{G}}_n^0$ by the following formula (cf. Ref. 21)

$$T_{ij} = t - (x_j - tx_i)[ij]. \tag{45}$$

It follows from (39) and (44), that

$$T_{ij}^2 = (t-1)T_{ij} + t.$$

Our next goal is to check that the elements $\tilde{T}_i := T_{i,i+1}$, $1 \le i \le n-1$, satisfy the Coxeter relations.

Proposition 2. *We have the following relation in the algebra* $\widetilde{\mathcal{G}}_n^0$:

$$T_{ab}T_{bc}T_{ab} = T_{bc}T_{ab}T_{bc}, \quad a < b < c. \tag{46}$$

Proof. Direct computation based on (39) and the following relation in the algebra $\widetilde{\mathcal{G}}_n^0$:

$$[ab][bc][ab] = [bc][ab][bc], \quad 1 \le a < b < c \le n. \qquad \square$$

As a corollary, we see that the elements w, $\tilde{T}_0 = w\tilde{T}_1 w^{-1}$, $\tilde{T}_1, \ldots, \tilde{T}_{n-1}$ generate the representation of the extended affine Hecke algebra $H(\widehat{W})$. In this representation the quasiclassical limit $q \to 1$ with rescaling $t = q^\beta$ of the Dunkl element $t^{-n+2j-1}Y_j$ is equal to

$$x_j \frac{\partial}{\partial x_j} + \beta \sum_{i<j} x_j[ij] - \beta \sum_{k>j} [jk]x_k. \tag{47}$$

The element (47) does not belong to the algebra $\widetilde{\mathcal{G}}_n$, but its further extension $\widetilde{\widetilde{\mathcal{G}}}_n$, which is an analog of the double affine Hecke algebra, introduced by I. Cherednik. It is an interesting problem to understand some connections between the Schubert calculus and the Dunkl–Cherednik operators.

Conjecture 1 (Nonnegativity conjecture; [10, Conjecture 8.1]).
For any $w \in S_n$, the Schubert polynomial $\mathfrak{S}_w$ evaluated at the Dunkl elements $\tilde{\theta}_1, \ldots, \tilde{\theta}_n$ belongs to the positive cone $\tilde{\mathcal{G}}_n^+$, where $\tilde{\mathcal{G}}_n^+$ is the cone of all nonnegative integer linear combinations of all (noncommutative in general) monomials in the generators x_i, $1 \le i \le n$, and $[ij]$, $1 \le i < j \le n$, of the algebra $\tilde{\mathcal{G}}_n^0$.

Acknowledgments: This work was done during my stay at the CRM, Université de Montréal. I would like to thank all my colleagues for discussions and support. I would like also to thank C. Dunkl, T. Maeno, M. Noumi, and A. N. Varchenko for fruitful and stimulating discussions. My special thanks to Arun Ram, who explained to me during my stay at the University of Wisconsin (1995) the importance of the Jucys–Murphy elements in the representation theory, and N. A. Liskova for inestimable help in different stages of this paper and constructive criticism.

7   REFERENCES

1. V. I. Arnold, *The cohomology ring of the group of dyed braids*, Mat. Zametki **5** (1969), 227–231 (Russian).

2. D. Bar-Natan, *Bibliography of Vassiliev invariants*, http://www.ma.huji.ac.il/~drorbn/VasiBib/VasiBib.html.

3. E. Brieskorn, *Automorphic sets and braids and singularities*, Braids (Santa Cruz, CA, 1986) (J. S. Birman and A. Libgober, eds.), Contemp. Math., Vol. 78, Amer. Math. Soc., Providence, RI, 1988, pp. 45–115.

4. I. Cherednik, *Double affine Hecke algebras and Macdonald's conjectures*, Ann. of Math. **141** (1995), No. 1, 191–216.

5. _____, *Lectures on affine quantum Knizhnik–Zamolodchikov equations, quantum many body problems, Hecke algebras, and Macdonald theory*, Lectures given at the International Institute for Advances Studies, Kyoto, Memoirs of the Mathematical Society of Japan, Vol. 1, Math. Soc. Japan, Tokyo, Kyoto, 1995, 1998.

6. V. G. Drinfeld, *Quasi-Hopf algebras*, Leningrad Math. J. **1** (1990), 1419–1457.

7. _____, *On quasitriangular quasi-Hopf algebras and on a group that is closely connected with* Gal($\bar{\mathbb{ii}}/\mathbb{ii}$), Leningrad Math. J. **2** (1991), No. 4, 829–860.

8. C. F. Dunkl, *Differential-difference operators associated to reflection groups*, Trans. Amer. Math. Soc. **311** (1989), No. 1, 167–183.

9. _____, *Harmonic polynomials and peak sets of reflection groups*, Geom. Dedicata **32** (1989), No. 2, 157–171.

10. S. Fomin and A. N. Kirillov, *Quadratic algebras, Dunkl elements, and Schubert calculus*, Advances in Geometry (J.-L. Brylinski, R. Brylinski, V. Nistor, B. Tsygan, and P. Xu, eds.), Progr. Math., Vol. 172, Birkhäuser, Boston, MA, 1999, pp. 147–182.

11. L. Funar, *Vassiliev invariants. I. Braid groups and rational homotopy theory*, Rev. Roumaine Math. Pures Appl. **42** (1997), No. 3-4, 245–272.

12. A. J. Hahn, *Quadratic algebras, Clifford algebras, and arithmetic Witt groups*, Universitext, Springer, New York, 1994.

13. G. J. Heckman, *Dunkl operators*, Séminaire Bourbaki. Vol. 1996/97, Astérisque, Vol. 245, Soc. Math. France, Paris, 1997, pp. 223–246.

14. H. Hiller, *Geometry of Coxeter Groups*, Res. Notes Math., Vol. 54, Pitman, Boston–London, 1982.

15. Yu. I. Manin, *Quantum Groups and Noncommutative Geometry*, Centre de recherches mathématiques, Montréal, 1988.

16. A. A. Jucys, *On the Young operators of symmetric groups*, Litovsk. Fiz. Sb. **6** (1966), 163–180 (Russian).

17. A. A. Kirillov, Jr., *Lectures on affine Hecke algebras and Macdonald's conjectures*, Bull. Amer. Math. Soc. **34** (1997), No. 3, 351–292.

18. A. N. Kirillov, *Cauchy identities for universal Schubert polynomials*, q-alg/9703047.

19. _____, *On some quadratic algebras*, Tech. Report CRM-2478, Centre de recherches mathématiques, 1997.

20. A. N. Kirillov and T. Maeno, *Bracket algebras and quantum Schubert calculus for Coxeter groups*, in preparation.

21. A. N. Kirillov and M. Noumi, *Affine Hecke algebras and raising operators for Macdonald polynomials*, Duke Math. J. **93** (1998), No. 1, 1–39.

22. T. Kohno, *Série de Poincaré–Koszul associée aux groupes de tresses pures*, Invent. Math. **82** (1985), No. 1, 57–75.

23. _____, *Linear representations of braid groups and classical Yang–Baxter equations*, Braids (Santa Cruz, CA, 1986) (J. S. Birman and A. Libgober, eds.), Contemp. Math., Vol. 78, Amer. Math. Soc., Providence, RI, 1988, pp. 339–363.

24. _____, *Vassiliev invariants and de Rham complex on the space of knots*, Symplectic Geometry and Quantization (Yokohama, 1993) (Y. Maeda, H. Omori, and A. Weinstein, eds.), Contemp. Math., Vol. 179, Amer. Math. Soc., Providence, RI, 1994, pp. 123–138.

25. X.-S. Lin, *Braid algebras, trace modules and Vassiliev invariants*, preprint, 1993.

26. I. G. Macdonald, *Affine Hecke algebras and orthogonal polynomials*, Séminaire Bourbaki. Vol. 1994/95, Astérisque, Vol. 237, Soc. Math. France, Paris, 1996, pp. 189–207.

27. G. E. Murphy, *A new construction of Young's seminormal representation of the symmetric groups*, J. Algebra **69** (1981), No. 2, 287–297.

28. A. Ram, *Seminormal representations of Weyl groups and Iwahori–Hecke algebras*, Proc. London Math. Soc. **75** (1997), No. 1, 99–133.

17

Elliptic Solutions to Difference Nonlinear Equations and Nested Bethe Ansatz Equations

I. Krichever

ABSTRACT We outline an approach to a theory of various generalizations of the elliptic Calogero–Moser (CM) and Ruijsenaars–Schneider (RS) systems based on a special inverse problem for linear operators with elliptic coefficients. Hamiltonian theory of such systems is developed with the help of universal symplectic structure proposed by D. H. Phong and the author. Canonically conjugated variables for spin generalizations of the elliptic CM and RS systems are found.

1 Introduction

The elliptic nested Bethe ansatz equations is a system of algebraic equations

$$\prod_{j\neq i} \frac{\sigma(x_i^n - x_j^{n+1})\sigma(x_i^n - \eta - x_j^n)\sigma(x_i - x_j^{n-1} + \eta)}{\sigma(x_i^n - x_j^{n-1})\sigma(x_i^n + \eta - x_j^n)\sigma(x_i - x_j^{n+1} - \eta)} = -1 \qquad (1)$$

for N unknown functions $x_i = x_i^n$, $i = 1, \ldots, N$, of a discrete variable n [13]. (Here and below $\sigma(x) = \sigma(x \mid \omega_1, \omega_2)$, $\zeta(z) = \zeta(z \mid \omega, \omega')$, and $\wp(x) = \wp(x \mid \omega, \omega')$ are the Weierstrass σ-, ζ-, and $\wp$-functions corresponding to the elliptic curve with periods 2ω, $2\omega'$.) This system is an example of a whole family of integrable systems that have attracted renewed interest for years. The most recent burst of interest is due to unexpected connections of these systems with the Seiberg–Witten solution to $N = 2$ supersymmetric gauge theories [27]. It turns out that the low-energy effective theory for SU(N) model with matter in the adjoint representation (identified first in Ref. 6 with SU(N) Hitchin systems) is isomorphic to the elliptic CM system. This connection quantum was used to find the order parameters in Ref. 5.

The elliptic Calogero–Moser (CM) system [3, 20] is a system of N identical particles on a line interacting with each other via the potential $V(x) = \wp(x)$. Its equations of motion have the form

$$\ddot{x}_i = 4 \sum_{j\neq i} \wp'(x_i - x_j). \qquad (2)$$

The CM system is completely integrable Hamiltonian system; i.e., it has N independent integrals H_k in involution [24, 25]. The second integral H_2 is the Hamiltonian of (2).

In Ref. 1 a remarkable connection of the CM system with a theory of elliptic solutions to the KdV equations was revealed. It was shown that the elliptic solutions of the KdV equations have the form $u(x, t) = 2 \sum_{i=1} \wp(x - x_i(t))$ and the poles x_i of the solutions satisfy the constraint $\sum_{j \neq i} \wp'(x_i - x_j) = 0$, which is the *locus* of stationary points of the CM system. Moreover, it turns out that a dependence of the poles with respect to t coincides with the Hamiltonian flow corresponding to the third integral H_3 of the system. In Refs. 14 and 4 it was found that this connection becomes an isomorphism in the case of the elliptic solutions to the Kadomtsev–Petviashvilii equation. Since then, the theory of the CM system and its various generalizations is inseparable from the theory of the elliptic solutions to the soliton equations.

In Ref. 22 system (1) was revealed as the pole system corresponding to the elliptic solutions of completely discretized version of the KP equation on the lattice. It was noticed that equations (1) have the form of the Bethe ansatz equations for the spin-$\frac{1}{2}$ Heisenberg chain with impurities. Its connection with the nested ansatz equations for the A_k-lattice models was established in [13].

As shown in Ref. 19, an intermediate discretization of the KP equation that is the $2D$ Toda lattice equations leads to the Ruijesenaars–Schneider system [26]:

$$\ddot{x}_i = \sum_{s \neq i} \dot{x}_i \dot{x}_s \big(V(x_i - x_s) - V(x_s - x_i) \big), \quad V(x) = \zeta(x) - \zeta(x + \eta). \quad (3)$$

which is a relativistic version of (2).

The main goal of this paper is to present a general approach to a theory of the CM-type many body systems that is based on a special inverse problem for *linear* operators with the elliptic coefficients. This approach originated in Ref. 15 and developed in Refs. 13, 16, and 19 clarifies the connection of these systems with the soliton equations for which the corresponding linear operator is the Lax operator. We formulate the inverse problem in the next section and show that its solution is equivalent to a finite-dimensional integrable system. We discuss an algebraic-geometric interpretation of the corresponding systems.

The advantage of our approach is that it generates the finite-dimensional system simultaneously with its Lax representation. Until recently, among its disadvantages was missing connection to the Hamiltonian theory. For example, in Ref. 19 spin generalization of the RS systems was proposed and explicitly solved in terms of the Riemann theta-functions of the auxiliary spectral curves. At the same time, all direct attempts to show that this system is Hamiltonian have failed, so far.

One of the existing general frameworks to the Hamiltonian theory of the CM-type systems is based on their geometric interpretation as reductions of geodesic flows on symmetric spaces [24]. Equivalently, these models can be obtained from free dynamics in a larger phase space possessing a rich symmetry by means of the Hamiltonian reduction [12]. A generalization to infinite-dimensional phase spaces (cotangent bundles to current algebras and groups) was suggested in Refs. 10 and 9. The infinite-dimensional gauge symmetry allows one to make a reduction to finite degrees of freedom. Among systems having appeared this way, there are Ruijsenaars–Schneider-type models and the elliptic Calogero–Moser model.

A further generalization of this approach consists in considering dynamical systems on cotangent bundles to moduli spaces of stable holomorphic vector bundles on Riemann surfaces. Such systems were introduced by Hitchin [11], where their integrability was proved. An attempt to identify the known many body integrable systems in terms of the abstract formalism developed by Hitchin was made in Ref. 21. To do this, it is necessary to consider vector bundles on algebraic curves with singular points. It turns out that the class of integrable systems corresponding to the Riemann sphere with marked points includes spin generalizations of the Calogero–Moser model as well as integrable Gaudin magnets [8] (see also Ref. 7).

Unfortunately, a geometric interpretation of the spin generalization of the elliptic RS system has not been found yet. Recently, such realization and consequently the Hamiltonian theory of the rational degeneration of that system were found in Ref. 2.

In Section 3 we develop Hamiltonian theory of the CM-type systems in the framework of a new approach to the Hamiltonian theory of soliton equations proposed in Refs. 18 and 17. It can be applied evenly to any equation having the Lax representation. The symplectic structure is constructed in terms of the Lax operator, only. We discuss three basic examples: spin generalizations of the CM and the RS systems, and the nested Bethe ansatz equations. We would like to emphasize that the universal form of the symplectic structure provides a universal and direct way to a canonical transformation to the action-angle type variables.

We would like to refer to Ref. 5 for analysis of connections of the elliptic CM system to Seiberg–Witten theory of $N = 2$ supersymmetric gauge theories.

2 Generating Problem

Let $\mathcal{L}$ be a linear differential or difference operator in two variables x, t with coefficients that are scalar or matrix elliptic functions of the variable x (i.e., meromorphic double-periodic functions with the periods $2\omega_\alpha$, $\alpha = 1, 2$). We do not assume any special dependence of the coefficients with respect to the second variable. Then it is natural to introduce a notion of *double-Bloch* solutions of the equation

$$\mathcal{L}\Psi = 0. \tag{4}$$

We call a *meromorphic* vector-function $f(x)$ that satisfies the following monodromy properties:

$$f(x + 2\omega_\alpha) = B_\alpha f(x), \quad \alpha = 1, 2, \tag{5}$$

a *double-Bloch function*. The complex numbers B_α are called *Bloch multipliers*. (In other words, f is a meromorphic section of a vector bundle over the elliptic curve.)

In the most general form a problem that we are going to address is to *classify* and to *construct* all the operators $\mathcal{L}$ such that Eq. (4) has *sufficient* double-Bloch solutions.

It turns out that existence of the double-Bloch solutions is so restrictive that only in exceptional cases do such solutions exist. A simple and general explanation of that is due to the Riemann–Roch theorem. Let D be a set of points x_i, $i = 1, \ldots, m$, on the elliptic curve Γ_0 with multiplicites d_i and let $V = V(D; B_1, B_2)$ be a linear space of the double-Bloch functions with the Bloch multipliers B_α that have poles at x_i of order less or equal to d_i and holomorphic outside of D. Then the dimension of D is equal to

$$\dim D = \deg D = \sum_i d_i.$$

Now let x_i depend on the variable t. Then for $f \in D(t)$ the function $\mathcal{L}f$ is a double-Bloch function with the same Bloch multipliers, but in general with higher orders of poles because taking derivatives and multiplication by the elliptic coefficients increase orders. Therefore, the operator $\mathcal{L}$ defines a linear operator

$$\mathcal{L}|_D : V(D(t); B_1, B_2) \mapsto V(D'(t); B_1, B_2), \quad N' = \deg D' > N = \deg D,$$

and (4) is *always* equivalent to an *over-determined* linear system of N' equations for N unknown variables that are the coefficients $C_i = C_i(t)$ of an expansion of $\Psi \in V(t)$ with respect to a basis of functions $f_i(t) \in V(t)$. With some exaggeration one may say that in the soliton theory a representation of a system in the form of a compatibility condition of an over-determined system of the linear problems is considered as equivalent to integrability.

In all of the examples that we are going to discuss $N' = 2N$ and the over-determined system of equations has the form

$$LC = kC, \quad \partial_t C = MC, \tag{6}$$

where L and M are $N \times N$ matrix functions depending on a point z of the elliptic curve as on a parameter. A compatibility condition of (6) has the standard Lax form $\partial_t L = [M, L]$ and is equivalent to a finite-dimensional integrable system.

A basis in the space of the double-Bloch functions can be written in terms of the fundamental function $\Phi(x, z)$ defined by the formula

$$\Phi(x, z) = \frac{\sigma(z + x)}{\sigma(z)\sigma(x)} e^{-\zeta(z)x}. \tag{7}$$

Note, that $\Phi(x, z)$ is a solution of the Lamé equation:

$$\left(\frac{d^2}{dx^2} - 2\wp(x)\right) \Phi(x, z) = \wp(z)\Phi(x, z). \tag{8}$$

From the monodromy properties it follows that Φ considered as a function of z is double-periodic:

$$\Phi(x, z + 2\omega_\alpha) = \Phi(x, z),$$

though it is not elliptic in the classical sense owing to an essential singularity at $z = 0$ for $x \neq 0$.

As a function of x the function $\Phi(x, z)$ is double-Bloch function, i.e.,

$$\Phi(x + 2\omega_\alpha, z) = T_\alpha(z)\Phi(x, z), \quad T_\alpha(z) = \exp(2\zeta(\omega_\alpha)z - 2\omega_\alpha\zeta(z)).$$

In the fundamental domain of a lattice defined by $2\omega_\alpha$ the function $\Phi(x, z)$ has a unique pole at the point $x = 0$:

$$\Phi(x, z) = x^{-1} + O(x). \tag{9}$$

The gauge transformation

$$f(x) \mapsto \tilde{f}(x) = f(x)e^{ax},$$

where a is an arbitrary constant, does not change poles of any functions and transform double Bloch-function into double-Bloch function. If B_α are Bloch multipliers for f, then the Bloch multipliers for $\tilde{f}$ are equal to

$$\tilde{B}_1 = B_1 e^{2a\omega_1}, \quad \tilde{B}_2 = B_2 e^{2a\omega_2}. \tag{10}$$

We call two pairs of Bloch multipliers equivalent if they are connected with each other through the relation (10) for some a. Note that for all equivalent pairs of Bloch multipliers the product $B_1^{\omega_2} B_2^{-\omega_1}$ is a constant depending on the equivalence class, only.

From (9) it follows that a double-Bloch function $f(x)$ with simple poles x_i in the fundamental domain and with Bloch multipliers B_α (such that at least one of them is not equal to 1) may be represented in the form

$$f(x) = \sum_{i=1}^{N} c_i \Phi(x - x_i, z)e^{kx}, \tag{11}$$

where c_i is a residue of f at x_i and z, k are parameters related by

$$B_\alpha = T_\alpha(z)e^{2\omega_\alpha k}. \tag{12}$$

(Any pair of Bloch multipliers may be represented in the form (12) with an appropriate choice of parameters z and k.)

For a proof of (11) it is enough to note that as a function of x the difference of the left- and right-hand sides is holomorphic in the fundamental domain. It is a double-Bloch function with the same Bloch multipliers as the function f. But a nontrivial double-Bloch function with at least one of the Bloch multipliers that is not equal to 1 has at least one pole in the fundamental domain.

Now we are in a position to present a few examples of the generating problem.

Example 1 (The elliptic CM system [15]). Let us consider the equation

$$\mathcal{L}\Psi = \left(\partial_x - \partial_x^2 + u(x,t)\right)\Psi = 0, \tag{13}$$

where $u(x,t)$ is an elliptic function. Then, as shown in Ref. 15, Eq. (13) has N linear independent double-Bloch solutions with equivalent Bloch multipliers and N simple poles at points $x_i(t)$ if and only if $u(x,t)$ has the form

$$u(x,t) = 2\sum_{i=1}^{N} \wp\big(x - x_i(t)\big) \tag{14}$$

and $x_i(t)$ satisfy the equation of motion of the elliptic CM system (2).

The assumption that their exist N linear independent double-Bloch solutions with equivalent Bloch multipliers implies that they can be written in the form

$$\Psi = \sum_{i=1}^{N} c_i(t,k,z)\Phi(x - x_i(t), z)e^{kx+k^2 t}, \tag{15}$$

with the same z but different values of the parameter k.

Let us substitute (15) into (13). Then (13) is satisfied if and only f we get a function holomorphic in the fundamental domain. First of all, we conclude that u has poles at x_i only. The vanishing of the triple poles $(x - x_i)^{-3}$ implies that $u(x,t)$ has the form (14). The vanishing of the double poles $(x - x_i)^{-2}$ gives the equalities that can be written as a matrix equation for the vector $C = (c_i)$:

$$(L(t,z) - kI)C = 0, \tag{16}$$

where I is the unit matrix and the Lax matrix $L(t, z)$ is defined as follows:

$$L_{ii} = -\frac{1}{2}\dot{x}_i, \quad L_{ij}(t, z) = -\Phi(x_i - x_j, z), \ i \neq j. \tag{17}$$

Finally, the vanishing of the simple poles gives the equations

$$(\partial_t - M(t, z))C = 0, \tag{18}$$

where

$$M_{ii} = \wp(z) - 2\sum_{j \neq i} \wp(x_i - x_j),$$

$$M_{ij}(t, z) = -2\Phi'(x_i - x_j, z), \ i \neq j. \tag{19}$$

The existence of N linear independent solutions for (13) with equivalent Bloch multipliers implies that (16) and (18) have N independent solutions corresponding to different values of k. Hence, as a compatibility condition we get the Lax equation $\dot{L} = [M, L]$ which is equivalent to (2). Note that the last system does not depend on z. Therefore, if (16) and (18) are compatible for some z, then they are compatible for all z. As a result we conclude that if (13) has N linear independent double-Bloch solutions with equivalent Bloch multipliers then it has infinitely many of them. All the double-Bloch solutions are parameterized by points of an algebraic curve Γ defined by the characteristic equation

$$R(k, z) \equiv \det(kI - L(z)) = k^N + \sum_{i=1}^{N} r_i(z)k^{N-i} = 0. \tag{20}$$

Equation (20) can be seen as a dispersion relation between two Bloch multipliers and defines Γ as N-sheet cover of Γ_0.

From (7) and (17) it follows that

$$L = G\tilde{L}G^{-1}, \quad G_{ij} = e^{-\zeta(z)x_i}\delta_{ij}. \tag{21}$$

At $z = 0$ we have the form

$$\tilde{L} = z^{-1}(I - F) + O(1), \quad F_{ij} = 1. \tag{22}$$

The matrix F has zero eigenvalue with multiplicity $N - 1$ and a simple eigenvalue $(-N)$. Therefore, in a neighborhood of $z = 0$ the characteristic polynomial (20) has the form

$$R(k, z) = \prod_{i=1}^{N}(k - \nu_i z^{-1} - h_i + O(z)), \quad \nu_1 = 1 - N, \ \nu_i = 1, \ i > 1. \tag{23}$$

We call the sheet of the covering Γ at $z = 0$ corresponding to the branch $k = z^{-1}(1 - N) + O(1)$ by upper sheet and mark the point P_1 on this sheet among the preimages of $z = 0$. From (23) it follows that in general position when the curve Γ is smooth its genus equals N.

The coefficient $r_i(z)$ in (20) is an elliptic function with a pole at $z = 0$ of order i. The coefficients $I_{i,s}$ of the expansion

$$r_i(z) = I_{i,0} + \sum_{s=0}^{i-2} I_{i,s}\partial_z^s \wp(z) \tag{24}$$

are integrals of motion. From (23) it is easy to show that there are only N independent among them. Recently, a remarkably explicit representation for the characteristic Eq. (20) was found [5]:

$$R(k,z) = R^*(k + \zeta(z), z), \quad R^*(k,z) = \frac{1}{\sigma(z)}\sigma\left(z - \frac{\partial}{\partial k}\right) H(k), \tag{25}$$

where $H(k)$ is a polynomial. Note that (25) may be written as

$$R^* = \frac{1}{\sigma(z)} \sum_{n=1}^{N} \frac{1}{n!}\partial_z^n \sigma(z)\left(-\frac{\partial}{\partial k}\right)^n H(k).$$

The coefficients of the polynomial $H(k)$ are free parameters of the spectral curve of the CM system.

After the Lax representation and the corresponding algebraic curve Γ are constructed, the next step is to consider analytical properties of the eigenvectors of the Lax operator L on Γ.

Let L be a matrix of the form (17). Then the components of the eigenvector $C(P) = (c_i(P))$, $P = (k, z) \in \Gamma$, normalized by the condition

$$\sum_{i=1}^{N} c_i(P)\Phi(-x_i, z) = 1, \tag{26}$$

are meromorphic functions on Γ outside the preimages P_i of $z = 0$. They have N poles $\gamma_1, \ldots, \gamma_N$ (in a general position γ_s are distinct). At the points P_j the coordinates $c_i(P)$ have the expansion

$$\begin{aligned} c_i &= z\left(c_i^1 + O(z)\right)e^{-\zeta(z)x_i}; \\ c_i &= \left(c_i^j + O(z)\right)e^{-\zeta(z)x_i}, \quad i > 1. \end{aligned} \tag{27}$$

If we denote the diagonal elements of L by $-p_i/2$, then the above constructed correspondence

$$p_i, \ x_i \mapsto \{\Gamma, \mathcal{D} = \{\gamma_s\}\} \tag{28}$$

is an isomorphism (on the open set).

Now let $x_i(t)$ be a solution of (2), then the divisor $\mathcal{D}$ corresponding to $p_i = \dot{x}_i(t)$, $x_i(t)$ depends on t, $\mathcal{D} = \mathcal{D}(t)$. It turns out that under the Abel transform this dependence becomes linear on the Jacobian $J(\Gamma)$ of the spectral curve. The final result is as follows.

Theorem 1. *The coordinates of the particles $x_i(t)$ are roots of the equation*

$$\theta(Ux + Vt + Z_0) = 0, \tag{29}$$

where $\theta(\xi) = \theta(\xi \mid B)$ is the Riemann theta-function corresponding to matrix of b-periods of holomorphic differentials on Γ; the vectors U and V are the vectors of b-periods of normalized meromorphic differentials on Γ with poles of order 2 and 3 at the point P_1; the vector Z_0 is the Abel transform of the divisor $\mathcal{D}_0 = \mathcal{D}(0)$.

This result can be reformulated in the following more geometric way. Let $J(\Gamma)$ be the Jacobain (N-dimensional complex torus) of a smooth genus N algebraic curve Γ. Abel transform defines imbedding of Γ into $J(\Gamma)$. A point $P \in \Gamma$ defines a vector U in $J(\Gamma)$ that is the tangent vector to the image of Γ at the point. Let us consider a class of curves having the following property: there exists a point on the curve such that the complex linear subspace generated by the corresponding vector U *is compact*; i.e., it is an elliptic curve Γ_0. This means that there exist two complex numbers $2\omega_\alpha$, $\mathrm{Im}\,\omega_2/\omega_1 > 0$, such that $2\omega_\alpha U$ belongs to the lattice of periods of holomorphic differentials on Γ. From a purely algebraic-geometrical point of view the problem of the description of such curves is transcendental. It turns out that this problem has an explicit solution and algebraic equations that define such curves can be written as the characteristic equation for the Lax operator corresponding to the CM system. Moreover, it turns out that in general position Γ_0 intersects theta-divisor at N points x_i and if we move Γ_0 in the direction that is defined by the vector V of the second jet of Γ at P then the intersections of Γ_0 with the theta-divisor move according to the CM dynamics.

Example 2 (Spin generalization of the elliptic CM system [16]).
Let $\mathcal{L}$ be an operator of the same form (13) as in the previous case but now $u = u_\alpha^\beta(x, t)$ is an elliptic $(l \times l)$ *matrix* function of the variable x. We slightly reformulate results of Ref. 16 to a form that would be used later.

Equation (13) has $N \geq l$ linear independent double-Bloch solutions with N simple poles at points $x_i(t)$ and such that $(l \times N)$ matrix formed by its residues at the poles has rank l if and only if

  i) the potential u has the form

$$u = \sum_{i=1}^{N} a_i(t) b_i^+(t) \wp(x - x_i(t)), \tag{30}$$

where $a_i = (a_{i,\alpha})$ are l-dimensional vectors and $b_i^+ = (b_i^\alpha)$ are l-dimensional co-vectors;

 ii) $x_i(t)$ satisfy the equations

$$\ddot{x}_i = \sum_{j \neq i} (b_i^+ a_j)(b_j^+ a_i) \wp'(x_i - x_j); \tag{31}$$

iii) the vectors a_i and co-vectors b_i^+ satisfy the constraints

$$b_i^+ a_i = \sum_{\alpha=1}^{l} b_i^{\alpha}(t) a_{i\alpha}(t) = 2 \qquad (32)$$

and a system of the equations

$$\dot{a}_i = -\sum_{j\neq i} a_j (b_j^+ a_i) \wp(x_i - x_j) - \lambda_i a_i,$$

$$\dot{b}_i = \sum_{j\neq i} b_j (b_i^+ a_j) \wp(x_i - x_j) + \lambda_i b_i^+, \qquad (33)$$

where $\lambda_i = \lambda_i(t)$ are scalar functions.

In order to get these results we represent a double-Bloch vector function Ψ in the form

$$\Psi = \sum_{i=1}^{N} s_i(t, k, z) \Phi(x - x_i(t), z) e^{kx + k^2 t}, \qquad (34)$$

where s_i are l-dimensional vectors and substitute it into (13). The vanishing of the triple pole $(x - x_i)^{-3}$ gives that u has the form (30), the vectors s_i are proportional to a_i, i.e., $s_i = c_i a_i$, where c_i are scalars, and that the constraints (32) are fulfilled.

The vanishing of the coefficients in front of $(x - x_i)^{-2}$ implies the equation (16) for the vector C with the coordinates c_i, where the Lax matrix has the form

$$L_{ii} = -\frac{1}{2}\dot{x}_i, \quad L_{ij} = -\frac{1}{2}(b_i^+ a_j)\Phi(x_i - x_j, z), \; i \neq j. \qquad (35)$$

The vanishing of the coefficients in front of $(x - x_i)^{-1}$ implies equations (33) for the vectors a_i and the second Lax equation (18). The matrix M has the form

$$M_{ii} = (\lambda_i + \wp(z)), \quad M_{ij} = -(b_i^+ a_j)\Phi'(x_i - x_j) \qquad (36)$$

The Lax equation for these matrices implies (31) and equations (33) for b_i^+ (here we use the assumption that a_i span the whole l-dimensional space).

System (33) after the gauge transformation

$$a_i \rightarrow a_i q_i, \; b_i^+ \rightarrow b_i q_i^{-1}, \quad q_i = \exp\left(\int^t \lambda_i(t)\, dt\right), \qquad (37)$$

which does not effect (31) and (32), becomes

$$\dot{a}_i = -\sum_{j\neq i} a_j(b_j^+ a_i)\wp(x_i - x_j), \quad \dot{b}_i = \sum_{j\neq i} b_j(b_i^+ a_j)\wp(x_i - x_j). \qquad (38)$$

Equations (32) and (38) are invariant under transformations

$$a_i \to \lambda_i^{-1} a_i, \quad b_i^+ \to \lambda_i b_i^+, \quad a_i \to W^{-1} a_i, \quad b_i^+ \to b_i^+ W, \qquad (39)$$

where λ_i are constants and W is a constant $(l \times l)$ matrix. A factorization with respect to these transformations leaves us with a reduced phase space $\mathcal{M}$ of the dimension $\dim \mathcal{M} = 2Nl - l(l-1)$. Let us introduce a canonical system of coordinates on $\mathcal{M}$. First of all, for any set of a_i, b_i^+ we define the matrix

$$S_\alpha^\beta = \sum_{i=1}^N a_{i\alpha} b_i^\beta. \qquad (40)$$

and diagonalize it with the help of a matrix $W_{0\alpha}^j$, which leaves the co-vector $(1, \dots, 1)$ invariant, i.e.,

$$S_\alpha^\beta W_{0\beta}^j = \kappa_j W_{0\alpha}^j, \quad \sum_\alpha W_{0\alpha}^j = 1, \ j = 1, \dots, l. \qquad (41)$$

Then we define

$$A_i = W_0^{-1} a_i, \quad B_i^+ = b_i^+ W_0. \qquad (42)$$

The vectors A_i and co-vectors B_i satisfy the conditions

$$\sum_{i=1}^N A_{i\alpha} B_i^\beta = \kappa_\alpha \delta_\alpha^\beta, \qquad (43)$$

which destroy the second half of the gauge transformations (39).

At $z = 0$ we have

$$L = G \tilde{L} G^{-1}, \quad \tilde{L} = z^{-1}(1 - F) + O(1), \quad F_{ij} = (b_i^+ a_j) = (B_i^+ A_j), \qquad (44)$$

where $\tilde{L}$ is given by (21). The matrix F has rank l. Its null subspace is a subspace of vectors such that

$$\sum_{i=1}^N A_{i,\alpha} c_i = 0. \qquad (45)$$

Note that eigenvectors of F corresponding to nonzero eigenvalues are identified with B_i^j.

From (44) it follows that

$$R(k, z) = \prod_{i=1}^N (k - \nu_i z^{-1} - h_i + O(z)), \quad \nu_i = 1, \ i > l. \qquad (46)$$

As shown in Ref. 16 expansion (46) implies that the spectral curve defined
by (20) has (in general position) genus $g = Nl - l(l + 1) + 1$. At the same
time (46) implies that a number of independent integrals given by (20) is
equal to $\frac{1}{2} \dim \mathcal{M}$.

The angle-type variables of our reduced system are the divisor of poles
of the solution of (16) with the following normalization:

$$\sum_{\alpha=1}^{l} \sum_{i=1}^{N} A_{i\alpha} c_i \Phi(-x_i, z) = 1. \qquad (47)$$

At the points P_j the components of C have the form

$$c_i = z(c_i^j + O(z))e^{-\zeta(z)x_i}, \ i \le l; \ c_i = (c_i^j + O(z))e^{-\zeta(z)x_i}, \ j > l, \qquad (48)$$

where

$$\sum_{i=1}^{N} A_{i,\alpha} c_i^j = \delta_\alpha^j, \ j \le l, \ \sum_{i=1}^{N} A_{i,\alpha} c_i^j = 0, \ j > l. \qquad (49)$$

The last formulae identify the normalization (47) with a canonical normalization used in the soliton theory. The coordinates

$$\Psi_\alpha = \sum_{i=1}^{N} A_{i\alpha} c_i \Phi(x - x_i, z)^{kx}$$

of the corresponding Baker–Akhiezer function Ψ are meromorphic outside
the punctures P_j, $j \le l$. At P_j they have the form

$$\Psi_\alpha = \left(\delta_\alpha^j + O(z) \right) e^{\kappa_j x}$$

(see details in Ref. 16).

Outside the punctures P_j the canonically normalized vector C has $g +$
$l - 1$ poles $\gamma_1, \ldots, \gamma_{g-l+1}$. This number is equal to $\frac{1}{2} \dim \mathcal{M}$. Note that
these poles are independent on the transformations (39). Therefore, we
have constructed an algebraic-geometric correspondence

$$\mathcal{M} = \{x_i, p_i, A_i, B_i\} \mapsto \{\Gamma, \mathcal{D} = (\gamma_s)\}, \qquad (50)$$

which is an isomorphism on the open set. The reconstruction formulae for
solutions of the elliptic CM system (31), (32), (38) can be found in Ref. 16.
It turns out that the coordinates of x_i are defined by the same equation
(29). The only difference is a set of the corresponding algebraic curves
that are defined by the characteristic equation for the Lax matrix L of the
form (35). In pure algebraic-geometric form they can be described in a way
similar to that in the previous example. Namely, they are curves such that
their exists a set of l points on it with the following property. A linear space
span by the tangent vectors to the curve at these points in the Jacobian
contains a vector U that spans in $J(\Gamma)$ an elliptic curve Γ_0.

Example 3 (Spin generalization of the elliptic RS system [19]).
Let us consider now the differential-difference equation

$$\mathcal{L}\Psi = \partial_t \Psi(x,t) - \Psi(x+\eta,t) - v(x,t)\Psi(x,t) = 0, \qquad (51)$$

where η is a complex number and $v(x,t)$ is an elliptic $(l \times l)$ matrix function. It has $N \geq l$ linear independent double-Bloch solutions with N simple poles at points $x_i(t)$ and such that $(l \times N)$ matrix formed by its residues at the poles has rank l if and only if

i) the potential u has the form

$$v = \sum_{i=1}^{N} a_i(t) b_i^+(t) V(x - x_i(t)), \qquad (52)$$
$$V = \zeta(x - x_i + \eta) - \zeta(x - x_i),$$

where $a_i = (a_{i,\alpha})$ are l-dimensional vectors and $b_i^+ = (b_i^\alpha)$ are l-dimensional co-vectors;

ii) $x_i(t)$ satisfy the equations

$$\ddot{x}_i = \sum_{j \neq i} (b_i^+ a_j)(b_j^+ a_i) V(x_i - x_j) - V(x_j - x_i); \qquad (53)$$

iii) the vectors $a_i = (a_\alpha^i)$ and co-vectors $b_i = (b_i^\alpha)$ satisfy the constraints

$$b_i^+ a_i = \sum_{\alpha=1}^{l} b_i^\alpha(t) a_{i\alpha}(t) = \dot{x}_i \qquad (54)$$

and the system of equations

$$\dot{a}_i = \sum_{j \neq i} a_j (b_j^+ a_i) V(x_i - x_j) + \lambda_i a_i,$$
$$\dot{b}_i = -\sum_{j \neq i} b_j (b_i^+ a_j) V(x_j - x_i) + \lambda_i b_i^+, \qquad (55)$$

where $\lambda_i = \lambda_i(t)$ are scalar functions.

The gauge transformation (39) allows us to eliminate λ_i in (55). The corresponding system was introduced in Ref. 19 and is a spin generalization of the elliptic RS model, which coincides with (53), (54) for $l = 1$.

Construction of integrals of system (53)–(55) and the angle-type coordinates is parallel to the previous case but requires some technical modification because $\mathcal{L}$ is a difference operator in x. First of all, we choose a different basis in a space of double-Bloch functions. The proper choice is defined by the formula

$$\Phi(x, z) = \frac{\sigma(z + x + \eta)}{\sigma(z + \eta)\sigma(x)} \left[\frac{\sigma(z - \eta)}{\sigma(z + \eta)}\right]^{x/2\eta}. \tag{56}$$

It satisfies the difference analog of the Lamé equation (8):

$$\Phi(x + \eta, z) + c(x)\Phi(x - \eta, z) = E(z)\Phi(x, z),$$

where

$$c(x) = \frac{\sigma(x - \eta)\sigma(x + 2\eta)}{\sigma(x + \eta)\sigma(x)}, \quad E(z) = \frac{\sigma(2\eta)}{\sigma(\eta)} \frac{\sigma(z)}{(\sigma(z - \eta)\sigma(z + \eta))^{1/2}}.$$

The Riemann surface $\widehat{\Gamma}_0$ of the function $E(z)$ is a twofold covering of the initial elliptic curve Γ_0 with periods $2\omega_\alpha$, $\alpha = 1, 2$. Its genus is equal to 2.

As a function of x, the function $\Phi(x, z)$ is a double-Bloch function. In the fundamental domain of the lattice defined by $2\omega_\alpha$ the function $\Phi(x, z)$ has a unique pole at the point $x = 0$:

$$\Phi(x, z) = x^{-1} + A + O(x), \quad A = \zeta(z + \eta) + \frac{1}{2\eta} \ln \frac{\sigma(z - \eta)}{\sigma(z + \eta)}. \tag{57}$$

Therefore, we may represent a double-Bloch solution Ψ of (51) in the form

$$\Psi = \sum_{i=1}^{N} s_i(t, k, z)\Phi(x - x_i(t), z)k^{x/\eta}, \tag{58}$$

substitute this ansatz into the equation, and proceed as before. We get (53)–(55). The corresponding Lax operators have the form

$$L_{ij}(t, z) = (b_i^+ a_j)\Phi(x_i - x_j - \eta, z) \tag{59}$$

$$M_{ij}(t, z) = (\lambda_i - (\zeta(\eta) - A)\dot{x}_i)\delta_{ij} + (1 - \delta_{ij})(b_i^+ a_j)\Phi(x_i - x_j, z). \tag{60}$$

Explicit formulae in terms of the Riemann theta functions are the same as for spin generalization of the CM system. The only difference is due to the different family of the spectral curves. In the case $l = 1$, they may be defined as a class of curves having the following property: there exists a pair of points on the curve such that the complex linear subspace span by the corresponding vector U is an elliptic curve Γ_0. If we move Γ_0 in the direction that is defined by the vector V^+ (V^-) tangent to $\Gamma \in J(\Gamma)$ at the point P^+ (P^-), then the intersections x_i of Γ_0 with the theta-divisor move according to the RS dynamics. The spectral curves for $l > 1$ are characterized by an existence of two sets of points $P_i^\pm$, $i = 1, \ldots, l$ such that in the linear subspace spanned by the vectors corresponding to each pair there exists a vector U with the same property as above.

Example 4 (The nested Bethe ansatz equations [13]). Let us consider two-dimensional difference equation

$$\Psi(x, m+1) = \Psi(x+\eta) + v(x, m)\Psi(x, m) \tag{61}$$

with an elliptic in x coefficient of the form

$$v(x, m) = \prod_{i=1}^{N} \frac{\sigma(x - x_i^m)\sigma(x - x_i^{m+1} + \eta)}{\sigma(x - x_i^{m+1})\sigma(x - x_i^m + \eta)} \tag{62}$$

It has N linear independent double-Bloch solutions with equivalent Bloch multipliers if and only if the functions x_i^m of the discrete variable m satisfy the nested Bethe ansatz equations. The Lax representation for this system has the form

$$L(m+1)M(m) = M(m)L(m+1), \tag{63}$$

where

$$
\begin{aligned}
L_{ij}(m) &= \lambda_i(m)\Phi(x_i^m - x_j^m - \eta, z), \\
M_{ij}(m) &= \mu_i(m)\Phi(x_i^{m+1} - x_j^m, z)
\end{aligned} \tag{64}
$$

and

$$\lambda_i(m) = \frac{\prod_{s=1}^{M} \sigma(x_i^m - x_s^m - \eta)\sigma(x_i^m - x_s^{m+1})}{\prod_{s=1,\neq i}^{M} \sigma(x_i^m - x_s^m)\prod_{s=1}^{N}\sigma(x_i^m - x_s^{m+1} - \eta)}, \tag{65}$$

$$\mu_i(m) = \frac{\prod_{s=1}^{M}\sigma(x_i^{m+1} - x_s^{m+1} + \eta)\sigma(x_i^{m+1} - x_s^m)}{\prod_{s=1,\neq i}^{M}\sigma(x_i^{m+1} - x_s^{m+1})\prod_{s=1}^{N}\sigma(x_i^{m+1} - x_s^m + \eta)}. \tag{66}$$

A class of the spectral curves is the same as for the RS system. The solution x_i^m of (1) corresponding to the spectral curve and the divisor is defined by the equation

$$\theta(Ux + Vm + Z) = 0. \tag{67}$$

Here V is the vector from the puncture P_+ to the third point Q on Γ. When this point tends to P_+ the vector V becomes the tangent vector to the curve and we come to the RS system as a continuous limit of (1).

3 Hamiltonian Theory of the CM-Type Systems

As we have seen, various generating problems lead to various integrable finite-dimensional systems that can be explicitly solved via the spectral transform of a phase space $\mathcal{M}$ to algebraic-geometric data. On this way we do not use Hamiltonian description of the system. Moreover, a priori it's not clear why all the systems that can be constructed with the help of the generating scheme are Hamiltonian. In this section we clarify this problem using the approach to the Hamiltonian theory of soliton equations proposed in Ref. 18 and developed in Ref. 17.

First of all, let us outline a framework that was presented in the previous section. The direct spectral transform identifies a space of the solutions with a bundle over a space of the corresponding spectral curves. The fiber over a curve Γ is a symmetric power $S^{g+l-1}\Gamma$ (i.e., an unordered set of points $\gamma_s \in \Gamma$). Dimension of the space of the spectral curves equals $g+l-1 = \frac{1}{2}M$. The spectral curves are realized as N-sheet covering of the elliptic curve Γ_0.

Let us consider the case of spin generalization of the CM system. Entries of $L(z)$ are explicitly defined as functions on $\mathcal{M}$. Therefore, $L(z)$ can be seen as an operator-valued function and its external differential δL as an operator-valued one-form on $\mathcal{M}$. Canonically normalized eigenfunction $\Psi(z,k)$ of $L(z)$ is the vector-valued function on $\mathcal{M}$. Hence, its differential is a vector-valued one-form. Let us defined a two-form on $\mathcal{M}$ by the formula

$$\omega = \sum_{i=1}^{N} \mathrm{res}_{P_i}(\langle \Psi^*(z,k)(\delta L(z) + \delta k) \wedge \delta \Psi(z) \rangle)\, dz, \qquad (68)$$

where $\Psi^*(z,k)$ is an eigen-covector (row vector) of $L(z)$, i.e., the solution of the equation $\Psi^* L = k\Psi^*$, normalized by the condition $\langle \Psi^*(z,k)\Psi(z,k) \rangle = 1$. The form ω can be rewritten as

$$\omega = \mathrm{res}_{z=0} \mathrm{Tr}(\hat{\Psi}^{-1}(z)\delta L(z) \wedge \delta \hat{\Psi}(z) - \hat{\Psi}^{-1}(z)\delta \hat{\Psi}(z) \wedge \delta \hat{k}\wedge) \qquad (69)$$

where $\hat{\Psi}(z)$ is a matrix with columns $\Psi(z,k_j)$; $k_j = k_j(z)$ are different eigenvalues of $L(z)$ and $\hat{k}(z)$ is the diagonal matrix $k_j(z)\delta_{ij}$.

Note that Ψ^* are rows of the matrix $\hat{\Psi}^{-1}$. That implies that Ψ^* as a function on the spectral curve is meromorphic outside the punctures, has poles at the branching points of the spectral curve and zeros at the poles γ_s of Ψ. These analytical properties are used in the proof of the following theorem.

Theorem 2. *The two-form ω equals*

$$\omega = 2 \sum_{s=1}^{g+l-1} \delta z(\gamma_s) \wedge \delta k(\gamma_s). \qquad (70)$$

The meaning of the right-hand side of this formula is as follows. The spectral curve by definition arises with a meromorphic function $k(Q)$ and multi-valued holomorphic function $z(Q)$. Their evaluations $k(\gamma_s)$, $z(\gamma_s)$ at the points γ_s defines functions on the space $\mathcal{M}$ and the wedge product of their external differentials is a two-form on $\mathcal{M}$. Formula (70) identifies $k_s = k(\gamma_s)$, $z_s = z(\gamma_s)$ as Darboux coordinates for ω.

Remark 1. The right-hand side of (70) can be identified with a particular case of universal algebraic-geometric symplectic forms proposed in Ref. 18. They are defined on the generalized Jacobian bundles over a proper subspaces of the moduli spaces of Riemann surfaces with punctures. In the case of families of hyperelliptic curves that form was pioneered by Novikov and Veselov [23].

Remark 2. Equations (31), (32), and (38) are linearized by generalized Abel transform

$$A\colon S^{g+l-1}\Gamma \mapsto J(\Gamma) \times C^{l-1}. \tag{71}$$

This transform is defined with the help of a basis of the normalized holomorphic differentials $d\omega_i$, $i \leq g$, and with the help of normalized meromorphic differentials $d\Omega_j$, $j \leq (l-1)$ of the third kind with residues 1 and -1 at the points P_j and P_l, respectively. Let Q be a point of Γ. Then we define $(g+l-1)$-dimensional vector $A_k(Q)$ with the coordinates

$$A_i(Q) = \int^Q d\omega_i, \quad A_{g+j}(Q) = \int^Q d\Omega_j.$$

The isomorphism (71) is given by

$$\phi_k = \sum_{s=1}^{g+l-1} A_k(\gamma_s).$$

The action variables I_k canonically conjugated to ϕ, i.e., such that

$$\omega = \sum_{k=1}^{g+l-1} \delta\phi_k \wedge \delta I_k,$$

are equal

$$I_i = \oint_{a_i^0} k\,dz, \quad I_{g+j} = \mathrm{res}_{P_j} k\,dz = \nu_j,$$

where a_i^0 are a-cycles of a basis of cycles on Γ with the canonical matrix of intersections.

Let us outline a proof Theorem 2. The differential $\Omega = \langle \Psi^* \delta L \wedge \delta \Psi \rangle\, dz$ is a meromorphic differential on the spectral curve (the essential singularities of the factors cancel each other at the punctures). Therefore, a sum of its residues at the punctures is equal to a sum of other residues with negative sign. There are poles of two types. First of all, Ω has poles at the poles γ_s of Ψ. Note that $\delta\Psi$ has pole of the second order at γ_s. Taking into account that Ψ^* has zero at γ_s we obtain

$$\mathrm{res}_{\gamma_s}\Omega = \langle \Psi^* \delta L \Psi \rangle \wedge \delta z(\gamma_s) = \delta k(\gamma_s) \wedge \delta z(\gamma_s). \tag{72}$$

The last equality follows from the standard formula for a variation of the eigenvalue of an operator. The second term in (68) has the same residue at γ_s.

The second set of poles of Ω is a set of the branching points q_i of the cover. At q_i a pole of Ψ^* cancels with zero of the differential dz, $dz(q_i) = 0$ considered as differential on Γ. The function Ψ is holomorphic at q_i. If we take an expansion of Ψ in the local coordinate $(z - z(q_i))^{1/2}$ (in general position when the branching point is simple) and consider its variation we get that

$$\delta\Psi = -\frac{d\Psi}{dz}\delta z(q_i) + O(1). \tag{73}$$

Therefore, $\delta\Psi$ has simple pole at q_i. In the similar way we obtain

$$\delta k = -\frac{dk}{dz}\delta z(q_i). \tag{74}$$

Equalities (73) and (74) imply that

$$\operatorname{res}_{q_i} \Omega = \operatorname{res}_{q_i} \left[\langle \Psi^* \delta L d\Psi \rangle \wedge \frac{\delta k \, dz}{dk} \right] \tag{75}$$

Due to skew-symmetry of the wedge product, we may replace δL in (75) by $(\delta L - \delta k)$. Then using identities $\Psi^*(\delta L - \delta k) = \delta\Psi^*(k - L)$ and $(k - L)d\Psi = (dL - dk)\Psi$ we obtain

$$\operatorname{res}_{q_i} \Omega = -\operatorname{res}_{q_i} \langle \delta\Psi^* \Psi \rangle \wedge \delta k \, dz = \operatorname{res}_{q_i} \langle \Psi^* \delta\Psi \rangle \wedge \delta k \, dz \tag{76}$$

This term cancels with a residue of the second term in the sum (68). The theorem is proved.   □

Now let us express ω in terms of the coordinates (42) on $\mathcal{M}$. Using the gauge transformation

$$L = G\tilde{L}G^{-1}, \quad \Psi = GC \tag{77}$$

where G is given by (21), we obtain

$$\omega = \sum_j \operatorname{res}_{z=0} \operatorname{Tr}[\delta\tilde{L} \wedge \delta h + \hat{C}^{-1}(\delta\tilde{L} \wedge \delta\hat{C} + [\delta h, \tilde{L}] \wedge \delta\hat{C} - \delta h\hat{C} \wedge \delta\hat{k})]. \tag{78}$$

where $\delta h = \delta G G^{-1}$, $\delta h = \operatorname{diag}(-\delta x_i \zeta(z))$. From the relation $\tilde{L}\hat{C} = \hat{C}\hat{k}$ it follows that

$$\operatorname{Tr}(C^{-1}[\delta h, L] \wedge \delta C) = \operatorname{Tr}(C^{-1}\delta h \wedge (L\delta C - \delta C\delta\hat{k}))$$
$$= -\operatorname{Tr}(C^{-1}\delta h \wedge (\delta L\hat{C} - \hat{C}\delta\hat{k})).$$

Therefore,

$$\omega = \operatorname{res}_{z=0} \operatorname{Tr}(2\delta\tilde{L} \wedge \delta h + \hat{C}^{-1}\delta L \wedge \delta C). \tag{79}$$

The first term equals $\sum_i \delta p_i \wedge \delta x_i$. The last term equals

$$-\frac{1}{2} \operatorname{Tr} C_0^{-1} \delta(B_i^+ A_j) \wedge \delta C_0$$

$$= -\frac{1}{2} \operatorname{Tr} \left(C_0^{-1}(\delta B_j A_j) \wedge \delta C_0 + \delta C_0^{-1}(B_j^+ \wedge \delta A_j) C_0 \right), \quad (80)$$

where C_0 is the matrix c_i^j of leading coefficients of the expansions (48). From (49) it follows that

$$\sum_{j=1}^{N} \left(A_{j\alpha} \delta c_j^k + \delta A_{j,\alpha} c_j^k \right) = 0, \quad \sum_{j=1}^{N} \left(c_{ki}^* \delta B_j^\alpha + \delta c_{ki}^* B_j^\alpha \right) = 0,$$

where c_{jk}^* are matrix elements of C_0^{-1}. Substitution of the last formulae into (80) completes the proof of the following theorem.

Theorem 3. *The symplectic form ω given by (68) equals*

$$\omega = \sum_{i=1}^{N} \left(\delta p_i \wedge \delta x_i + \sum_{\alpha=1}^{l} \delta B_i^\alpha \wedge \delta A_{i\alpha} \right). \quad (81)$$

After the symplectic structure ω is identified with a standard one it can be directly checked that equations (31), (38) are Hamiltonian with respect to ω and with the Hamiltonian

$$H = \frac{1}{2} \sum_{i=1}^{N} p_i^2 + \frac{1}{2} \sum_{i \neq j} (b_i^+ a_j)(b_j^+ a_i) \wp(x_i - x_j). \quad (82)$$

Nevertheless, we would like to show that the existence of the Hamiltonian for system (31), (38) can be proven in the framework of our approach. That makes it applicable even for cases when we cannot write the Hamiltonian explicitly.

By definition a vector field ∂_t on a symplectic manifold is Hamiltonian if the contraction $i_{\partial_t}\omega(X) = \omega(X, \partial_t)$ of the symplectic form is an exact one-form $dH(X)$. The function H is the Hamiltonian corresponding to the vector field ∂_t. The equations $\partial_t L = [M, L]$, $\partial_t k = 0$, and the equation

$$\partial_t \Psi(t, P) = M \Psi(t, P) + \mu(P, t) \Psi(t, P), \quad (83)$$

where $\mu(P, t)$ is a scalar function, imply

$$i_{\partial_t}\omega = \sum_{i=1}^{N} \operatorname{res}_{P_i} \left(\langle \Psi^*(\delta L + \delta k(z))(M + f)\Psi(z, k) \rangle \right.$$
$$\left. - \langle \Psi^*(z, k)[M, L]\delta\Psi \rangle \right) dz. \quad (84)$$

Note that if a matrix $\Lambda(z)$ is holomorphic outside P_j then the differential $\langle \Psi^* \Lambda \Psi \rangle \, dz$ is holomorphic outside P_j, as well. Therefore, a sum of its residues at P_j is equal to zero. Using that and the relations

$$\langle \Psi^* [M, L] \delta \Psi \rangle = \langle \Psi^* M (L - k) \delta \Psi \rangle = \langle \Psi^* (z, k) M (\delta k - \delta L) \Psi \rangle$$

we get

$$i_{\partial_t} \omega = \sum_{i=1}^{N} \mathrm{res}_{P_i} \left(\langle \Psi^* (\delta L + \delta k) \Psi \rangle \mu(P, t) \right) dz = 2 \sum_{i=1}^{N} \mathrm{res}_{P_i} \, \delta \mu(P, t) \, dz. \quad (85)$$

A singular part of the function $\mu(P, t)$ is equal to the singular part of the eigenvalues of the second Lax operator

$$\mu_j(z) = -z^{-2} + z^{-1} 2(k_j(z)) + O(1), \quad (86)$$

where $k_j(z)$ is an expansion of $k(z)$ at P_j (see (4.8) in Ref. 16). Hence,

$$i_{\partial_t} \omega = 2 \, \mathrm{res}_{z=0} \, \mathrm{Tr}(z^{-2} \delta \hat{k} + z^{-1} \delta \hat{k}^2) \, dz = 2 \, \mathrm{Tr} \, \hat{k}^2 = 2 \, \mathrm{Tr} \, L^2 = H. \quad (87)$$

Now let us define a symplectic structure for spin generalization of the elliptic RS system by the formula

$$\omega = \sum_{i=1}^{N} \mathrm{res}_{P_i} (\langle \Psi^* (z, k)(\delta L(z) L^{-1} + \delta \ln k) \wedge \delta \Psi(z) \rangle) dz, \quad (88)$$

where L and Ψ are the Lax operator (59) and its eigenvector.

Theorem 4. *The two-form (88) equals*

$$\omega = 2 \sum_{s=1}^{g+l-1} \delta z(\gamma_s) \wedge \delta \ln k(\gamma_s). \quad (89)$$

Equations (53)–(55) are Hamiltonian with respect to this symplectic structure with Hamiltonian

$$H = \sum_{i=1}^{N} (b_i^+ a_i). \quad (90)$$

Note that (89) implies that ω is closed and does define a symplectic structure on $\mathcal{M}$. A proof of Theorem 4 is almost identical to the previous case.

At this moment, we do not know for $l > 1$ an explicit expression for ω in the original coordinates on $\mathcal{M}$. Formula (88) contains the inverse matrix L^{-1}. For $l = 1$ it can be written explicitly. Namely, let $\hat{L}$ is the matrix

$$\hat{L}_{ij} = f_i \frac{\sigma(z + x_i - x_j)}{\sigma(x_i - x_j - \eta)},$$

which is (up to gauge transformation (21) and scalar factor) equal to (59) for $l = 1$. Then entries of the inverse matrix equal

$$(L^{-1})_{jm} = f_m^{-1} \frac{\sigma(z + x_j - x_m - (N-2)\eta)}{\sigma(z+\eta)\sigma(z-(N-1)\eta)}$$

$$\times \frac{\prod_{k \neq m} \sigma(x_j - x_k + \eta) \prod_k \sigma(x_j - x_k - \eta)}{\prod_{k \neq j} \sigma(x_j - x_k) \prod_{k \neq m} \sigma(x_m - x_k)}$$

This formula has been used for a proof of the following theorem.

Theorem 5. *Let L be a matrix $L_{ij} = f_i \Phi(x_i - x_j - \eta, z)$, where Φ is given by (56). Then the form ω defined by (88) equals*

$$\omega = \sum_i \delta \ln f_i \wedge \delta x_i + \sum_{i,j} V(x_i - x_j)\delta x_i \wedge \delta x_j, \tag{91}$$

where $V(x)$ is defined in (52).

   Our approach is evenly applicable to the nested Bethe ansatz equations. It gives a Hamiltonian version of a proof proposed in Ref. 22 with the help of a Lagrangian interpretation of (1) that discrete evolution x_i^n is a canonical transform of the RS symplectic structure (91).

4   REFERENCES

1. H. Airault, H. P. McKean, and J. Moser, *Rational and elliptic so-lutions of the Korteweg–de Vries equation and a related many-body problem*, Commun. Pure Appl. Math. **30** (1977), No. 1, 95–148.

2. G. E. Arutyunov and S. A. Frolov, *On the Hamiltonian structure of the spin Ruijsenaars–Schneider system*, J. Phys. A **31** (1998), No. 18, 4203–4216, hep-th/9703119.

3. F. Calogero, *Exactly solvable one-dimensional many-body problems*, Lett. Nuovo Cimento **13** (1975), No. 11, 411–416.

4. D. V. Chudnovsky and G. V. Chudnovsky, *Pole expansions of nonlin-ear partial differential equations*, Nuovo. Cimento **40B** (1977), No. 2, 339–350.

5. E. D'Hoker and D. H. Phong, *Calogero–Moser systems in* SU(*n*) *Seiberg–Witten theory*, Nucl. Phys. B **513** (1998), No. 1-2, 405–444, hep-th/970905.

6. R. Donagi and E. Witten, *Supersymmetric Yang–Mills theory and integrable systems*, Nucl. Phys. B **460** (1996), No. 2, 299–334.

7. B. Enriquez and V. N. Rubtsov, *Hitchin systems, higher Gaudin op-erators and R-matrices*, Math. Res. Lett. **3** (1996), No. 3, 343–357.

8. M. Gaudin, *Diagonalisation d'une classe d'hamiltoniens de spin*, J. Phys. (Paris) **37** (1976), No. 10, 1089–1098.

9. A. Gorsky and N. Nekrasov, *Elliptic Calogero–Moser system from two-dimensional current algebra*, hep-th/9401021.

10. A. Gorsky and N. Nekrasov, *Hamiltonian systems of Calogero-type, and two-dimensional Yang–Mills theory*, Nucl. Phys. B **414** (1994), No. 1-2, 213–238.

11. N. Hitchin, *Stable bundles and integrable systems*, Duke Math. J. **54** (1987), No. 1, 91–114.

12. D. Kazhdan, B. Kostant, and S. Sternberg, *Hamiltonian group actions and dynamical systems of Calogero type*, Commun. Pure Appl. Math. **31** (1978), No. 4, 481–507.

13. I. Krichever, O. Lipan, P. Wiegmann, and A. Zabrodin, *Quantum integrable models and discrete classical Hirota equations*, Commun. Math. Phys. **188** (1997), No. 2, 267–304.

14. I. M. Krichever, *Rational solutions of the Kadomtsev–Petviashvili equation and integrable systems of N particles on a line*, Funct. Anal. Appl. **12** (1978), No. 1, 59–61.

15. _____ , *Elliptic solutions of the Kadomtsev–Petviashvili equation and integrable systems of particles*, Funct. Anal. Appl. **14** (1980), No. 4, 282–290.

16. I. M. Krichever, O. Babelon, E. Billey, and M. Talon, *Spin generalization of the Caloger–Moser system and the matrix KP equation*, Topics in Topology and Mathematical Physics (A. B. Sossinsky, ed.), Amer. Math. Soc. Transl. Ser. 2, Vol. 170, Amer. Math. Soc., Providence, RI, 1995, pp. 83–119, hep-th/9411160.

17. I. M. Krichever and D. H. Phong, *Symplectic forms in the theory of solitons*, hep-th/9708170.

18. _____ , *On the integrable geometry of soliton equations and N = 2 supersymmetric gauge theories*, J. Differential Geom. **45** (1997), No. 2, 349–389.

19. I. M. Krichever and A. Zabrodin, *Spin generalization of the Ruijsenaars–Schneider model, non-Abelian 2D Toda chain and representations of Sklyanin algebra*, Russian Math. Surveys **50** (1995), No. 6, 1101–1150, hep-th/9505039.

20. J. Moser, *Three integrable Hamiltonian systems connected to isospectral deformations*, Adv. Math. **16** (1975), 197–220.

21. N. Nekrasov, *Holomorphic bundles and many-body systems*, Commun. Math. Phys. **180** (1996), No. 3, 587–603.

22. F. W. Nijhoff, O.Ragnisco, and V. B. Kuznetsov, *Integrable time-discretization of the Ruijsenaars–Schneider model*, Commun. Math. Phys. **176** (1996), No. 3, 681–700.

23. S. P. Novikov and A. Veselov, *Poisson brackets that are compatible with the algebraic geometry and the dynamics of the Korteweg–de Vries equations on the set of finite-gap potentials*, Siberian Math. J. **26** (1982), No. 2, 357–362.

24. M. A. Olshanetsky and A. M. Perelomov, *Classical integrable finite-dimensional systems related to Lie algebras*, Phys. Rep. **71** (1981), No. 5, 313–400.

25. A. M. Perelomov, *Completely integrable classical systems connected with semisimple Lie algebras*. III, Lett. Math. Phys. **1** (1975/77), No. 6, 531–534.

26. S. N. M. Ruijsenaars and H. Schneider, *A new class of integrable systems and its relation to solitons*, Ann. Phys. (NY) **170** (1986), No. 2, 370–405.

27. N. Seiberg and E. Witten, *Monopoles, duality and chiral symmetry breaking in $N = 2$ supersymmetric QCD*, Nucl. Phys. B **431** (1994), No. 2, 484–550.

18

Creation Operators for the Calogero–Sutherland Model and Its Relativistic Version

Luc Lapointe
Luc Vinet

ABSTRACT The Jack and Macdonald symmetric polynomials enter, respectively, in the wave functions of the quantum integrable Calogero–Sutherland (CS) and Ruijsenaars–Schneider (RS) models. This paper is devoted to recent advances in the combinatorial theory of these functions. The emphasis is put on the creation operators that allow us to construct the excited state wave functions of the CS and RS models from their ground state wave functions.

1 Introduction

The Calogero–Sutherland model and its generalizations are of great help in the understanding of quantum many-body physics. Moreover, the wave functions of these quantum integrable systems are typically given in terms of symmetric functions that are central in algebraic combinatorics (in the case of the Calogero–Sutherland (CS) model, the energy eigenfunctions involve the Jack polynomials and in the case of the Ruijsenaars–Schneider relativistic version of the CS model, the wave functions are expressed in terms of the Macdonald polynomials). It would hence be of considerable interest to provide an algebraic description of these models and of the associated symmetric functions. Advances toward this goal have been made recently and will be the subject of this paper. We shall in particular describe how the excited state wave functions of the CS model and its generalization can be obtained, as in the case of harmonic oscillators, from the action of creation operators on the ground state wave functions. From the mathematical point of view, this entails formulas of Rodrigues' type for the Jack and Macdonald polynomials from which the integrality conjectures of Macdonald for these functions simply follow.

2   Notations and Definitions [12]

Symmetric polynomials are labeled by partition λ of their degree n, that
is sequences $\lambda = (\lambda_1, \lambda_2, \ldots)$ of nonnegative integers in decreasing order
$\lambda_1 \geq \lambda_2 \geq \ldots$ such that $|\lambda| = \lambda_1 + \lambda_2 + \cdots = n$. The number of nonzero
parts in λ is denoted $\ell(\lambda)$. For instance, $\lambda = (5, 4, 4, 1)$ is a partition of
14 with $\ell(\lambda) = 4$. Let λ and μ be two partitions of n. In the dominance
ordering, $\lambda \geq \mu$ if $\lambda_1 + \lambda_2 + \cdots + \lambda_i \geq \mu_1 + \mu_2 + \cdots + \mu_i$ for all i. Note that
this is only a partial ordering since, for example, $(5, 2, 2)$ and $(4, 4, 1)$ are
incomparable in this ordering. We can associate a diagram to each partition
λ. The diagram is made out of $\ell(\lambda)$ rows, labeled by the integer i, with λ_i
squares in each one of them. The squares are identified by the coordinates
$(i, j) \in \mathbb{Z}^2$ with i, the row index, increasing as one goes downward and j,
the column index, increasing as one goes from left to right. The diagram
of $(5, 4, 4, 1)$ is

For each square $s = (i, j)$ in the diagram of a partition λ, let $\ell'(s)$, $\ell(s)$,
$a(s)$ and $a'(s)$ be, respectively, the number of squares in the diagram of λ
to the north, south, east, and west of the square s. Choosing the square
$s = (2, 1)$ in the preceding diagram,

we get $\ell'(s) = 1$, $\ell(s) = 2$, $a(s) = 3$, and $a'(s) = 0$. By $\lambda \supset \mu$ it is meant
that the diagram λ contains the diagram μ, that is, that $\lambda_i \geq \mu_i$ for all
$i \geq 1$. The set-theoretic difference $\theta = \lambda - \mu$ is called a skew diagram. For
example, if $\lambda = (5, 4, 4, 1)$ and $\mu = (4, 3, 2)$, $\lambda - \mu$ is made of the dotted
squares in the picture

A skew diagram θ is a vertical m-strip (horizontal m-strip) if $|\theta| = m$ and
$\theta_i \leq 1$ for each $i \geq 1$ ($\theta_j \leq 1$ for each $j \geq 1$). In other words, a vertical
(horizontal) strip has at most one square in each row (column). Finally, to
every partition λ corresponds the conjugate partition λ'. The diagram of λ'

is obtained from the diagram of λ by transposing with respect to the main diagonal. In the case of $\lambda = (5, 4, 4, 1)$, the following diagram is associated to λ':

that is $\lambda' = (4, 3, 3, 3, 1)$.

Let Λ_N denote the ring of symmetric functions in the variables z_1, z_2, $\ldots$, z_N. Three standard bases for the space of symmetric functions are

i) the power sum symmetric functions p_λ, which in terms of the power sums

$$p_i = \sum_k z_k^i, \tag{1}$$

are given by

$$p_\lambda = p_{\lambda_1} p_{\lambda_2} \cdots ; \tag{2}$$

ii) the monomial symmetric functions m_λ, which are

$$m_\lambda = \sum_{\substack{\text{distinct} \\ \text{permutations}}} z_1^{\lambda_1} z_2^{\lambda_2} \cdots ; \quad \text{and} \tag{3}$$

iii) the elementary symmetric functions e_λ, which in terms of the ith elementary function

$$e_i = \sum_{j_1 < j_2 < \cdots < j_i} x_{j_1} x_{j_2} \cdots z_{j_i} = m_{(1^i)}, \tag{4}$$

are given by

$$e_\lambda = e_{\lambda_1} e_{\lambda_2} \cdots . \tag{5}$$

The Macdonald and Jack polynomials on which this paper will focus can now be presented as follows. To the partition λ with $m_i(\lambda)$ parts equal to i, we associate the number

$$z_\lambda = 1^{m_1} m_1! \, 2^{m_2} m_2! \cdots . \tag{6}$$

Let q and t be parameters and $\mathbb{Q}(q, t)$ the field of all rational functions of q and t with rational coefficients and define a scalar product $\langle \ , \ \rangle_{q,t}$ on $\Lambda_N \otimes \mathbb{Q}(q, t)$ by

$$\langle p_\lambda, p_\mu \rangle_{q,t} = \delta_{\lambda\mu} z_\lambda \prod_{i=1}^{\ell(\lambda)} \frac{1 - q^{\lambda_i}}{1 - t^{\lambda_i}}. \tag{7}$$

The Macdonald polynomials $J_\lambda(z; q, t) \in \Lambda_N \otimes \mathbb{Q}(q, t)$ are uniquely specified by

$$\langle J_\lambda, J_\mu \rangle_{q,t} = 0, \quad \text{if } \lambda \neq \mu, \tag{8}$$

$$J_\lambda = \sum_{\mu \leq \lambda} v_{\lambda\mu}(q, t) m_\mu, \tag{9}$$

$$v_{\lambda\lambda}(q, t) = c_\lambda(q, t), \tag{10}$$

where

$$c_\lambda(q, t) = \prod_{s \in \lambda} (1 - q^{a(s)} t^{\ell(s)+1}). \tag{11}$$

The Jack polynomials are obtained from the Macdonald polynomials in the limit $q = t^\alpha$, $t \to 1$. More precisely, in this limit, with $\mathbb{Q}(\alpha)$ defined as the field of rational functions of α, (7) becomes

$$\langle p_\lambda, p_\mu \rangle_\alpha = \delta_{\lambda\mu} z_\lambda \alpha^{\ell(\lambda)} \tag{12}$$

and the Jack polynomials $J_\lambda(z; \alpha) \in \Lambda_N \otimes \mathbb{Q}(\alpha)$ are given by

$$J_\lambda(z; \alpha) = \lim_{\substack{q=t^\alpha \\ t \to 1}} \frac{J_\lambda(z; q, t)}{(1 - t)^{|\lambda|}}, \tag{13}$$

with $|\lambda| = \lambda_1 + \lambda_2 + \cdots$. They are thus uniquely specified by

$$\langle J_\lambda, J_\mu \rangle_\alpha = 0, \quad \text{if } \lambda \neq \mu, \tag{14}$$

$$J_\lambda = \sum_{\mu \leq \lambda} v_{\lambda\mu}(\alpha) m_\mu, \tag{15}$$

$$v_{\lambda\lambda}(\alpha) = c_\lambda(\alpha), \tag{16}$$

where

$$c_\lambda(\alpha) = \lim_{\substack{q=t^\alpha \\ t \to 1}} \frac{c_\lambda(q, t)}{(1 - t)^{|\lambda|}} = \prod_{s \in \lambda} (\alpha a(s) + \ell(s) + 1). \tag{17}$$

Since the ordering on the partitions is not a total order, the Gram-Schmidt orthogonalization procedure cannot be used to prove the existence of the Macdonald and Jack polynomials. Nevertheless, one can prove their existence by replacing conditions (8) and (14) by equivalent conditions using self-adjoint operators acting triangularly on the monomial symmetric functions.

3   Jack Polynomials and the Calogero–Sutherland Model

The Calogero–Sutherland model [1, 17, 18] describes a system of N particles on a circle interacting pairwise through long-range interaction. We shall denote by L the perimeter of that circle and by x_i, $i = 1, \ldots, N$; $0 \le x_i \le L$, the positions of the particles. The quantum Hamiltonian H_{CS} is

$$H_{CS} = -\sum_{j=1}^{N} \frac{\partial^2}{\partial x_j^2} + 2\beta(\beta - 1)\sum_{j<k} \frac{1}{d^2(x_j - x_k)}, \tag{18}$$

where β is a constant and

$$d(x_j - x_k) = \frac{L}{\pi} \sin \frac{\pi}{L}(x_j - x_k), \tag{19}$$

is the chord length between the positions of the particles j and k. The momentum operator is $P_{CS} = -i\sum_{j=1}^{N} \partial/\partial x_j$. It can be shown that H is positive and that its ground state (the state with the lowest eigenvalue) is explicitly given by

$$\psi_0(x) = \prod_{j<k}\left[\sin \frac{\pi}{L}(x_j - x_k)\right]^{\beta}, \tag{20}$$

with eigenvalue $E_0 = (\pi/L)^2 \beta^2 N(N^2 - 1)/3$.

The wave functions of the excited states can be written in the form $\psi(x) = \phi(x)\psi_0(x)$. In the coordinates

$$z_j = e^{2\pi i x_j/L}, \tag{21}$$

the ground state wave function is $\Delta^{\beta}(z) = \prod_{j<k}(z_j - z_k)^{\beta} \prod_k z_k^{-\beta(N-1)/2}$ and the Schrödinger equation $H_{CS}\psi = E\psi$ is transformed after conjugation with the ground state into the following equation for ϕ:

$$H\phi = \left(\frac{L}{2\pi}\right)^2 (E - E_0)\phi, \tag{22}$$

where

$$H = \left(\frac{L}{2\pi}\right)^2 \Delta^{-\beta}(H_{CS} - E_0)\Delta^{\beta} = H_1 + \beta H_2, \tag{23}$$

$$H_1 = \sum_{j=1}^{N}\left(z_j \frac{\partial}{\partial z_j}\right)^2, \tag{24a}$$

$$H_2 = \sum_{j<k}\left(\frac{z_j + z_k}{z_j - z_k}\right)\left(z_j \frac{\partial}{\partial z_j} - z_k \frac{\partial}{\partial z_k}\right). \tag{24b}$$

The momentum operator becomes $P = \Delta^{-\beta} P_{CS} \Delta^{\beta} = 2\pi/L \sum_j z_j \partial/\partial z_j$, the Euler operator, and commutes with H. The action of H_1 and H_2 on the monomial basis is easily found to be of the form

$$H_1 m_\lambda = \left(\sum_{j=1}^{N} \lambda_j^2 \right) m_\lambda \qquad (25)$$

and

$$H_2 m_\lambda = \beta \left[\sum_{j=1}^{N} (N + 1 - 2j)\lambda_j \right] m_\lambda + \sum_{\mu < \lambda} d_{\lambda\mu} m_\mu, \qquad (26)$$

with the $d_{\lambda\mu}$'s, some coefficients.

Since H is S_N-invariant, the symmetric polynomials of degree n are stable under its action. The wave functions $\phi(z)$ can thus be labeled by partitions $(\lambda_1, \ldots, \lambda_N)$ of n. In fact, they are found to be the Jack polynomials evaluated at $\alpha = 1/\beta$,

$$\phi_\lambda(z) = J_\lambda(z; 1/\beta), \qquad (27)$$

with energy given, using (9), (25), and (26), by $E_\lambda = \sum_{j=1}^{N} [\lambda_j^2 + \beta(N + 1 - 2j)\lambda_j]$. Moreover, we have that $P\phi_\lambda = 2\pi n \phi_\lambda / L$.

The Jack polynomials being solution of the CS model are also orthogonal under the norm

$$(\phi_1, \phi_2) = \int \frac{d\theta_1}{2\pi} \cdots \frac{d\theta_N}{2\pi} \prod_{j<k} |(z_j - z_k)|^{2\beta} \phi_1(z) \overline{\phi_2(z)}, \quad z_j = e^{i\theta_j}, \quad (28)$$

induced from the original quantum mechanical problem. This second scalar product is proportional to the one introduced in (12).

4   Goal and Outline

We now aim to describe creation operators that allow one to obtain the wave functions of the CS model excited states from the wave function of the ground state. In view of the fact that the CS model wave functions are of the form $\psi_\lambda = \phi_\lambda \psi_0$ (see (20)), it clearly suffices to find operators that give J_λ upon acting on $J_0 = 1$. Suppose $\ell(\lambda) \leq k$; what we want are operators B_k such that $B_k J_\lambda(z) = J_{\lambda+(1^k)}$, in other words operators whose effect on J_λ is to give the Jack polynomials associated to the partition where one column of length k has been added to λ:

$$k \left\{ \begin{matrix} \\ \\ \\ \\ \end{matrix} \right. \qquad \longrightarrow \qquad$$

$$(1^k) \qquad + \qquad \lambda$$

If such B_k can be found, it then follows that for arbitrary λ, the Jack polynomials are given by $J_\lambda = B_N^{\lambda_N} \cdots B_2^{\lambda_2 - \lambda_3} B_1^{\lambda_1 - \lambda_2} \cdot 1$. The remainder of this paper is organized as follows. Sections 5 and 6 are devoted to the description of these operators, their properties, and (some of) their applications. Their generalization to the case of Macdonald polynomials is briefly discussed in Section 9. Results related to the algebraic properties of the CS model creation operators are collected in Section 7, while the creation operators $\overline{B}_\lambda = \overline{B}_{\lambda_1} \overline{B}_{\lambda_2} \cdots$ that would build the Jack polynomials part by part are considered in Section 8.

5   Dunkl Operators

The Dunkl operators ∇_i [3], are the basic building blocks of the CS model creation operators. They are defined as follows:

$$\nabla_i = \alpha \frac{\partial}{\partial z_i} + \sum_{\substack{j=1 \\ j \neq i}}^{N} \frac{1}{(z_i - z_j)}(1 - K_{ij}), \qquad (29)$$

where $K_{ij} = K_{ji}$, $K_{ij}^2 = 1$, is the operator that permutes the variables z_i and z_j:

$$K_{ij} z_j = z_i K_{ij}. \qquad (30)$$

The Dunkl operators ∇_i are easily found to have the following properties:

$$[\nabla_i, \nabla_j] = 0, \qquad (31a)$$

$$K_{ij} \nabla_j = \nabla_i K_{ij}, \qquad (31b)$$

$$[\nabla_i, z_j] = \delta_{ij} \left(\alpha + \sum_{l=1}^{N} K_{il} \right) - K_{ij}. \qquad (31c)$$

Let

$$D_i \equiv z_i \nabla_i, \qquad (32)$$

and let $J = \{j_1, j_2, \ldots, j_\ell\}$ be a set of cardinality $|J| = \ell$ made of integers such that

$$j_\kappa \in \{1, 2, \ldots, N\}, \qquad j_1 < j_2 < \cdots < j_\ell. \qquad (33)$$

Introduce the operators $D_{J,\omega}$ labeled by such sets J and by a parameter ω:

$$D_{J,\omega} = (D_{j_1} + \omega)(D_{j_2} + \omega + 1) \cdots (D_{j_\ell} + \omega + \ell - 1). \qquad (34)$$

In the case where $|J| = 0$, we set $D_{J,\omega} = 1$.

We now list a few properties that will prove useful in the sequel:

1. In terms of Dunkl operators, the operator H given in (23) and (24) takes the simple form:

$$H = \mathrm{Res} \sum_{i=1}^{N} D_i^2. \tag{35}$$

Here and in the following, $\mathrm{Res}\, X$ indicates that the action of X is restricted to symmetric function of the variables $z_1, \ldots, z_N$.

As a matter of fact, H belongs to a set of mutually commuting operators. The CS system is completely integrable and admits N functionally independent constants of motion in involution. Since the quantities

$$L_j = \sum_{i=1}^{N} (D_i)^j, \quad j = 1, \ldots, N, \tag{36}$$

mutually commute, $[L_k, L_j] = 0$ [14], the N differential operators $\mathrm{Res}\, L_j$ are precisely the functionally independent constants of motion. It can be, moreover, shown that these quantities are diagonal on the Jack polynomials.

2. With $z_J = \prod_{k \in J} z_k$, we have

$$[D_i, z_J] = -z_i \sum_{j \in J} z_{J \setminus \{j\}} K_{ij}, \qquad i \notin J,$$

$$[D_i, z_J] = z_J \left(\alpha + \sum_{\substack{j \notin J \\ j \in \{1, \ldots, N\}}} K_{ij} \right), \quad i \in J. \tag{37}$$

Note that z_i always appear on the right-hand side

3. From (31c) and (32), we determine that the operators D_i satisfy the commutation relations

$$[D_i, D_j] = (D_j - D_i) K_{ij}. \tag{38}$$

4. $\mathrm{Res}\, D_{J,\omega}$ is invariant under the exchange of the variables z_{j_κ}, $j_\kappa \in J$. This is seen from the fact that, using (38), we have, for any δ,

$$\mathrm{Res}(D_i + \delta)(D_j + \delta + 1) = \mathrm{Res}(D_j + \delta)(D_i + \delta + 1). \tag{39}$$

5. $\mathrm{Res}\, D_{J,\omega}$ only depends on the variables z_{j_κ} with $j_\kappa \in J$. To see this, apply $\mathrm{Res}\, D_{J,\omega}$ on a symmetric function ϕ_S. Obviously

$$D_{j_\ell} \phi_S = \left[\alpha z_{j_\ell} \frac{\partial}{\partial z_{j_\ell}} + \sum_{\substack{i_\ell = 1 \\ i_\ell \neq j_\ell}}^{N} \frac{z_{j_\ell}}{z_{j_\ell} - z_{i_\ell}} (1 - K_{j_\ell i_\ell}) \right] \phi_S \alpha z_{j_\ell} \frac{\partial}{\partial z_{j_\ell}} \phi_S, \tag{40}$$

and we thus have

$$
\begin{aligned}
D_{J,\omega}\phi_S \\
= \cdots \left(\alpha z_{j_{\ell-1}} \frac{\partial}{\partial z_{j_{\ell-1}}} + \sum_{\substack{i_{\ell-1}=1 \\ i_{\ell-1} \neq j_{\ell-1}}}^{N} \frac{z_{j_{\ell-1}}}{z_{j_{\ell-1}} - z_{i_{\ell-1}}} (1 - K_{j_{\ell-1} i_{\ell-1}}) + \omega + \ell - 2 \right) \\
\times \left(\alpha z_{j_\ell} \frac{\partial}{\partial z_{j_\ell}} + \omega + \ell - 1 \right) \phi_S \\
= \cdots \left(\alpha z_{j_{\ell-1}} \frac{\partial}{\partial z_{j_{\ell-1}}} + \frac{z_{j_{\ell-1}}}{z_{j_{\ell-1}} - z_{j_\ell}} (1 - K_{j_{\ell-1} j_\ell}) + \omega + \ell - 2 \right) \\
\times \left(\alpha z_{j_\ell} \frac{\partial}{\partial z_{j_\ell}} + \omega + \ell - 1 \right) \phi_S. \quad (41)
\end{aligned}
$$

Iterating the last step proves the assertion.

6.

$$
(\mathrm{Res}\, D_{J,\omega}) J_\lambda(z(J); \alpha) = \left[\prod_{i=1}^{\ell} (\alpha \lambda_i + \omega + \ell - i) \right] J_\lambda(z(J); \alpha), \quad (42)
$$

with $z(J) = \{x_i \mid i \in J\}$. To show that this result holds, one needs simply to use the fact that $\mathrm{Res}\, D_{J,\omega}$ only depends on the variables z_{j_κ}, $j_\kappa \in J$ and is symmetric in these variables. Since it is a symmetric polynomial in the operators $D_i^{(J)}$, with the superscript J indicating that only the variables z_{j_κ}, $j_\kappa \in J$ appear in the operators, $\mathrm{Res}\, D_{J,\omega}$ is thus expandable in the operators $L_i^{(J)} = \sum_{m=1}^{\ell} (D_m^{(J)})^i$ and hence diagonal on $J_\lambda(z(J); \alpha)$. Its eigenvalues are found as follows. With

$$
J_\lambda \sim m_\lambda + \sum_{\mu < \lambda} d_{\lambda\mu} m_\mu \sim z_{j_1}^{\lambda_1} \cdots z_{j_\ell}^{\lambda_\ell} + \cdots + z_{j_1}^{\lambda_\ell} \cdots z_{j_\ell}^{\lambda_1} + \sum_{\mu < \lambda} d_{\lambda\mu} m_\mu, \quad (43)
$$

one checks [8] that the term $z_{j_1}^{\lambda_\ell} \cdots z_{j_\ell}^{\lambda_1}$ of $(\mathrm{Res}\, D_{J,\omega}) J_\lambda(z(J); \alpha)$ can only be obtained from the action of $(\alpha z_{j_1} \partial/\partial z_{j_1} + \omega) \cdots (\alpha z_{j_\ell} \partial/\partial z_{j_\ell} + \omega + \ell - 1)$ on $z_{j_1}^{\lambda_\ell} \cdots z_{j_\ell}^{\lambda_1}$. The result follows by symmetry.

7.

$$
(\mathrm{Res}\, D_{J,\omega}) z_J^\rho = z_J^\rho (\mathrm{Res}\, D_{J,\omega+\alpha\rho}). \quad (44)
$$

From property 5, we have $\mathrm{Res}\, D_{J,\omega} = \mathrm{Res}\, D_{J,\omega}^{(J)}$. The result then follows from (37), which gives $D_i^{(J)} z_J^\rho = z_J^\rho (D_i^{(J)} + \alpha\rho)$.

6 Rodrigues Formula

We now give three expressions $B_k^{(i)}$, $i = 1, 2, 3$; $k = 1, \ldots, N$ for the creation operators. They will be derived in the remainder of this section.

Expression 1.

$$B_k^{(1)} = \left(\prod_{j=k+1}^{N} \frac{1}{(-\alpha + k - 1 - j)} \right) D_{\{1,\dots,N\},k+1-N-\alpha} e_k = M_k e_k. \quad (45)$$

Expression 2.

$$B_k^{(2)} = \sum_{|I|=k} z_I \sum_{m=0}^{N-k} \frac{(-1)^m}{(\alpha + N - k - m)_m} \sum_{\substack{I' \subseteq I^c \\ |I'|=m}} D_{I \cup I', 1-m}, \quad (46)$$

where $I^c = \{1, \dots, N\} \backslash I$ and $(a)_n = (a) \dots (a + n - 1), (a)_0 \equiv 1$ is the Pochhammer symbol.

Expression 3.

$$B_k^{(3)} = \sum_{|I|=k} z_I D_{I,1}. \quad (47)$$

For applications and in particular for proving the conjecture of Macdonald and Stanley, it is Expression 3, that is the most useful. Nevertheless as will be shown, it is most easily proved starting from Expression 1, which follows from Pieri's formula and using, as intermediate step, Expression 2. We thus need to prove that these three sets of operators are such that

$$B_k^{(i)} J_\lambda(z) = J_{\lambda + (1^k)}(z), \quad (48)$$

if $\ell(\lambda) \leq k$. Given (48) and $J_0 = 1$, as was mentioned already, the following Rodrigues formulas for the Jack polynomials associated to any partition λ are easily seen to hold

$$J_\lambda(z; \alpha) = (B_N^{(i)})^{\lambda_N} (B_{N-1}^{(i)})^{\lambda_{N-1} - \lambda_N} \cdots (B_1^{(i)})^{\lambda_1 - \lambda_2} \cdot 1$$
$$= B_{\lambda_1'}^{(i)} B_{\lambda_2'}^{(i)} \cdots \cdot 1 \equiv B_{\lambda'}^{(i)} \cdot 1, \quad i = 1, 2, 3. \quad (49)$$

That the first expression has the property (48) will follow from the Pieri formulas which give the action of e_k and of J_k on J_λ. Let $P_\lambda = 1/c_\lambda(\alpha) J_\lambda$ be a renormalization of J_λ. The formulas read

$$e_k P_\lambda = \sum_\mu \Psi_{\mu/\lambda} P_\mu \quad (50)$$

and

$$J_k P_\lambda = \sum_\mu \Phi_{\mu/\lambda} P_\mu, \quad (51)$$

with the sum over partition $\mu \supset \lambda$ (of length $\leq N$) such that $\mu - \lambda$ is a vertical k-strip in the e_k case and a horizontal k-strip in the J_k case. The coefficients $\Psi_{\mu/\lambda}$ and $\Phi_{\mu/\lambda}$ are

$$\Psi_{\mu/\lambda} = \prod_{\substack{s \in C_{\mu/\lambda} \\ s \notin R_{\mu/\lambda}}} \frac{b_\mu(s)}{b_\lambda(s)} \tag{52}$$

and

$$\Phi_{\mu/\lambda} = \alpha^k k! \prod_{s \in C_{\mu/\lambda}} \frac{b_\mu(s)}{b_\lambda(s)}, \quad k = |\mu| - |\lambda|, \tag{53}$$

where

$$b_\lambda(s) = \begin{cases} \frac{\alpha a(s) + \ell(s) + 1}{\alpha(a(s)+1) + \ell(s)} & \text{if } s \in \lambda \\ 1 & \text{otherwise} \end{cases} \tag{54}$$

and where $C_{\mu/\lambda}$ (respectively $R_{\mu/\lambda}$) denotes the union of the columns (resp. rows) that intersect $\mu - \lambda$. For example, with $\mu = (4,2,2)$ and $\lambda = (3,2,1)$ we have

t	i	m	e	s
$\times$			

so the only s in $C_{\mu/\lambda}$ but not in $R_{\mu/\lambda}$ is in position (2,2). With $\lambda + 1$ representing the partition $(\lambda_1 + 1, \ldots, \lambda_N + 1)$, since $P_{\lambda+1} = e_N P_\lambda$, we have

$$\Psi_{\mu+1/\lambda+1} = \Psi_{\mu/\lambda}. \tag{55}$$

Theorem 1. *For any partition λ with $\ell(\lambda) \leq k$, the operators $B_k^{(1)}$ are such that on the Jack polynomials $J_\lambda(z; \alpha)$:*

$$B_k^{(1)} J_\lambda(z) = J_{\lambda+(1^k)}(z). \tag{56}$$

Proof. First, we need the following lemma obtained from the Pieri formula (50).

Lemma 1. *The action of the elementary function e_k on P_λ, with λ a partition such that $\ell(\lambda) \leq k$, is given by*

$$e_k P_\lambda = P_{\lambda+(1^k)} + \sum_{\mu \neq \lambda+(1^k)} \Psi_{\mu/\lambda} P_\mu, \tag{57}$$

where all the μ's in the sum are such that $\mu_{k+1} = 1$.

Proof. This is seen from the fact that μ must be a partition that contains λ and $\mu - \lambda$ a vertical k-strip. Hence the only way to construct a μ with $\mu_{k+1} \neq 1$ is to add a 1 in each of the first k entries of λ. $\qquad\square$

From Lemma 1, we then have

$$B_k P_\lambda = M_k P_{\lambda+(1^k)} + \sum_{\mu \neq \lambda+(1^k)} \Psi_{\mu/\lambda} M_k P_\mu, \qquad (58)$$

with all the terms in the sum vanishing since, from (42), the eigenvalue of the operator $D_{\{1,\dots,N\},k+1-N-\alpha}$ on P_μ, given by $\prod_{i=1}^{N}(\alpha\mu_i + k + 1 - \alpha - i)$, is equal to 0 when $\mu_{k+1} = 1$. Thus, we are left with

$$B_k P_\lambda = M_k P_{\lambda+(1^k)}$$
$$= \left(\frac{\prod_{j=1}^{k}(\alpha(\lambda_j + 1) + k + 1 - \alpha - j)\prod_{j=k+1}^{N}(k+1-\alpha-j)}{\prod_{j=k+1}^{N}(k+1-\alpha-j)} \right) P_{\lambda+(1^k)}$$
$$= \left(\prod_{j=k+1}^{N}(k+1-\alpha-j) \right) P_{\lambda+(1^k)}. \qquad (59)$$

In order to prove that formula (59) is in fact equivalent to $B_k J_\lambda = J_{\lambda+(1^k)}$, we need one more lemma.

Lemma 2. *If λ is a partition with $\ell(\lambda) \leq k$,*

$$\frac{c_{\lambda+(1^k)}}{c_\lambda} = \prod_{j=k+1}^{N}(k+1-\alpha-j). \qquad (60)$$

Proof. When going from the diagram of λ to the one of $\lambda + (1^k)$, what we actually do is add a column at the west of the diagram of λ. For instance, $\lambda = (5,4,4,1) + (1^4)$ is obtained from $\lambda = (5,4,4,1)$ by adding a column of length 4.

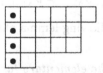

From the fact that c_λ (see Eq. (17)) only involves $a(s)$ and $\ell(s)$, which do not depend on the number of squares to the west of s, the contribution in $c_{\lambda+(1^k)}$ of the squares that have been shifted to the east is exactly c_λ. Hence, $c_{\lambda+(1^k)}/c_\lambda$ is given by the contribution in $c_{\lambda+(1^k)}$ of the first column of $\lambda + (1^k)$, which is $\prod_{j=k+1}^{N}(k+1-\alpha-j)$. $\qquad\square$

From Lemma 2 we have

$$B_k P_\lambda = \frac{c_{\lambda+(1^k)}}{c_\lambda} P_{\lambda+(1^k)}, \tag{61}$$

which gives, when passing from P_λ to J_λ, $B_k J_\lambda = J_{\lambda+(1^k)}$. This completes the proof of Theorem 1. □

Now that we have proved that $B_k^{(1)}$ is a creation operator, we can show that $B_k^{(2)}$ (see (46)) is also a creation operator.

Theorem 2.

$$\text{Res } B_k^{(1)} = \text{Res } B_k^{(2)}$$

$$\equiv \sum_{|I|=k} z_I \sum_{m=0}^{N-k} \frac{(-1)^m}{(\alpha+N-k-m)_m} \sum_{\substack{I'\subseteq I^c \\ |I'|=m}} \text{Res } D_{I\cup I',1-m}; \tag{62}$$

hence $B_k^{(2)}$ is also such that $B_k^{(2)} J_\lambda = J_{\lambda+(1^k)}$ for any partition λ such that $\ell(\lambda) \le k$.

This is shown using the two following lemmas.

Lemma 3.

$$\text{Res } B_k^{(1)} = \left(\prod_{j=k+1}^{N} \frac{1}{(-\alpha+k+1-j)} \right) \sum_{|I|=k} z_I \, \text{Res } D_{I^c,k+1-N-\alpha} D_{I,1}). \tag{63}$$

Lemma 4.

$$D_{\{1,\dots,\ell\},-\ell+1-\alpha} = \sum_{m=0}^{\ell} (-1)^{\ell-m} (\alpha)_{\ell-m} \sum_{\substack{I'\subseteq\{1,\dots,\ell\} \\ |I'|=m}} D_{I',1-m}. \tag{64}$$

Using the result of Lemma 4, rewritten in the form

$$D_{I^c,k+1-N-\alpha} = \sum_{m=0}^{N-k} (-1)^{N-k-m} (\alpha)_{N-k-m} \sum_{\substack{I'\subseteq I^c \\ |I'|=m}} D_{I',1-m}, \tag{65}$$

in (63), we get

$$\text{Res } B_k^{(1)} = \sum_{|I|=k} z_I \sum_{m=0}^{N-k} \frac{(-1)^m}{(\alpha+N-k-m)_m} \sum_{\substack{I'\subseteq I^c \\ |I'|=m}} \text{Res}(D_{I\cup I',1-m} D_{I,1})$$

$$= \sum_{|I|=k} z_I \sum_{m=0}^{N-k} \frac{(-1)^m}{(\alpha+N-k-m)_m} \sum_{\substack{I'\subseteq I^c \\ |I'|=m}} \text{Res } D_{I\cup I',1-m}, \tag{66}$$

since $D_{I',1-m}D_{I,1} = D_{I' \cup I,1-m}$ when $|I'| = m$. We thus only have to prove Lemmas 3 and 4 to complete the proof of Theorem 2.   □

Proof of Lemma 3. We want to switch to the left the variables at the right of the Dunkl operators. We have

$$\mathrm{Res}\, D_{\{1,\ldots,N\},k+1-N-\alpha}e_k$$
$$= \mathrm{Res}\big(D_{\{1,\ldots,N-k\},k+1-N-\alpha}\, \mathrm{Res}(D_{\{N-k+1,\ldots,N\},1-\alpha})\big)e_k. \quad (67)$$

Since this expression is symmetric, we can focus on the part involving only the variables z_j, $j \in \{N - k + 1, \ldots, N\}$ to the left of the Dunkl operators and symmetrize in order to get the whole expression. First we note, using (37), that $[D_i, z_j]$ contains z_i to the left. Hence, commuting any products of variables with $D_{\{1,\ldots,N-k\},k+1-N-\alpha}$ will give at least one variable z_j with $j \notin \{N - k + 1, \ldots, N\}$ to the left of this operator. We can thus consider $D_{\{1,\ldots,N-k\},k+1-N-\alpha}$ as a constant for our purposes. Next, since $\mathrm{Res}\, D_{\{N-k+1,\ldots,N\},\omega}$ only depends on the variables z_i, $i \in \{N - k + 1, \ldots, N\}$, all the terms containing initially some variables z_j with $j \notin \{N - k + 1, \ldots, N\}$ would still, after commutation, contain these variables. We are thus left with only the term $z_{N-k+1} \cdots z_N$ of e_k to consider. We will denote by $\mathrm{Res}\, D_{\{1,\ldots,N\},k+1-N-\alpha}e_k|_{z_{N-k+1}\cdots z_N}$ the term with $z_{N-k+1} \cdots z_N$ to the left in expression (67). From (44), we have

$$\mathrm{Res}\big(D_{\{N-k+1,\ldots,N\},1-\alpha}\big) z_{N-k+1} \cdots z_N$$
$$= z_{N-k+1} \cdots z_N\, \mathrm{Res}\, D_{\{N-k+1,\ldots,N\},1}. \quad (68)$$

This gives

$$\mathrm{Res}\, D_{\{1,\ldots,N\},k+1-N-\alpha}e_k\big|_{z_{N-k+1}\cdots z_N}$$
$$= \mathrm{Res}(D_{\{1,\ldots,N-k\},k+1-N-\alpha}\, \mathrm{Res}(D_{\{N-k+1,\ldots,N\},1-\alpha})$$
$$\times z_{N-k+1} \cdots z_N)\big|_{z_{N-k+1}\cdots z_N}$$
$$= \mathrm{Res}\big(D_{\{1,\ldots,N-k\},k+1-N-\alpha} z_{N-k+1} \cdots z_N$$
$$\times \mathrm{Res}(D_{\{N-k+1,\ldots,N\},1})\big)\big|_{z_{N-k+1}\cdots z_N}$$
$$= z_{N-k+1} \cdots z_N\, \mathrm{Res}(D_{\{1,\ldots,N-k\},k+1-N-\alpha}D_{\{N-k+1,\ldots,N\},1}). \quad (69)$$

Symmetrizing this result, we get

$$\mathrm{Res}\, B_k^{(1)} = \bigg(\prod_{j=k+1}^{N} \frac{1}{(-\alpha + k + 1 - j)} \bigg) \sum_{|I|=k} z_I\, \mathrm{Res}\, D_{I^c,k+1-N-\alpha}D_{I,1}, \quad (70)$$

which is the content of Lemma 3.   □

Proof of Lemma 4. One proceeds by induction. It is easily checked that the lemma holds in the case $\ell = 1$. Assuming then that

$$D_{\{2,\ldots,\ell+1\},-\ell+1-\alpha} = \sum_{m=0}^{\ell}(-1)^{\ell-m}(\alpha)_{\ell-m}\sum_{\substack{I'\subseteq\{2,\ldots,\ell+1\}\\|I'|=m}}D_{I',1-m}, \quad (71)$$

we have

$$D_{\{1,\ldots,\ell+1\},-\ell-\alpha}$$
$$= \left(D_1 + (-\alpha-\ell)\right)D_{\{2,\ldots,\ell+1\},-\ell+1-\alpha}$$
$$= \sum_{m=0}^{\ell}(-1)^{\ell-m}(\alpha)_{\ell-m}\left(D_1 - m + (-\alpha-\ell+m)\right)\sum_{\substack{I'\subseteq\{2,\ldots,\ell+1\}\\|I'|=m}}D_{I',1-m}$$
$$= \sum_{m=0}^{\ell}(-1)^{\ell-m+1}(\alpha)_{\ell-m+1}\sum_{\substack{I'\subseteq\{1,\ldots,\ell+1\}\\1\notin I';|I'|=m}}D_{I',1-m}$$
$$+ \sum_{m=0}^{\ell}(-1)^{\ell-m}(\alpha)_{\ell-m}\sum_{\substack{I'\subseteq\{1,\ldots,\ell+1\}\\1\in I';|I'|=m+1}}D_{I,-m}, \quad (72)$$

which finally gives

$$D_{\{1,\ldots,\ell+1\},-\ell-\alpha} = \sum_{m=0}^{\ell+1}(-1)^{\ell-m+1}(\alpha)_{\ell-m+1}\sum_{\substack{I'\subseteq\{1,\ldots,\ell+1\}\\|I'|=m}}D_{I',1-m}, \quad (73)$$

to complete the proof. $\qquad\square$

The connection between $B_k^{(2)}$ and $B_k^{(3)}$ is now readily obtained.

Theorem 3. *For any partition λ, such that $\ell(\lambda) \leq k$, the actions of $B_k^{(2)}$ and $B_k^{(3)}$ on the Macdonald polynomials $J_\lambda(z)$ coincide,*

$$B_k^{(3)}J_\lambda(z) = B_k^{(2)}J_\lambda(z) = J_{\lambda+(1^k)}(z). \quad (74)$$

This is shown to be true with the help of the following lemma

Lemma 5. *Let $|I| = k$ and $|I'| = m$, $I' \subseteq I^c$.*

$$D_{I\cup I',1-m}J_\lambda(z;\alpha) = 0, \quad (75)$$

if $\ell(\lambda) \leq k$ and $m > 0$.

Proof. The Jack polynomials are known [12] to enjoy the property according to which

$$J_\lambda\big(z(I), z(I^c)\big) = \sum_{\mu,\nu}\tilde{f}_{\mu\nu}^\lambda J_\mu\big(z(I)\big)J_\nu\big(z(I^c)\big), \quad (76)$$

with $\tilde{f}^{\lambda}_{\mu\nu} = 0$ unless $\mu \subset \lambda$ and $\nu \subset \lambda$ and in particular if $\ell(\mu)$ or $\ell(\nu)$ is greater than k.

Since $\mathrm{Res}\, D_{I\cup I',1-m}$ is a differential operator depending only on the variables z_i, $i \in I \cup I'$, we see from (76) that

$$D_{I\cup I',1-m} J_{\lambda}(z)$$
$$= \sum_{\mu,\nu} \tilde{f}^{\lambda}_{\mu\nu} J_{\mu}\big(z((I\cup I')^c)\big)(\mathrm{Res}\, D_{I\cup I',1-m}) J_{\nu}\big(z(I\cup I')\big). \quad (77)$$

The proof of Lemma 5 is then completed by observing from (42) that

$$\mathrm{Res}(D_{I\cup I',1-m}) J_{\nu}\big(z(I\cup I')\big)$$
$$= \left[\prod_{i=1}^{k+m} (\alpha\nu_i + k + 1 - i)\right] J_{\nu}\big(z(I\cup I')\big) = 0, \quad (78)$$

whenever $m > 0$, since $\nu_{k+1} = 0$. □

Theorem 3 is thus an immediate consequence of this lemma. Indeed, by (46), we have that

$$\left(\sum_{|I|=k} z_I \sum_{\substack{I' \subseteq I^c \\ |I'|=m}} D_{I\cup I',1-m}\right) J_{\lambda}(z;\alpha) = 0, \quad (79)$$

if $\ell(\lambda) \le k$ and $m > 0$. This only leaves the $m = 0$ term of $B_k^{(2)}$ which coincides with $B_k^{(3)}$ (see Eqs. (46) and (47)).

The operators $B_k^{(3)}$ are by far the most interesting. Some of their properties will be discussed in the next section. But first, we will show one of their applications. Recall that

$$J_{\lambda} = \sum_{\mu \le \lambda} v_{\lambda\mu}(\alpha) m_{\mu}. \quad (80)$$

Macdonald and Stanley had conjectured [12, 16] that the $v_{\lambda\mu}(\alpha)$'s are polynomials in α with positive integer coefficients. A straightforward corollary of Theorem 7 is that they are polynomials in α with integer coefficients, thereby proving that a week form of the conjecture is true [7]. Note that the positivity part of the conjecture has also been shown to hold afterward by Knop and Sahi using different methods [5].

7   Additional Results

We extend the definition of a partition to allow real entries:

$$P_{(\beta_1,\ldots,\beta_N)} = e_N^{\beta_N} P_{(\beta_1-\beta_N,\ldots,\beta_{N-1}-\beta_N,0)} = e_N^{\beta_N} P_{\beta-\beta_N} \quad (81)$$

$\forall \beta_N \in \mathbb{R}$ and $\beta_i - \beta_{i+1}$ integers ≥ 0, $i = 1, \ldots, N - 1$. Let $F(\kappa)$ be the operator acting as follows on P_β:

$$F(\kappa) P_{(\beta_1, \ldots, \beta_N)} = \prod_{i=1}^{N} \prod_{j=1}^{\infty} (\alpha(\beta_i - j) + \kappa + 1 - i) P_{(\beta_1, \ldots, \beta_N)}. \tag{82}$$

From this definition and the equality $P_{\lambda+1} = e_N P_\lambda$, one has

$$F(\kappa) e_N^\rho = e_N^\rho F(\kappa + \alpha \rho). \tag{83}$$

We now form the operators

$$F_{m,\kappa} = F(\kappa) e_m F(\kappa)^{-1}, \tag{84}$$

to see that these have on P_β actions that only involve a finite number of products. These are given by

$$F_{m,\kappa} P_\beta = \sum_\delta \Psi_{\delta/\beta} F_{\delta/\beta}(\kappa) P_\delta, \tag{85}$$

with $\delta - \beta$ m-vertical strips, and

$$F_{\delta/\beta}(\kappa) = \left[\prod_i (\alpha \beta_i + \kappa + 1 - i) \right], \tag{86}$$

where the product is taken over the i's such that $\delta_i - \beta_i = 1$. Note that by the argument given in (55), we also have that $\Psi_{\delta/\beta} = \Psi_{\delta - \beta_N / \beta - \beta_N}$. Hence, the action of $F_{m,\kappa}$ on the Jack polynomials is very similar to that of the elementary symmetric function e_m on these functions: the action of $F_{m,\kappa}$ differs from that of e_m by the presence of m additional factors in front of the coefficients $\Psi_{\delta/\beta}$. Taking $N = 4$ and $m = 2$, we have, for example,

$$\begin{aligned}
F_{2,\kappa} P_{(1,1,-1,-1)} = \ &\Psi_{(3,3)/(2,2)} (\alpha + \kappa)(\alpha + \kappa - 1) P_{(2,2,-1,-1)} \\
&+ \Psi_{(3,2,1)/(2,2)} (\alpha + \kappa)(-\alpha + \kappa - 2) P_{(2,1,0,-1)} \\
&+ \Psi_{(2,2,1,1)/(2,2)} (-\alpha + \kappa - 2)(-\alpha + \kappa - 3) P_{(1,1,0,0)}. \tag{87}
\end{aligned}$$

Remarkably the creation operators $B_k^{(3)}$ can be identified with a subset of the operators $F_{m,\kappa}$.

Theorem 4. *The creation operators $B_k^{(3)}$ can be written in the form*

$$\text{Res}\, B_k^{(3)} = F(k) e_k F(k)^{-1} = F_{k,k} \tag{88}$$

The proof of this theorem, which appeared first under the form of a conjecture in Ref. 9, can be found in in Ref. 6. This result immediately provides, through (60), the action of $B_k^{(3)}$ on arbitrary Jack polynomials.

In the framework of this theorem, the Hermitian conjugate $B_k^{(3)\dagger}$ of $B_k^{(3)}$ with respect to the scalar product defined in (28) is readily obtained from the fact that $F(\kappa)$ is Hermitian under this scalar product. We thus have

$$F_{m,\kappa}^\dagger = F(\kappa)^{-1} e_m^\dagger F(\kappa), \qquad (89)$$

with $e_m^\dagger = e_N^{-1} e_{N-m}$, which implies that

$$\operatorname{Res} B_k^{(3)\dagger} = F(k)^{-1} e_k^\dagger F(k). \qquad (90)$$

It is striking that the set of operators $F_{m,\kappa}$ contains a one-parameter family of N-dimensional Abelian algebras. From (83) we see that $F_{m,\kappa}$ can be generated from $F_{m,0}$ by conjugating with powers of e_N. More precisely, one has

$$F_{m,\kappa} = e_N^{-\kappa/\alpha} F_{m,0} e_N^{\kappa/\alpha}. \qquad (91)$$

Noting that

$$[F_{m,\kappa}, F_{n,\kappa}] = F(\kappa)[e_m, e_n]F(\kappa)^{-1} = 0 \qquad (92)$$

immediately shows that we can construct, by proper conjugation of $B_k^{(3)}$ with e_N, the following set of commuting operators:

$$\{F_{m,\kappa} = e_N^{(m-\kappa)/\alpha} B_k^{(3)} e_N^{(\kappa-m)/\alpha}, m = 1, \dots, N; \kappa \in \mathbb{R}\} \qquad (93)$$

and thus obtain this way, for each value of κ, a N-dimensional Abelian algebra.

Using

$$D_{J,\omega} e_N^\rho = e_N^\rho D_{J,\omega+\alpha\rho}, \qquad (94)$$

which follows from (44), we can furthermore provide a realization of the operators $F_{m,\kappa}$:

$$F_{m,\kappa} = \sum_{|I|=m} z_I \operatorname{Res} D_{I,\kappa-m+1}. \qquad (95)$$

8   Another Set of Creation Operators

We have been able to construct operators such that

$$J_\lambda(z;\alpha) = B_{\lambda'} \cdot 1. \qquad (96)$$

The question that one might ask is whether it is possible to obtain operators that would build the Jack polynomials part by part, that is

$$J_\lambda(z;\alpha) = \overline{B}_\lambda \cdot 1 = \overline{B}_{\lambda_1} \overline{B}_{\lambda_2} \dots \cdot 1. \qquad (97)$$

By analogy with the previous case, one is led to try using the other Pieri formula, the one involving J_k instead of e_k (see Eq. (51)). The next proposition uses the operators introduced in the previous section to build the operators $\overline{B}_k$.

Proposition 1. *By definition, the operators*

$$\overline{B}_k = F(-\alpha k)\frac{J_k}{(-\alpha)^k k!}F(-\alpha k)^{-1}, \tag{98}$$

are such that for any partition λ with $\lambda_1 \leq k$, $\overline{B}_k J_\lambda = J_{(k,\lambda)}$, where (k,λ) is the partition $(k, \lambda_1, \lambda_2, \dots)$ and where $F(-\alpha k)$ is as defined in (82).

Proof. The proof is quite similar to the one used to prove Theorem 1.

Lemma 6. *The action of J_k on P_λ with λ a partition such that $\lambda_1 \leq k$ is given by*

$$J_k P_\lambda = \Phi_{(k,\lambda)/\lambda} P_{(k,\lambda)} + \sum_{\mu \neq (k,\lambda)} \Phi_{\mu/\lambda} P_\mu, \tag{99}$$

where all the μ's in the sum are such that $\mu'_{k+1} = 1$, with the prime refering to the conjugate partition μ'.

Proof. This is seen from the fact (see Eq. (51)) that μ must be a partition that contains λ and $\mu - \lambda$ a horizontal k-strip. Hence the only way to construct a μ with $\mu'_{k+1} \neq 1$ is to add a row of length k to the diagram of λ.

The action of $F(\kappa)J_m F(\kappa)^{-1}$ on P_β is given, similarly to (85) and (86), by

$$F(\kappa)J_m F(\kappa)^{-1} P_\beta = \sum_\delta \Phi_{\delta/\beta} F'_{\delta/\beta}(\kappa) P_\delta, \tag{100}$$

with $\delta - \beta$ m-horizontal strips and where

$$F'_{\delta/\beta}(\kappa) = \left[\prod_i (\kappa - \beta'_i + (i-1)\alpha) \right], \tag{101}$$

with the product over the i's such that $\delta'_i - \beta'_i = 1$. We thus have

$$\overline{B}_k P_\lambda = \sum_\rho \Phi_{\rho/\lambda} \frac{F_{\rho/\lambda}(-\alpha k)}{(-\alpha)^k k!} P_\rho$$

$$= \Phi_{(k,\lambda)/\lambda} \frac{F_{(k,\lambda)/\lambda}(-\alpha k)}{(-\alpha)^k k!} P_{(k,\lambda)} + \sum_\mu \Phi_{\mu/\lambda} \frac{F_{\mu/\lambda}(-\alpha k)}{(-\alpha)^k k!} P_\mu, \tag{102}$$

where all the terms in the sum cancel since, from (101), $F'_{\mu/\lambda}(-\alpha k) = 0$ when $\mu'_{k+1} - \lambda'_{k+1} = 1$ and $\lambda'_{k+1} = 0$, which is actually the case from Lemma 6. $\qquad \square$

292   Luc Lapointe and Luc Vinet

With

$$F_{(k,\lambda)/\lambda}(-k\alpha) = \prod_{j=1}^{k}(-k\alpha + (j-1)\alpha - \lambda_i'), \tag{103}$$

$$\Phi_{(k,\lambda)/\lambda} = \frac{\prod_{j=1}^{k}((k-j)\alpha + \lambda_i' + 1)\alpha^k k!}{\prod_{j=1}^{k}(k\alpha - (j-1)\alpha + \lambda_i')} \tag{104}$$

and

$$\frac{c_{(k,\lambda)}}{c_\lambda} = \prod_{j=1}^{k}((k-j)\alpha + \lambda_i' + 1), \tag{105}$$

expression (102) becomes

$$\overline{B}_k P_\lambda = \frac{c_{(k,\lambda)}}{c_\lambda} P_{(k,\lambda)}, \tag{106}$$

which gives, when passing from P_λ to J_λ, $\overline{B}_k J_\lambda = J_{(k,\lambda)}$. This proves Proposition 1. $\qquad\square$

In order to get a realization of the operators involved in the proposition, we expand J_k in terms of elementary symmetric functions [19]. This reads

$$J_k = \sum_{|\lambda|=k} d_{k\lambda} e_\lambda = \sum_{|\lambda|=k} \frac{(1/\alpha)_{\ell(\lambda)}(-\alpha)^k k!}{(-1)^{\ell(\lambda)}(m_1(\lambda)!m_2(\lambda)!\dots)} e_\lambda, \tag{107}$$

where $m_i(\lambda)$ stands for the number of parts equals to i in the partition λ. Inserting this expansion in (98), we get

$$\overline{B}_k = \sum_{|\lambda|=k} \frac{(1/\alpha)_{\ell(\lambda)}(-1)^{\ell(\lambda)}}{(m_1(\lambda)!m_2(\lambda)!\dots)} \prod_{i=1}^{\ell(\lambda)}[F(-k\alpha)e_{\lambda_i}F(-k\alpha)^{-1}. \tag{108}$$

Using (95), a result following from Theorem 4, we thus get a realization of the operators $\overline{B}_k$.

Theorem 5.

$$\overline{B}_k = \sum_{|\lambda|=k} \frac{(1/\alpha)_{\ell(\lambda)}(-1)^{\ell(\lambda)}}{(m_1(\lambda)!m_2(\lambda)!\dots)} \prod_{i=1}^{\ell(\lambda)} F_{\lambda_i,-k\alpha}, \tag{109}$$

with

$$F_{\lambda_i,-k\alpha} = \sum_{|I|=\lambda_i} z_I D_{I,-k\alpha-\lambda_i+1}, \tag{110}$$

is such that for any partition λ with $\lambda_1 \le k$, $\overline{B}_k J_\lambda = J_{(k,\lambda)}$.

Note that the order in which the operators $F_{\lambda_i,-k\alpha}$, $i = 1, \dots, \ell(\lambda)$, appear in (109) is irrelevant since from (92), a result also following from Theorem 4, these operators are in involution.

9   Relativistic Generalization: The Macdonald Polynomials

The Calogero–Sutherland model has a relativistic generalization, the Ruijsenaars–Schneider model [15]. In this case, the wave functions are given in terms of Macdonald polynomials $J_\lambda(z; q, t)$. They are eigenfunctions of a set of N commuting operators, the Macdonald operators, denoted by

$$M_N^r = \sum_I t^{(N-r)r + r(r-1)/2} \tilde{A}_I(z; t) \prod_{i \in I} \tau_i, \quad r = 1, \dots, N, \tag{111}$$

where

$$\tilde{A}_I(z; t) = \prod_{\substack{i \in I \\ j \in I^c}} \frac{z_i - t^{-1} z_j}{z_i - z_j}, \tag{112}$$

and where τ_i, the shift operator, is such that

$$\tau_i f(z_1, \dots, z_N) = f(z_1, \dots, q z_i, \dots, z_N) \tag{113}$$

for any polynomial $f \in \Lambda_N$, now defined as the ring of polynomials in the variables $z_1, \dots, z_N$. From these operators, one constructs

$$M_N(X; q, t) = \sum_{r=0}^{N} M_N^r X^r, \tag{114}$$

with X an arbitrary parameter. Its action on $J_\lambda(z; q, t)$ with $\ell(\lambda) \leq N$ is given by

$$M_N(X; q, t) J_\lambda(z; q, t) = a_\lambda(X; q, t) J_\lambda(z; q, t), \tag{115}$$

where

$$a_\lambda(X; q, t) = \prod_{i=1}^{N} (1 + X q^{\lambda_i} t^{N-i}). \tag{116}$$

From (114) we see that the eigenvalue of M_N^r on $J_\lambda(z; q, t)$ is the coefficient of X^r in the polynomial (116).

The creation operators $B_k^{(1)}$, $B_k^{(2)}$, and $B_k^{(3)}$ have relativistic extensions. One way to obtain such extensions is to use the Dunkl–Cherednik operators constructed out of operators realizing an affine Hecke algebra [2, 13].

First, we need some definitions. The Weyl group $W \cong S_N$ is generated by the transpositions s_i, $i = 1, \dots, N$. On $z^\lambda = z_1^{\lambda_1} \cdots z_N^{\lambda_N}$ their action is such that

$$s_i z^\lambda = z^{s_i \lambda} s_i, \tag{117}$$

where $s_i\lambda = (\lambda_1, \ldots, \lambda_{i-1}, \lambda_{i+1}, \lambda_i, \lambda_{i+2}, \ldots, \lambda_N)$. We denote by Λ_N^W, the subring of all symmetric polynomials. We can extend the action of the Weyl group W on Λ_N to one of the affine Weyl group $\tilde{W}$ by introducing the elements s_0 and $\omega^{\pm 1}$ realized by

$$s_0 = s_{N-1} \ldots s_2 s_1 s_2 \ldots s_{N-1} T_1 T_N^{-1},$$
$$\omega = s_{N-1} \ldots s_1 T_1 = T_N s_{N-1} \ldots s_1. \tag{118}$$

The generators of $\tilde{W}$ obey the following fundamental relations:

i) $s_i^2 = 1$, $i = 0, 1, \ldots, N-1$,

ii) $s_i s_j = s_j s_i$, $i - j| \geq 2$,

iii) $s_i s_j s_i = s_j s_i s_j$, $|i - j| = 1$,

iv) $\omega s_i = s_{i-1}\omega$, $i = 0, 1, \ldots, N-1$.

where the indices $0, 1, \ldots, N-1$ are understood as elements of $\mathbb{Z}_N = \mathbb{Z}/N\mathbb{Z}$.

The operators

$$T_i = 1 + \frac{1 - t^{-1}z_{i+1}/z_i}{1 - z_{i+1}/z_i}(s_i - 1), \tag{119}$$

for $i = 1, \ldots, N-1$ and

$$T_0 = 1 + \frac{1 - t^{-1}q^{-1}z_1/z_N}{1 - q^{-1}z_1/z_N}(s_0 - 1), \tag{120}$$

and $\omega^{\pm 1}$ realize on Λ_N the affine Hecke algebra $H(\tilde{W})$ of $\tilde{W}$; that is they verify the defining relations

i) $(T_i - 1)(T_i + t^{-1}) = 0$, $i = 0, 1, \ldots, N-1$,

ii) $T_i T_j = T_j T_i$, $|i - j| \geq 2$,

iii) $T_i T_j T_i = T_j T_i T_j$ $|i - j| = 1$,

iv) $\omega T_i = T_{i-1}$, $i = 0, 1, \ldots, N-1$,

where again the indices are understood as elements of $\mathbb{Z}_N = \mathbb{Z}/N\mathbb{Z}$. The Dunkl–Cherednik operators $Y_1, \ldots, Y_N$ are constructed as follows from the generators of $H(\tilde{W})$:

$$Y_i = T_i \ldots T_{N-1}\omega T_1^{-1} \ldots T_{i-1}^{-1}. \tag{121}$$

They form an Abelian algebra: $[Y_i, Y_j] = 0$, $\forall i, j = 1, \ldots, N$. They also satisfy the following commutation relations with the T_i's:

$$T_i Y_{i+1} T_i = Y_i,$$
$$T_i Y_j = Y_j T_i \quad j \neq i, i+1. \tag{122}$$

Let $J = \{j_1, j_2, \ldots, j_\ell\}$ denote sets of cardinality $|J| = \ell$ made out of integers $j_\kappa \in \{1, \ldots, N\}$, $1 \leq \kappa \leq \ell$ such that $j_1 < j_2 < \cdots < j_\ell$. We introduce the operators

$$Y_{J,u} = (1 - ut^{\ell-1}Y_{j_1})\ldots(1 - utY_{j_{\ell-1}})(1 - uY_{j_\ell}), \qquad (123)$$

associated to such sets and labeled by a real number u. If $|J| = 0$, we define $Y_{J,u} = 1$.

It is known (see for instance Ref. 4) that the Macdonald operators can be rewritten in terms of Dunkl–Cherednik operators. In particular, we have that

$$\operatorname{Res} Y_{\{1,\ldots,N\},u} = M_N(-u; q, t). \qquad (124)$$

We now give the expressions $B_k^{(i)}$ $i = 1, 2, 3$; $k = 1, \ldots, N$ of the relativistic analogs of the creation operators obtained for the Jack polynomials (the proofs can be found in Ref. 10).

Expression 1.

$$B_k^{(1)} = \frac{1}{(q^{-1}; t^{-1})_{N-k}} Y_{\{1,\ldots,N\}, t^{k+1-N}q^{-1}} e_k, \qquad (125)$$

where for n positive integer, $(a; q)_n = (1 - a)(1 - qa)\ldots(1 - q^{n-1}a)$ and $(a; q)_0 \equiv 1$.

Expression 2.

$$B_k^{(2)} = \sum_{|I|=k} z_I \sum_{m=0}^{N-k} \sum_{\substack{I' \subseteq I^c \\ |I'|=m}} \frac{q^{-m}}{(t^{k-N+1}q^{-1}; t)_m} t^{-d(I', I^c)} Y_{I \cup I', t^{1-m}}. \qquad (126)$$

The quantity $d(J, I)$ entering in the above expression depends on nested subsets $J \subseteq I$ of $\{1, \ldots, N\}$ and is defined as follows. Order the elements of I so that $I = \{i_0, \ldots, i_{\ell-1}\}$ with $i_0 < i_1 < \cdots < i_{\ell-1}$. Let $J = \{i_{j_1}, \ldots, i_{j_m}\} \subseteq I$ with its elements ordered so that $(i_{j_1}, \ldots, i_{j_m})$ is a subsequence of $(i_0, \ldots, i_{\ell-1})$, $0 \leq j_\kappa \leq \ell - 1$; $k = 1, \ldots, m$. We then define

$$d(J, I) = \sum_{k=1}^{m} j_k - m(m-1)/2. \qquad (127)$$

Note that the sum is over the indices that identify the elements of J in the reference set I. If $|J| = 0$, then $d(J, I) \equiv 0$.

Expression 3.

$$B_k^{(3)} = \sum_{|I|=k} z_I Y_{I,t}. \qquad (128)$$

296   Luc Lapointe and Luc Vinet

The last expression has also been introduced in Ref. 4.

The Macdonald–Stanley conjecture generalizes in the context of the Macdonald polynomials. In this case, Macdonald has conjectured that the expansion coefficients of the Macdonald polynomials on a special basis (the big Schur basis) are polynomials in q and t with positive integer coefficients. Using the operators $B_k^{(3)}$, one can show that the expansion coefficients are polynomials in q,t with integer coefficients (see Refs. 4, 11, and 10). Showing the positivity of these coefficients is still an open problem. Finally, note that Lemma 6 and Theorem 5 have also their relativistic counterparts [6].

Acknowledgments: This work has been supported in part through funds provided by NSERC (Canada) and FCAR (Québec). L. Lapointe holds a FCAR postgraduate scholarship.

10 REFERENCES

1. F. Calogero, *Solution of a three-body problem in one dimension*, J. Math. Phys. **10** (1969), 2191–2196.

2. I. Cherednik, *Double affine Hecke algebras and Macdonald's conjectures*, Ann. of Math. **141** (1995), No. 1, 191–216.

3. C. F. Dunkl, *Differential-difference operators associated to reflection groups*, Trans. Amer. Math. Soc. **311** (1989), No. 1, 167–183.

4. A. N. Kirillov and M. Noumi, *Affine Hecke algebras and raising operators for Macdonald polynomials*, Duke Math. J. **93** (1998), No. 1, 1–39.

5. F. Knop and S. Sahi, *A recursion and a combinatorial formula for the Jack polynomials*, Invent. Math. **128** (1997), No. 1, 9–22.

6. L. Lapointe, *Modèles de Calogero et Sutherland, fonctions spéciales et symétries*, Ph.D. thesis, Université de Montréal, 1997.

7. L. Lapointe and L. Vinet, *A Rodrigues formula for the Jack polynomials and the Macdonald–Stanley conjecture*, Internat. Math. Res. Notices (1995), No. 9, 419–424.

8. _____ , *Exact operator solution of the Calogero–Sutherland model*, Commun. Math. Phys. **178** (1996), No. 2, 425–452, hep-th/9507073.

9. _____ , *Creation operators for the Macdonald and Jack polynomials*, Lett. Math. Phys. **40** (1997), No. 3, 469–486.

10. _____ , *Rodrigues formulas for the Macdonald polynomials*, Adv. Math. **130** (1997), No. 2, 261–279.

11. _____ , *A short proof of the integrality of the Macdonald (q, t)-Kostka coefficients*, Duke Math. J. **91** (1998), No. 1, 205–214.

12. I. G. Macdonald, *Symmetric Functions and Hall Polynomials*, 2nd ed., Oxford Math. Monogr., Oxford Univ. Press, New York, 1995.

13. _____ , *Affine Hecke algebras and orthogonal polynomials*, Séminaire Bourbaki. Vol. 1994/95, Astérisque, Vol. 237, Soc. Math. France, Paris, 1996, pp. 189–207.

14. A. P. Polychronakos, *Exchange operator formalism for integrable systems of particles*, Phys. Rev. Lett. **69** (1992), No. 5, 703–705.

15. S. N. M. Ruijsenaars, *Complete integrability of relativistic Calogero–Moser systems and elliptic function identities*, Commun. Math. Phys. **110** (1987), No. 2, 191–213.

16. R. P. Stanley, *Some combinatorial properties of Jack symmetric functions*, Adv. Math. **77** (1989), No. 1, 76–115.

17. B. Sutherland, *Exact results for a quantum many-body problem in one dimension*, Phys. Rev. A **4** (1971), 2019–2021.

18. _____ , *Exact results for a quantum many-body problem in one dimension. II*, Phys. Rev. A **5** (1972), 1372–1376.

19. N. J. Vilenkin and A. U. Klimyk, *Representation of Lie Groups and Special Functions. Recent Advances*, Math. Appl., Vol. 316, Kluwer Acad. Publ., Dordrecht, 1995.

19

New Exact Results for Quantum Impurity Problems

F. Lesage, H. Saleur, and P. Simonetti

1 Introduction

The main goal of these notes is to explain, in simple terms, how recent developments in integrable quantum field theories can be used to solve certain types of impurity problems consisting of a quantum mechanical degree of freedom, localized in space, interacting with extended, gapless excitations. The different Kondo models (single or multichanneled), 1D quantum wires, dissipative quantum mechanics, the quantum Hall effect subject to point-contact, and quantum optics are among the physical realizations of this type of problem.

The interest in these problems is manifold: although they are very simple they exhibit behaviors characteristic of strongly interacting systems. They are also simple enough that nontrivial information about correlators and dynamical properties can be obtained. Moreover, most of them have experimental applications and the analytical results can hopefully be compared with the data found there.

The key ingredient in the following study is that all these systems can be represented by one-dimensional field theories. This opens the door for very powerful techniques such as conformal field theory and integrability in one dimension using Yang-Baxter equations.

Our starting point will be the following models:

$$H^{(\mathrm{I})} = \frac{1}{2} \int_{-\infty}^{0} dx \left[8\pi g \Pi^2 + \frac{1}{8\pi g}(\partial_x \phi)^2 \right] + \mathcal{B}^{(\mathrm{I})} \qquad (1)$$

and

$$H^{(\mathrm{II})} = \frac{1}{2} \int_{-\infty}^{0} dx \sum_{i=1}^{2} \left[8\pi g_i \Pi_i^2 + \frac{1}{8\pi g_i}(\partial_x \phi_i)^2 \right] + \mathcal{B}^{(\mathrm{II})} \qquad (2)$$

where we have defined the theories on the half-line $x < 0$ and there is some interaction at the boundary $x = 0$. We see that the constants g, g_1, g_2 control the dimensions of the Vertex operators and although they could be scaled away in the free theory, they are crucial when the interaction is present. Our choice for the boundary interaction takes a peculiar form that, a priori, does not seem to be related to the physical problems at hand, we choose

$$B^{(I)} = \lambda[S_+ e^{i\phi(0)/2} + S_- e^{-i\phi(0)/2}] \qquad (3)$$

and

$$B^{(II)} = \lambda[S_+ e^{i\phi_1(0)/2} + S_- e^{-i\phi_1(0)/2}] \cos\left(\frac{\phi_2(0)}{2}\right). \qquad (4)$$

In these expressions, the $S_\pm$ and another generator S_z form a su(2)$_q$ algebra

$$[S_+, S_-] = \frac{q^{S_z} - q^{-S_z}}{q - q^{-1}}, \quad [S_z, S_\pm] = \pm S_\pm \qquad (5)$$

with $q = e^{i\pi g}$ for I and $q = e^{i\pi g_1}$ for II. It is this choice of algebra that seems a mere mathematical generalization that does not look physically motivated. The next section is devoted to show how this form relates to the physical problems mentioned before.

Our approach to solving the impurity models is largely based on their integrability. Both models presented in the introduction can be though as massless limits of massive relativistic field theories which are integrable. The massive models can be diagonalized in a basis of Bethe states and it is these same states, of which we will follow the trace in the massless limit, that will be used to provide a solution of the impurity problem. This is very akin to the original solution of the Kondo problem [3, 18].

2   Connection with Kondo Models and 1D Wires

The models presented in the introduction have a formal form and this section is devoted sketch the relationship with the physical models.[1] The key to the relation between models I and II and their physical counterpart is bosonization and representation theory of su(2)$_q$.

Model I

Let us start with the simplest example: the one-channel Kondo problem that describes dilute $s = \frac{1}{2}$ impurities in a metal. Since the impurities are dilute, we can restrict to the case where there is one impurity and the

---

[1]Not all the steps of the relationship are given restricting to the key ingredients. References useful for the reader wanting more explicit details are [1, 5, 7, 15]

interaction between the bulk electrons and the impurity takes the general form

$$\sum_{i=x,y,z} \lambda_i S_i \cdot (\psi_\alpha^\dagger \sigma_{\alpha\beta}^i \psi_\beta)(\vec{r}_0) \tag{6}$$

with $\vec{r}_0$ the position of the impurity and λ_i coupling constants. By expanding the electron field on spherical harmonics one observes that only s-waves interact with the impurity reducing the problem to one dimension. Upon bosonization of the Fermi fields at low energies, the impurity is seen to interact not with both fields but rather with a combination describing the spin fluctuations $\phi = \phi_\uparrow - \phi_\downarrow$. This leads to a boundary interaction of the form

$$\lambda \left(S_+ e^{i\sqrt{4\pi}\phi(0)} + S_- e^{-i\sqrt{4\pi}\phi(0)} \right) + \lambda_z \partial_x \phi \tag{7}$$

where we have chosen $\vec{r}_0 = 0$ and $\lambda_x = \lambda_y = \lambda$. Already this is very close to the form I of the introduction, the difference being that the spins are in a $s = \frac{1}{2}$ representation of su(2) and there is an extra term proportional to S_z. This last term can be removed by a unitary transformation of the form $U = e^{i\alpha S_z \phi(0)}$ (with α chosen appropriately) and doing so changes the dimension of the vertex operators at the boundary. This will be described by g in I. The algebra su(2)$_q$ has many representations, among them the $s = \frac{1}{2}$ representation in terms of Pauli matrices reduces the algebra to the ordinary su(2) algebra. This completes the equivalence between the models. In the notations of I the points $g = \frac{1}{2}, 1$ correspond to the Toulouse and isotropic limits of the Kondo model, respectively. It has been shown that the dissipative quantum mechanics problem is equivalent to this anisotropic Kondo model in Ref. 15 and the same for the optics problem of a two-level atom interacting with radiation in Ref. 14.

Another class of problems related to I are 1D quantum wires. There, following the standard analysis of Kane and Fisher [13], nonrelativistic spinless electrons are described, at low energies and long wavelengths, by the excitations in the Fermi sea around the Fermi momentum. Since the dispersion relation for the fermions is essentially relativistic in that region, one can again use the tools of bosonization to describe the system in terms of charge density fluctuations or bosonic fields. When describing a wire without impurity we need to take care of the electrons interactions. The constant g in I comes from the coulomb interaction, a four Fermi term, which under bosonization becomes quadratic in the currents, and amounts essentially to a rescaling of the kinetic term in this case. When adding a charge impurity, $\Psi^\dagger \Psi(x = 0)$, to this system, a term of the form

$$\lambda \cos(\sqrt{4\pi}\phi(0)) \tag{8}$$

is generated under bosonization. There are two remaining steps to understand how to relate to model I, the first is to "fold" the system by a nonlocal change of field basis to get a model on the half line. The end result is a model on the half line with a boundary interaction of the form (8). This would be of the form I if we could choose a representation of the algebra su(2)$_q$ where $S_\pm$ act like 1. When q is a k-root of unity, there are representations of the algebra called cyclic, where there are no highest nor lowest weight but rather the weights are distributed on a circle. These representations are described by a complex number δ such that

$$S_\pm |m\rangle = \frac{q^{\delta+m} - q^{-\delta-m}}{q - q^{-1}}. \qquad (9)$$

If we choose δ large imaginary, then we can find a limit in which the commutator of $S_\pm$ can be neglected and in perturbation the traces of all spin monomials are the same. It is clear that this is essentially the case where $S_\pm$ act like 1. The previous argument relied on q's being a root of unity, though it is generalizable to any q by using oscillator algebras [4].

Model II

Let us now come to model II, which, when the spins are chosen to be in a $s = \frac{1}{2}$ representation describes the two-channel Kondo model. This model is a simple generalization of the one-channel case where the impurity inter- acts with two types of electrons instead of one. This describe for example metals having two bands of conduction electrons in which there are dilute impurities. The bosonization sketched for the previous case is similar ex- cept that we have more bosonic fields corresponding to the presence of more flavors of electrons (two in our case). Some of these fields will decouple and only two bosonic densities at the end will interact with the boundary. This then leads to the interaction found in II with $g_2 = \frac{1}{2}$ [5].

The generalization of the 1D wires to the case of electrons with spin is similar, now we will bosonize both the spin up and the spin down electron. There are thus two types of bosons describing the two different electrons densities. Taking the so-called charge (+) and spin (−) combinations: $\phi_\pm = \phi_\uparrow \pm \phi_\downarrow$ we end up with the free Hamiltonian II. The constants g_i are related to the spin-spin and e-e electron interaction. A charge impurity added to this wire, of the form $\Psi_\uparrow^\dagger \Psi_\uparrow + \Psi_\downarrow^\dagger \Psi_\downarrow$, results in a boundary interaction of the form

$$\lambda \cos\left(\frac{\phi_+(0)}{2}\right) \cos\left(\frac{\phi_-(0)}{2}\right). \qquad (10)$$

The only remaining step is similar to I, where we have to take the bound- ary spins in II in a cyclic representation to get back the interaction term correctly.

The method of solution for these problems will be exposed in the follow- ing using mainly the model I since its spectra is simpler. We will explore the thermodynamic properties using a perturbative expansion and compare this with an approach based on integrability. Although they are interesting, thermodynamic properties are not the most valuable information for these previous models. Rather we would like to know about transport properties such as the conductivity in the 1D quantum wires. Having set up the framework, we will present more recent results on model II using these methods. More precisely we will concentrate on the problem of the 1D quantum wires with spin last mentioned, and present exact results for the I(V) characteristics of such a wire in the presence of an impurity.

3   Perturbation Theory and Jack Polynomials

Let us for the time being concentrate on model I. The simplest thing one can do is to look at the free energy of the model in perturbation theory. At finite temperature, the propagator on the cylinder is

$$\langle \phi(0, \tau) \phi(0, \tau') \rangle = -2g \log \left| \frac{\kappa}{\pi T} \sin \pi T (\tau - \tau') \right| \tag{11}$$

with τ imaginary time and κ the UV cut-off. For the case when the boundary is $\lambda \cos(\phi(0)/2)$ or, equivalently, the $S_\pm$ are taken in a cyclic representation, it is simple to compute the partition function that takes the form

$$\mathcal{Z} = \sum_{n=0}^{\infty} x^{2n} Z_{2n}, \quad x = \frac{\lambda}{T} \left(\frac{2\pi T}{\kappa} \right)^g. \tag{12}$$

The coefficients Z_{2n} are given by the $2n$ dimensional integrals

$$Z_{2n} = \frac{1}{(n!)^2} \int_0^{2\pi} du_1 \, du_1' \cdots \int_0^{2\pi} du_n \, du_n'$$
$$\left| \frac{\prod_{i<j}^n 4 \sin([u_i - u_j]/2) \sin([u_i' - u_j']/2)}{\prod_{i,j}^n 2 \sin([u_i - u_j']/2)} \right|^{2g}. \tag{13}$$

By changing variables to $z_k = e^{iu_k}$ the previous expression can be integrated completely by expanding the denominator on Jack polynomials. The result is sum of squared norms over all partitions of length less than n. An explicit evaluation gives

$$Z_{2n} = \frac{1}{\Gamma(g)^{2n}} \sum_\lambda \prod_{i=1}^n \left(\frac{\Gamma[\lambda_i + g(n - i + 1)]}{\Gamma[\lambda_i + g(n - i) + 1]} \right)^2. \tag{14}$$

Sadly, this sum, apart for $n = 1$, does not seem to have a simple expression. An open problem on this subject is that of duality: it has been shown from another approach that the expansion of the grand canonical partition

function at small λ and given g is related to the expansion of the same
function at large λ with $1/g$. A similar duality between Jack polynomials
exist

$$\prod_{ij}(1 + z_i w_j) = \sum_\lambda J_\lambda(z_i; g) J_{\lambda'}(w_i; 1/g), \tag{15}$$

and it would be interesting to find if there is a connection between these
two results.

Similar manipulations can be done for the case when $S_\pm$ are in a $s = \frac{1}{2}$
representation. Everything follows except that there are extra conditions
on the integrals coming from the fact that $S_\pm^2 = 0$. The result is exactly
the same as (13) apart from the condition on the integrals

$$0 < u_1 < u_1' < u_2 < u_2' < \cdots < u_n < u_n' < 2\pi;$$

that is, the integrals are ordered. The previous integration technique is no
longer available. Still the integrable approach presented in the next section
provides us with a way to compute the partition function in both cases,
leading to intriguing relations between ordered and nonordered integrals
[10].

Although explicit, this approach has its limitations; notably the perturbative sums are (up to now) not analytically known. It is possible to
compute the free energy in a completely different fashion, using the Bethe
ansatz; this is what is presented in the following.

4   Integrable Basis

To use the integrability of the models it is useful to think of them slightly
differently. Instead of a massless theory with interaction at the boundary,
let us think of it in terms of the following limiting procedure (for model I):

$$H^{(I)} = \frac{1}{2} \int_{-\infty}^0 dx \left[8\pi g \Pi^2 + \frac{1}{8\pi g}(\partial_x \phi)^2 \right]$$
$$+ \lim_{\Lambda \to 0} \int_{-\infty}^0 dx \, \Lambda \cos \phi(x) + \mathcal{B}^{(I)}. \tag{16}$$

First, let us forget about the boundary. Then the model above is simply the
sine-Gordon model (on the half-line) and its spectrum consists of solitons,
anti-solitons, and bound states (the number of which depends on g) called
breathers. The theory is completely described by the scattering between
these particles given by a factorized, elastic S-matrix

$$S(\theta)_{\epsilon_1 \epsilon_2}^{\epsilon_1' \epsilon_2'} \tag{17}$$

In this notation $\epsilon_i = \pm, 1, 2, \ldots$ indicates whether the particle is a soliton, antisoliton, or mth breather. The S-matrix is a function of the rapidity defined as

$$E = m\cosh\theta, \quad P = m\sinh\theta \tag{18}$$

satisfying the relativistic dispersion relation. The coupling Λ is related to the mass of these particles m through $m \propto \Lambda^{1-g}$. The idea is that in the massless or $\Lambda \to 0$ limit, we choose waves having solitonic profiles to describe the excitations instead of plane waves. With the above description the massless limit $\Lambda \to 0$ of these excitations is understood in the following way: take the limit $m \to 0$ and $\theta = \pm\theta_0 \pm \beta$ with $\theta_0 \to \infty$ such that the energy is kept finite, ie

$$E = \pm P = \mu e^{\beta}, \quad \mu = \frac{m}{2}e^{\theta_0} = \text{finite}. \tag{19}$$

A few comments are in order: in this description μ is a massless scale that can be chosen at will; there is no scale in a massless theory. Also we have a doubling of excitations corresponding to the right and left movers. One might wonder what sense does this limit have in the S matrices since it is difficult to understand what the scattering of two particles traveling at the speed of light means. This scattering can be understood in the classical limit as a monodromy of the classical solution [11]. In this new description we are left with a $R - R$ and $L - L$ scattering S matrix identical with the massive one (as seen by taking the limit) but a $R - L$ scattering which becomes a constant phase and will be taken to be one in the following. Thus this description in terms of massless right and left moving excitations forms an alternative description of the free bosonic theory of type I.

When the boundary interaction is present, the conserved quantities remain conserved (more precisely half of them). Since the integrable basis forms states that are eigenstates of these charges, we get strong constraint on the way the solitons are reflected at the boundary. In fact, as shown in the study [12], this suggest that the scattering at the boundary is elastic and described by a simple reflection matrix of the form

$$R(\theta - \theta_B)_{\epsilon}^{\epsilon'}. \tag{20}$$

Here θ_B is a boundary scale that is related to the coupling constant λ. The matrix has been determined explicitly in [12] using the bootstrap approach for the case where the perturbation is $\lambda\cos(\phi(0)/2)$. The last point remaining to complete the picture of model I is to understand what is the role of the representations used for the algebra $\mathrm{su}(2)_q$.

In the bootstrap approach, the boundary reflection matrix is found by analyzing the so-called boundary Yang-Baxter equation and finding its most general solution. On the other hand it is known that the S matrix of the sine-Gordon model is intimately related to the universal $\mathcal{R}$-matrix of $\mathrm{su}(2)_q$

in the $s = \frac{1}{2}$ representation. The reflection matrix of solitons scattering off an impurity in a spin $j/2$ representation is simply related to the $\mathcal{R}$ matrix intertwining the representations $\mathcal{R}_{1/2,j/2}$. In that case the boundary is thought of as an immobile particle in a spin $j/2$ representation. These solutions were found long-ago in the study of the Kondo model. When the boundary particle is taken in a cyclic representation, it is possible to red-erive [7] the solution found by the bootstrap approach, providing a unifying picture of model I. With this picture, we see that the differences between the various physical realizations of model I amounts to different reflection matrices. The excitation description is otherwise the same for all cases.

Having this picture of solitons, anti-solitons, and breathers scattering among each other and with a certain amplitude at the boundary should allow a complete description of the model I. Many results can be obtained by using the aforementioned properties; for example, using form factors of the sine-Gordon model, dynamical correlation functions of the current and spin operators were computed. Results for the spin-spin susceptibility in the Kondo problem or the AC conductivity in the Hall edge system were found in [17] that are valid for the whole RG flow from the UV to the IR. The exact $I(V)$ curve for the Hall problem was also computed in Ref. 8 using a Boltzmann-type equation and the thermodynamical Bethe ansatz. In the rest of these notes a more recent computation of the $I(V)$ characteristics is described for 1D wires with spin following Ref. 16.

The model II also has a description in terms of an integrable basis. There are some restrictions though on the parameters g_1 and g_2 arising from the integrability. An obvious choice would be $g_1 = g_2$: in that case by a simple change of variables in the fields we can show that the theory is equivalent to a sum of two sine-Gordon models. When one of the g_i is zero, we also have an integrable theory described by a free boson and a single sine-Gordon model for the other boson. In the general case, the model for generic g_i's is not integrable. The previous choices covered the trivial lines along which it is integrable, there are other lines of integrability that are more interesting though: when one of the g's for example $g_2 = 1$, then the theory is very close to that of a two-channel Kondo model. The only difference is the fact that the boundary spin now lives in $\mathrm{su}(2)_q$ instead of $\mathrm{su}(2)$. To understand the scattering description better, it is useful to review the basic features of the model.

The isotropic two-channel Kondo model describes two types of electrons interacting with a spin impurity. When the reduction to one dimension is done, the interaction with the impurity reduces to a term of the form

$$\lambda \sum_{i=1,2} \psi_i^\dagger(0)\vec{\sigma}\psi_i(0) \cdot \vec{S}, \qquad (21)$$

with i describing the type of electron and $\vec{S}$ the impurity spin. As mentioned at the beginning, at low energies the dispersion relation of the electrons is relativistic and we can use Abelian bosonization to find term at the boundary similar to what is found in model II. On the other hand, the two-channel Kondo model can be bosonized using non-Abelian bosonization. This will lead to a different description in terms of the charge, spin, and flavor currents

$$H = \int_{-\infty}^{0} JJ + J^{\lambda}J^{\lambda} + J^{i}J^{i} + \lambda J^{\lambda}S^{\lambda}(0), \qquad (22)$$

with J, J^{i}, J^{λ} the charge, flavor, and spin currents, respectively. From this it is clear that the theory is effectively given in the bulk by a WZW model based on

$$\mathrm{u}(1) \otimes \mathrm{su}(2)_2 \otimes \mathrm{su}(2)_2 \qquad (23)$$

with only the spin current interacting at the boundary. From this picture we see that the charge and flavor degrees of freedom decouples and only a $c = 3/2$ theory remains describing the spin degrees of freedom.

The scattering theory of the massless $\mathrm{su}(2)_2$ theory is known to be described by a basis akin to the sine-Gordon one used previously but with more structure. The particles have an extra quantum number: they are also a kink of an RSOS model. In our case the bulk S matrix is given by

$$S = S_{\mathrm{SG}} \otimes (\mathrm{RSOS})_3 \qquad (24)$$

with the subscript 3 indicating that this is the RSOS solution having three wells (or three possible states). The dispersion relation for the excitations of the system are massless and again the RR and LL scattering is given by that S matrix, but the RL scattering is just a constant phase that we set to one.

The alert reader will notice that there is a discrepancy when we compare the Abelian bosonization, leading to model II having $c = 2$ and the non-Abelian one with $c = 3/2$. It turns out that the non-Abelian version exhibits the symmetries more clearly but we can also recover the aforementioned scattering theory using the Abelian version. The trick is to realize that the boundary term involving the second field ϕ_2 can be fermionized

$$\cos(\phi_2(0)/2) \to \Psi\overline{\Psi}(0) \qquad (25)$$

with Ψ a *Majorana* fermions. This is possible because the cosine has the right dimension since we consider $g_2 = 1$ and also because at the boundary $\phi_2^{L}(0) = \phi_2^{R}(0)$. In the bulk, the boson ϕ_2 is fermionized with two Majorana fermions, Ψ, χ but only one of them interacts at the boundary. This leads to a boundary-interacting theory having $c = 3/2$ [16]. In fact, when the boundary is taken in a cyclic representation, this is just a $N = 1$ supersymmetric version of the sine-Gordon model. Anisotropy in the original

Kondo model is then equivalent to reestablishing the dependence on g_1 in the supersymmetric sine-Gordon model.

If we are interested in describing the boundary interaction, we need to understand how the RSOS kinks are scattered at the boundary. This is an involved problem, but the answer was found by Fendley [6] and depends on the representation in which the impurity spin is taken. In the standard two-channel Kondo model the reflection matrix depends on whether the bulk electrons "screen" the impurity. In the case of a cyclic representation the answer is

$$R = R_{\text{SG}} \otimes 1. \tag{26}$$

This is the case of interest for the 1D wires since the impurity involves no spin. Only the physical amplitude will be needed in the following; it is given by

$$|R_+^+|^2 = \frac{1}{1 + e^{-2(\theta - \ln T_B)/\gamma}}, \quad |R_+^-|^2 = 1 - |R_+^+|^2. \tag{27}$$

Having the reflection and scattering matrices, we have, in principle, all the quantum information needed about the theory. In the next section we show how to use that data to compute properties of the system.

5   Transport Properties for the 1D Wires

Let us now show how transport properties for the 1D wires can be computed using this integrable basis. In practice, such a wire will be subject to a voltage, or be at finite temperature. This means that the ground state is the one found by minimizing the free energy in the presence of either voltage or temperature. This can be done by using the Bethe ansatz, but first we need to specify how the voltage couples to the fields in the model II.

In the quantum wires description, the fields ϕ_1 and ϕ_2 correspond to the charge and spin densities, respectively. If a voltage is added to the system, it can be described by adding a chemical potential term to the Hamiltonian of the form[2]

$$H_V = V \int_{-\infty}^{0} dx \, \partial_x \phi_1. \tag{28}$$

It is also conventional to use the parameters $g_\rho = 2g_1$ and $g_\sigma = 2g_2 = 2$ in the 1D wires literature to describe the effect of the charge and spin interactions in the wire and we will use these in the following.

---

[2] There is a subtlety regarding the nature of the reservoirs that we do not discuss. A recent article about this problem is Ref. 2

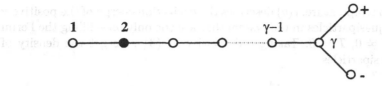

FIGURE 19.1. TBA diagram

Thus, the effect of a positive finite voltage is to give a chemical potential to the particles having positive charge in the system. From the Hamiltonian of model II these are easily identified as the ones having topological charge one in the sine-Gordon sector. The Bethe ansatz leads to a quantization of the rapidities in the system

$$e^{-iL\mu e^{\theta_i}} \Lambda(\theta_i \mid \theta_1, \dots, \theta_N) = 1. \tag{29}$$

In this expression Λ is a large matrix describing the nondiagonal scattering of the particles among themselves. The diagonalization of the matrix necessitate the use of a higher-level Bethe ansatz, which won't be exposed here but is well explained in [16]. The end result is given by the set of integral equations for γ an integer

$$\epsilon_j(\theta, V/T) = \delta_{j,2}\frac{me^\theta}{T} + K \star \sum_k N_{jk} \ln\left[1 + e^{-\mu_k/T} e^{\epsilon_k(\theta,V/T)}\right], \tag{30}$$

where $g_\rho/4 = \gamma/[2(\gamma + 2)]$ and the kernel $K(\theta) \equiv \gamma/(2\pi \cosh \gamma\theta)$, and N_{jk} is the incidence matrix of the diagram of Figure 19.1 (which is found by "gluing" the RSOS and the nondiagonal sine-Gordon TBA diagrams) that is $N_{jk} = 1$ if the nodes j and k are connected, and 0 otherwise (in particular $N_{jj} = 0$). The chemical potential vanishes, except for the end nodes, where $\mu_\pm = \pm V/2$, V the applied voltage. Here, all the nodes but the one with $j = 2$ correspond to pseudoparticles, which are necessary to diagonalize the S-matrix.

Let us now concentrate on the zero temperature, finite voltage solution. At finite chemical potential, there will be positively charged particles filling the sea up to a Fermi rapidity A in a compact way. In that case, the set of equations for the ϵ_j can be recast in a single equation for the node $j = 2$

$$\epsilon(\theta) - \int_{-\infty}^A \Phi(\theta - \theta')\epsilon(\theta') \, d\theta' = \frac{V}{2} - \frac{m}{2}e^\theta. \tag{31}$$

Here, Φ is a kernel with Fourier transform

$$\tilde\Phi = 1 - \frac{\sinh(\gamma + 2)\pi\omega/2}{4\cosh^2 \pi\omega/2 \sinh \gamma\pi\omega/2} = 1 - \frac{1}{N(\omega)N(-\omega)}. \tag{32}$$

The Fermi rapidity can be determined by the condition that $\epsilon(A) = 0$.

At zero temperature, $\epsilon(\theta)$ describes the excitation energy of the positively charged quasiparticles in the system that are the only ones filling the Fermi sea at $V \neq 0$, $T = 0$. Taking a derivative of (31), we get the density of these quasiparticles

$$\rho(\theta) - \int_{-\infty}^{A} d\theta' \, \rho(\theta')\Phi(\theta - \theta') = \frac{1}{2h}e^{\theta}. \qquad (33)$$

The current then follows from a Boltzmann equation describing the one-by-one scattering of these charged excitation on the boundary. There are $\rho(\theta)d\theta$ such particles in $[\theta, \theta + d\theta]$ and the the probability that a particle of type α is reflected as one of type α' is $\left| R_{\alpha}^{\alpha'} \left(\theta - \log(T_B) \right) \right|^2$. This leads, following the arguments of Ref. 9, to the current

$$I(V, T_B) = e \int_{-\infty}^{A} d\theta \, \rho(\theta) \left| R_u^u \left(\theta - \log(T_B) \right) \right|^2. \qquad (34)$$

After a few manipulations we get the exact expression

$$I(V, T_B) = \frac{e}{2h} e^A N(-i) \int_{-\infty}^{A} d\theta \int_{-\infty}^{\infty} \frac{d\omega}{2\pi} \frac{e^{-i\omega\theta}}{(1+i\omega)} \frac{N(\omega)}{1 + e^{-2/\gamma(\theta + A - \log T_B)}}. \qquad (35)$$

From these two expressions we can check that there is an exact duality between the UV expansion and the IR expansion under the exchange $g_\rho \to 4/g_\rho$. This duality is more transparent through the relation

$$I(V, T_B', g_\rho) = \frac{e^2 V}{h} g_\rho - \frac{g_\rho^2}{4} I(T_B', V, 4/g_\rho). \qquad (36)$$

where

$$T_B' = T_B \frac{2\pi}{2 - g_\rho}. \qquad (37)$$

This concludes the computation of the zero T $I(V)$ characteristics of the 1D wire.

6   Conclusion

The goal of these notes was to provide an overview of the integrable solution of certain impurity problems and give an explicit example in the case of quantum wires with spin. At the center of the method is a description of free theories by integrable bases that are interacting. Although complicated if one is only interested in the free theory, these bases lead to nonperturbative results when certain types of boundary interactions are added.

The integrable subspace given for model II is not unique: other values of the parameters g_i lead to integrability. One of the most interesting cases is

$$g_1 + g_2 = 1 \tag{38}$$

and a boundary interaction of the form $\lambda \cos\phi_1/2 \cos\phi_2/2$. It is possible to show the existence of nonlocal currents commuting to form a $su(2)_{q_1} \otimes su(2)_{q_2}$ algebra and from them deduce the scattering matrices. The interesting feature of this model is that it is not classically integrable. This leads to new previously unknown IR fixed points [16].

7 REFERENCES

1. I. Affleck, *Conformal field theory approach to the Kondo effect*, Acta Phys. Pol. **26** (1995), No. 12, 1869–1932.

2. A. Yu. Alekseev, V. V. Cheianov, and J. Fröhlich, *Comparing conductance quantization in quantum wires and quantum Hall systems*, cond-mat/9607144.

3. N. Andrei, K. Furuya, and J. Lowenstein, *Solution of the Kondo problem*, Rev. Mod. Phys. **55** (1983), No. 5, 331–402.

4. V. V. Bazhanov, S. L. Lukyanov, and A. B. Zamolodchikov, *Integrable structure of conformal field theory, quantum KdV theory and thermodynamic Bethe ansatz*, Commun. Math. Phys. **177** (1996), No. 2, 381–398.

5. V. J. Emery and S. A. Kivelson, *Mapping of the two-channel Kondo problem to a resonant-level model*, Phys. Rev. B **46** (1992), 10812–10817.

6. P. Fendley, *Kinks in the Kondo problem*, Phys. Rev. Lett. **71** (1993), 2485–2489.

7. P. Fendley, F. Lesage, and H. Saleur, *A unified framework for the Kondo problem and for an impurity in a Luttinger liquid*, J. Stat. Phys. **85** (1996), No. 1-2, 211–249.

8. P. Fendley, A. W. W. Ludwig, and H. Saleur, *Exact conductance through point contacts in the $\nu = \frac{1}{3}$ quantum Hall effect*, Phys. Rev. Lett. **74** (1995), 3005–3009.

9. _____, *Exact nonequilibrium transport through point contacts in quantum wires and fractional quantum Hall devices*, Phys. Rev. B **52** (1995), 8934.

10. P. Fendley and H. Saleur, *Exact perturbative solution of the Kondo problem*, Phys. Rev. Lett. **75** (1995), 4492–4496.

11. P. Fendley, H. Saleur, and N. P. Warner, *Exact solution of a massless scalar field with a relevant boundary interaction*, Nucl. Phys. B **430** (1994), No. 3, 577–596.

12. S. Ghoshal and A. B. Zamolodchikov, *Boundary S matrix and boundary state in two-dimensional integrable quantum field theory*, Internat. J. Modern Phys. A **9** (1994), No. 21, 3841–3885.

13. C. L. Kane and M. P. A. Fisher, *Transmission through barriers and resonant tunneling in an interacting one-dimensional electron gas*, Phys. Rev. B **46** (1992), 15233–15262.

14. A. LeClair, F. Lesage, S. Lukyanov, and H. Saleur, *The Maxwell–Bloch theory in quantum optics and the Kondo model*, Phys. Lett. A **235** (1997), 203–208.

15. A. J. Leggett, S. Chakravarty, A. T. Dorsey, M. P. A. Fisher, A. Garg, and W. Zwerger, *Dynamics of the dissipative two-state system*, Rev. Mod. Phys. **59** (1987), 1–85.

16. F. Lesage, H. Saleur, and P. Simonetti, *Tunneling in quantum wires. I. Exact solution of the spin isotropic case*, 1997, cond-mat/9703220, p. 7598.

17. F. Lesage, H. Saleur, and S. Skorik, *Form factors approach to current correlations in one-dimensional systems with impurities*, Nucl. Phys. B **474** (1996), No. 3, 602–640.

18. A. M. Tsvelick and P. B. Wiegmann, *Exact results in the theory of magnetic alloys*, Adv. Phys. **32** (1983), 453–708.

20

Painlevé–Calogero Correspondence

A. M. Levin
M. A. Olshanetsky

ABSTRACT It is proved that the Painlevé VI equation $\mathrm{PVI}_{\alpha,\beta,\gamma,\delta}$ for the special values of constants $\alpha = \nu^2/4$, $\beta = -\nu^2/4$, $\gamma = \nu^2/4$, $\delta = \frac{1}{2} - \nu^2/4$ is a reduced Hamiltonian system. Its phase space is the set of flat $\mathrm{SL}(2,\mathbb{C})$ connections over elliptic curves with a marked point, and time of the system is given by the elliptic module. This equation can be derived via reduction procedure from the free infinite Hamiltonian system. The phase space of the latter is the affine space of smooth connections, and the times are the Beltrami differentials. This approach allows us to define the associate linear problem, whose isomonodromic deformations are provided by the Painlevé equation and the Lax pair. In addition, it leads to a description of solutions by a linear procedure. This scheme can be generalized to G bundles over Riemann curves with marked points, where G is a simple complex Lie group. In some special limit such Hamiltonian systems convert into the Hitchin systems. In particular, for SL bundles over elliptic curves with a marked point, we obtain in this limit the elliptic Calogero N-body system. Relations to the classical limit of the Knizhnik–Zamolodchikov–Bernard equations are discussed.

1 Introduction

We learned from Yu. Manin's lectures in MPI (Bonn, 1996) about the elliptic form of the famous Painlevé VI equation (PVI) [21]. In this representation PVI looks very similar to the elliptic Calogero–Inozemtsev–Treibich–Verdier (CITV) rank one system [4, 12, 27]. Namely, both equations are Hamiltonian with the same symplectic structure for two dynamical variables and the same Hamiltonians. The only difference is that the time in the PVI system is nothing like the elliptic module. Therefore, it is a nonautonomous Hamiltonian system, while the CITV Hamiltonian is independent of time. This similarity is not accidental and is based on the very closed geometric origin of both systems, which we will elucidate in this talk. It should be confessed from the very beginning that at the present time our approach is to cover only the one parametric family of $\text{PVI}_{\alpha,\beta,\gamma,\delta}$. This family corresponds to the standard two-body elliptic Calogero equation.

1.1 Painlevé VI and Calogero Equations

The Painlevé VI $\text{PVI}_{\alpha,\beta,\gamma,\delta}$ equation depends of four free parameters and has the form

$$\frac{d^2X}{dt^2} = \frac{1}{2}\left(\frac{1}{X} + \frac{1}{X-1} + \frac{1}{X-t}\right)\left(\frac{dX}{dt}\right)^2 - \left(\frac{1}{t} + \frac{1}{t-1} + \frac{1}{X-t}\right)\frac{dX}{dt}$$
$$+ \frac{X(X-1(X-t)}{t^2(t-1)^2}\left(\alpha + \beta\frac{t}{X^2} + \gamma\frac{t-1}{(X-1)^2} + \delta\frac{t(t-1)}{X-t)^2}\right). \quad (1)$$

It is a Hamiltonian systems [23], but we will write the symplectic form and the Hamiltonian below in another variable. Among other distinguishing features of this equation we are interested in its relation to the isomonodromic deformations of linear differential equations. This approach was investigated by Fuchs [7], while $\text{PVI}_{\alpha,\beta,\gamma,\delta}$ first was written down by Gambier [8]. The equation has a lot of different applications (see Ref. 20). We shortly present $\text{PVI}_{\alpha,\beta,\gamma,\delta}$ in terms of elliptic functions [21].

Let $\wp(u \mid \tau)$ be the Weiershtrass function on the elliptic curve $T_\tau^2 = \mathbb{C}/(\mathbb{Z}+\mathbb{Z}\tau)$, and $e_i = \wp(T_i/2 \mid \tau)$, $(i = 1, 2, 3)$ $(T_0,\ldots,T_3) = (0, 1, \tau, 1+\tau)$. Consider instead of (t, X) in (1), the new variables

$$(\tau, u) \to \left(t = \frac{e_3 - e_1}{e_2 - e_1}, X = \frac{\wp(u \mid \tau) - e_1}{e_2 - e_1}\right). \quad (2)$$

Then $\text{PVI}_{\alpha,\beta,\gamma,\delta}$ takes the form

$$\frac{d^2u}{d\tau^2} = \partial_u U(u \mid \tau), \quad U(u \mid \tau) = \frac{1}{(2\pi i)^2}\sum_{j=0}^{3}\alpha_j\wp\left(u + \frac{T_j}{2}\,\Big|\,\tau\right), \quad (3)$$

$(\alpha_0,\ldots,\alpha_3) = (\alpha, -\beta, \gamma\frac{1}{2} - \delta)$. As usual in the nonautonomous case, the equations of motion (3) are derived from the variations of the degenerated symplectic form

$$\omega = \delta v\delta u - \delta H\delta\tau, \quad H = \frac{v^2}{2} + U(u \mid \tau), \quad (4)$$

which is defined over the extended phase space $\mathcal{P} = \{v, u, \tau\}$. The semidirect product of $\mathbb{Z} + \mathbb{Z}\tau$ and the modular group act on the dynamical variables (v, u, τ) preserving (4).

Let us introduce the new parameter κ (the level) and instead of (4) consider

$$\omega = \delta v\delta u - \frac{1}{\kappa}\delta H\delta\tau. \quad (5)$$

It corresponds to the overall rescaling of constants $\alpha_j \to \frac{\alpha_j}{\kappa^2}$. Put $\tau = \tau_0 + \kappa t^H$ and consider the system in the limit $\kappa \to 0$, which is called the critical level. We come to the equation

$$\frac{d^2 u}{(dt^H)^2} = \partial_u U(u \mid \tau_0), \tag{6}$$

It is just the rank one CITV$_{\alpha,\beta,\gamma,\delta}$ equation. Thus, we have in this limit PVI$_{\alpha,\beta,\gamma,\delta} \xrightarrow{\kappa \to 0}$ CI$_{\alpha,\beta,\gamma,\delta}$.

Consider one-parametric family PVI$_{\nu^2/4,-\nu^2/4,\nu^2/4,1/2-\nu^2/4}$. The potential (3) takes the form

$$U(u \mid \tau) = \frac{1}{(4\pi i)^2} \nu^2 \wp(2u \mid \tau). \tag{7}$$

We will prove that (5) with the potential (7) describe the dynamic of flat connections of SL$(2, \mathbb{C})$ bundles over elliptic curves T_τ with one marked point $\Sigma_{1,1}$. In fact, u lies on the Jacobian of T_τ, (v, u) defines a flat bundle, and τ defines a point in the moduli space $\mathcal{M}_{1,1} = \{\Sigma_{1,1}\}$. The choice of the polarization of connections, in other words, v and u, depends on the complex structure of $\Sigma_{1,1}$. The extended phase space $\mathcal{P}$ includes besides the dynamical variables v and u, the "time" τ. It is the bundle over $\mathcal{M}_{1,1}$ with the fibers $\mathcal{R} = \{v, u\}$, which is endowed by the degenerated symplectic structure ω (5). This system is derived by a reduction procedure from some free, but infinite Hamiltonian system. In this way we obtain the Lax equations, the linear system that monodromies preserve by (6) with $U(u \mid \tau)$ (7) and the explicit solutions of the Cauchy problem via the so-called projection method. The discrete symmetries of (4) are nothing else but the remnant gauge symmetries. On the critical level it is just the two-body elliptic Calogero system. The corresponding quantum system is identified with the KZB equation [2, 3, 16] for the one-vertex correlator on T_τ. In a similar way PVI$_{\nu^2/4,-\nu^2/4,\nu^2/4,1/2-\nu^2/4}$ is the classical limit of the KZB for $\kappa \neq 0$.

This particular example has far-reaching generalizations. Consider a phase space that is the moduli space of flat connections $\mathcal{A}$ of G bundle over Riemann curve $\Sigma_{g,n}$ of genus g with n marked points, where G is a complex simple Lie group. While the flatness is the topological property of bundles, the polarization of connections $\mathcal{A} = (A, \bar{A})$ depends on the choice of complex structure on $\Sigma_{g,n}$. Therefore, we consider a bundle $\mathcal{P}$ over the moduli space $\mathcal{M}_{g,n}$ of curves with flat connections $\mathcal{R} = (A, \bar{A})$ as fibers. The fibers are supplemented by elements of coadjoint orbits $\mathcal{O}_a$ in the marked points x_a. There exists a closed degenerate two-form ω on $\mathcal{P}$, which is nondegenerate on the fibers. The equations of motions are defined as variations of the dynamical variables along the null-leaves of this symplectic form. We call them the *hierarchies of isomonodromic deformations* (HID). They are attended by the *Whitham hierarchies*, which are discussed in Ref. 18 as a result of the averaging procedure, and also in Ref. 5 as the classical limit of "string equations." Our approach is close to the Hitchin construction of integrable systems, living on the cotangent bundles to the moduli space of holomorphic G bundles [10], generalized for singular curves in Refs. 6, 26, and 22. Namely, the connection $\bar{A}$ plays the same role as in the Hitchin scheme, while A replaces the Higgs field. Essentially, our construction is local—we work over a vicinity of some fixed curve $\Sigma_{g,n}$ in $\mathcal{M}_{g,n}$. The coordinates of tangent vector to $\mathcal{M}_{g,n}$ in this case play the role of times, while the Hitchin times have nothing to do with the moduli space. The Hamiltonians are the same quadratic Hitchin Hamiltonians, but now they are time dependent. There is a free parameter κ (*the level*) in our construction. On the critical level ($\kappa = 0$), after rescaling the times, our systems convert into the Hitchin systems. In concrete examples our work is based essentially on Ref. 22, which deals with the same systems on the critical level.

Like Hitchin systems, HID can be derived by the symplectic reduction from a free infinite Hamiltonian system. In our case the upstairs extended phase space is the space of the affine connections and the Beltrami differentials. We consider its symplectic quotient with respect to gauge action on the connections. In addition, to arrive at the moduli space $\mathcal{M}_{g,n}$ we need the subsequent factorization under the action of the diffeomorphisms, which effectively acts on the Beltrami differentials only. Apart from the last step, this derivation resembles the construction of the KZB systems in Ref. 1, where they are derived as a quantization of the very similar symplectic quotient. Our approach allows us to write down the Lax pairs; prove that the HID are consistency conditions of the isomonodromic deformations of the linear Lax equations; and, therefore, justify the notion of HID. Moreover, we describe solutions via linear procedures (the projection method). HID are the quasiclassical limit of the KZB equations for $\kappa \neq 0$, as the Hitchin systems are the quasiclassical limit of the KZB equations on the critical level [13, 22]. The quantum counterpart of the Whitham hierarchy is the flatness condition, which is discussed in Ref. 11 within the context of derivation of KZB. The interrelations between quantizations of isomonodromic deformations and the KZB equations are discussed in Refs. 9, 17, and 26.

For genus zero our procedure leads to Schlesinger equations. We restrict ourselves to the simplest cases with only simple poles of connections. Therefore, we don't include in the phase space the Stokes parameters. This phenomenon was investigated in the rational case in detail in Refs. 14 and 15. For genus one we obtain a particular case of the Painlevé VI equation (for $SL(2, \mathbb{C})$ bundles with one marked point), generalization of this case on arbitrary simple groups and arbitrary number of marked points.

2 Symplectic Reduction

2.1 Upstairs Extended Phase Space

Let $\Sigma_{g,n}$ be a Riemann curve of genus g with n marked points. Let us fix the complex structure of $\Sigma_{g,n}$ defining local coordinates $(z, \bar{z})$ in open maps covering $\Sigma_{g,n}$. Assume that the marked points $(x_1, \ldots, x_n)$ are in the generic positions. The deformations of the basic complex structure are determined by the Beltrami differentials μ, which are smooth $(-1, 1)$ differentials on $\Sigma_{g,n}$, $\mu \in \mathcal{A}^{(-1,1)}(\Sigma_{g,n})$. We identify this set with the space of times $\mathcal{N}'$. The Beltrami differentials can be defined in the following way. Consider the chiral diffeomorphisms of $\Sigma_{g,n}$

$$w = z - \epsilon(z, \bar{z}), \quad \bar{w} = \bar{z} \tag{8}$$

and the corresponding one-form dw. Up to the conformal factor $1 - \partial \epsilon(z, \bar{z})$, it is equal

$$dw = dz - \mu d\bar{z}, \quad \mu = \frac{\bar{\partial}\epsilon(z, \bar{z})}{1 - \partial\epsilon(z, \bar{z})}. \tag{9}$$

The new holomorphic structure is defined by the deformed antiholomorphic operator annihilating dw, while the antiholomorphic structure is kept unchanged:

$$\partial_{\bar{w}} = \bar{\partial} + \mu\partial, \quad \partial_w = \partial.$$

In addition, assume that μ vanishes in the marked points $\mu(z, \bar{z})|_{x_a} = 0$. We consider small deformations of the basic complex structure $(z, \bar{z})$. It allows us to replace (9) by

$$\mu = \bar{\partial}\epsilon(z, \bar{z}). \tag{10}$$

Let $\mathcal{E}$ be a principal stable G bundle over a Riemann curve $\Sigma_{g,n}$. Assume that G is a complex simple Lie group. The phase space $\mathcal{R}'$ is recruited from the following data:

i) the affine space $\{\mathcal{A}\}$ of the Lie(G)-valued connection on $\mathcal{E}$. It has the following component description:

   a) C^∞ connection $\{\bar{A}\}$, corresponding to the $d\bar{w} = d\bar{z}$ component of $\mathcal{A}$;

   b) the dual to the previous space $\{A\}$ of dw components of $\mathcal{A}$. A can have simple poles in the marked points.

   Moreover, assume that $\bar{A}\mu$ is a C^∞ object;

ii) cotangent bundles $T^*G_a = \{(p_a, g_a), p_a \in \text{Lie}^*(G_a), g_a \in G_a\}$ ($a = 1, \ldots, n$) in the points $(x_1, \ldots, x_n)$.

There is the canonical symplectic form on $\mathcal{R}'$

$$\omega_0 = \int_\Sigma \langle \delta A, \delta \bar{A} \rangle + 2\pi i \sum_{a=1}^n \delta \langle p_a, g_a^{-1} \delta g_a \rangle, \tag{11}$$

where $\langle \ , \ \rangle$ denotes the Killing form on $\mathrm{Lie}(G)$.

Consider the bundle $\mathcal{P}'$ over $\mathcal{N}'$ with $\mathcal{R}'$ as the fibers. It plays role of the extended phase space. There exists the degenerate form on $\mathcal{P}'$:

$$\omega = \omega_0 - \frac{1}{\kappa} \int_\Sigma \langle \delta A, A \rangle \delta \mu. \tag{12}$$

Thus, we deal with the infinite set of Hamiltonians $\langle A, A \rangle (z, \bar{z})$, parametrized by points of $\Sigma_{g,n}$ and the corresponding set of times $\mu(z, \bar{z})$. The equations of motion take the form

$$\frac{\partial A}{\partial \mu}(z, \bar{z}) = 0, \quad \kappa \frac{\partial \bar{A}}{\partial \mu}(z, \bar{z}) = A(z, \bar{z}), \quad \frac{\partial p_b}{\partial \mu} = 0, \quad \frac{\partial g_b}{\partial \mu} = 0. \tag{13}$$

We will apply the formalism of Hamiltonian reduction to these systems.

2.2 Symmetries

The form ω (12) is invariant with respect to the action of the group $\mathcal{G}_0$ of diffeomorphisms of $\Sigma_{g,n}$, which are trivial in vicinities $\mathcal{U}_a$ of marked points:

$$\mathcal{G}_0 = \{z \to N(z, \bar{z}), \bar{z} \to \overline{N}(z, \bar{z}), N(z, \bar{z}) = z + o(|z - x_a|), z \in \mathcal{U}_a\}. \tag{14}$$

Another infinite gauge symmetry of the form (12) is the group $\mathcal{G}_1 = \{f(z, \bar{z}) \in C^\infty(\Sigma_g, G)\}$ that acts on the dynamical fields as

$$A \to f(A + \kappa \partial)f^{-1}, \quad \bar{A} \to f(\bar{A} + \bar{\partial} + \mu \partial)f^{-1},$$
$$(\bar{A}' \to f(\bar{A}' + \bar{\partial})f^{-1}), \tag{15}$$
$$p_a \to f_a p_a f_a^{-1}, \quad g_a \to g_a f_a^{-1}, \quad (f_a = \lim_{z \to x_a} f(z, \bar{z})), \quad \mu \to \mu.$$

In other words, the gauge action of $\mathcal{G}_1$ does not touch the base $\mathcal{N}'$ and transforms only the fibers $\mathcal{R}'$. The whole gauge group is the semidirect product $\mathcal{G}_1 \oslash \mathcal{G}_0$.

2.3 Symplectic Reduction with Respect to $\mathcal{G}_1$

Since the symplectic form (12) is closed (though it is degenerated) one can consider the symplectic quotient of the extended phase space $\mathcal{P}'$ under the action of the gauge transformations (15). They are generated by the moment constraints

$$F_{A,\bar{A}}(z, \bar{z}) - 2\pi i \sum_{a=1}^n \delta^2(x_a)p_a = 0, \tag{16}$$

where $F_{A,\bar{A}} = (\bar{\partial} + \partial\mu)A - \kappa\partial\bar{A} + [\bar{A}, A]$. This means that we deal with the flat connection everywhere on $\Sigma_{g,n}$ except the marked points. The holonomies of $(A, \bar{A})$ around the marked points are conjugate to $\exp 2\pi i p_a$.

Let $(L, \bar{L})$ be the gauge-transformed connections

$$\bar{A} = f(\bar{L} + \bar{\partial} + \mu\partial)f^{-1}, \qquad A = f(L + \kappa\partial)f^{-1}, \tag{17}$$

Then (16) takes the form

$$(\bar{\partial} + \partial\mu)L - \kappa\partial\bar{L} + [\bar{L}, L] = 2\pi i \sum_{a=1}^{n} \delta^2(x_a)p_a. \tag{18}$$

Remark 1. The gauge fixing allows us to choose $\bar{A}$ in a such a way that $\partial\bar{L} = 0$. Then (18) takes the form

$$(\bar{\partial} + \partial\mu)L + [\bar{L}, L] = \sum_{a=1}^{n} \delta^2(x_a)p_a. \tag{19}$$

It coincides with the moment equation for the Hitchin systems on singular curves [22].

We can rewrite (19) as

$$\partial_{\bar{w}}L + [\bar{L}, L] = 2\pi i \sum_{a=1}^{n} \delta^2(x_a)p_a. \tag{20}$$

Anyway, by choosing $\bar{L}$ we can somehow fix the gauge in the generic case. There is additional gauge freedom h_a in the points x_a, which acts on T^*G_a as $p_a \to p_a$, $g_a \to h_a g_a$. It allows us to fix p_a on some coadjoint orbit $p_a = g_a^{-1}p_a^{(0)}g_a$ and obtain the symplectic quotient $\mathcal{O}_a = T^*G_a//G_a$. Thus in (18) or (19) p_a are elements of $\mathcal{O}_a$.

Let $\mathcal{I}_{g,n}$ be the equivalence classes of the connections $(A, \bar{A})$ with respect to the gauge action (17)—the moduli space of stable flat G bundles over $\Sigma_{g,n}$. It is a smooth finite-dimensional space. Fixing the conjugacy classes of holonomies $(L, \bar{L})$ around marked points (18) amounts to choosing a symplectic leave $\mathcal{R}$ in $\mathcal{I}_{g,n}$. Thereby we come to the symplectic quotient

$$\mathcal{R} = \mathcal{R}'//\mathcal{G}_1 = \mathcal{J}_1^{-1}(0)/\mathcal{G}_1 \subset \mathcal{I}_{g,n}.$$

The connections $(L, \bar{L})$ in addition to $\mathbf{p} = (p_1, \ldots, p_n)$ depend on a finite even number of free parameters $2r$ $(\mathbf{v}, \mathbf{u})$, $\mathbf{v} = (v_1, \ldots, v_r)$, $\mathbf{u} = (u_1, \ldots, u_r)$.

$$r = \begin{cases} 0 & g = 0, \\ \operatorname{rank} G, & g = 1, \\ (g-1)\dim G, & g \geq 2. \end{cases}$$

The fibers $\mathcal{R}$ are symplectic manifolds with the nondegenerate symplectic form, which is the reduction of (12):

$$\omega_0 = \int_\Sigma \langle \delta L, \delta \bar{L} \rangle + 2\pi i \sum_{a=1}^n \delta \langle p_a, g_a \delta g_a^{-1} \rangle. \tag{21}$$

At this stage we come to the bundle $\mathcal{P}''$ with the finite-dimensional fibers $\mathcal{R}$ over the infinite-dimensional base $\mathcal{N}'$ with the symplectic form

$$\omega = \omega_0 - \frac{1}{\kappa} \int_\Sigma \langle L, \delta L \rangle \delta \mu. \tag{22}$$

2.4   Factorization with Respect to the Diffeomorphisms $\mathcal{G}_0$

We can utilize invariance of ω with respect to $\mathcal{G}_0$ and reduce $\mathcal{N}'$ to the finite-dimensional space $\mathcal{N}$, which is isomorphic to the moduli space $\mathcal{M}_{g,n}$. The crucial point is that for the flat connections the action of diffeomorphisms $\mathcal{G}_0$ on the connection fields is generated by the gauge transforms $\mathcal{G}_1$. But we already have performed the symplectic reduction with respect to $\mathcal{G}_1$. Therefore, we can push ω (22) down on the factor space $\mathcal{P}''/\mathcal{G}_0$. Since $\mathcal{G}_0$ acts on $\mathcal{N}'$ only, it can be done by fixing the dependence of μ on the coordinates in the Teichmüller space $\mathcal{T}_{g,n}$. According to (10) μ is represented as

$$\mu = \sum_{s=1}^l \mu_s. \tag{23}$$

The Beltrami differential (23) defines the tangent vector $\mathbf{t} = (t_1, \ldots, t_l)$, to the Teichmüller space $\mathcal{T}_{g,n}$ at the fixed point of $\mathcal{T}_{g,n}$.

We specify the dependence of μ on the positions of the marked points in the following way. Let $\mathcal{U}'_a \supset \mathcal{U}_a$ be two vicinities of the marked point x_a such that $\mathcal{U}'_a \cap \mathcal{U}'_b = \varnothing$ for $a \neq b$, and $\chi_a(z, \bar{z})$ is a smooth function:

$$\chi_a(z, \bar{z}) = \begin{cases} 1, & z \in \mathcal{U}_a \\ 0, & z \in \Sigma_{g,n} \setminus \mathcal{U}'_a. \end{cases}$$

Introduce times related to the positions of the marked points $t_a = x_a - x_a^0$. Then

$$\mu_a = t_a \bar{\partial} n_a(z, \bar{z}), \quad n_a(z, \bar{z}) = \left(1 + c_a(z - x_a^0)\right)\chi_a(z, \bar{z}). \tag{24}$$

The action of $\mathcal{G}_0$ on the phase space $\mathcal{P}''$ reduces the infinite-dimensional component $\mathcal{N}'$ to $\mathcal{T}_{g,n}$. After the reduction we come to the bundle with base $\mathcal{T}_{g,n}$. The symplectic form (22) is transformed as follows:

$$\omega = \omega_0(\mathbf{v}, \mathbf{u}, \mathbf{p}, \mathbf{t}) - \frac{1}{\kappa}\sum_{s=1}^l \delta H_s(\mathbf{v}, \mathbf{u}, \mathbf{p}, \mathbf{t})\delta t_s, \quad H_s = \int_\Sigma \langle L, L \rangle \partial_s \mu, \tag{25}$$

where ω_0 is defined by (21).

In fact, we still have a redundant discrete symmetry, since ω is invariant under the mapping class group $\pi_0(\mathcal{G}_0)$. Eventually, we come to the moduli space $\mathcal{M}_{g,n} = \mathcal{T}_{g,n}/\pi_0(\mathcal{G}_0)$.

The extended phase space $\mathcal{P}$ is the result of the symplectic reduction with respect to the $\mathcal{G}_1$ action and subsequent factorization under the $\mathcal{G}_0$ action. We can write symbolically $\mathcal{P} = (\mathcal{P}''//\mathcal{G}_1)/\mathcal{G}_0$. It is endowed with the symplectic form (25).

2.5   The Hierarchies of the Isomonodromic Deformations

The equations of motion (HID) can be extracted from the symplectic form (25). In terms of the local coordinates they take the form

$$\kappa\partial_s \mathbf{v} = \{H_s, \mathbf{v}\}_{\omega_0}, \quad \kappa\partial_s \mathbf{u} = \{H_s, \mathbf{u}\}_{\omega_0}, \quad \kappa\partial_s \mathbf{p} = \{H_s, \mathbf{p}\}_{\omega_0}. \quad (26)$$

The Poisson bracket $\{\cdot, \cdot\}_{\omega_0}$ is the inverse tensor to ω_0. We also have the accompanying Whitham hierarchy (26):

$$\partial_s H_r - \partial_r H_s + \{H_r, H_s\}_{\omega_0} = 0. \quad (27)$$

There exists the one form on $\mathcal{M}_{g,n}$ defining *the tau function* of the hierarchy of isomonodromic deformations:

$$\delta \log \tau = \delta^{-1}\omega_0 - \frac{1}{\kappa}\sum H_s dt_s. \quad (28)$$

The following three statements are valid for the HID (26):

Proposition 1. *There exists the consistent system of linear equations*

$$(\kappa\partial + L)\Psi = 0, \quad (29)$$
$$(\partial_s + M_s)\Psi = 0, \quad (s = 1, \ldots, l = \dim \mathcal{M}_{g,n}) \quad (30)$$
$$(\bar{\partial} + \mu\partial + \bar{L})\Psi = 0, \quad (31)$$

where M_s is a solution to the linear equation

$$\partial_{\bar{w}} M_s - [M_s, \bar{L}] = \partial_s \bar{L} - \frac{1}{\kappa}L\partial_s\mu. \quad (32)$$

Proposition 2. *The linear conditions (30) provide the isomonodromic deformations of the linear system (29), (31) with respect to change the "times" on $\mathcal{M}_{g,n}$.*

Therefore, the HID (26) are the monodromy-preserving conditions for the linear system (29), (31).

The presence of a derivative with respect to the spectral parameter $w \in \Sigma_{g,n}$ in the linear equation (29) is a distinguishing feature of the monodromy-preserving equations. It hinders the application of the inverse scattering method to these types of systems. Nevertheless, in our case we have in some sense the explicit form of the solutions.

Proposition 3 (The projection method). *The solution of the Cauchy problem of (26) for the initial data* $\mathbf{v}^0, \mathbf{u}^0, \mathbf{p}^0$ *at the time* $\mathbf{t} = \mathbf{t}^0$ *is defined in terms of the elements* $L^0, \bar{L}^0$ *as the gauge transform*

$$\bar{L}(\mathbf{t}) = f^{-1}(L^0(\mu(\mathbf{t}) - \mu(\mathbf{t}^0)) + (\bar{L}^0))f + f^{-1}(\bar{\partial} + \mu(\mathbf{t})\partial)f, \qquad (33)$$

$$L(\mathbf{t}) = f^{-1}(\partial + L^0)f, \quad \mathbf{p}(\mathbf{t}) = f^{-1}(\mathbf{p}^0)f, \qquad (34)$$

where f *is a smooth* G-*valued function on* $\Sigma_{g,n}$ *fixing the gauge.*

3   Relations to the Hitchin Systems and the KZB Equations

3.1   Scaling Limit

Consider the HID in the limit $\kappa \to 0$. We will prove that in this limit we come to the Hitchin systems, which are living on the cotangent bundles to the moduli space of holomorphic G-bundles over $\Sigma_{g,n}$ [10]. The critical value $\kappa = 0$ looks singular (see Eqs. (12) and (25)). To get around this we rescale the times $\mathbf{t} = \kappa \mathbf{t}^H$, where $\mathbf{t}^H$ are the "Hitchin times." Therefore, $\delta\mu(\mathbf{t}) = \kappa \sum_s \partial_s \mu(\mathbf{t}^0)\delta t^H$. After this rescaling, the forms (12), (25) become regular in the critical limit. The rescaling procedure means that we blow up a vicinity of the fixed point corresponding to ($\mu = 0$), and the whole dynamic of the Hitchin systems is developed in this vicinity.[1] Denote $\partial_s \mu_o = \partial_s \mu(\mathbf{t})|_{\mathbf{t}=\mathbf{t}^0}$. Then we have instead of (12),

$$\omega = \int_\Sigma \langle \delta A, \delta \bar{A} \rangle + 2\pi i \sum_{a=1}^n \delta\langle p_a, g_a^{-1}\delta g_a \rangle - \sum_s \int_\Sigma \langle \delta A, A \rangle \partial_s \mu(\mathbf{t}^0)\delta t^H. \quad (35)$$

If $\kappa = 0$ the connection A behaves as the one-form $A \in \mathcal{A}^{(1,0)}(\Sigma_{g,n}, \mathrm{Lie}(G))$ (see Eq. (15)). It is so called the Higgs field. An important point is that the Hamiltonians now become the times independent. The form (35) is the starting point in the derivation of the Hitchin systems via the symplectic reduction [10, 22]. Essentially, it is the same procedure described above. Namely, we obtain the same moment constraint (19) and the same gauge fixing (17). But now we are sitting in a fixed point $\mu(\mathbf{t}^0) = 0$ of the moduli space $\mathcal{M}_{g,n}$ and don't need the factorization under the action of the diffeomorphisms. This only difference between the solutions L and $\bar{L}$ in the Hitchin systems and the hierarchies of isomonodromic deformations.

Propositions 1 and 3 are valid for the Hitchin systems in a slightly modified form.

---

[1] We are grateful to A. Losev for elucidating this point.

Proposition 4. *There exists the consistent system of linear equations*

$$(\lambda + L)\Psi = 0, \qquad \lambda \in \mathbb{C} \tag{36}$$

$$(\partial_s + M_s)\Psi = 0, \qquad \partial_s = \partial_{t_s^H}, (s = 1, \dots, l = \dim \mathcal{M}_{g,n}) \tag{37}$$

$$(\bar{\partial} + \bar{L})\Psi = 0, \qquad \bar{\partial} = \partial_{\bar{z}}, \tag{38}$$

where M_s is a solution to the linear equation

$$\bar{\partial}M_s - [M_s, \bar{L}] = \partial_s \bar{L} - L\partial_s \mu(\mathbf{t}^0). \tag{39}$$

The parameter λ in (36) can be considered as the symbol of ∂ (compare with Eq. (29)).

When L and M can be found explicitly, the simplified form of (19) allows us to apply "the inverse scattering method" to find solutions of the Hitchin hierarchy as was done for SL holomorphic bundles over $\Sigma_{1,1}$ [19], corresponding to the elliptic Calogero system with spins. We present an alternative way to describe the solutions:

Proposition 5 (The projection method).

$$\bar{L}(t_s) = f^{-1}(L^0 \partial_s \mu_o(t_s - t_s^0) + \bar{L}^0)f + f^{-1}\bar{\partial}f,$$

$$L(t_s) = f^{-1}L^0 f, \quad p_a(t_s) = f^{-1}(p_a^0)f.$$

The degenerate version of these expressions has been known for a long time [24, 25].

3.2 About KZB

The Hitchin systems are the classical limit of the KZB equations on the critical level [13, 22]. The latter has the form of the Schrödinger equations, which is the result of geometric quantization of the moduli of flat G bundles [1, 11]. The conformal blocks of the WZW theory on $\Sigma_{g,n}$ with vertex operators in marked points, are ground-state wave functions

$$\widehat{H}_s F = 0, \quad (s = 1, \dots, l = \dim \mathcal{M}_{g,n}).$$

The classical limit means that one replaces operators and their symbols and finite-dimensional representations in the vertex operators by the corresponding coadjoint orbits. The level κ plays the role of the Planck constant, but in contrast with the limit considerd before, we don't adjust the moduli of complex structures.

Generically, for $\kappa \neq 0$ the KZB equations can be written in the form of the nonstationary Schrödinger equations:

$$(\kappa \partial_s + \widehat{H}_s)F = 0, \quad (s = 1, \dots, l = \dim \mathcal{M}_{g,n}).$$

The flatness of this connection (see Ref. 11) is the quantum counterpart of the Whitham equations (27). The classical limit above-sense described leads to the HID. Summarizing, we arrange these quantum and classical systems in the diagram below. The vertical arrows indicates the classical limit, while the limit $\kappa \to 0$ (the horizontal arrows) also includes the rescaling of the moduli of complex structures. The examples in the bottom of the diagram will be considered in following sections.

$$
\left\{
\begin{array}{l}
\text{KZB Eqs.,} (\kappa, \mathcal{M}_{g,n}, G) \\
(\kappa \partial_{t_a} + \hat{H}_a) F = 0, \\
(a = 1, \ldots, \dim \mathcal{M}_{g,n})
\end{array}
\right\}
\xrightarrow{\kappa \to 0, t = \kappa t^H}
\left\{
\begin{array}{l}
\text{KZB Eqs. on critical level,} \\
(\mathcal{M}_{g,n}, G), \ (\hat{H}_a) F = 0, \\
(a = 1, \ldots, \dim \mathcal{M}_{g,n})
\end{array}
\right\}
$$

$$\Big\downarrow \kappa \to 0 \qquad\qquad\qquad\qquad\qquad \Big\downarrow \kappa \to 0$$

$$
\left\{
\begin{array}{l}
\text{Hierarchies of isomonodromic} \\
\text{deformations on } \mathcal{M}_{g,n}
\end{array}
\right\}
\xrightarrow{\kappa = 0, t = \kappa t^H}
\left\{\text{Hitchin systems}\right\}
$$

EXAMPLES

$$
\left\{
\begin{array}{l}
\text{Schlesinger Eqs.} \\
\text{Painlevé type Eqs.} \\
\text{Elliptic Schlesinger Eqs.}
\end{array}
\right\}
\xrightarrow{\kappa \to 0, t = \kappa t^H}
\left\{
\begin{array}{l}
\text{Classical Gaudin Eqs.} \\
\text{Calogero Eqs.} \\
\text{Elliptic Gaudin Eqs.}
\end{array}
\right\}
$$

4 Genus Zero—Schlesinger's Equation

Consider $\mathbb{C}P^1$ with n punctures $(x_1, \ldots, x_n \mid x_a \neq x_b)$. The Beltrami differential μ is related only to the positions of marked points (24). On $\mathbb{C}P^1$ the gauge transform (17) allows us to choose $\bar{L}$ to be identically zero. Let $A = f(L + \kappa \partial_w) f^{-1}$. Then the moment equation takes the form

$$(\bar{\partial} + \partial \mu) L = 2\pi i \sum_{a=1}^{n} \delta^2(x_a) p_a. \tag{40}$$

It allows us to find L:

$$L = \sum_{a=1}^{n} \frac{p_a}{w - x_a}. \tag{41}$$

On the symplectic quotient ω (25) takes the form

$$\omega = \delta \sum_{a=1}^{n} \langle p_a g_a^{-1} \delta g_a \rangle - \frac{1}{\kappa} \sum_{b=1}^{n} (\delta H_{b,1} + \delta H_{b,0}) \delta x_b. \tag{42}$$

$$H_{a,1} = \sum_{b \neq a} \frac{\langle p_a, p_b \rangle}{x_a - x_b}, \quad H_{2,a} = c_a \langle p_a, p_a \rangle.$$

$H_{1,a}$ are precisely the Schlesinger's Hamiltonians. Note that we still have a gauge freedom with respect to the G action. The corresponding moment constraint means that the sum of residues of L vanishes:

$$\sum_{a=1}^{n} p_a = 0. \tag{43}$$

While $H_{2,a}$ are Casimir's and lead to trivial equations, the equations of motion for $H_{1,a}$ are the Schlesinger equations:

$$\kappa \partial_b p_a = \frac{[p_a, p_b]}{x_a - x_b}, \ (a \neq b), \quad \kappa \partial_a p_a = -\sum_{b \neq a} \frac{[p_a, p_b]}{x_a - x_b}.$$

As a by-product, we obtain by this procedure the corresponding linear problem (29) and (30) with L (41) and

$$M_{a,1} = -\frac{p_a}{w - x_a}$$

as a solution to (32). The tau-function for the Schlesinger equations has the form [14, 15]

$$\delta \log \tau = \sum_{c \neq b} \langle p_b, p_c \rangle \delta \log(x_c - x_b).$$

5   Genus One—Elliptic Schlesinger, Painlevé VI, Etc.

5.1   *Deformations of Elliptic Curves*

In addition to the moduli coming from the positions of the marked points, there is an elliptic module τ, $\mathrm{Im}\,\tau > 0$ on $\Sigma_{1,n}$. As in (23) and (24) we take the Beltrami differential in the form $\mu = \sum_{a=1}^{n} \mu_a + \mu_\tau$ $(\mu_s = t_s \bar{\partial} n_s)$, where $n_a(z, \bar{z})$ is the same as in (24) and

$$n_\tau = (\bar{z} - z)\left(1 - \sum_{a=1}^{n} \chi_a(z, \bar{z})\right). \tag{44}$$

Then

$$\mu_\tau = \tilde{\mu}_\tau \left(1 - \sum_{a=1}^{n} \chi_a(z, \bar{z})\right), \quad \left(\tilde{\mu}_\tau = \frac{t_\tau}{\tau - \tau_0}, t_\tau = \tau - \tau_0\right). \tag{45}$$

Here τ_0 defines the reference comlex structure on the curve

$$T_0^2 = \{0 < x \leq 1, 0 < y \leq 1, z = x + \tau_0 y, \bar{z} = x + \bar{\tau}_0 y\}.$$

5.2   Flat Bundles on a Family of Elliptic Curves

Note first that $\bar{A}$ can be considered as a connection of holomorphic G bundle $\mathcal{E}$ over T_τ^2. For stable bundles $\bar{A}$ can be gauge transformed by (17) to the Cartan z-independent form $\bar{A} = f(\bar{L} + \bar{\partial} + \mu\partial)f^{-1}$, $\bar{L} \in \mathcal{H}$—Cartan subalgebra of Lie(G). Therefore, stable bundle $\mathcal{E}$ is decomposed into the direct sum of line bundles $\mathcal{E} = \bigoplus_{k=1}^r \mathcal{L}_k$, $r = \text{rank}\,G$. The set of gauge equivalent connections represented by $\{\bar{L}\}$ can be identified with the r power of the Jacobian of T_τ^2, factorized by the action of the Weyl group W of G. Put

$$\bar{L} = 2\pi i \frac{1 - \bar{\mu}_\tau}{\rho}\mathbf{u}, \quad \mathbf{u} \in \mathcal{H}, (\rho = \tau_0 - \bar{\tau}_0). \tag{46}$$

The moment constraints (20) leading to the flatness condition take the form

$$\partial_{\bar{w}} L + [\bar{L}, L] = 2\pi i \sum_{a=1}^n \delta^2(x_a,)p_a. \tag{47}$$

Let $R = \{\alpha\}$ be the root system of Lie $(G) = \mathcal{G}$ and $\mathcal{G} = \mathcal{H} \oplus_{\alpha \in R} \mathcal{G}_\alpha$ be the root decomposition. Impose the vanishing of the residues in (47):

$$\sum_{a=1}^n (p_a)_\mathcal{H} = 0, \tag{48}$$

where $p_a|_\mathcal{H}$ is the Cartan component and we have identified $\mathcal{G}$ with its dual space $\mathcal{G}^*$.

We will parametrize the set of its solutions by two elements $\mathbf{v}$, $\mathbf{u} \in \mathcal{H}$. Define the solutions L to the moment equation (47), which is double periodic on the deformed curve T_τ^2. Let $E_1(w)$ be the Eisenstein function

$$E_1(z \mid \tau) = \partial_z \log \theta(z \mid \tau),$$

where

$$\theta(z \mid \tau) = q^{\frac{1}{8}} \sum_{n \in \mathbf{Z}} (-1)^n e^{\pi i(n(n+1)\tau + 2nz)}.$$

It is connected with the Weirstrass zeta-function as

$$\zeta(z \mid \tau) = E_1(z \mid \tau) + 2\eta_1(\tau)z, \quad (\eta_1(\tau) = \zeta(\tfrac{1}{2})).$$

Another function we need is

$$\phi(u, z) = \frac{\theta(u + z)\theta'(0)}{\theta(u)\theta(z)} = \exp(-2\eta_1 uz)\frac{\sigma(u + z)}{\sigma(y)\sigma(z)},$$

where $\sigma(z)$ is the Weierstrass sigma function.

Lemma 1. *The solutions of the moment constraint equation have the form*

$$L = P + X, \quad P \in \mathcal{H}, X = \sum_{\alpha \in R} X_\alpha. \tag{49}$$

$$P = 2\pi i \left(\frac{\mathbf{v}}{1 - \tilde{\mu}_\tau} - \kappa \frac{\mathbf{u}}{\rho} + \sum_{a=1}^{n} (p_a)_{\mathcal{H}} E_1(w - x_a) \right), \tag{50}$$

$$X_\alpha = \frac{2\pi i}{1 - \tilde{\mu}_\tau} \sum_{a=1}^{n} (p_a)_\alpha \exp 2\pi i \left\{ \frac{(w - x_a) - (\bar{w} - \tilde{x}_a)}{\tau - \bar{\tau}_0} \alpha(u) \right\}$$
$$\times \phi(\alpha(u), w - x_a). \tag{51}$$

5.3   Symmetries

The remnant gauge transforms preserve the chosen Cartan subalebra $\mathcal{H} \subset G$. These transformations are generated by the Weyl subgroup W of G and elements $f(w, \bar{w}) \in \text{Map}(T_\tau^2, \text{Cartan}(G))$. Let Π be the system of simple roots, $R^\vee = \{\alpha^\vee = 2\alpha/(\alpha \mid \alpha)\}$, is the dual root system, and $\mathbf{m} = \sum_{\alpha \in \Pi} m_\alpha \alpha^\vee$ be the element from the dual root lattice $\mathbb{Z}R^\vee$. Then the Cartan valued harmonics

$$f_{\mathbf{m},\mathbf{n}} = \exp 2\pi i \left(\mathbf{m} \frac{w - \bar{w}}{\tau - \tau_0} + \mathbf{n} \frac{\tau \bar{w} - \bar{\tau}_0 w}{\tau - \tau_0} \right), \quad (\mathbf{m}, \mathbf{n} \in R^\vee) \tag{52}$$

generate the basis in the space of Cartan gauge transformations. In terms of the variables $\mathbf{v}$ and $\mathbf{u}$ they act as

$$\mathbf{u} \to \mathbf{u} + \mathbf{m} - \mathbf{n}\tau, \quad \mathbf{v} \to \mathbf{v} - \kappa \mathbf{n}, \quad (p_a)_\alpha \to \varphi(m_\alpha, n_\alpha)(p_a)_\alpha. \tag{53}$$

Here $\varphi(m_\alpha, n_\alpha) = \exp 4\pi i[(m_\alpha - n_\alpha \bar{\tau}_0 x_a^0) - (m_\alpha - n_\alpha \tau_0 \bar{x}_a^0)]/\rho$.

The whole discrete gauge symmetry is the semidirect product $\widehat{W}$ of the Weyl group W and the lattice $\mathbb{Z}R^\vee \oplus \tau \mathbb{Z}R^\vee$. It is the Bernstein–Schvartsmann complex crystallographic group. The factor space $\mathcal{H}/\widehat{W}$ is the genuine space for the "coordinates" $\mathbf{u}$.

According to (15) the transformations (52) also act on $p_a \in \mathcal{O}_a$. This action leads to the symplectic quotient $\mathcal{O}_a//H$ and generates the moment equation (48).

The modular group $\text{PSL}_2(\mathbb{Z})$ is a subgroup of the mapping class group for the Teichmüller space $\mathcal{T}_{1,n}$. Its action on τ is the Möbius transform. We summarise the action of symmetries on the dynamical variables in Table 20.1.

5.4   Symplectic Form

The set $(\mathbf{v}, \mathbf{u} \in \mathcal{H}, \mathbf{p} = (p_1, \ldots, p_n))$ of dynamical variables along with the times $\mathbf{t} = (t_\tau, t_1, \ldots, t_n)$ describe the local coordinates in the bundle

TABLE 20.1.

	W={s}	$\mathbb{Z}R^\vee \oplus \tau\mathbb{Z}R^\vee$	$\mathrm{PSL}_2(\mathbb{Z})$
$\mathbf{v}$	$s\mathbf{v}$	$\mathbf{v} - \kappa\mathbf{n}$	$\mathbf{v}(c\tau + d) - \kappa c\mathbf{u}$
$\mathbf{u}$	$s\mathbf{u}$	$\mathbf{u} + \mathbf{m} - \mathbf{n}\tau$	$\mathbf{u}(c\tau + d)^{-1}$
$(p_a)_{\mathcal{H}}$	$s(p_a)_{\mathcal{H}}$	$(p_a)_{\mathcal{H}}$	$(p_a)_{\mathcal{H}}$
$(p_a)_\alpha$	$(p_a)_{s\alpha}$	$\varphi(m_\alpha, n_\alpha)(p_a)_\alpha$	$(p_a)_\alpha$
τ	τ	τ	$\frac{a\tau+b}{c\tau+d}$
x_a	x_a	x_a	$\frac{x_a}{c\tau+d}$

$\mathcal{P}$. According to the general prescription, we can define the Hamiltonian system on this set.

The main statement, formulated in terms of the theta-functions and the Eisenstein functions, is

$$E_2(z \mid \tau) = -\partial_z E_1(z \mid \tau) = \partial_z^2 \log \theta(z \mid \tau) = \wp(z \mid \tau) + 2\eta_1(\tau). \quad (54)$$

It takes the following form:

Proposition 6. *The symplectic form ω (25) on $\mathcal{P}$ is*

$$\frac{1}{4\pi^2}\omega = (\delta\mathbf{v}, \delta\mathbf{u}) + \sum_{a=1}^n \delta\langle p_a, g_a^{-1}\delta g_a\rangle$$
$$- \frac{1}{\kappa}\left(\sum_{a=1}^n \delta H_{2,a} + \delta H_{1,a}\right)\delta t_a - \frac{1}{\kappa}\delta H_\tau \delta\tau, \quad (55)$$

with the Hamiltonians

$$H_{2,a} = c_a\langle p_a, p_a\rangle;$$

$$H_{1,a} = 2\left(\frac{\mathbf{v}}{1 - \tilde{\mu}_\tau} - \kappa\frac{\mathbf{u}}{\rho}, p_a|_{\mathcal{H}}\right) + \sum_{b\neq a}(p_a|_{\mathcal{H}}, p_b|_{\mathcal{H}})E_1(x_a - x_b)$$
$$+ \sum_{b\neq a}\sum_\alpha (p_a|_\alpha, p_b|_{-\alpha})\frac{\theta(-\alpha(\mathbf{u}) + x_a - x_b)\theta'(0)}{\theta(\alpha(\mathbf{u}))\theta(x_a - x_b)};$$

$$H_\tau = \frac{(\mathbf{v}, \mathbf{v})}{2} + \sum_{a=1}^n\sum_\alpha (p_a|_\alpha, p_a|_{-\alpha})E_2(\alpha(\mathbf{u}))$$
$$+ \sum_{a\neq b}^n (p_a|_{\mathcal{H}}, p_b|_{\mathcal{H}})(E_2(x_a - x_b) - E_1^2(x_a - x_b))$$
$$+ \sum_{a\neq b}\sum_\alpha (p_a|_\alpha, p_b|_{-\alpha})\frac{\theta(-\alpha(\mathbf{u}) + x_a - x_b)\theta'(0)}{\theta(\alpha(\mathbf{u}))\theta(x_a - x_b)}$$
$$\times \big(E_1(\alpha(\mathbf{u})) - E_1(x_b - x_a + \alpha(\mathbf{u})) - E_1(x_b - x_a)\big).$$

Example 1. Consider $SL(2, \mathbb{C})$ bundles over the family of $\Sigma_{1,1}$. Then (46) takes the form

$$\bar{L} = 2\pi i \frac{1 - \tilde{\mu}_\tau}{\rho} \operatorname{diag}(u, -u). \tag{56}$$

In this case the position of the marked point is no longer the module and we put $x_1 = 0$. Since $\dim \mathcal{O} = 2$ the orbit degrees of freedom can be gauged away by the Hamiltonian action of the diagonal group. We assume that $p = \nu[(1, 1)^T \otimes (1, 1) - \operatorname{Id}]$. Then we have from (49)–(51)

$$L = 2\pi i \begin{pmatrix} \dfrac{v}{1 - \tilde{\mu}_\tau} - \kappa \dfrac{u}{\rho} & x(u, w, \bar{w}) \\ x(-u, w, \bar{w}) & -\dfrac{v}{1 - \tilde{\mu}_\tau} + \kappa \dfrac{u}{\rho} \end{pmatrix} \tag{57}$$

$$x(u, w, \bar{w}) = \frac{\nu}{1 - \tilde{\mu}_\tau} \exp 4\pi i \left\{ (w - \bar{w}) u \frac{1 - \tilde{\mu}_\tau}{\rho} \right\} \phi(2u, w).$$

The symplectic form (55) is

$$\frac{1}{4\pi^2} \omega = (\delta v, \delta u) - \frac{1}{\kappa} \delta H_\tau \delta \tau,$$

and

$$H_\tau = v^2 + U(u \mid \tau), \quad U(u \mid \tau) = -\nu^2 E_2(2u \mid \tau).$$

It leads to the equation of motion:

$$\frac{\partial^2 u}{\partial \tau^2} = \frac{2\nu^2}{\kappa^2} \frac{\partial}{\partial u} E_2(2u \mid \tau). \tag{58}$$

In fact, owing to (54) we can use $\wp(2u \mid \tau)$ instead of $E_2(2u \mid \tau)$ and after rescaling the coupling constant come to (3) for special values of constants as in (7). Equation (58) represents the isomonodromic deformation conditions for the linear system (29) and (31) with L (57) and $\bar{L}$ (56). The Lax pair is given by L (57) and M_τ

$$M_\tau = \begin{pmatrix} 0 & y(u, w, \bar{w}) \\ y(-u, w, \bar{w}) & 0 \end{pmatrix},$$

where $y(u, w, \bar{w})$ is defined as the convolution integral on T_τ^2:

$$y(u, w, \bar{w}) = -\frac{1}{\kappa} x * x(u, w, \bar{w}).$$

The projection method determines solutions of (58) as a result of diagonalization of L (57) by the gauge transform on the deformed curve T_τ^2. On the critical level ($\kappa = 0$) we come to the two-body elliptic Calogero system.

Example 2. For flat G bundles over $\Sigma_{1,1}$ we obtain Painlevé-type equations, related to the arbitrary root system. They are described by the system of differential equations for the $\mathbf{u} = (u_1, \ldots, u_r)$,$(r = \mathrm{rank}\, G)$ variables. In addition there are the orbit variables $p \in \mathcal{O}(G)$ satisfying the Euler top equations. For SL bundles the most degenerate orbits $\mathcal{O} = T^*\mathbb{C}P^{N-1}$ have dimension $2N - 2$. These variables are gauged away by the diagonal gauge transforms, as in the previous example. On the critical level this Painlevé-type system degenerates into N-body elliptic Calogero system. For generic orbits we obtain the generalized Calogero–Euler systems.

Acknowledgments: We are thankful to Yu. Manin—his lectures and discussions with him concerning PVI stimulated our interest in these problems. We are grateful to the Max-Planck-Institut für Mathematik in Bonn for the hospitality, where this work was started. We would like to thank our colleagues V. Fock, A. Losev, A. Morozov, N. Nekrasov, and A. Rosly for fruitful discussions while we worked on this subject. The work is supported in part by by Award No. RM1-265 of the US Civilian Research & Development Foundation (CRDF) for the Independent States of the Former Soviet Union, and grant 96-15-96455 for support of scientific schools (A.L); grants RFBR-96-02-18046, Award No. RM2-150 of the US Civilian Research & Development Foundation (CRDF) for the Independent States of the Former Soviet Union, INTAS 930166 extension, and grant 96-15-96455 for support of scientific schools (M.O).

6   References

1. S. Axelrod, S. Della Pietra, and E. Witten, *Geometric quantization of the Chern–Simons gauge theory*, J. Differential Geom. **33** (1991), No. 3, 787–902.

2. D. Bernard, *On the Wess–Zumino–Witten models on Riemann surfaces*, Nucl. Phys. B **309** (1988), No. 1, 145–174.

3. ———, *On the Wess–Zumino–Witten models on the torus*, Nucl. Phys. B **303** (1988), No. 1, 77–93.

4. F. Calogero, *On a functional equation connected with integrable many-body problems*, Lett. Nuovo Cimento **16** (1976), No. 3, 77–80.

5. R. Dijkgraaf, H. Verlinde, and E. Verlinde, *Notes on topological string theory and 2D quantum gravity*, String Theory and Quantum Gravity (Trieste, 1990) (M. Green, R. Iengo, S. Randjbar-Daemi, E. Sezgin, and H. Verlinde, eds.), World Sci. Publishing, River Edge, NJ, 1991, pp. 91–156.

6. B. Enriquez and V. N. Rubtsov, *Hitchin systems, higher Gaudin operators and R-matrices*, Math. Res. Lett. **3** (1996), No. 3, 343–357.

7. R. Fuchs, *Über lineare homogene differentialgleichungen zweiter ordnung mit im endlich gelegne wesentlich singulären stellen*, Math. Ann. **63** (1907), 301–323.

8. B. Gambier, *Sur les équations différentielles du second ordre et du premier degré dont l'intégrale générale a ses points critiques fixes*, Acta Math. Ann. **33** (1910), 1–55.

9. J. Harnard, *Quantum isomonodromic deformations and the Knizhnik–Zamolodchikov equations*, Symmetries and Integrability of Difference Equations (Estérel, QC, 1994) (D. Levi, L. Vinet, and P. Winternitz, eds.), CRM Proc. Lecture Notes, Vol. 9, Amer. Math. Soc. Providence, RI, 1996, pp. 155–161, hep-th/9511087.

10. N. Hitchin, *Stable bundles and integrable systems*, Duke Math. J. **54** (1987), No. 1, 91–114.

11. _____, *Flat connections and geometric quantization*, Commun. Math. Phys. **131** (1990), No. 2, 347–380.

12. V. Inozemtsev, *Lax representation with spectral parameter on a torus for integrable particle systems*, Lett. Math. Phys. **17** (1989), No. 1, 11–17.

13. D. Ivanov, *Knizhnik–Zamolodchikov–Bernard equations as a quantization of nonstationary Hitchin system*, hep-th/9610207.

14. M. Jimbo and T. Miwa, *Monodromy preserving deformation of linear ordinary differential equations with rational coefficients. II*, Physica **2D** (1981), No. 3, 407–448.

15. M. Jimbo, T. Miwa, and K. Ueno, *Monodromy preserving deformation of linear ordinary differential equations with rational coefficients. I. General theory and τ-function*, Physica **2D** (1981), No. 2, 306–352.

16. V. G. Knizhnik and A. B. Zamolodchikov, *Current algebra and Wess–Zumino model in two dimensions*, Nucl. Phys. B **247** (1984), No. 1, 83–103.

17. D. A. Korotkin and J. A. H. Samtleben, *On the quantization of isomonodromic deformations on the torus*, Internat. J. Modern Phys. A **12** (1997), No. 11, 2013–2030.

18. I. Krichever, *The τ-function of the universal Whitham hierarchy, matrix models and topological field theories*, Commun. Pure Appl. Math. **47** (1994), No. 4, 437–475.

19. I. M. Krichever, O. Babelon, E. Billey, and M. Talon, *Spin generaliza-tion of the Caloger–Moser system and the matrix KP equation*, Topics in Topology and Mathematical Physics (A. B. Sossinsky, ed.), Amer. Math. Soc. Transl. Ser. 2, Vol. 170, Amer. Math. Soc., Providence, RI, 1995, pp. 83–119, hep-th/9411160.

20. D. Levi and P. Winternitz (eds.), *Painlevé Transcedents. Their Asymptotics and Physical Applications*, NATO Adv. Sci. Inst. Ser. B Phys., Vol. 278, Plenum, New York, 1992.

21. Yu. I. Manin, *Sixth Painlevé equation, universal elliptic curve, and mirror of p^2*, Tech. Report MPI-1996-114, Max-Planck-Institut für Mathematik, Bonn, 1996.

22. N. Nekrasov, *Holomorphic bundles and many-body systems*, Commun. Math. Phys. **180** (1996), No. 3, 587–603.

23. K. Okamoto, *Isomonodromic deformation and the Painlevé equations and the Garnier system*, J. Fac. Sci. Univ. Tokyo Sect. IA Math. **33** (1986), No. 3, 575–618.

24. M. A. Olshanetsky and A. M. Perelomov, *Explicit solutions of some completely integrable systems*, Lett. Nuovo Cimento **17** (1976), No. 3, 97–101.

25. _____ , *Explicit solutions of the Calogero models in the classical case and geodsic flows on symmetric spaces of zero curvature*, Lett. Nuovo Cimento **16** (1976), No. 11, 333–339.

26. N. Yu. Reshetikhin, *The Knizhnik–Zamolodchikov system as a defor-mation of the isomonodromy problem*, Lett. Math. Phys. **26** (1992), No. 3, 167–177.

27. A. Treibich and J.-L. Verdier, *Revêtements tangentiels et sommes de 4 nombres triangulaires*, C. R. Acad. Sci. A **311** (1990), No. 1, 51–54.

21

Yangian Symmetry in WZW Models

Z. Maassarani
P. Mathieu

ABSTRACT Yangian symmetry has been unraveled recently in a particular class of conformal field theories with SU(N) invariance (the spinon models). Such a symmetry is expected to be at the core of the integrable structure of the corresponding WZW models. However, if we try to characterize the integrability in terms of factorized S-matrices (in particular, S-matrices inherited from massive current-algebras theories), this Yangian symmetry is not the one that fixes these matrices; that is, its generators do not have the expected commutation relations with the S-matrix elements. This hints at the existence of a hidden Yangian symmetry in WZW models. After a brief review of the spinon-type Yangian symmetry in these models, we display a new Yangian algebra whose structure is generic and level independent. This Yangian is the natural extension at the conformal point of the one present in massive theories with current algebras.

1 Conformal Field Theory: The "Field Description"

Two-dimensional conformal field theories are primarily characterized by a factorization of the Hilbert space into two independent sectors (holomorphic and antiholomorphic) and the defining commutation relation of the Laurent coefficients of the holomorphic energy-momentum tensor $T(z) = \sum L_n z^{-n-2}$:

$$[L_m, L_n] = (m-n)L_{n+m} + \frac{c}{12}(m^3 - m)\delta_{n+m,0}. \quad (1)$$

and a similar relation for the antiholomorphic modes $\bar{L}_n$). This is the Virasoro algebra. The parameter c is the central charge, which characterizes the theory to a large extent.

An important part of the analysis of a specific conformal field theory can be rephrased into a problem in the representation theory of the Virasoro algebra. The natural language at this stage is that of Verma modules. The different fields in the theory are separated into two classes: the primary fields and their descendants. All the information is coded in the properties of the primary fields and the conformal symmetry (i.e., the Virasoro algebra). To each primary field $\phi(z)$ there corresponds a highest-weight state $|\phi\rangle$ such that

$$L_0|\phi\rangle = h|\phi\rangle \quad \text{and} \quad L_n|\phi\rangle = 0 \quad \text{for } n > 0. \tag{2}$$

The descendant fields correspond to the states obtained from this highest-weight state by the actions of the L_n modes with $n < 0$. These towers generate the Verma modules. Generically, they form irreducible representations. For particular values of the central charge, there singular vectors appear in the modules: these are both highest-weight states and descendants. In such cases, the irreducible modules are obtained by quotienting these states together with their descendants. It is only in such situations that there can be a finite number of primary fields in the theory. This occurs, for instance, when c takes the values

$$c = 1 - \frac{(p'-p)^2}{pp'}, \tag{3}$$

which correspond to the famous Belavin–Polyakov–Zamolodchikov minimal models [3].

An important class of conformal field theories are the WZW models, which are invariant under a Lie group symmetry G. Hence, in addition to the energy-momentum tensor, there are new conserved currents: $J^a(z)$ with $a = 1, \ldots, \dim G$ (and their antiholomorphic counterparts). Their modes $(J^a(z) = \sum J_n^a z^{-n-1})$ generate a current algebra

$$[J_m^a, J_n^b] = if^{abc}J_{n+m}^c + km\delta_{n+m,0}, \tag{4}$$

which is the defining commutation relation of the affine Lie algebra $\hat{g}$ at level k. For such models, the primary fields are in one-to-one correspondence with the integrable representations of the $\hat{g}_k$ algebra. The simplest example is the $\widehat{su}(2)_1$-WZW model, which has only two primary fields.

2   Integrable Quantum Field Theory: The "Particle Description"

A conformal field theory can always be viewed as the massless limit of a quantum field theory. Moreover, since every minimal (i.e., solvable) model has an integrable deformation, a stronger statement holds, which is that a conformal field theory can always be viewed as the massless limit of an *integrable* quantum field theory.

From this perspective, it is thus of interest to try to understand the integrable structure of conformal field theories. But for that purpose, the

Verma module or field description is not appropriate. The natural language is that of particles and factorized S-matrices. These S-matrices can be determined from the structure of the local conservation laws $\{H_n\}$. We recall that an integrable quantum field theory is characterized by the existence of an infinite number of local conservation laws all commuting among themselves. From various examples worked out over the past years (see, e.g., Ref. 8), it appears that most integrable quantum field theories are also characterized by an infinite number of nonlocal conserved charges $\{Q_n\}$, which, in contrast to the local ones, do not commute among themselves; rather, they generate a non-Abelian algebra, the dynamical algebra of the model. Understanding the structure of these nonlocal charges is expected to be an essential part of the understanding of integrability at criticality. The Yangian symmetry of the $\widehat{su}(2)_1$-WZW model, briefly reviewed in the next section, provides a neat example of such a structure.

3   The Yangian Symmetry of the $\widehat{su}(2)_1$-WZW Model

This symmetry is inherited from the dense limit of the Haldane–Shastry model, which is a spin chain with long-range interraction. Its defining Hamiltonian is [17, 22]

$$H = \sum_{\substack{,j=1 \\ i \neq j}}^{N} \sum_{a=1}^{3} \frac{z_i z_j}{(z_i - z_j)^2} \sigma_i^a \sigma_j^a \tag{5}$$

with $z_j = \exp(2\pi i j/N)$ and the σ_i^a's are the usual Pauli spin matrices. This spin chain is completely integrable since it has an infinite number of local conservation laws $\{H_n\}$ (with $H_2 = H$). In addition, there is an infinite number of nonlocal conservation laws $\{Q_n\}$, all recursively generated from the first two:

$$Q_0^a = \sum_{i=1}^{N} \sigma_i^a, \quad Q_1^a = \epsilon^{abc} \sum_{\substack{i,j=1 \\ i \neq j}}^{N} \frac{z_i + z_j}{z_i - z_j} \sigma_i^b \sigma_j^c \tag{6}$$

The commutator of these charges generate the su(2) Yangian algebra.

For su(N) with structure constant f^{abc}, the Yangian defining relations are [15]

Y(1) $$[Q_0^a, Q_0^b] = i f^{abc} Q_0^c \tag{7}$$

Y(2) $$[Q_0^a, Q_1^b] = i f^{abc} Q_1^c \tag{8}$$

In addition, it is required that the level n generator Q_n^a, defined recursively by $i f^{abc}[Q_1^b, Q_{n-1}^c]$, be unique; that is, it should not depend on the chosen sequence of multiple commutators of Q_1^a and Q_0^b used to reach it. For su($N > 2$), consistency at level 2 is sufficient to ensure this uniqueness property at all levels. This consistency condition translates into the following Serre relation:

$$Y(3) \quad [Q_1^a, [Q_1^b, Q_0^c]] - [Q_0^a, [Q_1^b, Q_1^c]] = \kappa A^{abc;def} \{Q_0^d, Q_0^e, Q_0^f\}_s, \quad (9)$$

where

$$A^{abc;def} = f^{adp} f^{beq} f^{cfr} f^{pqr}. \quad (10)$$

$\{a, b, c\}_s$ stand for the complete symmetrization of the three objects a, b, c (which incorporates a factor $1/6$) and κ is a constant related to the normalization of Q_1^a. For su(2), it turns out that Y(3) is trivial and a fourth relation needs to be checked. This is the level-3 consistency requirement that express as the fact that $[[Q_1^a, Q_1^b], [Q_0^c, Q_0^d]] + [[Q_1^d, Q_1^c], [Q_0^b, Q_0^a]]$ should be expressed solely in terms of $\{Q_1^k, Q_0^l, Q_0^m\}_s$.

To complete the description of the Yangian, we must stress that its fundamental property is its comultiplication, which gives the action (Δ) of its generators on two copies of the Hilbert space:

$$\Delta Q_0^a = Q_0^a \otimes 1 + 1 \otimes Q_0^a$$
$$\Delta Q_1^a = Q_1^a \otimes 1 + 1 \otimes Q_1^a + i\sqrt{\kappa} f^{abc} Q_0^b \otimes Q_0^c. \quad (11)$$

The constant κ is that same that appears in the relation Y(3).

All these relations are satisfied by the charges (6) with f^{abc} replaced by ϵ^{abc}. Let us now consider the low-energy dense limit ($N \to \infty$) of this spin chain model. This is obtained somewhat heuristically by replacing $\sum_i$ by $\oint dz_i/2\pi i$, σ_i^a by $J^a(z_i)$ (the $J^a(z_i)$ are the $\widehat{su}(2)_1$ generators) and

$$\sum f(z_i, z_j, \ldots, z_k) \sigma_i^a \sigma_j^b \cdots \sigma_k^c$$
$$\to \oint \frac{dz}{2\pi i} \oint \frac{dw}{2\pi i} \cdots \oint \frac{d\zeta}{2\pi i} f(z, w, \ldots, \zeta) J^a(z) J^b(w) \cdots J^c(\zeta) \quad (12)$$

with $|z| > |w| > \cdots > |\zeta|$ (i.e., field products are assumed to be radially ordered). With these prescriptions, the resulting expression for $\tilde{Q}_0^a$, $\tilde{Q}_1^a$, $\{\tilde{H}_n\}$ (where the tildes indicate the continuum limit) are found to satisfy the same properties as in the chain model [18]; in particular,

$$[\tilde{Q}_0^a, \tilde{H}_m] = [\tilde{Q}_1^a, \tilde{H}_m] = 0. \quad (13)$$

Moreover, L_0 (the zero-mode Virasoro generator) is nothing but the continuum limit of the crystal momentum H_1. This has the following implication. Each irreducible representation can be decomposed into a direct sum of L_0 subspace. Since L_0 commutes with the Yangian generators, each

L_0 subspace can be reorganized into Yangian representations. Quite remarkably, this decomposition codes a generalized exclusion principle [18]. This in turn implies that the characters can be reinterpreted in terms of quasiparticles. These are the spinons, related to the basic excitations of the chain model. The irreducible representations, which look quite complicated when viewed from the chiral algebra point of view owing to the presence of singular vectors, become very simple when reinterpreted from the quasiparticle perspective since they are nothing but free Fock modules on which no restriction is imposed other than the generalized exclusion principle [10, 11].

This transcription of the field formalism in terms of quasiparticles provides a concrete example of the kind of reinterpretation that we would like to find for generic conformal field theories. However, the limitation of the above results should be stressed. Although they generalize to $\widehat{su}(N)_1$-WZW models [13, 21], there are no explicit descriptions of the Yangian for higher levels [12]. This reflects the nonexistence of "higher spin" analogs of the Haldane–Shastry model (built from matrices in representations other than the fundamental one)—at least none has been constructed thus far. Moreover, this construction seems to use in an essential way the existence of a symmetric Casimir operator of degree three in the underlying algebra, a property that uniquely selects the su(N) models.

In addition to these limitations, we should stress that the commutativity of L_0 with the Yangian generators is not a property to be expected from the S-matrix description of the theory. More precisely, if we insist on an S-matrix formulation at criticality (these matrices describe thus the scattering of massless particles) and assume that these matrices would be fixed by the dynamical symmetry (generated by the nonlocal charges), then L_0 should not commute with Q_1^a. This argument is reviewed in the next section.

4 Nonlocal Charges and S-Matrices

For an integrable quantum field theory, the factorized S-matrices are fixed by the conservation laws. If there are nonlocal charges generating a Yangian symmetry, the S-matrices must then be Yangian invariant. The Yangian multiplets are then formed by the n-asymptotic particle states. These asymptotic particle states are also eigenstates of an infinite set of conservation laws. In particular, they must be eigenstates of the momentum operator. Hence, the Yangian symmetry that fixes the structure of the S-matrices must commute with the momentum operator.

All this can be made more explicit by considering the actual action of the nonlocal charges on the asymptotic states (see, e.g., [4, 5] and [2]). That first requires the introduction of asymptotic fields ϕ_λ associated to a representation λ of the Lie algebra symmetry. These create the asymptotic particles. The antiparticles are created by ϕ_{λ^*}, where λ^* indicates the conjugate of λ. The asymptotic currents J^a needed to evaluate the action of the nonlocal charges on asymptotic states are expressible as normal ordered products of ϕ_λ and ϕ_{λ^*} (since the OPE of the latter contain the identity, the ancestor of the currents J^a). It is sufficient to consider the action on one-particle states only (the action on multiparticle sates being fixed from those and the comultiplication rule). In this way, the action of Q_0^a and Q_1^a on a one-particle state of rapidity θ (denoted by $|\theta\rangle$) is found to have the following generic structure:

$$Q_0^a|\theta\rangle = -t^a|\theta\rangle$$
$$Q_1^a|\theta\rangle = \gamma\theta t^a|\theta\rangle, \tag{14}$$

where t^a is some matrix representation of the Lie algebra and γ is a constant). The second relation implies the following commutation with the boost operator B (whose on-shell representation is $\partial/\partial\theta$):

$$[B, Q_1^a] = -\gamma Q_0^a. \tag{15}$$

Moreover, translation invariance requires

$$[P_\mu, Q_0^a] = [P_\mu, Q_1^a] = 0. \tag{16}$$

For massive current algebra and sigma models (with or without topological term) there is a Yangian symmetry satisfying the above conditions for generic affine Lie algebra at any positive level [4, 5, 19, 24, 25]. These Yangian generators are nothing but the direct quantum extensions of the classical nonlocal charges of the sigma or current algebra models [1, 14, 20].

All these models flow to WZW models in the ultraviolet limit. Moreover, the S-matrices of the WZW model are known to be Yangian invariant [5]. Note finally that in the massless limit, P_μ and B are given by

$$P_\mu \sim L_{-1} + \bar{L}_{-1} \quad B \sim L_0 + \bar{L}_0, \tag{17}$$

so that the previous commutation relations translate into the requirements

$$[L_{-1}, Q_1^a] = 0 \quad [L_0, Q_1^a] = \gamma' Q_0^a \tag{18}$$

(where γ' is some constant.) This is in sharp contradistinction with the situation observed in the spinon description, where $[L_0, Q_1^a] = 0$.

These facts suggest that for any $\hat{g}_k$-WZW model, there should exist a Yangian symmetry with a structure quite different from that of the spinon models. This new Yangian symmetry, found in Ref. 9, is presented in the next section.

5 Nonlocal Currents in WZW Models

The classical equations of motion for the WZW models at the conformally invariant points can be written as the conservation laws, $\partial_\mu J_\mu^\pm = 0$, for two chiral currents $J_\mu^\pm$, respectively defined by

$$J_\mu^+ = g^{-1}\partial_\mu g - \epsilon_{\mu\nu} g^{-1}\partial_\nu g, \quad J_\mu^- = \partial_\mu g g^{-1} + \epsilon_{\mu\nu}\partial_\nu g g^{-1}, \quad (19)$$

where $g(x)$ is a matrix field valued in the Lie group maniflold G. As a consequence, there exist classically two nonlocal conserved currents defined by

$$\mathcal{J}_\mu^{\pm a}(x) = i f^{abc} \int^x dv^\sigma\, \epsilon^{\sigma\rho} J_\rho^{\pm b}(v) J_\mu^{\pm c}(x) \quad (20)$$

Similarly, in most nonlinear sigma models, the equations of motion can be written as a conservation law for a curl-free current taking values in a finite dimensional Lie algebra. This ensures the existence of nonlocal currents [1, 14, 20] given by an expression similar to (20). In favorable cases these currents persist in the quantum theory. They are then the generators of a Yangian algebra [4, 5, 19, 24, 25].

The structure of the nonlocal conserved currents in massive theories suggests the following form for the nonlocal density of the Yangian generator in the ultraviolet limit

$$Y^a(z; P) = \lim_{y \to z}\left\{ i f^{abc} \int_P^z dv\, J^b(v) J^c(y) + 2g J^a(y) \ln\left(\frac{z-y}{P-y}\right)\right\}, \quad (21)$$

Recall that the OPE version of the current-algebra commutation relations between the modes is

$$J^a(z) J^b(w) \sim \frac{k\delta^{ab}}{(z-w)^2} + \frac{i f^{abc} J^c(w)}{z-w} \quad (22)$$

The structure constants f^{abc} are assumed to be completely antisymmetric and normalized according to

$$f^{abc} f^{abd} = 2g\delta^{cd}, \quad (23)$$

where g stands for the dual Coxeter number and as before k is the level. The first integration contour in Eq. (21), $\int_P^z dv\, J^b(v)$, is the same as the contour used to define the second logarithmic term, $\ln([z-y]/[P-y]) = \int_P^z dv/(v-y)$.

The main observation that needs to be made from the expression of this nonlocal current is the dependence on a certain point P. This is a novelty brought by the conformal invariance, which prevents us from simply letting $P \to \infty$ from the onset. Many of the complications to follow are rooted in this simple fact. In the massive regime, P can be sent to ∞ since correlation functions decrease exponentially; mathematically said, the points are no longer all conformally equivalent.

Notice that the conservation of this nonlocal current is completely trivial: it is manifestly $\bar{z}$ independent. The real issue is to show that the corresponding charge is a level-one Yangian generator.

To establish this we first need to evaluate the effect of inserting nonlocal currents in correlation functions, that is, to calculate $\langle Y^a(z;P) \prod_j \phi_j(\zeta_j) \rangle$. Here $\phi_j(\zeta_j)$ stands for a WZW primary field:

$$J^a(z)\phi_j(w) \sim \frac{-t_j^a}{z-w}\phi(w) \tag{24}$$

where t_j^a is a matrix representation (specified by j) of the symmetry algebra: $[t_j^a, t_k^b] = i\delta_{jk}f^{abc}t_j^c$. A direct calculation yields

$$\left\langle Y^a(z;P) \prod_j \phi_j(\zeta_j) \right\rangle = if^{abc} \sum_{j,k} \frac{t_j^b t_k^c}{z-\zeta_k} \ln\left(\frac{z-\zeta_j}{P-\zeta_j}\right) \left\langle \prod_j \phi_j(\zeta_j) \right\rangle. \tag{25}$$

The nonlocal structure of Y^a manifests itself through logarithmic terms. The nonlocality of $Y(z;P)$ is encoded in the monodromy acquired by these integrals upon contour deformations.

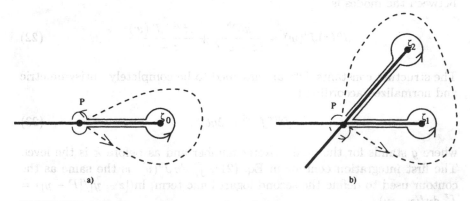

FIGURE 21.1. a) The contour γ_0 consists of an open contour of vanishing radius, starting below and to the left of P, and of a contour encircling ζ_0 (dashed line). The second contour can be deformed into the contour drawn with a solid line. b) The contour (broken line) circling around ζ_1 and ζ_2 is deformed into the contours drawn with solid lines.

6   Nonlocal Charges in WZW Models and Comultiplication

The action of the charge Q_1^a—associated to the current $Y^a(z; P)$—on a field $\phi_0(\zeta_0)$ is defined as follows [4, 5]:

$$\left\langle Q_1^a(\phi_0(\zeta_0)) \prod_{j\geq 1} \phi_j(\zeta_j) \right\rangle = \oint_{\gamma_0} dz \left\langle Y^a(z; P) \prod_{j\geq 0} \phi_j(\zeta_j) \right\rangle \qquad (26)$$

where the contour γ_0 is defined as in Figure 21.1a.

Essentially, the contour circles around ζ_0 but excludes all other ζ_i's. However, the presence of the logarithmic cut calls for a more careful specification. The cut, which extends from ζ_0 to infinity is chosen to pass through the point P and the contour includes P, although it remains slightly open owing to the cut. The point z starts then just below (and at the left of) P, hence just below the cut, follows the cut up to ζ_0, which is then encircled and returns to P above the cut (and again at the left of P). The phase of $(z - \zeta_0)$ above the cut differs by $2\pi i$ from its value just below it.[1]

The next step is to define the action of Q_1^a on two fields $\phi_1(\zeta_1)$ and $\phi_2(\zeta_2)$. This is still defined by a contour of integration:

$$\left\langle Q_1^a(\phi_1(\zeta_1)\phi_2(\zeta_2)) \prod_{j\geq 3} \phi_j(\zeta_j) \right\rangle = \oint_{\Gamma} dz \left\langle Y^a(z; P) \prod_{j\geq 0} \phi_j(\zeta_j) \right\rangle \qquad (27)$$

but with a different integration contour, which is illustrated in Figure 21.1b. This contour can be deformed in two contours γ_1 and γ_2. However, after the first contour a shift of $2\pi i$ occurs, which is responsible for the following comultiplication rule:

$$\Delta Q_1^a = Q_1^a \otimes 1 + 1 \otimes Q_1^a - 2\pi f^{abc} Q_0^b \otimes Q_0^c \qquad (28)$$

This is indeed the comultiplication of Yangians.

On the other hand, the action of the local charge Q_0^a on ϕ_0 is defined as usual in terms of a small contour around ζ_0. This results in $Q_0^a(\phi_0) = -t_0^a \phi_0$. Q_0^a acts additively on a product of fields.

Before going on, we pause to show that $Q_1^a(\phi_0)$ belongs to the same representation as ϕ_0.

From the correlator obtained at the end of the previous section we derive

---

[1]For subsequent applications of the charge Q_1^a, we mention that, in Figure 21.1a, the open contour around P has vanishing radius. Thus, for this part of γ_0, the logarithmic terms do not contribute, while the pole contributions are kept (which can arise for some of the terms in a correlation function), *as if*, in the absence of cuts, the contour was completely closed at the left of P.

the following Ward identity:

$$\sum_{j\geq0} t_j^b \left\langle Y^a(z;P) \prod_{j\geq0} \phi_j(\zeta_j) \right\rangle - if^{bac} \left\langle Y^c(z;P) \prod_{j\geq0} \phi_j(\zeta_j) \right\rangle = 0 \qquad (29)$$

Equation (29) implies that

$$\sum_{j\geq0} t_j^b \left\langle Q_1^a(\phi_0) \prod_{j\geq1} \phi_j(\zeta_j) \right\rangle = 0 \qquad (30)$$

7 Yangian Relations

In Ref. 9 we have checked that the charges Q_0^a and Q_1^a are the first two generators of a Yangian symmetry by verifying Y(1)–Y(3) for the action of Q_0^a and Q_1^a on a set of primary fields. (Actually, this check provides only a partial proof of the Yangian relations since acting with the Yangian generators produces fields that are not generated by products of primary fields; i.e., nonlocal fields are produced.) As is well known, the action of Q_0^a obtained previously implies Y(1) directly. Having obtained the Yangian comultiplication, it is enough to verify the basic relation Y(2) and the Serre relation Y(3) on a single field, since this will imply that they are also verified on an arbitrary number of primary fields. Hence, it suffices to evaluate the following expectation values:

$$\left\langle \left[Q_i^a(Q_1^b(\phi_0)) - Q_1^b(Q_i^a(\phi_0)) \right](\zeta_0) \prod_{j\geq1} \phi_j(\zeta_j) \right\rangle \qquad (31)$$

with $i = 0, 1$. These calculations turn out to be rather involved and they are described in detail in Ref 9. Note that in Y(3), the constant κ is found to be

$$\kappa = -4\pi^2 \qquad (32)$$

exactly the required value for this relation to be compatible with the co-multiplication law.

8 Yangians Densities as Logarithmic Operators

The OPE of the energy-momentum tensor and the nonlocal current is easily found to be

$$T(w)Y^a(z;P) \sim \frac{Y^a(z;P)}{(w-z)^2} + \frac{\partial_z Y^a(z;P)}{w-z} - \frac{J^a(z)}{(w-z)^2}$$

$$+ \frac{J^a(z)}{(w-z)(w-P)} - \frac{G^a(z;P)}{w-P} \qquad (33)$$

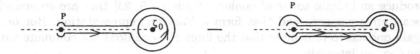

FIGURE 21.2. For the first term, we carry out the w-integration on the inner circle first and then the z-integration on the outer contour. For the second term the z-integration is carried out on the inner contour and the w-integration on the outer contour.

where $G^a(z; P) = if^{abc} J^b(P) J^c(z)$. Inserted in a correlation, this can be used to evaluate the commutator $[L_{-1}, Q_1^a]$ acting on the field $\phi_0(\zeta_0)$ using the contours in Figure 21.2.

The result is simply

$$[L_{-1}, Q_1^a] = -2if^{abc} \sum_{k \neq 0} \frac{t_0^b t_k^c}{P - \zeta_k} \left\langle \prod_j \phi_j(\zeta_j) \right\rangle, \qquad (34)$$

which vanishes only when $P \to \infty$. Thus, if $Q_1^a|_\infty$ refers to the charge associated to $Y^a(z; \infty)$, we have

$$[L_{-1}, Q_1^a]|_\infty = 0. \qquad (35)$$

This commutation relation has to be used carefully since it holds only in the limit $P \to \infty$ and with the limit taken after having performed the integration. The fact that the Yangian charges commute with the momentum operator L_{-1} only once the limit $P \to \infty$ has been taken is the analog of the well known fact that the quantum group symmetry of integrable lattice models only emerges once the infinite lattice limit as been taken.

When using this commutation relation, one has to remember that taking the limit $P \to \infty$ and integrating along the contour γ_0 are operations that do not commute. This noncommutativity of the limits prevents us from writing down directly the Ward indentities for the charges $Q_a^1|_\infty$. A similar phenomenon for nonlocal symmetry on finite lattices has been pointed out in Ref. 6.

If the point P could be set to ∞, the OPE of T with $Y^a(z; \infty)$ would simply be

$$T(w)Y^a(z; \infty) \sim \frac{Y^a(z; \infty)}{(w - z)^2} + \frac{\partial_z Y^a(z; \infty)}{w - z} - \frac{J^a(z)}{(w - z)^2} \qquad (36)$$

It reveals the standard Jordan cell structure observed in Ref. 16, which characterizes the presence of logarithmics operators. More precisely, it arises if at least two operators with the same dimension appear in the OPE of two other operators. Here the nonlocal operator $Y^a(z; \infty)$ would be the logarithmic partner of the operator $J^a(z)$. More generally, by acting recursively on a local field of the WZW model with $Y^a(z; \infty)$ one will produce an infinite tower of nonlocal fields [4, 5, 23] that are recursively logarithmic partners and that form a Yangian representation. But once again one has to remember that the limit $P \to \infty$ does not commute with the contour integrals.

9 Conclusions

The new WZW Yangian symmetry described here is completely general in that it holds for an arbitrary affine Lie algebra at any level. It provides the natural ultraviolet limit of the Yangian symmetry observed in the massive theories with current algebras. These massive theories can be described as perturbed conformal theories with the perturbation $J^a(z)\bar{J}^a(\bar{z})$. In general, we expect that any integrable perturbation that preserves the group symmetry will have a Yangian symmetry. It would be interesting to display further Yangians in WZW models from that point of view.

On the other hand, the present results raise a number of questions. To undertsand better the representations associated to the action of the Yangian charges on the fields clearly remains an open problem. Since the logarithmic Yangian acts on-shell on the n-particle states, to introduce these representations will be necessary if one wants to preserve the particle-field duality for massless theories. It would be interesting to see whether this action could be intertwined by transfer matrices associated to inhomogeneous spin chains. Another problem is to write the Yangian Ward identities in a compact form, the difficulty being related to coutour deformations in presence of logarithmic cuts. Note finally that in view of the S-matrices of the massive (WZW current-current perturbations) theories, which are products of Yangian matrices and R-matrices of the RSOS type, the logarithmic Yangian is expected to commute with the generators of the (restricted) affine quantum group symmetries $\mathcal{U}_q(\mathcal{G})$ that are associated to the RSOS factor [7].

Acknowledgments: We thank D. Bernard for his collaboration in Ref. 9 and the organizers of the workshop for the invitation to present this work. This work was supported by NSERC (Canada) and FCAR (Québec).

10 References

1. E. Abdalla, M. C. B. Abdalla, J. C. Brunelli, and A. Zadra, *The algebra of nonlocal charges in nonlinear sigma models*, Commun. Math. Phys. **166** (1994), No. 2, 379–396.

2. E. Abdalla, M. C. B. Abdalla, and K. D. Rothe, *Nonperturbative Methods in Two-Dimensional Quantum Field Theory*, World Sci.

Publ., River Edge, NJ, 1991.

3. A. A. Belavin, A. M. Polyakov, and A. B. Zamolodchikov, *Infinite conformal symmetry in two-dimensional quantum field theory*, Nucl. Phys. B **241** (1984), No. 2, 333–380.

4. D. Bernard, *Hidden Yangians in 2D massive current algebras*, Commun. Math. Phys. **137** (1991), No. 1, 191–208.

5. _____, *Quantum symmetries in 2D massive field theory*, New Symmetry Principles in Quantum Field Theory (Cargèse, 1991) (J. Fröhlich, G. 't Hooft, A. Jaffe, G. M. andP. K. Mitter, and R. Stora, eds.), NATO Adv. Sci. Inst. Ser. B Phys., Vol. 295, Plenum, New York, 1992, pp. 1–35, hep-th/9109058.

6. D. Bernard and G. Felder, *Quantum group symmetries in two-dimensional lattice quantum field theory*, Nucl. Phys. B **365** (1991), No. 1, 98–120.

7. D. Bernard and A. LeClair, *The fractional supersymmetric sine-Gordon models*, Phys. Lett. B **247** (1990), 309–316.

8. _____, *Quantum group symmetries and nonlocal currents in 2D QFT*, Commun. Math. Phys. **142** (1991), No. 1, 99–138.

9. D. Bernard, Z. Maassarani, and P. Mathieu, *Logarithmic Yangians in WZW models*, Modern Phys. Lett. **A12** (1997), No. 8, 535–544, hep-th/9612217.

10. D. Bernard, V. Pasquier, and D. Serban, *Spinons in conformal field theory*, Nucl. Phys. B **428** (1994), No. 3, 612–628, hep-th/9406020.

11. P. Bouwknegt, A. W. W. Ludwig, and K. Schoutens, *Spinon bases, Yangian symmetry and fermionic representations of Virasoro characters in conformal field theory*, Phys. Lett. B **338** (1994), No. 4, 448–456, hep-th/9406020.

12. P. Bouwknegt, A. W. W. Ludwig, and K. Schoutens, *Spinon basis for higher level* SU(2) *WZW models*, Phys. Lett. B **359** (1995), No. 3-4, 304–312, hep-th/941210.

13. P. Bouwknegt and K. Schoutens, *The* $\widehat{SU}(n)_1$ *WZW models. Spinon decomposition and Yangian structure*, Nucl. Phys. B **482** (1996), No. 1-2, 345–372, hep-th/9607064.

14. E. Brézin, C. Itzykson, J. Zinn-Justin, and J.-B. Zuber, *Remarks about the existence of nonlocal charges in two-dimensional models*, Phys. Lett. B **82** (1979), 442–445.

15. V. G. Drinfeld, *Hopf algebras and the quantum Yang–Baxter equation,* Soviet Math. Dokl. **32** (1985), No. 1, 254–258.

16. V. Gurarie, *Logarithmic operators in conformal field theory,* Nucl. Phys. B **410** (1993), No. 3, 535–549.

17. F. D. M. Haldane, *Exact Jastrow–Gutzwiller resonating-valence-bond ground state of the spin-$\frac{1}{2}$ antiferromagnetic Heisenberg chain with $1/r^2$ exchange,* Phys. Rev. Lett. **60** (1988), 635–638.

18. F. D. M. Haldane, Z. N. C. Ha, J. C. Talstra, D. Bernard, and V. Pasquier, *Yangian symmetry of integrable quantum chains with long-range interactions and a new description of states in conformal field theory,* Phys. Rev. Lett. **69** (1992), No. 14, 2021–2025.

19. M. Lüscher, *Quantum nonlocal charges and absence of particle production in the two-dimensional nonlinear σ-model,* Nucl. Phys. B **135** (1978), 1–19.

20. M. Lüscher and K. Pohlmeyer, *Scattering of massless lumps and nonlocal charges in the two-dimensional classical nonlinear σ-model,* Nucl. Phys. B **137** (1978), No. 1-2, 46–54.

21. K. Schoutens, *Yangian symmetry in conformal field theory,* Phys. Lett. B **331** (1994), No. 3-4, 335–341, hep-th/9401154.

22. B. S. Shastry, *Exact solution of an $S = \frac{1}{2}$ Heisenberg antiferromagnetic chain with long-ranged interactions,* Phys. Rev. Lett. **60** (1988), 639–642.

23. F. Smirnov, *Dynamical symmetries of massive integrable models. i, ii,* IJMA (1992), 813–858.

24. H. J. de Vega, H. Eichenherr, and J.-M. Maillet, *Classical and quantum algebras of nonlocal charges in σ models,* Commun. Math. Phys. **92** (1984), No. 4, 507–524.

25. _____ , *Yang–Baxter algebras of monodromy matrices in integrable quantum field theories,* Nucl. Phys. B **240** (1984), No. 3, 377–399.

22

The Quantized Knizhnik–Zamolodchikov Equation in Tensor Products of Irreducible sl$_2$-Modules

E. Mukhin
A. Varchenko

ABSTRACT We consider the quantized Knizhnik–Zamolodchikov difference equation (qKZ) with values in a tensor product of irreducible sl$_2$-mod-
ules, the equation defined in terms of rational R-matrices. We solve the
equation in terms of multidimensional q-hypergeometric integrals. We identify the space of solutions of the qKZ equation with the tensor product of
the corresponding modules over the quantum group U_q sl$_2$. We compute the
monodromy of the qKZ equation in terms of the trigonometric R-matrices.

1   Introduction

In this paper we solve the rational quantized Knizhnik–Zamolodchikov difference equation (qKZ) with values in a tensor product of irreducible highest weight sl$_2$-modules. The rational qKZ equation is a system of difference equations for a function $\Psi(z_1, \ldots, z_n)$ with values in a tensor product
$M_1 \otimes \cdots \otimes M_n$ of sl$_2$-modules. The system of equations has the form

$$\Psi(z_1, \ldots, z_m + p, \ldots, z_n)$$
$$= R_{m,m-1}(z_m - z_{m-1} + p) \cdots R_{m,1}(z_m - z_1 + p)e^{-\mu h_m}$$
$$\times R_{m,n}(z_m - z_n) \cdots R_{m,m+1}(z_m - z_{m+1})\Psi(z_1, \ldots, z_n),$$

$m = 1, \ldots, n$, where p, μ are complex parameters of the qKZ equation, h is
a generator of the Cartan subalgebra of sl$_2$, h_m is the operator h acting in
the mth factor, $R_{i,j}(x)$ is the rational R-matrix $R_{M_i M_j}(x) \in \operatorname{End}(M_i \otimes M_j)$
acting in the ith and jth factors of the tensor product. In this paper we
consider only steps p with negative real part.

The qKZ equation is an important system of difference equations. The
qKZ equation was introduced in Ref. 6 as an equation for matrix elements
of vertex operators of a quantum affine algebra. An important special case

of the qKZ equation had been introduced earlier in Ref. 9 as equations for form factors in integrable quantum field theory. Later, the qKZ equation was derived as an equation for correlation functions in lattice integrable models, cf. Ref. 7 and references therein.

Solutions of the rational qKZ equation with values in a tensor product of sl_2 Verma modules $V_{\lambda_1} \otimes \cdots \otimes V_{\lambda_n}$ with generic highest weights $\lambda_1, \ldots, \lambda_n$ were constructed in Ref. 11. The solutions have the form

$$\Psi(z) = \sum_{k_1,\ldots,k_n} I_{k_1,\ldots,k_n}(z) f^{k_1} v_1 \otimes \cdots \otimes f^{k_n} v_n,$$

where $\{f^{k_1} v_1 \otimes \cdots \otimes f^{k_n} v_n\} \in V_{\lambda_1} \otimes \cdots \otimes V_{\lambda_n}$ is the standard basis in the tensor product of sl_2 Verma modules, and the coefficients $I_{k_1,\ldots,k_n}(z)$ are given by suitable multidimensional q-hypergeometric integrals.

The space of solutions of the qKZ equation with values in a tensor product of sl_2 Verma modules with generic highest weights was described in Ref. 11 in terms of the representation theory of the quantum group $U_q(sl_2)$ with $q = e^{\pi i/p}$. Namely, consider the tensor product $V^q_{\lambda_1} \otimes \cdots \otimes V^q_{\lambda_n}$ of $U_q(sl_2)$ Verma modules, where $V^q_{\lambda_j}$ is the deformation of the sl_2 Verma module V_{λ_j}. It was shown in Ref. 11 that there is a natural isomorphism of the space $\mathcal{S}$ of meromorphic solutions of the qKZ equation with values in the tensor product $V_{\lambda_1} \otimes \cdots \otimes V_{\lambda_n}$ and the space $V^q_{\lambda_1} \otimes \cdots \otimes V^q_{\lambda_n} \otimes F$, where F is the space of meromorphic functions in $z_1, \ldots, z_n$, p-periodic with respect to each of the variables,

$$V^q_{\lambda_1} \otimes \cdots \otimes V^q_{\lambda_n} \otimes F \simeq \mathcal{S}.$$

This isomorphism was used in Ref. 11 to compute asymptotic solutions of the qKZ equation with values in a tensor product of sl_2 Verma modules with generic highest weights and to compute the transition functions between asymptotic solutions in terms of the trigonometric R-matrices acting in $V^q_{\lambda_1} \otimes \cdots \otimes V^q_{\lambda_n}$.

Assume that the Verma module V_{λ_j} is reducible and vectors $\{f^k v_j\}_{k \geq N_j}$ generate a proper submodule $S_{\lambda_j} \subset V_{\lambda_j}$. Then the vectors $\{f^k v_j\}_{k < N_j}$ form a basis in the irreducible module $L_{\lambda_j} = V_{\lambda_j}/S_{\lambda_j}$. If V_{λ_j} is irreducible, then we set $N_j = \infty$.

Return to the sum $\Psi(z)$. If the Verma modules of the tensor product $V_{\lambda_1} \otimes \cdots \otimes V_{\lambda_n}$ become reducible, then some of the coefficients in the sum become divergent. In this paper we show that the restricted sum

$$\Psi^0(z) = \sum_{\substack{k_1 < N_1 \\ \vdots \\ k_n < N_n}} I_{k_1,\ldots,k_n}(z) f^{k_1} v_1 \otimes \cdots \otimes f^{k_n} v_n$$

remains well defined even when some of the Verma modules become reducible.

Moreover, we show that the sum $\Psi^0(z)$ defines a solution of the qKZ equation with values in the tensor product $L_{\lambda_1} \otimes \cdots \otimes L_{\lambda_n}$ of irreducible sl$_2$-modules, and under certain conditions all solutions have this form.

These results allow us to describe the space of solutions to the qKZ equation with values in $L_{\lambda_1} \otimes \cdots \otimes L_{\lambda_n}$ in terms of representation theory of the quantum group $U_q(\text{sl}_2)$. Namely, consider the tensor product $L_{\lambda_1}^q \otimes \cdots \otimes L_{\lambda_n}^q$ of U_q sl$_2$-modules, where $L_{\lambda_j}^q$ is the deformation of the sl$_2$ irreducible module L_{λ_j}. We show that there is a natural isomorphism of the space S of meromorphic solutions of the qKZ equation with values in the tensor product $L_{\lambda_1}^q \otimes \cdots \otimes L_{\lambda_n}$ and the space $L_{\lambda_1}^q \otimes \cdots \otimes L_{\lambda_n}^q \otimes F$,

$$L_{\lambda_1}^q \otimes \cdots \otimes L_{\lambda_n}^q \otimes F \simeq S.$$

We compute asymptotic solutions of the qKZ equation with values in a tensor product of irreducible sl$_2$ Verma modules and the transition functions between the asymptotic solutions. The transition functions are given in terms of the trigonometric R-matrices acting in $L_{\lambda_1}^q \otimes \cdots \otimes L_{\lambda_n}^q$.

In this paper we consider the rational qKZ equation associated with sl$_2$. There are other types of the qKZ equation: the trigonometric qKZ equation [6, 12] and the elliptic qKZB equation [2–5]. The trigonometric qKZ equation with values in a tensor product of $U_q(\text{sl}_2)$ Verma modules with generic highest weights and the elliptic qKZB equation with values in a tensor product of $E_{\tau,\eta}(\text{sl}_2)$ Verma modules with generic highest weights were solved in Refs. 4 and 12, respectively. Here $E_{\tau,\eta}(\text{sl}_2)$ is the elliptic quantum group associated to sl$_2$. In the next paper we will extend our results to the trigonometric qKZ and elliptic qKZB equations and describe solutions of these equations with values in irreducible finite dimensional modules.

The paper is organized as follows.

In Section 2 we recall some facts about sl$_2$, $U_q(\text{sl}_2)$ and their representations. We define the rational R-matrix and the qKZ equation. In Section 3 we describe integral representations of solutions of the qKZ equation with values in a tensor product of sl$_2$ Verma modules. The statements of results are given in Section 4. The proofs are collected in Section 5.

2   General Definitions and Notations

2.1   The Lie Algebra sl$_2$

Let e, f, h be generators of the Lie algebra sl$_2$ such that

$$[h, e] = e, \quad [h, f] = -f, \quad [e, f] = 2h.$$

For an sl$_2$-module M, let M^* be its restricted dual with an sl$_2$-module structure defined by

$$\langle e\varphi, x \rangle = \langle \varphi, fx \rangle, \quad \langle f\varphi, x \rangle = \langle \varphi, ex \rangle, \quad \langle h\varphi, x \rangle = \langle \varphi, hx \rangle$$

for all $x \in M$, $\varphi \in M^*$. The module M^* is called the *dual* module.

For $\lambda \in \mathbb{C}$, denote V_λ the sl_2 Verma module with highest weight λ. Then $V_\lambda = \bigoplus_{i=0}^\infty \mathbb{C}f^i v$, where v is a highest-weight vector. Denote L_λ the irreducible module with highest weight λ.

Let $\Lambda^+ = \{0, \frac{1}{2}, 1, 3/2, 2, \dots\}$ be the set of dominant weights. If $\lambda \in \Lambda^+$; then L_λ is a $(2\lambda + 1)$-dimensional module and

$$L_\lambda \simeq V_\lambda / S_\lambda,$$

where $S_\lambda = \bigoplus_{i=2\lambda+1}^\infty \mathbb{C}f^i v \subset V_\lambda$ is the maximal proper submodule. The vectors $f^i v$, $i = 0, \dots, 2\lambda$, generate a basis in L_λ.

For $\lambda \notin \Lambda^+$, $L_\lambda = V_\lambda$. It is convenient to introduce S_λ to be the zero submodule of V_λ, then $L_\lambda \simeq V_\lambda / S_\lambda$, as we have for $\lambda \in \Lambda^+$.

For an sl_2-module M with highest weight λ, denote by $(M)_l$ the subspace of weight $\lambda - l$, by $(M)^{\mathrm{sing}}$ the kernel of the operator e, and by $(M)_l^{\mathrm{sing}}$ the subspace $(M)_l \cap (M)^{\mathrm{sing}}$.

2.2   The Algebra $U_q \mathrm{sl}_2$

Let q be a complex number different from ± 1. Let e_q, f_q, q^h, q^{-h} be generators of $U_q \mathrm{sl}_2$ such that

$$q^h q^{-h} = q^{-h} q^h = 1, \qquad\qquad [e_q, f_q] = \frac{q^{2h} - q^{-2h}}{q - q^{-1}},$$

$$q^h e_q = q e_q q^h, \qquad\qquad q^h f_q = q^{-1} f_q q^h.$$

A comultiplication $\Delta: U_q \mathrm{sl}_2 \to U_q \mathrm{sl}_2 \otimes U_q \mathrm{sl}_2$ is given by

$$\Delta(q^h) = q^h \otimes q^h, \qquad\qquad \Delta(q^{-h}) = q^{-h} \otimes q^{-h},$$

$$\Delta(e_q) = e_q \otimes q^h + q^{-h} \otimes e_q, \qquad\qquad \Delta(f_q) = f_q \otimes q^h + q^{-h} \otimes f_q.$$

The comultiplication defines a module structure on tensor products of $U_q \mathrm{sl}_2$-modules.

For $\lambda \in \mathbb{C}$, denote V_λ^q the $U_q \mathrm{sl}_2$ Verma module with highest weight q^λ. Then $V_\lambda^q = \bigoplus_{i=0}^\infty \mathbb{C}f_q^i v^q$, where v^q is a highest weight vector.

For $\lambda \in \Lambda^+$, $S_\lambda^q = \bigoplus_{i=2\lambda+1}^\infty \mathbb{C}f_q^i v^q$ is a submodule in V_λ^q. Denote L_λ^q the quotient module $V_\lambda^q / S_\lambda^q$. The module L_λ^q is the $(2\lambda + 1)$-dimensional highest-weight module with highest weight q^λ. The vectors $f_q^i v^q$, $i = 0$, $\dots, 2\lambda$, generate a basis in L_λ^q.

For $\lambda \notin \Lambda^+$, let $L_\lambda^q = V_\lambda^q$. It is convenient to introduce S_λ^q to be the zero submodule of V_λ^q; then $L_\lambda^q \simeq V_\lambda^q / S_\lambda^q$, as we have for $\lambda \in \Lambda^+$.

For an $U_q \mathrm{sl}_2$-module M^q with highest weight q^λ, denote by $(M^q)_l$ the subspace of weight $q^{\lambda-l}$, by $(M^q)^{\mathrm{sing}}$ the kernel of the operator e_q, and by $(M^q)_l^{\mathrm{sing}}$ the subspace $(M^q)_l \cap (M^q)^{\mathrm{sing}}$.

2.3 The Rational R-Matrix

The *Yangian* $Y(\mathrm{sl}_2)$ is a Hopf algebra that has a family of homomorphisms to the universal enveloping algebra of sl_2, $Y(\mathrm{sl}_2) \to U(\mathrm{sl}_2)$, depending on a complex parameter. Therefore, each sl_2-module M carries a Yangian module structure $M(x)$ depending on a parameter (see Refs. 1 and 11).

For sl_2 irreducible highest-weight modules L_{λ_1}, L_{λ_2} and generic numbers $x, y \in \mathbb{C}$, the Yangian modules $L_{\lambda_1}(x) \otimes L_{\lambda_2}(y)$ and $L_{\lambda_2}(y) \otimes L_{\lambda_1}(x)$ are irreducible and isomorphic. There exists a unique intertwiner that sends $v_1 \otimes v_2$ to $v_2 \otimes v_1$, where v_i is a highest vector in L_{λ_i}, $i = 1$, 2. This intertwiner has the form $P R_{L_{\lambda_1} L_{\lambda_2}}(x - y)$, where P is the operator of permutation of factors. The operator $R_{L_{\lambda_1} L_{\lambda_2}}(x) \in \mathrm{End}(L_{\lambda_1} \otimes L_{\lambda_2})$ is called the *rational R-matrix*, see details in Ref. 11. The rational R-matrix commutes with the sl_2 action on the tensor product $L_{\lambda_1} \otimes L_{\lambda_2}$.

Let λ_1, $\lambda_2 \notin \Lambda^+$. Let

$$V_{\lambda_1} \otimes V_{\lambda_2} = \bigoplus_{l=0}^{\infty} V_{\lambda_1 + \lambda_2 - l}$$

be the decomposition of the tensor product of sl_2 Verma modules into the direct sum of irreducibles, and let $\Pi^{(l)}$ be the projector on $V_{\lambda_1 + \lambda_2 - l}$ along the other summands.

There is a formula for the rational R-matrix,

$$R_{V_{\lambda_1} V_{\lambda_2}}(x) = \sum_{l=0}^{\infty} \Pi^{(l)} \prod_{s=0}^{l-1} \frac{x + \lambda_1 + \lambda_2 - s}{x - \lambda_1 - \lambda_2 + s}$$

(see Refs. 8 and 10).

Let λ_1, $\lambda_2 \in \Lambda^+$. Let

$$L_{\lambda_1} \otimes L_{\lambda_2} = \bigoplus_{l=0}^{2\min\{\lambda_1, \lambda_2\}} L_{\lambda_1 + \lambda_2 - l}$$

be the decomposition of the tensor product of finite dimensional irreducible sl_2-modules into the direct sum of irreducibles, and let $\Pi^{(l)}$ be the projector on $L_{\lambda_1 + \lambda_2 - l}$ along the other summands.

There is a formula for the rational R-matrix,

$$R_{L_{\lambda_1} L_{\lambda_2}}(x) = \sum_{l=0}^{2\min\{\lambda_1, \lambda_2\}} \Pi^{(l)} \prod_{s=0}^{l-1} \frac{x + \lambda_1 + \lambda_2 - s}{x - \lambda_1 - \lambda_2 + s}$$

(see Refs. 8 and 10).

The vector spaces $V_{\lambda_1} \otimes V_{\lambda_2}$ for different values of λ_1, λ_2 are identified by distinguished bases $\{ f^{l_1} v_1 \otimes f^{l_2} v_2 \mid l_1, l_2 \in \mathbb{Z}_{\geq 0} \}$.

Theorem 1. 1. *The rational R-matrix $R_{V_{\lambda_1} V_{\lambda_2}}(x) \in \text{End}(V \otimes V)$ is a meromorphic function of x, λ_1, λ_2. The poles of $R_{V_{\lambda_1} V_{\lambda_2}}(x)$ have the form $x - \lambda_1 - \lambda_2 + s = 0$, where $s \in \mathbb{Z}_{\geq 0}$.*

2. *Let x be generic. Then the rational R-matrix $R_{V_{\lambda_1} V_{\lambda_2}}(x)$ preserves $S_{\lambda_1} \otimes V_{\lambda_2} + V_{\lambda_1} \otimes S_{\lambda_2}$.*

3. *Let $V_{\lambda_1} \otimes V_{\lambda_2} \to L_{\lambda_1} \otimes L_{\lambda_2}$ be the canonical factorization map. Then for generic x, the rational R-matrix $R_{V_{\lambda_1} V_{\lambda_2}}(x)$ can be factorized to an operator $R(x) \in \text{End}(L_{\lambda_1} \otimes L_{\lambda_2})$ and, moreover, $R(x) = R_{L_{\lambda_1} L_{\lambda_2}}(x)$.*

An analog of Theorem 1 for the elliptic R-matrix is proved in Theorems 8 and 31 in Ref. 5. The same proof works for the rational R-matrix. We give an alternative proof in Section 5.

2.4   The Trigonometric R-Matrix

Let q be a complex number, and not a root of unity. The *quantum affine algebra* $\widehat{U_q \, \text{sl}_2}$ is a Hopf algebra that has a family of homomorphisms, $\widehat{U_q \, \text{sl}_2} \to U_q \, \text{sl}_2$, depending on a complex parameter. Therefore, each $U_q \, \text{sl}_2$-module M^q carries a $\widehat{U_q \, \text{sl}_2}$-module structure $M^q(x)$ depending on a parameter (see Refs. 10 and 1).

For $U_q \, \text{sl}_2$ irreducible highest weight modules $L_{\lambda_1}^q$, $L_{\lambda_2}^q$ and generic numbers $x, y \in \mathbb{C}$, the $\widehat{U_q \, \text{sl}_2}$-modules $L_{\lambda_1}^q(x) \otimes L_{\lambda_2}^q(y)$ and $L_{\lambda_2}^q(y) \otimes L_{\lambda_1}^q(x)$ are isomorphic. There exists a unique intertwiner that sends $v_1^q \otimes v_2^q$ to $v_2^q \otimes v_1^q$, where v_i^q is a highest vector in $L_{\lambda_i}^q$, $i = 1, 2$. This intertwiner has the form $P R_{L_{\lambda_1}^q L_{\lambda_2}^q}^q(x/y)$, where P is the operator of permutation of factors. The operator $R_{L_{\lambda_1}^q L_{\lambda_2}^q}^q(x) \in \text{End}(L_{\lambda_1}^q \otimes L_{\lambda_2}^q)$ is called the *trigonometric R-matrix* (see details in Ref. 11). The trigonometric R-matrix preserves the weight decomposition.

Let λ_1, $\lambda_2 \notin \Lambda^+$. Let

$$V_{\lambda_1}^q \otimes V_{\lambda_2}^q = \bigoplus_{l=0}^{\infty} V_{\lambda_1 + \lambda_2 - l}^q$$

be the decomposition of the tensor product of $U_q \, \text{sl}_2$ Verma modules into the direct sum of irreducible modules, and let $\Pi^{(l)}$ be the projector on $V_{\lambda_1 + \lambda_2 - l}^q$ along the other summands.

There is a formula for the trigonometric R-matrix:

$$R_{V_{\lambda_1}^q V_{\lambda_2}^q}^q(x) = R_{\lambda_1, \lambda_2}^q(0) \sum_{l=0}^{\infty} \Pi^{(l)} \prod_{s=0}^{l-1} \frac{1 - x q^{2s - 2\lambda_1 - 2\lambda_2}}{1 - x q^{2\lambda_1 + 2\lambda_2 - 2s}},$$

where

$$R^q_{\lambda_1,\lambda_2}(0) = q^{2\lambda_1\lambda_2 - 2h\otimes h} \sum_{k=0}^{\infty} (q^2-1)^{2k} \prod_{s=1}^{k}(1-q^{2s})^{-1}(q^h f_q \otimes q^{-h}e_q)^k$$

(see Refs. 1 and 10).

Let $\lambda_1, \lambda_2 \in \Lambda^+$. Let

$$L^q_{\lambda_1} \otimes L^q_{\lambda_2} = \bigoplus_{l=0}^{2\min\{\lambda_1,\lambda_2\}} L^q_{\lambda_1+\lambda_2-l}$$

be the decomposition of the tensor product of finite-dimensional irreducible U_q sl$_2$-modules into the direct sum of irreducibles, and let $\Pi^{(l)}$ be the projector on $L^q_{\lambda_1+\lambda_2-l}$ along the other summands.

There is a formula for the trigonometric R-matrix,

$$R^q_{L^q_{\lambda_1} L^q_{\lambda_2}}(x) = R^q_{\lambda_1,\lambda_2}(0) \sum_{l=0}^{2\min\{\lambda_1,\lambda_2\}} \Pi^{(l)} \prod_{s=0}^{l-1} \frac{1-xq^{2s-2\lambda_1-2\lambda_2}}{1-xq^{2\lambda_1+2\lambda_2-2s}}$$

(see Refs. 1 and 10). The vector spaces $V^q_{\lambda_1} \otimes V^q_{\lambda_2}$ for different values of λ_1, λ_2 are identified by distinguished bases $\{f^{l_1}v^q_1 \otimes f^{l_2}v^q_2 \mid l_1, l_2 \in \mathbb{Z}_{\geq 0}\}$.

Theorem 2. *Let $q \in \mathbb{C}$ be not a root of unity.*

1. *The trigonometric R-matrix $R^q_{V^q_{\lambda_1} V^q_{\lambda_2}}(x) \in \mathrm{End}(V \otimes V)$ is a meromorphic function of x, λ_1, λ_2. The poles of $R^q_{V^q_{\lambda_1} V^q_{\lambda_2}}(x)$ have the form $x = q^{-2\lambda_1-2\lambda_2+2s}$, where $s \in \mathbb{Z}_{\geq 0}$.*

2. *Let x be generic. Then the trigonometric R-matrix $R^q_{V^q_{\lambda_1} V^q_{\lambda_2}}(x)$ preserves*
$$S^q_{\lambda_1} \otimes V^q_{\lambda_2} + V^q_{\lambda_1} \otimes S^q_{\lambda_2}.$$

3. *Let $V^q_{\lambda_1} \otimes V^q_{\lambda_2} \to L^q_{\lambda_1} \otimes L^q_{\lambda_2}$ be the canonical factorization map. Then for generic x, the trigonometric R-matrix $R^q_{V^q_{\lambda_1} V^q_{\lambda_2}}(x)$ can be factorized to an operator $R^q(x) \in \mathrm{End}(L^q_{\lambda_1} \otimes L^q_{\lambda_2})$ and, moreover, $R^q(x) = R^q_{L^q_{\lambda_1} L^q_{\lambda_2}}(x)$.*

An analog of Theorem 1 for the elliptic R-matrix is proved in Theorems 8 and 31 in Ref. 5. The same proof works for the trigonometric R-matrix.

2.5 The qKZ Equation

The *rational quantized Knizhnik-Zamolodchikov equation* (qKZ) associated to sl$_2$ is the following system of linear difference equations for a function $\Psi(z_1, \ldots, z_n)$ with values in a tensor product $M_1 \otimes \cdots \otimes M_n$ of sl$_2$-modules:

$$\Psi(z_1, \ldots, z_m + p, \ldots, z_n)$$
$$= R_{M_m,M_{m-1}}(z_m - z_{m-1} + p) \cdots R_{M_m,M_1}(z_m - z_1 + p)e^{-\mu h_m}$$
$$\times R_{M_m,M_n}(z_m - z_n) \cdots R_{M_m,M_{m+1}}(z_m - z_{m+1})\Psi(z_1, \ldots, z_n),$$

for $m = 1, \ldots, n$. Here p, μ are complex parameters, μ is chosen so that $0 \le \mathrm{Im}\,\mu < 2\pi$; h_m is the operator $h \in \mathrm{sl}_2$ acting in the mth representation, $R_{M_i M_j}(x) \in \mathrm{End}(M_i \otimes M_j)$ is the rational R-matrix acting in the ith and jth factors (see Ref. 6). The linear operators on the right-hand side of the equations are called the *qKZ operators*.

The qKZ operators commute with the action of the operator $h \in \mathrm{sl}_2$ in the tensor product $M_1 \otimes \cdots \otimes M_n$. Therefore, in order to construct all solutions of the qKZ equation, it is enough to solve the qKZ equation with values in weight spaces $(M_1 \otimes \cdots \otimes M_n)_l$.

If the parameter μ of the equation is equal to zero, then the qKZ operators commute with the sl_2 action in the tensor product $M_1 \otimes \cdots \otimes M_n$, and in order to construct all solutions of the qKZ equation in this case, it is enough to solve the equation with values in singular weight spaces $(M_1 \otimes \cdots \otimes M_n)_l^{\mathrm{sing}}$.

Let $\pi\colon V_{\lambda_1} \otimes \cdots \otimes V_{\lambda_n} \to L_{\lambda_1} \otimes \cdots \otimes L_{\lambda_n}$ be the canonical projection map.

Lemma 1. *Let $\Psi(z)$ be a solution of the qKZ equation with values in $V_{\lambda_1} \otimes \cdots \otimes V_{\lambda_n}$. Then $\pi \circ \Psi(z)$ is a solution of the qKZ equation with values in $L_{\lambda_1} \otimes \cdots \otimes L_{\lambda_n}$.*

Lemma 1 follows from Theorem 1.

3 The Hypergeometric Pairing [11]

3.1 The Phase Function

Let $z = (z_1, \ldots, z_n) \in \mathbb{C}^n$, $\lambda = (\lambda_1, \ldots, \lambda_n) \in \mathbb{C}^n$, $t = (t_1, \ldots, t_l) \in \mathbb{C}^l$. The *phase function* is defined by the following formula:

$$\Phi_l(t, z, \lambda) = \exp\left(\mu \sum_{i=1}^{l} \frac{t_i}{p}\right) \prod_{i=1}^{n} \prod_{j=1}^{l} \frac{\Gamma((t_j - z_i + \lambda_i)/p)}{\Gamma((t_j - z_i - \lambda_i)/p)} \prod_{\substack{i,j=1 \\ i<j}}^{l} \frac{\Gamma((t_i - t_j - 1)/p)}{\Gamma((t_i - t_j + 1)/p)}.$$

3.2 Actions of the Symmetric Group

Let $f = f(t_1, \ldots, t_l)$ be a function. For a permutation $\sigma \in \mathbb{S}^l$, define the functions $[f]_\sigma^{\mathrm{rat}}$ and $[f]_\sigma^{\mathrm{trig}}$ via the action of the simple transpositions $(i, i+1) \in \mathbb{S}^l$, $i = 1, \ldots, l-1$, given by

$$[f]_{(i,i+1)}^{\mathrm{rat}}(t_1, \ldots, t_l) = f(t_1, \ldots, t_{i+1}, t_i, \ldots, t_l)\frac{t_i - t_{i+1} - 1}{t_i - t_{i+1} + 1},$$

$$[f]_{(i,i+1)}^{\mathrm{trig}}(t_1, \ldots, t_l) = f(t_1, \ldots, t_{i+1}, t_i, \ldots, t_l)\frac{\sin(\pi(t_i - t_{i+1} - 1)/p)}{\sin(\pi(t_i - t_{i+1} + 1)/p)}.$$

If for all $\sigma \in \mathbb{S}^l$, $[f]_\sigma^{\text{rat}} = f$, we will say that the function is *symmetric with respect to the rational action*. If for all $\sigma \in \mathbb{S}^l$, $[f]_\sigma^{\text{trig}} = f$, we will say that the function is *symmetric with respect to the trigonometric action*.

This definition implies the following important remark.

Remark 1. If $w(t_1, \ldots, t_l)$ is $\mathbb{S}^l$ symmetric with respect to the rational action and $W(t_1, \ldots, t_l)$ is $\mathbb{S}^l$ symmetric with respect to the trigonometric action, then $\Phi_l w W$ is a symmetric function of $t_1, \ldots, t_l$ (in the usual sense).

3.3 Rational Weight Functions

Fix natural numbers n, l. Set $\mathcal{Z}_l^n = \{l = (l_1, \ldots, l_n) \in \mathbb{Z}_{\geq 0}^n \mid \sum_{i=1}^n l_i = l\}$.
For $l \in \mathcal{Z}_l^n$ and $m = 0, 1, \ldots, n$, set $l^m = \sum_{i=1}^m l_i$.
For $l \in \mathcal{Z}_l^n$, define the *rational weight function* w_l by

$$w_l(t, z, \lambda) = \sum_{\sigma \in \mathbb{S}^l} \left[\prod_{m=1}^n \frac{1}{l_m!} \prod_{j=l^{m-1}+1}^{l^m} \left(\frac{1}{t_j - z_m - \lambda_m} \prod_{k=1}^m \frac{t_j - z_k + \lambda_k}{t_j - z_k - \lambda_k} \right) \right]_\sigma^{\text{rat}}.$$

For fixed $z, \lambda \in \mathbb{C}^n$, the space spanned over $\mathbb{C}$ by all rational weight functions $w_l(t, z, \lambda)$, $l \in \mathcal{Z}_l^n$, is called the *hypergeometric rational space specialized at* z, λ and is denoted $\mathfrak{F}(z, \lambda) = \mathfrak{F}_l^n(z, \lambda)$. This space is a space of functions of variable t.

For generic values of z, λ, the space $\mathfrak{F}(z, \lambda)$ can be identified with the space $(V_{\lambda_1}^* \otimes \cdots \otimes V_{\lambda_n}^*)_l$ by the map

$$\mathfrak{a}(z, \lambda) \colon (f^{l_1} v_1)^* \otimes \cdots \otimes (f^{l_n} v_n)^* \mapsto w_l(t, z, \lambda).$$

Here $\{(f^l v_i)^* \mid l \in \mathbb{Z}_{\geq 0}\}$ is the basis of $V_{\lambda_i}^*$, dual to the standard basis of V_{λ_i} given by $\{f^l v_i \mid l \in \mathbb{Z}_{\geq 0}\}$, see Lemma 4.5 and Corollary 4.8 in Ref. 11.

3.4 Trigonometric Weight Functions

Fix natural numbers n, l. For $l \in \mathcal{Z}_l^n$, define the *trigonometric weight function* W_l by

$$W_l(t, z, \lambda) = \sum_{\sigma \in \mathbb{S}^l} \left[\prod_{m=1}^n \prod_{s=1}^{l_m} \frac{\sin(\pi/p)}{\sin(\pi s/p)} \prod_{j=l^{m-1}+1}^{l^m} \frac{\exp(\pi i (z_m - t_j)/p)}{\sin(\pi(t_j - z_m - \lambda_m)/p)} \right.$$

$$\left. \times \prod_{k=1}^m \frac{\sin(\pi(t_j - z_k + \lambda_k)/p)}{\sin(\pi(t_j - z_k - \lambda_k)/p)} \right]_\sigma^{\text{trig}}.$$

A function $W(t, z, \lambda)$ is said to be a *holomorphic trigonometric weight function* if

$$W(t, z, \lambda) = \sum_{m_1 + \cdots + m_n = l} a_m(\lambda, e^{2\pi i z_1/p}, \ldots, e^{2\pi i z_n/p}) W_m(t, z, \lambda), \quad (1)$$

where $a_m(\lambda, u)$ are holomorphic functions of parameters $\lambda, u \in \mathbb{C}^n$. We denote $\mathfrak{G}$ the space of all holomorphic trigonometric weight functions. This space is a space of functions of variables of t, z, λ.

For a permutation $\sigma \in \mathbb{S}^n$, define the *trigonometric weight functions* W_l^σ by $W_l^\sigma(t, z, \lambda) = W_{\sigma l}(t, \sigma z, \sigma \lambda)$, where

$$\sigma l = (l_{\sigma_1}, \ldots, l_{\sigma_n}), \quad \sigma z = (z_{\sigma_1}, \ldots, z_{\sigma_n}), \quad \sigma \lambda = (\lambda_{\sigma_1}, \ldots, \lambda_{\sigma_n}).$$

For fixed λ, $z \in \mathbb{C}^n$ and $\sigma \in \mathbb{S}^n$, the space spanned over $\mathbb{C}$ by all trigonometric weight functions $W_l^\sigma(t, z, \lambda)$, $l \in \mathcal{Z}_l^n$, is called the *hypergeometric trigonometric space specialized at z, λ* and is denoted $\mathfrak{G}^\sigma(z, \lambda) = \mathfrak{G}_l^{n,\sigma}(z, \lambda)$. This space is a space of functions of variable t.

For generic z, λ, the space $\mathfrak{G}^\sigma(z, \lambda)$ does not depend on σ (see Section 2 in Ref. 11), we denote it as $\mathfrak{G}(z, \lambda)$.

For $l \in \mathcal{Z}_l^{n-1}$, define the *singular trigonometric weight function* W_l^{sing} by

$$W_l^{\text{sing}}(t, z, \lambda)$$

$$= \sum_{\sigma \in \mathbb{S}^l} \left[\prod_{m=1}^{n-1} \prod_{s=1}^{l_m} \frac{\sin(\pi/p)}{\sin(\pi s/p)} \sin\left(\frac{\pi(z_m - \lambda_m - z_{m+1} - \lambda_{m+1} + s - 1)}{p} \right) \right.$$

$$\times \prod_{j=l^{m-1}+1}^{l^m} \frac{1}{\sin(\pi(t_j - z_m - \lambda_m)/p) \sin(\pi(t_j - z_{m+1} - \lambda_{m+1})/p)}$$

$$\left. \times \prod_{k=1}^m \frac{\sin(\pi(t_j - z_k + \lambda_k)/p)}{\sin(\pi(t_j - z_k - \lambda_k)/p)} \right]_\sigma^{\text{trig}}.$$

For fixed z, $\lambda \in \mathbb{C}^n$, the space spanned over $\mathbb{C}$ by all singular trigonometric weight functions $W_l(t, z, \lambda)$, $l \in \mathcal{Z}_l^n$, is called the *singular hypergeometric trigonometric space specialized at z, λ* and is denoted $\mathfrak{G}^{\text{sing}}(z, \lambda) = \mathfrak{G}_l^{sing,n}(z, \lambda)$.

We have $\mathfrak{G}^{\text{sing}}(z, \lambda) \subset \mathfrak{G}(z, \lambda)$ (see Lemma 2.29 in Ref. [11]).

A function $W(t, z, \lambda)$ is said to be a *holomorphic singular trigonometric weight function* if $W(t, z, \lambda)$ is a holomorphic trigonometric weight function, and for all $z, \lambda \in \mathbb{C}^n$, the function $W(t, z, \lambda)$ belongs to $\mathfrak{G}^{\text{sing}}(z, \lambda)$. We denote as $\mathfrak{G}^{\text{sing}}$ the space of all holomorphic singular trigonometric weight functions. This is a space of functions of variables t, z, λ.

Lemma 2. *For any $l \in \mathcal{Z}_l^{n-1}$, the function $W_l^{\text{sing}}(t, z, \lambda)$ belongs to $\mathfrak{G}^{\text{sing}}$.*

Proof. By Lemmas 2.28 and 2.29 of Ref. 11, we have the following decomposition (1):

$$W_l^{\text{sing}}(t, z, \lambda) = \sum_{m_1 + \cdots + m_n = l} a_m(\lambda, e^{2\pi i z_1/p}, \ldots, e^{2\pi i z_n/p}) W_m(t, z, \lambda).$$

Here $a_m(\lambda, u)$ are meromorphic functions of λ, $u \in \mathbb{C}^n$. The functions $a_m(\lambda, u)$ are holomorphic for the following reason.

Suppose, for some $m_0 \in \mathcal{Z}_l^n$ and some Z, $\Lambda \in \mathbb{C}^n$ the function $a_m(\lambda, u)$ has a pole at Z, Λ. The functions $W_m(t, z, \lambda)$, $m \in \mathcal{Z}_l^n$ are linearly independent for generic t, z, λ by Lemma 2.28 in Ref. 11. Moreover, for generic t, the functions $W_m(t, z, \lambda)$, $m \in \mathcal{Z}_l^n$, do not have poles at Z, Λ. Hence, for generic t, the function $W_l^{\text{sing}}(t, z, \lambda)$ has a pole at Z, Λ. But this is not so. $\square$

For $l \in \mathcal{Z}_l^n$, define the *weight coefficient* $c_l(\lambda)$ by

$$c_l(\lambda) = \prod_{m=1}^{n} \prod_{s=0}^{l_m-1} \frac{\sin(\pi(s+1)/p)\sin(\pi(2\lambda_m - s)/p)}{\sin(\pi/p)}.$$

Let $q = e^{\pi i/p}$. The weight space $(V_{\lambda_{\sigma_1}}^q \otimes \cdots \otimes V_{\lambda_{\sigma_n}}^q)_l$ is mapped to the trigonometric hypergeometric space $\mathfrak{G}^\sigma(z, \lambda)$ by

$$\mathfrak{b}_\sigma(z, \lambda) \colon f_q^{l_{\sigma_1}} v_{\sigma_1}^q \otimes \cdots \otimes f_q^{l_{\sigma_n}} v_{\sigma_n}^q \mapsto c_l(\lambda) W_l^\sigma(t, z, \lambda). \qquad (2)$$

If z, λ are generic and q is not a root of unity, then the map $\mathfrak{b}_\sigma(z, \lambda)$ is an isomorphism of vector spaces, see Lemma 4.17 and Corollary 4.20 in Ref. 11. Moreover, the singular hypergeometric space is identified with the subspace of singular vectors,

$$\mathfrak{G}^{\text{sing}}(z, \lambda) = \mathfrak{b}_\sigma(z, \lambda)\big((V_{\lambda_{\sigma_1}}^q \otimes \cdots \otimes V_{\lambda_{\sigma_n}}^q)_l^{\text{sing}}\big),$$

see Corollary 4.21 in Ref. 11.

The composition maps

$$\mathfrak{b}_{\sigma,\sigma'}(z, \lambda) \colon (V_{\lambda_{\sigma_1'}}^q \otimes \cdots \otimes V_{\lambda_{\sigma_n'}}^q)_l \to (V_{\lambda_{\sigma_1}}^q \otimes \cdots \otimes V_{\lambda_{\sigma_n}}^q)_l,$$

$$\mathfrak{b}_{\sigma,\sigma'}(z, \lambda) = (\mathfrak{b}_\sigma(z, \lambda))^{-1} \circ \mathfrak{b}_{\sigma'}(z, \lambda),$$

are called the *transition functions*.

Theorem 3 ([11, Theorem 4.22]). *Let q be not a root of unity. For any $\sigma \in \mathbb{S}^n$ and any transposition $(m, m+1)$, $m = 1, \ldots, n-1$, the transition function*

$$\mathfrak{b}_{\sigma,\sigma\circ(m,m+1)}(z, \lambda) \colon (V_{\lambda_{\sigma_1}}^q \otimes \cdots \otimes V_{\lambda_{\sigma_{m+1}}}^q \otimes V_{\lambda_{\sigma_m}}^q \otimes \cdots \otimes V_{\lambda_{\sigma_n}}^q)_l$$
$$\to (V_{\lambda_{\sigma_1}}^q \otimes \cdots \otimes V_{\lambda_{\sigma_n}}^q)_l$$

equals the operator $P_{V_{\lambda_{\sigma_{m+1}}}^q V_{\lambda_{\sigma_m}}^q} R_{V_{\lambda_{\sigma_{m+1}}}^q V_{\lambda_{\sigma_m}}^q}(e^{2\pi i(z_{\sigma_{m+1}} - z_{\sigma_m})/p})$ *acting in the mth and $(m+1)$st factors. Here* $P_{V_{\lambda_{\sigma_{m+1}}}^q V_{\lambda_{\sigma_m}}^q}$ *is the operator of permutation of the factors* $V_{\lambda_{\sigma_{m+1}}}^q$ *and* $V_{\lambda_{\sigma_m}}^q$.

Corollary 1. *Let q be not a root of unity. For any $\sigma \in \mathbb{S}^n$ and any transposition $(m, m+1)$, $m = 1, \ldots, n-1$, the transition function $\mathfrak{b}_{\sigma,\sigma'}(z,\lambda)$ can be factorized to a transition function*

$$\mathfrak{B}_{\sigma,\sigma'}(z,\lambda) \colon (L^q_{\lambda_{\sigma'_1}} \otimes \cdots \otimes L^q_{\lambda_{\sigma'_n}})_l \to (L^q_{\lambda_{\sigma_1}} \otimes \cdots \otimes L^q_{\lambda_{\sigma_n}})_l.$$

Moreover, for any $\sigma \in \mathbb{S}^n$ and any transposition $(m, m+1)$, $m = 1, \ldots, n-1$, the induced transition function

$$\mathfrak{B}_{\sigma,\sigma\circ(m,m+1)}(z,\lambda) \colon (L^q_{\lambda_{\sigma_1}} \otimes \cdots \otimes L^q_{\lambda_{\sigma_{m+1}}} \otimes L^q_{\lambda_{\sigma_m}} \otimes \cdots \otimes L^q_{\lambda_{\sigma_n}})_l$$
$$\to (L^q_{\lambda_{\sigma_1}} \otimes \cdots \otimes L^q_{\lambda_{\sigma_n}})_l$$

equals the operator $P_{L^q_{\lambda_{\sigma_{m+1}}} L^q_{\lambda_{\sigma_m}}} R_{L^q_{\lambda_{\sigma_{m+1}}} L^q_{\lambda_{\sigma_m}}} (e^{2\pi i(z_{\sigma_{m+1}} - z_{\sigma_m})/p})$ acting in the mth and $(m+1)$th factors.

Corollary 1 follows from Theorems 2 and 3.

Thus, the map $\mathfrak{b}_\sigma(z,\lambda) \colon (V^q_{\lambda_{\sigma_1}} \otimes \cdots \otimes V^q_{\lambda_{\sigma_n}})_l \to \mathfrak{G}(z,\lambda)$ can be factorized to a map

$$\mathfrak{B}_\sigma(z,\lambda) \colon (L^q_{\lambda_{\sigma_1}} \otimes \cdots \otimes L^q_{\lambda_{\sigma_n}})_l \to \mathfrak{G}(z,\lambda), \qquad (3)$$

defined by the same formula (2). Moreover, the image of $\mathfrak{B}_\sigma(z,\lambda)$ does not depend on σ. Transition functions $(\mathfrak{B}_\sigma(z,\lambda))^{-1} \circ \mathfrak{B}_{\sigma'}(z,\lambda)$ are well defined, and $(\mathfrak{B}_\sigma(z,\lambda))^{-1} \circ \mathfrak{B}_{\sigma'}(z,\lambda) = \mathfrak{B}_{\sigma,\sigma'}(z,\lambda)$.

3.5 Hypergeometric Integrals

Fix $p \in \mathbb{C}$, $\operatorname{Re} p < 0$. Let $\operatorname{Im} \mu \neq 0$. Assume that the parameters $z, \lambda \in \mathbb{C}^n$ satisfy the condition $\operatorname{Re}(z_i + \lambda_i) < 0$ and $\operatorname{Re}(z_i - \lambda_i) > 0$ for all $i = 1, \ldots, n$. For a rational weight function $w_l(t,z,\lambda)$, $l \in \mathcal{Z}_l^n$, and a trigonometric weight function $W(t,z,\lambda) \in \mathfrak{G}$, define the *hypergeometric integral* $I(w,W)(z,\lambda)$ by the formula

$$I(w,W)(z,\lambda) = \int_{\substack{\operatorname{Re} t_i = 0, \\ i = 1, \ldots, l}} \Phi_l(t,z,\lambda) w(t,z,\lambda) W(t,z,\lambda) \, d^l t, \qquad (4)$$

where $d^l t = dt_1 \cdots dt_l$.

The hypergeometric integral for generic z, λ and an arbitrary step p with a negative real part is defined by analytic continuation with respect to z, λ and p. This analytic continuation makes sense since the integrand is meromorphic in z, λ, and p. The poles of the integrand are located at the union of hyperplanes

$$t_i = z_k \pm (\lambda_k + sp), \qquad t_i = t_j \pm (1 - sp), \qquad (5)$$

$i, j = 1, \ldots, l$, $k = 1, \ldots, n$, $s \in \mathbb{Z}_{\geq 0}$. We move the parameters z, λ, and p in such a way that the topology of the complement in $\mathbb{C}^l$ to the union of hyperplanes (5) does not change. We deform accordingly the integration cycle (the imaginary subspace) in such a way that it does not intersect the hyperplanes (5) at any moment of the deformation. The deformed integration cycle is called the *deformed imaginary subspace* and is denoted $\mathfrak{I}(z, \lambda)$, $\mathfrak{I}(z, \lambda) \subset \mathbb{C}^l$. Then the analytic continuation of integral (4) is given by

$$I(w, W)(z, \lambda) = \int_{\mathfrak{I}(z,\lambda)} \Phi_l(t, z, \lambda) w(t, z, \lambda) W(t, z, \lambda) \, d^l t \qquad (6)$$

(see Section 5 in Ref. 11).

Theorem 4 ([11, Theorem 5.7]). *Let* $\operatorname{Im} \mu \neq 0$. *For any* $w_l(t, z, \lambda)$, $l \in \mathcal{Z}_l^n$, *and any* $W(t, z, \lambda) \in \mathfrak{G}$, *the hypergeometric integral* (6) *is a univalued meromorphic function of variables* p, z, λ *holomorphic on the set*

$$
\begin{aligned}
&\operatorname{Re} p < 0, \quad \{1, \ldots, l\} \not\subset p\mathbb{Z}, \\
&2\lambda_m - s \notin p\mathbb{Z}, \qquad\qquad m = 1, \ldots, n, s = 1 - l, \ldots, l - 1, \\
&z_k \pm \lambda_k - z_m \pm \lambda_m - s \notin p\mathbb{Z}, \quad k, m = 1, \ldots, n, k \neq m, \\
&\qquad\qquad\qquad\qquad s = 1 - l, \ldots, l - 1,
\end{aligned} \qquad (7)
$$

where we allow arbitrary combinations of $\pm$.

The case $\operatorname{Im} \mu = 0$ is treated in the same manner.

Theorem 5 ([11, Theorem 5.8]). *Let* $\operatorname{Im} \mu = 0$. *For any* $w_l(t, z, \lambda)$, $l \in \mathcal{Z}_l^n$, *and any* $W(t, z, \lambda) \in \mathfrak{G}^{\mathrm{sing}}$, *the hypergeometric integral* (6) *is a univalued meromorphic function of variables* p, z, λ *holomorphic on the set* (7).

For a function $W(t, z, \lambda) \in \mathfrak{G}$, let $\Psi_W(z, \lambda)$ be the following $V_{\lambda_1} \otimes \cdots \otimes V_{\lambda_n}$-valued function

$$\Psi_W(z, \lambda) = \sum_{l_1 + \cdots + l_n = l} I(w_l, W)(z, \lambda) f^{l_1} v_1 \otimes \cdots \otimes f^{l_n} v_n. \qquad (8)$$

Theorem 6 ([11, Corollaries 5.25 and 5.26]). *Let* p, z, λ *be such that conditions* (7) *are satisfied.*

i) *Let* $\operatorname{Im} \mu \neq 0$. *Then for any function* $W \in \mathfrak{G}$, *the function* $\Psi_W(z, \lambda)$ *is a solution of the qKZ equation with values in* $(V_{\lambda_1} \otimes \cdots \otimes V_{\lambda_n})_l$.

ii) *Let* $\operatorname{Im} \mu = 0$. *Then for any function* $W \in \mathfrak{G}^{\mathrm{sing}}$, *the function* $\Psi_W(z, \lambda)$ *is a solution of the qKZ equation with values in* $(V_{\lambda_1} \otimes \cdots \otimes V_{\lambda_n})_l^{\mathrm{sing}}$.

The solutions of the qKZ equation defined by (8) are called the *hypergeometric solutions*.

4 Main Results

Fix natural numbers n, l. We will often assume the following restrictions on the parameters p, z, λ:

$$\operatorname{Re} p < 0, \quad 1 \notin p\mathbb{Z}, \tag{9}$$
$$\{s \mid s \in \mathbb{Z}_{>0}, s < 2\max\{\operatorname{Re}\lambda_1,\ldots,\operatorname{Re}\lambda_n\}, s \leq l\} \cap \{p\mathbb{Z}\} = \varnothing, \tag{10}$$
$$\{2\lambda_m - s \mid s \in \mathbb{Z}_{\geq 0}, s < 2\operatorname{Re}\lambda_m, s < l\} \cap \{p\mathbb{Z}\} = \varnothing,$$
$$m = 1,\ldots,n, \tag{11}$$
$$z_k - z_m \pm (\lambda_k + \lambda_m) + s \notin \{p\mathbb{Z}\}, \quad k,m = 1,\ldots,n, k \neq m,$$
$$s = 1 - l,\ldots,l - 1. \tag{12}$$

Sometimes, in addition we will make the following assumption. Let $\lambda \in \mathbb{C}^n$. For each $i \in \{1,\ldots,n\}$ such that $\lambda_i \notin \Lambda^+$, assume

$$\{1,\ldots,l\} \cap \{p\mathbb{Z}\} = \varnothing, \tag{13}$$
$$\{2\lambda_i - s \mid s = 0,1,\ldots,l-1\} \cap \{p\mathbb{Z}\} = \varnothing. \tag{14}$$

Notice that if the parameters p, z, λ satisfy conditions (9)–(14), then they also satisfy conditions (7).

4.1 Analytic Continuation. The Case $\operatorname{Im}\mu \neq 0$

Let $\Lambda = (\Lambda_1,\ldots,\Lambda_n) \in \mathbb{C}^n$, $l = (l_1,\ldots,l_n) \in \mathbb{Z}_{\geq 0}^n$. An ith coordinate of l is called Λ-*admissible* if either $\Lambda_i \notin \Lambda^+$ or $\Lambda_i \in \Lambda^+$ and $l_i \leq 2\Lambda_i$. The index l is called Λ-*admissible* if all its coordinates are Λ-admissible. Denote $B_\Lambda(l) \subset \{1,\ldots,n\}$ the set of all non-Λ-admissible coordinates of l.

Remark 2. Let p be generic, $\Lambda \in \mathbb{C}^n$, $l \in \mathcal{Z}_l^n$. Then the weight coefficient $c_l(\Lambda)$ is not equal to zero if and only if the index l is Λ-admissible.

Let $m \in \mathcal{Z}_l^n$ and $B_\Lambda(l) \subseteq B_\Lambda(m)$. Then the function $c(\lambda) = c_m(\lambda)/c_l(\lambda)$ is holomorphic at Λ. Moreover, $c(\Lambda) = 0$ if $B_\Lambda(l) \neq B_\Lambda(m)$, and $c(\Lambda)$ is not zero if $B_\Lambda(l) = B_\Lambda(m)$.

Theorem 7. *Let* $\operatorname{Im}\mu \neq 0$. *Let* p *satisfy condition* (9). *Let* $Z, \Lambda \in \mathbb{C}^n$ *satisfy conditions* (10)–(12). *Let* $l, m \in \mathcal{Z}_l^n$. *Assume that for all* $i = 1, \ldots,$ n *either the ith coordinate of* l *or the ith coordinate of* m *is* Λ-admissible, *that is* $B(l) \cap B(m) = \varnothing$. *Then the hypergeometric integral* $I(w_l, W_m)(z,\lambda)$ *is holomorphic at* Z, Λ. *Moreover, there exists a contour of integration* $\mathfrak{I}(Z,\Lambda) \subset \mathbb{C}^l$ *independent on* l, m, *such that for all* z, λ, *in a small neighborhood of* Z, Λ *we have*

$$I(w_l, W_m)(z,\lambda) = \int_{\mathfrak{I}(Z,\Lambda)} \Phi_l(t,z,\lambda) w_l(t,z,\lambda) W_m(t,z,\lambda)\, d^l t.$$

A contour $\mathfrak{I}(Z,\Lambda)$ with properties indicated in Theorem 7 is called an *integration contour associated to Z, Λ.*

Theorem 7 is proved in Section 5.

For $l, m \in \mathcal{Z}_l^n$, introduce a function

$$J_{l,m}(z,\lambda) = c_m(\lambda)I(w_l, W_m)(z,\lambda). \tag{15}$$

Theorem 8. *Let* $\operatorname{Im}\mu \neq 0$. *Let p satisfy condition* (9). *Let $Z, \Lambda \in \mathbb{C}^n$ satisfy conditions* (10)–(12). *Let $l, m \in \mathcal{Z}_l^n$. Then the meromorphic function $J_{l,m}(z,\lambda)$ is holomorphic at Z, Λ. Moreover, if $B_\Lambda(l) \subset B_\Lambda(m)$ and $B_\Lambda(l) \neq B_\Lambda(m)$, then $J_{l,m}(Z,\Lambda) = 0$.*

Theorem 8 is proved in Section 5.

Let $\Lambda \in \mathbb{C}^n$, $l, m \in \mathcal{Z}_l^n$ and $B_\Lambda(l) = B_\Lambda(m)$. Denote as B the set $B_\Lambda(l) = B_\Lambda(m)$. Introduce $\Lambda'(B) = (\Lambda_1', \ldots, \Lambda_n') \in \mathbb{C}^n$ by the rule $\Lambda_i' = \Lambda_i$ if $i \notin B$ and $\Lambda_i' = -\Lambda_i - 1$ if $i \in B$.

Let $l'(B) = (l_1', \ldots, l_n')$, where $l_i' = l_i$ if $i \notin B$ and $l_i' = l_i - 2\Lambda_i - 1$ if $i \in B$. Similarly, let $m'(B) = (m_1', \ldots, m_n')$, where $m_i' = m_i$ if $i \notin B$ and $m_i' = m_i - 2\Lambda_i - 1$ if $i \in B$.

We have $l_1' + \cdots + l_n' = m_1' + \cdots + m_n' = l - 2\sum_{i \in B}\Lambda_i - |B|$. We call this number $l'(B)$.

Theorem 9. *Let* $\operatorname{Im}\mu \neq 0$. *Let p satisfy condition* (9). *Let $Z, \Lambda \in \mathbb{C}^n$ satisfy conditions* (10)–(12). *Let $l, m \in \mathcal{Z}_l^n$ and $B_\Lambda(l) = B_\Lambda(m) = B$. Then*

$$J_{l,m}(Z,\Lambda) = \mathfrak{C}_\Lambda(Z)J_{l'(B),m'(B)}(Z,\Lambda'(B)),$$

where $\mathfrak{C}_\lambda(z)$ is a nonzero holomorphic function at Z given below.

Theorem 9 is proved in Section 5.

Notice that

$$J_{l'(B),m'(B)}(z,\Lambda'(B)) = c_{m'(B)}(\Lambda'(B))I(w_{l'(B)}, W_{m'(B)})(z,\Lambda'(B)),$$

where $I(w_{l'(B)}, W_{m'(B)})(z,\lambda)$ is an $l'(B)$-dimensional hypergeometric integral. The indices $l'(B)$, $m'(B)$ are $\Lambda'(B)$-admissible. By Theorem 7 the integral $I(w_{l'(B)}, W_{m'(B)})(z,\Lambda'(B))$ is well defined. Theorem 9 connects l- and $l'(B)$-dimensional hypergeometric integrals.

Now, we describe the function $\mathfrak{C}_\lambda(z)$. For $k \in \mathbb{Z}_{\geq 0}$, let

$$\psi_k = \frac{-1}{\pi^{k+2}\,k!\,(k+1)!}\Gamma(-(k+1)/p)\Gamma(1/p)^{k+1}\prod_{j=1}^{k+1}\sin(j\pi/p).$$

For $z, \lambda \in \mathbb{C}^n, j \in \{1, \ldots, n\}, k \in \mathbb{Z}_{\geq 0}$, let

$$\phi_{\lambda,j,k}(z)$$

$$= \prod_{s=0}^{k}\left(\prod_{i=0}^{j-1}\frac{\Gamma((z_i - z_j + \lambda_i + \lambda_j - s)/p)}{\Gamma((z_i - z_j - \lambda_i - \lambda_j + s)/p)}\prod_{i=j+1}^{n}\frac{\Gamma((z_j - z_i + \lambda_j + \lambda_i - s)/p)}{\Gamma((z_j - z_i - \lambda_j - \lambda_i + s)/p)}\right).$$

Then

$$\mathfrak{C}_\lambda(z) = \frac{l!}{(l'(B))!} \prod_{j \in B} e^{\mu(2\lambda_j+1)z_j \pi i/p} \psi_{2\lambda_j} \phi_{\lambda,j,2\lambda_j}(z).$$

Consider a square matrix

$$J^l(z, \lambda) = \{J_{l,m}(z, \lambda)\}_{l,m \in \mathcal{Z}_l^n}. \tag{16}$$

Recall that for generic p, we have $c_m(\lambda) \neq 0$ if m is λ-admissible. According to Theorem 4, the matrix (16) is holomorphic if conditions (7) hold. According to Theorem 8, the matrix (16) is holomorphic on a larger set of parameters (9)–(12).

According to Theorem 8, if some of λ_1, ..., λ_n become nonnegative half-integers, then the matrix (16) becomes upper block triangular in the following sense.

Let $\Lambda \in \mathbb{C}^n$. Divide the set of indices $\mathcal{Z}_l^n$ into subsets labeled by subsets of $\{1, \ldots, n\}$. The subset of $\mathcal{Z}_l^n$ corresponding to a subset $B \subset \{1, \ldots, n\}$ consists of all $l \in \mathcal{Z}_l^n$ such that $B_\Lambda(l) = B$. Since the rows and columns of matrix (16) are labeled by elements of $\mathcal{Z}_l^n$, matrix (16) is divided into blocks labeled by pairs $B_1, B_2 \subset \{1, \ldots, n\}$.

Choose any order $<$ on the set of all subsets of $\{1, \ldots, n\}$ such that for any $B_1, B_2 \subset \{1, \ldots, n\}$, $B_1 \subset B_2$, we have $B_1 < B_2$. Theorem 8 says that matrix (16) is upper block triangular at (z, Λ) with respect to this order. Namely, the block corresponding to a pair B_1, B_2 is equal to zero if $B_1 < B_2$.

Theorem 9 describes the diagonal blocks of matrix (16). The diagonal block of $J^l(z, \Lambda)$ corresponding to a subset $B \subset \{1, \ldots, n\}$ coincides (up to multiplication from the left by the nondegenerate diagonal matrix $\{\mathfrak{C}_{\Lambda,m}(z)\}_{m \in \mathcal{Z}_l^n}$) with the diagonal block of the matrix $J^{l'(B)}(z, \Lambda'(B))$ corresponding to the empty subset of $\{1, \ldots, n\}$.

4.2 Analytic Continuation

The case $\operatorname{Im}\mu = 0$ *is treated similarly.*

A function $W \in \mathfrak{G}^{\mathrm{sing}}$ is called Λ-*admissible* if for all non-Λ-admissible $m \in \mathcal{Z}_l^n$, the functions $a_m(\lambda, u)$ in decomposition (1) are equal to zero.

Theorem 10. *Let* $\operatorname{Im}\mu = 0$. *Let* p *satisfy condition* (9). *Let* $Z, \Lambda \in \mathbb{C}^n$ *satisfy conditions* (10)–(12). *Let* $l \in \mathcal{Z}_l^n$. *Let* $W \in \mathfrak{G}^{\mathrm{sing}}$ *be* Λ-*admissible. Then the hypergeometric integral* $I(w_l, W)(z, \lambda)$ *is holomorphic at* Z, Λ. *Moreover, there exists a contour of integration* $\mathfrak{I}(Z, \Lambda) \subset \mathbb{C}^l$ *independent on* l *and* W, *such that for all* z, λ, *in a small neighborhood of* Z, Λ *we have*

$$I(w_l, W)(z, \lambda) = \int_{\mathfrak{I}(Z,\Lambda)} \Phi_l(t, z, \lambda) w_l(t, z, \lambda) W(t, z, \lambda) \, d^l t.$$

A contour $\mathfrak{I}(Z, \Lambda)$ with properties indicated in Theorem 10 is called an *integration contour associated to Z, Λ*.

The proof of Theorem 10 is similar to the proof of Theorem 7.

It follows from the proof that a contour of integration associated to Z, Λ with respect to Theorem 7 is also an integration contour associated to Z, Λ with respect to Theorem 10 and vice versa.

Theorem 11. *Let* $\operatorname{Im} \mu = 0$. *Let* p *satisfy condition* (9). *Let* Z, $\Lambda \in \mathbb{C}^n$ *satisfy conditions* (10)–(12). *Let* $l \in \mathcal{Z}_l^n$, $W \in \mathfrak{G}^{\mathrm{sing}}$. *Then the meromorphic function* $c_l(\lambda) I(w_l, W)(z, \lambda)$ *is holomorphic at* Z, Λ. *Moreover, if* l *is not* Λ-*admissible and* W *is* Λ-*admissible, then the function* $c_l(\lambda) I(w_l, W)(z, \lambda)$ *is equal to zero at* (Z, Λ).

The proof of Theorem 11 is similar to the proof of Theorem 8.

Notice that in Theorem 11 we consider the function $c_l(\lambda) I(w_l, W)(z, \lambda)$ and not the function $c_m(\lambda) I(w_l, W_m)(z, \lambda)$ as in Theorem 8.

4.3 Solutions of qKZ with Values in Irreducible Representations

For $m \in \mathcal{Z}_l^n$, consider a $V_{\lambda_1} \otimes \cdots \otimes V_{\lambda_n}$-valued function

$$\Psi_m(z, \lambda) = \sum_{l_1 + \cdots + l_n = l} J_{l,m}(z, \lambda) f^{l_1} v_1 \otimes \cdots \otimes f^{l_n} v_n. \qquad (17)$$

Corollary 2. *Let* $\operatorname{Im} \mu \neq 0$. *Let* $p \in \mathbb{C}$ *and* $\Lambda \in \mathbb{C}^n$ *satisfy conditions* (9)–(11). *Then the function* $\Psi_m(z, \Lambda)$ *given by* (17) *is a meromorphic solution of the qKZ equation with values in* $(V_{\Lambda_1} \otimes \cdots \otimes V_{\Lambda_n})_l$, *holomorphic for all* z *satisfying condition* (12).

Corollary 2 follows from Theorems 1, 6, and 8.

Let $\Lambda \in \mathbb{C}^n$. For any Λ-admissible $m \in \mathcal{Z}_l^n$, consider a function

$$\Psi_m(z, \Lambda) = \sum I(w_l, W_m)(z, \lambda) f^{l_1} v_1 \otimes \cdots \otimes f^{l_n} v_n, \qquad (18)$$

where the sum is over all Λ-admissible $l \in \mathcal{Z}_l^n$.

Corollary 3. *Let* $\operatorname{Im} \mu \neq 0$. *Let* $p \in \mathbb{C}$ *and* $\Lambda \in \mathbb{C}^n$ *satisfy conditions* (9)–(11). *Then for any* Λ-*admissible* $m \in \mathcal{Z}_l^n$, *the function* $\Psi_m(z, \Lambda)$ *given by* (18) *is a meromorphic solution of the qKZ equation with values in* $(L_{\Lambda_1} \otimes \cdots \otimes L_{\Lambda_n})_l$, *holomorphic for all* z *satisfying condition* (12).

Notice that vectors $f^{l_1} v_1 \otimes \cdots \otimes f^{l_n} v_n$ with Λ-admissible $l \in \mathcal{Z}_l^n$ form a basis in $(L_{\Lambda_1} \otimes \cdots \otimes L_{\Lambda_n})_l$.

Corollary 3 follows from Corollary 2 and Lemma 1.

Let $\operatorname{Im} \mu = 0$. For $W \in \mathfrak{G}^{\mathrm{sing}}$, consider a $V_{\lambda_1} \otimes \cdots \otimes V_{\lambda_n}$-valued function $\Psi_W(z, \lambda)$ given by (8).

Corollary 4. *Let* $\operatorname{Im}\mu = 0$. *Let* $p \in \mathbb{C}$ *and* $\Lambda \in \mathbb{C}^n$ *satisfy conditions* (9)–(11). *Let* $W \in \mathfrak{G}^{\mathrm{sing}}$ *be* Λ-*admissible. Then the function* $\Psi_W(z, \Lambda)$ *given by* (8) *is a meromorpfic solution of the qKZ equation with values in* $(V_{\Lambda_1} \otimes \cdots \otimes V_{\Lambda_n})_l^{\mathrm{sing}}$, *holomorphic for all* z *satisfying condition* (12).

Corollary 4 follows from Theorems 1, 6, and 10.

Let $\Lambda \in \mathbb{C}^n$, $W \in \mathfrak{G}^{\mathrm{sing}}$. Consider a function

$$\Psi_W(z, \Lambda) = \sum I(w_l, W)(z, \lambda) f^{l_1} v_1 \otimes \cdots \otimes f^{l_n} v_n, \tag{19}$$

where the sum is over Λ-admissible $l \in \mathcal{Z}_l^n$.

Corollary 5. *Let* $\operatorname{Im}\mu = 0$. *Let* $p \in \mathbb{C}$ *and* $\Lambda \in \mathbb{C}^n$ *satisfy conditions* (9)–(11). *Let* $W \in \mathfrak{G}^{\mathrm{sing}}$ *be* Λ-*admissible. Then the function* $\Psi_W(z, \Lambda)$ *given by* (19) *is a meromorphic solution of the qKZ equation with values in* $(L_{\Lambda_1} \otimes \cdots \otimes L_{\Lambda_n})_l^{\mathrm{sing}}$, *holomorphic for all* z *satisfying condition* (12).

Corollary 5 follows from Corollary 4 and Lemma 1.

The solutions of the qKZ equation defined by formulas (18), and (19) are called the *hypergeometric solutions*.

4.4 Determinant of the Hypergeometric Pairing

Let $\operatorname{Im}\mu \neq 0$. The determinant of the matrix $J^l(z, \lambda)$ defined by (16) is given by

$$\det\big(J_l(z, \lambda)\big) = (2i)^{l\binom{n+l-1}{n-1}} (l!)^{\binom{n+l-1}{n-1}} (e^\mu - 1)^{-2\sum_{m=1}^n \lambda_m \binom{n+l-1}{n}/p + 2n\binom{n+l-1}{n+1}/p}$$

$$\times \exp\left(\mu \sum_{m=1}^n \frac{z_m}{p} \binom{n+l-1}{n} \right)$$

$$\times \exp\left((\mu + \pi i)\left(\sum_{m=1}^n \frac{\lambda_m}{p} \binom{n+l-1}{n} - \frac{n}{p}\binom{n+l-1}{n+1} \right) \right)$$

$$\times \prod_{s=0}^{l-1} \left[\Gamma\left(1 + \frac{1}{p}\right)^n \Gamma\left(1 + \frac{s+1}{p}\right)^{-n} \prod_{m=1}^n \frac{\pi}{\Gamma(1 - (2\lambda_m - s)/p)} \right.$$

$$\times \left. \prod_{\substack{k,m=1 \\ k<m}}^n \frac{\Gamma((z_k + \lambda_k - z_m + \lambda_m - s)/p)}{\Gamma((z_k - \lambda_k - z_m - \lambda_m + s)/p)} \right]^{\binom{n+l-s-2}{n-1}},$$

cf. Theorem 5.14 of Ref. [11].

For $\Lambda \in \mathbb{C}^n$, consider a matrix

$$J_{\mathrm{adm}}^l(z, \lambda) = \{J_{l,m}(z, \lambda)\}, \tag{20}$$

where $l, m \in \mathcal{Z}_l^n$ run through the set of all Λ-admissible indices.

By Theorem 7, the matrix $J_{\mathrm{adm}}^l(z, \lambda) = J_{\mathrm{adm}}^l(z, \lambda, p)$ is holomorphic at (z, Λ) if parameters p, z, Λ satisfy conditions (9)–(12).

Theorem 12. *Let* $\operatorname{Im}\mu \neq 0$. *Let* $p \in \mathbb{C}$ *and* $z, \Lambda \in \mathbb{C}^n$ *satisfy conditions* (9)–(14). *Then the matrix* $J^l_{\mathrm{adm}}(z, \Lambda)$ *is nondegenerate. Moreover, for generic* p,

$$\det\left(J^l_{\mathrm{adm}}(z, \Lambda)\right) = \prod_{A \subset B(\Lambda)} \left(C_\Lambda(A)(z) \det\left(J^{l'(A)}(z, \Lambda'(A))\right)\right)^{(-1)^{|A|}}, \quad (21)$$

where $B(\Lambda)$ *is the set of all* $i \in \{1, \dots, n\}$ *such that* $\Lambda_i \in \Lambda^+$, *and the function* $C_\Lambda(A)(z)$ *is given by*

$$C_\Lambda(A)(z) = \prod_{\substack{m, \\ B(m)\,\sup\,seteq A}} \prod_{i \in A} \left(\frac{l!}{(l - 2\Lambda_i - 1)!} \psi_{2\Lambda_i} \phi_{\Lambda, i, 2\Lambda_i}(z)\right).$$

Theorem 12 is proved in Section 5.

Consider a pairing

$$s_\mu(z) \colon (L^q_{\Lambda_1} \otimes \cdots \otimes L^q_{\Lambda_n})_l \otimes \left((L_{\Lambda_1} \otimes \cdots \otimes L_{\Lambda_n})_l\right)^* \to \mathbb{C},$$

defined by

$$(f^{m_1} v^q_1 \otimes \cdots \otimes f^{m_n} v^q_n) \otimes (f^{l_1} v_1 \otimes \cdots \otimes f^{l_n} v_n)^* \mapsto J_{l,m}(z, \Lambda) \in \mathbb{C},$$

where $l, m \in \mathcal{Z}^n_l$ are Λ-admissible.

Corollary 6. *Let* $\operatorname{Im}\mu \neq 0$. *Let* $p \in \mathbb{C}$ *and* $z, \Lambda \in \mathbb{C}^n$ *satisfy conditions* (9)–(14). *Then the pairing* $s_\mu(z)$ *is well defined and nondegenerate.*

Corollary 6 follows from Theorem 12.

Corollary 7. *Let* $\operatorname{Im}\mu \neq 0$. *Let* $p \in \mathbb{C}$ *and* $\Lambda \in \mathbb{C}^n$ *satisfy conditions* (9)–(11), (13) *and* (14). *Then any solution of the qKZ equation with values in* $(L_{\Lambda_1} \otimes \cdots \otimes L_{\Lambda_n})_l$ *is a linear combination of the hypergeometric solutions* (18) *with* p-*periodic coefficients.*

Corollary 7 follows from Corollary 6.

Similarly, for a permutation $\sigma \in \mathbb{S}^n$, consider a pairing

$$s^\sigma_\mu(z) \colon (L^q_{\Lambda_{\sigma_1}} \otimes \cdots \otimes L^q_{\Lambda_{\sigma_n}})_l \otimes \left((L_{\Lambda_1} \otimes \cdots \otimes L_{\Lambda_n})_l\right)^* \to \mathbb{C},$$

defined by

$$(f^{m_{\sigma_1}} v^q_{\sigma_1} \otimes \cdots \otimes f^{m_{\sigma_n}} v^q_{\sigma_n}) \otimes (f^{l_1} v_1 \otimes \cdots \otimes f^{l_n} v_n)^*$$
$$\mapsto c_m(\Lambda) I(w_l, W^\sigma_m)(z, \Lambda) \in \mathbb{C},$$

where $l, m \in \mathcal{Z}^n_l$ are Λ-admissible.

Corollary 8. *Let* $\operatorname{Im}\mu \neq 0$. *Let* $p \in \mathbb{C}$ *and* $z, \Lambda \in \mathbb{C}^n$ *satisfy conditions* (9)–(14). *Then for any permutation* $\sigma \in \mathbb{S}^n$, *the pairing* $s^\sigma_\mu(z)$ *is well defined and nondegenerate.*

Corollary 8 follows from Corollaries 6 and 1.

For a permutation $\sigma \in \mathbb{S}^n$ and a Λ-admissible index $m \in Z_l^n$, consider a function

$$\Psi_m^\sigma(z, \Lambda) = \sum I(w_l, W_m^\sigma)(z, \lambda) f^{l_1} v_1 \otimes \cdots \otimes f^{l_n} v_n, \qquad (22)$$

where the sum is over all Λ-admissible $l \in Z_l^n$.

Corollary 9. *Let* $\operatorname{Im}\mu \neq 0$. *Let* $\sigma \in \mathbb{S}^n$ *be a permutation. Let* $p \in \mathbb{C}$ *and* $\Lambda \in \mathbb{C}^n$ *satisfy conditions* (9)–(11), *and* (13) *and* (14). *Then any solution of the qKZ equation with values in* $(L_{\Lambda_1} \otimes \cdots \otimes L_{\Lambda_n})_l$ *is a linear combination of the hypergeometric solutions* (22) *with p-periodic coefficients.*

Corollary 9 follows from Corollary 8.

The map $s_\mu^\sigma(z)$ induces a map

$$\tilde{s}_\mu^\sigma(z) \colon (L_{\Lambda_{\sigma_1}}^q \otimes \cdots \otimes L_{\Lambda_{\sigma_n}}^q)_l \to (L_{\Lambda_1} \otimes \cdots \otimes L_{\Lambda_n})_l. \qquad (23)$$

This map is an isomorphism of vector spaces for all z satisfying (12). For a fixed vector $v^q \in (L_{\Lambda_{\sigma_1}}^q \otimes \cdots \otimes L_{\Lambda_{\sigma_n}}^q)_l$, the $(L_{\Lambda_1} \otimes \cdots \otimes L_{\Lambda_n})_l$-valued function $\tilde{s}_\mu^\sigma(z) v^q$ is a solution of the qKZ equation. This construction identifies the space $\mathcal{S}$ of all meromorphic solutions to the qKZ equation with the space $(L_{\Lambda_{\sigma_1}}^q \otimes \cdots \otimes L_{\Lambda_{\sigma_n}}^q)_l \otimes F$ where F is the field of scalar meromorphic functions in $z_1, \ldots, z_n$, p-periodic with respect to each of the variables,

$$\iota_\sigma \colon (L_{\Lambda_{\sigma_1}}^q \otimes \cdots \otimes L_{\Lambda_{\sigma_n}}^q)_l \otimes F \to \mathcal{S}.$$

Let now $\operatorname{Im}\mu = 0$. Consider a pairing

$$s_\mu(z) \colon (L_{\Lambda_1}^q \otimes \cdots \otimes L_{\Lambda_n}^q)_l^{\operatorname{sing}} \otimes ((L_{\Lambda_1} \otimes \cdots \otimes L_{\Lambda_n})_l)^* \to \mathbb{C},$$

defined by

$$v^q \otimes (f^{l_1} v_1 \otimes \cdots \otimes f^{l_n} v_n)^* \mapsto I(w_l, \mathfrak{B}_{\mathrm{id}}(z, \Lambda) v^q)(z, \Lambda) \in \mathbb{C},$$

where $l \in Z_l^n$ is admissible, $v^q \in (L_{\Lambda_1}^q \otimes \cdots \otimes L_{\Lambda_n}^q)_l^{\operatorname{sing}}$. The map $\mathfrak{B}_{\mathrm{id}}(z, \Lambda)$ is defined in (3), the integral $I(w_l, \mathfrak{B}_{\mathrm{id}}(z, \Lambda) v^q)(z, \Lambda)$ is taken over an integration contour associated to (z, Λ).

The map $s_\mu(z)$ induces a map

$$\tilde{s}_\mu(z) \colon (L_{\Lambda_1}^q \otimes \cdots \otimes L_{\Lambda_n}^q)_l^{\operatorname{sing}} \to (L_{\Lambda_1} \otimes \cdots \otimes L_{\Lambda_n})_l. \qquad (24)$$

By Corollary 5, the image of $\tilde{s}_\mu(z)$ belongs to $(L_{\Lambda_1} \otimes \cdots \otimes L_{\Lambda_n})_l^{\operatorname{sing}}$.

Theorem 13. *Let* $\operatorname{Im}\mu = 0$. *Let* $p \in \mathbb{C}$ *and* $z, \Lambda \in \mathbb{C}^n$ *satisfy conditions* (9)–(14). *Let* $\sum_{m=1}^n 2\Lambda_m - s \notin \{p\mathbb{Z}_{<0}\}$ *for* $s = l-1, \ldots, 2l-2$. *Then the map*

$$\tilde{s}_\mu(z) \colon (L_{\Lambda_1}^q \otimes \cdots \otimes L_{\Lambda_n}^q)_l^{\operatorname{sing}} \to (L_{\Lambda_1} \otimes \cdots \otimes L_{\Lambda_n})_l^{\operatorname{sing}}$$

is an isomorphism of vector spaces.

Theorem 13 is proved similarly to Theorem 12 using Theorem 5.15 in Ref. 11.

Corollary 10. *Let* $\operatorname{Im}\mu = 0$. *Let* $p \in \mathbb{C}$ *and* $\Lambda \in \mathbb{C}^n$ *satisfy conditions* (9)– (11) *and* (13)–(14). *Let* $\sum_{m=1}^{n} 2\Lambda_m - s \notin \{p\mathbb{Z}_{<0}\}$ *for* $s = l - 1, \ldots, 2l - 2$. *Then any solution of the qKZ equation with values in* $(L_{\Lambda_1} \otimes \cdots \otimes L_{\Lambda_n})_l^{\text{sing}}$ *is a linear combination of the hypergeometric solutions* (19) *with p-periodic coefficients.*

Corollary 10 follows from Theorem 13.

4.5   Asymptotic Solutions

Let U be a domain in $\mathbb{C}^n$ and let $M_1, \ldots, M_n$ be sl$_2$-modules. A basis Ψ_1, $\ldots, \Psi_N$ of solutions to the qKZ equation with values in $(M_1 \otimes \cdots \otimes M_n)_l$, is called an *asymptotic solution* in the domain U if

$$\Psi_j(z) = \exp\left(\sum_{m=1}^{n} a_{mj} \frac{z_m}{p} \right) \prod_{\substack{m,k=1 \\ m<k}}^{n} (z_k - z_m)^{b_{jkm}} \left(v_j + o(1) \right),$$

where a_{mj} and b_{jkm} are suitable numbers, $v_1, \ldots, v_N$ are vectors that form a basis in $(M_1 \otimes \cdots \otimes M_n)_l$ and $o(1)$ tends to zero as z tends to infinity in U. The domain U is called an *asymptotic zone*.

Let $\Psi_1^1(z), \ldots, \Psi_N^1(z)$ and $\Psi_1^2(z), \ldots, \Psi_N^2(z)$ be asymptotic solutions in asymptotic zones $U_1 \in \mathbb{C}^n$ and $U_2 \in \mathbb{C}^n$, respectively. Then

$$\Psi_i^1(z) = \sum_{j=1}^{N} T_i^j(z) \Psi_j^2(z), \quad i = 1, \ldots, N$$

for suitable meromorphic functions $T_i^j(z)$. The matrix-valued function $T(z) = (T_i^j(z))$ is p-periodic with respect to variables $z_1, \ldots, z_n$ and has a nonzero determinant for generic z. The function $T(z)$ is called *the transition function between the asymptotic solutions* Ψ^1 *and* Ψ^2.

We describe the asymptotic zones, asymptotic solutions, and transition functions of the qKZ equation with values in $(L_{\lambda_1} \otimes \cdots \otimes L_{\lambda_n})_l$. Our results are parallel to the results of [11], where the asymptotic zones, asymptotic solutions and transition functions are described for the qKZ equation with values in $(V_{\lambda_1} \otimes \cdots \otimes V_{\lambda_n})_l$.

For a permutation $\sigma \in \mathbb{S}^n$, define an asymptotic zone in $\mathbb{C}^n$,

$$U_\sigma = \{z \in \mathbb{C}^n \mid \operatorname{Re} z_{\sigma_1} \ll \cdots \ll \operatorname{Re} z_{\sigma_n}\}. \tag{25}$$

Say that $z \to \infty$ in U_σ if $\operatorname{Re}(z_{\sigma_m} - z_{\sigma_{m+1}}) \to -\infty$ for all $m = 1, \ldots, n-1$.

For each permutation $\sigma \in \mathbb{S}^n$, let $\Psi^\sigma(\lambda)$ be the set of solutions $\{\Psi_m^\sigma\}$ of the qKZ equation with values in $(L_{\lambda_1} \otimes \cdots \otimes L_{\lambda_n})_l$ given by (22). Here $m \in \mathcal{Z}_l^n$ runs through the set of all λ-admissible indices.

Theorem 14. *Let* $\mathrm{Im}\,\mu \neq 0$. *Let* $p \in \mathbb{C}$, $\Lambda \in \mathbb{C}^n$ *satisfy the conditions* (9)–(11). *Then for any permutation* $\sigma \in \mathbb{S}^n$, *the set of solutions* $\Psi^\sigma(\Lambda)$ *forms a basis of solutions. Moreover, the basis of solutions* $\Psi^\sigma(\Lambda)$ *is an asymptotic solution of the qKZ equation with values in* $(L_{\Lambda_1} \otimes \cdots \otimes L_{\Lambda_n})_l$ *in the asymptotic zone* U_σ. *Namely,*

$$\Psi_l^\sigma(z) = \Xi_l \exp\left(\mu \sum_{m=1}^n l_m \frac{z_m}{p} \right) \prod_{\substack{k,m=1 \\ k<m}}^n \left(\frac{z_{\sigma_k} - z_{\sigma_m}}{p} \right)^{2(l_{\sigma_k}\Lambda_{\sigma_m} + l_{\sigma_m}\Lambda_{\sigma_k} - l_{\sigma_k}l_{\sigma_m})/p}$$
$$\times \left(f^{l_1}v_1 \otimes \cdots \otimes f^{l_n}v_n + o(1) \right)$$

as $z \to \infty$ *in* U_σ *so that at any moment condition* (12) *is satisfied. Here* $|\arg((z_{\sigma_k} - z_{\sigma_m})/p)| < \pi$ *and* Ξ_l *is a constant independent of the permutation* σ *and given by*

$$\Xi_l = (2i)^l \, |\,\Gamma(-1/p)^{-l} \prod_{m=1}^n \left[(e^\mu - 1)^{(l_m(l_m-1) - 2l_m\Lambda_m)/p} \right.$$
$$\times \exp\left(\frac{(\mu + \pi i)(l_m\Lambda_m - l_m(l_m - 1)/2)}{p} \right)$$
$$\left. \times \prod_{s=1}^{l_m-1} \Gamma\left(\frac{2\Lambda_m - s}{p} \right) \Gamma\left(-\frac{s+1}{p} \right) \right],$$

where $0 \leq \arg(e^\mu - 1) < 2\pi$.

The proof of Theorem 14 is the same as the proof of Theorem 6.4 in Ref. 11.

Notice that the asymptotics of the basis of solutions $\Psi^\sigma(\Lambda)$ determine the basis uniquely. Namely, if a basis of solutions meromorphically depends on parameters μ, z, λ and has asymptotics in U_σ described in Theorem 14, then such a basis coincides with the basis $\Psi^\sigma(\Lambda)$; see the remark after Theorem 6.4 in Ref. 11.

Using the isomorphism $\iota_\sigma \colon (L_{\Lambda_{\sigma_1}}^q \otimes \cdots \otimes L_{\Lambda_{\sigma_n}}^q)_l \otimes F \to \mathcal{S}$, defined in Section 4.4, we can identify the basis of solutions $\Psi^\sigma(\Lambda)$ with a basis of $(L_{\Lambda_{\sigma_1}}^q \otimes \cdots \otimes L_{\Lambda_{\sigma_n}}^q)_l \otimes F$, considered as a vector space over F. Then the transition function between two asymptotic solutions $\Psi^\sigma(\Lambda)$ and $\Psi^\nu(\Lambda)$ is identified with an F-linear map $(L_{\Lambda_{\nu_1}}^q \otimes \cdots \otimes L_{\Lambda_{\nu_n}}^q)_l \otimes F \to (L_{\Lambda_{\sigma_1}}^q \otimes \cdots \otimes L_{\Lambda_{\sigma_n}}^q)_l \otimes F$.

Corollary 11. *Let* $\mathrm{Im}\,\mu \neq 0$. *Let* $p \in \mathbb{C}$, $\Lambda \in \mathbb{C}^n$ *satisfy the conditions* (9)–(12), (13) *and* (14). *Then for any permutation* $\sigma \in \mathbb{S}^n$ *and a simple transposition* $(m, m+1)$, *the transition function between asymptotic solutions* $\Psi^\sigma(\Lambda)$ *and* $\Psi^{\sigma\circ(m,m+1)}(\Lambda)$,

$$(L_{\Lambda_{\sigma_1}}^q \otimes \cdots \otimes L_{\Lambda_{\sigma_{m+1}}}^q \otimes L_{\Lambda_{\sigma_m}}^q \otimes \cdots \otimes L_{\Lambda_{\sigma_n}}^q)_l \otimes F$$
$$\to (L_{\Lambda_{\sigma_1}}^q \otimes \cdots \otimes L_{\Lambda_{\sigma_n}}^q)_l \otimes F,$$

equals the operator $P_{L^q_{\lambda_{\sigma_{m+1}}} L^q_{\lambda_{\sigma_m}}} R_{L^q_{\lambda_{\sigma_{m+1}}} L^q_{\lambda_{\sigma_m}}} (e^{2\pi i (z_{\sigma_{m+1}} - z_{\sigma_m})/p})$ *acting in the mth and* $(m+1)$*th factors.*

Corollary 11 follows from Theorem 14 and Corollary 1.

Theorem 14 allows us to obtain another formula for the determinant of the hypergeometric pairing (cf. formula (21)).

Theorem 15 (Joint with V. Tarasov). *Let* $\operatorname{Im}\mu \neq 0$. *Let* $p \in \mathbb{C}$ *and* z, $\Lambda \in \mathbb{C}^n$ *satisfy conditions* (9)–(14). *Then the matrix* $J^l_{\mathrm{adm}}(z, \Lambda)$ *defined in* (20) *is nondegenerate. Moreover, for generic* p,

$$\det(J^l_{\mathrm{adm}}(z, \Lambda)) = \mathcal{D}_l(\Lambda) \exp\left(\mu \sum_{m=1}^n D_m(\Lambda) \frac{z_m}{p}\right)$$

$$\times \prod_{\substack{k,m=1 \\ k<m}}^n \prod_{s=0}^{d_{km}} \left(\frac{\Gamma((z_k + \lambda_k - z_m + \lambda_m - s)/p)}{\Gamma((z_k - \lambda_k - z_m - \lambda_m - s)/p)}\right)^{E_{km}(s,\Lambda)},$$

where $d_{km} = \min\{l, \dim L_{\Lambda_k} - 1, \dim L_{\Lambda_m} - 1\}$, *the functions* $\mathcal{D}_l(\Lambda)$, $D_m(\Lambda)$ *and* $E_{km}(s, \Lambda)$ *are given below.*

For $m = 1, \ldots, n$ and $\Lambda \in \mathbb{C}^n$, set

$$D_m(\Lambda) = \sum_{r=1}^{\min\{l, \dim L_{\Lambda_m} - 1\}} (r \dim(L_{\Lambda_1} \otimes \cdots \otimes \widehat{L_{\Lambda_m}} \otimes \cdots \otimes L_{\Lambda_n})_{l-r}).$$

For $k, m = 1, \ldots, n$, $s \in \mathbb{Z}_{\geq 0}$ and $\Lambda \in \mathbb{C}^n$, set

$$E_{km}(s, \Lambda) = \sum_r (\dim(L_{\Lambda_k} \otimes L_{\Lambda_m})_r - s - 1)$$
$$\times \dim(L_{\Lambda_1} \otimes \cdots \otimes \widehat{L_{\Lambda_k}} \otimes \cdots \otimes \widehat{L_{\Lambda_m}} \otimes \cdots \otimes L_{\Lambda_n})_{l-r},$$

where the sum is over $r = s+1, \ldots, \min\{l, \dim L_{\Lambda_k} + \dim L_{\Lambda_k} - 1 - s\}$.
Finally, for $\Lambda \in \mathbb{C}^n$, set

$$\mathcal{D}_l(\Lambda) = \prod_l c_l \Xi_l,$$

where the product is over Λ-admissible indices $l \in \mathcal{Z}^n_l$.

The proof of Theorem 15 is the same as the proof of Theorem 6.4 in Ref. 11.

5   Proofs

In this section we collected the proofs of statements from Section 4.

5.1 Proof of Theorem 7

Let $\mathrm{Im}\,\mu \neq 0$. Fix a rational weight function $w = w_l$, and a trigonometric weight function $W = W_m \in \mathfrak{G}$, such that $B(l) \cap B(m) = \varnothing$.

For $z, \lambda \in \mathbb{C}^n$ such that $\mathrm{Re}(z_i + \lambda_i) < 0$ and $\mathrm{Re}(z_i - \lambda_i) > 0$ for all $i = 1, \dots, n$, the integral $I(w, W)(z, \lambda)$ is defined by formula (4),

$$I(w, W)(z, \lambda) = \int_{\substack{\mathrm{Re}\,t_i = 0, \\ i = 1, \dots, l}} f(t, z, \lambda)\, d^l t,$$

where $f(t, z, \lambda) = \Phi_l(t, z, \lambda) w(t, z, \lambda) W(t, z, \lambda)$. The integrand f has simple poles at

$$t_i = z_j \pm (\lambda_j + kp), \quad k \in \mathbb{Z}_{\geq 0}, i = 1, \dots, l, j = 1, \dots, n, \qquad (26)$$

and at

$$t_i - t_j = \pm(1 - kp), \quad k \in \mathbb{Z}_{\geq 0}, i, j = 1, \dots, l, i < j. \qquad (27)$$

On the complex line the poles of the first type and the integration contour can be represented as in Figure 22.1.

The integral $I(w, W)(z, \lambda) = I(w, W)(z, \lambda, p)$ is a meromorpic function of z, λ, p.

Fix $Z, \Lambda \in \mathbb{C}^n$, $P \in \mathbb{C}$, satisfying conditions (9)–(12). Our goal is to prove that $I(w, W)(Z, \Lambda, P)$ is well defined and is given by the integral over a suitable cycle.

We fix parameters z^0, λ^0, p^0 in such a way that $\mathrm{Re}\,\lambda_k^0 < 0$, p^0 is a negative number with large absolute value and $|\mathrm{Im}(z_k^0 + \Lambda_k - z_m^0 - \Lambda_m)|$ is large for k, $m = 1, \dots, n$ and $k \neq m$. Namely, let $p^0 = -2\sum_{j=1}^n |\Lambda_j| - 1$, $z_k^0 = i(-\mathrm{Im}\,\Lambda_k + 3kA)$, $\lambda_k^0 = -1 + i\,\mathrm{Im}\,\Lambda_k$, and A is a large real number such that $A > 2|\Lambda_k|$, $k = 1, \dots, n$.

The proof is done according to the following plan. First, we represent the integral $I(w, W)(z^0, \lambda^0, p^0)$ as a sum of new integrals of different dimensions. Then we analytically continue the function $I(w, W)(z, \lambda, p)$ from z^0, λ^0, p^0 to z^0, Λ, p^0. Then we continue the function $I(w, W)(z, \Lambda, p)$ analytically from z^0, Λ, p^0 to z^0, Λ, P. In the last step of the proof we analytically continue the function $I(w, W)(z, \Lambda, P)$ from z^0, Λ, P to Z, Λ, P.

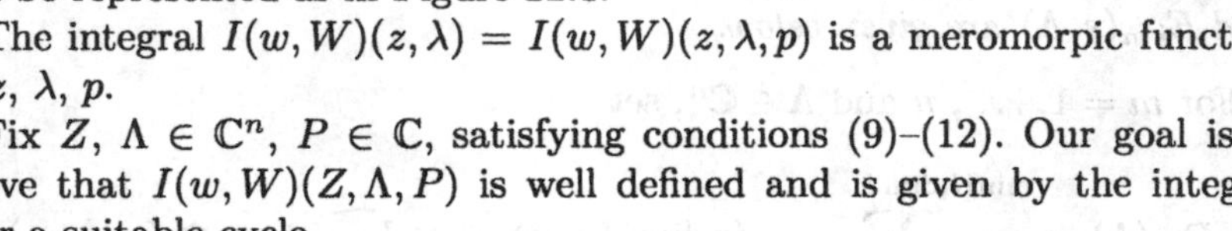

| $z_1+\lambda_1+2p$ | $z_1+\lambda_1+p$ | $z_1+\lambda_1$ | | $z_1-\lambda_1$ | $z_1-\lambda_1-p$ | $z_1-\lambda_1-2p$ |

| $z_2+\lambda_2+2p$ | $z_2+\lambda_2+p$ | $z_2+\lambda_2$ | | $z_2-\lambda_2$ | $z_2-\lambda_2-p$ | $z_2-\lambda_2-2p$ |

$\cdots$ $t_1,\dots,t_l$ $\cdots$

FIGURE 22.1.

For $k \in \mathbb{Z}_{\geq 0}^n$, set $k = k_1 + \cdots + k_n$ and define the multiple residues of f by the formula

$$\mathrm{res}_k \, f = \mathrm{res}_{t_k = z_n + \lambda_n - k_n + 1} \cdots \mathrm{res}_{t_{k_1 + k_2} = z_2 + \lambda_2 - k_2 + 1} \cdots \mathrm{res}_{t_{k_1 + 1} = z_2 + \lambda_2}$$

$$\mathrm{res}_{t_{k_1} = z_1 + \lambda_1 - k_1 + 1} \cdots \mathrm{res}_{t_2 = z_1 + \lambda_1 - 1} \, \mathrm{res}_{t_1 = z_1 + \lambda_1} \, f. \quad (28)$$

If $k > l$, then we set $\mathrm{res}_k \, f = 0$.

The function $\mathrm{res}_k \, f$ is a function of variables $t_{k+1}, \ldots, t_l; z, \lambda$. In particular, for $k = (0, \ldots, 0)$, we have $\mathrm{res}_k \, f = f$.

Lemma 3. *Let k, $k' \in \mathbb{Z}_{\geq 0}^n$. If k' can be obtained by a permutation of components of k then $\mathrm{res}_k \, \bar{f} = \mathrm{res}_{k'} \, f$.*

Lemma 3 follows from the symmetry of the integrand f, see Remark 1 in Section 3.2.

Lemma 4. *For all $i, j = 1, \ldots, l$, and for any $a \in \mathbb{C}$, we have*

$$\mathrm{res}_{t_j = a} \, \mathrm{res}_{t_i = a} \, f = 0.$$

Proof. Recall that $f = \Phi_l w W$. On the hyperplane $t_i = t_j$, any function that is symmetric with respect to either rational or trigonometric action is identically equal to zero. In particular, w is equal to zero at the hyperplane $t_i = t_j$. Since all poles of f are at most of first order, we have the statement of the lemma. $\qquad \square$

Lemma 5. *Let $k \in \mathbb{Z}_{\geq 0}^n$. The function $\mathrm{res}_k \, f$ is holomorpic for all t_{k+1}, $\ldots, t_l; z, \lambda, p$ satisfying conditions (9) and such that*

$$t_i \neq z_j \pm (\lambda_j + sp), \qquad s \in \mathbb{Z}_{\geq 0}, i = k+1, \ldots, l, j = 1, \ldots, n,$$
$$t_i - t_j \neq \pm(1 - sp), \qquad s \in \mathbb{Z}_{\geq 0}, i, j = k+1, \ldots, l, i < j,$$
$$t_i \neq z_j + \lambda_j - m \pm (1 - sp), \qquad s \in \mathbb{Z}_{\geq 0}, i = k+1, \ldots, l, j = 1, \ldots, n,$$
$$m = 0, \ldots, k_j,$$
$$z_i + \lambda_i - z_j \pm (\lambda_j + sp) \neq m, \qquad s \in \mathbb{Z}_{\geq 0}, i, j = 1, \ldots, k, m = 0, \ldots, k, i \neq j,$$
$$2\lambda_i + sp \neq m, \qquad s \in \mathbb{Z}_{\geq 0}, i = 1, \ldots, k, m = 0, \ldots, k_i.$$

Lemma 5 follows from formulas (26) and (27) for the poles of f, and formula (28) for the function $\mathrm{res}_k \, f$.

For $u \in \mathbb{R}^n$, define a curve $\mathcal{C}_u = \{\mathcal{C}_u(x) \in \mathbb{C} \mid x \in \mathbb{R}\}$ with the following properties. The curve $\mathcal{C}_u$ consists of $2n + 1$ line segments and none of the line segments is horizontal (i.e., no line-segment is parallel to the real line). There are two line segments (rays) which go to $+i\infty$ and $-i\infty$, these two rays are parts of the imaginary line. There are n vertical segments with real coordinates $u_1, \ldots, u_n$ and imaginary coordinates close to $\mathrm{Im}(z_1^0 + \lambda_1^0)$, $\ldots, \mathrm{Im}(z_n^0 + \lambda_n^0)$.

The precise formulas defining $\mathcal{C}_u$ are as follows. For $x \leq A$ and for $x \geq (3n - 1)A$, let $\mathcal{C}_u(x) = ix$. For $(3m - 1)A \leq x \leq (3m + 1)A$, $m = 1$,

..., n, let $C_u(x) = u_m + ix$, and this segment is called the mth *vertical segment*. For $m = 0, \ldots, n$, let $\{C_u(x), (3m+1)A \leq x \leq (3m+2)A\}$ be a parametrization of the line segment connecting $C_u((3m+1)A)$ and $C((3m+2)A)$.

See an example of such a curve on Figure 22.2.

Lemma 6. *Let $, \Lambda, z^0, \lambda^0, p^0$ be as above. For any sufficiently small ϵ,*

$$I(w, W)(z^0, \lambda^0, p^0)$$

$$= \sum_k \frac{l!}{n!, k_1! \cdots k_n!(l-k)!} \int_{\substack{t_j \in C, \\ j=k+1,\ldots,l}} \mathrm{res}_k \, f(t, z^0, \lambda^0, p^0) \, d^{l-k}t, \quad (29)$$

where $d^{l-k}t = dt_{k+1} \cdots dt_l$. Here C denotes the curve C_u with $u_m = \min\{-\mathrm{Re}(2\Lambda_m - \lambda_m^0) - \epsilon, 0\}$, $m = 1, \ldots, n$, and the sum is over all $k \in \mathbb{Z}_{\geq 0}^n$ such that $k \leq l$, and $k_m \leq 2\,\mathrm{Re}\,\Lambda_m + 1$ for $m = 1, \ldots, n$.

Proof. The integrand decays exponentially with respect to all $t_1, \ldots, t_l$. We move the contour of integration with respect to each variable t_i to the left avoiding the poles at $t_i - t_j = (1 - kp)$.

Initially, each of the variables runs through the imaginary line. We deform the contours of integration with respect to $t_1, \ldots, t_l$ in such a way that at every moment the contours of integration with respect to $t_1, \ldots, t_l$ are curves C_{u^i} with different parameters $u^i \in \mathbb{R}^n$.

First, we separate the contours of integration with respect to different variables by small distances as follows.

$$I(w, W)(z^0, \lambda^0, p^0) = \int_{\substack{\mathrm{Re}\,t_j=0, \\ j=i,\ldots,l}} f \, d^l t = \int_{\substack{t_j \in C_{u^j}, \\ j=1,\ldots,l}} f \, d^l t,$$

where $u^j = ((j-l-1)\rho, \ldots, (j-l-1)\rho) \in \mathbb{R}^n$, and ρ is a small positive number.

We deform the integration cycle $\{t \in \mathbb{C}^l \mid t_j \in C_{u^j}, j = 1, \ldots, l\}$ changing $u^j, j = 1, \ldots, l$. Namely, we move the vertical segments of contours of integrations $C_{u^j}, j = 1, \ldots, n$, to the left, until we reach the vertical segments of the curve C, described in the lemma, and get the integration cycle $\{t \in \mathbb{C}^l \mid t_j \in C, j = 1, \ldots, l\}$. During the deformation, we keep the same small distances between the contours of integrations C_{u^j} with respect to different variables. At every moment, when one of the vertical segments of the contour of integration goes through a pole of the integrand, we add another integral of lower dimension that is the integral with respect to remaining variables of the residue. For example, the first such event leads to the following decomposition.

$$\int_{\substack{t_j \in \mathcal{C}_{u^j}, \\ j=1,\ldots,l}} f\, d^l t = \int_{\substack{t_j \in \mathcal{C}_{(\operatorname{Re}\lambda_1^0 + j\rho, u_2^j, \ldots, u_n^j)}, \\ j=1,\ldots,l}} f\, d^l t$$

$$= \int_{\substack{t_1 \in \mathcal{C}_{(\operatorname{Re}\lambda_1^0 - \rho/2, u_2^1, \ldots, u_n^1)} \\ t_j \in \mathcal{C}_{(\operatorname{Re}\lambda_1^0 + (j-3/2)\rho, u_2^j, \ldots, u_n^j)}, \\ j=2,\ldots,l}} f\, d^l t \quad + \int_{\substack{t_j \in \mathcal{C}_{(\operatorname{Re}\lambda_1^0 + (j-3/2)\rho, u_2^j, \ldots, u_n^j)}, \\ j=2,\ldots,l}} \operatorname{res}_{t_1 = z_1 + \lambda_1} f\, d^{l-1} t.$$

Recall that $\operatorname{Re}\lambda_1^0 = -1$. In the first term of the resulting decomposition, moving the first vertical segment of the contour $\mathcal{C}_{(\operatorname{Re}\lambda_1^0 + 1/2\rho, u_2^2, \ldots, u_n^2)}$ corresponding to t_2, through the point $z_1^0 + \lambda_1^0$, we also get a lower dimensional integral. This integral is equal to the second term of the above decomposition by Lemma 3. Note that owing to Lemma 4, the integrand of the second term of the decomposition does not have poles at $t_j = z_1 + \lambda_1$ for all $j = 2, \ldots, l$. Thus, we can move the first vertical segments of the contours corresponding to t_j, $j = 2, \ldots, l$, through $z_1^0 + \lambda_1^0$ without getting new residues.

Finally, we have

$$I(w, W)(z^0, \lambda^0, p^0) = \int_{\substack{t_j \in \mathcal{C}_{(\operatorname{Re}\lambda_1^0 + (j-l-1)\rho, u_2^j, \ldots, u_n^j)}, \\ j=1,\ldots,l}} f\, d^l t \quad + l \int_{\substack{t_j \in \mathcal{C}_{(\operatorname{Re}\lambda_1^0 + (j-l-1)\rho, u_2^j, \ldots, u_n^j)}, \\ j=2,\ldots,l}} \operatorname{res}_{t_1 = z_1 + \lambda_1} f\, d^{l-1} t.$$

In the first summand we move the contour of integration to

$$\{t \in \mathbb{C}^l \mid t_j \in \mathcal{C}_{(\min\{-\operatorname{Re}(2\Lambda_1 - \lambda_1^0) - \epsilon, 0\}, u_2^j, \ldots, u_n^j)}, j = 1, \ldots, l\}$$

and the contour does not meet any other poles of the integrand.

The integrand of the second summand has a simple pole at $t_j = z_1 + \lambda_1 - 1$, since f had a pole at $t_1 - t_j = 1$. It also may have a pole at $2\lambda_1 = 0$ coming from the pole at $t_1 = z_1 - \lambda_1$, see Lemma 5. So, as before, we have

$$\int_{\substack{t_j \in \mathcal{C}_{(\operatorname{Re}\lambda_1^0 + (j-l-1)\rho, u_2^j, \ldots, u_n^j)}, \\ j=2,\ldots,l}} \operatorname{res}_{t_1 = z_1 + \lambda_1} f\, d^{l-1} t = \int_{\substack{t_j \in \mathcal{C}_{(\operatorname{Re}\lambda_1^0 - 1 + (j-1)\rho, u_2^j, \ldots, u_n^j)}, \\ j=2,\ldots,l}} \operatorname{res}_{t_1 = z_1 + \lambda_1} f\, d^{l-1} t$$

$$= \int_{\substack{t_j \in \mathcal{C}_{(\operatorname{Re}\lambda_1^0 - 1 + (j-l-1)\rho, u_2^j, \ldots, u_n^j)}, \\ j=2,\ldots,l}} \operatorname{res}_{t_1 = z_1 + \lambda_1} f\, d^{l-1} t$$

$$+ (l-1) \int_{\substack{t_j \in \mathcal{C}_{(\operatorname{Re}\lambda_1^0 - 1 + (j-l-1))\rho, u_2^j, \ldots, u_n^j)}, \\ j=3,\ldots,l}} \operatorname{res}_{t_2 = z_1 + \lambda_1 - 1} \operatorname{res}_{t_1 = z_1 + \lambda_1} f\, d^{l-2} t.$$

Again, we move the contour of integration in the first summand to

$$\{(t_2, \ldots, t_l) \in \mathbb{C}^{l-1} \mid t_j \in \mathcal{C}_{(\min\{-\operatorname{Re}(2\Lambda_1 - \lambda_1^0) - \epsilon, 0\}, u_2^j, \ldots, u_n^j)}, j = 2, \ldots, l\}$$

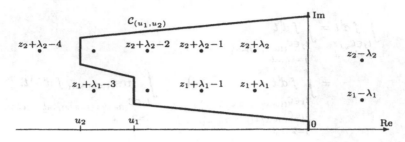

FIGURE 22.2.

without meeting any poles of the integrand, and the integrand of the second summand has simple poles at $t_j = z_1 + \lambda_1 - 2$ coming from the poles at $t_2 - t_j = 1$. It may also have poles at $2\lambda_1 = 1$ coming from $t_2 = z_1 - \lambda_1$.

Continuing the process we prove Lemma 6.   □

Lemma 7. *The analytic continuation of the function* $I(w, W)(z, \lambda, p)$ *from* z^0, λ^0, p^0 *to a neighborhood of the point* z^0, Λ, p^0 *is given by*

$$
I(w, W)(z^0, \lambda, p^0)
$$
$$
= \sum_{k} \frac{l!}{n! k_1! \cdots k_n! (l - k)!} \int_{\substack{t_i \in \mathcal{C}, \\ i = k+1, \ldots, l}} \operatorname{res}_k f(t, z^0, \lambda, p^0)\, d^{l-k} t, \quad (30)
$$

where $k = k_1 + \cdots + k_n$, $d^{l-k} t = dt_{k+1} \cdots dt_l$. *Here* $\mathcal{C}$ *denotes the curve* $\mathcal{C}_u$ *with* $u_m = \min\{-\operatorname{Re} \Lambda_m - \epsilon, 0\}$, $m = 1, \ldots, n$, ϵ *is a sufficiently small positive number, and the sum is over all* $k \in \mathbb{Z}_{\geq 0}^n$ *such that* $k \leq l$, *and* $k_m \leq 2 \operatorname{Re} \Lambda_m + 1$ *for* $m = 1, \ldots, n$. *Moreover, if* $B(l) \cap B(m) = \varnothing$, *then sum* (30) *is well defined at* z^0, Λ, p^0.

Proof. We analytically continue each summand in (29) from z^0, λ^0, p^0 to z^0, Λ, p^0. By Lemma 6, the poles of the integrand of a summand and the contour of integration can be pictured as in Figure 22.2.

Here $z_j = z_j^0$, $\lambda_j = \lambda_j^0$, $j = 1, \ldots, n$.

Note that there are also poles of the integrand of the type $t_i = z_j^0 \pm (\lambda_j + p^0 s)$, $s = 1, 2, \ldots$. These poles are far away from our picture according to our choice of p^0. Note also that all poles are simple.

We move the parameters λ_m from λ_m^0 to Λ_m. The contour of integration of each summand in Lemma 6 $\{t \in \mathbb{C}^l \mid t_j \in \mathcal{C} = \mathcal{C}_{u(\lambda)}, j = k+1, \ldots, l\}$, depends on λ. At every moment of deformation of parameters λ_m, we define $I(w, W)(z^0, \lambda, p^0)$ by

$$
I(w, W)(z^0, \lambda, p^0)
$$
$$
= \sum_{k} \frac{l!}{n! k_1! \cdots k_n! (l - k)!} \int_{\substack{t_i \in \mathcal{C}_{u(\lambda)}, \\ i = k+1, \ldots, l}} \operatorname{res}_k f(t, z^0, \lambda, p^0)\, d^{l-k} t,
$$

where $u(\lambda) = \big(u_1(\lambda_1), \ldots, u_n(\lambda_n)\big) \in \mathbb{R}^n$, $u_m(\lambda_m) = \min\{-\operatorname{Re}(2\Lambda_m - \lambda_m) - \epsilon, 0\}$, $m = 1, \ldots, n$, and the sum is over all $k \in \mathbb{Z}_{\geq 0}^n$ such that $k \leq l$, and $k_m \leq 2\operatorname{Re}\Lambda_m$ for $m =, \ldots, n$.

If $\Lambda_m \notin \Lambda^+$ for all $m = 1, \ldots, n$, then at every moment of the deformation the integrand is holomorphic by Lemma 5. On the drawing, the points $z_i - \lambda_i$, $i = 1, \ldots, n$, move to the left and the rest of the picture moves to the right. Note that the points $z_i - \lambda_i$, $i = 1, \ldots, n$, encounter neither other points on the picture nor the curve of integration $\mathcal{C}_{u(\lambda)}$.

If, for some $i \in \{1, \ldots, n\}$, $\Lambda_i \in \Lambda^+$, then by Lemma 5, the function $\operatorname{res}_k(z, \lambda, p)$ can have a pole at $\lambda_i = \Lambda_i$ for k such that $k_i = 2\Lambda_i$. In this case, in the picture, in the last moment of the deformation the point $z_i - \lambda_i$ coincides with the point $z_i + \lambda_i - k$.

However, if $w = w_l$, $W = W_m$ and $B(l) \cap B(m) = \varnothing$, then the pole is trivial by the following two lemmas.

Lemma 8. *Let $i \in \{1, \ldots, n\}$ and $\Lambda_i \in \Lambda^+$. Let $W = W_m$ and $m_i \leq 2\Lambda_i$. Then for any $w = w_l$,*

$$\operatorname{res}_{\lambda_i = \Lambda_i} \operatorname{res}_{t_{2\Lambda_i + 1} = z_i + \lambda_i - 2\Lambda_i} \cdots \operatorname{res}_{t_2 = z_i + \lambda_i - 1} \ \operatorname{res}_{t_1 = z_i + \lambda_i} \Phi_l w W = 0.$$

Proof. Let $i = 1$. We prove that

$$\operatorname{res}_{t_{2\Lambda_1 + 1} = z_1 + \lambda_1 - 2\Lambda_1} \cdots \operatorname{res}_{t_2 = z_1 + \lambda_1 - 1} \operatorname{res}_{t_1 = z_1 + \lambda_1} \ \Phi_l w W = 0.$$

Recall that W has the form $W = \sum_{\sigma \in S^l} [\ldots]_\sigma^{\mathrm{trig}}$. The residue is nontrivial only for the terms corresponding to permutations σ, which satisfy the following condition:

$$m_1 \geq (\sigma^{-1})_1 > (\sigma^{-1})_2 > \cdots > (\sigma^{-1})_{2\Lambda + 1}.$$

If $m_1 \leq 2\Lambda$ then there are no such permutations.

The cases $i = 2, \ldots, n$ are proved similarly. $\square$

Lemma 9. *Let $i \in \{1, \ldots, n\}$ and $\Lambda_i \in \Lambda^+$. Let $w = w_l$ and $l_i \leq 2\Lambda_i$. Then for any $W = W_m$,*

$$\operatorname{res}_{\lambda_i = \Lambda_i} \operatorname{res}_{t_{2\Lambda_i + 1} = z_i + \lambda_i - 2\Lambda_i} \cdots \operatorname{res}_{t_2 = z_i + \lambda_i - 1} \operatorname{res}_{t_1 = z_i + \lambda_i} \Phi_l w_l W_m = 0.$$

Proof. Let $i = 1$. We prove that

$$\operatorname{res}_{t_{2\Lambda_1 + 1} = z_1 + \lambda_1 - 2\Lambda_1} \cdots \operatorname{res}_{t_2 = z_1 + \lambda_1 - 1} \operatorname{res}_{t_1 = z_1 + \lambda_1} \ \Phi_l w_l \tilde{W}_m = 0.$$

Recall that w_l has the form $w_l = \sum_{\sigma \in S^l} [\ldots]_\sigma^{\mathrm{rat}}$. The residue is nontrivial only for the terms corresponding to permutations σ that satisfy the condition

$$l_1 \geq (\sigma^{-1})_1 > (\sigma^{-1})_2 > \cdots > (\sigma^{-1})_{2\Lambda + 1}.$$

Since $1 \notin B(l)$, we have $l_1 \leq 2\Lambda$ and there are no such permutations.
The cases $i = 2, \ldots, n$ are proved similarly. $\square$

Lemma 7 is proved.   □

Note that a residue of a meromorphic function can be described as an integral over a small circle.

For $B \in \mathbb{C}$ and $\epsilon > 0$, denote $\mathcal{D}_B \subset \mathbb{C}$ the circle with center B and radius ϵ. Denote $\mathcal{D}$ the curve $\mathcal{C}_u$ with $u_m = \min\{-\operatorname{Re}\Lambda_m - \epsilon, 0\}$.

Lemma 10. *The analytic continuation of the function $I(w, W)$ from z^0, Λ, p^0 to z^0, Λ, P is well defined. Moreover, for sufficiently small $\epsilon > 0$,*

$$I(w, W)(z^0, \Lambda, P) = \sum_{m=0}^{l} \sum_{a_1, \ldots, a_m} \frac{l!}{(l-m)!}$$

$$\times \int_{\substack{t_i \in \mathcal{D}, \\ i = m+1, \ldots, l}} \left(\sum_{\sigma \in \mathbb{S}^m} \int_{\substack{t_i \in \mathcal{D}_{a_i}, \\ i = 1, \ldots, m}} f(t, z^0, \Lambda, P)\, dt_{\sigma_1} \ldots dt_{\sigma_m} \right) dt_{m+1} \ldots dt_l, \quad (31)$$

where the second sum is over all $a = (a_1, \ldots, a_m) \in \mathbb{C}^m$ such that for each $i = 1, \ldots, m$, there exist $j \in \{1, \ldots, n\}$, $k, s \in \mathbb{Z}_{>0}$, such that $a_i = z_j + \Lambda_j - k + sP$; a_i is located to the right from $\mathcal{D}$ and $a_i \neq a_k$ for all $k, i = 1, \ldots, m$, $k \neq i$.

Note that formula (31) is a generalization of formula (30); however, in sum (31) there are many additional zero terms.

Proof. We analytically continue each summand in (30) with respect to p from p^0 to P preserving z^0, Λ. We move p and preserve the contour of integration $\mathcal{I}$ as long as the integrand is holomorphic for all $t \in \mathcal{I}$ at z, Λ, p (cf. Lemma 5). When a pole of the integrand goes through the contour of integration, we do the same procedure as in the proof of Lemma 6.

An individual integral on the right-hand side of (31) is over the cycle

$$\{t \mid t_i \in \mathcal{D}_{a_i}, i = 1, \ldots, m; t_i \in \mathcal{D}, i = m+1, \ldots, l\}.$$

The points a_i are the poles of the integrand that were on the left of $\mathcal{D}$ at the beginning of the deformation and are on the right of $\mathcal{D}$ at the final moment of the deformation. All poles of the integrand that were on the right of $\mathcal{D}$ at the beginning of the deformation remain on the right at every moment of the deformation. The new poles a_i on the right of $\mathcal{D}$ do not coincide with the old poles on the right of $\mathcal{D}$ owing to conditions (9)–(11) on P and Λ.

Note that the procedure of analytic continuation, described in Lemma 6, gives decomposition (31) in which the sum is over $a = (a_1, \ldots, a_m)$, where some of coordinates a_i could be equal. However, such integrals are equal to zero. In fact, let $a_i = a_j$ for some $i < j$. When we integrate over the variables $t_1, \ldots, t_{j-1}$, the resulting integrand could have a pole at $t_j = a_j$ of at most first order. Then the reason of Lemma 4 shows that the resulting integrand is holomorphic at $t_j = a_j$.   □

Let, $\mathcal{D}'$ be the line $\{x \in \mathbb{C} \mid \operatorname{Re} x = \min\{\operatorname{Re}(Z_i - \Lambda_i), i = 1, \ldots, n\} - \epsilon\}$.

Lemma 11. *The analytic continuation from* z^0, Λ, P *to* Z, Λ, P *is well defined. Moreover, for sufficiently small* $\epsilon > 0$,

$$I(w,W)(Z, \Lambda, P) = \sum_{m=0}^{l} \sum_{b_1, \ldots, b_m} \frac{l!}{(l-m)!}$$

$$\times \int_{\substack{t_i \in \mathcal{D}', \\ i=m+1,\ldots,l}} \left(\sum_{\sigma \in \mathbb{S}^m} \int_{\substack{t_i \in \mathcal{D}_{b_i}, \\ i=1,\ldots,m}} f(t, Z, \Lambda, P) \, dt_{\sigma_1} \cdots dt_{\sigma_m} \right) dt_{m+1} \cdots dt_l, \quad (32)$$

where the second sum is over all $b = (b_1, \ldots, b_m) \in \mathbb{C}^m$ *such that for each* $i = 1, \ldots, m$, *there exist* $j \in \{1, \ldots, n\}$, $k, s \in \mathbb{Z}_{>0}$, *such that* $b_i = Z_j + \Lambda_j - k + sP$; *and* b_i *is located to the right from* $\mathcal{D}'$.

Proof. First, we move the parameters z_m from z_m^0 to $z_m^0 + \operatorname{Re} Z_m$. The contour of integration of a summand in (31) has the form

$$\{t \in \mathbb{C}^l \mid t_i \in \mathcal{D}_{a_i(z)}, i = 1, \ldots, m; \, t_i \in \mathcal{D} = \mathcal{C}_{u(z)}, i = m+1, \ldots, l\},$$

where $u(z) = \operatorname{Re}(z_m - \Lambda_m) - \epsilon$, and depends on z. At every moment of the deformation of parameters z_m, we define $I(w, W)(z, \Lambda, P)$ by

$$I(w,W)(z, \Lambda, P) = \sum_{m=0}^{l} \sum_{a_1, \ldots, a_m} \frac{l!}{(l-m)!}$$

$$\times \int_{\substack{t_i \in \mathcal{C}_{u(z)}, \\ i=m+1,\ldots,l}} \left(\sum_{\sigma \in \mathbb{S}^m} \int_{\substack{t_i \in \mathcal{D}_{a_i}, \\ i=1,\ldots,m}} f(t, z, \Lambda, P) \, dt_{\sigma_1} \ldots dt_{\sigma_m} \right) dt_{m+1} \ldots, dt_l,$$

where the second sum is over all $a = (a_1, \ldots, a_m) \in \mathbb{C}^m$ such that for each $i = 1, \ldots, m$, there exist $j \in \{1, \ldots, n\}$, $k, s \in \mathbb{Z}_{>0}$, such that $a_i = z_j + \Lambda_j - k + sP$; and a_i is located to the right from $\mathcal{C}_{u(z)}$.

Consider the summand related to $a \in \mathbb{C}^m$. We move the contour of integration with respect to $t_{m+1}, \ldots, t_l$ from $\mathcal{C}_{u(z^0 + \operatorname{Re} Z)}$ to $\mathcal{D}'$. We use the method described in Lemma 6.

Finally, we move parameters $z_1, \ldots, z_n$ from $z_1^0 + \operatorname{Re} Z_1, \ldots, z_n^0 + \operatorname{Re} Z_n$ to $Z_1, \ldots, Z_n$ in the same way as we moved $\lambda_1, \ldots, \lambda_n$ in Lemma 7.

An individual integral in the RHS of (32) is over the cycle

$$\{t \mid t_i \in \mathcal{D}_{b_i}, i = 1, \ldots, m; t_i \in \mathcal{D}', i = m+1, \ldots, l\}.$$

The points b_i are the poles of the integrand that were on the left of $\mathcal{D}$ at the beginning of the deformation and are on the right of $\mathcal{D}'$ at the final moment of the deformation. All poles of the integrand that were on the right of $\mathcal{D}$ at the beginning of the deformation remain on the right of $\mathcal{D}'$ at the every moment of the deformation. The new poles b_i on the right of $\mathcal{D}'$ do not coincide with the old poles on the right of $\mathcal{D}'$ owing to conditions (12) on Z. $\square$

Theorem 7 is proved. $\square$

5.2   Proof of Theorems 8 and 9

Let $\mathrm{Im}\,\mu \neq 0$. Fix a rational weight function $w = w_l$, and a trigonometric weight function $W = W_m \in \mathfrak{G}$. Fix $\Lambda \in \mathbb{C}^n$.

We fix parameters z, p, λ^0 in such a way that $\mathrm{Re}\,\lambda_m^0 < 0$, p is a negative number with large absolute value and $|\,\mathrm{Im}(z_k + \Lambda_k - z_m - \Lambda_m)|$ is large for k, $m = 1, \ldots, n$, $k \neq m$. Namely, let $p = -2\sum_{j=1}^{n} |\Lambda_j| - 1$, $\lambda_k^0 = -1 + i\,\mathrm{Im}\,\Lambda_k$, $z_k^0 = i(-\mathrm{Im}\,\Lambda_k + 3kA)$, where A is a large real number such that $A > 2|\Lambda_k|$, $k = 1, \ldots, n$.

By Lemma 7,

$$I(w, W)(z, \lambda, p) = \sum_k \frac{l!}{n! k_1! \cdots k_n! (l-k)!} \int_{\substack{t_i \in \mathcal{C}, \\ i=k+1,\ldots,l}} \mathrm{res}_k f(t, z, \lambda, p)\, d^{l-k} t, \quad (33)$$

where $k = k_1 + \cdots + k_n$, $\mathcal{C}$ denotes the curve $\mathcal{C}_u$ with $u_m = \min\{-\mathrm{Re}(2\Lambda_m - \lambda_m) - \epsilon, 0\}$, $m = 1, \ldots, n$, ϵ is a sufficiently small positive number, and the sum is over all $k \in \mathbb{Z}_{\geq 0}^n$ such that $k \leq l$, and $k_m \leq 2\,\mathrm{Re}\,\Lambda_m + 1$ for $m = 1, \ldots, n$.

Let $i \in \{1, \ldots, n\}$. We have the following cases.

1) If i does not belong to $B_\Lambda(m)$, then the analytic continuation of the function $I(w, W)(z, \lambda, p)$ with respect to λ_i from λ_i^0 to Λ_i is well defined; see the proof of Lemma 7.

2) If i belongs to $B_\Lambda(m)$ and i does not belong to $B_\Lambda(l)$, then the analytic continuation of the function $I(w, W)(z, \lambda, p)$ with respect to λ_i from λ_i^0 to Λ_i is well defined; see the proof of Lemma 7. In this case $c_m(\lambda, p)$ is equal to zero at $\lambda_i = \Lambda_i$, so $J_{l,m}(z, \lambda, p) = c_m(\lambda, p) I(w, W)(z, \lambda, p)$ is equal to zero at $\lambda_i = \Lambda_i$.

3) If i belongs to $B_\Lambda(m)$ and to $B_\Lambda(l)$, then in decomposition (33) all the summands are well defined at $\lambda_i = \Lambda_i$ except for the summands corresponding to k such that $k_i = 2\Lambda_i + 1$. Such summands have a simple pole at $\lambda_i = \Lambda_i$; see the proof of Lemma 7. In this case $c_m(\lambda, p)$ is equal to zero at $\lambda_i = \Lambda_i$, so the function $J_{l,m}(z, \lambda, p) = c_m(\lambda, p) I(w, W)(z, \lambda, p)$ is well defined at $\lambda_i = \Lambda_i$.

This proves Theorem 8. $\square$

Moreover, if $B_\Lambda(l) = B_\Lambda(m) = B$; then we have

$$J_{l,m}(z, \Lambda, p)$$

$$= \sum_k \frac{l!}{n!\, k_1! \ldots k_n!(l-k)!} \int\limits_{\substack{t_i \in \mathcal{C}, \\ i=k+1,\ldots,l}} (c_m \operatorname{res}_k f)(t_{k+1}, \ldots, t_l, z, \Lambda, p)\, d^{l-k}t,$$

where $\mathcal{C}$ denotes the curve $\mathcal{C}_u$ with $u_m = \min\{-\operatorname{Re}\Lambda_m - \epsilon, 0\}$, $m = 1$, $\ldots$, n, and the sum is over all $k \in \mathbb{Z}_{\geq 0}^n$ such that $k \leq l$, $k_m \leq 2\operatorname{Re}\Lambda_m + 1$ for $m = 1, \ldots, n$, and $k_m = 2\Lambda_m + 1$ for all $m \in B$.

Let $j \in B$. We compute the residues at points $z_j + \lambda_j$, $z_j + \lambda_j - 1$, $\ldots$, $z_j + \lambda_j - 2\Lambda_j$ explicitly using the following lemma.

Lemma 12. *Let $B_\Lambda(l) = B_\Lambda(m) = B$ and let $j \in B$, so that $\Lambda_j \in \Lambda^+$, l, $m \in \mathcal{Z}_l^n$ and l_j, $m_j > 2\Lambda_j$. Let $k = 2\Lambda_j$,*

$$(c_m \operatorname{res}_{t_{k+1}=z_j+\lambda_j-k} \cdots \operatorname{res}_{t_2=z_j+\lambda_j-1} \operatorname{res}_{t_1=z_j+\lambda_j} \Phi_l w_l W_m)(t_{k+2}, \ldots, t_l, z, \Lambda)$$

$$= (k+1)!\, e^{\mu(k+1)z_j \pi i/p} \psi_{m_j,k} \phi_{\Lambda,j,k}(z)$$

$$\times (c_{m^{(j)}} \Phi_{l-k-1} w_{l^{(j)}} W_{m^{(j)}})(t_{k+2}, \ldots, t_l, z, \Lambda^{(j)}),$$

where

$$l^{(j)} = (l_1, \ldots, l_{j-1}, l_j - k - 1, l_{j+1}, \ldots, l_n),$$
$$m^{(j)} = (m_1, \ldots, m_{j-1}, m_j - k - 1, m_{j+1}, \ldots, m_n),$$
$$\Lambda^{(j)} = (\Lambda_1, \ldots, \Lambda_{j-1}, -\Lambda_j - 1, \Lambda_{j+1}, \ldots, \Lambda_n).$$

Proof. Consider the case $j = 1$. For a function $g(t, z, \lambda)$, the coefficient in the Laurent expansion with respect to variables $t_1, \ldots, t_{2\Lambda_1+1}$ at $t_i = z_1 + \lambda_1 - i + 1$, $i = 1, \ldots, 2\Lambda_1 + 1$, computed at $\lambda_1 = \Lambda_1$, that is, the coefficient of

$$\prod_{i=1}^{2\Lambda_1+1} \frac{1}{t_i - z_1 - \lambda_1 + i - 1},$$

is called the *main coefficient*. We compute the main coefficients of the functions Φ_l, w_l, W_m.

The main coefficient of the function $\Phi_l(t, z, \lambda)$ is

$$e^{\mu(2\Lambda_1+1)z_j\pi i/p} a_1(\Lambda) \phi_{\Lambda^{(1)},1,2\Lambda_1}(z) \prod_{j=2\Lambda_1+2}^{l} \prod_{i=1}^{2\Lambda_1+1} \frac{\Gamma((z_1 + \Lambda_1 - i + 1 - t_j - 1)/p)}{\Gamma((z_1 + \Lambda_1 - i + 1 - t_j + 1)/p)}$$

$$\times \prod_{j=2\Lambda_1+2}^{l} \prod_{i=1}^{n} \frac{\Gamma((t_j - z_i + \lambda_i)/p)}{\Gamma((t_j - z_i - \lambda_i)/p)} \prod_{2\Lambda_1+1 < i < j \leq l} \frac{\Gamma((t_i - t_j - 1)/p)}{\Gamma((t_i - t_j + 1)/p)}.$$

Here $a_1(\Lambda_1) \in \mathbb{C}$ is some explicitly computable constant. Note that $\phi_{\Lambda^{(1)},1,2\Lambda_1}(z) = \phi_{\Lambda,1,2\Lambda_1}(z)$. The main coefficient of the function $\Phi_l(t, z, \lambda)$ is equal to

$$\prod_{j=2\Lambda_1+2}^{l} \frac{t_j - z_1 - \Lambda_1}{t_j - z_1 + \Lambda_1} \frac{\sin((t_j - z_1 - \Lambda_1)\pi i/p)}{\sin((t_j - z_1 + \Lambda_1)\pi i/p)}$$

$$\times \frac{t_j - z_1 + \Lambda_1'}{t_j - z_1 - \Lambda_1'} \frac{\sin((t_j - z_1 + \Lambda_1')\pi i/p)}{\sin((t_j - z_1 - \Lambda_1')\pi i/p)} a_1(\Lambda) e^{\mu(2\Lambda_1+1)z_j \pi i/p}$$

$$\times \phi_{\Lambda,1,2\Lambda_1}(z)\Phi_{l-2\Lambda_1-1}(t_{2\Lambda_1+1},\ldots,t_l,z,\Lambda^{(1)}),$$

where $\Lambda_1' = -\Lambda_1 - 1$.

The function $w_l(t, z, \lambda)$ has the form $w_l = \sum_{\sigma \in \mathbb{S}^l} [\ldots]_\sigma^{\mathrm{rat}}$. Consider the sum over $\sigma \in \mathbb{S}^l$ such that $\sigma(i) = i$ for all $i = l_1 + 1, \ldots, l$. We have

$$w_l(t, z, \lambda) = a_2(\Lambda_1) \sum_{\sigma \in \mathbb{S}^l} \left[\prod_{\substack{i,j=1 \\ i<j}}^{l_1} \frac{t_i - t_j}{t_i - t_j + 1} \prod_{i=1}^{l} \frac{1}{t_i - z_1 - \lambda_1} \right.$$

$$\left. \times \prod_{i=l_1+1}^{l} (t_j - z_1 + \lambda_1) w_{(l_2,\ldots,l_n)}(t_{l_1+1},\ldots,t_l,z,\lambda) \right]_\sigma^{\mathrm{rat}}.$$

Here, again, $a_2(\Lambda_1) \in \mathbb{C}$ is an easily computable constant.

Let $\sigma \in \mathbb{S}^l$ be a permutation and assume there exists $i \leq 2\Lambda_1 + 1$ such that $\sigma(i) > l_1$. Then the term corresponding to σ does not contribute to the main coefficient. Hence, this coefficient is

$$\tilde{a}_2(\Lambda_1) \sum_{\sigma \in \mathbb{S}^{l-2\Lambda_1-1}} \left[\prod_{j=2\Lambda_1+2}^{l_1} \prod_{i=1}^{2\Lambda+1} \frac{z_1 + \Lambda_1 - i + 1 - t_j - 1}{z_1 + \Lambda_1 - i + 1 - t_j + 1} \right.$$

$$\times \prod_{i=2\Lambda_1+2}^{l} \frac{1}{t_i - z_1 - \lambda_1} \prod_{i=l_1+1}^{l} (t_j - z_1 + \lambda_1)$$

$$\left. \times \prod_{\substack{i,j=2\Lambda_1+2 \\ i<j}}^{l_1} \frac{t_i - t_j}{t_i - t_j + 1} w_{(l_2,\ldots,l_n)}(t_{l_1+1},\ldots,t_l,z,\Lambda_1,\lambda_2,\ldots,\lambda_n) \right]_\sigma^{\mathrm{rat}},$$

where $\tilde{a}_2(\Lambda_1) \in \mathbb{C}$ is another easily computable constant and the group $\mathbb{S}^{l-2\Lambda_1-1}$ permutes the variables $t_{2\Lambda_1+1}, \ldots, t_l$. Simplifying, we get the main coefficient of the function $w_l(t, z, \lambda)$, given by

$$\tilde{a}_2(\Lambda_1) \prod_{j=2\Lambda+1}^{l} \frac{t_j - z_1 + \Lambda_1}{t_j - z_1 - \Lambda_1} \frac{t_j - z_1 - \Lambda_1'}{t_j - z_1 + \Lambda_1'} w_{(l_1',l_2\ldots,l_n)}.$$

The main coefficient of the function $W_m(t, z, \lambda)$ is computed similarly and is equal to

$$a_3(\Lambda_1) \prod_{j=2\Lambda_1+1}^{l} \frac{\sin((t_j - z_1 + \Lambda_1)\pi i/p)}{\sin((t_j - z_1 - \Lambda_1)\pi i/p)} \frac{\sin((t_j - z_1 - \Lambda_1')\pi i/p)}{\sin((t_j - z_1 + \Lambda_1')\pi i/p)} W_{(m_1', m_2 \ldots, m_n)},$$

where $a_3(\Lambda_1) \in \mathbb{C}$ is some easily computable comstant.

Multiplying the above main coefficients we get the statement of the Lemma. We have

$$a_1(\Lambda_1)\tilde{a}_2(\Lambda_1)a_3(\Lambda_1)c_m(\Lambda) = (k+1)!\, c_m^{(1)}(\Lambda^{(1)})\, \psi_{2\Lambda_1}.$$

The cases $j = 2, \ldots, n$ are proved similarly.   □

We have

$$J_{l,m}(z, \Lambda, p)$$

$$= \sum_k \frac{l!}{n! k_1! \ldots k_n! (l-k)!} \int_{\substack{t_i \in \mathcal{C}, \\ i=k+1+l'(B),\ldots,l}} \mathfrak{C}_{\Lambda,m}\big(c_{m'(B)} \operatorname{res}_k \Phi_{l'(B)} w_{l'(B)} W_{m'(B)}\big)$$

$$\times (t_{k+1+l'(B)}, \ldots, t_l, z, \Lambda, p)\, d^{l-k-l'(B)}t, \quad (34)$$

where $\mathcal{C}$ denotes the curve $\mathcal{C}_u$ with $u_m = \min\{-\operatorname{Re}\Lambda_m - \epsilon, 0\}$, $m = 1, \ldots,$ n, and the sum is over all $k \in \mathbb{Z}_{\geq 0}^n$ such that $k \leq l - l'(B)$, $k_m \leq 2\operatorname{Re}\Lambda_m + 1$ for $m = 1, \ldots, n$, and $k_m = 0$ for all $m \in B$.

Let $\mathcal{C}'$ be the curve $\mathcal{C}_u$ with $u_m = \min\{-\operatorname{Re}\Lambda_m' - \epsilon, 0\}$, $m = 1, \ldots,$ n. Then the integrand of each integral in sum (34) does not have poles of the type $t_j = a$, where $a \in \mathbb{C}$ is located between $\mathcal{C}'$ and $\mathcal{C}$, see Lemma 5. Thus, we can move the contour of integration in each summand to $\{t \in \mathbb{C}^{l-k-l'(B)} \mid t_i \in \mathcal{C}', i = k+1+l'(B), \ldots, l\}$.

We proved Theorem 9 for our choice of z, p. Theorem 9 holds for all z, p, since both left- and right-hand sides of the formula of Theorem 9 are meromorphic functions of parameters z, p.   □

5.3 Proof of Theorem 12

Consider a matrix

$$I^l(z, \lambda, p) = \left\{ \prod_{i \in B_\Lambda(m)} (\lambda_i - \Lambda_i) I_{l,m}(z, \lambda, p) \right\}_{l,m \in \mathcal{Z}_l^n}.$$

The matrix $I^l(z, \lambda, p)$ is obtained from the matrix $J^l(z, \lambda) = J^l(z, \lambda, p)$ by multiplication the mth row by $(\prod_{i \in B_\Lambda(m)}(\lambda_i - \Lambda_i))/c_m(\lambda)$ for all $m \in \mathcal{Z}_l^n$. For generic p, these factors are well defined and not equal to zero. All entries of the matrix $I^l(z, \lambda, p)$ are well defined and the matrix $I^l(z, \lambda)$ is upper block triangular at $\lambda = \Lambda$.

It follows from formula 5.14 in Ref. 11 that the matrix $I^l(z, \lambda)$ is nondegenerate for all z, λ, p satisfying conditions (9)–(14).

Hence, the matrix

$$I^l_{\text{adm}}(z, \lambda, p) = \left\{ \prod_{i \in B_\Lambda(m)} (\lambda_i - \Lambda_i) I_{l,m}(z, \lambda, p) \right\},$$

where l, $m \in \mathcal{Z}_l^n$ run through the set of all Λ-admissible indices, is also nondegenerate for all z, λ, p satisfying conditions (9)–(14).

For all p, Λ satisfying conditions (9)–(14) and Λ-admissible indices m, the factors $(\prod_{i \in B_\Lambda(m)} (\lambda_i - \Lambda_i))/c_m(\lambda)$ are well defined and nonzero. Hence, the matrix $J^l_{\text{adm}}(z, \lambda, p)$ is also nondegenerate for all z, λ, p satisfying conditions (9)–(14).

The matrix $J^l(z, \lambda)$ is upper block triangular at $\lambda = \Lambda$. Now, formula (21) follows from the Inclusion-Exclusion Principle and Theorem 9.

The theorem is proved.   □

5.4   Proof of Theorem 1

The rational R-matrix $R_{\lambda_1 \lambda_2}(x) \in \text{End}(V_{\lambda_1} \otimes V_{\lambda_2})$ is a meromorphic function of λ_1, λ_2, defined explicitly in Section 2.3 for $\lambda_1, \lambda_2 \notin \Lambda^+$.

Consider the qKZ equation with $n = 2$, $\mu = 2$, $p = -2(|\lambda_1| + |\lambda_2|) - 1$. Consider the functions $\{\Psi_m(z, \lambda) \mid m \in \mathcal{Z}_l^2\}$ given by

$$\Psi_m(z, \lambda) = \sum_{l_1 + l_2 = l} J_{l,m}(z, \lambda) f^{l_1} v_1 \otimes f^{l_2} v_2,$$

where the function $J_{l,m}$ is defined in (15).

The functions $\{\Psi_m(z, \lambda) \mid m \in \mathcal{Z}_l^2\}$ are solutions of the qKZ equation. Moreover, for generic λ, the values of these functions span $V_{\lambda_1} \otimes V_{\lambda_2}$ over $\mathbb{C}$ by Theorem 5.14 in Ref. 11. By Theorem 8, these functions are well defined for all λ_1, λ_2. Moreover, they span $V_{\lambda_1} \otimes V_{\lambda_2}$, for all λ_1, λ_2 and generic z; see Theorem 5.14 in Ref. 11 and Section 4.4. The first statement of theorem 1 follows, now, from the equality of meromorphic functions:

$$\Psi_m(z_1 + p, z_2, \lambda_1, \lambda_2) = e^{-\mu h_1} R_{\lambda_1 \lambda_2}(z_1 - z_2) \Psi_m(z_1, z_2, \lambda_1, \lambda_2)$$

for all $m \in \mathcal{Z}_l^2$.

For all complex λ_1, λ_2, the R-matrix $R_{\lambda_1 \lambda_2}(x)$ commutes with the action of sl_2.

If $\lambda_1, \lambda_2 \notin \Lambda^+$, then $S_{\lambda_1} \otimes V_{\lambda_2} + V_{\lambda_1} \otimes S_{\lambda_2}$ is the zero submodule and the second statement of the Theorem is trivial. Otherwise, let $\lambda \in \{\lambda_1, \lambda_2\}$ be the minimal dominant weight. Namely, if $\lambda_1 \in \Lambda^+$ and $\lambda_2 \notin \Lambda^+$, set

$\lambda = \lambda_1$. If $\lambda_2 \in \Lambda^+$ and $\lambda_1 \notin \Lambda^+$, set $\lambda = \lambda_2$. If $\lambda_1 \in \Lambda^+$ and $\lambda_2 \in \Lambda^+$, set $\lambda = \min\{\lambda_1, \lambda_2\}$. Then the submodule $S_{\lambda_1} \otimes V_{\lambda_2} + V_{\lambda_1} \otimes S_{\lambda_2}$ is generated by all singular vectors in $V_{\lambda_1} \otimes V_{\lambda_2}$ of weights less than $\lambda_1 + \lambda_2 - 2\lambda$. Hence, the R-matrix $R_{\lambda_1 \lambda_2}(x)$ preserves $S_{\lambda_1} \otimes V_{\lambda_2} + V_{\lambda_1} \otimes S_{\lambda_2}$.

The third statement of the theorem follows from the first two statements.
$\square$

Acknowledgments: The authors thank V. Tarasov for useful discussions.

6   REFERENCES

1. V. Chari and A. Pressley, *A Guide to Quantum Groups*, Cambrige Univ. Press, Cambridge, 1994.

2. G. Felder, *Conformal field theory and integrable systems associated to elliptic curves*, Proc. International Congress of Mathematicians (Zürich, 1994) (S. D. Chatterji, ed.), Birkhäuser, Basel, 1994, pp. 1247–1255, hep-th/9407154.

3. _____, *Elliptic quantum groups*, XIth International Congress of Mathematical Physics (Paris, 1994) (D. Iagolnitzer, ed.), Internat. Press, Cambridge, MA, 1995, pp. 211–218.

4. G. Felder, V. O. Tarasov, and A. N. Varchenko, *Monodromy of solutions of the elliptic quantum Knizhnik–Zamolodchikov–Bernard difference equation*, q-alg/9705017.

5. _____, *Solutions of the elliptic qKZB equations and Bethe ansatz.* I, Topics in Singularity Theory: V. I. Arnold's 60th Anniversary Collection (A. Khovanskii, A. Varchenko, and V. Vassiliev, eds.), Amer. Math. Soc. Transl. Ser. 2, Vol. 180, Amer. Math. Soc., Providence, RI, 1997, pp. 45–76, q-alg/9606005.

6. I. Frenkel and N. Yu. Reshetikhin, *Quantum affine algebras and holonomic difference equations*, Commun. Math. Phys. **146** (1992), No. 1, 1–60.

7. M. Jimbo and T. Miwa, *Algebraic Analysis of Solvable Lattice Models*, CBMS Regioal Conf. Ser. in Math., Vol. 85, Conf. Board Math. Sci., Washington, DC, 1995.

8. P. P. Kulish, N. Yu. Reshetikhin, and E. K. Sklyanin, *Yang–Baxter equation and representation theory.* I, Lett. Math. Phys. **5** (1981), No. 5, 393–403.

9. F. A. Smirnov, *Form Factors in Completely Integrable Models of Quantum Field Theory*, Adv. Ser. Math. Phys., World Sci. Publ., River Edge, NJ, 1992.

10. V. O. Tarasov, *Irreducible monodromy matrices for the R-matrix of the XXZ–model and lattice local Hamiltonians*, Theoret. and Math. Phys. **63** (1985), No. 2, 440–454.

11. V. O. Tarasov and A. N. Varchenko, *Geometry of q-hypergeometric functions as a bridge between Yangians and quantum affine algebras*, Invent. Math. **128** (1997), No. 3, 501–588.

12. _____ , *Geometry of q-Hypergeometric Functions, Quantum Affine Algebras and Elliptic Quantum Groups*, Astérisque, Vol. 246, Soc. Math. France, Paris, 1997, q-alg/9703044.

23
Gauge Fields and Interacting Particles

N. Nekrasov

ABSTRACT In this short survey we discuss integrable many-body systems and their relation to gauge theories. One aspect of such a relation is the Hamiltonian reduction, which produces the model out of a simple dynamical system. The phase spaces of original simple systems are constructed using the infinite-dimensional current algebras. We briefly discuss dualities, relating integrable systems. Finally, we outline the applications to the studies of moduli spaces of vacua of supersymmetric gauge theories in four and five dimensions and present derivations of some of these results using string theory.

1 Introduction

These notes describe our current understanding of relations between the integrable systems of certain kinds and gauge theories in various dimensions. There are at least two ways the integrable system can describe the dynamics of a gauge theory. The first one is relevant for the low-dimensional gauge theories with no propagating degrees of freedom, like the Yang–Mills theory in two dimensions or the Chern–Simons theory in three dimensions. There the nontrivial dynamics occurs if the topology of space-time is nontrivial. Then the evolution of global gauge invariant observables turns out to be that of an integrable system of particles [11, 12, 15, 24].

Once the discussion is switched to the normal field theories, the relation to integrable models is tricky. Indeed, the gauge field in three and higher dimensions has propagating degrees of freedom and the corresponding phase space is infinite-dimensional. The information stored in the infinity of degrees of freedom of the gauge model is hardly of immediate interest for the low-energy observer. The idea is to get rid of the majority of the degrees of freedom by integrating out the high-energy modes. This approach (created essentially by K. Wilson) produces the low-energy effective action. In the context of supersymmetric gauge theories this action is highly constrained by holomorphy and duality. This allows us to determine the actions for a large class of theories. It turns out that the low-energy action for the vector multiplets in $\mathcal{N} = 2$ supersymmetric gauge theories in four dimensions is governed by an integrable systems in holomorphic sense [5]. The vacuum expectation values of scalars in the vector multiplet are identified with the integrals of motion, and the Liouville tori determine the gauge coupling. This is the second way the integrable system can encode the dynamics of the gauge theory.

In our review we merely present a few examples of both situations and will speculate at the end on the possible unification of the two approaches.

The paper is organized as follows. In Section 2 we list the many-body systems of our interest. In Section 3 we explain their equivalence to the low-dimensional gauge theories. In Section 4 we describe the low-energy effective actions of four- and five-dimensional supersymmetric gauge theories and the corresponding integrable systems. In Section 5 we derive the statement about the five-dimensional gauge theory using summation over world-sheet instantons in IIA string theory. In Section 6 we present our conclusions and speculations.

2   The Cast

The main characters of our play are the integrable systems describing the interaction of the collections of identical particles. The phase space of the many-body system has the form $P = X/\Gamma$, where X is the phase space of the system of the distinguishable particles and Γ is the group of allowed permutations. In all our examples X has the form $Y \times \cdots \times Y$ (N factors) and Γ can be identified with the symmetric group S_N. One may distinguish the following cases:

Nonrelativistic cases The space Y is the cotangent bundle to the space of configurations of the single particle: $Y = T^*Z$.

Relativistic case The space Y is the torus bundle over the space of configurations of the single particle:

$$F \longrightarrow Y$$
$$\downarrow$$
$$Z. \tag{1}$$

There are also further subdivisions according to the following classification:

Rational case Z is the line $\mathbb{R}^1$ or $\mathbb{C}^1$.

Trigonometric case Z is the torus S^1 or $\mathbb{C}^*$.

Elliptic case Z is the elliptic curve E. One fixes the symplectic form $\omega = dp \wedge dq$ on Y, which determines a symplectic form on X: $\Omega =$

$\pi_* \sum_{i=1}^{N} p_i^* \omega$, where $\pi: Y^{\times N} \to X$ is the projection-factorization; π_* is the corresponding pushforward map; and $p_i: Y^{\times N} \to Y$ is the projection onto the ith factor.

The typical system has a basic Hamiltonian H, which gives rise to the time evolution of the model. One has also the higher Hamiltonians, which Poisson-commute with H and give rise to the higher flows. These Hamiltonians provide the complete set of Poisson-commuting integrals of motion.

The basic Hamiltonian H_{nonrel} in the nonrelativistic case is the sum $T+U$ of the kinetic and potential energies. The kinetic energy is the quadratic form in momenta $T = \sum_{i=1}^{N} \frac{1}{2} p_i^2$, while the potential energy is the sum of the pairwise potentials: $U = \sum_{i,j} U(q_i, q_j)$, $q_i \in Z$, $i = 1, \ldots, N$. The potentials $U(q_i, q_j)$ can be associated to the root systems (more generally, Coxeter groups). For the root system of A_N type the potential has the form

$$U(x,y) = \begin{cases} \frac{g}{(x-y)^2} & \text{rational} \\ \frac{g}{\sin^2(x-y)} & \text{trigonometric} \\ g\wp(x-y) & \text{elliptic} \end{cases} \tag{2}$$

where g is the coupling (a complex number). The Weierstrass function is defined as the following series:

$$\wp(x) = \frac{1}{x^2} + \sum_{(m,n)\in\mathbb{Z}^2\backslash(0,0)} \frac{1}{(m\tau+n+x)^2} - \frac{1}{(m\tau+n)^2}.$$

In the limit $\text{Im}\,\tau \to \infty$ $\wp(x)$ reduces to $\pi^2/\sin^2(\pi x) - \pi^2/3$. The system with rational potential is referred to as Calogero–Moser system [4], the one with trigonometric potential is the Sutherland system [31].

The Hamiltonian of the relativistic elliptic model has the form [26–28]:

$$H_{\text{rel}} = \sum_{i=1}^{N} \cos\beta\theta_i \prod_{j\neq i} f(q_{ij}), \quad q_{ij} = q_i - q_j \tag{3}$$

with

$$f(q) = \sqrt{1 - \frac{\wp(q)}{\wp(\beta\nu)}}$$

This model is called *elliptic Ruijsenaars–Schneider* or *relativistic Calogero–Moser* system (see Ref. 3 for discussion of the appropriateness of the latter name). It has the following degenerations:

Nonrelativistic limit $\beta \to 0$. In this limit $H_{\text{rel}} = N + \beta^2 H_{\text{nonrel}} + \cdots$, where H_{nonrel} is the Hamiltonian of elliptic Calogero–Moser system.

Trigonometric limit $\text{Im}\,\tau \to \infty$. The function $f(q)$ goes to: $f(q) = (1 - \sin^2\beta\nu/\sin^2(q))^{1/2}$.

Relativistic Toda limit In this limit the elliptic curve degenerates and
the parameter ν is tuned in such a way that $\gamma = \exp(2\pi i(\nu + \tau/N))$ is
kept finite while $\operatorname{Im}\tau \to \infty$. The Hamiltonian becomes (after a shift
of variables: $q_k \to q_k - k\tau/N$):

$$H_{\text{Toda}} = \sum_{i=1}^{N} \cos\beta\theta_i \sqrt{(1 - \gamma^2 e^{q_{i+1} - q_i})(1 - \gamma^2 e^{q_i - q_{i-1}})},$$

$$q_{N+1} = q_1, q_0 = q_N. \quad (4)$$

3 The Hamiltonian Reduction and Gauge Theories

In this section we explain the origin of the integrable systems described in
the previous chapter. All of them can be obtained as a result of the reduction of simple "free motion" systems defined on somewhat bigger (infinite-
dimensional) phase spaces. The reduction involved is the Hamiltonian reduction with respect to the action of a certain natural symmetry group.
This is a symplectic version of the general idea of "projection method", suggested by M. Olshanetsky and A. Perelomov in Ref. 23 (see also Ref. 15).
In fact, the Hamiltonian reduction is the procedure that is familiar to the
gauge theorists. In the gauge theory setup one starts with the model with
a larger phase space then one actually needs with an action of a symmetry
group. The symmetry group is generated by Gauss law (moment map in a
more sophisticated language). One imposes the Gauss law constraints (fixes
the level of moment map) and reduces with respect to the group acton. As a
result the phase space reduces to a smaller physical one. Our original phase
spaces P (we call them "big" phase spaces) are the cotangent bundles to a
certain manifolds M_A that are acted on by the Lie group A (for example,
M_A may be the Lie algebra $\mathfrak{a}$): $P = T^* M_A$. The group A acts on P and
gives rise to the moment map $\mu: P \to \mathfrak{a}^*$. Explicitly, if $V_a \in \operatorname{Vect}(M_A)$
is the vector field on M_A corresponding to the element $\xi_a \in \mathfrak{a}$ of the Lie
algebra of A, then $\mu_a(x, p) = \langle p, V_a(x) \rangle$, where $p \in T_x^* M_A$ and $\langle \, , \, \rangle$ is the
canonical pairing between $T_x M_A$ and $T_x^* M_A$. We proceed with imposing
the moment map equations. To this end we choose a certain coadjoint orbit
$\mathcal{O} \in \mathfrak{a}^*$ and consider its preimage $\mu^{-1}(\mathcal{O}) \subset P$. It is invariant under the A
action and (formally) the quotient $P_{\mathcal{O}} = \mu^{-1}(\mathcal{O})/A$ is defined. It carries the
induced symplectic form. In our problems the quotients may have orbifold
singularities. The Hamiltonian system on P gives rise to the system on $P_{\mathcal{O}}$
if the restriction of the Hamiltonian on $\mu^{-1}(\mathcal{O})$ is A-invariant. To illustrate
this construction we shall prove the equivalence of Sutherland model to the
two-dimensional Yang–Mills theory on a cylinder with the time-like Wilson
line.

We take as a "big" space P the cotangent bundle to the affine Lie algebra
$\hat{\mathfrak{g}}$. The latter is the central extension of the Lie algebra $L\mathfrak{g}$ of loops in the
finite-dimensional Lie algebra $\mathfrak{g}$. An element of this algebra is the pair

(ϕ, c), where $\phi\colon S^1 \to \mathfrak{g}$ is the loop and c is a number. The dual space can be identified with the space of the first-order differential operators on the circle which have the form of connections $\nabla = k\partial + A$.

The symmetry group is the loop group LG acting in the adjoint representation:

$$A \to g^{-1}Ag + kg^{-1}\partial g \qquad \phi \to g^{-1}\phi g$$

$$c \to c + \int_{S^1} \mathrm{Tr}(\phi \partial g g^{-1}) \quad k \to k. \tag{5}$$

The phase space has a canonical LG-invariant symplectic structure:

$$i\Omega = \delta c \wedge \delta k + \int_{S^1} \mathrm{Tr}\, \delta\phi \wedge \delta A.$$

The corresponding moment map sends (∇, ϕ, c) to

$$\mu = k\partial\phi + [A, \phi]. \tag{6}$$

At this point we fix $G = \mathrm{SU}(N)$. The orbit $\mathcal{O}$ in this case is going to be the conjugacy class of the element $J \in \hat{\mathfrak{g}}^*$, which is the distribution $J(t) = \delta(t)j$, where $j_{kl} = i\nu(\delta_{kl} - e_k e_l^*)$, $k, l = 1, \ldots, N$, $e_i \in \mathbb{C}$, t is the coordinate on the circle and $\delta(t)$ is the unit delta function with the support at the marked point $t = 0$. We identify t and $t + 2\pi R$.

The moment equation $\mu = J$ can be solved in the particular gauge:

$$A = \mathrm{diag}(a_1, \ldots, a_N), \tag{7}$$

with $\sum a_i = 0$. The eigenvalues a_i's are not very well defined. There are two kinds of gauge transformations that preserve (7): the diagonal transformations

$$g(t) = \exp\frac{it}{R}\, \mathrm{diag}(n_1, \ldots, n_N), \quad n_i \in \mathbb{Z} \tag{8}$$

(they shift a_i by ikn_i/R) and the permutations of a_i's. Altogether they form the affine Weyl group $\widehat{W}$ of $\mathrm{SU}(N)$. The reduced phase space can be identified with the quotient $T^*\mathfrak{t}/\widehat{W}$. Indeed, let us plug (7) into (6) and $\mu = J$:

$$k\partial\phi_{ii} = \nu(1 - |e_i|^2)\delta(t)$$
$$k\partial\phi_{ij} + (a_i - a_j)\phi_{ij} = -\nu e_i^* e_j \delta(t), \quad i \neq j. \tag{9}$$

The equation is solved by the following ansatz:

$$\phi_{ii} = p_i = \mathrm{const}_t \quad |e_i| = 1$$

$$\phi_{ij}(t) = \exp\left(-\frac{t}{k}(a_i - a_j)\right)\phi_{ij}(0) \tag{10}$$

$$\phi_{ij}(2\pi R) - \phi_{ij}(0) = \frac{\nu}{k}e_i^* e_j.$$

The last equation follows from

$$\partial f = l\delta(t) \implies f(+0) - f(-0) = l$$

and from the periodicity of $\phi_{ij}(t)$. The equation $|e_i| = 1$ allows us to set all e_i's to be equal to 1 by the constant diagonal gauge transformation. Finally, we get

$$\phi_{ij}(t) = \frac{\nu e^{-t a_{ij}/k}/k}{e^{-2\pi R a_{ij}/k} - 1},$$

where $a_{ij} = a_i - a_j$.

The Hamiltonian H comes from the quadratic casimir on the big phase space

$$H = \frac{1}{4\pi R} \int_{S^1} dt \, \mathrm{Tr}\, \phi^2,$$

which is equal to

$$H = \sum_{i=1}^{N} \frac{1}{2} p_i^2 - \sum_{i<j} \frac{\nu^2}{4k^2 \sinh^2(\pi R a_{ij}/k)}$$

on the solutions of the moment equation. The symplectic form equals

$$\Omega = -i\delta k \wedge \delta c - i \sum_{i=1}^{N} 2\pi R \delta a_i \wedge \delta p_i.$$

The variable, canonically conjugate to p_i (neglecting the subtlety associated with the $\sum p_i = \sum a_i = 0$ conditions) is

$$q_i = 2\pi i R a_i$$

In terms of canonical variables we get the Sutherland Hamiltonian:

$$H = \sum_{i=1}^{N} \frac{1}{2} p_i^2 + \sum_{i<j} \frac{\nu^2}{4k^2 \sin^2\left(q_{ij}/(2k)\right)} \tag{11}$$

Notice that the effect of the "level" k is to set the period of the potential to be $2\pi k$. One can rewrite the final potential in the form of infinite sum of rational potentials:

$$U(x) = \sum_{n \in \mathbb{Z}} \frac{\nu^2}{(x + 2\pi n k)^2}.$$

In the limit $k \to \infty$ it reduces to the rational potential. Now, to see that this construction has anything to do with two-dimensional Yang–Mills let us write down the Lagrangian of the system together with constraint $\mu = J$ imposed by means of a Lagrange multiplier A_0:

$$\mathcal{L} = \int d^2t \operatorname{Tr}(\phi\partial_0 A_1 + A_0(\mu - J) + \phi^2),$$

where $A_1 = A$, $t^1 = t$ and t^0 represents time. In the quantum theory one studies the path integral over the field configurations:

$$[D\phi DA DA_t De] \exp - \mathcal{L}.$$

One can integrate out ϕ and also replace the integration over e_i's by the insertion of a Wilson line:

$$\langle v_1 | P \exp \int A_0 \, dt^0 | v_2 \rangle,$$

which leaves the action

$$S = \int d^2t \operatorname{Tr} F_{01}^2$$

of Yang–Mills theory.

One can elaborate further and get the relativistic as well as elliptic systems with the help of certain gauge theories in two/three dimensions (by studying the Hamiltonian reduction of the curved spaces with infinite-dimensional current groups as symmetries).

It turns out that elliptic Calogero–Moser system is a special case of the more general Hitchin system associated to a curve with punctures.

One of the applications of the approach involving Hamiltonian reduction is the understanding of the phenomenon of *duality*.

The point is that the explicit form of the reduced Hamiltonian system requires a choice of gauge, that is, the specific way of parameterizing the quotient. Clearly, one may imagine different gauge choices. They might lead to the Hamiltonian systems, defined on the same phase space, but looking differently in the different coordinates. In particular, one may imagine a situation where the "coordinates" of one system coincide with "action" variables of the other and vice versa. Such a situation has been called "action-coordinate" duality in Ref. 6. It was shown there that rational Calogero–Moser system is self-dual, trigonometric Calogero–Moser (= Sutherland) system is dual to rational Ruijsenaars–Schneider model and the trigonometric Ruijsenaars–Schneider system is self-dual as well.

One may also encounter a situation where one set of Hamiltonians look differently in different coordinate systems. In this case the systems are simply equivalent. It was shown in Ref. 21 that such identification is possible for the Sutherland model and the Calogero model (it is the model with particles interactiong pairwise with the inverse square potential and confined by a external oscillator potential).

The lesson here is that the reduction procedure broadens our point of view on the symmetries and properties of the system and allows us to find the relations between the different models.

4 Supersymmetric Gauge Theories in Four Dimensions

The supersymmetry relates bosonic fields to the fermionic ones. This suggests that the field content of the susy gauge theory involves fermionic partners of the gauge fields. In addition, in the theories with extended supersymmetry the vector multiplet (the representation of super-Poincare algebra, containing vector field) contains scalars. All the fields entering one susy multiplet belong to the same representation of the gauge group. The supersymmetry $\mathcal{N} = 2$ implies that the gauge field is augmented with the complex scalar in the adjoint representation.

Let us consider pure $G = \mathrm{SU}(N)$ $\mathcal{N} = 2$ theory. Its microscopic Lagrangian has the form

$$L = \frac{1}{g_0^2} \int d^4x \, \mathrm{Tr}(F_{mn}^2 + |D_m\phi|^2 + [\phi, \bar\phi]^2) + \cdots + \theta \int \mathrm{Tr} \, F \wedge F, \quad (12)$$

where ... denote the fermionic terms. In the low-energy limit the potential $[\phi, \bar\phi]^2$ vanishes, breaking the gauge group down to the normalizer $N(\mathbf{T})$ of the maximal torus $\mathbf{T}$. One may fix a gauge:

$$\phi = \mathrm{diag}(\phi_1, \ldots, \phi_N), \quad \sum_i \phi_i = 0.$$

The effective Abelian theory contains among other particles the W-bosons, which are charged under the torus $\mathbf{T}$. Their loops produce the running of the effective couplings $1/g^2 \sim \log \phi$. The whole low-energy effective action, containing not more then two derivatives, can be encoded in one function $\mathcal{F}(\phi_k)$, which must be holomorphic:

$$L = \frac{1}{4\pi} \mathrm{Im} \left[\int d^4\theta \, \frac{\partial \mathcal{F}}{\partial A^i} \bar{A}^i + \frac{1}{2} \int d^2\theta \, \frac{\partial^2 \mathcal{F}}{\partial A^i \partial A^j} W_\alpha^i W^{\alpha, j} \right]. \quad (13)$$

Here we used the $\mathcal{N} = 1$ notations, A^i denote the chiral multiplets, whose scalar components contain ϕ_i (we are not careful here about the imposing $\sum \phi_i = 0$ condition). The main applications of the abovementioned constructions are in the field of supersymmetric gauge theories. One can see that the second derivatives of the function $\mathcal{F}$ determine the gauge couplings:

$$L \sim \tau_{ij} F_-^i \wedge F_-^j - \bar\tau_{ij} F_+^i \wedge F_+^j,$$

with $\tau_{ij} = \partial^2 \mathcal{F}/\partial A^i \partial A^j$. The problem is to find τ as a function of ϕ.

Vacuum expectation values of ϕ_k form a complex manifold $\mathcal{M}$, called a moduli space of vacua, since each value of ϕ determines a vacuum of the effective theory. The good coordinates on the moduli space are the invariant polynomials of ϕ_k, or, the coefficients of the characteristic polynomial:

$$\mathcal{P}(\lambda) = \prod_k (\lambda - \phi_k) = \sum_l u_l \lambda^l$$

The solution, found for $N = 2$ in Refs. 29 and 30 and then extended in Refs. 1 and 16 to the general case makes use of a family of hyperelliptic curves:

$$y^2 = \mathcal{P}(\lambda)^2 - \Lambda^{2N} \tag{14}$$

It turns out that the period matrix of such curve is precisely τ_{ij}. Moreover, the curve comes equipped with a certain meromorphic one-differential η, whose periods determine the values of the possible central charges of the BPS (short massive) representations of $\mathcal{N} = 2$ superalgebra, corresponding to the particles in the theory. It has been noticed in Ref. 9 that the family of curves (14) together with the one-differential η can be interpreted with the help of integrable system. Consider the periodic Toda chain:

$$H = \sum_{i=1}^{N} \frac{1}{2} p_i^2 + \Lambda^2 \sum_i e^{q_i - q_{i+1}}, \qquad q_{N+1} = q_1 \tag{15}$$

It is integrable system, it has N integrals I_k, $k = 1, \ldots, N$ in involution and the common level set of the integrals I_k is the Abelian variety, which can be identified with the Jacobian of the spectral curve (14) provided that $I_k = u_k$. Moreover the periods of differential η are the periods of the one-form $\sum_k p_k \, dq_k$ along the cycles on the Jacobian.

This picture was explained on the general grounds of $\mathcal{N} = 2$ supersymmetry in Ref. 5 where it was shown that the Coulomb branch of $\mathcal{N} = 2$ gauge theory with massive adjoint hypermultiplet is described in a similar vein by the space of integrals of motion of the elliptic Calogero–Moser system (in fact, the description of the system, presented in Ref. 5, uses the auxilliary three-dimensional gauge theory, which is equivalent to elliptic Calogero–Moser system owing to the construction of Ref. 10):

$$H = \sum_{i=1}^{N} \frac{1}{2} p_i^2 + m^2 \sum_{i \neq j} \wp(q_{ij}), \tag{16}$$

where m is the mass of the hypermultiplet and the elliptic modulus τ entering the $\wp$-function is the ultraviolet bare coupling (which is finite in such a theory). The flow to the pure $\mathcal{N} = 2$ theory proceeds by the tuning m and τ in such a way that $\text{Im}\,\tau \to \infty$, and $\Lambda^{2N} = m^{2N} \exp(2\pi i \tau)$ is finite.

The main property of both the systems (15) and (16)) is the polynomiality of the integrals of motion in the momenta of the particles. Asymptotically the integrals behave like

$$\sum_k \lambda^k I_k \sim \prod_i (\lambda - p_i),$$

which allows to identify the asymptotic momenta with the asymptotic eigenvalues of the Higgs field ϕ.

The natural origin of $\mathcal{N} = 2$ supersymmetry in four dimensions is the minimal supersymmetry in five dimensions. One may imagine that five-dimensional supersymmetric gauge theory exists (or it exists effectively in some intermediate energy regime) and consider compactifying it on a circle of finite radius R.

Then by going to small energies in the four-dimensional sense one gets a modification of the previous picture in the following sense. The scalar ϕ of the four-dimensional theory is in fact a sum of the real scalar φ of five-dimensional theory and the fifth component of the gauge field:

$$\phi = A_5 + i\varphi.$$

The remnant of the five-dimensional gauge invariance is the shift of A_5 by a discrete multiple of $1/R$, exactly like in (8). It implies that good coordinates on $\mathcal{M}$ are no longer polynomials of ϕ_k but rather trigonometric functions, like $\mathrm{Tr}\exp(R\phi)$. Given the relation between the asymptotic momenta and the eigenvalues of the Higgs field, we conclude that the five-dimensional theories compactified on a circle are described by the relativistic integrable systems, since it is there that the momenta enter the Hamiltonians in a trigonometric fashion. The argument led to the conjecture [22] that the pure supersymmetric theory is described by the relativistic Toda system, while the theory with adjoint hypermultiplet corresponds to elliptic Ruijsenaars–Schneider model. The radius R of the fifth dimension is identified with the parameter β (inverse speed of light).

5   Stringy Derivation

In this section, following Ref. 18, we derive the last statement using instanton calculus. We consider SU(2) five-dimensional theory, compactified on circle. This theory can be realized in the context of M-theory compactified on a *noncompact* Calabi–Yau manifold CY times the same circle (for detailed introduction into the ideology of *geometrical engeneering* see Ref. 14 and references there; toric mirrors are presented in Ref. 2 and mathematical attempts to derive them were made in Refs. 7, 8, and 17). The relevant CY is the total space of the line bundle $p_1^* \mathcal{O}(-2) \otimes p_2^* \mathcal{O}(-2)$ over $\mathbb{P}^1 \times \mathbb{P}^1$. The sizes t_1, t_2 of the spheres are the scalars in the two vector multiplets. We are interested in the dependence of the coupling corresponding to one of the spheres (in the limit where the sphere shrinks to zero the five-dimensional theory develops an enhancement of the gauge symmetry).

It turns out that one can compute the corrections to the central charge, corresponding to the W-boson, coming from the $M2$-brane, wrapping one of these spheres. The answer is

$$Z = t_1 + \sum_{m,n} e^{-nt_1 - mt_2} \frac{(2m + 2n - 1)!}{m!^2 n!^2} \qquad (17)$$

where (m, n) is the bidegree of a worldsheet instanton (one uses the equivalence of M-theory compactified on $CY \times \mathbf{S}^1$ to the IIA string compactified on CY). The formula (17) is a result of the expansion of the generating function of the equivariant Euler characters of the moduli spaces of holomorphic maps of $\mathbb{P}^1$ to $\mathbb{P}^1 \times \mathbb{P}^1$. The numerator $(2m + 2n - 1)!$ corresponds roughly to the obstruction bundle, while the denominators $m!^2$ and $n!^2$ come from the integration over $\mathbb{P}^{2m+1} \times \mathbb{P}^{2n+1}$, which is the toric compactification of the space of parameterized maps, as suggested by Refs. 19 and 20 and elaborated further in Refs. 7 and 8. The connection to the relativistic Toda becomes transparent when one notices that (17) is actually one of the periods of a differential $\eta = \log wdz/z$ on a curve

$$\sqrt{q_1}(w + w^{-1}) + \sqrt{q_2}(z + z^{-1}) = 1, \qquad (18)$$

which is equivalent to the spectral curve of $N = 2$ relativistic Toda (4) provided that the change of variables ($\theta_1 + \theta_2 = 0$ to exclude the center of mass) is made: $q_1 = 1/H^2$, $q_2 = i\gamma^4/H^2$ and

$$w = e^{\beta\theta_1} \sqrt{\frac{1 - \gamma^2 e^{q_2 - q_1}}{1 - \gamma^2 e^{q_1 - q_2}}} \quad z = we^{q_1 - q_2}. \qquad (19)$$

6   Conclusions and Speculations

To summarize, we presented two interesting sources of the appearence of integrable systems in the gauge theories. The system may be equivalent to the gauge theory and this is manifested in its origin as coming by the reduction-projection from the simple system with bigger phase space. The system may encode the information about effective degrees of freedom of the gauge theory through the geometry of its phase space. It is tempting to speculate that there exists a unifying description of both situations. A step in this direction is the work of A. Hanany and E. Witten [13], who explained the coincidence of the moduli space of the three-dimensional $\mathcal{N} = 4$ SU(N) gauge theory and the moduli space of charge N BPS monopoles in the SU(2) theory broken down to U(1). Their explanation involves the realization of both gauge theory and the monopole configurations using the

D- and NS-branes of IIB string theory. The monopole moduli space is the analog of the phase space of integrable model, obtained by the Hamiltonian reduction (in fact, in this case there exist two reductions leading to the same moduli space, as a consequence of Nahm' duality). Finally, the integrable system may tell us something interesting about the gauge theory. For example, the solution of the five-dimensional theory with massive adjoint hypermultiplet identifies the coupling ν with the mass m. The solution exhibits a symmetry $m \rightarrow m + i(n_1 + n_2\tau)/(2R)$, n_1, $n_2 \in \mathbb{Z}$ (see Ref. 22), which suggests a presence of a hidden six-dimensional tower of states. Recently this six-dimensional structure was used by M. Rozali in the attempt to see the U-duality group of seven dimensional compactification of M-theory in the Matrix theory description [25].

Acknowledgments: I would like to thank my collaborators V. Fock, A. Gorsky, A. Lawrence, A. Losev, G. Moore, A. Rosly, V. Rubtsov, and S. Shatashvili for their insights. My thanks to R. Donagi, A. Gerasimov, E. Martinec, A. Polyakov, N. Seiberg, and A. Turbiner for useful discussions.

The research was supported in part by Harvard Society of Fellows, by NSF grant PHY-92-18167, by RFFI grant 96-02-18046 and grant 96-15-96455 for support of scientific schools.

The paper was completed during my visit to LPTHE, Université Pierre et Marie Curie. I am grateful to L. Baulieu for his hospitality.

7   References

1. P. C. Argyres and A. E. Farragi, *The vacuum structure and spectrum of $N = 2$ supersymmetric* SU(n) *gauge theory*, Phys. Rev. Lett. **74** (1995), 3931–3934, hep-th/9411057.

2. V. V. Batyrev, *Variations of the mixed Hodge structure of affine hypersurfaces in algebraic tori*, Duke Math. J. **69** (1993), No. 2, 349–409.

3. H. W. Braden and R. Sasaki, *The Ruijsenaars–Schneider model*, Progr. Theor. Phys. **97** (1997), No. 6, 1003–1017, hep-th/9702182.

4. F. Calogero, *Solution of the one-dimensional N-body problems with quadratic and/or inversely quadratic pair potentials*, J. Math. Phys. **12** (1971), 419–436.

5. R. Donagi and E. Witten, *Supersymmetric Yang–Mills theory and integrable systems*, Nucl. Phys. B **460** (1996), No. 2, 299–334.

6. V. Fock, A. Gorsky, N. Nekrasov, and V. Rubtsov, *Duality in integrable systems and gauge theories*, hep-th/9906235.

7. A. Givental, *Equivariant Gromov–Witten invariants*, Internat. Math. Res. Notices (1996), No. 13, 613–663, alg-geom/ 9603021.

8. _____ , *A mirror theorem for complete intersections*, Topological Field Theory, Primitive Forms and Related Topics (Kyoto,1996) (M. Kashiwara, A. Matsuo, K. Saito, and I. Satake, eds.), Progr. Math., Vol. 160, Birkhäuser, Boston, MA, 1998, pp. 141–175, alg-geom/9701016.

9. A. Gorsky, I. M. Krichever, A. Marshakov, A. Morozov, and A. Mironov, *Integrablity and Seiberg–Witten exact solution*, Phys. Lett. B **355** (1995), No. 3-4, 466–474, hep-th/9505035.

10. A. Gorsky and N. Nekrasov, *Elliptic Calogero–Moser system from two-dimensional current algebra*, hep-th/9401021.

11. A. Gorsky and N. Nekrasov, *Hamiltonian systems of Calogero-type, and two-dimensional Yang–Mills theory*, Nucl. Phys. B **414** (1994), No. 1-2, 213–238.

12. A. Gorsky and N. Nekrasov, *Relativistic Calogero–Moser model as gauged WZW theory*, Nucl. Phys. B **436** (1995), No. 3, 582–608, hep-th/9401017.

13. A. Hanany and E. Witten, *Type IIB superstrings, BPS monopoles, and three-dimensional gauge theories*, Nucl. Phys. B **492** (1997), No. 1-2, 152–190, hep-th/9611230.

14. S. Katz, P. Mayr, and C. Vafa, *Mirror symmetry and exact solution of 4D N = 2 gauge theories. I*, Adv. Theor. Math. Phys. **1** (1997), No. 1, 53–114, hep-th/9706110.

15. D. Kazhdan, B. Kostant, and S. Sternberg, *Hamiltonian group actions and dynamical systems of Calogero type*, Commun. Pure Appl. Math. **31** (1978), No. 4, 481–507.

16. A. Klemm, W. Lerche, S. Theisen, and S. Yankielowicz, *Simple singularities and N = 2 supersymmetric Yang–Mills theory*, Phys. Lett. B **344** (1995), No. 1-4, 169–175, hep-th/9411048.

17. M. Kontsevich, *Enumeration of rational curves via torus actions*, The Moduli Space of Curves (Texel, 1994) (R. Dijkgraaf, C. Faber, and G. van der Geer, eds.), Progr. Math., Vol. 129, Birkhäuser, Boston, MA, 1995, pp. 335–368, hep-th/9405035.

18. A. Lawrence and N. Nekrasov, *Instanton sums and five-dimensional gauge theories*, Nucl. Phys. B **513** (1998), No. 1-2, 239–265, hep-th/9706025.

19. D. R. Morrison and M. R. Plesser, *Summing up the instantons: quantum cohomology and mirror symmetry in toric varieties*, Nucl. Phys. B **440** (1995), No. 1-2, 279–354, hep-th/9412136.

20. _____, *Towards mirror symmetry as duality for two-dimensional Abelian gauge theories*, S-duality and Mirror Symmetry (Trieste, 1995) (E. Gava, K. S. Narain, and C. Vafa, eds.), Nuclear Phys. B Proc. Suppl., Vol. 46, North-Holland, Amsterdam, 1997, pp. 177–186, hep-th/9508107.

21. N. Nekrasov, *On a duality in Calogero–Moser systems*, Tech. Report ITEP-TH-16/97, ITEP, Moscow, 1997.

22. _____, *Five-dimensional gauge theories and relativistic integrable systems*, Nucl. Phys. B **531** (1998), No. 1-3, 323–344, hep-th/9609219.

23. M. A. Olshanetsky and A. M. Perelomov, *Completely integrable Hamiltonian systems connected with semisimple Lie algebras*, Invent. Math. **37** (1976), No. 2, 93–108.

24. _____, *Classical integrable finite-dimensional systems related to Lie algebras*, Phys. Rep. **71** (1981), No. 5, 313–400.

25. M. Rozali, *Matrix theory and U-duality in seven dimensions*, Phys. Lett. B **400** (1997), No. 3-4, 260–264, hep-th/9702136.

26. S. N. M. Ruijsenaars, *Complete integrability of relativistic Calogero–Moser systems and elliptic function identities*, Commun. Math. Phys. **110** (1987), No. 2, 191–213.

27. _____, *Finite-dimensional soliton systems*, Integrable and Superintegrable Systems (B. A. Kupershmidt, ed.), World Scientific, Singapore, 1990, pp. 165–206.

28. S. N. M. Ruijsenaars and H. Schneider, *A new class of integrable systems and its relation to solitons*, Ann. Phys. (NY) **170** (1986), No. 2, 370–405.

29. N. Seiberg and E. Witten, *Electric-magnetic duality, monopole condensation, and confinement in $N = 2$ supersymmetric Yang–Mills theory*, Nucl. Phys. B **426** (1994), No. 1, 19–52, Errata, **430** (1994), no. 2, 485–486.

30. _____, *Monopoles, duality and chiral symmetry breaking in $N = 2$ supersymmetric QCD*, Nucl. Phys. B **431** (1994), No. 2, 484–550.

31. B. Sutherland, *Exact results for a quantum many-body problem in one dimension. II*, Phys. Rev. A **5** (1972), 1372–1376.

24

Generalizations of Calogero Systems

Alexios P. Polychronakos

ABSTRACT We point out some directions for potential generalizations of Calogero-type systems. In particular, we demonstrate that a many-matrix model gives rise, upon Hamiltonian reduction, to a multidimensional version of the Calogero–Sutherland–Moser model and its spin generalizations. Some simple solutions of these models are demonstrated by solving the corresponding matrix equations. We also show that a supersymmetric system of spinless particles in which supersymmetry is realized through exchange operators exhibits reflectionless two-body scattering for arbitrary prepotential. The exchange-Calogero system is the simplest example, but it is conjectured that appropriate three-body forces would make all such systems integrable.

The quest for integrable (nonrelativistic) particle systems in more than one spatial dimension is often frustrating. In general, no nontrivial such systems exist (with the exception of some isolated few-body cases), that is, systems with a nonquadratic potential that are not a repackaging of one-dimensional degrees of freedom. (See F. Calogero's chapter for some demonstrations of such repackagings, and a physical definition of true multidimensional models.) In one dimension things are different. The celebrated Calogero model and its various generalizations (also known as the Calogero–Sutherland–Moser systems) [2–5, 12, 15, 20–22], which is the topic of this book, is an integrable class of models that has rewarded us with much fun and many analytical results. The question then is, firstly, to what extent these models remain solvable, if at all, in higher dimensions and, secondly, if there is any more general class of models that are integrable in one dimension.

In this chapter, we choose to look at techniques that have proved convenient and fruitful for the Calogero model and try to push them further to explore uncharted territory. The results are not conclusive, but they are to some degree encouraging that as yet uncovered integrable models exist and are within our reach.

One such approach to Calogero-like systems is though matrix models, in which the particle positions are regained as the eigenvalues of some appropriate matrix [10, 16, 17]. The integrability, as well as the solutions of the equations of motion, are simpler to obtain this way. It would seem, then, that this is the most promising route to systems of higher dimension. We will show that, indeed, appropriate matrix models give rise, under Hamiltonian reduction, to multidimensional many-body systems of the Calogero type. These matrix models are not in general integrable as they stand, and therefore we do not expect the corresponding particle systems to be integrable either. The hope is, nevertheless, that some integrable version may exist, and at any rate the (inherently simpler) matrix modes dynamics will allow for a better study of the dynamics of the particle systems.

The starting point will be a many-matrix model consisting of d time-dependent Hermitian $N \times N$ matrices M_i, $i = 1, \ldots, d$, which we will also represent as a vector of matrices $\mathbf{M}$. The action will be the usual kinetic term for each matrix plus some potential invariant under simultaneous unitary conjugation of the matrices. The eigenvalues of M_i then will be interpreted as the i-component of the position vectors of N particles moving in flat d-dimensional space. For this interpretation to be possible, however, the matrices must be simultaneously diagonalizable, else there is no invariant association of the n-th eigenvalue of the matrices as the coordinates of the *same* particle. The eigenvalues of all M_i can be simultaneously permuted with a common unitary transformation, which corresponds to having identical particles. So we write the Lagrangian

$$L = \mathrm{tr}\left\{ \sum_i \frac{1}{2}\dot{M}_i^2 + i\sum_{ij} \frac{1}{2}\Lambda_{ij}[M_i, M_j] \right\} - V(\mathbf{M}), \qquad (1)$$

where overdot stands for time derivative. $\Lambda_{ij} = -\Lambda_{ji}$ is an antisymmetric set of $d(d-1)/2$ Hermitian matrices serving as Lagrange multipliers for the commutativity constraint between the M_i. The potential $V(\mathbf{M}) = V(U^{-1}\mathbf{M}U)$ can be any real function of the M_i invariant under simultaneous unitary transformations of the M_i. The form $V = \mathrm{tr}\,V(\mathbf{M})$, where $V(\mathbf{x})$ is some real scalar function on R^d, will be assumed in what follows, which leads to an external potential $V(\mathbf{x})$ for the particles. The harmonic oscillator potential $V(\mathbf{x}) = \frac{1}{2}\omega^2 \mathbf{x}^2$, corresponding to the matrix potential $V = \frac{1}{2}\omega^2 \sum_i \mathrm{tr}\, M_i^2$, is the simplest example.

For a central potential the model is invariant under $SO(d)$ rotations R_{ij}:

$$M_i \to R_{ij}M_j, \quad \Lambda_{ij} \to R_{ik}R_{jl}\Lambda_{kl}. \qquad (2)$$

As a result, there is a conserved angular momentum

$$J_{ij} = \mathrm{tr}\{M_i\dot{M}_j - M_j\dot{M}_i\}. \qquad (3)$$

Time translation invariance implies the conservation of energy:

$$E = \mathrm{tr}\left\{ \sum_i \frac{1}{2}\dot{M}_i^2 + V(\{M_i\}) \right\}. \qquad (4)$$

The invariance of (1) under simultaneous conjugation of all matrices by a time-independent unitary matrix implies the existence of a conserved matrix "angular momentum":

$$K = i \sum_i [M_i, \dot{M}_i], \tag{5}$$

where K is a traceless Hermitian matrix. Choosing it to have the "minimal" form where all its eigenvalues are equal but one, that is,

$$K_{mn} = \ell(\delta_{mn} - u_m^* u_n), \quad \sum_m |u_m|^2 = N, \tag{6}$$

where u is a fixed N-vector, will lead to Calogero dynamics for the eigenvalues of M_i, just as in the one-dimensional case,

The equations of motion are

$$\ddot{M}_i + V_{,i}(\mathbf{M}) + i \sum_j [\Lambda_{ij}, M_j] = 0 \tag{7}$$

plus the constraint

$$[M_i, M_j] = 0. \tag{8}$$

The constraint implies that M_i can all be diagonalized with a common time-dependent unitary rotation $U(t)$:

$$M_i = U^{-1} X_i U, \quad X_i = diag(x_{i,1}, \ldots x_{i,N}). \tag{9}$$

In terms of (9) the equations of motion acquire the form

$$\ddot{X}_i + 2[\dot{X}_i, A] + [X_i, \dot{A}] + [[X_i, A], A] + V_{,i}(\mathbf{X}) + i \sum_j [\tilde{\Lambda}_{ij}, X_j] = 0, \tag{10}$$

where $\tilde{\Lambda} = U\Lambda U^{-1}$ and $A = \dot{U} U^{-1}$ is the "gauge potential" generated by the time variance of U.

We now recall that the commutator of a diagonal matrix with any matrix has zero diagonal elements, since

$$[D, B]_{mn} = (d_m - d_n) B_{mn}, \quad \text{if } D = diag(d_1, \ldots d_N). \tag{11}$$

Therefore, isolating the diagonal terms in (10), only the first, fourth, and fifth term contribute, and we have

$$\ddot{x}_{i,m} + \sum_n 2(x_{i,m} - x_{i,n}) A_{mn} A_{nm} + V_{,i}(\mathbf{x}_m) = 0 \tag{12}$$

Plugging the form (6) and (9) in (5), on the other hand we have

$$i \sum_i (x_{i,m} - x_{i,n})^2 A_{mn} = \ell(\delta_{mn} - \tilde{u}_m^* \tilde{u}_n) \tag{13}$$

where $\tilde{u} = Uu$. For $m \neq n$ and $m = n$ we obtain the relations for A_{mn} and $\tilde{u}_m$, respectively,

$$A_{mn} = \frac{i\ell \tilde{u}_m^* \tilde{u}_n}{(\mathbf{x}_m - \mathbf{x}_n)^2}, \quad |\tilde{u}_m|^2 = 1. \tag{14}$$

Plugging these in (12) and calling $(x_{1,m}, \ldots x_{d,m}) = \mathbf{x}_m$ we finally obtain

$$\ddot{\mathbf{x}}_m - 2\ell^2 \sum_{n \neq m} \frac{\mathbf{x}_m - \mathbf{x}_n}{(\mathbf{x}_m - \mathbf{x}_n)^4} + \nabla V(\mathbf{x}_m) = 0, \tag{15}$$

which is the equation of motion for the positions of particles $\mathbf{x}_m$ in an external potential $V(\mathbf{x})$ and interacting through a d-dimensional two-body inverse square potential ℓ^2/x^2, that is, a d-dimensional generalization of the Calogero model. The key elements in the derivation are that the Lagrange multiplier term does not influence the eigenvalue equations of motion and that the angular part reproduces the rotationally invariant d-dimensional inverse-square potential. Note that the energy (4) and angular momentum (3) become the corresponding quantities of the particle system in the constraint subspace, that is,

$$E = \sum_m \frac{1}{2}\dot{\mathbf{x}}_m^2 + \sum_{m \neq n} \frac{\ell^2}{(\mathbf{x}_m - \mathbf{x}_n)^2} + \sum_m V(\mathbf{x}_m) \tag{16}$$

$$J_{ij} = x_i \dot{x}_j - x_j \dot{x}_i. \tag{17}$$

If the "angular momentum" K is not in the "minimal" form (6), it will enter the equations for the eigenvalues in a nontrivial way and will give rise to multidimensional generalizations of the "spin-Calogero" model [7, 9, 13, 24]. To see this, we point out that the restriction of the Hamiltonian in the constraint subspace $[M_i, M_j] = 0$ takes the form

$$H = \sum_m \frac{1}{2}\dot{\mathbf{x}}_m^2 + \sum_{m \neq n} \frac{\tilde{K}_{mn}\tilde{K}_{nm}}{(\mathbf{x}_m - \mathbf{x}_n)^2} + \sum_m V(\mathbf{x}_m), \tag{18}$$

where $\tilde{K} = UKU^{-1}$. As usual, $\tilde{K}_{mn}$ Poisson-commute to the SU(N) algebra and can be recast into internal degrees of freedom ("spin") for the particles [14]:

$$\tilde{K}_{mn} = \sum_{a=1}^p S_m^a S_n^a. \tag{19}$$

To study the matrix equations of motion we specify to the minimum nontrivial dimensions $d = 2$ and to the rotationally invariant harmonic external potential $V(\mathbf{x}) = \frac{1}{2}\omega^2 \mathbf{x}^2$. Defining the non-Hermitian matrix $M = M_1 + iM_2$, the equations of motion and constraint become

$$\ddot{M} + [\Lambda, M] + \omega^2 M = 0, \quad [M, M^\dagger] = 0, \tag{20}$$

while the "angular momentum" K takes the form

$$K = i[M^\dagger, \dot{M}] \tag{21}$$

in the constraint subspace. Solving the two-dimensional Calogero model amounts to finding solutions of the above matrix equations for M with the form (6) for K.

The simplest possible class of solutions is the one with $\Lambda = 0$. It can be shown, however, that these solutions correspond to linear motion of the particles and the system reduces to the one-dimensional Calogero model. The solution to the equations of motion is

$$M = Ae^{i\omega t} + B^\dagger e^{-i\omega t}, \tag{22}$$

where the matrices A, B, to satisfy the commutativity and "angular momentum" constraints, must obey

$$[A, B] = 0, \quad [A, A^\dagger] = [B, B^\dagger] = \frac{K}{2\omega} \tag{23}$$

with K as in (6). In terms of the new matrices $Q = A + B^\dagger$ and $P = i\omega(A - B^\dagger)$ (representing the position and velocity matrices at $t = 0$) relations (23) become

$$[Q, Q^\dagger] = [P, P^\dagger] = 0, \quad [Q, P^\dagger] = -iK \tag{24}$$

This tells us that P and Q, although non-Hermitian, can each be diagonalized with a unitary rotation, with complex eigenvalues (representing the initial positions and velocities of the particles on the complex plane). Choosing a basis where Q is diagonal with eigenvalues q_m, we deduce from the last relation in (24) that the matrix elements of P are

$$P_{mn} = ip_m \delta_{mn} + \frac{i\ell}{q_m^* - q_n^*}(1 - \delta_{mn}) \tag{25}$$

where we used $|u_m| = 1$ in the Q-diagonal basis and further chose the phases of the states such that $u_m = 1$. From $[P, P^\dagger] = 0$ now we obtain

$$\frac{p_m - p_n}{q_m - q_n} = \text{real}, \quad \sum_{k \neq m,n} \frac{1}{(q_m - q_k)(q_m^* - q_n^*)} = \text{real} \tag{26}$$

By using the invariance of the equations by a shift of Q and P by a multiple of the unit matrix (which is related to the fact that the center-of-mass motion decouples from the relative motion) we can always choose $q_1 = p_1 = 0$. Then the first relation above implies that all q_m are colinear (i.e., $q_m/q_n = \text{real}$) unless $p_m = aq_m$ for some real a. The second relation, however, is satisfied only if the q_m are colinear. By the first relation, p_m

will also be colinear with q_m. Therefore, we see that the one-dimensional Calogero model is included in the $\Lambda = 0$ sector of the general model.

The other case in which the equations of motion have an obvious solution is when $\Lambda = $ constant. Choosing a basis in which Λ is diagonal, with (real) eigenvalues λ_n, we have

$$\ddot{M}_{mn} + \omega_{mn}^2 M_{mn} = 0, \quad \omega_{mn}^2 = \omega^2 + \lambda_m - \lambda_n, \tag{27}$$

which has as solutions (we assume $\omega_{mn}^2 > 0$)

$$M_{mn} = A_{mn} e^{i\omega_{mn}t} + B_{mn}^\dagger e^{-i\omega_{mn}t}. \tag{28}$$

The task of finding the most general A_{mn}, B_{mn} that satisfy the commutativity and "angular momentum" constraints is not trivial. We demonstrate here a particularly simple solution, namely,

$$A_{mn} = A_m \delta_{mn} + a_n \delta_{m,n+1}, \quad B_{mn} = B_n \delta_{mn},$$
$$|a_n| = a, \quad \omega_{n+1,n+2} - \omega_{n,n+1} = \frac{\ell}{a^2}. \tag{29}$$

The last constraint for $\omega_{n,n+1}$ translates into $N-1$ algebraic equations for the $N-1$ variables $\lambda_n - \lambda_{n+1}$ (clearly Λ can be shifted by any multiple of the unit matrix). It is obvious that the diagonal part A_m, B_m represents a general motion of the decoupling center of mass. The eigenvalues of the off-diagonal part of A are the Nth roots of $a_1 a_2 \cdots a_N$. So the off-diagonal part of M has eigenvalues

$$z_m = x_m + iy_m = ae^{i(2\pi m/N + \omega_r t)},$$
$$\omega_r = \sum_n \omega_{n,n+1} = \omega_{12} + (N-1)\frac{\ell}{a^2}. \tag{30}$$

Therefore the relative motion is one in which the particles are regularly positioned on a circle of radius $|a|$ and rotate with constant angular velocity ω_r.

The above model can be generalized to one with unitary matrices. Omitting the details of the calculation, we simply state the result. The Lagrangian of the model is

$$L = \text{tr}\left\{ \sum_i \frac{1}{2} R_i^2 \dot{U}_i^\dagger \dot{U}_i + i \sum_{ij} (\Lambda_{ij}[U_i, U_j] + \Lambda_{ij}^\dagger [U_i^\dagger, U_j^\dagger]) \right\}$$
$$- V(\{U_i\}), \tag{31}$$

where again V is some real conjugation-invariant potential. The Lagrange multiplier matrices $\Lambda_{ij} = -\Lambda_{ji}$ are not Hermitian, but the constraints arising from the variation of Λ_{ij} and $\Lambda_{ij}^\dagger$ are compatible (in fact, equivalent) for unitary U_i. The eigenvalues of U_i, written as $\exp(ix_{i,m}/R_i)$ represent coordinates of particles on a d-dimensional torus of radii R_i. Upon choosing the "angular momentum"

$$i \sum_i R_i^2 [U_i^\dagger, U_i] = \ell(1 - uu^\dagger) \tag{32}$$

as before, the $\mathbf{x}_n$ move like particles on the torus in an external potential $V(\mathbf{x}_n)$ and interacting through a periodic generalization of the d-dimensional two-body inverse-square potential

$$V(\mathbf{x}_n - \mathbf{x}_m) = \frac{\ell}{\sum_i \pi^2 R_i^2 \sin^2([x_{i,m} - x_{i,n}]/\pi R_i)}. \tag{33}$$

Similarly, the d-dimensional generalization of the inverse-sinh model can be obtained by taking $R_i \to iR_i$.

Clearly, there are a lot of unanswered questions in the above. Firstly, the physical meaning of the Lagrange multiplier matrices Λ_{ij} and their role at classifying the types of solutions are yet to be understood. The matrix equations of motion have barely been touched in the general case, and their solutions are unknown. Even in the case $\Lambda = $ constant the general solution has not been fully studied. Generalizations of these models involving the Weierstrass function potential could be sought, where instead of a matrix model one would have to deal with an appropriate topological model [8]. In fact, it would be interesting to consider what type of model would give rise to Calogero-type dynamics on a manifold of more general geometry and/or topology. Finally, the quantization of the model, being a constrained system, is a nontrivial issue.

To tackle, now, the question of finding more general models in one dimension, we turn to another useful approach in the study of these systems, that of exchange operators [18]. Using a generalization of the momentum involving operators exchanging the particles (similar to the so-called Dunkl operators [6]), one can streamline the proof of the quantum integrability of these systems. The energy wavefunctions can also, in some cases, be given in a ladder operator form [1, 11]. One interesting feature of these exchange-operator models is that they allow the particles to penetrate perfectly each other, with no backscattering. This makes them very close to being free, and provides simple physical intuition on the degeneracies of their spectrum [19]. It would seem, then, that systems with this property are prime candidates for integrability or solvability. We will show that a generic class of these systems can be obtained through a supersymmetric construction.

Let us first look at supersymmetric quantum mechanics in one dimension. They are characterized by the existence of a set of Hermitian operators, Q_a, called supersymmetry charges, and of a grading operator G (also written $(-)^F$, where F is the fermion operator), satisfying the algebra

$$\{Q_a, Q_b\} = \delta_{ab} 2H, \quad \{Q_a, G\} = 0, \quad G^2 = 1. \tag{34}$$

H is the Hamiltonian of the system. It is realized in the standard way in the Hilbert space of a one-dimensional spin-$\frac{1}{2}$ particle [23], where there are $N = 2$ SUSY generators. We will present another realization here, not requiring spin and leading to $N = 2$. If M is the parity operator for the particle, we have

$$Q_1 = p - iMW, \quad Q_2 = iMp + W, \quad G = M, \tag{35}$$

where p is the momentum operator and $W(x)$ a prepotential. Hermiticity and supersymmetry restrict $W(x)$ to be a real antisymmetric function. The Hamiltonian, then, is

$$H = \frac{1}{2}(p^2 + W^2 + MW'). \tag{36}$$

It is parity invariant, and contains a parity-dependent term. For parity even (odd) states the potential is $W^2 \pm W'$, respectively.

Let us assume that W goes to a constant at spatial infinity, that is,

$$W(\pm\infty) = \pm c. \tag{37}$$

The potential, then, goes to a constant at infinity and the Hamiltonian supports scattering. For a fixed momentum the scattering matrix is a 2×2 matrix defined through the relation

$$\begin{pmatrix} A' \\ B' \end{pmatrix} = S \begin{pmatrix} A \\ B \end{pmatrix}, \tag{38}$$

where A and B are the amplitudes of the incoming asymptotic plane waves from the left and right, respectively, and A', B' the amplitudes of the corresponding outgoing waves (A' to the right and B' to the left). S is expressed in terms of transmission and reflection amplitude coefficients as

$$S = \begin{pmatrix} T_L & R_R \\ R_L & T_R \end{pmatrix} \tag{39}$$

in an obvious notation. Unitarity and parity invariance imply

$$S^\dagger S = 1, \quad S = \sigma_1 S \sigma_1; \tag{40}$$

that is,

$$R_L = R_R = R, \quad T_L = T_R = T, \quad R^*T + T^*R = 0. \tag{41}$$

Supersymmetry, on the other hand, means that if Ψ is a (scattering) energy eigenstate, so are $Q_1\Psi$ and $Q_2\Psi$. Since $Q_2 = iMQ_1$, and we have already considered parity, only $Q_1\Psi$ need be considered. Using (35) and (37), we see hat the action of Q_1 on the asymptotic part of the wavefunction is

$$Q_1 \begin{pmatrix} A \\ B \end{pmatrix} = \begin{pmatrix} k & -ic \\ ic & -k \end{pmatrix} \begin{pmatrix} A \\ B \end{pmatrix}, \quad Q_1 \begin{pmatrix} A' \\ B' \end{pmatrix} = \begin{pmatrix} k & ic \\ -ic & -k \end{pmatrix} \begin{pmatrix} A' \\ B' \end{pmatrix}. \tag{42}$$

Implementing the above in terms of the matrix elements of S, we finally obtain

$$R = -\frac{ic}{k}T,$$ (43)

which is our main result. This also implies for the reflection and transmission probabilities

$$|T|^2 = 1 - |R|^2 = \frac{k^2}{c^2 + k^2};$$ (44)

that is, they are *independent* of the details of the potential and only depend on its asymptotic value! In the special case where the potential goes to zero at infinity, that is, $c = 0$, we obtain the result

$$R = 0.$$ (45)

So the potential becomes transparent in that case. This result holds also for an arbitrary, non-antisymmetric prepotential W, in which case there is no parity invariance and no supersymmetry, provided that it goes to zero at spatial infinity.

We now consider a many-particle system with only two-body interactions of the form $W(x_i - x_j)^2 + M_{ij}W'(x_i - x_j)$, where M_{ij} is the operator that exchanges particles i and j. From the previous discussion it is clear that the two-body scattering matrix will have the property (43). In the particular case of a potential falling off to zero, (45) means that the particles go through each other without backscattering, the wavefunction simply picking up a momentum-dependent phase. This guarantees that an asymptotic Bethe-ansatz (ABA) solution is trivially self-consistent.

In general, for ABA to be consistent, the two-body scattering matrix must satisfy the Yang–Baxter relation:

$$S_{12}S_{13}S_{23} = S_{23}S_{13}S_{12}.$$ (46)

S_{ij} is the two-body scattering matrix in the Hilbert space of particles i and j; that is, $S_{ij} = T(p_{ij}) + M_{ij}R(p_{ij})$ where T and R are the previously defined transmission and reflection amplitudes as functions of the relative momentum p_{ij} of particles i and j. Equation (46) guarantees that the wavefunction after scattering is independent of the order of scattering. (Alternatively, it can be thought of as the condition that the transfer matrix between different asymptotic regions of the wavefunction is independent of the order of transition, which we interpret as an absence of non-Abelian "Aharonov–Bohm"-type effects from three-particle coincidence points. See E. Gutkin's chapter for a discussion of this point of view.) It can, now, be checked that any two T and R satisfying (43) will automatically satisfy the Yang–Baxter relation. Therefore, the models presented above admit ABA-type solutions and are good candidates for integrability.

The Yang–Baxter relation is only a necessary condition, and does not guarantee integrability. In other words, although three-particle scattering regions do not introduce global obstructions to the ABA solution, they may still produce local effects that destroy integrability. A detailed analysis of the dynamics of these models is, then, necessary to decide this question. To this end, we propose a somewhat modified version of the models, with Hamiltonian

$$H = \sum_i \frac{1}{2} \pi_i^2, \quad \text{where } \pi_i = p_i + i \sum_{j \neq i} M_{ij} W(x_i - x_j). \tag{47}$$

Upon expanding the squares of the operators π_i, we recover the terms of the previous Hamiltonians, plus some three-body terms. In fact, the condition that the three-body interactions reduce to two-body ones leads to the usual class of Calogero–Sutherland–Moser systems [18]. The commutator $[\pi_i, \pi_j]$ is also expressed in terms of the resulting three-body potential, and in the special case of the CSM model the proof of involubility of $I_n = \sum_i \pi_i^n$ is quite straightforward [18].

In the case of arbitrary (odd) prepotential $W(x)$ with the asymptotic properties assumed in the previous discussion, it is clear that the asymptotic momenta should be conserved quantities, if the system is integrable. The corresponding permutation-invariant quantities I_n should therefore be as above, with the possible addition of terms that fall off to zero at asymptotic regions. The question of the involubility of the above I_n is, however, algebraically quite hard to decide for generic $W(x)$ and remains an open issue.

We are faced with a situation where our computational power (or at least that of the author) is not up to the task. Perhaps we need to develop new tools and techniques that can deal with this problem in a more efficient way. At any rate, the challenge of uncovering a yet wider class of integrable models, be it in one or more dimensions, is still with us.

REFERENCES

1. L. Brink, T. H. Hansson, and M. A. Vassiliev, *Explicit solution to the N-body Calogero problem*, Phys. Lett. B **286** (1992), No. 1-2, 109–11.

2. F. Calogero, *Ground state of a one-dimensional N-body system*, J. Math. Phys. **10** (1969), 2197–2200.

3. _____, *Solution of a three-body problem in one dimension*, J. Math. Phys. **10** (1969), 2191–2196.

4. _____, *Solution of the one-dimensional N-body problems with quadratic and/or inversely quadratic pair potentials*, J. Math. Phys. **12** (1971), 419–436.

5. _____, *Exactly solvable one-dimensional many-body problems*, Lett. Nuovo Cimento **13** (1975), No. 11, 411–416.

6. C. F. Dunkl, *Differential-difference operators associated to reflection groups*, Trans. Amer. Math. Soc. **311** (1989), No. 1, 167–183.

7. J. Gibbons and T. Hermsen, *A generalisation of the Calogero–Moser system*, Physica **11D** (1984), No. 3, 337–348.

8. A. Gorsky and N. Nekrasov, *Quantum integrable systems of particles as gauge theories*, Theoret. and Math. Phys. **100** (1994), No. 1, 874–878.

9. Z. N. C. Ha and F. D. M. Haldane, *On models with inverse-square exchange*, Phys. Rev. B **46** (1992), 9359–9368.

10. D. Kazhdan, B. Kostant, and S. Sternberg, *Hamiltonian group actions and dynamical systems of Calogero type*, Commun. Pure Appl. Math. **31** (1978), No. 4, 481–507.

11. L. Lapointe and L. Vinet, *Exact operator solution of the Calogero–Sutherland model*, Commun. Math. Phys. **178** (1996), No. 2, 425–452, hep-th/9507073.

12. C. Marchioro, F. Calogero, and O. Ragnisco, *Exact solution of the classical and quantal one-dimensional many-body problems with the two-body potential $V_a(x) = g^2 a^2 / \sinh^2(ax)$*, Lett. Nuovo Cimento **13** (1975), No. 10, 383–387.

13. J. A. Minahan and A. P. Polychronakos, *Integrable systems for particles with internal degrees of freedom*, Phys. Lett. B **302** (1993), No. 2-3, 265–270.

14. _____, *Interacting fermion systems from two-dimensional QCD*, Phys. Lett. B **336** (1994), 288–294.

15. J. Moser, *Three integrable Hamiltonian systems connected to isospectral deformations*, Adv. Math. **16** (1975), 197–220.

16. M. A. Olshanetsky and A. M. Perelomov, *Classical integrable finite-dimensional systems related to Lie algebras*, Phys. Rep. **71** (1981), No. 5, 313–400.

17. _____, *Quantum integrable systems related to Lie algebras*, Phys. Rep. **94** (1983), No. 6, 313–404.

18. A. P. Polychronakos, *Exchange operator formalism for integrable systems of particles*, Phys. Rev. Lett. **69** (1992), No. 5, 703–705.

19. B. S. Shastry and B. Sutherland, *Solution of some integrable one-dimensional quantum systems*, Phys. Rev. Lett. **71** (1993), No. 1, 5–8.

20. B. Sutherland, *Exact results for a quantum many-body problem in one dimension*, Phys. Rev. A **4** (1971), 2019–2021.

21. _____ , *Exact results for a quantum many-body problem in one dimension*. II, Phys. Rev. A **5** (1972), 1372–1376.

22. _____ , *Exact ground-state wave function for a one-dimensional plasma*, Phys. Rev. Lett. **34** (1975), 1083–1085.

23. E. Witten, *Constraints on supersymmetry breaking*, Nucl. Phys. B **202** (1982), No. 2, 253–316.

24. S. Wojciechowski, *An integrable marriage of the Euler equations with the Calogero–Moser system*, Phys. Lett. A **111** (1985), No. 3, 101–103.

25

Three-Body Generalizations of the Sutherland Problem

C. Quesne

ABSTRACT The three-particle Hamiltonian obtained by replacing the two-body trigonometric potential of the Sutherland problem by a three-body one of a similar form is shown to be exactly solvable. When written in appropriate variables, its eigenfunctions can be expressed in terms of Jack symmetric polynomials. The exact solvability of the problem is explained by a hidden $sl(3, \mathbb{R})$ symmetry. A generalized Sutherland three-particle problem including both two- and three-body trigonometric potentials and internal degrees of freedom is then considered. It is analyzed in terms of three first-order noncommuting differential-difference operators, which are constructed by combining SUSYQM upercharges with the elements of the dihedral group D_6. Three alternative commuting operators are also introduced.

1 Introduction

In 1974, Calogero and Marchioro [8], on one hand, and Wolfes [21], on the other, extended the Calogero problem [5–7] for three particles on a line interacting via inverse-square two-body potentials (and harmonic forces in the case of bound states) to a problem where there is an additional three-body potential of a similar form. Later on, it was pointed out by Olshanetsky and Perelomov [11] that the Calogero–Marchioro–Wolfes (CMW) problem is related to the root system of the exceptional Lie algebra G_2, and to the Weyl group of the latter, namely, the dihedral group D_6. More recently, the Brink et al. [4] and Polychronakos [12] exchange operator formalism was extended to the CMW problem to deal with particles with internal degrees of freedom, thereby leading to a D_6-extended Heisenberg algebra [13].

In this chapter, we present some results for similar generalizations of the Sutherland problem [18–20], wherein trigonometric potentials are considered instead of inverse-square ones [14, 15]. The Hamiltonian considered here is

$$H = -\sum_{i=1}^{3} \partial_i^2 + ga^2 \sum_{\substack{i,j=1 \\ i\neq j}}^{3} \csc^2\big(a(x_i - x_j)\big)$$
$$+ 3fa^2 \sum_{\substack{i,j,k=1 \\ i\neq j\neq k\neq i}}^{3} \csc^2\big(a(x_i + x_j - 2x_k)\big)), \quad (1)$$

where x_i, $i = 1, 2, 3$, $0 \leq x_i \leq \pi/a$, denote the particle coordinates, $\partial_i \equiv \partial/\partial x_i$, and g, f are assumed not to vanish simultaneously and to be such that $g > -\frac{1}{4}$, $f > -\frac{1}{4}$. In the case where $g \neq 0$ and $f = 0$, Hamiltonian (1) reduces to the Sutherland Hamiltonian [18–20], while for $a \to 0$, it goes over into the CMW Hamiltonian [8, 21].

Hamiltonian (1) is invariant under translations of the centre-of-mass, whose coordinate will be denoted by $R = (x_1 + x_2 + x_3)/3$. In other words, H commutes with the total momentum $P = -i\sum_{i=1}^{3} \partial_i$, which may be simultaneously diagonalized. It proves convenient to use two different systems of relative coordinates, namely, $x_{ij} \equiv x_i - x_j$, $i \neq j$, and $y_{ij} \equiv x_i + x_j - 2x_k$, $i \neq j \neq k \neq i$, where in the latter, we suppressed index k as it is entirely determined by i and j.

Since the potentials are singular and crossing is therefore not allowed, in the case of distinguishable particles the wave functions in different sectors of configuration space are disconnected, while for indistinguishable particles, they are related by a symmetry requirement.

For distinguishable particles in a given sector of configuration space, the unnormalized ground-state wave function of Hamiltonian (1) is given by

$$\psi_0(\boldsymbol{x}) = \prod_{\substack{i,j=1 \\ i\neq j}}^{3} |\sin(ax_{ij})|^\kappa |\sin(ay_{ij})|^\lambda, \quad (2)$$

where $\kappa \equiv (1+\sqrt{1+4g})/2$ or 0, and $\lambda \equiv (1+\sqrt{1+4f})/2$ or 0, according to whether $g \neq 0$ or $g = 0$, and $f \neq 0$ or $f = 0$, respectively (or, equivalently, $g = \kappa(\kappa - 1)$, $f = \lambda(\lambda - 1)$). The corresponding eigenvalues of H and P are $E_0 = 8a^2(\kappa^2 + 3\kappa\lambda + 3\lambda^2)$, and $p_0 = 0$ [14].

In Section 2, we will prove that the Hamiltonian (1) with pure three-body interactions, that is, for $g = 0$, is exactly solvable, and we will derive its energy spectrum and eigenfunctions. In Section 3, we will propose an extension of (1) for a system of three particles with internal degrees of freedom, and introduce the corresponding exchange operator formalism.

2   Exact Solvability of the Pure Three-Body Problem

Let us assume that $g = 0$ (hence $\kappa = 0$), and $f \neq 0$ in Eq. (1). For distinguishable particles in a given sector of configuration space, the simultaneous solutions of the eigenvalue equations $H\psi(\boldsymbol{x}) = E\psi(\boldsymbol{x})$, and $P\psi(\boldsymbol{x}) = p\psi(\boldsymbol{x})$ can be found by setting $\psi(\boldsymbol{x}) = \psi_0(\boldsymbol{x})\varphi(\boldsymbol{x})$. The functions $\varphi(\boldsymbol{x})$ satisfy the equations $h\varphi(\boldsymbol{x}) = \epsilon\varphi(\boldsymbol{x})$, and $P\varphi(\boldsymbol{x}) = p\varphi(\boldsymbol{x})$, where $h \equiv (\psi_0(\boldsymbol{x}))^{-1}(H - E_0)\psi_0(\boldsymbol{x})$, and $\epsilon \equiv E - E_0$. In terms of the new variables $z_i \equiv \exp(2ia(x_i - 2x_j + 4x_k)/3)$, where $(ijk) = (123)$, the gauge-transformed Hamiltonian h becomes

$$h = 12a^2\left(\sum_i (z_i\partial_{z_i})^2 + \lambda\sum_{\substack{i,j \\ i\neq j}} \frac{z_i + z_j}{z_i - z_j}z_i\partial_{z_i}\right) - \frac{8}{3}a^2\left(\sum_i z_i\partial_{z_i}\right)^2, \quad (3)$$

while $P = 2a\sum_i z_i\partial_{z_i}$.

It can be easily proved [15] that the eigenfunctions and eigenvalues of h and P are given by

$$\varphi_{\{k\}}(\boldsymbol{x}) = \exp(6iaqR)J_{\{\mu\}}(\boldsymbol{z}; \lambda^{-1}), \quad (4)$$

and

$$\epsilon_{\{k\}} = 4a^2\left[3\sum_i k_i^2 - \frac{2}{3}\left(\sum_i k_i\right)^2 - 6\lambda^2\right],$$

$$p_{\{k\}} = 2a\sum_i k_i = 2a\left(\sum_i \mu_i + 3q\right), \quad (5)$$

where $J_{\{\mu\}}(\boldsymbol{z}; \lambda^{-1})$ denotes the Jack (symmetric) polynomial in the variables z_i, $i = 1, 2, 3$, corresponding to the parameter λ^{-1}, and the partition $\{\mu\} = \{\mu_1\mu_2\}$ into not more than two parts [17]. In Eqs. (4) and (5), $k_1 = q - \lambda$, $k_2 = \mu_2 + q$, $k_3 = \mu_1 + q + \lambda$, and $q \in \mathbb{R}$. In Table 25.1, the explicit form of $J_{\{\mu\}}(\boldsymbol{z}; \lambda^{-1})$ is given for $\mu_1 + \mu_2 \leq 4$.

The eigenfunctions of h can be separated into centre-of-mass and relative functions as follows:

$$\varphi_{\{k\}}(\boldsymbol{x}) = \exp\left[2ia\left(\sum_i k_i\right)R\right]P_{\{\mu\}}(\boldsymbol{\zeta}; \lambda^{-1}), \quad (6)$$

where $P_{\{\mu\}}(\boldsymbol{\zeta}; \lambda^{-1})$ is the polynomial in $\zeta_1 \equiv \sum_i v_i$ and $\zeta_2 \equiv \sum_{i<j} v_iv_j$, $v_i \equiv \exp(-2iax_{jk}) = z_i \exp(-2iaR)$ for $(ijk) = (123)$, obtained from the corresponding Jack polynomial $J_{\{\mu\}}(\boldsymbol{v}; \lambda^{-1})$ by making the change of variables $v_i \to \zeta_1, \zeta_2$. It satisfies the eigenvalue equation

$$h^{\text{rel}}P_{\{\mu\}}(\boldsymbol{\zeta}; \lambda^{-1}) = \epsilon_{\{\mu\}}^{\text{rel}}P_{\{\mu\}}(\boldsymbol{\zeta}; \lambda^{-1}), \quad (7)$$

where

$$h^{\text{rel}} = 8a^2[(\zeta_1^2 - 3\zeta_2)\partial_{\zeta_1}^2 + (\zeta_1\zeta_2 - 9)\partial_{\zeta_1\zeta_2}^2 + (\zeta_2^2 - 3\zeta_1)\partial_{\zeta_2}^2$$
$$+ (3\lambda + 1)(\zeta_1\partial_{\zeta_1} + \zeta_2\partial_{\zeta_2})], \quad (8)$$

$$\epsilon_{\{\mu\}}^{\text{rel}} = 8a^2(\mu_1^2 - \mu_1\mu_2 + \mu_2^2 + 3\lambda\mu_1). \quad (9)$$

TABLE 25.1. Jack polynomials $J_{\{\mu\}}(z;\lambda^{-1})$ for $\mu_1 + \mu_2 \leq 4$.

$\{\mu\}$	$J_{\{\mu\}}(z;\lambda^{-1})$
$\{0\}$	1
$\{1\}$	$\sum_i z_i$
$\{1^2\}$	$\sum_{i<j} z_i z_j$
$\{2\}$	$\sum_i z_i^2 + \frac{2\lambda}{\lambda+1}\sum_{i<j} z_i z_j$
$\{21\}$	$\sum_{i\neq j} z_i^2 z_j + \frac{6\lambda}{2\lambda+1} z_1 z_2 z_3$
$\{3\}$	$\sum_i z_i^3 + \frac{3\lambda}{\lambda+2}\sum_{i\neq j} z_i^2 z_j + \frac{6\lambda^2}{(\lambda+1)(\lambda+2)} z_1 z_2 z_3$
$\{2^2\}$	$\sum_{i<j} z_i^2 z_j^2 + \frac{2\lambda}{\lambda+1} z_1 z_2 z_3 \sum_i z_i$
$\{31\}$	$\sum_{i\neq j} z_i^3 z_j + \frac{2\lambda}{\lambda+1}\sum_{i<j} z_i^2 z_j^2 + \frac{\lambda(5\lambda+3)}{(\lambda+1)^2} z_1 z_2 z_3 \sum_i z_i$
$\{4\}$	$\sum_i z_i^4 + \frac{4\lambda}{\lambda+3}\sum_{i\neq j} z_i^3 z_j + \frac{6\lambda(\lambda+1)}{(\lambda+2)(\lambda+3)}\sum_{i<j} z_i^2 z_j^2$ $+ \frac{12\lambda^2}{(\lambda+2)(\lambda+3)} z_1 z_2 z_3 \sum_i z_i$

TABLE 25.2. Eigenvalues $\epsilon_{\{\mu\}}^{\text{rel}}$ and eigenfunctions $P_{\{\mu\}}(\zeta;\lambda^{-1})$ of h^{rel} for $\mu_1 + \mu_2 \leq 4$.

$\{\mu\}$	$\epsilon_{\{\mu\}}^{\text{rel}}/(8a^2)$	$P_{\{\mu\}}(\zeta;\lambda^{-1})$
$\{0\}$	0	1
$\{1\}$	$3\lambda+1$	ζ_1
$\{1^2\}$	$3\lambda+1$	ζ_2
$\{2\}$	$2(3\lambda+2)$	$\zeta_1^2 - \frac{2}{\lambda+1}\zeta_2$
$\{21\}$	$3(2\lambda+1)$	$\zeta_1\zeta_2 - \frac{3}{2\lambda+1}$
$\{3\}$	$9(\lambda+1)$	$\zeta_1^3 - \frac{6}{\lambda+2}\zeta_1\zeta_2 + \frac{6}{(\lambda+1)(\lambda+2)}$
$\{2^2\}$	$2(3\lambda+2)$	$\zeta_2^2 - \frac{2}{\lambda+1}\zeta_1$
$\{31\}$	$9\lambda+7$	$\zeta_1^2\zeta_2 - \frac{2}{\lambda+1}\zeta_2^2 - \frac{3\lambda+1}{(\lambda+1)^2}\zeta_1$
$\{4\}$	$4(3\lambda+4)$	$\zeta_1^4 - \frac{12}{\lambda+3}\zeta_1^2\zeta_2 + \frac{12}{(\lambda+2)(\lambda+3)}(\zeta_2^2 + 2\zeta_1)$

The relative energies are similar to those obtained with pure two-body interactions, that is for $g \neq 0$ and $f = 0$. In Table 25.2, they are listed for $\mu_1 + \mu_2 \leq 4$, together with the corresponding eigenfunctions $P_{\{\mu\}}(\zeta; \lambda^{-1})$. On the results displayed in the table, it can be checked that $P_{\{\mu\}}(\zeta; \lambda^{-1})$ belongs to the space $V_{\mu_1}(\zeta)$, where $V_n(\zeta)$, $n \in \mathbb{N}$, is defined as the space of polynomials in ζ_1 and ζ_2 that are of degree less than or equal to n (hence, $\dim V_n = (n+1)(n+2)/2$).

In Ref. 15, the degeneracies of the relative energy spectrum (9) were obtained for both distinguishable and indistinguishable (either bosonic or fermionic) particles on the line interval $(0, \pi/a)$, interacting via pure two-body or three-body potential. It was shown that although the results do not depend on the nature of interactions for distinguishable particles, they do for indistinguishable ones. Such a property is due to the fact that both the configuration space sectors and the variables the relative wave functions depend on have different transformation properties under particle permutations for the problems with pure two-body or pure three-body potential.

The exact solvability of H for $g = 0$ and $f \neq 0$, or equivalently of h^{rel}, defined in Eq. (8), can be easily explained by a hidden $sl(3, \mathbb{R})$ symmetry [15]. The Hamiltonian h^{rel} can indeed be rewritten as a quadratic combination

$$h^{\mathrm{rel}} = 8a^2 \big[E_{11}^2 + E_{11}E_{22} + E_{22}^2 - 3E_{12}E_{32} - 3E_{21}E_{31}$$
$$- 9E_{31}E_{32} + 3\lambda\,(E_{11} + E_{22}) \big] \quad (10)$$

of the operators

$$E_{11} = \zeta_1 \partial_{\zeta_1}, \quad E_{22} = \zeta_2 \partial_{\zeta_2}, \quad E_{33} = n - \zeta_1 \partial_{\zeta_1} - \zeta_2 \partial_{\zeta_2},$$
$$E_{21} = \zeta_2 \partial_{\zeta_1}, \quad E_{12} = \zeta_1 \partial_{\zeta_2},$$
$$E_{31} = \partial_{\zeta_1}, \quad E_{13} = n\zeta_1 - \zeta_1^2 \partial_{\zeta_1} - \zeta_1 \zeta_2 \partial_{\zeta_2}, \quad (11)$$
$$E_{32} = \partial_{\zeta_2}, \quad E_{23} = n\zeta_2 - \zeta_1 \zeta_2 \partial_{\zeta_1} - \zeta_2^2 \partial_{\zeta_2},$$

satisfying $gl(3, \mathbb{R})$ commutation relations $[E_{ij}, E_{kl}] = \delta_{kj} E_{il} - \delta_{il} E_{kj}$, together with the constant trace condition $\sum_i E_{ii} = n$ for any real n value. Whenever n is a non-negative integer, the operators E_{ij} preserve the space $V_n(\zeta)$. Hence, h^{rel} preserves an infinite flag of spaces, $V_0(\zeta) \subset V_1(\zeta) \subset V_2(\zeta) \subset \cdots$. Its representation matrix is therefore triangular in the basis wherein all spaces $V_n(\zeta)$ are naturally defined, so that h^{rel} is exactly solvable. This result is similar to that previously obtained for the pure two-body trigonometric potential [16].

3 Two- and Three-body Problem with Internal Degrees of Freedom

Let us now assume that both g and f are nonvanishing. From the ground-state wave function (2) of Hamiltonian (1), one can construct SUSYQM supercharge operators $\widehat{Q}^+$, $\widehat{Q}^- = (\widehat{Q}^+)^\dagger$, whose matrix elements can be expressed in terms of six differential operators $Q_i^\pm = \mp\partial_i - \partial_i \ln\psi_0(\boldsymbol{x})$, $i = 1, 2, 3$ [1, 2]. The latter are given by [14]

$$Q_i^\pm = \mp\partial_i - \kappa a \sum_{j\neq i} \cot(ax_{ij})$$
$$- \lambda a \left(\sum_{j\neq i} \cot(ay_{ij}) - \sum_{\substack{j,k \\ i\neq j\neq k\neq i}} \cot(ay_{jk}) \right). \quad (12)$$

The corresponding supersymmetric Hamiltonian is

$$\widehat{H} = \mathrm{diag}(H^{(0)}, H^{(1)}, H^{(2)}, H^{(3)}),$$

where $H^{(0)} = H - E_0 = \sum_i Q_i^+ Q_i^-$, $H^{(1)}$ and $H^{(2)}$ contain matrix potentials, while $H^{(3)} = \sum_i Q_i^- Q_i^+$ only differs from $H^{(0)}$ by the replacement in H of $g = \kappa(\kappa - 1)$, $f = \lambda(\lambda - 1)$ by $g = \kappa(\kappa + 1)$, $f = \lambda(\lambda + 1)$, respectively.

In the case of the CMW problem, it was shown in Ref. 13 that the corresponding operators Q_i^- can be transformed into three commuting differential-difference operators D_i, the so-called Dunkl operators of the mathematical literature [9], by inserting in appropriate places some finite-group elements K_{ij} and $L_{ij} \equiv K_{ij}I_r$. Here K_{ij} are particle permutation operators, while I_r is the inversion operator in relative-coordinate space. In the centre-of-mass coordinate system, they satisfy the relations

$$K_{ij} = K_{ji} = K_{ij}^\dagger, \quad K_{ij}^2 = 1, \qquad K_{ij}K_{jk} = K_{jk}K_{ki} = K_{ki}K_{ij},$$
$$K_{ij}I_r = I_rK_{ij}, \qquad I_r = I_r^\dagger, \qquad\qquad I_r^2 = 1, \qquad\qquad (13)$$
$$K_{ij}x_j = x_iK_{ij}, \qquad K_{ij}x_k = x_kK_{ij}, \qquad I_rx_i = -x_iI_r,$$

for all $i \neq j \neq k \neq i$. The operators 1, K_{ij}, $K_{ijk} \equiv K_{ij}K_{jk}$, I_r, L_{ij}, and $L_{ijk} \equiv K_{ijk}I_r$, where i, j, k run over the set $\{1, 2, 3\}$, are the 12 elements of the dihedral group D_6.

By proceeding in a similar way in the present problem, we find the three differential-difference operators [14]

$$D_i = \partial_i - \kappa a \sum_{j\neq i} \cot(ax_{ij})K_{ij}$$
$$- \lambda a \left(\sum_{j\neq i} \cot(ay_{ij})L_{ij} - \sum_{\substack{j,k \\ i\neq j\neq k\neq i}} \cot(ay_{jk})L_{jk} \right), \quad (14)$$

where $i = 1, 2, 3$. From their definition and Eq. (13), it is obvious that such operators are both anti-Hermitian and D_6-covariant, that is, $D_i^\dagger = -D_i$, $K_{ij}D_j = D_iK_{ij}$, $K_{ij}D_k = D_kK_{ij}$, and $I_rD_i = -D_iI_r$, for all $i \neq j \neq k \neq i$, but that they do not commute among themselves. Their commutators are indeed given by

$$[D_i, D_j] = -a^2(\kappa^2 + 3\lambda^2 - 4\kappa\lambda I_r) \sum_{k \neq i,j} (K_{ijk} - K_{ikj}), \quad i \neq j, \quad (15)$$

and only vanish in the $a \to 0$ limit, that is, for the CMW problem.

The operators D_i may be used to construct a generalized Hamiltonian with exchange terms

$$\begin{aligned}
H_{\text{exch}} &\equiv -\sum_i \partial_i^2 + a^2 \sum_{\substack{i,j \\ i \neq j}} \csc^2(ax_{ij})\kappa(\kappa - K_{ij}) \\
&\qquad\qquad + 3a^2 \sum_{\substack{i,j \\ i \neq j}} \csc^2(ay_{ij})\lambda(\lambda - L_{ij}) \\
&= -\sum_i D_i^2 + 6a^2(\kappa^2 + 3\lambda^2) \\
&\qquad\qquad + a^2(\kappa^2 + 3\lambda^2 + 12\kappa\lambda I_r)(K_{123} + K_{132}). \quad (16)
\end{aligned}$$

In those subspaces of Hilbert space wherein $(K_{ij}, L_{ij}) = (1,1)$, $(1,-1)$, $(-1,1)$, or $(-1,-1)$, the latter reduces to Hamiltonian (1) corresponding to $(g, f) = \big(\kappa(\kappa - 1), \lambda(\lambda - 1)\big)$, $\big(\kappa(\kappa - 1), \lambda(\lambda + 1)\big)$, $\big(\kappa(\kappa + 1), \lambda(\lambda - 1)\big)$, or $\big(\kappa(\kappa + 1), \lambda(\lambda + 1)\big)$, respectively.

From the operators D_i and the elements K_{ij}, L_{ij} of D_6, it is also possible to construct an alternative set of three Dunkl operators, that is, three anti-Hermitian, commuting, albeit non-covariant, differential-difference operators

$$\widehat{D}_i = D_i + i\kappa a \sum_{j \neq i} \alpha_{ij} K_{ij} + i\lambda a \left(\sum_{j \neq i} \beta_{ij} L_{ij} - \sum_{\substack{j,k \\ i \neq j \neq k \neq i}} \beta_{jk} L_{jk} \right),$$
$$\alpha_{ij}, \beta_{ij} \in \mathbb{R}, \quad (17)$$

in terms of which the generalized Hamiltonian with exchange terms, defined in Eq. (16), can be rewritten as $H_{\text{exch}} = -\sum_i \widehat{D}_i^2$. By choosing for the α_{ij}'s the values previously considered for the Dunkl operators of the pure two-body problem [3], namely $\alpha_{ij} = -\alpha_{ji} = -1$, $i < j$, one finds [14] that there are four equally acceptable choices for the remaining constants β_{ij}: $(\beta_{12}, \beta_{23}, \beta_{31}) = (-1, 1, 1)$, $(-1, 1, -1)$, $(-5/3, 1/3, 1/3)$, and $(-1/3, 5/3, -1/3)$.

The transformation properties under D_6 of the new operators $\widehat{D}_i$ are given by

$$\begin{aligned}
K_{ij}\widehat{D}_j - \widehat{D}_i K_{ij} &= -i\kappa a\left(2\alpha_{ij} + \sum_{k \neq i,j} (\alpha_{ik} - \alpha_{jk})K_{ijk} \right) \\
&\qquad - i\lambda a \sum_{k \neq i,j} (\beta_{ik} - \beta_{jk})I_r(K_{ijk} + 2K_{ikj}), \quad i \neq j,
\end{aligned}$$

$$[K_{ij}, \widehat{D}_k] = ia[\kappa(\alpha_{ik} - \alpha_{jk}) - \lambda(\beta_{ik} - \beta_{jk})I_r](K_{ijk} - K_{ikj}),$$
$$i \neq j \neq k \neq i, \quad (18)$$

$$\{I_r, \widehat{D}_i\} = 2i\kappa a \sum_{j \neq i} \alpha_{ij} L_{ij} + 2i\lambda a \left(\sum_{j \neq i} \beta_{ij} K_{ij} - \sum_{\substack{j,k \\ i \neq j \neq k \neq i}} \beta_{jk} K_{jk} \right).$$

The Hamiltonian with exchange terms H_{exch} can be related to a Hamiltonian $\mathcal{H}^{(\kappa, \lambda)}$ describing a one-dimensional system of three particles with SU(n) "spins" (or colours in particle physics language), interacting via spin-dependent two and three-body potentials [14],

$$\mathcal{H}^{(\kappa, \lambda)} = -\sum_i \partial_i^2 + a^2 \sum_{\substack{i,j \\ i \neq j}} \csc^2(ax_{ij})\kappa(\kappa - P_{ij})$$
$$+ 3a^2 \sum_{\substack{i,j \\ i \neq j}} \csc^2(ay_{ij})\lambda(\lambda - \widetilde{P}_{ij}). \quad (19)$$

Here each particle is assumed to carry a spin with n possible values, and P_{ij}, $\widetilde{P}_{ij} \equiv P_{ij}\widetilde{P}$ are some operators acting only in spin space. The operator P_{ij} is defined as the operator permuting the ith and jth spins, while $\widetilde{P}$ is a permutation-invariant and involutive operator, that is, $\widetilde{P}\sigma_i = \sigma_i^* \widetilde{P}$, for some σ_i^* such that $P_{jk}\sigma_i^* = \sigma_i^* P_{jk}$ for all i, j, k, and $\sigma_i^{**} = \sigma_i$. For $SU(2)$ spins for, instance, $\sigma_i = \pm 1/2$, $P_{ij} = (\sigma_i^a \sigma_j^a + 1)/2$, where σ^a, $a = 1$, 2, 3, denote the Pauli matrices, $\widetilde{P}$ may be taken as 1 or $\sigma_1^1 \sigma_2^1 \sigma_3^1$, and accordingly $\sigma_i^* = \sigma_i$ or $-\sigma_i$. The operators P_{ij} and $\widetilde{P}$ satisfy relations similar to those fulfilled by K_{ij} and I_r (cf. Eq. (13)), with x_i and $-x_i$ replaced by σ_i and σ_i^* respectively. Hence 1, P_{ij}, $P_{ijk} \equiv P_{ij}P_{jk}$, $\widetilde{P}$, $\widetilde{P}_{ij}$, and $\widetilde{P}_{ijk} \equiv P_{ijk}\widetilde{P}$ realize the dihedral group D_6 in spin space. Such a realization will be referred to as $D_6^{(s)}$ to distinguish it from the realization $D_6^{(c)}$ in coordinate space, corresponding to K_{ij} and I_r.

The Hamiltonian $\mathcal{H}^{(\kappa, \lambda)}$ remains invariant under the combined action of D_6 in coordinate and spin spaces (to be referred to as $D_6^{(cs)}$), since it commutes with both $K_{ij}P_{ij}$ and $I_r\widetilde{P}$. Its eigenfunctions corresponding to a definite eigenvalue therefore belong to a (reducible or irreducible) representation of $D_6^{(cs)}$. For indistinguishable particles that are bosons (respectively fermions), only those irreducible representations of $D_6^{(cs)}$ that contain the symmetric (respectively antisymmetric) irreducible representation of the symmetric group S_3 should be considered. There are only two such inequivalent representations, which are both one-dimensional and denoted by A_1 and B_1 (respectively A_2 and B_2) [10]. They differ in the eigenvalue of $I_r\widetilde{P}$, which is equal to $+1$ or -1, respectively.

In such representations, for an appropriate choice of the parameters κ, λ, $\mathcal{H}^{(\kappa, \lambda)}$ can be obtained from H_{exch} by applying some projection operators. Let indeed $\Pi_{B\pm}$ (respectively $\Pi_{F\pm}$) be the projection operators that consist in replacing K_{ij} and I_r by P_{ij} (respectively $-P_{ij}$) and $\pm\tilde{P}$, respectively, when they are at the right-hand side of an expression. It is obvious that $\Pi_{B\pm}(H_{\text{exch}}) = \mathcal{H}^{(\kappa,\pm\lambda)}$, and $\Pi_{F\pm}(H_{\text{exch}}) = \mathcal{H}^{(-\kappa,\pm\lambda)}$. If H_{exch} has been diagonalized on a basis of functions depending on coordinates and spins, then its eigenfunctions $\Psi(\boldsymbol{x},\boldsymbol{\sigma})$ are also eigenfunctions of $\mathcal{H}^{(\kappa,\pm\lambda)}$ (respectively $\mathcal{H}^{(-\kappa,\pm\lambda)}$) provided that $(K_{ij} - P_{ij})\Psi(\boldsymbol{x},\boldsymbol{\sigma}) = 0$ (respectively $(K_{ij} + P_{ij})\Psi(\boldsymbol{x},\boldsymbol{\sigma}) = 0$) and $(I_r \mp \tilde{P})\Psi(\boldsymbol{x},\boldsymbol{\sigma}) = 0$.

In conclusion, the three-body generalization of the Sutherland problem with internal degrees of freedom, corresponding to the Hamiltonian $\mathcal{H}^{(\kappa,\lambda)}$, is directly connected with the corresponding problem with exchange terms, governed by the Hamiltonian H_{exch}. The exchange operator formalism developed for the latter should therefore be relevant to a detailed study of the former.

Acknowledgments: The author is a Research Director of the National Fund for Scientific Research (Fonds National de la Recherche Scientifique), Belgium.

4 REFERENCES

1. A. A. Andrianov, N. V. Borisov, M. I. Eides, and M. V. Ioffe, *Supersymmetric origin of equivalent quantum systems*, Phys. Lett. A **109** (1985), No. 4, 143–148.

2. A. A. Andrianov, N. V. Borisov, and M. V. Ioffe, *The factorization method and quantum systems with equivalent energy spectra*, Phys. Lett. A **105** (1984), No. 1-2, 19–22.

3. D. Bernard, M. Gaudin, F. D. M. Haldane, and V. Pasquier, *Yang–Baxter equation in long-range interacting systems*, J. Phys. A **26** (1993), No. 20, 5219–5236, hep-th/9301084.

4. L. Brink, T. H. Hansson, and M. A. Vassiliev, *Explicit solution to the N-body Calogero problem*, Phys. Lett. B **286** (1992), No. 1-2, 109–11.

5. F. Calogero, *Ground state of a one-dimensional N-body system*, J. Math. Phys. **10** (1969), 2197–2200.

6. _____, *Solution of a three-body problem in one dimension*, J. Math. Phys. **10** (1969), 2191–2196.

7. _____, *Solution of the one-dimensional N-body problems with quadratic and/or inversely quadratic pair potentials*, J. Math. Phys. **12** (1971), 419–436.

8. F. Calogero and C. Marchioro, *Exact solution of a one-dimensional three-body scattering problem with two-body and/or three-body inverse-square potentials*, J. Math. Phys. **15** (1974), 1425–1430.

9. C. F. Dunkl, *Differential-difference operators associated to reflection groups*, Trans. Amer. Math. Soc. **311** (1989), No. 1, 167–183.

10. M. Hamermesh, *Group Theory and Its Application to Physical Problems*, Addison-Wesley Series in Physics, Addison-Wesley, Reading, MA, 1962.

11. M. A. Olshanetsky and A. M. Perelomov, *Quantum integrable systems related to Lie algebras*, Phys. Rep. **94** (1983), No. 6, 313–404.

12. A. P. Polychronakos, *Exchange operator formalism for integrable systems of particles*, Phys. Rev. Lett. **69** (1992), No. 5, 703–705.

13. C. Quesne, *Exchange operators and the extended Heisenberg algebra for the three-body Calogero–Marchioro–wolfes problem*, Modern Phys. Lett. **A10** (1995), No. 18, 1323–1330.

14. _____ , *Three-body generalization of the Sutherland model with internal degrees of freedom*, Europhys. Lett. **35** (1996), 407–412.

15. _____ , *Exactly solvable three-particle problem with three-body interaction*, Phys. Rev. A **55** (1997), No. 5, 3931–3934.

16. W. Rühl and A. Turbiner, *Exact solvability of the Calogero and Sutherland models*, Modern Phys. Lett. **A10** (1995), No. 29, 2213–2221.

17. R. P. Stanley, *Some combinatorial properties of Jack symmetric functions*, Adv. Math. **77** (1989), No. 1, 76–115.

18. B. Sutherland, *Exact results for a quantum many-body problem in one dimension*, Phys. Rev. A **4** (1971), 2019–2021.

19. _____ , *Exact results for a quantum many-body problem in one dimension. II*, Phys. Rev. A **5** (1972), 1372–1376.

20. _____ , *Exact ground-state wave function for a one-dimensional plasma*, Phys. Rev. Lett. **34** (1975), 1083–1085.

21. J. Wolfes, *On the three-body linear problem with three-body interaction*, J. Math. Phys. **15** (1974), 1420–1424.

26

On Relativistic Lamé Functions

S. N. M. Ruijsenaars

ABSTRACT To date, the quantum relativistic Calogero–Moser–Sutherland system with elliptic interactions is the most general one in the hierarchy of "A_{M-1}-symmetric" integrable M-particle systems. We present and discuss eigenfunctions for this system. More specifically, we only deal with the $M = 2$ case, but we handle a dense set in the relevant parameter space.

1 Introduction

The following serves to report on explicit eigenfunctions for quantum Calogero–Moser–Sutherland systems of the relativistic variety. More precisely, we restrict attention to eigenfunctions for the case of two particles, but we do treat the most general (elliptic) type of interaction for which integrability is known to persist. Our account is based on our papers [20–22]. In keeping with these papers, we emphasize quantum-mechanical/functional-analytic aspects. In particular, to ensure that the defining dynamics is at least formally self-adjoint, we restrict attention to real couplings and elliptic functions with a real and purely imaginary period.

We begin by recalling the Hamiltonians defining the nonrelativistic and relativistic Calogero–Moser–Sutherland systems for arbitrary particle number M. The nonrelativistic quantum dynamics is given by the PDO:

$$H_{\text{nr}} = -\frac{\hbar^2}{2m} \sum_{j=1}^{M} \left(\frac{\partial}{\partial x_j} \right)^2 + \frac{g(g - \hbar)}{m} \sum_{\substack{j,k=1 \\ j<k}}^{M} \wp(x_j - x_k). \quad (1)$$

Here, m is the particle mass, g is the coupling constant, $\hbar$ is Planck's constant and $\wp(x; \omega, \omega')$ is the Weierstrass $\wp$-function with half-periods ω and ω'. From now on we will set

$$\omega = \frac{\pi}{2r}, \quad \omega' = \frac{ia}{2}, \quad r, a \in (0, \infty). \quad (2)$$

This somewhat unusual parametrization anticipates our conventions for the relativistic level. Here we find it convenient to work with a close relative of the Weierstrass σ-function, viz.,

$$s(r,a;x) = \sigma\left(x; \frac{\pi}{2r}, \frac{ia}{2}\right) \exp\left(-\frac{\eta r x^2}{\pi}\right). \tag{3}$$

This function is entire, odd, and π/r-antiperiodic. It satisfies the analytic difference equation

$$\frac{s(r,a;x+ia/2)}{s(r,a;x-ia/2)} = -e^{-2irx}, \tag{4}$$

and has trigonometric and hyperbolic limits

$$\lim_{a\to\infty} s(r,a;x) = \frac{\sin rx}{r}, \quad \lim_{r\to 0} s(r,a;x) = \frac{a}{\pi} \text{sh}\left(\frac{\pi x}{a}\right), \tag{5}$$

uniformly on compact subsets of the complex x-plane.

The relativistic interaction involves the functions

$$f_\pm(x) = \left(\frac{s(r,a;x\pm ig/mc)}{s(r,a;x)}\right)^{1/2}, \tag{6}$$

where c is the speed of light; the defining quantum dynamics is then given by the analytic difference operator (henceforth AΔO):

$$H_{\text{rel}} = mc^2 \sum_{j=1}^{M} \prod_{k\neq j} f_-(x_j - x_k) \exp\left(-\frac{i\hbar\partial_j}{mc}\right) f_+(x_j - x_k). \tag{7}$$

The connection between H_{rel} and H_{nr} is given by the nonrelativistic limit $c \to \infty$: One clearly gets

$$H_{\text{rel}} = Mmc^2 + H_{\text{nr}} + C_M + O(c^{-2}), \quad c \to \infty. \tag{8}$$

(Here, C_M is a constant.) From now on, we work with the parameter $\beta = 1/mc$ and put $m = \hbar = 1$.

Next, we recall that the above quantum dynamics can be supplemented with $M-1$ independent and commuting PDOs and AΔOs, respectively; this is why H_{nr} and H_{rel} are viewed as quantum integrable systems. Background information on the nonrelativistic systems can be found in the surveys by Olshanetsky and Perelomov [14, 15]. More recent accounts including the relativistic versions are our survey [17] and lecture notes [23].

At the quantum level there are two basic problems associated with the above formal operators. First, one wants to find joint eigenfunctions for the whole commuting family of PDOs or AΔOs, in a form that is as explicit as possible. Second, one wants to redefine the operators as bona fide commuting self-adjoint operators on a Hilbert space.

We are mentioning the two problems in this order, since the second problem appears quite inaccessible without having explicit joint eigenfunctions available. More precisely, at the nonrelativistic level the Hilbert space aspects are relatively simple for $M = 2$, but already hard to control for $M > 2$, whereas at the relativistic level quite novel difficulties are present. In particular, it is not clear at face value that the Hamiltonian (7) can be defined as a symmetric operator on a dense subspace of the pertinent Hilbert space.

We continue by sketching the state of the art concerning the problems just mentioned, restricting attention to the elliptic settings. Beginning with the nonrelativistic level, there are explicit results only for integer g. For the two-particle case one is dealing with the Lamé operator

$$H_0 = -\frac{d^2}{dx^2} + g(g-1)\wp(x), \quad g \in \mathbb{R}. \tag{9}$$

(The center-of-mass motion may be ignored.) Eigenfunctions in product form were already found by Hermite in the last century; cf. the last pages of Whittaker and Watson [26]. (These functions will be detailed in Section 2.) For $M > 2$ the first results were obtained by Dittrich and Inozemtsev [3] (cf. also Ref. 9). More recently, Felder and Varchenko [4, 5] handled the arbitrary M, integer g case in a quite different, representation-theoretic and algebro-geometric setting, obtaining eigenfunctions without addressing their quantum-mechanical features.

At the relativistic level eigenfunctions for arbitrary M are only known when g equals an integer, just as in the nonrelativistic case. Such eigenfunctions were quite recently constructed by Billey [2] via a nested Bethe ansatz. See also papers by Hasegawa [8] and by Komori and Hikami [11], where it is shown (among others things) that the commuting A∆Os admit finite-dimensional invariant subspaces spanned by theta functions. Other results relevant to the arbitrary M case can be found in a recent paper by Komori [10]. He shows in particular that one can associate a symmetric Hilbert space operator to the A∆O (7).

Let us next specialize to the $M = 2$ case. Separating off a center-of-mass factor from H_{rel} (7), one winds up with the generalized Lamé operator:

$$H_\beta = \left(\frac{s(r, a; x - i\beta g)}{s(r, a; x)}\right)^{1/2} \exp\left(-i\beta\frac{d}{dx}\right)\left(\frac{s(r, a; x + i\beta g)}{s(r, a; x)}\right)^{1/2}$$
$$+ (i \to -i), \quad \beta > 0. \tag{10}$$

We obtained integer g eigenfunctions of H_β in 1988, announced this in Ref. 17, and presented details in our 1994 lecture notes [23]. Integer g eigenfunctions in a different guise (for $g > 2$) were then presented by Krichever and Zabrodin [12], who used them to study certain solutions to their spin generalizations of the relativistic elliptic systems. Their work emphasizes the finite-gap properties associated with these functions; roughly speaking, the integer g equals the number of bands in the spectrum of the operators arising from (9) and (10) when one shifts x over half the imaginary period.

In this connection we also point out that a close relative S_0 of the generalized Lamé operator H_β (10) was already introduced by Sklyanin in the early eighties. He studied quite special eigenfunctions of S_0 corresponding to the band edges in the finite-gap picture [24, 25].

In Ref. 7 Felder and Varchenko obtained integer g eigenfunctions in a form substantially equivalent to ours (cf. also Ref. 6). From their perspective, the functions arise via the algebraic Bethe ansatz, as a special case of their extensive work on representations of elliptic quantum groups.

In our paper [20] and in the present contribution as well, we are dealing with eigenfunctions for a dense set in the parameter space r, a, $\beta > 0$, $g \in \mathbb{R}$. As it turns out, these functions are in fact joint eigenfunctions of three independent commuting AΔOs. To handle Hilbert space aspects, however, the spectral variable must be discretized. Thus one ends up with *two* commuting generalized Lamé operators, namely H_β (10) and the AΔO obtained by interchanging β and a. (Note that from a physical point of view both $\hbar\beta$ and a have dimension of length.)

It so happens that the hyperbolic and trigonometric specializations can be treated in far more detail (cf. Refs. 21 and 22). The results obtained in these settings have their own flavor and are of independent interest. We will mostly deal with the elliptic case, however. In Section 2 we recall what is known about the two problems mentioned earlier for the integer g Lamé operator H_0. We summarize these results mainly to prepare the ground for Section 3, where we consider the problems for the relativistic generalization H_β, choosing again $g \in \mathbb{N}$.

In Section 4 we extend the results to parameters that are dense in the natural parameter domain. To bring out some remarkable symmetry properties, we adopt another normalization and notation. In particular, this enables us to handle at once the two generalized Lamé operators mentioned earlier. Section 5 contains several concluding remarks. In particular, we discuss the existence and features of interpolating eigenfunctions, and we briefly consider the hyperbolic specialization, where an explicit interpolation is known to exist [19, 23]. We also add some speculations about $M > 2$ eigenfunctions.

2   The Nonrelativistic Integer g Case

Let us consider the time-independent Schrödinger equation

$$-f''(x) + g(g-1)\wp(x)f(x) = Ef(x), \tag{11}$$

arising from the Hamiltonian H_0 (9). It is a second-order ODE to which standard existence and uniqueness results apply. Iterating the integral equation corresponding to it, one can obtain an infinite series representation for solutions on the interval $(0, \pi/r)$ (say), with arbitrary initial conditions $f(x_0)$, $f'(x_0)$ for $x_0 \in (0, \pi/r)$.

The ODE (11) has quite special features, however. The singularity at $x = 0$ is of the regular (Fuchsian) type, and the indicial equation has roots g and $1 - g$. For $g \notin \mathbb{Z}/2$ one therefore obtains two fractional power series solutions with behavior $f_1(x) \sim x^g$ and $f_2(x) \sim x^{1-g}$ as $x \to 0$. For $g \in \mathbb{N}^*/2$ there exists a power series solution behaving as x^g for $x \to 0$, but a second linearly independent power series solution need not exist. When it exists, however (as is the case for (11) with $g \in \mathbb{N}^*$), then it behaves once more as x^{1-g} for $x \to 0$.

The upshot is that in the nonrelativistic setting there is no difficulty concerning existence and uniqueness of solutions, and the solutions can actually be represented in two distinct forms. Even so, both formulas are not sufficiently explicit to get detailed information on the second problem mentioned above.

Of course, it is not a priori clear that more explicit formulas exist, but this turns out to be true for integer g. (These formulas can be derived in various ways; for a complete account, see the last chapter of Ref. 26.) To specify them, we put from now on

$$g = N + 1, \quad N \in \mathbb{N}^*. \tag{12}$$

(Note that the cases $g = 0, 1$ are trivial.) Then the eigenfunctions of H_0 are linear combinations of functions $\mathcal{F}(x, y)$ and $\mathcal{F}(-x, y)$ of the form

$$\mathcal{F}(x, y) = \prod_{j=1}^{N} \frac{s(x + z_j)}{s(x)} \cdot \exp[irx(N + 1) + ixy]. \tag{13}$$

Here, $s(x)$ stands for $s(r, a; x)$, and the spectral variable y reads

$$y = -(N + 1)r + i \sum_{j=1}^{N} \frac{s'(z_j)}{s(z_j)}. \tag{14}$$

(Clearly, the first term on the right-hand side can be absorbed in y; it is needed for later purposes, however.) The numbers $z_1, \ldots, z_N$ ("zeros") satisfy the constraint system

$$N \frac{s'(z_k)}{s(z_k)} + \sum_{\substack{j=1 \\ j \neq k}}^{N} \frac{s'(z_j - z_k)}{s(z_j - z_k)} - \sum_{j=1}^{N} \frac{s'(z_j)}{s(z_j)} = 0, \quad k = 1, \ldots, N, \tag{15}$$

and this system admits a solution curve. (Note that the sum of the N left-hand-sides vanishes identically, so that this is a priori plausible.) Adding the relation (14) between y and $z_1, \ldots, z_N$ to the system, one may view y as the curve parameter. For $y \in (K, \infty)$ with K sufficiently large, one can then choose $z_j = i\epsilon_j(y), \epsilon_j > 0$, with $\epsilon_j \downarrow 0$ for $y \uparrow \infty$.

It is not obvious, but true that $\mathcal{F}(x,y)$ thus defined is a solution to (11) with E given by

$$E = -(2N-1)\sum_{j=1}^{N} \wp(z_j).\tag{16}$$

Thus one gets $E = E(y) \uparrow \infty$ as $y \uparrow \infty$.

The functions $\mathcal{F}(\pm x, y)$ clearly become singular as $x^{-N} = x^{1-g}$ for $x \to 0$, in agreement with Fuchs theory. Moreover, the functions

$$\Phi(x,y) \equiv \mathcal{F}(x,y) - (-1)^N \mathcal{F}(-x,y)\tag{17}$$

vanish as x^{N+1} for $x \downarrow 0$. (Again this is not obvious at first sight. Note, however, that the leading singularity x^{-N} is taken out in Φ, so that its x^{N+1}-behavior follows from Fuchs theory.)

Next, note that $\mathcal{F}(x,y)$ satisfies the quasi-periodicity relation

$$\mathcal{F}\left(x + \frac{\pi}{r}, y\right) = (-1)^{N+1} \exp\left(\frac{i\pi y}{r}\right)\mathcal{F}(x,y).\tag{18}$$

Hence $\mathcal{F}(x, nr)$, $n \in \mathbb{N}$, is π/r-periodic or -antiperiodic. As a consequence, the functions

$$\Phi_n(x) \equiv \Phi(x, nr), \quad n \in \mathbb{N},\tag{19}$$

vanish not only at $x = 0$, but also at $x = \pi/r$. Thus they belong to the Hilbert space

$$\mathcal{H} \equiv L^2((0, \pi/r), dx)\tag{20}$$

of square-integrable functions on the interval $(0, \pi/r)$. For $n \to \infty$ one gets $E_n = E(nr) \uparrow \infty$, so the eigenvalues of H_0 on Φ_n are distinct for n large. It now follows from a standard argument that the functions Φ_n are pairwise orthogonal for n large.

More generally, from the well-developed self-adjointness theory for ordinary differential operators one readily deduces that H_0 is essentially self-adjoint on $C_0^\infty\left((0, \pi/r)\right)$ for $g \geq 3/2$. Moreover, the Weyl–Kodaira–Titchmarsh theory of eigenfunction expansions yields the existence of an orthonormal base of eigenfunctions. It is natural to expect that the latter is given by the (renormalized) functions $\{\Phi_n\}_{n=0}^\infty$, but to our knowledge this has not even been shown for $g = 2$.

In this connection it should be noted that the $g = 2$ case is particularly accessible, since the constraint system (15) is trivial for $N = 1$. From (14) one then sees that the sequence of values $y = 0, r, 2r, \ldots$ yields a sequence of distinct z_1-values $z_1(nr)$ in the interval $i(0, a/2)$. The corresponding energies E_n are obviously distinct, too (cf. (16)), so the functions Φ_0, Φ_1,

... are well-defined, nonzero, and pairwise orthogonal. (But it is not clear that they are *complete*.)

By contrast, the system (15) for $N > 1$ is quite inaccessible. Though results for generic $y \in \mathbb{C}$ can be gleaned from Refs. 26 and 5, the Hilbert space aspects involve the (nongeneric) values $y = nr$. (It is not clear, for example whether the functions Φ_n are nonzero for n small and whether the eigenvalues E_n are distinct.) Thus, even in the nonrelativistic integer g context the Hilbert space questions have not been completely elucidated.

On the other hand, for the trigonometric specialization we have

$$\lim_{a \to \infty} \wp\left(x; \frac{\pi}{2r}, \frac{ia}{2}\right) = \frac{r^2}{\sin^2(rx)} - \frac{r^2}{3}. \tag{21}$$

For this potential the Hilbert space theory is in great shape. Indeed, in the trigonometric case the above functions $\Phi_n(x)$ and eigenvalues E_n can be seen to be of the form

$$\Phi_n(x) = w(x)^{1/2} P_n(\cos rx), \quad n \in \mathbb{N}, \tag{22}$$

$$E_n = (n + N + 1)^2 r^2 - (N + 1) N \frac{r^2}{3}. \tag{23}$$

Here, the functions $P_n(u)$ are Gegenbauer polynomials, of degree n and parity $(-1)^n$, and

$$w(x) = (\sin rx)^{2N+2} \tag{24}$$

amounts to the weight function with regard to which they are orthogonal. Therefore, completeness is obvious, and this is one important reason to conjecture that for $a < \infty$ the functions Φ_0, Φ_1, ... are still complete. Unfortunately, no representation analogous to (22) is known for the elliptic case.

3   The Relativistic Integer g Case

Let us now turn to a consideration of the AΔO H_β (10). Here the time-independent Schrödinger equation is an analytic difference equation (AΔE), viz.,

$$\left(\frac{s(x - i\beta g)}{s(x)} \cdot \frac{s(x - i\beta + i\beta g)}{s(x - i\beta)}\right)^{1/2} F(x - i\beta) + (i \to -i) = EF(x). \tag{25}$$

The theory of such equations is far less developed than for ordinary differential and *discrete* difference equations. Though some existence results are known, the main problem is to single out solutions with special properties. Indeed, the key difference between the ODE (11) and the AΔE (25) is that the former has a two-dimensional solution space, whereas for the latter the existence of one nontrivial solution $F_E(x)$ already entails that the solution space is infinite-dimensional: For any meromorphic (say) multiplier $M(x)$ with period $i\beta$ the function $M(x)F_E(x)$ solves (25) as well.

The problem is, then, to find solutions with special properties, preferably such that they can be used to define the AΔO H_β (10) as a genuine self-adjoint operator on $\mathcal{H}$ (20). The point is that there is no obvious way to define H_β first as a symmetric operator on a dense subspace, by contrast to H_0, where for instance $C_0^\infty((0, \pi/r))$ serves this purpose. Since no general Hilbert space theory for AΔOs exists at the present time, one may instead try to find sufficiently explicit pairwise orthogonal eigenfunctions $\Phi_n \in \mathcal{H}$ with real eigenvalues E_n. Setting then $H\Phi_n \equiv E_n\Phi_n$, extending linearly, and taking the Hilbert space closure $\overline{H}$ of the symmetric operator H thus defined, one obtains a self-adjoint operator $\overline{H}$ on (a dense subspace of) the closed subspace spanned by the pertinent eigenfunctions.

As it turns out, this scenario can be realized to a large extent. We continue by describing the eigenfunctions that generalize the above eigenfunctions $\mathcal{F}(x, y)$ and that play the desired role in rigorously redefining H_β as a self-adjoint quantum dynamics. Choosing as before $g = N + 1$ with $N \in \mathbb{N}^*$, and requiring first

$$2N\beta \in (0, a), \tag{26}$$

they are of the form

$$\mathcal{F}(x, y) = \prod_{j=1}^{N} \frac{s(x + z_j)}{[s(x + ij\beta)s(x - ij\beta)]^{1/2}} \cdot \exp[irx(N+1) + ixy]. \tag{27}$$

Here, the spectral variable y is related to the zero functions via

$$y = -(N+1)r - \frac{1}{2\beta} \ln\left(\prod_{j=1}^{N} \frac{s(z_j - i\beta)}{s(z_j + i\beta)}\right), \tag{28}$$

and the latter obey the constraint system

$$s(z_k - iN\beta) \prod_{j \neq k} s(z_j - z_k - i\beta) \prod_j s(z_j + i\beta) - (\beta \to -\beta) = 0,$$
$$k = 1, \ldots, N. \tag{29}$$

It is clear that for $\beta \downarrow 0$ these equations yield the nonrelativistic counterparts (13)–(15). But in contrast to (15), it is by no means clear that one of the N equations for the N unknowns $z_1, \ldots, z_N$ is a consequence of the remaining $N - 1$ equations. This is, however, true, and it is important to understand the reason. Viewing (27) as an Ansatz for solving (25) with

$g = N + 1$, one obtains the function

$$E \equiv \frac{1}{s(x)} \left(s(x + iN\beta) \exp[(N+1)\beta r + \beta y] \prod_{j=1}^{N} \frac{s(x - i\beta + z_j)}{s(x + z_j)} \right.$$

$$\left. + (\beta \to -\beta) \right). \qquad (30)$$

Of course, this function depends on x when one gives $z_1, \ldots, z_N$ arbitrary values. But since the function is elliptic in x (with periods π/r, ia, cf. (4)), one need only require that the residues at N of its $N + 1$ (generically) simple poles in a period cell vanish to ensure that it is constant. Now the requirement that the residue at $x = 0$ vanish yields (28), whereas the residues at $x = -z_k$ give rise to (29). Thus we need only prove that the system (29) with $k = 2, \ldots, N$ (say) admits a solution curve to infer that all of the equations are solved.

Now it is obvious that all of the N equations are solved by choosing

$$z_j = ij\beta, \quad j = 1, \ldots, N. \qquad (31)$$

An application of the implicit function theorem then shows that the equations with $k = 2, \ldots, N$ have a unique holomorphic solution $z_k(z_1)$, $k = 2$, $\ldots$, N, near (31). Moreover, taking $z_1(t) = i\beta + it$ with $t \in [0, \epsilon)$, the functions $z_j(z_1(t))$ are real-analytic functions from $[0, \epsilon)$ to $i(0, \infty)$ for ϵ small enough. From (28) it is then clear that (eventually decreasing ϵ) $y = y(t)$ is real-analytic and real-valued on $(0, \epsilon)$, and that one has $y \uparrow \infty$ for $t \downarrow 0$.

As a consequence, one can trade t for y in a neighborhood (K, ∞) of ∞. Since we know very little about the minimal K satisfying various requirements, we may increase K as the need arises. In particular, we can choose it sufficiently large so that the functions

$$\Phi_n(x) \equiv \mathcal{F}(x, nr) - \mathcal{F}(-x, nr), \quad nr > K, \ n \in \mathbb{N}, \qquad (32)$$

are well defined and nonzero. Indeed, the above functions $z_j(z_1(t(y)))$ (denoted simply $z_j(y)$ from now on) satisfy

$$y \uparrow \infty \implies z_j(y) \to ij\beta, \ j = 1, \ldots, N, \qquad (33)$$

so that the summands on the right-hand side of (32) have distinct zeros for y large enough. (Recall our standing assumption (26).) Moreover, taking $x = iN\beta$ in (30) (which we may do, since E is x-independent), one deduces that an eventual increase of K ensures $E(y)$ is increasing on (K, ∞). Then H_β has distinct eigenvalues on (K, ∞).

The crux is now that all of the functions $\Phi_n(x)$ just defined belong to a dense subspace $\mathcal{A} \subset \mathcal{H}$ such that $H_\beta \mathcal{A} \subset \mathcal{H}$ and such that H_β is symmetric on $\mathcal{A}$. It is important to point out that the definition of $\mathcal{A}$ (which we do not present here) is not directly motivated by H_β, but rather by properties of the above (very special!) H_β-eigenfunctions. Since the eigenvalues E_n and E_m are distinct for $n \neq m$, it now follows from symmetry that Φ_n and Φ_m are orthogonal. Thus we obtain a self-adjoint operator (denoted again H_β) on the closed subspace $\mathcal{H}_K \subset \mathcal{H}$ spanned by the functions (32).

We expect that $\mathcal{H}_K$ equals $\mathcal{H}$ whenever K can be chosen negative. Put differently, we conjecture that for $K < 0$ the functions $\Phi_0, \Phi_1, \ldots$ are an orthogonal base for $\mathcal{H}$. In the special case $N = 1$ one can choose $K = -r$ (cf. Ref. 22, Eq. (2.35)), but completeness is still open, even in this simple case. More generally, we expect that the orthocomplement of $\mathcal{H}_K$ is spanned by functions $\Phi_0, \ldots, \Phi_{[K/r]}$ that are eigenfunctions of H_β with real eigenvalues.

Once again, the orthogonality and completeness problems are trivial for the trigonometric specialization, since one winds up with orthogonal polynomials in that case. Specifically, for $a = \infty$ the functions Φ_n are still of the form (22), with the weight function (24) now given by

$$w(x) = \sin^2(rx) \prod_{j=1}^{N} \sin r(x - ij\beta) \sin r(x + ij\beta). \qquad (34)$$

The associated orthogonal polynomials are then q-Gegenbauer polynomials (cf., for example, Ref. 1), with q given by

$$q = \exp(-2\beta r). \qquad (35)$$

Returning to the elliptic case, we recall that we have restricted β by (26) in the above account. But a substantial part of our results continues to be valid under the restriction

$$k\beta \notin \mathbb{N}a, \quad k = 1, \ldots, 2N. \qquad (36)$$

In particular, this suffices to infer the existence of eigenfunctions of the form (27)–(29). (Note that this more general restriction still guarantees that for y large the zeros $z_1(y), \ldots, z_N(y), -z_1(y), \ldots, -z_N(y)$ are distinct modulo the period ia, cf. (33).)

A key difference is, however, that for $N\beta > a$ the functions Φ_n (32) are most likely no longer pairwise orthogonal. More precisely, our symmetry proof breaks down for $N\beta > a$, and orthogonality is indeed violated in all cases where this could be tested. For $N = 1$ and $\beta > a$ a breakdown of orthogonality occurs in the strongest possible form: One has $(\Phi_n, \Phi_m) \neq 0$ for all $n \neq m$. (Here, we still assume (36); note that H_β becomes "free" for $N = 1$ and $\beta = la/2$, $l \in \mathbb{N}^*$.)

We prove the latter assertion in Ref. 22, which is concerned with the $g = 2$ case. To conclude this section we mention another remarkable result from this paper. Taking $\beta \uparrow a$ (the edge of the unitarity region) and simultaneously $a \downarrow 0$ in a certain way, the above eigenfunctions $\Phi_n(x)$ converge to the Lieb–Liniger eigenfunctions [13] for the ($M = 2$, center-of-mass) repulsive delta-function Bose gas. The role of the finite volume in [13] is played by the elliptic period $\pi/r < \infty$. This limiting transition generalizes the connection between the $g = 2$ *hyperbolic* relativistic $M = 2$ eigenfunctions and the *infinite-volume* delta-function eigenfunctions, which we pointed out at the end of [17].

4   Eigenfunctions for a Dense Parameter Set

Thus far, we have viewed the functions $\mathcal{F}(x, y)$ (27) as eigenfunctions of H_β (10). However, they also satisfy the quasi-periodicity relations (cf. (4))

$$\mathcal{F}(x + ia, y) = \exp[-2ir \sum_{j=1}^{N} z_j(y) - (N+1)ar - ay]\mathcal{F}(x, y), \tag{37}$$

$$\mathcal{F}\left(x + \frac{\pi}{r}, y\right) = -\exp\left(\frac{i\pi y}{r}\right)\mathcal{F}(x, y). \tag{38}$$

Thus, they can also be regarded as eigenfunctions of the AΔOs $T_{\pm ia}$ and $T_{\pm\pi/r}$, where we use the notation

$$(T_\alpha F)(x) \equiv F(x - \alpha), \quad \alpha \in \mathbb{C}. \tag{39}$$

Now this is true for $\mathcal{F}(-x, y)$, too, but then we obtain different eigenvalues. On the other hand, introducing the "extra" AΔO,

$$H_e \equiv T_{ia} + T_{-ia}, \tag{40}$$

and the "quasi-periodicity" AΔO,

$$Q = T_{\pi/r} + T_{-\pi/r}, \tag{41}$$

we obtain the *same* eigenvalues for $\mathcal{F}(x, y)$ and $\mathcal{F}(-x, y)$. Hence the functions $\mathcal{F}(\pm x, y)$ are joint eigenfunctions of the triple of independent AΔOs (H_β, H_e, Q).

As will now be detailed, we have found eigenfunctions $\mathcal{F}(\pm x, y)$ of H_β for a set that is dense in the parameter space r, β, $a > 0$, $g \in \mathbb{R}$. These eigenfunctions of H_β are once again eigenfunctions of Q with eigenvalue $-2\cos(\pi y/r)$ and of an extra AΔO H_e involving the shifts $T_{\pm ia}$. But the latter operator is no longer "free": It involves the functions (6) with an interchange of a and $\hbar\beta = \hbar/mc$. Within this more general setting, H_β and H_e are on the same footing from a mathematical viewpoint, and we take this into account by switching to notation that makes this symmetry manifest.

Specifically, we work from now on with parameters a_+, a_-, and b defined by

$$a_+ = \beta, \quad a_- = a, \quad b = \beta g. \tag{42}$$

The pertinent parameter domain at the elliptic level is then

$$\mathcal{E} \equiv \{(r, a_+, a_-, b) \in (0, \infty)^3 \times \mathbb{R}\}. \tag{43}$$

Introducing the notation

$$s_\delta(x) \equiv s(r, a_\delta; x), \quad \delta = +, -, \tag{44}$$

we now define the AΔOs

$$H_\delta \equiv e^{-br} \left(\left(\frac{s_\delta(x - ib)}{s_\delta(x)} \right)^{1/2} T_{ia_{-\delta}} \left(\frac{s_\delta(x + ib)}{s_\delta(x)} \right)^{1/2} + (i \to -i) \right),$$

$$\delta = +, -. \tag{45}$$

Comparing H_β (10), one sees that H_- arises from H_β via the substitutions (42), but for the prefactor $\exp(-br)$. Likewise, taking $g = N + 1$, the AΔO H_+ reduces to a positive multiple of H_e (40). The choice of prefactor in (45) ensures that the operators thus defined satisfy the invariance property

$$H_\delta(a_+ + a_- - b) = H_\delta(b), \quad \delta = +, -. \tag{46}$$

The assertions just made can be easily verified by using the AΔE (4). Similarly, this AΔE can be used to check that H_+ and H_- *commute*. Now this is in accord with the existence of joint eigenfunctions, but there are no general results to the effect that commutativity of two AΔOs *implies* the existence of joint eigenfunctions. (In this connection it is important to observe that when the two summands of H_+ are multiplied by meromorphic functions with period ia_+, the resulting AΔO still commutes with H_-.)

Even so, we have found joint eigenfunctions $\mathcal{F}(\pm x, y)$ of the three independent commuting AΔOs H_+, H_- and Q for a dense set $\mathcal{D}$ in $\mathcal{E}$ (43). For expository simplicity, we will specify these functions for a subset $\mathcal{D}_+ \cup \mathcal{D}_-$ of $\mathcal{D}$ that is already dense. The two sets $\mathcal{D}_\alpha$, $\alpha \in \{+, -\}$, are defined by

$$\mathcal{D}_\alpha \equiv \{(r, a_+, a_-, b) \in \mathcal{E} \mid b = (N_\alpha + 1)a_\alpha - N_{-\alpha}a_{-\alpha},$$
$$N_+, N_- \in \mathbb{N}, a_+/a_- \notin \mathbb{Q}\}. \tag{47}$$

Since the quotient a_+/a_- is allowed to be an arbitrary positive irrational number, the b-values occurring here are dense in $\mathbb{R}$. Hence each of the two (disjoint) sets $\mathcal{D}_+$ and $\mathcal{D}_-$ is dense in $\mathcal{E}$. To visualize the situation, it may be helpful to inspect Figure 26.1, where we have fixed a_- and drawn some of the pertinent lines in the (a_+, b)-plane.

The two sets $\mathcal{D}_+$, $\mathcal{D}_-$ are interchanged under the transformation $b \to a_+ + a_- - b$. Fixing $r > 0$, $a_+/a_- \notin \mathbb{Q}$ and $N_+, N_- \in \mathbb{N}$, we get a point in $\mathcal{D}_+$ by taking $b = (N_+ + 1)a_+ - N_-a_-$ and a point in $\mathcal{D}_-$ by taking

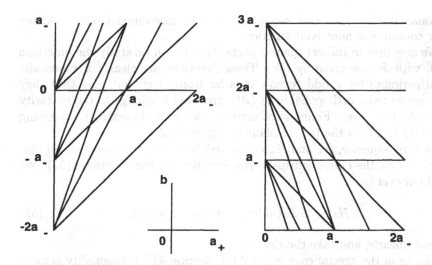

FIGURE 26.1. Some lines in the (a_+, b)-plane belonging to $\mathcal{D}_+$ (left) and $\mathcal{D}_-$ (right); the parameter a_- is fixed.

$b = (N_- + 1)a_- - N_+ a_+$. In agreement with the invariance property (46), the joint eigenfunctions $\mathcal{F}(\pm x, y)$ are the same in both points. They read

$$\mathcal{F}(x,y) = \prod_{\delta=+,-} \prod_{j=1}^{N_\delta} \frac{s_{-\delta}(x + z_j^\delta(y))}{[s_{-\delta}(x + ija_\delta)s_{-\delta}(x - ija_\delta)]^{1/2}}$$

$$\cdot \exp[irx(2N_+ N_- + N_+ + N_- + 1) + ixy]. \quad (48)$$

The zero functions z_j^δ, $j = 1, \ldots, N_\delta$, satisfy the constraint system (29) with $N \to N_\delta$, $\delta = +, -$, and the two systems are coupled via the spectral variable y in a somewhat involved way that we will not detail here.

We do specify the asymptotics of the zero functions and eigenvalues:

$$y \uparrow \infty \implies z_j^\delta(y) \to ija_\delta, \quad j = 1, \ldots, N_\delta, \ \delta = +, -, \quad (49)$$

$$y \uparrow \infty \implies E_\delta(y) \sim \exp(a_{-\delta}y), \frac{dE_\delta(y)}{dy} \sim a_{-\delta}\exp(a_{-\delta}y),$$

$$\delta = +, -. \quad (50)$$

In view of the eigenvalue asymptotics, we can choose K such that on (K, ∞) the two eigenvalues separate points:

$$K < y_1 < y_2 \implies (E_+(y_1), E_-(y_1)) \neq (E_+(y_2), E_-(y_2)). \quad (51)$$

Another important feature is that the joint eigenspace is two-dimensional. (Again, we can only prove this for $y \in (L, \infty)$ and sufficiently large $L \geq K$, cf. Appendix B in Ref. 20.) Notice that this is false when a_+/a_- is rational: In that case joint eigenspaces are infinite-dimensional whenever they contain one nontrivial function.

We now turn to Hilbert space aspects. To this end we study the functions (32), with $\mathcal{F}$ now given by (48). These functions are clearly π/r-periodic (-antiperiodic) for n odd (even). It is far from clear, but true that they are also pairwise orthogonal in $\mathcal{H}$ (20), provided b belongs to the unitarity interval $(0, a_+ + a_-$. Fixing once more a_-, we have depicted the resulting unitarity region in the (a_+, b)-plane in Figure 26.2.

As a consequence, H_+ and H_- can be redefined as commuting self-adjoint operators on the closed subspace $\mathcal{H}_K$ spanned by the functions (32): We need only set

$$H_\delta \Phi_n \equiv E_\delta(nr)\Phi_n, \quad nr > K, \ n \in \mathbb{N}, \tag{52}$$

extend linearly, and take the closure.

Just as in the special case $N_- = 0$ (cf. Section 3), orthogonality is most likely violated for $b < 0$ and $b > a_+ + a_-$. We also expect that for $b \in (0, a_+ + a_-)$ the subspace $\mathcal{H}_K^\perp$ has dimension $[K/r] + 1$ and is spanned by joint eigenfunctions $\Phi_0, \ldots, \Phi_{[K/r]}$ with real eigenvalues.

We conclude this section by explaining a key feature in the symmetry/orthogonality analysis. To this end we first rewrite $\mathcal{F}(x, y)$ as

$$\mathcal{F}(x,y) = \mathcal{H}(x,y) / \prod_{\delta=+,-} \prod_{j=1}^{N_\delta} [s_{-\delta}(x + ija_\delta)s_{-\delta}(x - ija_\delta)]^{1/2}, \tag{53}$$

so that we have

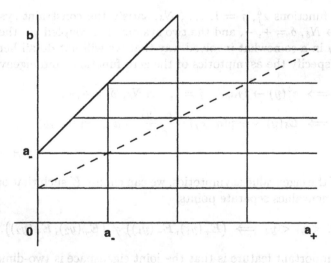

FIGURE 26.2. The unitarity region and the symmetry line $b = (a_+ + a_-)/2$ (dashed) in the (a_+, b)-plane.

$$\mathcal{H}(x,y) = \prod_{\delta=+,-} \prod_{j=1}^{N_\delta} s_{-\delta}\left(x + z_j^\delta(y)\right)$$
$$\times \exp[irx(2N_+N_- + N_+ + N_- + 1) + ixy]. \quad (54)$$

Note that this similarity transformation yields *holomorphic* eigenfunctions. By exploiting the transformed AΔEs it can now be proved that one has the identities

$$\mathcal{H}(ik_+a_+ + ik_-a_-, y) = \mathcal{H}(-ik_+a_+ - ik_-a_-, y),$$
$$k_\delta \in \{-N_\delta, \ldots, 0, \ldots, N_\delta\}, \ \delta = +, -. \quad (55)$$

Therefore, the function $\mathcal{H}(x,y) - \mathcal{H}(-x,y)$ has $(2N_+ + 1)(2N_- + 1)$ explicitly known zeros on the imaginary axis. Since the functions

$$\psi_n(x) \equiv \mathcal{H}(x, nr) - \mathcal{H}(-x, nr) \quad (56)$$

are π/r-periodic (n odd) or π/r-antiperiodic (n even), the above zeros repeat under a shift by π/r:

$$\psi_n\left(ik_+a_+ + ik_-a_- + \frac{l\pi}{r}\right) = 0,$$
$$k_\delta \in \{-N_\delta, \ldots, 0, \ldots, N_\delta\}, \ \delta = +, -, \ l \in \mathbb{Z}. \quad (57)$$

These explicit zeros of $\psi_n(x)$ are crucial in canceling poles arising from the squared denominator in (48). But only for $b \in (0, a_+ + a_-)$ one gets a pole-zero cancellation in a strip around the real x-axis that suffices to deduce pairwise orthogonality. For $b < 0$ and $b > a_+ + a_-$ one obtains instead a residue sum that has no reason to vanish (although in general its vanishing cannot be ruled out).

5   Concluding Remarks

In this final section we sketch some more results related to the above eigenfunctions. First, we point out that the functions $\mathcal{F}(\pm x, y)$ (48) can also be viewed as joint eigenfunctions of H_+ and H_- for the b-values $-N_+a_+ - N_-a_-$ and $(N_+ + 1)a_+ + (N_- + 1)a_-$ (in addition to $b = (N_\alpha + 1)a_\alpha - N_{-\alpha}a_{-\alpha}$, $\alpha = +, -$). This is because one has the identities

$$H_\delta(-N_+a_+ - N_-a_-) = H_\delta((N_+ + 1)a_+ + (N_- + 1)a_-)$$
$$= r_\delta H_\delta((N_+ + 1)a_+ - N_-a_-), \quad (58)$$

where

$$r_\delta \equiv \exp(2N_+ + 1)(2N_- + 1)a_{-\delta}r. \tag{59}$$

(The first equality follows from (46); the second can be checked using (4).) To obtain once again the eigenvalue asymptotics (50) for these new b-values, one should, however, shift the spectral variable y over a distance $(2N_+ + 1)(2N_- + 1)r$.

More generally, since we can handle a dense subset of $\mathcal{E}$ (43), a natural question concerns the existence of continuous interpolating joint eigenfunctions for all of $\mathcal{E}$. Clearly, such interpolations are uniquely determined up to multipliers depending solely on y and the parameters. Now a crucial feature of the above functions $\mathcal{F}(x, y)$ is that their $y \to \infty$ asymptotics is given in terms of a scattering function $u(r, a_+, a_-, b; x)$ that has a real-analytic extension to all of $\mathcal{E}$ (taking $x \in \mathbb{R}$), as will now be detailed.

First, we note that (49) entails

$$\mathcal{F}(x, y) \sim \zeta(N_+, N_-)[-\exp(2irx)u(x)]^{1/2} \exp(ixy), \quad y \to \infty, \tag{60}$$

where ζ is a suitable phase and where

$$u(x) = (-1)^{N_+ + N_-} \prod_{\delta=+,-} \prod_{j=1}^{N_\delta} \frac{s_\delta(x + ija_{-\delta})}{s_\delta(x - ija_{-\delta})}$$
$$\times \exp[2irx(2N_+N_- + N_+ + N_-)]. \tag{61}$$

To explain why this function extends to all of $\mathcal{E}$ (and for other purposes), it is convenient to introduce a function

$$c(r, a_+, a_-, b; x) = \frac{G(r, a_+, a_-; x - ib + i(a_+ + a_-)/2)}{G(r, a_+, a_-; x + i(a_+ + a_-)/2)}. \tag{62}$$

Here, $G(r, a_+, a_-; z)$ is the elliptic generalized gamma function introduced and studied in Ref. 18. It is meromorphic in z and in the parameters r, a_+ and a_- as long as ra_+ and ra_- stay in the right half-plane. Now the extension to $\mathcal{E}$ of the function $u(x) = u(r, a_+, a_-, (N_\alpha + 1)a_\alpha - N_{-\alpha}a_{-\alpha}; x)$ given by (61) reads

$$u(x) = -e^{-2irx} \frac{c(x)}{c(-x)}. \tag{63}$$

For real x this function is real-analytic on $\mathcal{E}$, as advertized. In particular, it is uniquely determined by (61). Thus the *asymptotics* of the joint eigenfunctions admits a unique real-analytic interpolation.

The similarity transformation (53) turning the two-valued eigenfunction $\mathcal{F}(x, y)$ into the holomorphic function $\mathcal{H}(x, y)$ is readily seen *not* to admit a continuous interpolation. But when we introduce a generalized weight function

$$w(x) = \frac{1}{c(x)c(-x)}, \tag{64}$$

then the similarity transformation to a new, meromorphic function

$$\Psi(x,y) \equiv \frac{\mathcal{F}(x,y)}{\zeta(N_+,N_-)w(x)^{1/2}}, \tag{65}$$

does admit an interpolation, since $w(x)$ does. The function Ψ has asymptotics

$$\Psi(x,y) \sim c(x)e^{ixy}, \quad y \to \infty, \tag{66}$$

and correspondingly $c(x)$ may be viewed as a generalization of the Harish-Chandra c-function for symmetric spaces of rank 1 (cf. Refs. 15 and 17).

Of course, this does not answer the question of whether the meromorphic joint eigenfunctions $\Psi(\pm x, y)$ of the similarity-transformed AΔOs

$$A_\delta \equiv w(x)^{-1/2} H_\delta w(x)^{1/2} = e^{-br} \frac{s_\delta(x-ib)}{s_\delta(x)} T_{ia_{-\delta}} + (i \to -i),$$

$$\delta = +, -, \tag{67}$$

admit an extension to $\mathcal{E}$. This appears to be a quite delicate issue. For the even combination

$$\chi(x,y) \equiv \Psi(x,y) + \Psi(-x,y) \tag{68}$$

the identities (55) give rise to pole-zero cancellations on the imaginary axis, but as before one needs to choose $y = nr$, $n \in \mathbb{N}$, to ensure the same cancellation on the lines $\mathrm{Re}\, x = k\pi/r$, $k \in \mathbb{Z}^*$.

These cancellations are not only crucial for the orthogonality issue, but they are also relevant for the question of meromorphic interpolations. Indeed, for convergence to points in $\mathcal{E}$ for which a_+/a_- is irrational and b not equal to $ka_+ + la_-$ with k,l integers (for example), one needs to let N_+ and N_- go to ∞; hence cancellations are needed to prevent poles from becoming dense.

Even so, it appears hard to *exclude* the existence of a meromorphic interpolation for $\Psi(x,y)$. (This is because poles of meromorphic functions can exhibit drastic changes under convergence.) At any rate, for Hilbert space purposes it would suffice to control the convergence to arbitrary points in $\mathcal{E}$ for the functions $\chi(x,nr)$, $n \in \mathbb{N}$.

Though we have discussed these questions in more detail in Ref. 20, we have obtained no clear-cut answers. By contrast, at the hyperbolic level an interpolation of $\chi(x,y)$ is explicitly known. More precisely, there exists a function $R(a_+,a_-,b;x,p)$ that is real-analytic in a_+, a_-, and b in the hyperbolic parameter domain $\{a_+, a_- > 0, b \in \mathbb{R}\}$ for x, p fixed, meromorphic in x, p for fixed parameters, and that satisfies

$$R(a_+, a_-, ka_+ + la_-; x, p)$$
$$= r_{kl}(a_+, a_-; p)\chi(a_+, a_-, ka_+ + la_-; x, \pi p/a_+a_-), \quad k, l \in \mathbb{Z}. \tag{69}$$

The renormalization of the dependence on the spectral variable ensures the self-duality property

$$R(a_+, a_-, b; x, p) = R(a_+, a_-, b; p, x). \tag{70}$$

Furthermore, one has the symmetry property

$$R(a_+, a_-, b; x, p) = R(a_-, a_+, b; x, p), \tag{71}$$

and the A_δ-eigenvalues take the simple form

$$A_\delta R(a_+, a_-, b; x, p) = 2 \operatorname{ch}\left(\frac{\pi p}{a_\delta}\right) R(a_+, a_-, b; x, p). \tag{72}$$

The latter properties are proved in Ref. 21 for the b-values occurring in (69). The interpolating function R is detailed in Ref. 23; we study a more general function in Ref. 19 and elsewhere.

We conclude this contribution with some conjectures concerning M-particle eigenfunctions of the M commuting AΔOs. We believe that such joint eigenfunctions exist not only at the hyperbolic, but also at the elliptic level. (At the trigonometric level the joint eigenfunctions needed for quantum-mechanical purposes amount to the A_{M-1} Macdonald polynomials, cf. Ref. 23, Section 6.2.) In the hyperbolic case we expect self-duality (invariance under $(x_1, \ldots, x_M) \leftrightarrow (p_1, \ldots, p_M)$), as the quantum generalization of the classical self-duality first proved in Ref. 16. Moreover, the parameter symmetry (71) should still hold true for $M > 2$, and the eigenvalues should be the obvious (scattering theory) generalizations of (72).

In the elliptic case we also expect the symmetry property (71). Note in this connection that $a_+ \leftrightarrow a_-$ symmetry would entail the joint eigenfunction property for the M AΔOs obtained by interchanging a_+ and a_-; This second AΔO family is readily seen to commute with the first one.

Finally, both at the elliptic and at the hyperbolic level we expect that the unitarity region is given by $b \in [0, a_+ + a_-]$ for all $M \geq 2$ and that the scattering (eigenfunction asymptotics) is factorized in terms of the (2-particle) u-function. To date, we are not aware of any evidence contradicting the above scenario, but the evidence supporting it is mainly circumstantial.

6   REFERENCES

1. R. Askey and M. E. H. Ismail, *A generalization of ultraspherical polynomials*, Studies in Pure Mathematics (Boston, MA) (P. Erdős, ed.), Birkhäuser, 1983, pp. 55–78.

2. E. Billey, *Algebraic nested Bethe ansatz for the elliptic Ruijsenaars model*, math.qa/9806068.

3. J. Dittrich and V. I. Inozemtsev, *On the structure of eigenvectors of the multidimensional Lamé operator*, J. Phys. A **26** (1993), No. 16, L753–L756.

4. G. Felder and A. Varchenko, *Integral representation of solutions of the elliptic Knizhnik–Zamolodchikov–Bernard equations*, Internat. Math. Res. Notices (1995), No. 5, 221–233.

5. _____, *Three formulas for eigenfunctions of integrable Schrödinger operators*, Compositio Math. **107** (1997), No. 2, 143–175.

6. _____, *Algebraic integrability of the two-body Ruijsenaars operator*, Funct. Anal. Appl. **32** (1998), No. 2, 81–92.

7. G. Felder and A. N. Varchenko, *Algebraic Bethe ansatz for the elliptic quantum group $E_{\tau,\eta}(\mathrm{sl}_2)$*, Nucl. Phys. B **480** (1996), No. 1-2, 485–503, q-alg/9605024.

8. K. Hasegawa, *Ruijsenaars' commuting difference operators as commuting transfer matrices*, Commun. Math. Phys. **187** (1997), No. 2, 289–325, q-alg/9512029.

9. V. I. Inozemtsev, *Solution to three-magnon problem for $S = \frac{1}{2}$ periodic quantum spin chains with elliptic exchange*, J. Math. Phys. **37** (1996), No. 1, 147–159.

10. Y. Komori, *Notes on the elliptic Ruijsenaars operators*, Lett. Math. Phys. **46** (1998), 147–155.

11. Y. Komori and K. Hikami, *Elliptic K-matrix associated with Belavin's symmetric R-matrix*, Nucl. Phys. B **494** (1997), 687–701.

12. I. M. Krichever and A. Zabrodin, *Spin generalization of the Ruijsenaars–Schneider model, non-Abelian 2D Toda chain and representations of Sklyanin algebra*, Russian Math. Surveys **50** (1995), No. 6, 1101–1150, hep-th/9505039.

13. E. H. Lieb and W. Liniger, *Exact analysis of an interacting Bose gas. I. The general solution and the ground state*, Phys. Rev. **130** (1963), 1605–1616.

14. M. A. Olshanetsky and A. M. Perelomov, *Classical integrable finite-dimensional systems related to Lie algebras*, Phys. Rep. **71** (1981), No. 5, 313–400.

15. _____, *Quantum integrable systems related to Lie algebras*, Phys. Rep. **94** (1983), No. 6, 313–404.

16. S. N. M. Ruijsenaars, *Action-angle maps and scattering theory for some finite-dimensional integrable systems. I. The pure soliton case*, Commun. Math. Phys. **115** (1988), No. 1, 127–165.

440   S. N. M. Ruijsenaars

17. _____, *Finite-dimensional soliton systems*, Integrable and Superintegrable Systems (B. A. Kupershmidt, ed.), World Scientific, Singapore, 1990, pp. 165–206.

18. _____, *First order analytic difference equations and integrable quantum systems*, J. Math. Phys. **38** (1997), No. 2, 1069–1146.

19. _____, *A generalized hypergeometric function satisfying four analytic difference equations of Askey-Wilson type*, Commun. Math. Phys. **206** (1999), 639–690.

20. _____, *Generalized Lamé functions. I. The elliptic case*, J. Math. Phys. **40** (1999), No. 3, 1595–1626.

21. _____, *Generalized Lamé functions. II. Hyperbolic and trigonometric specializations*, J. Math. Phys. **40** (1999), No. 3, 1627–1663.

22. _____, *Relativistic Lamé functions: the special case $g = 2$*, J. Phys. A **32** (1999), 1737–1772.

23. _____, *Systems of Calogero–Moser type*, Particles and Fields (Banff, 1994) (G. Semenoff and L. Vinet, eds.), CRM Series in Mathematical Physics, Springer, New York, 1999, pp. 251–352.

24. E. K. Sklyanin, *Some algebraic structure connected with the Yang–Baxter equation*, Funct. Anal. Appl. **16** (1982), No. 4, 263–270.

25. _____, *Some algebraic structures connected with the Yang–Baxter equation. Representations of quantum algebras*, Funct. Anal. Appl. **17** (1983), No. 4, 273–284.

26. E. T. Whittaker and G. N. Watson, *A Course of Modern Analysis*, Cambridge Univ. Press, Cambridge, 1927.

27
Exact Solution for the Ground State of a One-Dimensional Quantum Lattice Gas with Coulomb–Like Interaction

Bill Sutherland

1 Introduction

Over the last 25 years, there has been a steadily increasing interest in exactly soluble one-dimensional quantum many-body systems, interacting by long-ranged potentials. This vigorous interest is why we gathered in Montreal for a Workshop. At the Workshop, I presented a talk giving an overview of the field as it has developed over the last 25 years. The beginning of these investigations was the solution of the $1/r^2$ model by Calogero [1–3] and by myself [8–11] in the early 1970s. For our purposes, exactly soluble means either the exact determination of the spectrum of the Hamiltonian, or the exact determination of the ground-state wavefunction and its correlators. Both were achieved for the $1/r^2$ model.

Using the asymptotic Bethe ansatz, I evaluated exactly the spectrum and the thermodynamics of the $1/r^2$ model. In fact, the asymptotic Bethe ansatz [14] allows the determination of the spectrum for any integrable system that supports scattering. This then includes the inverse-sinh-squared potential, and the sinh-cosh potential, as well as an extensive elliptic potential. The nonextensive elliptic potential has been shown by Calogero [4, 6] to be integrable, but as yet, there is no determination of the spectrum.

In this early work, I also showed that the ground-state wavefunction for the periodic $1/r^2$ potential is of product form. A connection was then made with the eigenvalue distribution for random matrices, and various ground-state correlations were explicitly evaluated. (Recent mathematical developments have now allowed the calculation of time-dependent ground-state correlations.) Using this exact ground-state wavefunction, I then constructed all excited states as a set of orthogonal polynomials.

Soon afterwards, I found another elliptic potential, which is extensive, and whose ground-state wavefunction is of product form [12]. This was shown to exhibit long-ranged crystalline order [13]. However, it is generally believed that this system is not integrable.

A few years ago, interest in the $1/r^2$ system was renewed when Haldane [5] and Shastry [7] showed that the ground-state wavefunction for the continuum problem was also the ground-state wavefunction for a lattice gas, with a long-range one-body hopping operator. The correlations in the ground-state of the lattice problem could then be evaluated by modifying the techniques used for the continuum problem. (For a simple explanation of the Haldane and Shastry trick, as well as its limits, the interested reader might like to look at my interpretation [15].)

In discussing these early results with several participants of the Workshop —particularly with Philippe Choquard and Simon Ruijsenaars—I realized that the same trick of putting the continuum model onto a lattice might be made to work for the elliptic case. (I had attempted this earlier, but happily this time it worked.)

So, in this paper, rather than presenting the overview talk, I would instead like to share these new results with you, and return to the two early papers on the ground-state of a very-long range one-dimensional plasma with long-ranged crystalline order; we shall call them I [12] and II [13], respectively. We will show—as did Haldane and Shastry for the $1/r^2$ potential—that the ground-state wavefunction for the continuum problem is also the ground-state wavefunction for a lattice gas, with a long-range one-body hopping operator. However, the implementation is now very different, and rather subtle. In addition, we will explicitly evaluate the ground-state correlations, and verify that the ground-state has long-range crystalline order; it is a so-called Wigner crystal.

2   Summary of Paper I

In paper I, we introduced a wavefunction of product form,

$$\Psi_0 = C(r) \prod_{j>j'=1}^{N} \vartheta_1 \left(\frac{\pi(x_j - x_{j'})}{L} \,\middle|\, \frac{ir}{L} \right). \tag{1}$$

We always assume N to be odd. The functions $\vartheta_\alpha(z \mid \tau)$ are the Jacobi theta functions with parameter τ; we will make extensive use of the properties of these functions [16]. This wavefunction is antisymmetric, translationally invariant, and periodic in each of the x_j with period L. It has zeros only when $x_j = x_{j'}$, and so it is a suitable candidate for the ground-state wavefunction of a system of spinless, one-dimensional fermions. Then it was established that Ψ_0 satisfies the N-body Schrödinger equation

$$\frac{1}{\Psi_0} \sum_{j=1}^{N} \frac{\partial^2 \Psi_0}{\partial x_j^2} = V - E(r), \tag{2}$$

where the potential V is the sum of pair potentials of the form

$$V = \frac{N\pi^2}{L^2} \sum_{j>j'=1}^{N} \frac{\vartheta_1''(\pi(x_j - x_{j'})/L \mid ir/L)}{\vartheta_1(\pi(x_j - x_{j'})/L \mid ir/L)}, \tag{3}$$

and the ground-state energy is

$$E(r) = \frac{N(N-1)(N-2)}{3!} \frac{\pi^2}{L^2} \frac{\vartheta_1'''(0 \mid ir/L)}{\vartheta_1'(0 \mid ir/L)}. \tag{4}$$

This division of $V - E(r)$ into potential and ground-state energy is obviously arbitrary, and for the physical interpretation of this model, we modify it.

We have parameterized the system so as to easily take the limit $N, L \to \infty$ with $N/L \equiv d$ finite—the so-called *thermodynamic limit*. In this limit, since the parameter $r \to 0$, we can use the asymptotic expansions for the theta functions, correct to terms exponentially small in L. We emphasize that everything we do in this paper in exact for finite systems; however, we will often simplify the equations by showing this limit.

Thus, the pair potential for $L \geq x \geq -L$ becomes

$$v(x) \to d \left(\frac{2\pi}{r}\right)^2 \left[\frac{(|x| - L/2)^2}{L} - |x| \left[\coth\left(\frac{\pi|x|}{r}\right) - 1\right]\right] - \frac{2\pi d}{r}. \tag{5}$$

It is convenient to redefine the zero of $v(x)$, so $v(x) \to v(x) - \epsilon_0$. Then the energy is also redefined so $E \to E - N(N-1)\epsilon_0/2$. We choose $\epsilon_0 = N(\pi/r)^2 - 2\pi d/r$. Then the new pair potential becomes

$$v(x) \to d \left(\frac{2\pi}{r}\right)^2 \left[\frac{x^2}{L} - |x|\right] + d \left(\frac{2\pi}{r}\right)^2 |x| \left[1 - \coth\left(\frac{\pi|x|}{r}\right)\right]. \tag{6}$$

Consider the long-range part of this potential, $v_0(x) \equiv d(2\pi/r)^2[x^2/L - |x|]$. We interpret this as a one-dimensional Coulomb potential, which obeys Gauss's law, so that each particle has a charge density $\rho(x)$ given by $\rho(x) = -v_0''/4\pi q = q[\delta(x) - 1/L]$. Thus, each point particle of charge q is surrounded by a uniform cloud of background-charge $-q$. This neutral composite object gives rise to the periodic potential v_0. The relationship between the parameter r and the charge q is $q^2 = 2\pi d/r^2$.

We calculate the energy of interaction between these composite charges, and the self-energy of the composite charges. Adding these two contributions together, we have for the total electrostatic energy $V + N^3\pi^2/3r^2$. Since our potential is simply V, then we must add the additional term to the previous ground-state energy, so we now have our final result for the total ground-state energy E_0:

$$E_0 = E(r) - \frac{N(N-1)}{2}\epsilon_0 + \frac{N^3\pi^2}{3r^2} \to N \left[\frac{\pi^2}{3r^2} + \frac{2\pi(N-1)}{rL}\right]. \tag{7}$$

This is an extensive energy, and is correct for a finite system, up to terms exponentially small in the size of the system. In terms of the charge q,

$$\frac{E_0}{L} \to \frac{\pi q^2}{6} + \sqrt{2\pi} q d^{3/2} - \sqrt{2\pi} \frac{q d^{1/2}}{L}. \tag{8}$$

3   Summary of Paper II

In paper II, we established that if we choose $C(r)$ so that it satisfies the equation $4\pi d\, dC(r)/dr + E(r)C(r) = 0$, then Ψ_0 satisfies the N-particle diffusion equation,

$$\sum_{j=1}^{N} \frac{\partial^2 \Psi_0}{\partial x_j^2} = 4\pi d \frac{\partial \Psi_0}{\partial r}. \tag{9}$$

We now depart slightly from paper II. We define the center of mass coordinate $x = \sum_{j=1}^{N} x_j - L(N+1)/2, \bmod L$, and define a new wavefunction $\Phi_\alpha = \vartheta_3(\pi(x - \alpha N)/L \mid ir/L)\Psi_0$. This wavefunction is still antisymmetric and periodic in each of the x_j with period L. Further, $\Phi_{\alpha+L/N} = \Phi_\alpha$. Now because Ψ_0 is translationally invariant, and x is the center of mass coordinate,

$$\sum_{j=1}^{N} \frac{\partial^2 \Phi_\alpha}{\partial x_j^2} = \left[V + \frac{\pi^2}{L^2} \frac{\vartheta_3''(\pi(x - N\alpha)/L \mid ir/L)}{\vartheta_3(\pi(x - N\alpha)/L \mid ir/L)} - E(r) \right] \Phi_\alpha$$

$$= 4\pi d \frac{\partial \Phi_\alpha}{\partial r}. \tag{10}$$

Thus, Φ_α satisfies both the Schrödinger equation and the diffusion equation. However, there is an additional potential acting only on the center of mass coordinate. This is smaller than the pair potential, by a factor of $1/N$, and breaks translational invariance.

If we let $r \to 0$, then the wavefunction Φ_α localizes into a uniform lattice with lattice constant $1/d$. Thus, if $x_j = j/d + \alpha$, $j = 0, 1, \ldots, N-1$, and $0 \le \alpha < 1/d$—or permutations thereof—then the extra factor in the wavefunction simply localizes α, the location of the first particle, to be some fraction of a lattice constant.

In the same paper, we examined the wavefunction

$$\Phi_\alpha = \det_{1 \le j, j' \le N} \vartheta_3\left(\frac{pi(x_{j'} - j/d - \alpha)}{L} \,\middle|\, \frac{ir}{NL} \right). \tag{11}$$

This wavefunction also is antisymmetric, periodic in each x_j with period L, and satisfies the N-particle diffusion equation. Further, as $r \to 0$, the particles also become localized into the lattice $j/d + \alpha$, $j = 0, 1, \ldots, N-1$, and $0 \leq \alpha < 1/d$. Thus, this wavefunction must be a constant multiple of the previous one, and can be taken as identical, by a suitable choice of $C(0)$.

The important feature of this second form of the wavefunction is that it is a Slater determinant of one-body wavefunctions. Thus, it appears very much like an energy eigenstate for a system of free particles. This allows one to evaluate explicitly all correlation functions, and in particular, to exhibit long-range crystalline order.

We may construct total momentum eigenstates by

$$\Psi_k = e^{-\pi r k^2/L} e^{2\pi i k x/L} \Psi_0,$$

with k integer. It is easily verified that this is an eigenstate of the Schrödinger equation with the original potential V, and an energy $E(r) + N(2\pi k/L)^2$. This extra term is the familiar center of mass energy.

4 A Lattice Version

We now suppose that L is an integer, and that the coordinates x_j take only discrete integer values. We consider the complete set of L one-body lattice wavefunctions

$$\phi(x - y \mid r) \equiv \vartheta_3\left(\frac{\pi(x-y)}{L} \ \bigg| \ \frac{ir}{NL}\right), \quad y = 0, 1, \ldots, L-1. \quad (12)$$

Since x is an integer, we can simplify $\phi(x|r)$ as $\phi(x|r) = \sum_{s=0}^{L-1} \lambda_s(r) e^{2\pi i x s/L}$, with

$$\lambda_s(r) = \sum_{m=-\infty}^{\infty} e^{-\pi r(mL+s)^2/LN} = \sqrt{\frac{d}{r}} \vartheta_3\left(\frac{\pi s}{L} \ \bigg| \ \frac{id}{r}\right). \quad (13)$$

We see that $\lambda_{s+L} = \lambda_s$.

We now take

$$\frac{\partial \phi}{\partial r} = \sum_{s=0}^{L-1} \frac{\partial \lambda_s(r)}{\partial r} e^{2\pi i x s/L}. \quad (14)$$

What we want is to find a difference operator, which has the same effect on ϕ as $\partial/\partial r$. Thus, we seek $J(x)$ such that

$$\sum_{y=0}^{L-1} J(x-y)\phi(y - z \mid r) = 4\pi d \frac{\partial \phi(x-z)}{\partial r}. \quad (15)$$

This is the discrete diffusion equation. All that is needed is to Fourier transform the equation, so that with $\tilde{J}_s(r)$ the Fourier transform of J, then $\tilde{J}_s(r) = 4\pi d \partial \log \lambda_s(r)/\partial r$.

We now take the one-body operator $J = \sum_{j=1}^{N} J_j$, where J_j acts only on coordinate x_j. Let J act on the determinant wavefunction

$$\Phi = \det_{1 \leq j, j' \leq N} \phi(x_{j'} - y_j \mid r),$$

where the quantum numbers y_j are also integers. Then Φ satisfies the discrete diffusion equation $J\Phi = 4\pi d \partial \Phi / \partial r$.

Let N divide L, so $L = NM$. Thus, the density or *filling factor* is $d = N/L = 1/M$. Now this discrete wavefunction is identical to the continuous wavefunction, evaluated on the lattice, provided we choose $y_j = jM + \alpha$. Since the y's must be integers, this restricts α to $\alpha = 0, 1, \ldots, M - 1$. Then, we can write

$$J\Phi_\alpha = \left[V + \frac{\pi^2}{L^2} \frac{\vartheta_3''(\pi(x - N\alpha)/L \mid ir/L)}{\vartheta_3(\pi(x - N\alpha)/L \mid ir/L)} - E(r) \right] \Phi_\alpha = 4\pi d \frac{\partial \Phi_\alpha}{\partial r}. \quad (16)$$

Thus, Φ_α satisfies a discrete Schrödinger equation. The extra potential, acting on the center of mass, selects one of the M equivalent sublattices.

Suppose we take the momentum eigenstates $\Psi_k = \sum_{\alpha=0}^{M-1} e^{2\pi i k\alpha/M} \Phi_\alpha / M$, with $k/N = 0, 1, \ldots, M - 1$. Then, projecting out the appropriate momentum component of the Schrödinger equation, we find that $J\Psi_k = [V - E(r)]\Psi_k + v_k(x)\Psi_k$, where $v_k(x)$ is a translationally invariant potential which acts only on the center of mass x.

5   Ground-State Correlations

Let us take one of the localized states, say,

$$\Phi_0 = \det_{1 \leq i, j \leq N} \phi(x_i - jM \mid r) \equiv \det_{1 \leq i, j \leq N} \phi_j(x_i \mid r), \quad (17)$$

since the correlations in this state will be typical of the correlations in the ground-state of the translationally invariant problem. The one-particle orbitals $\phi_j(x_i)$ are localized about site $j = 0, 1, \ldots, N - 1$, and are not orthogonal.

We make a transformation to orthonormal Bloch functions $\psi_k(x) = \sum_{j=0}^{N-1} M_{kj} \phi_j(x)$, with $k = 0, 1, \ldots, N - 1$, so that

$$\delta_{kk'} = \sum_{x=0}^{L-1} \psi_k^*(x) \psi_{k'}(x) = \sum_{j,j'=0}^{N-1} M_{kj} N_{jj'} M_{j'k}^\dagger, \quad (18)$$

where

$$N_{jj'} = \sum_{x=0}^{L-1} \phi_{j'}(x)\, \phi_j(x) = L \sum_{k=0}^{N-1} e^{2\pi i k(j'-j)/N} \sum_{\beta=0}^{M-1} \lambda_{\beta N + k}^2(r). \quad (19)$$

Thus, the eigenvectors of the matrix N are $e^{2\pi i k j / N}$, and the eigenvalues are $L^2 \sum_{\beta=0}^{M-1} \lambda_{\beta N+k}^2(r) \equiv \gamma_k^2(r)$. We see that $\gamma_{k+N} = \gamma_k$, and so $M_{kj} = e^{2\pi i k j / N} / [\sqrt{L}\gamma_k(r)]$, and thus the Bloch functions are

$$\psi_k(x) = \frac{1}{M\gamma_k(r)} \sqrt{\frac{N}{2r}} e^{-\pi x^2 / rM} \vartheta_3 \left(\pi \left(\frac{k}{N} - \frac{ix}{r} \right) \mid \frac{iM}{r} \right). \quad (20)$$

We see that $\psi_k(x + M) = e^{2\pi i k / N} \psi_k(x)$, and $\psi_{k+N}(x) = \psi_k(x)$.

Because of the orthonormality of the Bloch functions, the normalized wavefunction Φ_0 is given as $\Phi_0 = \det_{1 \leq i,j \leq N} \psi_j(x_i \mid r) / \sqrt{N!}$. By Gram's result, the n-particle reduced density matrices $\rho_n(y; x)$ are given by

$$\rho_n(y; x) = \det_{1 \leq i,j \leq n} \sum_{k=0}^{N-1} \psi_k^*(y_i) \psi_k(x_j) = \det_{1 \leq i,j \leq n} \rho_1(y_i; x_j). \quad (21)$$

Thus, all higher reduced density matrices are given in terms of the one-particle density matrix $\rho_1(y; x) = \sum_{k=0}^{N-1} \psi_k^*(y) \psi_k(x)$.

The absolute ground-state (unnormalized) of the original translationally invariant problem, is the wavefunction Ψ_0 with zero total momentum, given by

$$\Psi_0 = \sum_{\alpha=0}^{M-1} \Phi_\alpha = \sum_{\alpha=0}^{M-1} \Phi_0(\ldots, x_j - \alpha, \ldots). \quad (22)$$

This state is not normalized, so let us first calculate

$$\Psi_0^\dagger \Psi_0 = N! \sum_{\alpha', \alpha=0}^{M-1} \det_{1 \leq j,j' \leq N} N_{jj'}(\alpha' - \alpha),$$

where $N_{jj'}(\alpha) = \sum_{x=0}^{L-1} \phi_j(x) \phi_{j'}(x + \alpha)$. This is very similar to previously, where $N_{jj'}$ becomes $N_{jj'}(0)$. The eigenvectors are again $e^{2\pi i k j / N}$, while the eigenvalues are now

$$\gamma_k^2(\alpha \mid r) = e^{2\pi i k \alpha / L} L^2 \sum_{\beta=0}^{M-1} \lambda_{\beta N+k}^2(r) e^{2\pi i \alpha \beta / M}. \quad (23)$$

This gives $\Psi_0^\dagger \Psi_0 = N! \sum_{\alpha', \alpha=0}^{M-1} \prod_{j=1}^{N-1} \gamma_k^2(\alpha' - \alpha \mid r)$. However, for a large system, the double summation over α', α will be dominated by the largest terms, with corrections exponentially small in N. These largest terms are all those with $\alpha' - \alpha = 0$. Thus, in the thermodynamic limit, $\Psi_0^\dagger \Psi_0 \rightarrow N N! \prod_{j=1}^{N-1} \gamma_k^2(r)$. Similarly, the correlations themselves can be expressed as a double summation, and once again, in the thermodynamic limit this summation is dominated by the maximum terms, so

$$\bar{\rho}_n(y; x) = \frac{1}{M} \sum_{\alpha=0}^{M-1} \rho_n(y - \alpha; x - \alpha).$$

Here, $\rho_n(y; x)$ are the previous reduced density matrices.

As an important example, let us consider the half-filled lattice, when $M = 2$. First, we explicitly evaluate the one-particle density matrix ρ_1 as

$$2\rho_1(y; x) = \delta_{y,x} + \frac{1}{\vartheta_3^2(0 \mid i/r)}\left[\frac{\sin(\pi(y - x)/2)}{\sinh((y - x)/2r)}(1 - \delta_{y,x}) \right.$$
$$\left. - \frac{\cos(\pi(y + x)/2)}{\cosh((y - x)/2r)}\right]. \quad (24)$$

Then, we can obtain the one-particle density matrix of the translationally invariant ground-state as

$$2\bar{\rho}_1(y; x) = 2\bar{\rho}_1(y - x)$$
$$= \delta_{y-x,0} + \frac{1}{\vartheta_3^2(0 \mid i/r)}\frac{\sin(\pi(y - x)/2)}{\sinh((y - x)/2r)}(1 - \delta_{y-x,0}). \quad (25)$$

The momentum distribution $n(k)$ is given by the Fourier transform as

$$n(k) = \frac{1}{2}\left[1 + \frac{\vartheta_2(0 \mid i/r)\vartheta_2(k \mid i/r)}{\vartheta_3(0 \mid i/r)\vartheta_3(k \mid i/r)}\right]. \quad (26)$$

From the diagonal elements of the two-body density matrix, we find the pair correlation function $g(x)$, defined as $\bar{\rho}_2(x_1, x_2; x_1, x_2) \equiv d^2 g(x_2 - x_1)$, to be

$$g(x) = 1 - \delta_{x,0} + \frac{1}{\vartheta_3^4(0 \mid i/r)}\left[(-1)^x - \frac{2}{\cosh(\pi x/r) + (-1)^x}\right]. \quad (27)$$

Thus, there is long-range crystalline order, with an order parameter given by $1/\vartheta_3^4(0 \mid i/r)$.

In a longer publication, we will consider the other cases, as well as the low-lying states, other fillings, and extensions based on powers of this ground- state.

6 References

1. F. Calogero, *Ground state of a one-dimensional N-body system*, J. Math. Phys. **10** (1969), 2197–2200.

2. ———, *Solution of a three-body problem in one dimension*, J. Math. Phys. **10** (1969), 2191–2196.

3. ———, *Solution of the one-dimensional N-body problems with quadratic and/or inversely quadratic pair potentials*, J. Math. Phys. **12** (1971), 419–436.

4. ———, *Exactly solvable one-dimensional many-body problems*, Lett. Nuovo Cimento **13** (1975), No. 11, 411–416.

5. F. D. M. Haldane, *Exact Jastrow–Gutzwiller resonating-valence-bond ground state of the spin-$\frac{1}{2}$ antiferromagnetic Heisenberg chain with $1/r^2$ exchange*, Phys. Rev. Lett. **60** (1988), 635–638.

6. C. Marchioro, F. Calogero, and O. Ragnisco, *Exact solution of the classical and quantal one-dimensional many-body problems with the two-body potential $V_a(x) = g^2 a^2 / \sinh^2(ax)$*, Lett. Nuovo Cimento **13** (1975), No. 10, 383–387.

7. B. S. Shastry, *Exact solution of an $S = \frac{1}{2}$ Heisenberg antiferromagnetic chain with long-ranged interactions*, Phys. Rev. Lett. **60** (1988), 639–642.

8. B. Sutherland, *Exact results for a quantum many-body problem in one dimension*, Phys. Rev. A **4** (1971), 2019–2021.

9. _____, *Quantum many-body problem in one dimension: ground state*, J. Math. Phys. **12** (1971), 246–250.

10. _____, *Quantum many-body problem in one dimension: thermodynamics*, J. Math. Phys. **12** (1971), 251–256.

11. _____, *Exact results for a quantum many-body problem in one dimension. II*, Phys. Rev. A **5** (1972), 1372–1376.

12. _____, *Exact ground-state wave function for a one-dimensional plasma*, Phys. Rev. Lett. **34** (1975), 1083–1085.

13. _____, *One-dimensional plasma as an example of a Wigner solid*, Phys. Rev. Lett. **35** (1975), 185–188.

14. _____, *A brief history of the quantum soliton with new results on the quantization of the Toda lattice*, Rocky Mountain J. Math. **8** (1978), No. 1-2, 413–428.

15. _____, *Exact solution of a lattice band problem related to an exactly soluble many-body problem: the missing-states problem*, Phys. Rev. B **38** (1988), No. 10, 6689–6692.

16. E. T. Whittaker and G. N. Watson, *A Course of Modern Analysis*, Cambridge Univ. Press, Cambridge, 1927.

28

Differential Operators that Commute with the r^{-2}-type Hamiltonian

Kenji Taniguchi[1]

1 Introduction

A quantum mechanical Hamiltonian H in n variables is said to be completely integrable if there exist n algebraically independent differential operators $I_1 = H$, I_2, $\ldots$, I_n commuting with each other. The Calogero–Moser–Sutherland type Hamiltonian is one of the most popular completely integrable systems.

In this paper, we investigate what kind of differential operator commutes with the CMS type Hamiltonian. This investigation is motivated by the following questions about the integrals of motion.

The first question is about the symmetry of integrals: The complete integrability of CMS type Hamiltonian is proved by constructing symmetric operators I_2, $\ldots$, I_n. Conversely, if a differential operator P commutes with H, then is it automatically symmetric? When the parameter of the potential function is special, we know that there exist nonsymmetric operators that commute with H when the parameter of the potential function is special (Ref. 5 and the references in it). But since the orders of such nonsymmetric operators are very high, it is expected that, under some order condition, the commutants of H are symmetric.

The second question is about commutativity of integrals: If both P_1 and P_2 commute with H, then do P_1 and P_2 commute mutually?

The third question is about the classification of integrals. Do there exist unknown integrals of motion? If so, what is the physical or mathematical meaning of them?

This chapter is organized as follows. In Section 2 we review the commutative families of Weyl group invariant differential operators, which contains the CMS models related to classical Lie algebras. In Section 3 we treat periodic (i.e., trigonometric or elliptic) potential cases, and in Section 4 rational cases. This is because the arguments and conclusions differ with

---

[1] JSPS Research Fellow.

the potential functions.

2 Commutative Families of Weyl Group Invariant Differential Operators

As a formulation of Calogero–Moser–Sutherland-type completely integrable systems, we adopt the following one, attributable to Oshima–Sekiguchi. For details, see Refs. 3 and 4.

We define commutative families of Weyl group invariant differential operators.

Let W be a Weyl group and Σ be the corresponding reduced root system. Σ is realized in $\mathbb{R}^n$, and then W acts on $\mathbb{R}^n$ and $\mathbb{C}^n = \mathbb{R}^n \otimes \mathbb{C}$.

Definition 1 (Oshima–Sekiguchi). We define C to be a commutative ring of differential operators satisfying the following conditions:

C1) Elements of C are holomorphic differential operators on some appropriate W-invariant open connected subset Ω of $\mathbb{C}^n$ with $0 \in \overline{\Omega}$.

C2) Elements of C commute mutually.

C3) Elements of C are W-invariant and their highest-order terms are of constant coefficient.

C4) C contains a second-order operator $H = \frac{1}{2} \sum_{i=1}^{n} \partial_{x_i}^2 + R(x)$.

C5) Highest-order terms of elements of C generate $\mathbb{C}[\partial_x]^W$.

When W is a classical Weyl group, Ochiai, Oshima and Sekiguchi determined the potential function $R(x)$ and constructed generators of C explicitly.

Theorem 1 ([3]). $R(x)$ can be expressed as

$$R(x) = \sum_{\substack{i,j=1 \\ i<j}}^{n} u(x_i - x_j), \quad \text{if } W \text{ is of type } A_{n-1}, \tag{1}$$

$$R(x) = \sum_{\substack{i,j=1 \\ i<j}}^{n} \left(u(x_i - x_j) + u(x_i + x_j)\right) + \sum_{i=1}^{n} v(x_i),$$
$$\text{if } W \text{ is of type } B_n, \tag{2}$$

$$R(x) = \sum_{\substack{i,j=1 \\ i<j}}^{n} \left(u(x_i - x_j) + u(x_i + x_j)\right), \quad \text{if } W \text{ is of type } D_n. \tag{3}$$

The above functions $u(t)$ and $v(t)$ are as follows:

If W is of type A_{n-1} with $n \geq 3$,

$$u(t) = C_1 \wp(t) + C_2. \tag{4}$$

If W is of type B_n with $n \geq 3$,

$$\begin{cases} u(t) = C_1 \wp(t) + C_2, \\ v(t) = \dfrac{C_3 \wp(t)^4 + C_4 \wp(t)^3 + C_5 \wp(t)^2 + C_6 \wp(t) + C_7}{\wp'(t)^2} \end{cases} \tag{5}$$

or

$$u(t) = C_1 t^{-2} + C_2 t^2 + C_3 \quad and \quad v(t) = C_4 t^{-2} + C_5 t^2 + C_6 \tag{6}$$

or

$$u(t) = C_1 \quad and \quad v(t) \ is \ any \ even \ function. \tag{7}$$

If W is of type D_n with $n \geq 4$, then u is (5) or (6).
If W is of type B_2 then $\bigl(u(t), v(t)\bigr)$ is (5) or (6) or (7) or

$$\begin{cases} u(t) = \dfrac{C_3 \wp(t/2)^4 + C_4 \wp(t/2)^3 + C_5 \wp(t/2)^2 + C_6 \wp(t/2) + C_7}{\wp'(t/2)^2}, \\ v(t) = C_1 \wp(t) + C_2 \end{cases} \tag{8}$$

or

$$\begin{cases} u(t) = C_1 \wp(t) + C_2 \dfrac{(\wp(t/2) - e_3)^2}{\wp'(t/2)^2} + C_3, \\ v(t) = C_4 \wp(t) + \dfrac{C_5}{\wp(t) - e_3} + C_6 \end{cases} \tag{9}$$

or

$$v(t) = C_1 \quad and \quad u(t) \ is \ any \ even \ function. \tag{10}$$

Here, C_i's are arbitrary complex numbers and $\wp(t)$ is the Weierstrass elliptic function $\wp(t \mid 2\omega_1, 2\omega_2)$ with primitive half-periods ω_1 and ω_2 which allowed to be infinity, and e_3 is a complex number satisfying $\wp'^2 = 4(\wp - e_1)(\wp - e_2)(\wp - e_3)$. Note that $\wp(t|\infty, \infty) = t^{-2}$, and $\wp(t|\sqrt{-1}\pi, \infty) = \sinh^{-2} t + 1/3$.

3   Periodic Potential Cases

At first, we introduce some notation.

Notation. 1) Let P be an mth-order differential operator on some open subset of $\mathbb{C}^n$. We write P as

$$
\begin{cases}
P = \displaystyle\sum_{k=0}^{m} P_k, \\
P_k := \displaystyle\sum_{|\mu|=k} a_\mu(x)\partial_x^\mu.
\end{cases}
$$

Here, $\partial_x = (\partial_{x_1}, \ldots, \partial_{x_n})$, μ is a multi-index $\mu = (\mu_1, \ldots, \mu_n)$ and $|\mu| := \sum_{i=1}^{n} \mu_i$.

2) Let

$$
\widetilde{P}_k := \sum_{|\mu|=k} a_\mu(x)\xi^\mu, \quad \xi = (\xi_1, \ldots, \xi_n),
$$

and call it the kth symbol of P.

3) For a root $\alpha \in \Sigma$ and $x = (x_1, \ldots, x_n)$, we denote $\langle \alpha, x \rangle$ to be their coupling. For example, if $\alpha = e_1 - e_2$, then $\langle \alpha, x \rangle = x_1 - x_2$. Analogously, we define $\langle \alpha, \partial_x \rangle$, $\langle \alpha, \xi \rangle$ and $\langle \alpha, \partial_\xi \rangle$.

4) We define a subring $\mathcal{R}$ of $\mathbb{C}[x, \xi]$ by

$$
\mathcal{R} := \mathbb{C}[\xi, x_j\xi_i - x_i\xi_j \ (1 \le i < j \le n)].
$$

We denote the second-order operator by

$$
H = \frac{1}{2} \sum_{i=1}^{n} \partial_{x_i}^2 + \sum_{\alpha \in \Sigma^+} u_\alpha(\langle \alpha, x \rangle).
$$

Here, (1) $u_\alpha(t) = u_{w\alpha}(t)$ for $\alpha \in \Sigma$ and $w \in W$, (2) $u_\alpha(t)$'s are given by the classification in Theorem 1. We shall investigate the equation

$$
[H, P] = 0.
$$

Lemma 1. 1) *If* $[H, P] = 0$, *then* $\widetilde{P}_m \in \mathcal{R}$.

2) *If* $\widetilde{Q}(x, \xi) \in \mathcal{R}$ *is symmetric with respect to* ξ, *taht is* $\widetilde{Q}(x, \sigma(\xi)) = \widetilde{Q}(x, \xi)$ *($\forall \sigma \in \mathfrak{S}_n$), then $\widetilde{Q}$ is constant with respect to x, that is $\widetilde{Q} \in \mathbb{C}[\xi]$.*

Lemma 2. *If* $u_\alpha(t)$ *is a (nontrivial) periodic function for any* $\alpha \in \Sigma$, *then* $\langle \alpha, \partial_\xi \rangle \widetilde{P}_m$ *is divisible by* $\langle \alpha, \xi \rangle$.

Sketch of proof. The $m+1$, m and $m-1$st-order terms of $[H, P] = 0$ imply that $\widetilde{P}_{m-2}$ can be expressed as

$$\widetilde{P}_{m-2} = \frac{1}{8}\left(\sum_{i=1}^{n} \partial_{x_i}\partial_{\xi_i}\right)^2 \widetilde{P}_m + \frac{1}{2}\left(\sum_{i=1}^{n} \partial_{x_i}\partial_{\xi_i}\right)\widetilde{Q}_{m-1} + \widetilde{Q}_{m-2}$$

$$+ \sum_{\alpha \in \Sigma^+}\left\{\frac{\langle \alpha, \partial_\xi\rangle\widetilde{P}_m}{\langle\alpha,\xi\rangle}u_\alpha(\langle\alpha,x\rangle) + \frac{\langle\alpha,\partial_x\rangle\widetilde{P}_m}{\langle\alpha,\xi\rangle^2}U_\alpha(\langle\alpha,x\rangle)\right\},$$

where, $\widetilde{P}_m, \widetilde{Q}_{m-1} \in \mathcal{R}$, $\widetilde{Q}_{m-2} \in \sum_{\alpha\in\Sigma^+}\frac{1}{\langle\alpha,\xi\rangle^2}\mathcal{R}$ and $U_\alpha(t)$ is a primitive function of $u_\alpha(t)$.

Since $\widetilde{P}_m$, $\widetilde{P}_{m-2}$ and $\widetilde{Q}_{m-1}$ are polynomials with respect to ξ, there exists a polynomial $\widetilde{Q}'_{m-2} \in \mathcal{R}$ such that

$$\lim_{\langle\alpha,\xi\rangle\to 0}\{\langle\alpha,\partial_x\rangle\widetilde{P}_m U_\alpha(\langle\alpha,x\rangle) + \widetilde{Q}'_{m-2}\} = 0.$$

Since $\langle\alpha,\partial_x\rangle\widetilde{P}_m$ and $\widetilde{Q}'_{m-2}$ are polynomial with respect to x but $U_\alpha(t)$ is not a rational function, this implies that $\langle\alpha,\partial_x\rangle\widetilde{P}_m$ is divisible by $\langle\alpha,\xi\rangle$.

Next, we have

$$\lim_{\langle\alpha,\xi\rangle\to 0}\left\{\langle\alpha,\partial_\xi\rangle\widetilde{P}_m u_\alpha(\langle\alpha,x\rangle) + \frac{\langle\alpha,\partial_x\rangle\widetilde{P}_m}{\langle\alpha,\xi\rangle}U_\alpha(\langle\alpha,x\rangle) + \langle\alpha,\xi\rangle\widetilde{Q}_{m-2}\right\} = 0.$$

If $\langle\alpha,\partial_\xi\rangle\widetilde{P}_m$ is not divisible by $\langle\alpha,\xi\rangle$, then there exist polynomials $f_\alpha(x) \neq 0$ and $g_\alpha(x)$ such that

$$f_\alpha(x)u_\alpha(\langle\alpha,x\rangle) + |\alpha|^{-2}(\langle\alpha,\partial_x\rangle f_\alpha)(x)U_\alpha(\langle\alpha,x\rangle) = g_\alpha(x).$$

This contradicts the periodicity of $u_\alpha(t)$. $\square$

Lemma 3. *Suppose that $\langle\alpha,\partial_\xi\rangle\widetilde{P}_m$ is divisible by $\langle\alpha,\xi\rangle$ and*

$$m \leq \begin{cases} n, & (W, \Sigma) \text{ is of type } A_{n-1} \text{ or } D_n\text{-type,} \\ 2n & (W, \Sigma) \text{ is of type } B_n\text{-type.} \end{cases}$$

Then $\widetilde{P}_m(x, \sigma(\xi)) = \widetilde{P}_m(x,\xi)$ for any $\sigma \in W$.

Proof. This can be proved by direct calculus. $\square$

We now explain the main theorem of periodic cases.

Theorem 2. *Assume that the following conditions are satisfied:*

1)

$$H = \frac{1}{2} \sum_{i=1}^{n} \frac{\partial^2}{\partial x_i{}^2} + \sum_{\alpha \in \Sigma^+} u_\alpha(\langle \alpha, x \rangle),$$

where $u_\alpha(t)$'s are nontrivial periodic functions in the classification in Theorem 1.

2) P is a holomorphic differential operator, defined on some connected open subset of the domain where H is defined.

3)

$$\operatorname{ord} P \leq \begin{cases} n & (W, \Sigma) \text{ is of type } A_{n-1} \text{ or } D_n\text{-type}, \\ 2n & (W, \Sigma) \text{ is of type } B_n\text{-type}. \end{cases}$$

Under these assumptions, $[H, P] = 0$ implies $P \in C$.

Proof. This theorem can be proved by induction on the order of P.   □

Remark 1. In the proof of this theorem, we only examined the $m + 1$, m and $m - $1st-order terms of $[H, P] = 0$. It is expected that we can weaken the order condition by investigating the lower-order terms of $[H, P] = 0$. On the other hand, when the parameters of potential functions are special, it is known that there exist higher-order operators that commute with H but are not W-invariant [5].

4   Rational Potential Cases

In the rational potential cases, the situation is a little more complicated than the periodic cases. In this section, we construct commutants of H in two ways.

At first, we restrict our interest to the A_{n-1}-type cases, and we construct a family of differential operators that commute with

$$H = \frac{1}{2} \sum_{i=1}^{n} \partial_{x_i}^2 + \lambda \sum_{\substack{i,j=1 \\ i<j}}^{n} (x_i - x_j)^{-2} \quad (\lambda \neq 0).$$

Definition 2. For a polynomial $p_0(x)$ of $x_1, \ldots, x_n$, we define differential operators $p_j(x, \partial_x)$ by

$$p_j(x, \partial_x) := (\operatorname{ad} H)^j \big(p_0(x)\big) \quad (j = 0, 1, 2, \ldots).$$

Here, $\operatorname{ad} H$ is an operator $[H, \cdot]$.

We can prove the following lemma by direct calculus.

Lemma 4 (Key lemma). *If $p_0(x)$ is a homogeneous symmetric polynomial of degree k, then $p_k(x, \partial_x) \in C$. Especially, $p_{k+1}(x, \partial_x) = 0$.*

Corollary 1. *For a homogeneous symmetric polynomial $p_0(x)$ of degree k, we define*

$$P := \sum_{j=0}^{k} (-1)^j p_j(x, \partial_x) \circ p_{k-j}(x, \partial_x).$$

Then $[H, P] = 0$.

Example 1. 1) If $p_0 = \sum_{i=1}^{n} x_i^2$, then

$$
\begin{cases}
p_1 = 2 \sum_{i=1}^{n} x_i \partial_{x_i} + n, \\
p_2 = 2 \sum_{i=1}^{n} \partial_{x_i}^2 + 4\lambda \sum_{\substack{i,j=1 \\ i<j}}^{n} (x_i - x_j)^{-2},
\end{cases}
$$

and

$$P = 4 \left\{ \sum_{\substack{i,j=1 \\ i<j}}^{n} (x_j \partial_{x_i} - x_i \partial_{x_j})^2 + 2\lambda \left(\sum_{i=1}^{n} x_i^2 \right) \sum_{\substack{i,j=1 \\ i<j}}^{n} (x_i - x_j)^{-2} \right\} + 4n - n^2.$$

2) If $q_0 = \sum_{\substack{i,j=1 \\ i<j}}^{n} x_i x_j$, then

$$
\begin{cases}
q_1 = \sum_{\substack{i,j=1 \\ i \neq j}}^{n} x_j \partial_{x_i}, \\
q_2 = 2 \sum_{\substack{i,j=1 \\ i<j}}^{n} \partial_{x_i} \partial_{x_j} - 2\lambda \sum_{\substack{i,j=1 \\ i<j}}^{n} (x_i - x_j)^{-2},
\end{cases}
$$

and

$$Q = -\left\{ \sum_{\substack{i,j=1 \\ i<j}}^{n} (x_j \partial_{x_i} - x_i \partial_{x_j})^2 + \sum_{\substack{ii,j,k=1 \\ i \neq j \neq k \neq i}}^{n} (x_j \partial_{x_i} - x_i \partial_{x_j})(x_k \partial_{x_i} - x_i \partial_{x_k}) \right.$$

$$\left. + 4\lambda \left(\sum_{\substack{i,j=1 \\ i<j}}^{n} x_i x_j \right) \sum_{\substack{i,j=1 \\ i<j}}^{n} (x_i - x_j)^{-2} \right\} + n(n-1).$$

Remark 2. Any holomorphic differential operator that commutes with H and of order at most two is a linear combination of

$$1, \quad \Delta_1 := \sum_{i=1}^{n} \partial_{x_i}, \quad \Delta_1^2, \quad H, \quad P, \quad Q, \quad \text{and} \quad [\Delta_1, P].$$

The second construction is based on the nature of the Dunkl differential difference operator.

Let (W, Σ) be a pair of a finite Coxeter group and the corresponding root system in $\mathbb{R}^n$. We choose a positive system Σ^+ of Σ and fix it. Let $k = (k_\alpha): \Sigma \to \mathbb{C}$ be a W-invariant function.

Definition 3. The Dunkl operator ∇_η ($\eta \in \mathbb{R}^n$) is the differential difference operator on $\mathbb{R}^n$ defined by

$$\nabla_\eta = \langle \eta, \partial_x \rangle - \sum_{\alpha \in \Sigma^+} k_\alpha \frac{\langle \alpha, \eta \rangle}{\langle \alpha, x \rangle} s_\alpha,$$

where s_α is the reflection operator $s_\alpha(x) = x - 2\langle \alpha, x \rangle \alpha / \langle \alpha, \alpha \rangle$ associated to α.

Lemma 5 ([1, 2]). *The Dunkl operator has the following properties.*

$$[\nabla_{\eta_1}, \nabla_{\eta_2}] = 0, \tag{11}$$

$$w \circ \nabla_\eta \circ w^{-1} = \nabla_{w(\eta)}, \tag{12}$$

$$[\nabla_{\eta_1}, \langle \eta_2, x \rangle] = \langle \eta_1, \eta_2 \rangle + 2 \sum_{\alpha \in \Sigma^+} \frac{k_\alpha \langle \alpha, \eta_1 \rangle \langle \alpha, \eta_2 \rangle}{\langle \alpha, \alpha \rangle} s_\alpha. \tag{13}$$

Let $\{e_i\}$ be an orthonormal basis of $\mathbb{R}^n$. We define $\nabla_i = \nabla_{e_i}$ and $\Delta = \sum_{i=1}^{n} \nabla_i^2$. Then

$$[\Delta, \langle \eta, x \rangle] = 2\nabla_\eta. \tag{14}$$

Here, η, η_1, η_2 are arbitrary elements of $\mathbb{R}^n$ and w is an arbitrary element of W.

Let $\mathcal{A}$ be the algebra generated by 1, x_i, ∇_j. For any element $D \in \mathcal{A}$, we denote by $\text{Res}\, D$ the differential operator that coincides on the space $C^\infty(\mathbb{R}^n - \bigcup_{\alpha \in \Sigma^+} \{x; \langle \alpha, x \rangle = 0\})^W$ of W-invariant functions. Note that

$$H := \text{Res}\, \Delta = \sum_{i=1}^{n} \partial_{x_i}^2 - \sum_{\alpha \in \Sigma^+} \frac{k_\alpha(k_\alpha - 1)\langle \alpha, \alpha \rangle}{\langle \alpha, x \rangle^2}$$

is the Calogero operator.

Lemma 6 ([1]). *If D_1 and $D_2 \in \mathcal{A}$ commute and are W-invariant, then $\text{Res}\, D_1$ and $\text{Res}\, D_2$ commute.*

By Lemmas 1, 5, and 6, we obtain the classification of W-invariant commutants of H.

Theorem 3. *For each element* $\widetilde{P}(\xi, x_j\xi_i - x_i\xi_j) \in \mathcal{R}^W$,

$$P = \operatorname{Res} \widetilde{P}(\nabla, x_j\nabla_i - x_i\nabla_j)$$

commutes with $H = \sum_{i=1}^n \partial_{x_i}^2 - \sum_{\alpha \in \Sigma^+} k_\alpha(k_\alpha - 1)\langle\alpha, \alpha\rangle/\langle\alpha, x\rangle^2$. *Moreover, such operators exhaust W-invariant commutants of H.*

Proof. The first assertion is proved by Lemma 6 and

$$[\Delta, x_j\nabla_i - x_i\nabla_j] = 2\nabla_j\nabla_i - 2\nabla_i\nabla_j = 0,$$

which is a consequence of Lemma 5.

If a mth-order W-invariant differential operator P commutes with H, then its principal symbol $\widetilde{P}_m$ is an element of $\mathcal{R}^W$ by Lemma 1(1). Since $P - \widetilde{P}_m(\nabla, x_j\nabla_i - x_i\nabla_j)$ commutes with H, the second assertion is proved by induction on the order of P.   □

5   REFERENCES

1. Yu. Berest and Yu. Molchanov, *Fundamental solutions for partial differential equations with reflection group invariance*, J. Math. Phys. **36** (1995), No. 8, 4324–4339.

2. C. F. Dunkl, *Differential-difference operators associated to reflection groups*, Trans. Amer. Math. Soc. **311** (1989), No. 1, 167–183.

3. H. Ochiai, T. Oshima, and H. Sekiguchi, *Commuting families of symmetric differential operators*, Proc. Japan Acad. Ser. A Math. Sci. **70** (1994), No. 2, 62–66.

4. T. Oshima and H. Sekiguchi, *Commuting families of differential operators invariant under the action of a Weyl group*, J. Math. Sci. Univ. Tokyo **2** (1995), No. 1, 1–75.

5. A. P. Veselov, K. L. Styrkas, and O. A. Chalykh, *Algebraic integrability for the Schrödinger equation and finite reflection groups*, Theoret. and Math. Phys. **94** (1993), No. 2, 182–197.

29

The Distribution of the Largest Eigenvalue in the Gaussian Ensembles: $\beta = 1, 2, 4$

Craig A. Tracy
Harold Widom

ABSTRACT The focus of this survey is on the distribution function $F_{N\beta}(t)$ for the largest eigenvalue in the finite N Gaussian Orthogonal Ensemble (GOE, $\beta = 1$), the Gaussian Unitary Ensemble (GUE, $\beta = 2$), and the Gaussian Symplectic Ensemble (GSE, $\beta = 4$) in the edge scaling limit of $N \to \infty$. These limiting distribution functions are expressible in terms of a particular Painlevé II function. Comparisons are made with finite N simulations and of the universality of these distribution functions is discussed.

1 Introduction

In the well-known Gaussian random matrix models of Wigner, Dyson, and Mehta [11, 13, 14], the probability density that the eigenvalues lie in infinitesimal intervals about the points $x_1, \ldots, x_N$ is given by

$$P_{N\beta}(x_1, \ldots, x_N) = C_{N\beta} e^{-\beta \sum x_i^2/2} \prod_{j<k} |x_j - x_k|^\beta,$$

where $C_{N\beta}$ is a normalization constant and

$$\beta := \begin{cases} 1 & \text{for GOE,} \\ 2 & \text{for GUE,} \\ 4 & \text{for GSE.} \end{cases}$$

We recall that for $\beta = 1$ the matrices are $N \times N$ real symmetric, for $\beta = 2$ the matrices are $N \times N$ complex Hermitian, and for $\beta = 4$ the matrices are $2N \times 2N$ self-dual Hermitian matrices. (For $\beta = 4$ each eigenvalue has multiplicity two.)

We are interested in

$$E_{N\beta}(0; J) := \int \cdots \int_{x_j \notin J} P_{N\beta}(x_1, \ldots, x_N) \, dx_1 \cdots dx_N$$

$$= \text{probability no eigenvalues lie in } J. \tag{1}$$

For $J = (t, \infty)$, $F_{N\beta}(t) := E_{N\beta}(0; (t, \infty))$ is the distribution function for the largest eigenvalue; that is,

$$\text{Prob}(\lambda_{\max} < t) = F_{N\beta}(t).$$

If we introduce the n–point correlations

$$R_{n\beta}(x_1, \ldots, x_n) := \frac{N!}{(N-n)!} \int_{\mathbb{R}} \cdots \int_{\mathbb{R}} P_{N\beta}(x_1, \ldots, x_N) \, dx_{n+1} \cdots dx_N,$$

and we denote by χ_J the characteristic function of the set J, then we may rewrite (1) as

$$E_{N\beta}(0; J) = \int_{\mathbb{R}} \cdots \int_{\mathbb{R}} P_{N\beta}(x_1, \ldots, x_N) \prod_i \left(1 - \chi_J(x_i)\right) dx_1 \cdots dx_N$$

$$= 1 - \int_J R_{1\beta}(x_1) \, dx_1 + \frac{1}{2!} \int_J \int_J R_{2\beta}(x_1, x_2) \, dx_1 \, dx_2$$

$$- \frac{1}{3!} \int_J \int_J \int_J R_{3\beta}(x_1, x_2, x_3) \, dx_1 \, dx_2 \, dx_3 + \cdots . \tag{2}$$

If we reinterpret $P_{N\beta}$ as the equilibrium Gibbs measure for N like charges interacting with a logarithmic Coulomb potential (confined to the real line) subject to a harmonic confining potential, then everything we have said so far is valid for arbitrary inverse temperature $\beta > 0$. In this interpretation $F_{N\beta}(t)$ is the probability, at inverse temperature β, that the interval (t, ∞) is free from charge.

2   Why $\beta = 2$ is the Simplest Case

For $\beta = 2$ the n–point functions take a particularly simple form [11]

$$R_{n2}(x_1, \ldots, x_n) = \det(K_N(x_i, x_j)|_{i,j=1}^n) \tag{3}$$

where

$$K_N(x, y) := \sum_{i=0}^{N-1} \varphi_i(x)\varphi_i(y) = \frac{\varphi(x)\psi(y) - \psi(x)\varphi(y)}{x - y},$$

$\varphi_i(x)$ are the orthonormal harmonic oscillator wave functions, i.e., $\varphi_k(x) := c_k e^{-x^2/2} H_k(x)$, H_k Hermite polynomials, and

$$\varphi(x) := \left(\frac{N}{2}\right)^{1/4} \varphi_N(x), \quad \psi(x) := \left(\frac{N}{2}\right)^{1/4} \varphi_{N-1}(x).$$

Using (3) in (2) we see that this expansion is the Fredholm expansion of the operator S whose kernel is $K_N(x, y)\chi_J(y)$; that is,

$$E_{N2}(0; J) = \det(I - S).$$

At this point we can use the general theory [15, 17] for Fredholm determinants of operators K whose kernel is of the form

$$\frac{\varphi(x)\psi(y) - \psi(x)\varphi(y)}{x - y}\chi_J(y),$$

where the φ and ψ are assumed to satisfy

$$\frac{d}{dx}\begin{pmatrix}\varphi \\ \psi\end{pmatrix} = \Omega(x)\begin{pmatrix}\varphi \\ \psi\end{pmatrix}$$

with $\Omega(x)$ a 2×2 matrix, trace zero, with rational entries in x. It is shown in [15, 17] that if

$$J := \bigcup_{j=1}^{m}(a_{2j-1}, a_{2j}),$$

then

$$\frac{\partial}{\partial a_j}\log\det(I - K), \quad j = 1, \ldots, 2m,$$

are expressible polynomially in terms of solutions to a total system of partial differential equations (a_j are the independent variables).

In the finite N GUE case

$$\Omega(x) = \begin{pmatrix} -x & \sqrt{2N} \\ -\sqrt{2N} & x \end{pmatrix},$$

and the theory gives for $J = (t, \infty)$

$$F_{N2}(t) = \exp\left(-\int_t^{\infty} R(x)\, dx\right),$$

where R satisfies ($' = d/dt$)

$$(R'')^2 + 4(R')^2(R' + 2N) - 4(tR' - R)^2 = 0.$$

This last differential equation is the σ version of Painlevé IV (P_{IV}) [8, 12]. What is remarkable about this result is that the size of the matrix, N, (equivalently the dimension of the integral (1)) enters only as a coefficient in the above second-order equation.

3 Edge Scaling Limit

The famous Wigner semicircle law states that if $\rho_N(x)$ is the density of eigenvalues in any of the three Gaussian ensembles, then

$$\lim_{N_\beta \to \infty} \frac{1}{2\sigma\sqrt{N_\beta}}\rho_N(2\sigma\sqrt{N_\beta}x) = \begin{cases} \sqrt{1 - x^2}/\pi & |x| < 1, \\ 0 & |x| > 1. \end{cases}$$

Here σ, $\sigma/\sqrt{2}$, $\sigma/\sqrt{2}$ (for $\beta = 1, 2, 4$, respectively) is the standard deviation of the Gaussian distribution in the off-diagonal elements and

$$N_\beta = \begin{cases} N, & \beta = 1, \\ N, & \beta = 2, \\ 2N + 1, & \beta = 4. \end{cases}$$

For the normalization here and in Ref. 11,

$$\sigma = \frac{1}{\sqrt{2}}.$$

Perhaps less well known is that the following is also true [1]:

$$F_{N\beta}(2\sigma\sqrt{N_\beta} + x) \to \begin{cases} 0 & \text{if } x < 0 \\ 1 & \text{if } x > 0 \end{cases}$$

as $N \to \infty$. The edge scaling variable, s, gives the scale on which to study fluctuations [3, 7, 16]:

$$t = 2\sigma\sqrt{N_\beta} + \frac{\sigma s}{N_\beta^{1/6}}. \qquad (4)$$

The edge scaling limit is the limit $N \to \infty$, s fixed, and in this limit [15, 16]

$$F_{N2}(t) \to \exp\left(-\int_s^\infty (x - s)q(x)^2\, dx\right) =: F_2(s),$$

where q is the solution to the P_{II} equation

$$q'' = sq + 2q^3 \qquad (5)$$

satisfying the condition

$$q(s) \sim \mathrm{Ai}(s) \quad \text{as } s \to \infty, \qquad (6)$$

with Ai the Airy function.

We note that $F_2(s)$ is the P_{II} τ–function [8, 12]. There is a rather complete description of the one-parameter family of solutions to P_{II} satisfying the condition

$$q(s; \lambda) \sim \lambda\, \mathrm{Ai}(s)$$

as $s \to \infty$ and an analysis of the corresponding connection problem for $s \to -\infty$ (see Refs. 4, 5 and references therein).

4   Cases $\beta = 1$ and 4

In Ref. 6 Dyson showed for the circular ensembles with $\beta = 1$ or 4 that the n-point correlations could be written as

$$\left(R_{n\beta}(x_1, \ldots, x_n)\right)^2 = \det(K_{N\beta}(x_i, x_j)|_{i,j=1}^n)$$

where now

$$K_{N\beta}(x, y) = 2 \times 2 \text{ matrix.}$$

Mehta generalized this result to the finite N Gaussian ensembles [11] and Mahoux and Mehta [10] gave a general method for invariant matrix models for both $\beta = 1$ and $\beta = 4$. Mehta's result implies that for $\beta = 1$ or 4 that

$$\left(E_{N\beta}(0; J)\right)^2 = \det(I - K_{N\beta}),$$

where $K_{N\beta}$ is an operator with 2×2 matrix kernel, or equivalently a 2×2 matrix with operator entries. Explicitly for $\beta = 1$ (and N even):

$$K_{N1} = \chi_J \begin{pmatrix} S + \psi \otimes \varepsilon\varphi & SD - \psi \otimes \varphi \\ \varepsilon S - \varepsilon + \varepsilon\psi \otimes \varepsilon\varphi & S + \varepsilon\varphi \otimes \psi \end{pmatrix} \chi_J,$$

where S, φ and ψ are as before for $\beta = 2$, and

$$\varepsilon f(x) := \int_{-\infty}^{\infty} \varepsilon(x - y)f(y)\, dy,$$

$$Df(x) := \frac{df(x)}{dx},$$

with

$$\varepsilon(x) := \begin{cases} \frac{1}{2} & \text{if } x > 0, \\ 0 & \text{if } x = 0, \\ -\frac{1}{2} & \text{if } x < 0. \end{cases}$$

We have used the notation $a \otimes b$ for the operator with kernel $a(x)b(y)$.
For $\beta = 4$, $\left(E_4(0; J/\sqrt{2})\right)^2$ is again a Fredholm determinant with

$$K_{N4} = \frac{1}{2}\chi_J \begin{pmatrix} S + \psi \otimes \varepsilon\varphi & SD - \psi \otimes \varphi \\ \varepsilon S + \varepsilon\psi \otimes \varepsilon\varphi & S + \varepsilon\varphi \otimes \psi \end{pmatrix} \chi_J.$$

In Ref. 18 these Fredholm determinants are related to integrable systems for general J. Although Ref. 18 treats exclusively the Gaussian ensembles, the methods appear quite general and should apply to other ensembles as well.

5   Idea of Proof and Results

Our derivation [18] rests on the fact that operator determinants may be manipulated much as scalar determinants, as long as one exercises some care. We first write

$$K_{N1} = \begin{pmatrix} \chi_J D & 0 \\ 0 & \chi_J \end{pmatrix} \begin{pmatrix} (S\varepsilon - \varepsilon\varphi \otimes \varepsilon\psi)\chi_J & (S + \varepsilon\varphi \otimes \psi)\chi_J \\ (S\varepsilon - \varepsilon - \varepsilon\varphi \otimes \varepsilon\psi)\chi_J & (S + \varepsilon\varphi \otimes \psi)\chi_J \end{pmatrix}.$$

This uses the trivial fact that $D\varepsilon = I$ and the commutator identities

$$[S, D] = \varphi \otimes \psi + \psi \otimes \varphi, \quad [\varepsilon, S] = -\varepsilon\varphi \otimes \varepsilon\psi - \varepsilon\psi \otimes \varepsilon\varphi.$$

By the general identity $\det(I - AB) = \det(I - BA)$ the determinant is unchanged if the factors are interchanged and so we may work instead with

$$\begin{pmatrix} (S\varepsilon - \varepsilon\varphi \otimes \varepsilon\psi)\chi_J D & (S + \varepsilon\varphi \otimes \psi)\chi_J \\ (S\varepsilon - \varepsilon - \varepsilon\varphi \otimes \varepsilon\psi)\chi_J D & (S + \varepsilon\varphi \otimes \psi)\chi_J \end{pmatrix}.$$

Applying a pair of row and column operations, which is justified by the identity $\det(I - A) = \det(I - BAB^{-1})$, we reduce this to

$$\begin{pmatrix} (S\varepsilon - \varepsilon\varphi \otimes \varepsilon\psi)\chi_J D + (S + \varepsilon\varphi \otimes \psi)\chi_J & (S + \varepsilon\varphi \otimes \psi)\chi_J \\ -\varepsilon\chi_J D & 0 \end{pmatrix}.$$

More row and column operations, which this time use $\det(I - A)(I - B) = \det(I - A)\det(I - B)$, followed by factoring out

$$I - S\chi_J.$$

These show that $E_{N1}(0; J)^2$ equals $E_{N2}(0; J)$ times a determinant of the general form

$$\det\left(I - \sum_{k=1}^n \alpha_k \otimes \beta_k\right),$$

whose value equals that of the scalar determinant

$$\det\big(\delta_{j,k} - (\alpha_j, \beta_k)\big)_{j,k=1,\dots,n}.$$

Thus the evaluation of $E_{N1}(0; J)$ is reduced to the evaluation of certain inner products and the result for $E_{N2}(0; J)$.

To evaluate the inner products differential equations are derived for them. For $J = (t, \infty)$ and in the edge scaling limit these differential equations can be solved with the result that

$$F_1(s)^2 = F_2(s)\exp\left(-\int_s^\infty q(x)\,dx\right), \tag{7}$$

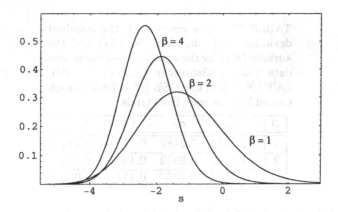

FIGURE 29.1. The probability densities $f_\beta(s) = dF_\beta/ds$, $\beta = 1, 2, 4$, for the position of the largest eigenvalue in the edge scaling limit.

TABLE 29.1. The mean (μ_β), the standard deviation (σ_β), the skewness (S_β) and the kurtosis (K_β) for the the densities f_β.

β	μ_β	σ_β	S_β	K_β
1	-1.20653	1.2680	0.2935	0.1653
2	-1.77109	0.9018	0.2241	0.0935
4	-2.30688	0.7195	0.1655	0.0492

where we recall

$$F_2(s) = \exp\left(-\int_s^\infty (x-s)\, q(x)^2\, dx \right).$$ (8)

Similarly for $\beta = 4$,

$$F_4\left(\frac{s}{\sqrt{2}}\right)^2 = F_2(s)\left(\frac{\exp(\frac{1}{2}\int_s^\infty q(x)\,dx) + \exp(-\frac{1}{2}\int_s^\infty q(x)\,dx)}{2} \right)^2.$$ (9)

We note that there are no adjustable parameters in F_β.

Using the known asymptotics of $q(s)$ as $s \to \pm\infty$, it is straightforward to solve (5) numerically and produce accurate numerical values for $F_\beta(s)$ and the corresponding densities $f_\beta(s) = dF_\beta/ds$. The densities for the largest eigenvalue in each of the three ensembles in the edge scaling limit are displayed in Figure 29.1. Observe that for higher "temperature" (lower β) the variance of $f_\beta(s)$ increases as one would expect from the Coulomb gas interpretation.

In Table 29.1 we give the mean (μ_β), standard deviation (σ_β), skewness

TABLE 29.2. The mean (μ_β), the standard deviation (σ_β), the skewness (S_β) and the kurtosis (K_β) for the scaled largest eigenvalue data from simulations in GOE ($N = 500$), GUE ($N = 100$) and GSE ($N = 100$). In each ensemble there were 5000 trials.

β	μ_β	σ_β	S_β	K_β
1	-1.2614	1.2182	0.273	0.171
2	-1.7772	0.8952	0.133	-0.089
4	-2.3154	0.7057	0.171	0.023

(S_β), and the kurtosis (K_β) of the densities f_β.[1]

6   How "Useful" Are These Scaling Functions?

Using a random number generator to simulate a real symmetric matrix of size $N \times N$ in the GOE, we calculate its largest eigenvalue $\lambda_{\max}$. For N large the expected value should be approximately

$$E(\lambda_{\max}) = 2\sigma\sqrt{N} + \frac{\sigma\mu_1}{N^{1/6}},$$

where μ_1 is the mean of f_1 (see Table 29.1). In one such simulation (with $\sigma = 1/\sqrt{2}$) with $N = 500$ and 5000 trials the mean largest eigenvalue is 31.3062, which should be compared with the theoretical prediction of 31.353.

For each largest eigenvalue $\lambda_{\max}$ we compute a scaled largest eigenvalue s from

$$\lambda_{\max} = 2\sigma\sqrt{N} + \frac{\sigma s}{N^{1/6}},$$

and form a histogram of s values. In Figure 29.2 we compare this histogram with the limiting density $f_1(s)$. In Table 29.2 we give the mean, standard deviation, skewness, and kurtosis of this same scaled largest eigenvalue data. This should be compared with the limiting theoretical values in the first row of Table 29.1. Similarly, in Figure 29.3 we give comparisons of finite N ($= 100$) simulations in GUE and GSE with the limiting densities f_2 and f_4, respectively. The descriptive statistics of this data are given in Table 29.2.

---

[1]The skewness of a density f is $\int ([x - \mu]/\sigma)^3 f(x)\, dx$ and the kurtosis is $\int ([x - \mu]/\sigma)^4 f(x)\, dx - 3$ where the -3 term makes the value zero for the normal distribution.

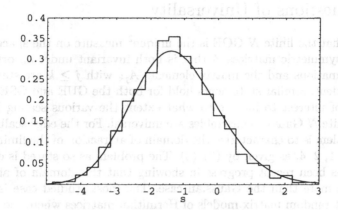

FIGURE 29.2. Histogram of scaled largest eigenvalues for $N = 500$, GOE for 5000 trials. Solid curve is the limiting density $f_1(s)$.

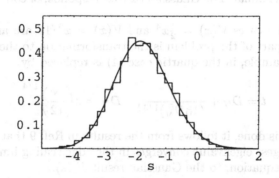

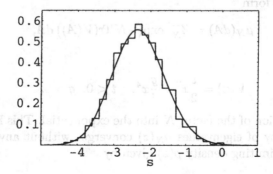

FIGURE 29.3. Histograms for the scaled largest eigenvalues in GUE and GSE each with $N = 100$ and 5000 trials. Solid curves are the limiting densities $f_2(s)$ and $f_4(s)$, respectively.

7 Questions of Universality

Recall that the finite N GOE is the unique[2] measure on the space of $N \times N$ real symmetric matrices A that is both invariant under all orthogonal transformations and the matrix elements A_{jk} with $j \geq k$ are statistically independent. Similar statements hold for both the GUE and GSE.

It is of interest to inquire to what extent the various scaling limits of these finite N Gaussian ensembles are universal. For the edge scaling limit, the problem is to characterize the domain of attraction of the limiting laws F_β, $\beta = 1, 2, 4$, as given by (7)–(9). The problem as so stated is open but there has been recent progress in showing that this domain of attraction contains more than the Gaussian case. The most studied case is that of invariant random matrix models of Hermitian matrices where the measure is of the form

$$\mu_N(dA) = Z_N^{-1} \exp\bigl(-\operatorname{Tr}(V(A))\bigr) dA$$

and V is a polynomial. The Gaussian case corresponds, of course, to $V(x) = \frac{1}{2}\beta x^2$.

In Ref. 9 the cases $V(x) = \frac{1}{2}x^4$ and $V(x) = x^6/12$ are analyzed. The nonuniversal part of the problem is the transformation to the scaling variable s. For example, in the quartic case (4) is replaced by

$$t = D_N + \frac{s}{(18D_N^5)^{1/3}}, \quad D_N = 2\left(\frac{N}{12}\right)^{1/4}.$$

But once this is done, it follows from the results in Ref. 9 that the distribution of the largest eigenvalue converges in the edge scaling limit, as defined by the above equation, to the Gaussian result $F_2(s)$.

In Ref. 2 there is a an account of the Hermitian matrix model with measure of the form

$$\mu_N(dA) = Z_N^{-1} \exp\bigl(-N\operatorname{Tr}(V(A))\bigr) dA,$$

with

$$V(x) = \frac{t}{2}x^2 + \frac{g}{4}x^4, \quad t < 0, \ g > 0.$$

Note the insertion of the factor N into the exponential. This has the effect that the density of eigenvalues $\rho_N(z)$ converges without any rescaling of variables to a limiting density $\rho(z)$ given by

$$\rho(z) = \frac{g|z|}{2\pi}\sqrt{(z^2 - z_1^2)(z_2^2 - z^2)}$$

[2]Uniqueness is up to centering and choice of standard deviation on, say, the off-diagonal matrix elements.

with

$$z_{1,2} = \left(\frac{-t \mp 2\sqrt{g}}{g} \right)^{1/2}.$$

In this setup the edge of the spectrum remains bounded and the transformation to the edge scaling variable s is now

$$z = z_2 + \frac{s}{cN^{2/3}}, \quad c = 2^{1/3}g^{1/2}z_2.$$

It is proved [2] that the distribution of the largest eigenvalue converges to the Gaussian result $F_2(s)$.

Physicists (see Ref. 3 and references therein) have heuristic arguments that suggest that the domain of attraction of the invariant matrix models for Hermitian matrices contains all those potentials V for which the density vanishes at the edge of the spectrum as does the Wigner semicircle, i.e., a square root. By tuning the potential one obtains densities that vanish faster than a square root—these models will be in a different universality class with regard to the edge scaling limit.

Acknowledgments: The first author wishes to thank Jan Felipe van Diejen and Luc Vinet for the invitation to speak and the kind hospitality at the Workshop on Calogero–Moser–Sutherland Models. This work was supported in part by the National Science Foundation (NSF) through grants DMS-9303413 (first author) and DMS-9424292 (second author).

8 REFERENCES

1. Z. D. Bai and Y. Q. Yin, *Necessary and sufficient conditions for almost sure convergence of the largest eigenvalue of a Wigner matrix*, Ann. Probab. **16** (1988), 1729–1741.

2. P. Bleher and A. Its, *Semiclassical asymptotics of orthogonal polynomials, Riemasnn–Hilbert problem, and universality in the matrix model*, preprint 97-2, IUPU, 1997.

3. M. J. Bowick and E. Brézin, *Universal scaling of the tail of the density of eigenvalues in random matrix models*, Phys. Lett. B **268** (1991), No. 1, 21–28.

4. P. A. Clarkson and J. B. McLeod, *A connection formula or the second Painlevé transcendent*, Arch. Rational Mech. Anal. **103** (1988), No. 2, 97–138.

5. P. A. Deift and X. Zhou, *Asymptotics for the Painlevé II equation*, Commun. Pure Appl. Math. **48** (1995), No. 3, 277–337.

6. F. J. Dyson, *Correlations between eigenvalues of a random matrix*, Commun. Math. Phys. **19** (1970), 235–250.

7. P. J. Forrester, *The spectrum edge of random matrix ensembles*, Nucl. Phys. B **402** (1993), No. 3, 709–728.

8. M. Jimbo and T. Miwa, *Monodromy preserving deformation of linear ordinary differential equations with rational coefficients. II*, Physica **2D** (1981), No. 3, 407–448.

9. E. Kanzieper and V. Freilikher, *Universality in invariant random-matrix models. Existence near the soft edge*, Phys. Rev. E **55** (1997), No. 3, part B, 3712–3715.

10. G. Mahoux and M. L. Mehta, *A method of integration over matrix variables. IV*, J. Phys. (Paris) **1** (1991), No. 8, 1093–1108.

11. M. L. Mehta, *Random Matrices*, 2nd ed., Academic Press, Boston, MA, 1991.

12. K. Okamoto, *Studies on the Painlevé equations. III. Second and fourth Painlevé equations, P_{II} and P_{IV}*, Math. Appl. **275** (1986), No. 2, 221–255.

13. C. E. Porter, *Statistical Theory of Spectra. Fluctuations*, Academic Press, New York, 1965.

14. C. A. Tracy and H. Widom, *Introduction to random matrices*, Geometric and Quantum Aspects of Integrable Systems (Scheveningen, 1992) (G. F. Helminck, ed.), Lecture Notes in Phys., Vol. 424, Springer, Berlin, 1993, pp. 103–130.

15. ———, *Fredholm determinants, differential equations and matrix models*, Commun. Math. Phys. **163** (1994), No. 1, 33–72.

16. ———, *Level-spacing distributions and the Airy kernel*, Commun. Math. Phys. **159** (1994), No. 1, 151–174.

17. ———, *Systems of partial differential equations for a class of operator determinants*, Partial Differential Operators and Mathematical Physics (Holzhau, 1994) (M. Demuth and B.-W. Schulze, eds.), Oper. Theory Adv. Appl, Vol. 78, Birkhäuser, Basel, 1995, pp. 381–388.

18. ———, *On orthogonal and symplectic matrix ensembles*, Commun. Math. Phys. **177** (1996), No. 1, 727–754.

30

Two-Body Elliptic Model in Proper Variables: Lie Algebraic Forms and Their Discretizations

Alexander Turbiner[1]

ABSTRACT Two Lie algebraic forms of the two-body elliptic Calogero model are presented and translation-invariant and dilatation-invariant discretizations of the model are obtained.

1 Introduction

The general N-body Lamé operator is given by

$$L \equiv \Delta - n(n+1) \sum_{i>j}^{N} \mathcal{P}(x_i - x_j), \qquad (1)$$

where n is positive integer number and the Weierstrass function $\mathcal{P}(x)$ is a double-periodic meromorphic function, which obeys the equation

$$\mathcal{P}'^2 = 4(\mathcal{P} - e_1)(\mathcal{P} - e_2)(\mathcal{P} - e_3) \equiv 4\mathcal{P}^3 - g_2\mathcal{P} - g_3.$$

Here e_1, e_2, e_3 such that $e_1 + e_2 + e_3 = 0$ are called the roots of the Weierstrass function. According to Olshanetsky–Perelomov (see, for example, Ref. 7) it is known that the problem (1) is completely integrable.

2 Lie Algebraic Analysis

For the 2-body case we need to make a slight modification in (1), in particular, adding an extra constant term

$$\tilde{L}^{(2)} \equiv \Delta^{(2)} - \frac{n(n+1)}{2} \left[\mathcal{P}\left(\frac{x_1 - x_2}{2} \right) + A \right], \qquad (2)$$

---

[1]On leave of absence from the Institute for Theoretical and Experimental Physics, Moscow 117259, Russia.

where $A(\omega, \omega')$ is a real constant with a condition that $A(\omega, \infty) = (\pi/2\omega)^2/3$; here ω, ω' are the half-periods of $\mathcal{P}$-function. In particular, one can choose $A(\omega, \omega') = (\pi/2\omega)^2/3$. Hermiticity of L requires us to consider one of the periods as real. Let it be ω. The extra term in (2) just changes the reference point for energies.

Hereafter we assume that center-of-mass coordinate $X = \sum x_i$ is separated and the remaining part of the Laplace operator is already written in Perelomov relative coordinates [8],

$$y_i = x_i - \frac{X}{2}, \quad i = 1, 2, \tag{3}$$

and, say, $y_1 = (x_1 - x_2)/2 = -y_2 \equiv y$; and, thus the Laplace operator is equal to

$$\Delta^{(2)} = 2\partial_{XX}^2 + \frac{1}{2}\partial_{yy}^2. \tag{4}$$

The operator $\tilde{L}^{(2)}$ after separation of the center-of-mass motion becomes

$$\tilde{L}_{\text{rel}}^{(2)} = \{2\tilde{L}^{(2)}\}|_{\text{rel}} = \partial_{yy}^2 - n(n+1)[\mathcal{P}(y) + A], \tag{5}$$

and has an important feature of coinciding either to the Hamiltonian of Calogero model (both periods tend to infinity, thus $A = 0$) or to the Sutherland model (if the second (complex) period tends to infinity while ω is kept fixed; in this case $A = 3g_3/2g_2$). This operator (5) is the Weierstrass form of Lamé operator from textbooks (see, for example, Refs. 1 and 5).

In further analysis we assume that n is even number. We study pure polynomial solutions (nonvanishing in the points, where $\mathcal{P}(y) = e_i$, $i = 1$, 2, 3). These solutions are called the Lamé polynomials of the first type (see Ref. 5). Implicitly, it means that we explore periodic-periodic solutions of the Lamé equation.

The well-known algebraic form of the Lamé operator (5) occurs if a new variable, $\xi = \mathcal{P}(y)$, is introduced:

$$\tilde{L}_{\text{rel}}^{(2)} = (4\xi^3 - g_2\xi - g_3)\partial_{\xi\xi}^2 + (6\xi^2 - \frac{g_2}{2})\partial_\xi - n(n+1)(\xi + A). \tag{6}$$

The operator (6) can be immediately rewritten in terms of the generators of the sl(2)-algebra:

$$J_k^+ = \rho^2 \frac{\partial}{\partial \rho} - k\rho, \quad J_k^0 = \rho \frac{\partial}{\partial \rho} - \frac{k}{2}, \quad J_k^- = \frac{\partial}{\partial \rho}, \tag{7}$$

and then the following expression for (6) emerges[2] [12]:

$$\tilde{L}_{\text{rel}}^{(2)} = J_{n/2}^+[4J_{n/2}^0 + (3n+2)] - g_2\left(J_{n/2}^0 + \frac{n+2}{4}\right)J_{n/2}^- \\ - g_3 J_{n/2}^- J_{n/2}^-. \tag{8}$$

---

[2] ρ is replaced by ξ.

in the representation (7) of the spin $n/2$. It is evident that the operator (6) has $n/2 + 1$ polynomial eigenfunctions.

It is worth mentioning that for even values n in (5) there exist also three different kinds of the eigenfunctions besides those proportional to $\eta_{ij} = (\xi - e_i)^{1/2}(\xi - e_j)^{1/2}$, $i \neq j = 1, 2, 3$. One can show that making the gauge rotation of (6): $\eta_{ij}^{-1}\tilde{L}_{rel}^{(2)}\eta_{ij}$ we get again the operator (6) with different coefficients in the part of it, which is a linear function in derivative. In particular, the only new term $\xi\partial_\xi$ occurs. This operator can be also rewritten in terms of the generators of the sl(2)-algebra (7) but with the spin of $n/2 - 1$ [12, 14, 15]. It proves that it contains $n/2$ polynomial eigenfunctions. They are nothing but the Lamé polynomials of the third type.

Now we pose a question: what are the proper coordinates whic occur in these *algebraic* and *Lie algebraic* forms of the rational, trigonometric, and elliptic models, if corresponding limits are taken? After some analysis we arrive at the following expression:

$$\rho = -\frac{1}{\mathcal{P}([x_1 - x_2]/2) + A} \equiv -\frac{1}{\mathcal{P}(y) + A}, \tag{9}$$

where, for the sake of convenience, we put

$$A \equiv \frac{\alpha^2}{12}.$$

In the trigonometric limit $\alpha = \frac{\pi}{4\omega}$. The variable ρ obeys the following equation:

$$(\rho')^2 = -4\rho - \alpha^2\rho^2 - (12A^2 - g_2)\rho^3 - (4A^3 - g_2A + g_3)\rho^4. \tag{10}$$

The trigonometric limit turns out to correspond to the disappearance of the ρ^3 and ρ^4 terms, while in the rational limit the ρ^2 term is also vanish. Then in these limits (5) becomes

$$\tilde{L}_{rel}^{(2)} = \partial_{yy}^2 - \frac{n(n+1)\alpha^2}{4\sin^2(\alpha y/2)},$$

or

$$= \partial_{yy}^2 - \frac{n(n+1)}{y^2},$$

correspondingly. It is worth mentioning the "embedding chain" rule of degeneration:

$$\rho(y) \to \frac{2(\cos\alpha y - 1)}{\alpha^2} = -\frac{4}{\alpha^2}\sin^2\left(\frac{\alpha y}{2}\right) \to -y^2. \tag{11}$$

Now we calculate the gauge rotated $\tilde{L}^{(2)}_{\text{rel}}$ operator in the relative coordinate $\rho(y)$:

$$h = \rho^{-n/2}\{\partial^2_{yy} - n(n+1)[\mathcal{P}(y) + A]\}\rho^{n/2}$$

$$= -\{4\rho + 12A\rho^2 + (12A^2 - g_2)\rho^3 + (4A^3 - g_2A + g_3)\rho^4\}\partial^2_\rho$$

$$+ \left\{2(2n-1) + 12(n-1)A\rho + \left(n - \frac{3}{2}\right)(12A^2 - g_2)\rho^2\right.$$

$$\left. + (n-2)(4A^3 - g_2A + g_3)\rho^3\right\}\partial_\rho$$

$$- 3An^2 - \frac{n(n-1)}{4}(12A^2 - g_2)\rho - \frac{n(n-2)}{4}(4A^3 - g_2A + g_3)\rho^2.$$

$$(12)$$

This expression also can be represented in terms of sl_2-generators (7) of spin $n/2$:

$$h = -(4A^3 - g_2A + g_3)J^+_{n/2}J^+_{n/2} - (12A^2 - g_2)J^+_{n/2}\left\{J^0_{n/2} - \frac{n(n-2)}{4}\right\}$$

$$- 12AJ^0_{n/2}\left(J^0_{n/2} - \frac{n}{2}\right) - 4\left(J^0_{n/2} - \frac{3n-2}{4}\right)J^-_{n/2} + \frac{3An^2}{4} \quad (13)$$

It is worth noting that the gauge factor $\rho^{n/2}$ in rational and trigonometric limits degenerates to the Vandermonde determinant factor or its trigonometric generalization, respectively, which we also gauged away in Calogero and Sutherland models [9].

The trigonometric limit corresponds to zeroing of two expressions: $12A^2 - g_2 = 0$ and $4A^3 - g_2A + g_3 = 0$, which leads to vanishing of the terms containing positive-root J^+-generator. Following a general criterion we conclude that the operator (12) preserves the infinite flag of spaces of polynomials and the corresponding problem is exactly solvable [14, 15]. The rational limit occurs if $A = g_2 = g_3 = 0$. Since in the rational limit the harmonic oscillator interaction is omitted there are no polynomial eigenfunctions. It is known that the eigenvalues corresponding to the polynomial eigenfunctions and the eigenfunctions themselves of (6) and (12) are the branches of $(n/2 + 1)$-sheeted Riemann surface in the variable $g_2(g_3)$ characterized by the square-root branch points only [12]. In the trigonometric limit the residues of these branch points vanish and this surface splits off into a separate sheets.

There is an intrinsic hierarchy of the algebraic form (12) of the Hamiltonian (5): elliptic (grading $+2$), trigonometric (grading 0), rational (grading -1). The elliptic case of the grading $+1$[3] has no meaning in our study, since there is no solution of (10) satisfying "embedding chain" rule (11). In

---

[3] $4A^3 - g_2A + g_3 = 0$ and hence rhs in (10) is the cubic polynomial in ρ.

principle, the variable (9) can be modified, allowing higher-order terms in (10) and insisting on fulfillment of (11). It would become a hyper-elliptic variable.

The above analysis is performed for periodic-periodic solutions of the Lamé equation, for the case of an even number of zones, $n = 2k$, $k = 1$, $2, 3, \ldots$. Similar studies can be carried out for other types of solutions (see discussion above) gauging away the factor $\rho^{n/2-1}$ in (5) in addition to the factor η_{ij}. Finally, we get algebraic and Lie algebraic forms of the Lamé equation similar to (12) and (13)). Similar results can be obtained for the case of odd number of zones.

As a certain line of possible development one can substitute the generators gl(2 | k)-superalgebra instead of sl$_2$ ones (see Ref. 2) and explore SUSY QM models those will occur. We will discuss this exercise elsewhere.

3   Translation-Invariant Discretization

The next problem we proceed is how to discretize the Lamé operator in translation-invariant way, simultaneously preserving the property of polynomiality of the eigenfunctions and also isospectrality. Again we restrict our analysis of the case to the periodic-periodic solutions and even number of zones, $n = 2k$, $k = 1, 2, 3, \ldots$, respectively.

Let us introduce a translationally covariant finite-difference operator

$$\mathcal{D}_+ f(x) = \frac{(e^{\delta \partial_x} - 1)}{\delta} f(x) = \frac{f(x + \delta) - f(x)}{\delta}, \tag{14}$$

where δ is the parameter. It is not difficult to find that the canonically conjugated operator to $\mathcal{D}_+ f(x)$ is of a form [10, 11]

$$x(1 - \delta \mathcal{D}_-) f(x) = x e^{-\delta \partial_x} f(x) = x f(x - \delta), \tag{15}$$

where $\mathcal{D}_+ \to \mathcal{D}_-$, if $\delta \to -\delta$. So, the operators (14) and (15) span the Heisenberg algebra, $[a, b] = 1$. One can easily show that using the operators (14) and (15) we can construct a representation of the sl$_2$ algebra in terms of finite-difference operators:

$$J_k^+ = x \left(\frac{x}{\delta} - 1 \right) e^{-\delta \partial_x} (1 - k - e^{-\delta \partial_x}),$$

$$J_k^0 = \frac{x}{\delta}(1 - e^{-\delta \partial_x}) - \frac{n}{2}, \quad J^- = \frac{1}{\delta}(e^{\delta \partial_x} - 1), \tag{16}$$

or, equivalently,

$$J_n^+ = x(1 - \frac{x}{\delta})(\delta^2 \mathcal{D}_- \mathcal{D}_- - (n + 1)\delta \mathcal{D}_- + k),$$

$$J_k^0 = x \mathcal{D}_- - \frac{k}{2}, \quad J^- = \mathcal{D}_+. \tag{17}$$

In the limit $\delta \to 0$, the representation (16) and (17) coincides with (7). The finite-dimensional representation space for (7) and for (16) and (17) occurring for the integer values of k is the same, being a linear space of polynomials of a degree not higher than k.

It is quite obvious that any operator written in terms of the generators of the algebra sl_2 has the same eigenvalues for both representation (7) and (16) and (17) (for the proof and discussion see [10, 11]). Let us take the operator (6) and construct the isospectral finite-difference operator. In order to do it we take the representation (8) of this operator and substitute the sl_2 generators in the form (16) or (17). Finally, we arrive at the following finite-difference operator:

$$\frac{4\xi^{(3)}}{\delta^2}e^{-3\delta\partial_\xi} - 2\frac{\xi^{(2)}}{\delta}\left(4\frac{\xi}{\delta}-5\right)e^{-2\delta\partial_\xi} + \left\{4\frac{\xi^{(3)}}{\delta^2}+6\frac{\xi^{(2)}}{\delta} - \left[\frac{g_2}{\delta^2}n(n+1)\right]\xi\right\}e^{-\delta\partial_\xi}$$

$$+ 2\frac{g_2}{\delta^2}\xi + \frac{g_2}{2\delta} - \frac{g_3}{\delta^2} - n(n+1)A$$

$$- \left(\frac{g_2}{\delta^2}\xi + \frac{g_2}{2\delta} - 2\frac{g_3}{\delta^2}\right)e^{\delta\partial_\xi} - \frac{g_3}{\delta^2}e^{2\delta\partial_\xi}, \quad (18)$$

The spectral problem corresponding to the operator (18) can be written as the 6-point finite-difference equation:

$$\frac{4\xi^{(3)}}{\delta^2}f(\xi-3\delta) - 2\frac{\xi^{(2)}}{\delta}\left(4\frac{\xi}{\delta}-5\right)f(\xi-2\delta)$$

$$+ \left\{4\frac{\xi^{(3)}}{\delta^2}+6\frac{\xi^{(2)}}{\delta} - \left[\frac{g_2}{\delta^2}n(n+1)\right]\xi\right\}f(\xi-\delta) + \left[2\frac{g_2}{\delta^2}\xi + \frac{g_2}{2\delta} - \frac{g_3}{\delta^2} - n(n+1)A\right]f(\xi)$$

$$- \left(\frac{g_2}{\delta^2}\xi + \frac{g_2}{2\delta} - 2\frac{g_3}{\delta^2}\right)f(\xi+\delta) - \frac{g_3}{\delta^2}f(\xi+2\delta) = \lambda f(\xi), \quad (19)$$

where λ is the spectral parameter and $\xi^{(n+1)} = \xi(\xi-\delta)(\xi-2\delta)\ldots(\xi-n\delta)$ is the so called "quasi-monomial."

Polynomial solutions of Eq. (19) are associated to the Lamé polynomials of the first type. They are constructed as

$$\tilde{P}_{n/2}(\xi) = \sum_{i=0}^{n/2} \ell_i \xi^{(i)}, \quad (20)$$

where ℓ_i are the coefficients of the Lamé polynomial of the first type. We call these polynomials "the 1-associated finite-difference Lamé polynomials of the first type." One can show that there exist slightly modified 6-point finite-difference equations of the same functional form as (19) having polynomial solutions of the form (20) associated with the Lamé polynomials of the 2nd, 3rd, and 4th types. All their eigenvalues corresponding to the polynomial eigenfunctions have no dependence on the parameter δ.

Now we proceed to the study of the translation-invariant discretization of the operator (12). We substitute the sl_2-generators of the form (16) and (17) into (13)) and arrive at

$$
-(4A^3 - g_2 A + g_3)\frac{\rho^{(4)}}{\delta^2} e^{-4\delta\partial_\rho}
$$

$$
+ \left[2(4A^3 - g_2 A + g_3)\left(\frac{\rho}{\delta} - \frac{n}{2} - 2\right) - (12A^2 - g_2)\right] \frac{\rho^{(3)}}{\delta} e^{-3\delta\partial_\rho}
$$

$$
- \left\{ (4A^3 - g_2 A + g_3)\delta^2 \left(\frac{\rho}{\delta} - \frac{n}{2} - 1\right)\left(\frac{\rho}{\delta} - \frac{n}{2} - 2\right) \right.
$$

$$
\left. - (12A^2 - g_2)\delta\left(2\frac{\rho}{\delta} - n - \frac{5}{2}\right) - 12A \right\} \frac{\rho^{(2)}}{\delta^2} e^{-2\delta\partial_\rho}
$$

$$
- \left\{ (12A^2 - g_2)\delta\left(\frac{\rho}{\delta} - \frac{n}{2} - \frac{1}{2}\right)\left(\frac{\rho}{\delta} - \frac{n}{2} - 1\right) \right.
$$

$$
\left. - 24A\left(\frac{\rho}{\delta} - \frac{n}{2} - \frac{1}{2}\right) + \frac{4}{\delta} \right\} \frac{\rho}{\delta} e^{-\delta\partial_\rho}
$$

$$
- \left[12A\left(\frac{\rho^2}{\delta^2} - n\frac{\rho}{\delta} + \frac{n^2}{4}\right) - 8\frac{\rho}{\delta^2} + \frac{2(2n-1)}{\delta} \right]
$$

$$
- \frac{4}{\delta}\left(\frac{\rho}{\delta} - n + \frac{1}{2}\right) e^{\delta\partial_\rho}. \quad (21)
$$

The spectral problem corresponding to the operator (21) can be written as the 6-point finite-difference equation

$$
-(4A^3 - g_2 A + g_3)\frac{\rho^{(4)}}{\delta^2} f(\rho - 4\delta)
$$

$$
+ \left[2(4A^3 - g_2 A + g_3)\left(\frac{\rho}{\delta} - \frac{n}{2} - 2\right) - (12A^2 - g_2)\right] \frac{\rho^{(3)}}{\delta} f(\rho - 3\delta)
$$

$$
- \left\{ (4A^3 - g_2 A + g_3)\delta^2 \left(\frac{\rho}{\delta} - \frac{n}{2} - 1\right)\left(\frac{\rho}{\delta} - \frac{n}{2} - 2\right) \right.
$$

$$
\left. - (12A^2 - g_2)\delta\left(2\frac{\rho}{\delta} - n - \frac{5}{2}\right) - 12A \right\} \frac{\rho^{(2)}}{\delta^2} f(\rho - 2\delta)
$$

$$
- \left\{ (12A^2 - g_2)\delta\left(\frac{\rho}{\delta} - \frac{n}{2} - \frac{1}{2}\right)\left(\frac{\rho}{\delta} - \frac{n}{2} - 1\right) \right.
$$

$$- 24A\left(\frac{\rho}{\delta} - \frac{n}{2} - \frac{1}{2}\right) + \frac{4}{\delta}\Big\}\frac{\rho}{\delta}f(\rho - \delta)$$

$$- \left[12A\left(\frac{\rho^2}{\delta^2} - n\frac{\rho}{\delta} + \frac{n^2}{4}\right) - 8\frac{\rho}{\delta^2} + \frac{2(2n-1)}{\delta}\right]f(\rho)$$

$$- \frac{4}{\delta}\left(\frac{\rho}{\delta} - n + \frac{1}{2}\right)f(x + \delta) = \lambda f(\rho). \quad (22)$$

Polynomial solutions of Eq. (22) are associated to the Lamé polynomials of the first type. They constructed as

$$\widetilde{P}_{n/2}(\rho) = \sum_{i=0}^{n/2} \ell_{n/2-i}\rho^{(i)}, \quad (23)$$

(cf. Eq. (20)), where ℓ_i are the coefficients of the Lamé polynomial of the first type. We call these polynomials "the 2-associated finite-difference Lamé polynomials of the first type." One can show that there exist slightly modified 6-point finite-difference equations of the same functional form as (22) having polynomial solutions of the form (23) associated with the Lamé polynomials of the 2nd, 3rd, and 4th types. All their eigenvalues corresponding to the polynomial eigenfunctions have no dependence on the parameter δ. In the trigonometric limit the coefficients before $f(\rho - 4\delta)$, $f(\rho - 3\delta)$ terms disappear and (22) becomes the 4-point finite-difference equation.

4   Dilatation-Invariant Discretization

Now we proceed to study a dilatationally invariant discretization of the Lamé operator (5). The prescription of the discretization we are going to use is a preservation of a polynomiality of the eigenfunctions under discretization.

Let us introduce a dilatationally invariant, finite-difference operator. One of the simplest operators possessing such a property is known in literature as the Jackson symbol (see, e.g., [3, 4]):

$$Df(x) = \frac{f(x) - f(qx)}{(1 - q)x}, \quad (24)$$

where $q \in C$. The Leibnitz rule for the operator D is

$$Df(x)g(x) = (Df(x))g(x) + f(qx)Dg(x).$$

One can construct the algebra of the first-order finite-difference operators

D possessing finite-dimensional irreps for generic q [6]:

$$\tilde{J}_n^+ = x^2 D - \{n\}x$$
$$\tilde{J}_n^0 = xD - \hat{n} \qquad\qquad (25)$$
$$\tilde{J}_n^- = D,$$

where $\{n\} = (1 - q^n)/(1 - q)$ is so called q-number and $\hat{n} \equiv \{n\} \times \{n+1\}/\{2n+2\}$. If n is a nonnegative integer number, the operators (25) possess a common finite-dimensional invariant subspace realized by the polynomials of the degree not higher than n similarly to what happens to the algebra (7). The operators (25) span the algebra sl_{2q} (see discussion in Refs. 14 and 15). In the limit $q \to 1$ the generators (25) become (7) and the sl_{2q} algebra reduces to the standard sl_2 one.

In Ref. 13 (see also Refs. 14 and 15) it has been proven that a linear differential (finite-difference) operator has a certain number of polynomial eigenfunctions if and only if it admits a representation in terms of the generators (7) of the sl_2-algebra (the generators (25) sl_{2q}-algebra). In order to fulfill the theorem it is sufficient to require that a deformation of the positive-grading part of the operator (7),

$$\tilde{L}_+^{(2)} = 4\xi^3 \partial_{\xi\xi}^2 + 6\xi^2 \partial_\xi - n(n+1)\xi, \qquad\qquad (26)$$

continues to map a polynomial of degree $n/2$ onto itself.[4] A minimal and perhaps natural deformation, which occurs after replacing $\partial_\xi \to D_\xi$, is given by

$$\tilde{L}_{+,\mathrm{def}}^{(2)} = 4\xi^3 D_{\xi\xi}^2 + 6\xi^2 D_\xi - 4\left\{\frac{n}{2}\right\}\left[\left\{\frac{n}{2} - 1\right\} + \frac{3}{2}\right]\xi. \qquad (27)$$

and it can be immediately rewritten as a linear combination of $\tilde{J}_{n/2}^+ \tilde{J}_{n/2}^0$ and $\tilde{J}_{n/2}^+$ terms. In general, this result can be treated in such a way that the q-deformation of the Lamé equation (5) is reduced to the q-deformation of its coupling constant:

$$n(n+1) \to 4\left\{\frac{n}{2}\right\}\left[\left\{\frac{n}{2} - 1\right\} + \frac{3}{2}\right]. \qquad\qquad (28)$$

A similar deformation can be carried out for other algebraic forms of the Lamé operator associated to the Lamé polynomials of the 2nd, 3rd, and 4th types.

---

[4]There are no restrictions on deformation of the numeric coefficients in the remaining part of the operator (7) – they can be deformed in any way we like.

In a similar way one can carry out a deformation of another algebraic form of the Lamé operator (12). Again a deformation of the only positive-grading part of (12),

$$
\begin{aligned}
h_+ = &-[(12A^2 - g_2)\rho^3 + (4A^3 - g_2 A + g_3)\rho^4]\partial_\rho^2 \\
&+ \left[\left(n - \frac{3}{2}\right)(12A^2 - g_2)\rho^2 + (n-2)(4A^3 - g_2 A + g_3)\rho^3\right]\partial_\rho \\
&- \frac{n(n-1)}{4}(12A^2 - g_2)\rho - \frac{n(n-2)}{4}(4A^3 - g_2 A + g_3)\rho^2,
\end{aligned} \quad (29)
$$

is essential. The simplest and perhaps the most natural deformation is given by the following expression:

$$
\begin{aligned}
h_{+,\mathrm{def}} = &-[(12A^2 - g_2)\rho^3 + (4A^3 - g_2 A + g_3)\rho^4]D_\rho^2 \\
&+ \left[\left[\left\{\frac{n}{2} - 1\right\} + \left\{\frac{n}{2} - \frac{1}{2}\right\}\right](12A^2 - g_2)\rho^2 + 2\left\{\frac{n}{2} - 1\right\}(4A^3 - g_2 A + g_3)\rho^3\right]D_\rho \\
&- \left\{\frac{n}{2}\right\}\left\{\frac{n}{2} - \frac{1}{2}\right\}(12A^2 - g_2)\rho \\
&- \left\{\frac{n}{2}\right\}\left\{\frac{n}{2} - 1\right\}(4A^3 - g_2 A + g_3)\rho^2.
\end{aligned} \quad (30)
$$

This expression can be immediately rewritten as a linear combination of $\tilde{J}_{n/2}^+ \tilde{J}_{n/2}^+$, $\tilde{J}_{n/2}^+ \tilde{J}_{n/2}^0$ and $\tilde{J}_{n/2}^+$ terms. A similar deformation can be carried out for other algebraic forms of the Lamé operator associated to the Lamé polynomials of the 2nd, 3rd, and 4th types.

In conclusion one should mention that another dilatation-invariant deformation of Lamé equation based on on different form of the dilatation-invariant finite-difference operator was studied in Ref. 16. This study was done in relation to the problem of Bloch electrons in a magnetic field (Azbel–Hofstadter problem).

5   References

1. F. M. Arscott, *Periodic Differential Equations. An Introduction to Mathieu, Lamé, and Allied Functions*, International Series of Monographs in Pure and Applied Mathematics, Vol. 66, Macmillan, New York, 1964.

2. L. Brink, A. V. Turbiner, and N. Wyllard, *Hidden algebras of the (super) Calogero and Sutherland models*, J. Math. Phys. **39** (1998), No. 3, 1285–1315.

3. H. Exton, *q-Hypergeometrical Functions and Applications*, Ellis Horwood Ser. Math. Appl., Ellis Horwood Ltd., Chichester; Halsted Press, New York, 1983.

4. G. Gasper and M. Rahman, *Basic Hypergeometric Series*, Encyclopedia Math. Appl., Vol. 35, Cambridge Univ. Press, Cambridge, 1990.

5. E. Kamke, *Differentialgleichungen. Lösungsmethoden und Lösungen. I. Gewöhnliche Differentialgleichungen*, Mathematik und ihre Anwendungen in Physik und Technik, Vol. 18, Geest & Portig, Leipzig, 1959.

6. O. Ogievetsky and A. V. Turbiner, $sl(2, \mathbb{R})_q$ *and quasi-exactly-solvable problems*, preprint CERN-TH: 6212/91, CERN, 1991.

7. M. A. Olshanetsky and A. M. Perelomov, *Quantum integrable systems related to Lie algebras*, Phys. Rep. **94** (1983), No. 6, 313–404.

8. A. M. Perelomov, *Algebraic approach to the solution of a one-dimensional model of N interacting particles*, Theoret. and Math. Phys. **6** (1971), 263–282.

9. W. Rühl and A. Turbiner, *Exact solvability of the Calogero and Sutherland models*, Modern Phys. Lett. **A10** (1995), No. 29, 2213–2221.

10. Y. F. Smirnov and A. V. Turbiner, *Lie algebraic discretization of differential equations*, Modern Phys. Lett. **A10** (1995), No. 24, 1795–1802, Errata, no. 40, 3139.

11. _____, *Hidden* sl_2-*algebra of finite-difference equations*, Proceedings of the IV Wigner Symposium (Guadalajara, 1995) (N. M. Atakishiyev, T. H. Seligman, and K. B. Wolf, eds.), World Sci. Publishing, River Edge, NJ, 1996, pp. 435–440.

12. A. V. Turbiner, *Lamé equation,* sl_2 *and isospectral deformation*, J. Phys. A **22** (1989), No. 1, L1–L3.

13. _____, *On polynomial solutions of differential equations*, J. Math. Phys. **33** (1992), No. 12, 3989–3994.

14. _____, *Lie algebras and linear operators with invariant subspace*, Lie Algebras, Cohomologies and New Findings in Quantum Mechanics (Springfield, MO, 1992) (N. Kamran and P. J. Olver, eds.), Contemp. Math., Vol. 160, Amer. Math. Soc., Providence, RI, 1994, pp. 263–310.

15. _____, *Quasi-exactly-solvable differential equations*, CRC Handbook of Lie Group Analysis of Differential Equations. Vol. 3. New Trends in Theoretical Developments and Computational, CRC Press, Boca Raton, FL, 1996, pp. 331–366, hep-th/9409068.

16. P. B. Wiegmann and A. V. Zabrodin, *Algebraization of difference eigenvalue equations related to* $U_q(sl_2)$, Nucl. Phys. B **451** (1995), No. 3, 699–724, cond-mat/9501129.

31

Yangian Gelfand–Zetlin Bases, $\mathfrak{gl}_N$-Jack Polynomials, and Computation of Dynamical Correlation Functions in the Spin Calogero–Sutherland Model

Denis Uglov

ABSTRACT We consider the $\mathfrak{gl}_N$-invariant Calogero–Sutherland models with $N = 1, 2, 3, \ldots$ in the framework of symmetric polynomials. The Hamiltonian of any such model admits a distinguished orthogonal eigenbasis characterized as the union of Yangian Gelfand–Zetlin bases of irreducible components with respect to the Yangian action on the space of states. We construct an isomorphism from the space of states into the space of symmetric Laurent polynomials that maps the eigenbasis into the basis of $\mathfrak{gl}_N$-Jack polynomials. These polynomials are defined as specializations of Macdonald polynomials where both parameters approach an Nth primitive root of unity. As an application of this isomorphism we compute two-point dynamical spin-density and density correlation functions in the $\mathfrak{gl}_2$-invariant Calogero–Sutherland model at integer values of the coupling constant.

1 Introduction

The subject of this work is the spin generalization of the Calogero–Sutherland model proposed by Cherednik [3] and Bernard et al. [2]. This model describes n quantum particles with coordinates $y_1, y_2, \ldots, y_n$ moving along a circle of length L ($0 \le y_i \le L$). Each particle carries a spin with N possible values. The dynamics of the particles are governed by the Hamiltonian

$$\widehat{H}_{\beta,N} = -\frac{1}{2}\sum_{i=1}^{n}\frac{\partial^2}{\partial y_i^2} + \frac{\pi^2}{2L^2}\sum_{\substack{i,j=1 \\ i \ne j}}^{n}\frac{\beta(\beta + P_{ij})}{\sin^2 \pi(y_i - y_j)/L}. \tag{1}$$

Here β is the coupling constant which we assume to be a positive integer, and the P_{ij} stands for the spin exchange operator of the ith and jth particles.

Following Ref. 2 it is convenient to make a gauge transformation by taking

$$W = \prod_{\substack{i,j=1 \\ i<j}}^{n} \sin \frac{\pi}{L}(y_i - y_j)$$

and defining the gauge-transformed Hamiltonian $H_{\beta,N}$ by

$$H_{\beta,N} = W^{-\beta} \widehat{H}_{\beta,N} W^{\beta}. \tag{2}$$

If we introduce the new coordinates $z_1, \ldots, z_n$ by $z_j = \exp(2\pi i y_j/L)$ then the Hamiltonian $H_{\beta,N}$ becomes a well-defined operator on the vector space $F_{N,n}$ which is the total antisymmetrization of the tensor product

$$\mathbb{C}[z_1^{\pm 1}, \ldots, z_n^{\pm 1}] \otimes (\mathbb{C}^N)^{\otimes n}. \tag{3}$$

In other words, if we let K_{ij} denote the permutation operator for coordinates z_i, z_j in $\mathbb{C}[z_1^{\pm 1}, \ldots, z_n^{\pm 1}]$, and let P_{ij} be the permutation of ith and jth factors in $(\mathbb{C}^N)^{\otimes n}$, then $F_{N,n}$ is the subspace of (3) spanned by vectors f such that $K_{ij}f = -P_{ij}f$ $(i, j = 1, 2, \ldots, n)$.

It is well known how to solve the scalar version of the model ($N = 1$). In this case the space of states $F_{1,n}$ is just the space of antisymmetric Laurent polynomials, which is obviously isomorphic, via division by the Vandermonde determinant, to the space of symmetric Laurent polynomials. There is an orthogonal with respect to the appropriate scalar product eigenbasis of the gauge-transformed Hamiltonian, which under this isomorphism is given by the Jack polynomials. The properties of the Jack polynomials known in mathematical literature can then be used to compute various quantities of interest, such as the two-point dynamical correlation functions [10, 11, 15].

To obtain a solution of the model for an arbitrary N is a more complicated problem. One approach to a solution was suggested in Ref. 2. It is based on properties of the nonsymmetric Jack polynomials [1, 4, 5, 17] and easily yields the energy spectrum and an eigenbasis of the Hamiltonian. It seems to be quite difficult, however, to use this approach effectively in order to compute the correlation functions. Still, progress has been achieved recently in Ref. 14, where a variation of the nonsymmetric polynomial-based way of treating the problem was employed.

In the present work we propose a different treatment of the problem. In our approach we stay entirely within the framework of symmetric (Laurent) polynomials for all N. It is rather easy to see that there is an isomorphism between the linear space $F_{N,n}$ and the space of symmetric Laurent polynomials for all N. This isomorphism is just a finite-particle version of the Fermion-Boson correspondence well-known in the representation theory of infinite-dimensional Lie algebras [12, 13]. Using the Yangian symmetry of the model [2], we define a particular eigenbasis of the Hamiltonian as the unique eigenbasis of the maximal commutative subalgebra of the Yangian. The main result which we present in this work, is the description of the image of this eigenbasis under the above-mentioned isomorphism. As with the scalar situation, we find that this image is given by specializations of symmetric Macdonald polynomials.

Let us recall (we use the notations of Macdonald's book [16]) that a Macdonald polynomial $P_\lambda(q, t)$ is labeled by a partition λ and depends on two parameters: q and t. Let γ be a positive real number, and let ω_N be an Nth primitive root of unity, that is: $\omega_N^N = 1$, $\omega_N^j \neq 1$ for $j = 1, 2, \ldots, N - 1$. We define a specialization $P_\lambda^{(\gamma, N)}$ of the Macdonald polynomial as

$$P_\lambda^{(\gamma, N)} = \lim_{p \to 1} P_\lambda(\omega_N p, \omega_N p^\gamma) \qquad (4)$$

and call it a $\mathfrak{gl}_N$-Jack polynomial. Obviously the $\mathfrak{gl}_1$-Jack polynomial and the usual Jack polynomial coincide.

The eigenbasis of the maximal commutative subalgebra of the Yangian is identified, under the isomorphism between $F_{N,n}$ and the space of symmetric Laurent polynomials, with the set of the $\mathfrak{gl}_N$-Jack polynomials $P_\lambda^{(N\beta+1, N)}$. To be more precise we need to extend the basis of the $\mathfrak{gl}_N$-Jack polynomials to a basis of the entire space of symmetric Laurent polynomials by certain trivial modifications—in the same way as in the scalar case.

By specialization of the orthogonality for Macdonald polynomials one easily finds that the $\mathfrak{gl}_N$-Jack polynomials $P_\lambda^{(N\beta+1, N)}$ are pairwise orthogonal with respect to the scalar product $\langle \cdot, \cdot \rangle_{\beta, N}$ defined for any two symmetric Laurent polynomials $f = f(x_1, \ldots, x_n)$ and $g = g(x_1, \ldots, x_n)$ as

$$\langle f, g \rangle_{\beta, N} = \frac{1}{n!} [f(x_1^{-1}, \ldots, x_n^{-1}) \Delta(\beta, N; x) g(x_1, \ldots, x_n)]_1 \qquad (5)$$

where

$$\Delta(\beta, N; x) = \prod_{\substack{i,j=1 \\ i \neq j}}^{n} (1 - x_i^N x_j^{-N})^\beta (1 - x_i x_j^{-1})$$

and for a Laurent polynomial h the $[h]_1$ stands for the constant term of h.

A number of properties of the $\mathfrak{gl}_N$-Jack polynomials are easily derived from the known properties of Macdonald polynomials by specialization. In particular, we get a normalization formula for these polynomials with respect to the scalar product (5), and a formula which gives an expansion of power-sums in the basis formed by these polynomials. These formulas are straightforward to apply, just as in the scalar case, to a computation of two-point dynamical density and spin-density correlation functions. We carry out this computation (without taking the thermodynamic limit) for the case $N = 2$.

Let us now make a remark on the scope of this chapter. Necessarily we omit proofs of all statements. A complete version of the paper appeared as the preprint [20], where the detailed exposition can be found.

2 Cherednik–Dunkl operators, Yangian, Yangian Gelfand–Zetlin Bases

The origin of solvability of the spin Calogero–Sutherland model lies in the relationship between this model and a representation in $\mathbb{C}[z_1^{\pm 1}, \dots, z_n^{\pm 1}]$ of the degenerate affine Hecke algebra given by the permutations K_{ij} and the pairwise commutative Cherednik–Dunkl operators [2, 3] defined as

$$d_i(\beta) = \beta^{-1} z_i \frac{\partial}{\partial z_i} - i + \sum_{i < j} \frac{z_j}{(z_j - z_i)}(K_{ij} - 1) - \sum_{i > j} \frac{z_i}{(z_i - z_j)}(K_{ij} - 1),$$

$$(i = 1, 2, \dots, n).$$

According to a result of Bernstein, symmetric polynomials in Cherednik–Dunkl operators commute with permutations K_{ij} and therefore are well-defined operators on the space $F_{N,n}$. In particular, the gauge-transformed Hamiltonian is an example of such an operator [2, 3]:

$$H_{\beta,N} = \frac{2\pi^2 \beta^2}{L^2} \sum_{i=1}^{n} \left(d_i(\beta) + \frac{n+1}{2} \right)^2, \tag{6}$$

where the equality is understood as that of operators on $F_{N,n}$.

The Yangian $Y(\mathfrak{gl}_N)$ is an associative unital algebra, dual to the degenerate affine Hecke algebra in the sense of Drinfeld [7–9]. The algebra $Y(\mathfrak{gl}_N)$ is generated by the unit 1 and elements $T_{ab}^{(s)}$ where $a, b \in \{1, \dots, N\}$ and $s = 1, 2, \dots$. In terms of the formal power series $T_{ab}(u)$ in a formal variable u^{-1} defined as

$$T_{ab}(u) = \delta_{ab} 1 + u^{-1} T_{ab}^{(1)} + u^{-2} T_{ab}^{(2)} + \cdots,$$

the relations of $Y(\mathfrak{gl}_N)$ are written as follows

$$(u - v)[T_{ab}(u), T_{cd}(v)] = T_{cb}(v)T_{ad}(u) - T_{cb}(u)T_{ad}(v).$$

The infinite-dimensional center of $Y(\mathfrak{gl}_N)$ is generated by coefficients of the series

$$A_N(u) = \sum_{w \in S_N} \mathrm{sgn}(w) T_{1w(1)}(u) T_{2w(2)}(u-1) \cdots T_{Nw(N)}(u - N + 1)$$

known as the quantum determinant of $Y(\mathfrak{gl}_N)$. The Yangian is equipped with a distinguished maximal commutative subalgebra, $A(\mathfrak{gl}_N)$, [6, 18].

This subalgebra is generated by coefficients of the series $A_1(u), \ldots, A_N(u)$ defined as quantum principal minors of the T-matrix $T_{ab}(u)$:

$$A_m(u) = \sum_{w \in S_m} \mathrm{sgn}(w) T_{1w(1)}(u) T_{2w(2)}(u-1) \cdots T_{mw(m)}(u-m+1),$$
$$(m = 1, 2, \ldots, N),$$

that is to say, by the centers of all algebras in the chain:

$$Y(\mathfrak{gl}_1) \subset Y(\mathfrak{gl}_2) \subset \cdots \subset Y(\mathfrak{gl}_N),$$

where for $m = 1, 2, \ldots, N-1$ the $Y(\mathfrak{gl}_m)$ is realized inside $Y(\mathfrak{gl}_N)$ as the subalgebra generated by coefficients of the series $T_{ab}(u)$ with a, $b = 1, 2, \ldots, m$.

The subalgebra $A(\mathfrak{gl}_N)$ is somewhat analogous to the Cartan subalgebra of an ordinary simple Lie algebra (note that the Cartan subalgebra of the Lie algebra $\mathfrak{gl}_N \subset Y(\mathfrak{gl}_N)$ is a subalgebra of $A(\mathfrak{gl}_N)$). The distinctive feature of the Yangian case is that action of $A(\mathfrak{gl}_N)$ on an irreducible Yangian module is not, in general, semisimple. The class of finite-dimensional $Y(\mathfrak{gl}_N)$ modules that admit a semisimple action of the subalgebra $A(\mathfrak{gl}_N)$ are called *tame* modules. Irreducible tame modules were classified in Ref. 18. Eigenbases of $A(\mathfrak{gl}_N)$ in these modules are called *Yangian Gelfand–Zetlin bases* [6, 18].

The space of states, $F_{N,n}$, of the gauge-transformed Hamiltonian admits a $Y(\mathfrak{gl}_N)$-action that is defined in the standard way [2] by the Drinfeld duality [7] from the representation of the degenerate affine Hecke algebra on $\mathbb{C}[z_1^{\pm 1}, \ldots, z_n^{\pm 1}]$ given by the coordinate permutations and the Cherednik–Dunkl operators. We denote by $T_{ab}(u; \beta)$ the T-matrix of generators of this $Y(\mathfrak{gl}_N)$-action, and by $A(\mathfrak{gl}_N; \beta)$—the corresponding action of the maximal commutative subalgebra $A(\mathfrak{gl}_N)$. It is known from the work [2] that the Hamiltonian $H_{\beta,N}$ and the momentum operator

$$P = \frac{2\pi}{L} \sum_{i=1}^{n} z_i \frac{\partial}{\partial z_i}$$

are among the operators that give the action of the Yangian center. Since the center is a subalgebra of $A(\mathfrak{gl}_N)$, these operators belong to the commutative family $A(\mathfrak{gl}_N; \beta)$.

3 The Wedge Basis and the Scalar Product

Let $V = \mathbb{C}^N$ with the basis $\{v_1, v_2, \ldots, v_N\}$. The affinization $V(z)$ of V defined as $V(z) = \mathbb{C}[z^{\pm 1}] \otimes V$ has a basis $\{z^m \otimes v_a \mid m \in \mathbb{Z}, a \in \{1, 2, \ldots, N\}\}$. For any integer k the integer m and $a \in \{1, 2, \ldots, N\}$ such that $k = a - Nm$ are defined uniquely. With these m and a we define for any $k \in \mathbb{Z}$: $u_k =$

$z^m \otimes v_a$. Then the set $\{u_k \mid k \in \mathbb{Z}\}$ is a basis of the infinite-dimensional space $V(z)$. The tensor product $\mathbb{C}[z_1^{\pm 1}, \dots, z_n^{\pm 1}] \otimes (\bigotimes^n V)$ is naturally identified with the tensor product $V(z)^{\otimes n}$.

For a sequence of n integers $(k_1, k_2, \dots, k_n)$ define the *wedge* vector as

$$u_{k_1} \wedge u_{k_2} \wedge \cdots \wedge u_{k_n} = \sum_{w \in S_n} \operatorname{sgn}(w) u_{k_{w(1)}} \otimes u_{k_{w(2)}} \otimes \cdots \otimes u_{k_{w(n)}}. \quad (7)$$

Observe that the linear space $F_{N,n} \subset V(z)^{\otimes n}$ is spanned by the wedges, and that in view of the antisymmetry relation $u_k \wedge u_l = -u_l \wedge u_k$, the set of all wedges with $k_1 > k_2 > \cdots > k_n$ is a basis of $F_{N,n}$. We call an element of this basis a *normally ordered wedge*.

Now we can define an appropriate scalar product on the space of states $F_{N,n}$ of the gauge-transformed Hamiltonian. The "free" scalar product $(\cdot, \cdot)_{0,N}$ is defined by requiring that the normally ordered wedges be orthonormal:

$$(u_{k_1} \wedge \cdots \wedge u_{k_n}, u_{l_1} \wedge \cdots \wedge u_{l_n})_{0,N} = \prod_{i=1}^{n} \delta_{k_i, l_i}.$$

The β-dependent scalar product $(\cdot, \cdot)_{\beta,N}$ is defined in terms of the symmetric Laurent polynomial (the weight function)

$$\Delta(z; \beta) = \prod_{\substack{i,j=1 \\ i \neq j}}^{n} (1 - z_i z_j^{-1})^{\beta}$$

by setting for any $f, g \in F_{N,n}$: $(f, g)_{\beta,N} = (f, \Delta(z; \beta)g)_{0,N}$. For an operator B acting on $F_{N,n}$ let B^* be its adjoint relative to the scalar product $(\cdot, \cdot)_{\beta,N}$ and for a series $B(u)$ with operator-valued coefficients we will use $B(u)^*$ to denote the series whose coefficients are adjoints of coefficients of the series $B(u)$. The generators of the Yangian action on $F_{N,n}$ satisfy the relations [19]: $T_{ab}(u; \beta)^* = T_{ba}(u; \beta)$, which imply that the action of the subalgebra $A(\mathfrak{gl}_N)$ is self-adjoint: that is, $A_m(u; \beta)^* = A_m(u; \beta)$ ($m = 1$, $2, \dots, N$). In particular the gauge-transformed Hamiltonian and the momentum operators are selfadjoint relative to the scalar product $(\cdot, \cdot)_{\beta,N}$.

4 Spin Calogero–Sutherland Model in the Framework of Symmetric Polynomials

For each integer M we define a map Ω_M from the space of states $F_{N,n}$ into the linear space of symmetric Laurent polynomials in variables $x_1, x_2, \dots, x_n$. In the basis of normally ordered wedges in $F_{N,n}$ the map Ω_M is given by

$$\Omega_M(u_{k_1} \wedge u_{k_2} \wedge \cdots \wedge u_{k_n}) = (x_1 x_2 \cdots x_n)^r s_\lambda(x_1, \dots, x_n)$$

where the integer r and the partition λ of length $l(\lambda) \leq n - 1$ are uniquely defined by the sequence $k_1 > k_2 > \cdots > k_n$ as follows: $k_i = \lambda_i + r + M - i + 1$ $(i = 1, 2, \ldots, n)$, and $s_\lambda(x_1, \ldots, x_n)$ is the Schur polynomial [16].

In the space of states there is a distinguished wedge vector (vacuum of charge M)

$$\mathrm{vac}(M) = u_M \wedge u_{M-1} \wedge \cdots \wedge u_{M-n+1},$$

which is mapped by Ω_M into the Laurent polynomial 1.

It is easy to see that each of the maps Ω_M is an isomorphism of linear spaces. Indeed an Ω_M is a finite-particle analogue of the Fermion-Boson correspondence [12, 13] well known in Conformal Field Theory.

Proposition 1. *For each $M \in \mathbb{Z}$ and any f, $g \in F_{N,n}$ we have*

$$(f, g)_{\beta,N} = \langle \Omega_M(f), \Omega_M(g) \rangle_{\beta,N}$$

That is, the isomorphism Ω_M is an isometry of the scalar product of the gauge-transformed spin Calogero–Sutherland model and the scalar product (5) for $\mathfrak{gl}_N$-Jack polynomials.

Theorem 1. A. 1. *In the space of states $F_{N,n}$ there exists a unique (up to normalization of eigenvectors) eigenbasis $\{X^{(\beta,N)}_{(k_1,k_2,\ldots,k_n)}\}$ of the commutative family $A(\mathfrak{gl}_N; \beta)$. Elements of this eigenbasis are parameterized by decreasing sequences of n integers: $k_1 > k_2 > \cdots > k_n$.*

2. *This eigenbasis is a disjoint union of Yangian Gelfand–Zetlin bases in (finite-dimensional) irreducible components of the Yangian action on $F_{N,n}$.*

3. *The spectrum of $A(\mathfrak{gl}_N; \beta)$ in this eigenbasis is simple.*

4. *The eigenbasis $\{X^{(\beta,N)}_{(k_1,k_2,\ldots,k_n)}\}$ is orthogonal with respect to the scalar product $(\cdot, \cdot)_{\beta,N}$.*

B. *Each of the isomorphisms Ω_M maps an element of this eigenbasis into a $\mathfrak{gl}_N$-Jack polynomial multiplied by a certain Laurent polynomial:*

$$\Omega_M(X^{(\beta,N)}_{(k_1,k_2,\ldots,k_n)}) = (x_1 x_2 \cdots x_n)^r P^{(N\beta+1,N)}_\lambda(x_1, \ldots, x_n),$$

where the integer r and the partition λ $(l(\lambda) \leq n - 1)$ are uniquely defined by the sequence $k_1 > k_2 > \cdots > k_n$ as follows: $k_i = \lambda_i + r + M - i + 1$ $(i = 1, 2, \ldots, n)$.

5 Dynamical Spin-Density and Density Correlation Functions for $N = 2$

It is straightforward to apply the proposition and the theorem of the preceding section in order to compute spin-density and density two-point dynamical correlation functions in the spin Calogero–Sutherland model. So as to avoid cumbersome technical details we consider only the case where the spin has two values, that is, $N = 2$ and the number of particles in the model n is an even number such that $n/2$ is odd. Under these assumptions the ground state (of the gauge-transformed Hamiltonian) is nondegenerate and is a Yangian singlet. It is identified with the vacuum vector $\text{vac}(n/2 + 1)$.

In order to derive formulas for the correlation functions, we need to introduce the eigenvalue spectra of the Hamiltonian and the momentum on the polynomial eigenvectors of the commutative family $A(\mathfrak{gl}_2; \beta)$. The polynomial eigenvectors are

$$X^{(\beta,2)}_{(k_1,k_2,\ldots,k_n)} \quad \text{with } k_1 > k_2 > \cdots > k_n \geq -n/2 + 1.$$

They are labeled by partitions $\lambda = (\lambda_1, \ldots, \lambda_n)$ of length less than or equal to n defined by $k_i = \lambda_i + n/2 + 2 - i$, and are mapped by the isomorphism $\Omega_{n/2+1}$ into polynomials $P^{(2\beta+1,2)}_\lambda(x_1, \ldots, x_n)$.

The corresponding eigenvalues of the momentum and the Hamiltonian are conveniently parameterized by a colored diagram of λ obtained by coloring the usual Young diagram by white and black colors in the checkerboard order so that the square $(1, 1)$ is colored white. An example of this coloring is given below, which represents the partition $\lambda = (6, 4, 4, 3, 1)$.

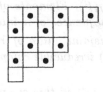

Let $W_\lambda(B_\lambda)$ be the subset of all white (black) squares in λ. Then the corresponding eigenvalue of the momentum is $\mathcal{P}(\lambda) = \frac{2\pi}{L}|W_\lambda|$, and the eigenvalue of the Hamiltonian $H_{\beta,N}$ minus energy of the ground state is

$$\mathcal{E}(\lambda) = \frac{2\pi^2}{L^2}\left(n_w(\lambda') - (2\beta + 1)n_w(\lambda) + \left(\frac{n}{2}(2\beta + 1) + \frac{1}{2}\right)|W_\lambda|\right).$$

In this formula λ' is the conjugate partition to λ and $n_w(\lambda) = \sum(i-1)w_i(\lambda)$, where $w_i(\lambda)$ is the number of white squares in the ith row of λ.

Let us consider the operator of spin-density:

$$s(0,0) = \sum_{i=1}^{n} \delta(y_i)\frac{\sigma_i^z}{2} = \frac{1}{L}\sum_{m\in\mathbb{Z}}\sum_{i=1}^{n} z_i^m \frac{\sigma_i^z}{2}.$$

The two-point correlation function

$$\langle s(x,t)s(0,0)\rangle$$
$$= \frac{(\text{vac}(n/2+1), e^{-itH_{\beta,2}-ixP}s(0,0)e^{itH_{\beta,2}+ixP}s(0,0)\,\text{vac}(n/2+1))_{\beta,2}}{(\text{vac}(n/2+1), \text{vac}(n/2+1))_{\beta,2}} \tag{8}$$

can be evaluated by using the isomorphism $\Omega_{n/2+1}$ together with the norm formulas and the formulas for expansion of power sums in the basis of $\mathfrak{gl}_2$-Jack polynomials in essentially the same fashion as in the scalar case.

For a square $s \in \lambda$, arm-length $a_\lambda(s)$, leg-length $l_\lambda(s)$, arm-colength $a'(s)$, and leg-colength $l'(s)$ are defined as the number of squares in the diagram of λ to the east, south, west, and north from s, respectively. Also for a parameter γ let

$$c(s; \gamma) = a'(s) - \gamma l'(s),$$
$$h_\lambda^*(s; \gamma) = a_\lambda(s) + \gamma l_\lambda(s) + 1, \quad h_*^\lambda(s; \gamma) = a_\lambda(s) + \gamma l_\lambda(s) + \gamma. \tag{9}$$

Then the formula for the correlation function reads

$$\langle s(x, t)s(0, 0)\rangle$$
$$= \frac{1}{2L^2} \sum_{\substack{\lambda \\ |W_\lambda| - |B_\lambda| = \pm 1 \\ |W_\lambda| = |H_2(\lambda)| + 1}} Z_\lambda(\beta, n) \frac{\prod_{s \in W_\lambda \setminus (1,1)} c(s; 2\beta + 1)^2}{\prod_{s \in H_2(\lambda)} h_\lambda^*(s; 2\beta + 1) h_*^\lambda(s; 2\beta + 1)} e^{it\mathcal{E}(\lambda)} \cos(x\mathcal{P}(\lambda))$$

where

$$Z_\lambda(\beta, n) = \prod_{s \in W_\lambda} \frac{a'(s) + (2\beta + 1)(n - l'(s))}{a'(s) + 1 + (2\beta + 1)(n - l'(s) - 1)}$$

and $H_2(\lambda)$ is the subset of squares of even hook-length [16] in λ.

A similar formula can be derived for the two-point correlation function of the density operator

$$\rho(0, 0) = \sum_{i=1}^{n} \delta(y_i) - \frac{n}{L} = \frac{1}{L} \sum_{m=1}^{\infty} \sum_{i=1}^{n} (z_i^m + z_i^{-m}). \tag{10}$$

The detailed derivations are contained in Ref. 20.

Acknowledgments: I would like to thank the organizers of the Workshop on Calogero–Moser–Sutherland models for the opportunity to present this chapter. I am grateful to Kouichi Takemura for collaboration. Some of the results summarized here are based on our joint work [19].

6 REFERENCES

1. T. H. Baker and P. J. Forrester, *Nonsymmetric Jack polynomials and integral kernels*, Duke Math. J. **95** (1998), No. 1, 1–50, q-alg/9612003.

2. D. Bernard, M. Gaudin, F. D. M. Haldane, and V. Pasquier, *Yang–Baxter equation in long-range interacting systems*, J. Phys. A **26** (1993), No. 20, 5219–5236, hep-th/9301084.

3. I. Cherednik, *Integration of quantum many-body problems by affine Knizhnik–Zamolodchikov equations*, Adv. Math. **106** (1994), No. 1, 65–95.

4. _____, *Double affine Hecke algebras and Macdonald's conjectures*, Ann. of Math. **141** (1995), No. 1, 191–216.

5. _____, *Nonsymmetric Macdonald polynomials*, Internat. Math. Res. Notices (1995), No. 10, 483–515.

6. I. V. Cherednik, *A new interpretation of Gelfand–Tzetlin bases*, Duke Math. J. **54** (1987), No. 2, 563–577.

7. V. G. Drinfeld, *Degenerate affine Hecke algebras and Yangians*, Funct. Anal. Appl. **20** (1986), No. 1, 62–64.

8. _____, *Quantum groups*, Proc. International Congress of Mathematicians (Berkeley, CA, 1986) (A. M. Gleason, ed.), Amer. Math. Soc., Providence, RI, 1987, pp. 798–820.

9. _____, *A new realization of Yangians and quantum affine algebras*, Soviet Math. Dokl. **36** (1988), No. 2, 212–216.

10. Z. N. C. Ha, *Exact dynamical correlation functions of Calogero–Sutherland model and one-dimensional fractional statistics*, Phys. Rev. Lett. **73** (1994), No. 13, 1574–1577.

11. _____, *Fractional statistics in one dimension: view from an exactly solvable model*, Nucl. Phys. B **435** (1995), No. 3, 604–636.

12. M. Jimbo and T. Miwa, *Solitons and infinite-dimensional Lie algebras*, Publ. Res. Inst. Math. Sci. **19** (1983), No. 3, 943–1001.

13. V. G. Kac and A. K. Raina, *Bombay Lectures on Highest Weight Representations of Infinite-Dimensional Lie Algebras*, Adv. Ser. Math. Phys., Vol. 2, World Sci. Publishing Co., Inc.,, Teaneck, NJ, 1987.

14. Y. Kato, *Green function of the Sutherland model with* SU(2) *internal symmetry*, preprint.

15. F. Lesage, V. Pasquier, and D. Serban, *Dynamical correlation functions in the Calogero–Sutherland model*, Nucl. Phys. B **435** (1995), No. 3, 585–603.

16. I. G. Macdonald, *Symmetric Functions and Hall Polynomials*, 2nd ed., Oxford Math. Monogr., Oxford Univ. Press, New York, 1995.

17. _____, *Affine Hecke algebras and orthogonal polynomials*, Séminaire Bourbaki. Vol. 1994/95, Astérisque, Vol. 237, Soc. Math. France, Paris, 1996, pp. 189–207.

18. M. Nazarov and V. Tarasov, *Representations of Yangians with Gelfand–Zetlin bases*, J. Reine Angew. Math. **496** (1998), 181–212.

19. K. Takemura and D. Uglov, *The orthogonal eigenbasis and norms of eigenvectors in the spin Calogero–Sutherland model*, J. Phys. A **1997** (30), No. 10, 3685–3717, solv-int/9611006.

20. D. Uglov, *Yangian Gelfand–Zetlin bases, $\mathfrak{gl}_N$-Jack polynomials and computation of dynamical correlation functions in the spin Calogero–Sutherland model*, Commun. Math. Phys. **191** (1998), No. 3, 663–696.

32

Thermodynamics of Moser–Calogero Potentials and Seiberg–Witten Exact Solution

K. L. Vaninsky

ABSTRACT We describe a recent attempt to compute thermodynamics for classical Moser–Calogero particles. Our approach is based on the description of the Gibbs states in the action-angle coordinates.

1 Introduction

It is generally believed in physics that Gibbs' distribution gives correct statistical description of mechanical system in equilibrium. At the present time, Gibbs' states have been constructed for infinite particle systems (see Ref. 2), and for Hamiltonian partial differential equations (see Refs. 1, 7, 9, and 13).

To perform the next step, we try to compute thermodynamics for classical 1-dimensional particles interacting via Moser–Calogero potential. Namely, we derive relations between macroscopical observables such as pressure and average energy as functions of density and temperature. Our main tool is the decomposition of the Gibbs' states proved previously for the Klein–Gordon (linear equation) [13], nonlinear Schrödinger equation (NLS) [10, 11]. Making some additional assumptions, we conjecture an explicit form of thermodynamic functions. Justification of our assumptions leads to interesting geometrical questions.

The construction which appears in our analysis is very similar to the one of Seiberg–Witten [3, 12] in their study of supersymmetric Yang–Mills theory. The direct relation (if there is any) between these two problems is unknown at the moment. In our analysis we essentially use results of Krichever and Phong [6] developed to study Seiberg–Witten type meromorphic differentials on Riemann surfaces.

We try to give an overview of our results here. For more details one should consult original papers.

2   Gibbs' States for NLS in Action-Angle Variables

For the cubic Schrödinger equation, the global action-angle variables $0 \leq I_k$, $0 \leq \phi_k < 2\pi$, $k = \ldots, -1, 0, 1, \ldots$, associated with the basic symplectic structure ω were constructed by McKean–Vaninsky, [10]. Now we want to express the measure

$$e^{-H} \, d\text{vol}, \quad \text{where } H = \int_0^1 |\psi'|^2 + 2|\psi|^4$$

in such coordinates. The main ingredient is the trace formula $H = \sum I'_k$, where the I''s are periods of some Abelian differential on the Riemann surface. They are the actions, but relative to some higher symplectic structure ω'; see Ref. 15. On a formal level,

$$e^{-H} \, d\text{vol} = e^{-\sum I'} \prod dI \, d\phi = e^{-\sum I'} \text{Jac} \prod dI' \, d\phi,$$

where Jac is the Jacobian between the variables I and I'. Therefore,

$$\text{Jac}^{-1} e^{-H} \, d\text{vol} = e^{-\sum I'} \prod dI' \, d\phi;$$

that is, I''s and ϕ's are independent, the I''s are exponential, and the ϕ's are uniform. Such a decomposition was proved by McKean–Vaninsky [11]. It can be useful in computing different thermodynamical quantities, as explained below.

3   Thermodynamics of the Moser–Calogero Potentials

Consider particles on the line interacting with the potential $U = 2x^{-2}$ (the rational case) or $2\sinh^{-2} x$ (the trigonometric case). To construct the Gibbs' state for an infinite system of such particles, we periodize the potential as in $U_{\omega_1}(\cdot) = \sum_k U(\cdot + k\omega_1)$. In our case, U_{ω_1} is simply the Weierstrass $2\wp$ with possibly infinite second pure imaginary period ω_2 (the rational case). Now N particles on the circle of the perimeter ω_1 are governed by the Hamiltonian

$$H(q,p) = \sum_{k=1}^{N} \frac{p_k^2}{2} + 2 \sum_{k,j=1}^{N} \wp(q_k - q_j).$$

The partition function is defined as

$$Z(N, \omega_1, \beta) = \frac{1}{N!} \int_M e^{-\beta H(q,p)} \, d^N q \, d^N p,$$

where the integral is taken over the phase space $M = ([0, \omega_1) \times R^1)^N$ and $\beta = T^{-1}$ is the reciprocal temperature. In the thermodynamic limit N and ω_1 tend to ∞, but their ratio ω_1/N tends to some constant: the specific volume $v > 0$. It is well known that, in the limit, under suitable assumptions on the decay of the potential, the specific free energy per particle

$$\psi(v, \beta) = \lim N^{-1} \log Z(N, \omega_1, \beta)$$

exists in the limiting ensemble. The average energy per particle and the pressure can be computed by simple differentiation of ψ:

$$\langle E \rangle = -\frac{\partial \psi(v, \beta)}{\partial \beta}, \quad p = T\frac{\partial \psi(v, \beta)}{\partial v}.$$

Let us introduce the function $h(v) \equiv \sum_{k \neq 0} U(kv)$ as the specific energy of one particle in the ground state of the limiting infinite system with the specific volume v. Now we are ready to formulate our basic conjecture.

Conjecture 1. *For Moser–Calogero potentials,*

$$\psi(v, \beta) = \log 2\pi + 1 - \beta h(v) - \log \beta - \log h(v).$$

In the rational case $h(v)$ can be computed explicitly:

$$h(v) = \sum_{k \neq 0} U(kv) = \sum_{k \neq 0} \frac{2}{k^2 v^2} = \frac{2\pi^2}{3v^2}.$$

Therefore,

$$p = -T\frac{\partial}{\partial v} \log h(v) - h'_v = \frac{2T}{v} + \frac{4\pi^2}{3v^3}, \quad \langle E \rangle = T + h(v) = T + \frac{2\pi^2}{3v^2}.$$

These equations are similar to the ideal gas for which $p = T/v$ and $\langle E \rangle = T/2$.

4   Sketch of the Computations

Assuming the possibility of decomposition of Gibbs' states described in Section 2 for particles with elliptic potentials, we compute $\psi(v, \beta)$ by some version of the method of stationary phase.

Obvious rotational symmetry allows one to reduce the dimension of the phase space. Indeed, according to the classical prescription,

$$I_1 = \frac{1}{2\pi} \int p \, dq = \frac{1}{2\pi} \int \sum p_k \, dq_k = \frac{\omega_1}{2\pi} \sum p_k = \frac{\omega_1}{2\pi} P,$$

where P is the total momentum. The corresponding angle φ_1 is just the usual on the circle. The change of variables $q_k \rightarrow q'_k$ and $p_k \rightarrow p'_k + P/N$

makes the total momentum P' vanish. The construction of the action-angle variables $I_2, \varphi_2, \ldots, I_N, \varphi_N$ on the reduced phase space is based on the KP equation (see Refs. 4 and 5). We do not need an explicit representation for them now.

In the computation of the partition function, the domain of integration is similarly reduced:

$$
\begin{aligned}
Z(N,\omega_1,\beta) &= \frac{1}{N!} \int_M e^{-\beta H(q,p)} \int_{R^1} \delta(P(q,p) - P_0)\, dP_0\, d^N q\, d^N p \\
&= \int_{R^1} dP_0 \frac{1}{N!} \int_M e^{-\beta H(q,p)} \delta(P(q,p) - P_0)\, d^N q\, d^N p.
\end{aligned}
$$

The canonical transformation $q_k \to q_k'$ and $p_k \to p_k' + P_0/N$ produces

$$
H(q,p) = H(q',p') + \frac{P_0^2}{2N}, \qquad P = P' + P_0
$$

and

$$
Z = \sqrt{2\pi NT} \frac{1}{N!} \int_{\substack{\text{range of } I_1,\ldots,I_N \\ \times [0,2\pi)^N}} e^{-\beta H(I_2,\ldots,I_N)} \delta\left(I_1 \frac{2\pi}{\omega_1}\right) dI_1\, d\varphi_1 \prod_{k=2}^{N} dI_k\, d\varphi_k.
$$

Integrating out I_1 and all the angles, we obtain

$$
Z = \sqrt{2\pi NT} \frac{\omega_1}{2\pi} (2\pi)^N \frac{1}{N!} \int_{\substack{\text{range of} \\ I_2,\ldots,I_N}} e^{-\beta H(I_2,\ldots,I_N)} \prod_{k=2}^{N} dI_k.
$$

To proceed farther we have to make few assumptions. The arguments in favor of them are supplied at the end of this section.

Assumption 1. The energy H can be expressed as $H = \sum_{k=2}^{N} I_k'$ ("trace formula"), where $I_2', \ldots, I_N'$ are the actions relative to higher symplectic structure ω'.

Assumption 2. The range of the variables $I_2', \ldots, I_N'$ is the rectangular domain $\{(I_2',\ldots,I_N') : I_{k\,\min}' \le I_k', k = 2,\ldots,N\}$, Then $H_{\min} = \sum I_{\min}'$ and in the thermodynamic limit, $H_{\min}/N \to h(v)$.

Assumption 3. The Jacobian between the variables $I_2, \ldots, I_N$ and $I_2', \ldots, I_N'$ in the thermodynamic limit has the asymptotics

$$
N^{-1} \log \operatorname{Jac}(I_2',\ldots,I_N',\omega_1,N) = \log N - \log h(v) + o(1)
$$

in probability relative to the Gibbs' ensemble. The nontrivial part of the asymptotics $\log h(v)$, we call *entropy*, by analogy with statistical mechanics.

Now we employ the trace formula:

$$Z(N, \omega_1, \beta) = \sqrt{2\pi N T} \omega_1 (2\pi)^{N-1} \frac{1}{N!} \int_{\substack{\text{range of} \\ I'_2, \dots, I'_N}} e^{-\beta \sum I'_k} \operatorname{Jac}(I'_2, \dots, I'_N) \prod_{k=2}^{N} dI'_k.$$

According to assumption 3 the Jac can be pulled out of the integral sign, and the rest is just elementary integration. For more details see Ref. 14.

Now we explain why we expect our assumptions to be true.

Assumption 1 The existence of the trace formula was proved recently in Ref. 16. We give description of this in Section 5 below.

Assumption 2 A general fact posited by Atiyah–Guillemin–Sternberg states that, for a compact symplectic phase space with a torus action, the image of the momentum map is a convex polytope. The shape of the momentum map here is suggested by the previous work on PDE and by the fact that the energy in the ground state is strictly positive.

Assumption 3 To explain why the $\operatorname{Jac}(I'_2, \dots, I'_N, \omega_1, N)$ becomes a constant in thermodynamic limit, we have to define all objects under consideration on one probability space. Consider a new phase space M for the infinite-particle system on the entire line. M contains all possible positions and velocities of particles, with the sole restriction that any finite region in the configuration space contains a finite number of particles (so-called locally finite).

Let $M_{\omega_1} \subset M$ comprise all spatially ω_1-periodic configurations. On M_{ω_1}, let $\mu_{\omega_1}(\cdot)$ be a Gibbs' state with periodic boundary conditions. Namely, consider projection of any configuration from M_{ω_1} into a finite volume, from 0 to ω_1. Any configuration is determined by the projection. Define on the projections a Gibbs' state with periodic boundary condition. It induces a measure $\mu_{\omega_1}(\cdot)$ on M_{ω_1} and on the big space M, too. A limiting Gibbs' state $\mu_{\infty}(\cdot)$ on M is obtained from $\mu_{\omega_1}(\cdot)$ by passing to the limit $\omega_1 \to \infty$.

On the big space M acts the one-parameter group of spatial translations τ_b, $b \in R^1$. Obviously, $\mu_{\omega_1}(\cdot)$ and M_{ω_1} itself are invariant under the action of τ. The limiting measure $\mu_{\infty}(\cdot)$ is also invariant under such translations and even ergodic.

For any such M_{ω_1} we can define a Jacobian $\operatorname{Jac}(I'_2, \dots, I'_N, \omega_1, N)$; it can be considered on the whole M as a function defined almost everywhere with respect to $\mu_{\omega_1}(\cdot)$. The Jacobians are τ-invariant functions. If they have a limit defined almost everywhere with respect $\mu_{\infty}(\cdot)$, then it is τ-invariant and must be a constant due to the ergodicity of this measure.

A similar formula for the ratio of two symplectic volumes for NLS was proved in [15]. For more details see Section 6 below.

5 Trace Formula for Particles with Elliptic Potential

In this section we will outline the result of Ref. 16. Consider the N-particle Hamiltonian on the line

$$H_N = \sum_{n=1}^{N} \frac{p_n^2}{2} - 2\sigma^2 \sum_{n,m=1}^{N} \wp(q_n - q_m).$$

The parameter $\sigma = 1$ corresponds to attractive particles and $\sigma = i^1$ corresponds to repulsive particles. The equations of motion can be written in the form

$$\ddot{q}_n = 4\sigma^2 \sum_{m \neq n} \wp'(q_n - q_m), \quad n = 1, \dots, N. \tag{1}$$

The key step in the embedding of the particle system into the class of elliptic solutions of the KP equation is the following theorem.

Theorem 1 ([[4]]). *The equations*

$$\left[\sigma \partial_y - \partial_x^2 + 2 \sum_{n=1}^{N} \wp(x - q_n(y))\right]\psi = 0,$$

$$\psi\dagger \left[\sigma \partial_y - \partial_x^2 + 2 \sum_{n=1}^{n} \wp(x - q_n(y))\right] = 0$$

have solutions of the form

$$\psi(x, y, k, z) = \sum_{n=1}^{N} a_n(y, k, z) \Phi(x - q_n, z) e^{kx + \sigma^{-1}k^2 y}$$

$$\psi^\dagger(x, y, k, z) = \sum_{n=1}^{N} a_n^\dagger(y, k, z) \Phi(-x + q_n, z) e^{-kx - \sigma^{-1}k^2 y},$$

where

$$\Phi(x, z) = \frac{\sigma(z - x)}{\sigma(z)\sigma(x)} e^{\zeta(z)x},$$

if and only if $q_n(y)$ satisfy Eq. (1).

The proof is obtained by requring that singularities of the form $(x - q_n)^{-2}$ and $(x - q_n)^{-1}$ vanish. This condition can be written in a compact form with the aid of $N \times N$ matrices L and M

$$L_{nm} = \sigma p_n \delta_{nm} + 2\Phi(q_n - q_m, z)(1 - \delta_{nm}),$$

$^1 i = \sqrt{-1}.$

$$M_{nm} = \left(-\wp(z) + 2\sum_{s \neq n} \wp(q_n - q_s)\right)\delta_{nm} + 2\Phi'(q_n - q_m, z)(1 - \delta_{nm}).$$

Lemma 1 ([4]). *The vectors $a(y, k, z)$ and $a^\dagger(y, k, z)$ satisfy the equations*

$$(L + 2k)a = 0 \qquad (\sigma\partial_y + M)a = 0$$

and

$$a^\dagger(L + 2k) = 0 \quad a^\dagger(\sigma\partial_y + M) = 0.$$

The existence of the nontrivial vector a: $(L + 2k)a = 0$ implies that $R_N(k, z) = \det(\tilde{L} + 2k)$ vanishes and this condition determines the curve.

The curve Γ_N is an N-sheeted covering of the elliptic curve. The function k is meromorphic on Γ_N. It has simple poles on all sheets above $z = 0$. The index "α" numbers the sheets of the curve Γ_N. The eigenvalues $k_\alpha(z)$ can be expanded in power series in z,

$$k_\alpha(z) = \frac{1}{z}k_\alpha^{(-1)} + k_\alpha^{(0)} + k_\alpha^{(1)}z + \cdots,$$

with $k_\alpha^{(-1)} = -1$ for $\alpha = 2, \ldots, N$ and $k_1^{(-1)} = N - 1$.

Theorem 2 ([16]). *On the "upper" sheet for $k_1(z)$ the following asymptotics hold:*

$$k_1(z) = \frac{N - 1}{z} - \frac{\sigma P_N}{2N} + z\left(\frac{\sigma^2 H_N}{2N^2} - \frac{\sigma^2 P_N^2}{4N^3}\right) + O(z^2),$$

where

$$H_N = K_N - \sigma^2 V_N = \sum_{n=1}^{N}\frac{p_n^2}{2} - 2\sigma^2\sum_{n,m=1}^{N}\wp(q_n - q_m),$$

and

$$P_N = \sum_{n=1}^{N} p_n.$$

This theorem is the key step to the main result.

Theorem 3 ([16]). *The following identity holds:*

$$\frac{\sigma^2}{2N^2}H_N = \sum_{\alpha=2}^{N} I'_\alpha,$$

where $I'_\alpha = -k_\alpha^{(1)}$. The variables I'_α are real for any configuration of particles.

Presumably I'_α are the actions relative some symplectic structure ω' on the phase space, but this is not proved (see Refs. [8] and [15] for the case of NLS). We will return to this issue elsewhere.

6   Symplectic Structures and Volume Elements for NLS

A similar question for the asymptotics of the ratio of two symplectic volumes was considered in Ref. [15].

The cubic nonlinear Schrödinger equation (NLS)

$$i\dot{\psi} = -\psi'' + 2|\psi|^2\psi$$

on the circle of perimeter $2l_x$ has an infinite series of conserved quantities—integrals of motion $H_1, H_2, \ldots$. The first three are "classical" integrals:

$$H_1 = \frac{1}{2}\int |\psi|^2 \, dx = \mathcal{N} = \text{number of particles},$$

$$H_2 = \frac{1}{2i}\int \psi'\bar{\psi}\, dx = \mathcal{P} = \text{momentum},$$

$$H_3 = \frac{1}{2}\int |\psi'|^2 + |\psi|^4 \, dx = \mathcal{H} = \text{energy}.$$

The others $H_4, H_5, \ldots$ do not have classical names. For sufficiently smooth ψ we have trace formulas [10],

$$\mathcal{N} = \sum_n I_n, \qquad I_n = \frac{l_x}{2\pi i}\int_{a_n} p(\lambda)d\lambda, \qquad (2)$$

$$\mathcal{P} = \sum_n I'_n, \qquad I'_n = \frac{l_x}{2\pi i}\int_{a_n} \lambda p(\lambda)d\lambda, \qquad (3)$$

$$\mathcal{H} = \sum_n I''_n, \qquad I''_n = \frac{l_x}{2\pi i}\int_{a_n} \lambda^2 p(\lambda)d\lambda, \qquad (4)$$

etc. The expression of the algebro-geometrical 2-form

$$\omega = \sum_{k=1}^{g+1} i\delta p(\gamma_k) \wedge \delta\lambda(\gamma_k)$$

entering into (2) in QP coordinates ($\psi = Q + iP$) is obtained using techniques of [6]. The symplectic structure can be written in the form

$$\omega = 4\langle \delta Q \wedge \delta P\rangle_x, \quad \psi = Q + iP.$$

Another algebro-geometrical 2-form entering (3),

$$\omega' = \sum_{k=1}^{g+1} i\delta p(\gamma_k) \wedge \delta\lambda^2(\gamma_k),$$

in QP coordinates is

$$\omega' = 2\rangle\delta Q' \wedge \delta Q + \delta P' \wedge \delta P + (\delta Q^2 + \delta P^2) \wedge \partial^{-1}(\delta Q^2 + \delta P^2)\langle_x,$$

restricted to submanifold

$$\int_{-l_x}^{l_x} Q^2 + P^2 = \text{const.} \tag{5}$$

The constant of integration in $\partial^{-1} = \int_{-l_x}^{x} dx'$ is irrelevant since 1-form $\langle Q\delta Q + P\delta P\rangle_x$ vanishes on the vector fields tangent to the sphere (5).

Now we will state a result about the ratio of these symplectic volumes. Consider a submanifold in the function space with $2N+1$ gaps open and all other closed. Denote as before by ω_N, ω_N' the restriction of the symplectic forms ω, ω' on this submanifold.

Theorem 4 ([15]). *In the limit of infinite genus the asymptotic identity holds*

$$\frac{1}{2N}\log \frac{\bigwedge^{2N} \omega_N'}{\bigwedge^{2N} \omega_N} = \log N + \left[\log\frac{\pi}{l_x} - 1\right] + \frac{\log 2\pi N}{2N}$$

$$+ \frac{1}{2N}\left[\log \int_{[0,1)^\infty} m_{12}(0)\, d^\infty\tilde\theta - \log l_x\right] + o\left(\frac{1}{N}\right).$$

Acknowledgments: The work is supported by NSF grant DMS-9501002.

7 REFERENCES

1. J. Bourgain, *Periodic nonlinear Schrödinger equation and invariant measures*, Commun. Math. Phys. **168** (1994), No. 1, 1–26.

2. R. L. Dobrushin, Ya. G. Sinai, and Yu. M. Suhov, *Dynamical systems of statistical mechanics*, Dynamical systems. II. Ergodic theory with applications to dynamical systems and statistical mechanics (Ya. G. Sinai, ed.), Encyclopaedia Math. Sci., Springer, Berlin, 1989, pp. 208–253.

3. R. Donagi and E. Witten, *Supersymmetric Yang-Mills theory and integrable systems*, Nucl. Phys. B **460** (1996), No. 2, 299–334.

4. I. M. Krichever, *Elliptic solutions of the Kadomtsev–Petviashvili equation and integrable systems of particles*, Funct. Anal. Appl. **14** (1980), No. 4, 282–290.

5. I. M. Krichever, O. Babelon, E. Billey, and M. Talon, *Spin generalization of the Caloger–Moser system and the matrix KP equation*, Topics in Topology and Mathematical Physics (A. B. Sossinsky, ed.), Amer. Math. Soc. Transl. Ser. 2, Vol. 170, Amer. Math. Soc., Providence, RI, 1995, pp. 83–119, hep-th/9411160.

506   K. L. Vaninsky

6. I. M. Krichever and D. H. Phong, *On the integrable geometry of soliton equations and N = 2 supersymmetric gauge theories*, J. Differential Geom. **45** (1997), No. 2, 349–389.

7. H. P. McKean, *Statistical mechanics of nonlinear wave equations. IV. Cubic Schrödinger*, Commun. Math. Phys. **168** (1995), No. 3, 479–491.

8. _____, *Trace formulas and the canonical 1-form*, Algebraic Aspects of Integrable Systems (A. S. Fokas and I. M. Gelfand, eds.), Progr. Nonlinear Differential Equations Appl., Vol. 26, Birkhäuser, Boston, MA, 1996, pp. 217–235.

9. H. P. McKean and K. L. Vaninsky, *Statistical mechanics of nonlinear wave equations*, Trends and Perspectives in Applied Mathematics (L. Sirovich, ed.), Appl. Math. Sci., Vol. 100, Springer, New York, 1994, pp. 239–264.

10. _____, *Action-angle variables for nonlinear Schrödinger equation*, Commun. Pure Appl. Math. **50** (1997), No. 6, 489–562.

11. _____, *Cubic Schrödinger: the petit canonical ensemble in action-angle variables*, Commun. Pure Appl. Math. **50** (1997), No. 7, 593–622.

12. N. Seiberg and E. Witten, *Electric-magnetic duality, monopole condensation, and confinement in N = 2 supersymmetric Yang–Mills theory*, Nucl. Phys. B **426** (1994), No. 1, 19–52, Errata, **430** (1994), no. 2, 485–486.

13. K. L. Vaninsky, *Invariant Gibbsian measures of the Klein–Gordon equation*, Stochastic analysis (Ithaca, NY, 1993) (M. C. Cranston and M. A. Pinsky, eds.), Proc. Sympos. Pure Math., Vol. 57, Amer. Math. Soc., Providence, RI, 1995, pp. 495–510.

14. _____, *Gibbs states for Moser–Calogero potentials*, Internat. J. Modern Phys. B **11** (1997), No. 1-2, 203–211, solv-int/9607008.

15. _____, *Symplectic structures and volume elements in the function space for the cubic Schrödinger equation*, Duke Math. J. **92** (1998), No. 2, 381–402.

16. _____, *Trace formula for a system of particles with elliptic potential*, 1999.

33

New Integrable Generalizations of the CMS Quantum Problem and Deformations of Root Systems

A. P. Veselov

Introduction

In 1976 M. Olshanetsky and A. Perelomov suggested some integrable generalisations of the Calogero–Moser system related to the root systems both in the classical and quantum case. In the quantum case one has the following Schrödinger operators (see Ref. 14):

$$L = -\Delta + \sum_{\alpha \in \Re_+} m_\alpha (m_\alpha + 1)(\alpha, \alpha) u((\alpha, x)), \tag{1}$$

where $\Re$ is a root system, $u(x) = x^{-2}$ (rational case), $\omega^2 \sin^{-2} \omega x$ (trigonometric case) or $\wp(x)$ (elliptic case), m_α are some parameters invariant under the action of the corresponding Weyl group. In the rational case one can consider any Coxeter group (not necessary crystallographic one).

The appearance of the Coxeter groups in this context is quite natural for many reasons, so it was believed that all such a generalisations should be related somehow to finite reflection groups. In 1988 the author made a conjecture that this is the case in the so-called algebraically integrable case (see Ref. 7). Although this conjecture has been later justified under some assumptions (see Ref. 19) in general it turned out not to be true! In 1995 O. Chalykh, M. Feigin, and the author proved that the following Scrödinger operator,

$$L = -\Delta + \sum_{i<j}^N \frac{2m(m+1)}{(x_i - x_j)^2} + \sum_{i=1}^N \frac{2(m+1)}{(x_i - \sqrt{m}x_{N+1})^2}, \tag{2}$$

is integrable for any m and algebraically integrable for any integer m (see Ref. 18). This operator is related to the set $\mathbf{A}_N(m)$, which consists of the vectors $\alpha = e_i - e_j$ $(i, j \le N)$ with multiplicities $m_\alpha = m$ and $\alpha = \pm(e_i - \sqrt{m}e_{N+1})$ $(i = 1, \ldots, N)$ with $m_\alpha = 1$ and may be considered as a nonsymmetric deformation of the root system $\mathbf{A}_N$. Notice that when $m = 1$ we have exactly this root system and a particular case of Calogero operator [3] considered in Ref. 7:

$$L = -\Delta + \sum_{i<j}^{N+1} \frac{4}{(x_i - x_j)^2}. \tag{3}$$

The same results are true also in the trigonometric case for the operator

$$L = -\Delta + \sum_{i<j}^{N} \frac{2m(m+1)\omega^2}{\sin^2 \omega(x_i - x_j)} + \sum_{i=1}^{N} \frac{2(m+1)\omega^2}{\sin^2 \omega(x_i - \sqrt{m}x_{N+1})}, \tag{4}$$

which is the deformation of the Sutherland operator [17],

$$L = -\Delta + \sum_{i<j}^{N+1} \frac{4\omega^2}{\sin^2 \omega(x_i - x_j)}. \tag{5}$$

We conjecture that the integrability holds also in elliptic case:

$$L = -\Delta + \sum_{i<j}^{N} 2m(m+1)\wp(x_i - x_j)$$
$$+ \sum_{i=1}^{N} 2(m+1)\wp(x_i - \sqrt{m}x_{N+1}), \tag{6}$$

but only partial results are found in this direction (see Ref. 6).

In this talk I will explain the way these generalizations appeared and the reasons why we think they might be in some sense unique and therefore important.

Notice that in the coordinates $q_1 = x_1, \ldots, q_N = x_N, q_{N+1} = \sqrt{m}x_{N+1}$ the operators (2), (4), (6) can be written as the following Schrödinger operator describing the interaction of $N + 1$ particles on the line with the masses $m_1 = \cdots = m_N = 1, m_{N+1} = m^{-1}$:

$$L = \sum_{j=1}^{N+1} \frac{\hat{p}_j^2}{2m_j} + m(m+1) \sum_{i<j}^{N+1} m_i m_j u(q_i - q_j), \tag{7}$$

where $u(x) = x^{-2}, \omega^2 \sin^{-2} \omega x$ or $\wp(x)$ and $\hat{p}_j = i\partial/\partial q_j$ $(j = 1, \ldots, N+1)$.

I would like to mention that a formal classical version of this problem, which is the Hamiltonian system with the Hamiltonian

$$H = \sum_{j=1}^{N+1} \frac{p_j^2}{2m_j} + m(m+1) \sum_{i<j}^{N+1} m_i m_j u(q_i - q_j),$$

seems to be nonintegrable (!) for any m different from 1. This means that Moser's results [13] can not be generalised for this Hamiltonian system.

Notice that the classical limit of our quantum system has a different form. To obtain such a limit we multiply the operator (2) by $\hbar^2$ and replace $i\hbar\partial/\partial x_k$ by p_k:

$$\sum_{k=1}^{N+1} p_k^2 + \sum_{i<j}^{N} 2\hbar^2 m(m+1)u(x_i - x_j) + \sum_{i=1}^{N} 2\hbar^2(m+1)u(x_i - \sqrt{m}x_{N+1}).$$

Then in the limit $\hbar \to 0$, $m \to \infty$, $m\hbar \to \mu$ we obtain formally the Hamiltonian of the classical N-body Calogero system with one additional noninteracting particle:

$$H = \sum_{k=1}^{N+1} p_k^2 + \sum_{i<j}^{N} 2\mu^2 u(x_i - x_j),$$

which is integrable for trivial reasons.

Another remark (by V. Inosemtsev) is that in the limit $\omega \to \infty$, $m+1 \to 0$, $\omega(m+1) \to$ const we formally have the following operator:

$$L = -\Delta + \sum_{k<l}^{N} \mu\delta(x_k - x_l) - \sum_{k=1}^{N} \mu\delta(x_k - ix_{N+1}),$$

or, in the coordinates $q_1 = x_1, \ldots, q_N = x_N$, $q_{N+1} = ix_{N+1}$,

$$L = -\partial_1^2 - \cdots - \partial_N^2 + \partial_{N+1}^2 + \sum_{k<l}^{N} \mu\delta(q_k - q_l) - \sum_{k=1}^{N} \mu\delta(q_k - q_{N+1}).$$

1 Basic Construction and Coxeter Root Systems

The following construction [7, 19] plays the fundamental role in our approach.

Let $\mathcal{A}$ be a finite set of noncolinear vectors α in $\mathbb{R}^n$ with multiplicities $m_\alpha \in \mathbb{N}$. This set determines the configuration Π of the hyperplanes Π_α given by the equations $(k, \alpha) = 0$ taken with multiplicities m_α.

Let us define the *Baker–Akhiezer function associated with the set $\mathcal{A}$* as the function $\psi(k, x)$, $k, x \in \mathbb{R}^n$ of the form

$$\psi = P(k, x)e^{(k,x)}, \tag{8}$$

where $P(k, x)$ is a polynomial in k with the highest-degree term $A(k) = \prod_{\alpha\in\mathcal{A}}(k, \alpha)^{m_\alpha}$, which satisfies the following condition of "quasi-invariance": for any $\alpha \in \mathcal{A}$ the difference $\psi(k, x) - \psi(s_\alpha k, x)$ is divisible by $(k, \alpha)^{2m_\alpha}$.

Here s_α as above is a reflection with respect to the hyperplane Π_α. The last condition means that all normal derivatives of ψ of any odd order less than $2m_\alpha$ vanish at the hyperplane Π_α.

The following basic result explains the important role of such a function in the theory of integrable Schrödinger operators in many dimensions.

Theorem 1 ([7, 19]). *If the Baker–Akhiezer function ψ exists then it is unique and satisfies the algebraically integrable Schrödinger equation $L\psi = -k^2\psi$, where*

$$L = -\Delta + \sum_{\alpha \in \mathcal{A}} \frac{m_\alpha(m_\alpha + 1)(\alpha, \alpha)}{(\alpha, x)^2}. \qquad (9)$$

Recall that a Schrödinger operator $L = -\Delta + u(x_1, \ldots, x_n)$ in n-dimensional space is *integrable*, if there exist n pairwise commuting differential operators $L_1 = L, L_2, \ldots, L_n$ with algebraically independent constant highest symbols $P_i(k_1, \ldots, k_n)$. L is called *algebraically integrable*, if there is at least one more operator L_{n+1}, which commutes with all L_i and such that its highest symbol $P_{n+1}(k_1, \ldots, k_n)$ takes different values at the solutions of the system of the equations $P_i(k) = c_i$ $(i = 1, \ldots, n)$ (see Refs. 7, 12, and 19). In the dimension $n = 1$ all algebraically integrable Schrödinger operators are described by the "finite-gap" theory (see, for example, Ref. 9). In the dimension more than 1 it is an open problem.

Our construction provide us with the examples of such operators in case the Baker–Akhiezer function exists.

Remark 1. One can show that such a function after division by $A(k)$ becomes symmetric with respect to k and x (see Ref. 19 and 5):

$$\phi(k, x) = \frac{\psi(k, x)}{A(k)}, \quad \phi(k, x) = \phi(x, k).$$

Remark 2. Very recently Yu. Berest discovered the following remarkable formula for the Baker–Akhiezer function under the assumption that it does exist:

$$\psi(k, x) = \frac{1}{2^M M!}(L - (k, k))^M [A(x)e^{(k,x)}],$$

$$A(x) = \prod_{\alpha \in \mathcal{A}}(x, \alpha)^{m_\alpha}, \quad M = \sum_{\alpha \in \mathcal{A}} m_\alpha. \qquad (10)$$

Thus the main problem is the existence of the Baker–Akhiezer function. At first we believed that all the configurations when ψ does exist are the Coxeter ones. Indeed, the following result has been proved.

Theorem 2 ([19]). *If the configuration Π consists of the hyperplanes of reflections ("mirrors") of any Coxeter group with the multiplicities, which are invariant under the action of this group, then the Baker–Akhiezer function does exist. If all $m_\alpha = 1$ then a converse statement is true: if ψ exists then Π is a configuration of mirrors of some Coxeter group.*

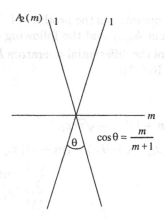

$$\cos\theta = \frac{m}{m+1}$$

FIGURE 33.1.

The proof of the direct statement is based on the results of Ch. Dunkl [10], G. Heckman [11], and E. Opdam [15]. The proof of the existence is effective in the sense that an explicit formula for the Baker–Akhiezer function is given in terms of the *Dunkl operators* [10] (see Ref. 19). The converse statement was in a good agreement with the conjecture that only Coxeter configurations are related to algebraically integrable Schrödinger operators [7].

Therefore it came as surprise that there exist Baker–Akhiezer functions associated with non-Coxeter configurations [18]!

2   Deformations of the Root Systems and New Integrable Quantum Problems

The following non-Coxeter configuration with Baker–Akhiezer function has been discovered by O. Chalykh et al. (see, Ref. 18).

Let $\mathcal{A} = \mathbf{A}_n^+(m)$ be the set of the following vectors in $\mathbb{R}^{n+1}$: $e_i - e_j$ ($i < j$; $i, j = 1, \ldots, n$) taken with multiplicity m and $e_i - \sqrt{m}e_{n+1}$ ($i = 1, \ldots, n$) with multiplicity 1. In the first nontrivial case we have the following configuration $\mathbf{A}_2(m)$ (see Figure 33.1).

Theorem 3 ([18]). *For the system $\mathbf{A}_n^+(m)$ there exists the Baker–Akhiezer function ψ. The corresponding algebraically integrable Schrödinger operator has a form*

$$L = -\Delta + \sum_{i<j}^{n} \frac{2m(m+1)}{(x_i - x_j)^2} + \sum_{i=1}^{n} \frac{2(m+1)}{(x_i - \sqrt{m}x_{n+1})^2}. \qquad (11)$$

The idea of the proof of this theorem is similar to the one we used in Ref. 7 when the Dunkl operator technique was not known. We write down the corresponding shift operator as the product of the shift operator related to the Coxeter subsystem $\mathbf{A}_{n-1}$ and the following operator $D_{n+1}^{1,\ldots,n}$.

Let's introduce a set of the differential operators $D_N^{i_1,\ldots,i_k}$ by the following recursion relations (cf. Ref. 7):

$$D_N = 1,$$
$$D_N^i = \partial_i - \sqrt{m}\partial_N - (m+1)(x_i - \sqrt{m}x_N)^{-1},$$
$$D_N^{i_1,\ldots,i_k} = D_N^{i_1,\ldots,i_{k-1}} \circ (\partial_{i_k} - \sqrt{m}\partial_N) - (m+1)(x_{i_k} - \sqrt{m}x_N)^{-1}D_N^{i_1,\ldots,i_{k-1}}$$
$$+ \sum_{s=1}^{k-1} \frac{m(m+1)}{(x_{i_s} - x_{i_k})^2} D_N^{i_1,\ldots,\widehat{i_s},\ldots,i_{k-1}}.$$

Here $\partial_i = \partial/\partial x_i$, $i_1,\ldots,i_k$ are different natural indeces less than N and the symbol $\widehat{i_s}$ means that the index $\widehat{i_s}$ must be omitted.

We have proved (see Refs. 18 and 6) that the Baker–Akhiezer function associated with the system $\mathbf{A}_n(m)$ has the form

$$\psi(k,x) = D_{n+1}^{1,\ldots,n}\psi_0(k,x),$$

where $\psi_0(k,x)$ is the Baker–Akhiezer function associated with the Coxeter subsystem $\mathbf{A}_{n-1}$, which consists of all vectors from $\mathbf{A}_n(m)$ with multiplicity m.

Notice that the shift operator $D = D_{n+1}^{1,\ldots,n}$ intertwines the operator (11) and

$$\widehat{\mathcal{L}} = -\partial_{N+1}^2 + \mathcal{L}, \tag{12}$$

where $\mathcal{L}$ is the standard Calogero operator in $\mathbb{R}^N$:

$$\mathcal{L} = -\Delta + \sum_{i<j}^N 2m(m+1)(x_i - x_j)^{-2}: \tag{13}$$

$$L \circ D = D \circ \widehat{\mathcal{L}}. \tag{14}$$

As a corollary we have that for any operator $\widehat{A}$ commuting with Calogero operator $\widehat{\mathcal{L}}$ the operator $A = D \circ \widehat{A} \circ D^*$ commutes with L, where D^* means the operator formally adjoint to D.

Indeed, since L and $\widehat{\mathcal{L}}$ are self-adjoint, it follows from (14) that $D^* \circ L = \widehat{\mathcal{L}} \circ D^*$ and therefore

$$AL = D\widehat{A}D^*L = D\widehat{A}\widehat{\mathcal{L}}D^* = D\widehat{\mathcal{L}}\widehat{A}D^* = LD\widehat{A}D^* = LA.$$

Now taking integrals for $\widehat{\mathcal{L}}$ as $\widehat{A}$, we obtain independent integrals for L. The same arguments as in [4] show that these integrals commute with each other.

Taking the symmetric integrals $\hat{A} = \mathcal{L}_i$ ($i = 1, \ldots, N$) and $\hat{A} = \partial_{N+1}$ we obtain $N + 1$ commuting integrals and therefore the integrability for the deformed Calogero system (11). Using the extra integrals of the system (13) found in Ref. 19 for integer m we construct in such a way the extra integrals for the system (11) to prove its algebraic integrability. Of course this fact also follows from the existence of Baker–Akhiezer function and the general Theorem 1.

We should note that not all the integrals have the form $D \circ \hat{A} \circ D^*$. For example, one can prove using the results of Ref. 18 the following:

Theorem 4 ([6]). *For the deformed Calogero operator (11) there exist commuting integrals $I_s(m)$ with the highest symbols*

$$p_s(k) = k_1^s + \cdots + k_N^s + m^{\frac{s-2}{2}} k_{N+1}^s.$$

When $m = 1$ these integrals coincide with the standard integrals L_s of Calogero system so $L_s(m)$ can be considered as their deformation.

Now having the rational case completed, it is natural to ask about a trigonometric version of these results. Everything works perfectly also in this case (see Refs. 6 and 18). In particular, the shift (intertwining) operator D is determined the following recursion formulas (cf. Ref. 8):

$$D_n = 1,$$
$$D_n^i = \partial_i - \sqrt{m}\partial_n - (m+1)\omega \cot \omega(x_i - \sqrt{m}x_n),$$
$$D_n^{i_1,\ldots,i_k} = D_n^{i_1,\ldots,i_{k-1}} \circ (\partial_{i_k} - \sqrt{m}\partial_n) - (m+1)\omega \cot \omega(x_{i_k} - \sqrt{m}x_n)D_n^{i_1,\ldots,i_{k-1}}$$
$$+ \sum_{s=1}^{k-1} \frac{m(m+1)\omega^2}{\sin^2 \omega(x_{i_s} - x_{i_k})}D_n^{i_1,\ldots,\hat{i}_s,\ldots,i_{k-1}},$$
$$D = D_{N+1}^{1,\ldots,N}.$$

It satisfies the relation

$$L \circ D = D \circ \hat{\mathcal{L}},$$

where

$$L = -\Delta + \sum_{i<j}^{N} 2m(m+1)\frac{\omega^2}{\sin^2 \omega(x_i - x_j)}$$

$$+ \sum_{i=1}^{N} 2(m+1)\frac{\omega^2}{\sin^2 \omega(x_i - \sqrt{m}x_{N+1})}, \quad (15)$$

and

$$\hat{\mathcal{L}} = -(\partial_1^2 + \cdots + \partial_{N+1}^2) + \sum_{i<j}^{N} 2m(m+1)\frac{\omega^2}{\sin^2 \omega(x_i - x_j)} \quad (16)$$

is the Sutherland operator [17].

Corresponding solution of the Schrödinger equation for (15): $L\psi = \lambda\psi$, has a form $\psi = D\psi_0$, where ψ_0 is the solution of the equation $\widehat{\mathcal{L}}\psi_0 = \lambda\psi_0$. In particular, for integer m we have the eigenfunction for L of the form

$$\psi = P(k, x)e^{(k, x)}, \tag{17}$$

where $P(k, x)$ is some polynomial on k with the highest terms

$$P(k) = \prod_{i<j}^{N}(k_i - k_j)^m \prod_{i=1}^{N}(k_i - \sqrt{m}k_{N+1}).$$

One can check that it satisfies the properties of the Baker–Akhiezer function in trigonometric case (see Refs. 18 and 19). This gives another proof of the algebraic integrability of the deformed Sutherland operator (15).

As an example let's consider the deformed three-particle Sutherland operator (15):

$$L = -\partial_1^2 - \partial_2^2 - m\partial_3^2 + 2m(m+1)u(q_1 - q_2)$$
$$+ 2(m+1)u(q_1 - q_3) + 2(m+1)u(q_2 - q_3), \tag{18}$$

where $u(x) = \omega^2 \sin^{-2}\omega x$.

In this case the shift operator D has the following form:

$$D = (\partial_1 - m\partial_3)(\partial_2 - m\partial_3) - (m+1)\omega\cot\omega(q_1 - q_3)(\partial_2 - m\partial_3)$$
$$- (m+1)\omega\cot\omega(q_2 - q_3)(\partial_1 - m\partial_3)$$
$$+ (m+1)^2\omega^2\cot\omega(q_1 - q_3)\cot\omega(q_2 - q_3)$$
$$+ m(m+1)\omega^2\sin^{-2}\omega(q_1 - q_2). \tag{19}$$

We have $L \circ D = D \circ \mathcal{L}$ with

$$\mathcal{L} = -\partial_1^2 - \partial_2^2 - m\partial_3^2 + 2m(m+1)\omega^2\sin^{-2}\omega(q_1 - q_2).$$

When m is integer we can intertwine $\mathcal{L}$ with $-\Delta = -\partial_1^2 - \partial_2^2 - m\partial_3^2$ by the following operator S:

$$\mathcal{L} \circ S = S \circ (-\Delta),$$
$$S = \left(\partial_1 - \partial_2 - 2m\omega\cot\omega(q_1 - q_2)\right) \circ \left(\partial_1 - \partial_2 - 2(m-1)\omega\cot\omega(q_1 - q_2)\right)$$
$$\circ \cdots \circ \left(\partial_1 - \partial_2 - 2\omega\cot\omega(q_1 - q_2)\right).$$

It follows from this that the operator $D \circ S$ intertwines L and $-\Delta$:

$$L \circ (D \circ S) = D \circ \mathcal{L} \circ S = (D \circ S) \circ (-\Delta).$$

The quantum integrals for L can be written explicitly:

$$L_1 = \partial_1 + \partial_2 + \partial_3,$$
$$L_2 = L,$$
$$L_3 = \partial_1 \partial_2 \partial_3 + \frac{1-m}{2}(\partial_1 + \partial_2)\partial_3^2 + \frac{(1-m)(1-2m)}{6}\partial_3^3$$
$$+ (m+1)u(q_2 - q_3)\partial_1 + (m+1)u(q_1 - q_3)\partial_2 + m(m+1)u(q_1 - q_2)\partial_3$$
$$+ \frac{1-m^2}{2}\left(u(q_1 - q_3) \circ \partial_3 + \partial_3 \circ u(q_1 - q_3)\right)$$
$$+ \frac{1-m^2}{2}\left(u(q_2 - q_3) \circ \partial_3 + \partial_3 \circ u(q_2 - q_3)\right),$$

where $u(x) = \omega^2 \sin^{-2} \omega x$.

While $\omega \to 0$ we obtain the analogous formulas for the deformed Calogero operator. It can be checked that the above formulas work also for the elliptic version of the operator (12) with $u(x) = \wp(x)$. This provides the integrability of the deformed elliptic Calogero–Moser system of three particles for the quantum case.

The intertwining operator exists also for the Schrödinger operator $L + \omega^2 x^2 = L_\omega$:

$$L_\omega = -\Delta + \sum_{i<j}^{N} \frac{2m(m+1)}{(x_i - x_j)^2} + \sum_{i=1}^{N} \frac{2(m+1)}{(x_i - \sqrt{m}x_{N+1})^2}$$
$$+ \omega^2(x_1^2 + \cdots + x_{N+1}^2). \quad (20)$$

It is defined by the following recursive formulas:

$$D_n = 1,$$
$$D_n^i = \partial_i - \sqrt{m}\partial_n - \left(\frac{m+1}{x_i - \sqrt{m}x_n} + \omega(x_i - \sqrt{m}x_n)\right),$$
$$D_n^{i_1,\dots,i_k} = D_n^{i_1,\dots,i_{k-1}} \circ (\partial_{i_k} - \sqrt{m}\partial_n)$$
$$- \left(\frac{m+1}{x_{i_k} - \sqrt{m}x_n} + \omega(x_{i_k} - \sqrt{m}x_n)\right)D_n^{i_1,\dots,i_{k-1}}$$
$$+ \sum_{s=1}^{k-1}\left(\frac{m(m+1)}{(x_{i_s} - x_{i_k})^2} - m\omega\right)D_n^{i_1,\dots,\widehat{i_s},\dots,i_{k-1}}.$$

The following result can be proven similar to Ref. 8.

Theorem 5. *For* $D_\omega = D_{N+1}^{1,\dots,N}$ *we have*

$$D_\omega \circ (L_\omega + 2N\omega) = L_\omega \circ D_\omega, \quad (21)$$

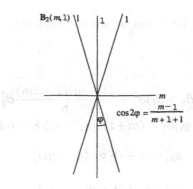

FIGURE 33.2.

where

$$\mathcal{L}_\omega = -\Delta + \sum_{i<j}^{N} \frac{2m(m+1)}{(x_i - x_j)^2} + \omega^2(x_1^2 + \cdots + x_{N+1}^2). \quad (22)$$

Corollary 1. $\mathcal{L}_\omega$ *is integrable for any* m.

Again, the integrals L_i can be obtained as

$$L_i^\omega = D_\omega \mathcal{L}_i D_\omega^*$$

where $\mathcal{L}_i$ $(i = 1, \ldots, N+1)$ are the integrals for the Calogero operator (22).

All these new integrable systems are related to the deformation $\mathbf{A}_N(m)$ of the root system $\mathbf{A}_N$. The natural question arises: Is it possible to deform the other root systems in a similar way? Until now we know only the following deformation of $\mathbf{B}_2$ case. It depends on two parameters l, m and consists of the vectors $\alpha_1 = \sqrt{2m+1}e_1$ with multiplicity l, $\alpha_2 = \sqrt{2l+1}e_1$ with multiplicity m and two vectors $\alpha_{3,4} = \frac{1}{2}(\alpha_1 \pm \alpha_2)$ with multiplicity 1. We denote this set as $\mathbf{B}_2(m,l)$. Corresponding configuration of the lines is shown in Figure 33.2.

The Schrödinger operator related to this configuration has the form

$$L = -\Delta + l(l+1)(2m+1)u(\sqrt{2m+1}x_1) + m(m+1)(2l+1)u(\sqrt{2l+1}x_2)$$
$$+ (l+m+1)u(\tfrac{1}{2}\sqrt{2m+1}x_1 + \tfrac{1}{2}\sqrt{2l+1}x_2)$$
$$+ (l+m+1)u(\tfrac{1}{2}\sqrt{2m+1}x_1 - \tfrac{1}{2}\sqrt{2l+1}x_2). \quad (23)$$

We have proved that this operator is integrable in rational $u = x^{-2}$ and trigonometric $u = \omega^2 \sin^{-2} \omega x$ cases for any m, n and algebraically integrable for any integer m, n (see Ref. 6). The proof of algebraic integrability is again based on the existence of the Baker–Akhiezer function. The conjecture is that this is true in the elliptic case $u = \wp(x)$ as well.

3   Concluding Remarks

In spite of existence of the "integrable" deformation of B_2 root system described above, we do not think that the same is true for all root systems. First of all the two-dimensional case is exceptional from many points of view. There exists a recent remarkable result in the theory of Huygens' principle proved by Yu. Berest [2], which gives under certain assumptions the description of all possible algebraically integrable Schrödinger operators in two dimensions, but to include all his examples we have to change the axiomatics of the Baker–Akhiezer function (see Ref. 5). Computer experiments made by Yu. Berest and I. Lutsenko in dimensions more than two did not show anything else but the deformation $A_N(m)$ of A_N we described, but of course this cannot pretend to be the proof of nonexistence of other "integrable" configurations. At the moment this investigation is still far from being completed. The relations with Huygens' principle (see, e.g., Ref. 1) make this problem of great interest.

Unfortunately we still have no understanding of the geometry behind the non-Coxeter configurations we found. The quasi-crystals and the theory of the quantum groups may give a natural setup for the answer. At the moment we can prove only that the subconfiguration of the hyperplanes with "large" multiplicities of any configuration with BA function is always a Coxeter one. This means that in the classical limit we will always have the Olshanetsky–Perelomov generalizations of the CMS system (see Ref. 5 and the Introduction).

Concerning our deformations of CMS we should mention that although we have given some formulas for the solutions of the corresponding Schrödinger equations the problem of finding the spectrum and nonsingular eigenfunctions, satisfying suitable boundary conditions, is still to be investigated. The lack of symmetry makes this problem nontrivial.

Another question is about different versions of our system. Such a version of the CMS quantum problem has been discovered by S. Ruijsenaars (see his review [16]). In the rational case Ruijsenaars's system can be described as dual system in the "bispectral sense" to the trigonometric CMS problem (Sutherland system). One can show that the same relation is valid for our generalizations of the Sutherland system (15) and certain different systems, which can be considered as deformations of Ruijsenaars's system (see Ref. 5).

Acknowledgments: I am grateful to J. Gibbons, A. Grünbaum, V. Inosemtsev, and especially to Yu. Berest for very useful discussions. Special thanks are to my collaborators: O. Chalykh, K. Styrkas, and M. Feigin: no need to say that their contribution to this work has been crucial.

4   REFERENCES

1. Yu. Yu. Berest and A. P. Veselov, *Huygens' principle and integrability*, Russian Math. Surveys **49** (1994), No. 6, 5–77.

2. H. W. Braden, *R-matrices and generalized inverses*, J. Phys. A **30** (1997), No. 15, L485–L493, q-alg/9706001, solv-int/9706001.

3. F. Calogero, *Solution of the one-dimensional N-body problems with quadratic and/or inversely quadratic pair potentials*, J. Math. Phys. **12** (1971), 419–436.

4. O. A. Chalykh, *Additional integrals for the generalized Calogero–Moser quantum problem*, Theoret. and Math. Phys. **109** (1996), No. 1, 1269–1273.

5. O. A. Chalykh, M. V. Feigin, and A. P. Veselov, *Multidimensional Baker–Akhiezer functions and Huygens' principle*, in preparation.

6. ———, *New integrable generalizations of Calogero–Moser quantum problem*, J. Math. Phys. **39** (1998), No. 2, 695–703.

7. O. A. Chalykh and A. P. Veselov, *Commutative rings of partial differential operators and Lie alrebras*, Commun. Math. Phys. **126** (1990), No. 3, 597–611.

8. ———, *Integrability in the theory of Schrödinger operator and harmonic analysis*, Commun. Math. Phys. **152** (1993), No. 1, 29–40.

9. B. A. Dubrovin, V. B. Matveev, and S. P. Novikov, *Nonlinear equations of Korteweg–de Vries type, finite-gap linear operators and Abelian varieties*, Russian Math. Surveys **31** (1976), No. 1, 51–125.

10. C. F. Dunkl, *Differential-difference operators associated to reflection groups*, Trans. Amer. Math. Soc. **311** (1989), No. 1, 167–183.

11. G. J. Heckman, *A remark on the Dunkl differential-difference operators*, Harmonic Analysis on Reductive Groups (Bowdoin, 1989) (W. Barker and P. Sally, eds.), Progr. Math., Vol. 101, Birkhäuser, Boston, MA, 1991, pp. 181–191.

12. I. M. Krichever, *Methods of algebraic geometry in the theory of nonlinear equations*, Russian Math. Surveys **32** (1977), No. 6, 185–213.

13. J. Moser, *Three integrable Hamiltonian systems connected to isospectral deformations*, Adv. Math. **16** (1975), 197–220.

14. M. A. Olshanetsky and A. M. Perelomov, *Quantum integrable systems related to Lie algebras*, Phys. Rep. **94** (1983), No. 6, 313–404.

15. E. Opdam, *Root systems and hypergeometric functions*. IV, Compositio Math. **67** (1988), No. 2, 191–209.

16. S. N. M. Ruijsenaars, *Systems of Calogero–Moser type*, Particles and Fields (Banff, 1994) (G. Semenoff and L. Vinet, eds.), CRM Series in Mathematical Physics, Springer, New York, 1999, pp. 251–352.

17. B. Sutherland, *Exact results for a quantum many-body problem in one dimension*, Phys. Rev. A **4** (1971), 2019–2021.

18. A. P. Veselov, M. V. Feigin, and O. A. Chalykh, *New integrable deformations of quantum Calogero–Moser problem*, Russian Math. Surveys **51** (1996), No. 3, 573–574.

19. A. P. Veselov, K. L. Styrkas, and O. A. Chalykh, *Algebraic integrability for the Schrödinger equation and finite reflection groups*, Theoret. and Math. Phys. **94** (1993), No. 2, 182–197.

34

The Calogero Model: Integrable Structure and Orthogonal Basis

Miki Wadati
Hideaki Ujino

ABSTRACT Integrability, algebraic structures, and orthogonal basis of the Calogero model are studied by the quantum Lax and Dunkl operator formulations. The commutator algebra among operators including conserved operators and creation-annihilation operators has the structure of the W-algebra. Through an algebraic construction of the simultaneous eigenfunctions of all the commuting conserved operators, we show that the Hi-Jack (hidden-Jack) polynomials, which are an multivariable generalization of the Hermite polynomials, form the orthogonal basis.

1 Introduction

In memory of the pioneering works in the 1970s [7, 12, 16, 17], a class of one-dimensional quantum many-body systems with inverse-square long-range interactions are generally called the Calogero–Moser–Sutherland models. The celebrated Hamiltonians are

$$\text{Calogero–Moser:} \quad H_{\text{CM}} = \frac{1}{2}\sum_{j=1}^{N} p_j^2 + \frac{1}{2}\sum_{\substack{j,k=1\\j\neq k}}^{N} \frac{a^2 - \hbar a}{(x_j - x_k)^2}, \tag{1}$$

$$\text{Calogero:} \quad \widehat{H}_{\text{C}} = \frac{1}{2}\sum_{j=1}^{N}(p_j^2 + \omega^2 x_j^2) + \frac{1}{2}\sum_{\substack{j,k=1\\j\neq k}}^{N} \frac{a^2 - \hbar a}{(x_j - x_k)^2}, \tag{2}$$

$$\text{Sutherland:} \quad \widetilde{H}_{\text{S}} = \frac{1}{2}\sum_{j=1}^{N} p_j^2 + \frac{1}{2}\sum_{\substack{j,k=1\\j\neq k}}^{N} \frac{a^2 - \hbar a}{\sin^2(x_j - x_k)}, \tag{3}$$

where the constants N, a, and ω are the particle number, the coupling parameter, and the strength of the external harmonic well, respectively. The momentum operator p_j is given by a differential operator, $p_j = -i\hbar\partial/\partial x_j$.

The Calogero and Sutherland models are a harmonic confinement and a periodic version of the Calogero–Moser model, respectively. Thus these two models have discrete energy spectra, whereas the other has continuous one. From now on, we set the Planck constant at unity, $\hbar = 1$.

The Lax formulation for the classical Calogero–Moser model was discovered by Moser [12]. Let us introduce two $N \times N$ Hermitian matrices:

$$L_{ij} = p_i \delta_{ij} + ia(1 - \delta_{ij}) \frac{1}{x_i - x_j},$$

$$M_{ij} = a\delta_{ij} \sum_{\substack{l=1 \\ l \neq i}}^{N} \frac{1}{(x_i - x_l)^2} - a(1 - \delta_{ij}) \frac{1}{(x_i - x_j)^2}.$$

We call them Lax pair. The classical Calogero–Moser Hamiltonian is given by Eq. (1) with $p_j = \mathrm{d}x_j/\mathrm{d}t$ and $\hbar = 0$. The time evolution of the L-matrix is expressed as the Lax equation,

$$\frac{\mathrm{d}L}{\mathrm{d}t} = \left\{ L, H_{\mathrm{CM}}^{\mathrm{cl}} \right\}_{\mathrm{P}} = [L, iM], \tag{4}$$

where the Poisson bracket is defined by $\{f, g\}_{\mathrm{P}} \overset{\text{def}}{=} \sum_j (\partial f/\partial x_j \partial g/\partial p_j - \partial g/\partial x_j \partial f/\partial p_j)$. Thanks to the trace identity for c-number-valued matrices, $\mathrm{Tr}\, AB = \mathrm{Tr}\, BA$, we can easily see that the trace of the power of the L-matrix, $I_n^{\mathrm{cl}} \overset{\text{def}}{=} \mathrm{Tr}\, L^n$, gives the conserved quantities: $\mathrm{d}I_n^{\mathrm{cl}}/\mathrm{d}t = \mathrm{Tr}[L^n, iM] = 0$. Using the classical matrix [5] or the generalized Lax equations for higher conserved quantities [6], we can show that the conserved quantities are Poisson-commutative, $\{I_n^{\mathrm{cl}}, I_m^{\mathrm{cl}}\}_{\mathrm{P}} = 0$, which proves the integrability of the classical Calogero–Moser model in Liouville's sense. Natural quantization of the Lax equation for the classical case (4) by the correspondence principle, $\{\clubsuit, \spadesuit\}_{\mathrm{P}} \rightarrow -i[\clubsuit, \spadesuit]$, gives an equality for the quantum Calogero–Moser Hamiltonian (1). However, the trace trick is not available to construct the commuting conserved operators for the quantum case because of the noncommutativity of the canonical conjugate variables. The initial motivation of our study was to find out a way to construct the conserved operators for the quantum models using the Lax formulation. The key of our idea is the sum-to-zero condition of the M-matrix:

$$\sum_{j=1}^{N} M_{jk} = 0, \quad \sum_{j=1}^{N} M_{kj} = 0, \quad \text{for } k = 1, 2, \dots, N.$$

This property tells us that the (commuting) conserved operators can be obtained by summing up all the matrix elements of the powers of the L-matrix instead of taking traces [20, 27],

$$I_n^{\mathrm{CM}} = \sum_{j,k=1}^{N} (L^n)_{jk} \overset{\text{def}}{=} \mathrm{T}_\Sigma L^n \implies [H_{\mathrm{CM}}, I_n^{\mathrm{CM}}] = \mathrm{T}_\Sigma[L^n, M] = 0,$$

which proves the quantum integrability, or the existence of sufficiently many conserved operators, of the Calogero–Moser model. Encouraged by the result, we further investigated the integrable structure of the Calogero model (2) through the quantum Lax formulation.

2 Integrability and Algebraic Structure

Let us start from the Lax equation of the Calogero Hamiltonian (2),

$$-i\frac{dL^\pm}{dt} = [\widehat{H}_C, L^\pm] = [L^\pm, M] \pm \omega L^\pm,$$

where the new matrices $L^\pm$ are defined by $L^\pm \overset{\text{def}}{=} L \pm Q$, $Q_{jk} \overset{\text{def}}{=} ix_j\delta_{jk}$. Using the sum-to-zero trick, we can get the conserved operators of the Calogero model as follows:

$$\widehat{I}_n \overset{\text{def}}{=} \mathrm{T}_\Sigma(L^+L^-)^n, \qquad\qquad [\widehat{H}_C, \widehat{I}_n] = \mathrm{T}_\Sigma[(L^+L^-)^n, M] = 0,$$

$$\widehat{I}_1 = 2\widehat{H}_C - N\omega(Na + (1-a)), \qquad \widehat{I}_n = \sum_{j=1}^N p_j^{2n} + \cdots.$$

Mutual commutativity of the above conserved operators is verified rather easily by the Dunkl operator formulation [8, 14]. Introducing the coordinate exchange operator,

$$(K_{lk}f)(\ldots, x_l, \ldots, x_k, \ldots) = f(\ldots, x_k, \ldots, x_l, \ldots),$$

we define the creation-annihilation like operators as

$$c_l^\dagger = p_l + ia\sum_{\substack{k=1 \\ k\neq l}}^N \frac{1}{x_l - x_k}K_{lk} + i\omega x_l,$$

$$c_l = p_l + ia\sum_{\substack{k=1 \\ k\neq l}}^N \frac{1}{x_l - x_k}K_{lk} - i\omega x_l.$$

(5)

Commutation relations among the creation-annihilation operators are

$$[c_l^\dagger, c_m^\dagger] = 0, \quad [c_l, c_m] = 0,$$

$$[c_l, c_m^\dagger] = 2\omega\delta_{lm}\left(1 + a\sum_{\substack{k=1 \\ k\neq l}}^N K_{lk}\right) - 2\omega a(1 - \delta_{lm})K_{lm},$$

which prove that the Hermitian operators, $I_n \overset{\text{def}}{=} \sum_{j=1}^N (c_j^\dagger c_j)^n$, are commuting operators. We denote the restriction of the operand to the space of symmetric functions by $|_{\mathrm{Sym}}$. Under the restriction, the conserved operators $\hat{I}_n$ and commuting Hermitian operators I_n are considered to be the same, $\hat{I}_n|_{\mathrm{Sym}} = \mathsf{I}_n|_{\mathrm{Sym}}$. Thus we have proved the quantum integrability of the Calogero model.

We can recursively construct generalized Lax equations for a family of operators O_m^p, m, $p = 1, 2, \ldots$, which reveal the W-algebraic structure of the Calogero model [21]. The operators are defined by the sum of all the matrix elements of the Weyl-ordered product of p L^+s and m L^-s:

$$O_m^p \overset{\text{def}}{=} \mathrm{T}_\Sigma [(L^+)^p (L^-)^m]_{\mathrm{W}},$$

$$[(L^+)^p (L^-)^m]_{\mathrm{W}} \overset{\text{def}}{=} \frac{p!\, m!}{(p+m)!} \sum_{\substack{\text{all possible} \\ \text{order}}} (L^+)^p (L^-)^m.$$

The generalized Lax equations for the operators O_m^p are

$$[O_m^p, L^\pm] = [L^\pm, M_m^p] + m\omega(1 \pm 1)[(L^+)^p (L^-)^{m-1}]_{\mathrm{W}}$$
$$- p\omega(1 \mp 1)[(L^+)^{p-1}(L^-)^m]_{\mathrm{W}}, \quad (6)$$

and the M_m^p matrices satisfy the sum-to-zero condition:

$$\sum_{j=1}^N (M_m^p)_{jk} = 0, \quad \sum_{j=1}^N (M_m^p)_{kj} = 0.$$

The Hamiltonian $\hat{H}_{\mathrm{C}}$ belongs to the operator family, $2\hat{H}_{\mathrm{C}} = O_1^1$. The operators O_n^n are conserved operators, though they do not commute each other. The family has two interesting subsets of commuting non-Hermitian operators,

$$B_n^\dagger \overset{\text{def}}{=} O_0^n, \quad B_n \overset{\text{def}}{=} O_n^0, \quad n = 1, 2, \ldots, \quad (7)$$

which we call power-sum creation-annihilation operators. They will play an important role in the algebraic construction of the energy eigenfunctions in the next section.

Let us introduce operators $W_n^{(s)}$:

$$W_n^{(s)} \overset{\text{def}}{=} \frac{1}{4\omega} O_{s+n-1}^{s-n-1}, \quad s \geq |n| + 1,$$

where the indices n and s are integer or half-odd integer, and respectively correspond to the Laurent mode and the conformal spin. The commutator among the operators above is

$$[W_n^{(s)}, W_m^{(t)}] = (n(t-1) - m(s-1)) W_{n+m}^{(s+t-2)} + \mathcal{P}_{n,m}^{(s,t)}(W_l^{(u)}),$$

where $\mathcal{P}_{n,m}^{(s,t)}\left(W_l^{(u)}\right)$ is a polynomial of $W_l^{(u)}$, $u \le s+t-3$, $l \le n+m$. The polynomial is generated while the products of $L^\pm$-matrices are rearranged into the Weyl-ordered products by replacements of L^+ and L^-,

$$[L^+, L^-] = 2\omega((a-1)\mathbf{1} - aT), \quad \mathbf{1}_{jk} = \delta_{jk}, \; T_{jk} = 1.$$

In terms of the W-operators, conserved operators and power-sum creation-annihilation operators are respectively expressed as $O_n^n \propto W_0^{n+1}$, $B_n^\dagger \propto W_{-n/2}^{(n/2+1)}$ and $B_n \propto W_{n/2}^{(n/2+1)}$.

For the classical case, the W-algebraic structure of the Calogero model was discovered by the collective field theory [2] and the classical r-matrix method [1]. The quantum collective field theory also possesses the W-algebraic structure [3, 4], though its relationship with the quantum Calogero model is not directly confirmed. An SU(ν) generalization of our approach is presented in Ref. 22.

3   Perelomov Basis

The eigenvalue problem of the Calogero model was first solved by Calogero [7]. Later, inspired by the simple form of its energy spectrum, Perelomov tried an algebraic construction of the energy eigenfunctions [13]. In what follows, we shall complete the Perelomov's approach [21]. The generalized Lax equations (6) for the power-sum creation operators (7) yield the following commutators:

$$\left[\widehat{H}_\mathrm{C}, B_n^\dagger\right] = n\omega B_n^\dagger, \; \left[B_n^\dagger, B_m^\dagger\right] = 0, \quad n, m = 1, 2, \dots, N. \qquad (8)$$

To construct all the eigenfunctions for the N-body Calogero model, we need N creation operators. By straightforward calculations of the commutators (8), Perelomov presented three creation operators, $B_n^\dagger$, $n = 2$, 3, 4. The quantum Lax formulation provides an easy way to make a sufficient number of such operators. By successive operations of the power-sum creation operators on the ground-state wave function, we can get all the excited states. The Calogero Hamiltonian is cast into the following form,

$$\widehat{H}_\mathrm{C} = \frac{1}{2}\,\mathrm{T}_\Sigma\,L^+L^- + \frac{1}{2}N\omega(Na + (1-a)) = \frac{1}{2}\sum_{j=1}^N h_j^\dagger h_j + E_\mathrm{g},$$

where the operators $h_j^\dagger$ and h_j are defined as

$$h_j^\dagger \stackrel{\text{def}}{=} \sum_{k=1}^N L_{kj}^+ = p_j + i\omega x_j - ia\sum_{\substack{k=1 \\ k \ne j}}^N \frac{1}{x_j - x_k},$$

$$h_j \overset{\text{def}}{=} \sum_{k=1}^{N} L_{jk}^- = p_j - i\omega x_j + ia \sum_{\substack{k=1 \\ k \neq j}}^{N} \frac{1}{x_j - x_k}.$$

Thus the differential equations, $h_j|0\rangle = 0$, $j = 1,2,\ldots,N$, are the sufficient conditions for the ground-state $|0\rangle$. In the coordinate representation, the solution is expressed as the (real) Laughlin wave function:

$$\langle x \mid 0 \rangle = \prod_{\substack{j,k=1 \\ j<k}}^{N} |x_j - x_k|^a \exp\left(-\frac{1}{2}\omega \sum_{j=1}^{N} x_j^2\right). \qquad (9)$$

As is similar to the free boson case, the excited states are labeled by the Young diagram,

$$\lambda = \{\lambda_1 \geq \lambda_2 \geq \cdots \geq \lambda_N \geq 0\} \in Y_N,$$

where λ_k, $k = 1,2,\ldots,N$, are nonnegative integers. Conventionally, we omit zeroes and use superscript for a sequence of the same numbers, for example, $\{4,1^2\} = \{4,1,1,0,\ldots,0\}$. The excited state labeled by the Young diagram λ is given by

$$|\lambda\rangle = \prod_{k=1}^{N} (B_k^\dagger)^{\lambda_k - \lambda_{k+1}}|0\rangle, \quad \lambda_{N+1} = 0,$$

$$\widehat{H}_C = (|\lambda|\omega + E_g)|\lambda\rangle \overset{\text{def}}{=} E(\lambda)|\lambda\rangle, \qquad (10)$$

where $|\lambda|$ denotes the weight of the Young diagram, $|\lambda| \overset{\text{def}}{=} \sum_{k=1}^{N} \lambda_k$. The above energy spectrum has the same form as that of noninteracting bosons confined in an external harmonic well up to the ground-state energy. In other words, the inverse-square interactions just shift the ground-state energy. The multiplicity of the nth energy level, $n\omega + E_g$, is equal to the number of the Young diagrams of the weight n, $\#\{\lambda \mid |\lambda| = n\}$. This is also the same as that of the noninteracting case. Thus we have algebraically constructed a basis of the eigenfunctions of the Calogero Hamiltonian.

4   Diagonalization of $\hat{I}_2$

The algebraic construction à la Perelomov generates a basis of the eigenfunctions of the Calogero Hamiltonian. Unfortunately, the basis is not orthogonal. To make an orthogonal basis from a basis, we usually try the Gram–Schmidt method. Here we take another way. As we have confirmed before, the Calogero model has a set of commuting conserved operators, which means existence of simultaneous eigenfunctions for them. The simultaneous eigenfunctions should be the orthogonal basis because they must be nondegenerate eigenfunctions of Hermitian operators. As the first step of our approach, we get some simultaneous eigenfunctions of the Hamiltonian and the second conserved operator $\hat{I}_2$, and observe their properties [23].

Since the Hamiltonian and $\hat{I}_2$ commute, the matrix representation of $\hat{I}_2$ on the Perelomov basis has a block-diagonalized form and each block consists of the wave functions of a weight (energy eigenvalue). By a straightforward calculation of commutators between $\hat{I}_2$ and $B_n^\dagger$, we calculated the first seven blocks, whose weights are from zero to six, of the matrix representation and their eigenvalues. The eigenvalues imply the general form of the eigenvalue of $\hat{I}_2$:

$$\widehat{E}_2(\lambda) = 4\omega^2 \sum_{k=1}^{N} \left((\lambda_k)^2 + a(N + 1 - 2k)\lambda_k\right).$$

Though a combination of $E(\lambda)$ and $\widehat{E}_2(\lambda)$ removes most of the degeneracies, there still remain degeneracies for the states whose weights larger than or equal to six. For example, the following two pairs of the Young diagrams with the weight six give such degeneracies:

$$\{4, 1^2\}, \{3^2\} \rightarrow \widehat{E}_2 = 4\omega^2\left(18 + 6a(N - 2)\right),$$

$$\{3, 1^3\}, \{2^3\} \rightarrow \widehat{E}_2 = 4\omega^2\left(12 + 6a(N - 3)\right).$$

It is interesting that the pairs have a common property. We can not compare two Young diagrams of each pair by the dominance order. The dominance order $\overset{D}{\leq}$, sometimes called the natural partial order, is defined as follows:

$$\mu \overset{D}{\leq} \lambda \iff |\mu| = |\lambda| \text{ and } \sum_{k=1}^{l} \mu_k \leq \sum_{k=1}^{l} \lambda_k \text{ for all } l.$$

We can readily confirm that the Young diagrams of each pair are incomparable in the dominance order,

$$\{4, 1^2\} \overset{D}{\not\leq} \{3^2\} \text{ and } \{3^2\} \overset{D}{\not\leq} \{4, 1^2\},$$

$$\{3, 1^3\} \overset{D}{\not\leq} \{2^3\} \text{ and } \{2^3\} \overset{D}{\not\leq} \{3, 1^3\}.$$

The specific observation above is, in fact, a general fact. We cannot define the dominance order between any pair of distinct Young diagrams λ and μ of a weight that share the common eigenvalue $\widehat{E}_2$ [15, 24], that is,

$$|\lambda| = |\mu| \text{ and } \widehat{E}_2(\lambda) = \widehat{E}_2(\mu) \implies \lambda \overset{D}{\not\leq} \mu \text{ and } \mu \overset{D}{\not\leq} \lambda.$$

We calculated the eigenvectors of the blocks with weights up to three in the matrix representation of $\hat{I}_2$. The eigenvectors correspond to seven simultaneous eigenfunctions of $\widehat{H}_C$ and $\hat{I}_2$. Since the eigenvalues E and $\widehat{E}_2$ for the seven functions have no degeneracy, they belong to the orthogonal basis and also to the simultaneous eigenfunctions of all the commuting conserved operators $\hat{I}_n$ of the Calogero model. The eigenfunction of the Calogero model is factorized into the ground-state wave function (9) and a symmetric polynomial. Symmetric polynomial parts of the seven simultaneous eigenfunctions, which we denote by $[\lambda]$, are

$$[0] = 1, \quad [1] = m_1, \quad [1^2] = m_{1^2} + \frac{a}{2\omega}\frac{N(N-1)}{2},$$

$$[2] = (1+a)m_2 + 2am_{1^2} - \frac{1}{2\omega}N(Na+1),$$

$$[1^3] = m_{1^3} + \frac{1}{2\omega}a\frac{(N-1)(N-2)}{2}m_1,$$

$$[2,1] = (2a+1)m_{2,1} + 6am_{1^3} - \frac{1}{2\omega}(1-a)(N-1)(Na+1)m_1,$$

$$[3] = (a^2+3a+2)m_3 + 3a(a+1)m_{2,1} + 6am_{1^3} - \frac{3}{2\omega}(a^2N^2+3aN+2)m_1,$$

where m_λ is the monomial symmetric polynomial defined by

$$m_\lambda(x_1,\ldots,x_N) = \sum_{\substack{\sigma \in S_N,\text{ distinct} \\ \text{permutations}}} (x_{\sigma(1)})^{\lambda_1}\cdots(x_{\sigma(N)})^{\lambda_N}.$$

Note that the sum over S_N is performed so that any monomial in the summand appears only once. In the above expressions, we notice that all the seven symmetric polynomials share a common property, triangularity. Namely, the seven polynomials $[\lambda]$ are expanded by the monomial symmetric polynomials m_μ whose Young diagram μ is smaller than or equal to the Young diagram λ in the weak dominance order $\overset{d}{\leq}$, that is,

$$\mu \overset{d}{\leq} \lambda \iff \sum_{k=1}^{l}\mu_k \leq \sum_{k=1}^{l}\lambda_k \text{ for all } l.$$

The observation means that we can uniquely identify the simultaneous eigenfunctions of the first two conserved operators of the Calogero model just by the first two eigenvalues and triangularity up to normalization. We shall confirm the existence of such functions by algebraically constructing them.

5   Hi-Jack Polynomials

Since our interest now concentrates on the symmetric polynomial parts of the simultaneous eigenfunctions, we modify some operators to make them suitable for the aim. A gauge transformation of the creation-annihilation-like operators (5) yields the following Dunkl operators:

$$\alpha_l^\dagger \stackrel{\text{def}}{=} \langle x \mid 0 \rangle \left(-\frac{i}{2\omega}\right) c_l^\dagger \frac{1}{\langle x \mid 0 \rangle}$$

$$= -\frac{i}{2\omega}\left(p_l + ia\sum_{\substack{k=1\\k\neq l}}^{N}\frac{1}{x_l - x_k}(K_{lk}-1) + 2i\omega x_l\right),$$

$$\alpha_l \stackrel{\text{def}}{=} \langle x \mid 0 \rangle i c_l \frac{1}{\langle x \mid 0 \rangle} = i\left(p_l + ia\sum_{\substack{k=1\\k\neq l}}^{N}\frac{1}{x_l - x_k}(K_{lk}-1)\right),$$

$$d_l \stackrel{\text{def}}{=} \alpha_l^\dagger \alpha_l. \tag{11}$$

The gauge transformation above removes the action on the ground-state wave function from the operators. Note that the definition of Hermiticity of such gauge-transformed operators is modified and different from the ordinary one. Using the d_l-operators, we define the normalized conserved operators:

$$I_n \stackrel{\text{def}}{=} \sum_{l=1}^{N}(d_l)^n\Big|_{\text{Sym}} = \left(\frac{1}{2\omega}\right)^n \langle x \mid 0 \rangle \hat{I}_n \frac{1}{\langle x \mid 0 \rangle}\Big|_{\text{Sym}},$$

$$\langle x \mid 0 \rangle \hat{H}_{\text{C}} \frac{1}{\langle x \mid 0 \rangle} = \omega I_1 + E_{\text{g}}. \tag{12}$$

We note that the Dunkl operators (11) reduce to those for the Sutherland model (3) in the limit, $\omega \to \infty$:

$$\alpha_l^\dagger \to z_l,$$

$$\alpha_l \to \nabla_l = i\left(p_{z_l} + ia\sum_{\substack{k=1\\k\neq l}}^{N}\frac{1}{z_l - z_k}(K_{lk}-1)\right), \quad p_{z_l} \stackrel{\text{def}}{=} -i\frac{\partial}{\partial z_l}, \tag{13}$$

$$d_l \to D_l = z_l \nabla_l.$$

We change the variables by

$$\exp 2i x_j = z_j, \quad j = 1, 2, \ldots, N,$$

and denote the ground-state wave function and the ground-state energy of the Sutherland model by

$$\tilde{\psi}_{\text{g}} = \prod_{\substack{j,k=1\\j<k}}^{N} |z_j - z_k|^a \prod_{j=1}^{N} z_j^{-\frac{1}{2}a(N-1)},$$

$$\epsilon_{\text{g}} = \frac{1}{6}a^2 N(N-1)(N+1).$$

Then the Sutherland Hamiltonian (3) is gauge-transformed to and related with the D-operator by

$$H_S - \epsilon_g = \tilde{\psi}_g^{-1}(\tilde{H}_S - \epsilon_g)\tilde{\psi}_g$$

$$= -2\sum_{j=1}^{N}(z_j p_{z_j})^2 + ia\sum_{\substack{j,k=1 \\ j\neq k}}^{N} \frac{z_j + z_k}{z_j - z_k}(z_j p_{z_j} - z_k p_{z_k})$$

$$= 2\sum_{l=1}^{N}(D_l)^2\big|_{\text{Sym}}. \tag{14}$$

Commutation relations among the Dunkl operators (11) and the action of α_l on 1 are

$$[\alpha_l, \alpha_m] = 0, \quad [\alpha_l^\dagger, \alpha_m^\dagger] = 0,$$

$$[\alpha_l, \alpha_m^\dagger] = \delta_{lm}\left(1 + a\sum_{\substack{k=1 \\ k\neq l}}^{N} K_{lk}\right) - a(1 - \delta_{lm})K_{lm},$$

$$[d_l, d_m] = a(d_m - d_l)K_{lm}, \quad \alpha_l \cdot 1 = 0.$$

We should remark that the above relations do not explicitly depend on the parameter ω, which implies the Dunkl operators for the Sutherland model (13) also satisfy the above relations. Hence the Calogero and Sutherland models share the same algebraic structure [19, 25, 26]. To put it another way, the theory of the Calogero model is a one-parameter deformation of that of the Sutherland model. Thus the simultaneous eigenfunction of the Calogero model is expected to be a one-parameter deformation of that of the Sutherland model, which is known to be the Jack polynomial [9]. In the following, we call the simultaneous eigenfunction of the Calogero model Hi-Jack (hidden-Jack) polynomial.

Using the normalized conserved operators (12), we define the Hi-Jack polynomials $j_\lambda(\boldsymbol{x};\omega,1/a)$ in a similar fashion to a definition of the Jack polynomials:

$$I_1 j_\lambda(\boldsymbol{x};\omega,1/a) = \sum_{k=1}^{N}\lambda_k j_\lambda(\boldsymbol{x};\omega,1/a) \overset{\text{def}}{=} E_1(\lambda)j_\lambda(\boldsymbol{x};\omega,1/a), \tag{15}$$

$$I_2 j_\lambda(\boldsymbol{x};\omega,1/a) = \sum_{k=1}^{N}(\lambda_k^2 + a(N+1-2k)\lambda_k)j_\lambda(\boldsymbol{x};\omega,1/a)$$

$$\overset{\text{def}}{=} E_2(\lambda)j_\lambda(\boldsymbol{x};\omega,1/a), \tag{16}$$

$$j_\lambda(\boldsymbol{x};\omega,1/a) = \sum_{\mu \overset{d}{\leq} \lambda} w_{\lambda\mu}(a,1/2\omega)m_\mu(\boldsymbol{x}), \tag{17}$$

$$w_{\lambda\lambda}(a,\omega) = 1. \tag{18}$$

We can prove the existence of the Hi-Jack polynomials by explicit construction. Following the results by Lapointe and Vinet on the Jack polynomials [10], we introduce the raising operators for the Hi-Jack polynomials,

$$b_k^+ = \sum_{\substack{J \subseteq \{1,2,\ldots,N\} \\ |J|=k}} \alpha_J^\dagger d_{1,J}, \text{ for } k = 1, 2, \ldots, N-1,$$

$$b_N^+ = \alpha_1^\dagger \alpha_2^\dagger \cdots \alpha_N^\dagger.$$

The operators, $\alpha_J^\dagger$ and $d_{1,J}$, stand for

$$\alpha_J^\dagger = \prod_{j \in J} \alpha_j^\dagger,$$

$$d_{1,J} = (d_{j_1} + a)(d_{j_2} + 2a) \cdots (d_{j_k} + ka),$$

where J is a subset of a set $\{1, 2, \ldots, N\}$ whose number of elements $|J|$ is equal to k, $J \subseteq \{1, 2, \ldots, N\}$, $|J| = k$. From Eq. (5), we can verify an identity,

$$(d_i + ma)(d_j + (m+1)a)\big|_{\text{Sym}}^{\{i,j\}} = (d_j + ma)(d_i + (m+1)a)\big|_{\text{Sym}}^{\{i,j\}}, \quad (19)$$

where m is some integer. The symbol $\big|_{\text{Sym}}^{J}$ where J is some set of integers means that the operands are restricted to the space that is symmetric with respect to the exchanges of any indices in the set J. This identity (19) guarantees that the operator $d_{1,J}$ does not depend on the order of the elements of a set J when it acts on symmetric functions and hence operation of the raising operators on symmetric functions yields symmetric functions. The function generated by the following Rodrigues formula,

$$j_\lambda(x; \omega, 1/a) = C_\lambda^{-1}(b_N^+)^{\lambda_N}(b_{N-1}^+)^{\lambda_{N-1}-\lambda_N} \cdots (b_1^+)^{\lambda_1 - \lambda_2} \cdot 1,$$

with the normalization constant C_λ given by

$$C_\lambda = \prod_{k=1}^{N-1} C_k(\lambda_1, \lambda_2, \ldots, \lambda_{k+1}; a),$$

where

$$C_k(\lambda_1, \lambda_2, \ldots, \lambda_{k+1}; a)$$
$$= (a)_{\lambda_k - \lambda_{k+1}}(2a + \lambda_{k-1} - \lambda_k)_{\lambda_k - \lambda_{k+1}} \cdots (ka + \lambda_1 - \lambda_k)_{\lambda_k - \lambda_{k+1}},$$

satisfies the definition of the Hi-Jack polynomial $j_\lambda(x; \omega, 1/a)$. The symbol $(\beta)_n$ in the above expression is the Pochhammer symbol, that is, $(\beta)_n = \beta(\beta + 1) \cdots (\beta + n - 1)$, $(\beta)_0 \overset{\text{def}}{=} 1$.

The first seven Hi-Jack polynomials are, for instance, given as follows:

$$j_0(x;\omega,1/a) = J_0(x;1/a) = m_0(x) = 1,$$

$$j_1(x;\omega,1/a) = J_1(x;1/a) = m_1(x),$$

$$j_{1^2}(x;\omega,1/a) = J_{1^2}(x;1/a) + \frac{a}{2\omega}\frac{N(N-1)}{2}J_0(x;1/a)$$

$$= m_{1^2}(x) + \frac{a}{2\omega}\frac{N(N-1)}{2}m_0(x),$$

$$(a+1)j_2(x;\omega,1/a) = (a+1)J_2(x;1/a) - \frac{1}{2\omega}N(Na+1)J_0(x;1/a)$$

$$= (a+1)m_2(x) + 2am_{1^2}(x) - \frac{1}{2\omega}N(Na+1)m_0(x),$$

$$j_{1^3}(x;\omega,1/a) = J_{1^3}(x;1/a) + \frac{1}{2\omega}a\frac{(N-1)(N-2)}{2}J_1(x;1/a)$$

$$= m_{1^3}(x) + \frac{1}{2\omega}a\frac{(N-1)(N-2)}{2}m_1(x),$$

$$(2a+1)j_{2,1}(x;\omega,1/a)$$

$$= (2a+1)J_{2,1}(x;1/a) - \frac{1}{2\omega}(1-a)(N-1)(Na+1)J_1(x;1/a)$$

$$= (2a+1)m_{2,1}(x) + 6am_{1^3}(x) - \frac{1}{2\omega}(1-a)(N-1)(Na+1)m_1(x),$$

$$(a^2+3a+2)j_3(x;\omega,1/a)$$

$$= (a^2+3a+2)J_3(x;1/a) - \frac{3}{2\omega}(a^2N^2+3aN+2)J_1(x;1/a)$$

$$= (a^2+3a+2)m_3(x) + 3a(a+1)m_{2,1}(x) + 6a^2m_{1^3}(x)$$

$$- \frac{3}{2\omega}(a^2N^2+3aN+2)m_1(x),$$

where the symbol $J_\lambda(x;1/a)$ denotes the Jack polynomial. The explicit forms also show the fact that the Hi-Jack polynomial reduces to the Jack polynomial in the limit, $\omega \to \infty$,

$$j_\lambda(x;\omega=\infty,1/a) = J_\lambda(x;1/a). \tag{20}$$

Besides the above relation, we have some other relations between the Hi-Jack and Jack polynomials [18, 19, 26]. While the Hi-Jack polynomial is a one-parameter deformation of the Jack polynomial, we can get the Hi-Jack polynomial from the Jack polynomial by the following formula,

$$J_\lambda(\alpha_1^\dagger,\alpha_2^\dagger,\ldots,\alpha_N^\dagger;1/a) \cdot 1 = j_\lambda(x;\omega,1/a),$$

which gives another relation between the Jack polynomials and the Hi-Jack polynomials. In the above expansion, we have an observation, and it is generally true, that increasing the order of $1/2\omega$ by one causes the weight of the symmetrized monomial to decrease by two. The fact yields an stronger form of the triangularity:

$$j_\lambda(\boldsymbol{x}; \omega, 1/a) = \sum_{\mu \overset{d}{\leq} \lambda, \, |\mu| \equiv |\lambda| \bmod 2} \left(\frac{1}{2\omega}\right)^{(|\lambda|-|\mu|)/2} w_{\lambda\mu}(a) m_\mu(\boldsymbol{x}),$$

(21)

$$w_{\lambda\lambda}(a) = 1.$$

Combining Eqs. (20) and (21), we have the following expansion form of the Hi-Jack polynomial with respect to the Jack polynomial:

$$j_\lambda(\boldsymbol{x}; \omega, 1/a) = J_\lambda(\boldsymbol{x}; \omega, 1/a) + \sum_{\substack{\mu \overset{d}{\leq} \lambda, |\mu| < |\lambda| \\ |\mu| \equiv |\lambda| \bmod 2}} \left(\frac{1}{2\omega}\right)^{(|\lambda|-|\mu|)/2} \mathsf{w}_{\lambda\mu}(a) J_\mu(\boldsymbol{x}; 1/a).$$

Relationship between the Hi-Jack polynomials and the Perelomov basis (10) is given as follows. The power-sum creation operator B_k^+ is cast into the power-sum of $\alpha_l^\dagger$-operators,

$$B_k^\dagger = (2\mathrm{i}\omega)^k \sum_{l=1}^{N} (\alpha_l^\dagger)^k \big|_{\mathrm{Sym}} \overset{\text{def}}{=} (2\mathrm{i}\omega)^k \mathsf{p}_k(\boldsymbol{\alpha}^\dagger) \big|_{\mathrm{Sym}},$$

and the Perelomov basis is expressed by the power-sum of the $\alpha_l^\dagger$-operators as

$$\frac{\langle \boldsymbol{x} \mid \lambda \rangle}{\langle \boldsymbol{x} \mid 0 \rangle} = (2\mathrm{i}\omega)^{|\lambda|} \prod_{k=1}^{N} \left(\mathsf{p}_k(\boldsymbol{\alpha}^\dagger)\right)^{\lambda_k - \lambda_{k+1}} \cdot 1 = (2\mathrm{i}\omega)^{|\lambda|} \mathsf{p}_\lambda(\boldsymbol{\alpha}^\dagger) \cdot 1.$$

Thus the transition matrix between the power-sums and Jack polynomials $M(J, \mathsf{p})$,

$$J_\lambda(\boldsymbol{x}, 1/a) = \sum_{\substack{\mu \\ |\mu|=|\lambda|}} M(J, \mathsf{p})_{\lambda\mu} \mathsf{p}_\mu(\boldsymbol{x}),$$

gives a relation between the Hi-Jack polynomials and the Perelomov basis,

$$j_\lambda(\boldsymbol{x}; \omega, 1/a) = (2\mathrm{i}\omega)^{-|\lambda|} \sum_{\substack{\mu \\ |\mu|=|\lambda|}} M(J, \mathsf{p})_{\lambda\mu} \frac{\langle \boldsymbol{x} \mid \mu \rangle}{\langle \boldsymbol{x} \mid 0 \rangle}.$$

We have introduced the Hi-Jack polynomials as the simultaneous eigenfunctions for the first two commuting conserved operators with the triangularity. As we shall see shortly, they are nondegenerate simultaneous eigenfunctions for all the commuting conserved operators of the Calogero model. From a calculation of the action of d_l operator on a symmetrized monomial of $\alpha_k^\dagger$'s, $m_\lambda(\alpha_1^\dagger, \ldots, \alpha_N^\dagger)$, we can prove the following expression:

$$I_n j_\lambda(\boldsymbol{x}; \omega, 1/a) = \sum_{\substack{D \\ \mu \leq \lambda \\ \text{or } |\mu| < |\lambda|}} w'_{\lambda,\mu}(a, 1/2\omega) m_\mu(\boldsymbol{x}). \tag{22}$$

This means that operation of the conserved operators on the Hi-Jack polynomials keeps their triangularity. Since the nth conserved operator commutes with the first and second conserved operators, $[I_1, I_n] = [I_2, I_n] = 0$, we can easily verify,

$$I_1 I_n j_\lambda(\boldsymbol{x}; \omega, 1/a) = E_1(\lambda) I_n j_\lambda(\boldsymbol{x}; \omega, 1/a), \tag{23}$$
$$I_2 I_n j_\lambda(\boldsymbol{x}; \omega, 1/a) = E_2(\lambda) I_n j_\lambda(\boldsymbol{x}; \omega, 1/a). \tag{24}$$

Equations (23), (24), and (22) for $I_n j_\lambda$ are, respectively, the same as (15), (16), and (17) for the Hi-Jack polynomial j_λ, which means $I_n j_\lambda$ satisfies the definition of the Hi-Jack polynomial except for normalization. Our definition of the Hi-Jack polynomial uniquely specifies the Hi-Jack polynomial. So we conclude that $I_n j_\lambda$ must coincide with j_λ up to normalization. Thus we confirm that the Hi-Jack polynomials j_λ simultaneously diagonalize all the commuting conserved operators I_n, $n = 1, \ldots, N$. The eigenvalues of the conserved operators,

$$I_n j_\lambda(\boldsymbol{x}; \omega, 1/a) = E_n(\lambda) j_\lambda(\boldsymbol{x}; \omega, 1/a),$$

are generally polynomials of the coupling parameter a:

$$E_n(a) = e_n^{(0)}(\lambda) + e_n^{(1)}(\lambda) a + \ldots.$$

It is easy to get the constant (a-independent) term $e_n^{(0)}(\lambda)$ because the term corresponds to the nth eigenvalue for N noninteracting bosons confined in an external harmonic well:

$$e_n^{(0)}(\lambda) = \sum_{k=1}^{N} (\lambda_k)^n.$$

It is clear that there is no degeneracy in the constant terms of the eigenvalues $\{e_n^{(0)}(\lambda) \mid n = 1, \ldots, N\}$. Since the conserved operators I_n are Hermitian operators concerning the inner product,

$$\langle j_\lambda, j_\mu \rangle = \int_{-\infty}^{\infty} \prod_{k=1}^{N} \mathrm{d}x_k |\langle \boldsymbol{x} \mid 0 \rangle|^2 j_\lambda j_\mu \propto \delta_{\lambda,\mu},$$

the Hi-Jack polynomials are the orthogonal symmetric polynomials with respect to the above inner product. From the explicit form of the weight function,

$$|\langle \boldsymbol{x} \mid 0 \rangle|^2 = \prod_{\substack{j,k=1 \\ j<k}}^{N} |x_j - x_k|^{2a} \exp\left(-\omega \sum_{l=1}^{N} x_l^2\right),$$

we conclude that the Hi-Jack polynomial is a multivariable generalization of the Hermite polynomial [11].

Acknowledgments: One of the authors (M.W.) thanks Professor L. Vinet and workshop organizers for their warm hospitality during the conference. The other author (H.U.) appreciates a Research Fellowship of the Japan Society for the Promotion of Science for Young Scientists.

6   REFERENCES

1. J. Avan, *Integrable extensions of the rational and trigonometric A_N Calogero–Moser potentials*, Phys. Lett. A **185** (1994), No. 3, 293–303.

2. J. Avan and A. Jevicki, *Classical integrability and higher symmetries of collective string field theory*, Phys. Lett. B **266** (1991), No. 1-2, 35–41.

3. ———, *Quantum integrability and exact eigenstates of the collective string field theory*, Phys. Lett. B **272** (1991), No. 1-2, 17–24.

4. ———, *Algebraic structures and eigenstates for integrable collective field theories*, Commun. Math. Phys. **150** (1992), No. 1, 149–166.

5. J. Avan and M. Talon, *Classical R-matrix structure for the Calogero model*, Phys. Lett. B **303** (1993), No. 1-2, 33–37.

6. G. Barucchi and T. Regge, *Conformal properties of a class of exactly solvable N-body problems in space dimension one*, J. Math. Phys. **18** (1977), 1149–1153.

7. F. Calogero, *Solution of the one-dimensional N-body problems with quadratic and/or inversely quadratic pair potentials*, J. Math. Phys. **12** (1971), 419–436.

8. C. F. Dunkl, *Differential-difference operators associated to reflection groups*, Trans. Amer. Math. Soc. **311** (1989), No. 1, 167–183.

9. H. Jack, *A class of symmetric polynomials with a parameter*, Proc. Royal Soc. Edinburgh Sect. A. **69** (1970/71), 1–18.

10. L. Lapointe and L. Vinet, *Exact operator solution of the Calogero–Sutherland model*, Commun. Math. Phys. **178** (1996), No. 2, 425–452, hep-th/9507073.

11. M. Lassalle, *Polynômes de Hermite généralisés*, C. R. Acad. Sci. A **313** (1991), No. 9, 579–582.

12. J. Moser, *Three integrable Hamiltonian systems connected to isospectral deformations*, Adv. Math. **16** (1975), 197–220.

13. A. M. Perelomov, *Algebraic approach to the solution of a one-dimensional model of N interacting particles*, Theoret. and Math. Phys. **6** (1971), 263–282.

14. A. P. Polychronakos, *Exchange operator formalism for integrable systems of particles*, Phys. Rev. Lett. **69** (1992), No. 5, 703–705.

15. R. P. Stanley, *Some combinatorial properties of Jack symmetric functions*, Adv. Math. **77** (1989), No. 1, 76–115.

16. B. Sutherland, *Exact results for a quantum many-body problem in one dimension*, Phys. Rev. A **4** (1971), 2019–2021.

17. _____, *Quantum many-body problem in one dimension: ground state*, J. Math. Phys. **12** (1971), 246–250.

18. H. Ujino, *Algebraic study on the quantum Calogero model*, Ph.D. thesis, Univ. of Tokyo, 1996.

19. _____, *Orthogonal symmetric polynomials associated with the Calogero model*, Extended and Quantum Algebras and their Applications in Mathematical Physics (Tianjin, 1996) (M.-L. Ge, Y. Saint-Aubin, and L. Vinet, eds.), Publications CRM, Montréal, to appear.

20. H. Ujino, K. Hikami, and M. Wadati, *Integrability of the quantum Calogero–Moser model*, J. Phys. Soc. Japan **61** (1992), No. 10, 3425–3427.

21. H. Ujino and M. Wadati, *The quantum Calogero model and the W-algebra*, J. Phys. Soc. Japan **63** (1994), No. 10, 3585–3597.

22. _____, *The SU(ν) quantum Calogero model and W_∞ algebra with SU(ν) symmetry*, J. Phys. Soc. Japan **64** (1995), No. 1, 39–56.

23. _____, *Orthogonal symmetric polynomials associated with the quantum Calogero model*, J. Phys. Soc. Japan **64** (1995), No. 8, 2703–2706.

24. _____, *Orthogonality of the Hi–Jack polynomials associated with the Calogero model*, J. Phys. Soc. Japan **66** (1995), No. 2, 345–350.

25. _____ , *Algebraic construction of the eigenstates for the second conserved operator of the quantum Calogero model*, J. Phys. Soc. Japan **65** (1996), No. 3, 653–656.

26. _____ , *Rodrigues formula for hi-Jack symmetric polynomials associated with the quantum Calogero model*, J. Phys. Soc. Japan **65** (1996), No. 8, 2423–2439.

27. H. Ujino, M. Wadati, and K. Hikami, *The quantum Calogero-Moser model: algebraic structures*, J. Phys. Soc. Japan **62** (1993), No. 9, 3035–3043.

35

The Complex Calogero–Moser and KP Systems

George Wilson

1   Introduction

The following remarkable fact has been known since about 1977 (see Refs. 1, 2, 14, and 15). Suppose we seek (complex) solutions of the Kadomtsev–Petviashvili (KP) equation,

$$\frac{3}{4}u_{yy} = \left\{u_t - \frac{1}{4}(u_{xxx} + 6uu_x)\right\}_x, \tag{1}$$

that are rational in x and vanish as $x \to \infty$; such solutions have the form

$$u(x, y, t) = -2\sum_1^n \{x - x_i(y, t)\}^{-2}$$

for some integer n. Then the dynamics with respect to y and t of the positions of the poles x_i is described (up to some constant factors) by the first two nontrivial flows of the rational n-particle Calogero–Moser hierarchy. In particular, this means that as functions of y, the x_i move like a classical particle system with inverse square potential, that is, with Hamiltonian

$$H = \frac{1}{2}\sum_1^n p_i^2 - \sum_{i<j}(x_i - x_j)^{-2}. \tag{2}$$

In this Chapter I want to uncover something of the geometry underlying this fact; the most essential new observation is that, in a sense that will be clarified in Section 2 below, the correspondence between the Calogero–Moser and KP systems extends even to the collision locus (of course, since the Calogero–Moser Hamiltonians are singular when any of the x_i coincide, the statement above makes sense at first only up to the point where some particles collide). The text follows closely my talk at the conference; it is based mainly on my paper [25], where the details can be found.

2   The Completed Phase Space for the Complex Calogero–Moser Flows [13]

V_n denote the vector space of all quadruples of matrices $(X, Z; v, w)$, where X and Z are $n \times n$, v is $n \times 1$, and w is $1 \times n$. Let the group $\mathrm{GL}(n, \mathbb{C})$ act on V_n by

$$g \circ (X, Z; w) = (gXg^{-1}, gZg^{-1}; gv, wg^{-1}). \tag{3}$$

If we regard V_n as a cotangent space by equipping it with the 2-form

$$\omega = \mathrm{tr}\{dZ \wedge dX + dv \wedge dw\},$$

then the action (3) of $\mathrm{GL}(n, \mathbb{C})$ is symplectic, and the moment map for it is

$$\mu(X, Z; v, w) = [X, Z] - vw.$$

Let $\widetilde{C}_n = \mu^{-1}(-I)$. It is easy to show that $\mathrm{GL}(n, \mathbb{C})$ acts freely on $\widetilde{C}_n$, so we can form the symplectic quotient $C_n = \widetilde{C}_n / \mathrm{GL}(n, \mathbb{C})$; it is a smooth irreducible affine algebraic variety with a holomorphic symplectic form induced from ω.

Over the dense open subset C'_n of C_n where X has distinct eigenvalues, the fibration $\widetilde{C}_n \to C_n$ is trivial: indeed, in each such $\mathrm{GL}(n, \mathbb{C})$-orbit we can find a representative such that

i) X is diagonal, say $X = \mathrm{diag}(x_1, x_2, \ldots, x_n)$;

ii) $v^T = w = (1, 1, \ldots, 1)$; and hence

iii) Z is a *Calogero–Moser matrix* with entries

$$Z_{ii} = p_i \text{ (say)}; \quad Z_{ij} = (x_i - x_j)^{-1} \text{ for } i \neq j;$$

this representative is unique up to simultaneous permutation of the parameters (p_i, x_i). In the induced symplectic structure on C'_n the coordinates (p_i, x_i) are canonically conjugate (as the notation implies); so we may identify C'_n with the phase space for a system of n indistinguishable and noncolliding particles: the positions of the particles are thus the eigenvalues of X. The crucial remark (see Refs. 17 and 18) that the Hamiltonian (2) is equal to $\frac{1}{2}\mathrm{tr}\,Z^2$ now implies that the Calogero–Moser flow on C'_n comes from the linear flow

$$X \mapsto X + tZ; \quad Z, v, \text{ and } w \text{ constant}$$

on the unreduced space $\widetilde{C}_n$. More generally, the various commuting flows of the hierarchy, with Hamiltonians $(-1)^{k-1}\mathrm{tr}\,Z^k$, come from similarly trivial flows

$$X \mapsto X + kt(-Z^{k-1}; \quad Z, v \text{ and } w \text{ constant}$$

on $\widetilde{C}_n$. These flows are clearly complete (that is, they exist for all complex time); it follows that the induced flows on C_n are also complete. Since the complement of C'_n in C_n is made up of the (classes of) quadruples $(X, Z; v, w)$ where X has a repeated eigenvalue, we may interpret C_n as giving a description of possible ways in which the Calogero–Moser particles may collide. More formally, C_n is one solution to the problem of finding a holomorphic symplectic manifold containing the classical phase space C'_n and such that all the (incomplete) Calogero–Moser flows on C'_n extend to complete ones on C_n.

Before discussing the connexion with the KP equations, I want to say a little about the topology of the space C_n, which is not at all trivial. We say above (using the parameters (p_i, x_i)) that C'_n is symplectically isomorphic to the cotangent bundle $T^*(\mathbb{C}^{(n)} \backslash \Delta)$, where $\mathbb{C}^{(n)}$ denotes the nth symmetric power of $\mathbb{C}$, and Δ is the "big diagonal" (consisting of (unordered) n-tuples of points of $\mathbb{C}$ that are not all distinct). The most obvious slightly larger symplectic manifold (cf. Ref. 23) is the contractible space $T^*\mathbb{C}^{(n)}$, and it might be natural to guess that this is our C_n; however, that is quite wrong. The space C_n arises (in a slightly disguised form) in recent work of H. Nakajima; I have to refer to Ref. 20 (see also Refs. 19 and 25) for a detailed account of this viewpoint, but the situation is as follows. First, the holomorphic symplectic structure that we have found on C_n is part of a hyper-Kähler structure. Indeed, that is not surprising; the origin of the hyper-Kähler structure is that we can make the complex vector space of quadruples $(X, Z; v, w)$ into a quaternionic vector space by letting the quaternion j act by

$$j \circ (X, Z; v, w) = (Z^*, -X^*; w^*, -v^*).$$

What is interesting, however, is that when we change the complex structure on C_n to a different one in the hyper-Kähler family, we get a fairly well studied space, namely, the Hilbert scheme parametrizing families of n points in the affine plane $\mathbb{C}^2$. This space has quite a lot of homology (all in even dimensions, see Refs. 10, 19, and 20): its Betti numbers are given by

$$b_{2i} = \text{(number of partitions of } n \text{ into } n - i \text{ parts)}.$$

3 Geometry of Rational KP Solutions

Coming back to the Calogero–Moser flows on C_n, if we denote the time variable for the ith flow by t_i, we may write down the simultaneous solution to all the flows of the hierarchy: it is given by

$$X(\mathbf{t}) = X + \sum_{i=1}^{\infty} i t_i (-Z)^{t-1}. \tag{4}$$

One formultation of the connexion between Calogero–Moser and KP is the following.

Theorem 1 (see Ref. 22). *The determinant* $\det X(\mathbf{t})$ *is a τ-function for the KP hierarchy.*

I shall not explain here exactly what that means; but it implies (among other things) that the function

$$u(t_1, t_2, t_3) = 2\frac{\partial^2}{\partial t_1^2} \log \det X(t_1, t_2, t_3, 0, \dots)$$

is a solution to the KP equation (1) if we indentify $t_1 = x$, $t_2 = y$, $t_3 = t$. It is clear that these solutions are rational (in all three variables) and that they vanish as $x \to \infty$; and also that the motion of the poles is described by the Calogero–Moser equations.

Rather than discuss the KP hierarchy, I confine myself here to describing the space that parametrizes the class of rational solutions that interests us. These solutions come (in the style of Sato, see Refs. 3–6, 6, 7, 12 and 21), from a certain "adelic" Grassmannian $\mathrm{Gr}^{\mathrm{ad}}$ that is constructed from the space $\mathbb{R}$ of rational functions in one variable z and its decomposition $\mathbb{R} = \mathbb{R}_+ \oplus \mathbb{R}_-$, where $\mathbb{R}_+$ is the space of polynomials in z amd $\mathbb{R}_-$ is the space of rational functions that vanish at infinity. The construction of $\mathrm{Gr}^{\mathrm{ad}}$ can be split into several simple steps, as follows.

Step 1. Fix a complex number λ and an integer $k \geq 0$, and let $\mathrm{Gr}_\lambda(k)$ be the space of all subspaces $W \subset \mathbb{R}$ such that

$$(z - \lambda)^k \mathbb{R}_+ \subset W \subset (z - \lambda)^{-k} \mathbb{R}_+, \tag{5}$$

and the codimensions of these inclusions are equal to k. Clearly, each $\mathrm{Gr}_\lambda(k)$ is isomorphic to the finite-dimensional Grassmanian of k-dimensional subspaces of $\mathbb{C}^{2k}$.

Step 2. Let Gr_λ be the union of all the $\mathrm{Gr}_\lambda(k)$ for all $k \geq 0$.

Each Gr_λ is isomorphic to the standard model for the classifying space BU, and is a very well understood space; for example, its Betti numbers are given by

$$b_{2i}(\mathrm{Gr}_\lambda) = \{\text{number of partitions of } i\}.$$

Step 3. Now choose a point of Gr_λ for each of a finite number of values of $\lambda \in \mathbb{C}$: this is a point of $\mathrm{Gr}^{\mathrm{ad}}$.

Although this description of $\mathrm{Gr}^{\mathrm{ad}}$ makes it very clear what sort of space it is, we need also a more concrete one that realizes each point of it as a space of rational functions (as was the case for the individual Gr_λ). From (5) it is clear that any $W \in \mathrm{Gr}_\lambda$ can be obtained in the following way: we choose some number $k = k(\lambda)$ of homogeneous linear conditions on the

Laurent series of a function at λ, and let W be the space of all rational functions that are regular on $\mathbb{C}$ except (possibly) for a pole of order $\leq k$ at λ, and that satisfy the k chosen conditions. Now, if we are given a point W of $\mathrm{Gr}^{\mathrm{ad}}$, that is, a point W_λ of Gr_λ for each of a finite number of $\lambda \in \mathbb{C}$, we can identify W with the space of rational functions that satisfy all the conditions defining each of the W_λ. An example should make the construction clear.

Example 1. Suppose that at each of a finite number of distinct points λ_1, $\ldots$, λ_n we let W_{λ_i} be the space of rational functions that are regular on $\mathbb{C}$ except for a simple pole at λ_i, and whose Laurent series λ_i satisfy the single condition

$$(\text{constant term}) + \alpha_i(\text{residue}) = 0,$$

where the α_i are some other chosen constants. Then the corresponding point W of $\mathrm{Gr}^{\mathrm{ad}}$ is the space of rational functions that are regular everywhere on $\mathbb{C}$ except for possible simple poles at each of the points λ_i, and whose Laurent series at each λ_i satisfy the condition above.

From W (or from the condtions that define W, cf. Ref. 14) we can construct its *Baker function* $\psi_W(\mathbf{t}, z)$: by definition, this is the unique function of the form

$$\psi_W(\mathbf{t}, z) = \exp\left(\sum_1^\infty t_i z^i\right)\left\{1 + \sum_1^\infty a_i(\mathbf{t}) z^{-i}\right\}$$

that belongs to a suitable closure of W for each value of $\mathbf{t} \equiv (t_1, t_2, \ldots)$. The function $-\partial/\partial t_1\big(a_1(t_1, t_2, t_3, 0, \ldots)\big)$ is then a solution of the KP equation (see Refs. [3–6, 6, 7, 12] and 21): more generally, translation of any of the t_i in the coefficients $a_r(\mathbf{t})$ corresponds to the so-called KP flows $W \longmapsto \exp(t_i z^i) W$ on $\mathrm{Gr}^{\mathrm{ad}}$. In the case when W is one of the spaces in example 1 above, parametrized by the numbers (λ_i, α_i), it is quite easy to calculate ψ_W: the answer is

$$\psi_W(\mathbf{t}, z) = \exp\left(\sum_1^\infty t_i z^i\right)\left\{1 - wX(\mathbf{t})^{-1}(Z_z I)^{-1} v\right\}, \qquad (6)$$

where $X(\mathbf{t})$ is given by the formula (4), but with the quadruple $(X, Z; v, w)$ belonging to the part C_n^d of C_n where Z (rather than X, as before) is diagonalizable: precisely, $Z = \mathrm{diag}(-\lambda_1, \ldots, -\lambda_n)$, v an w have all their entries equal to 1, and X has entries

$$X_{ii} = \alpha_i, \quad X_{ij} = (\lambda_i, \lambda_j)^{-1} \text{ for } i \neq j.$$

It follows that assigning to this quadruple the space W whose Baker function is given by (6) defines a map from C_n^d to $\mathrm{Gr}^{\mathrm{ad}}$, such that the Calogero–Moser flows on C_n^d correspond to the KP flows on Gr^{ad}.

Theorem 2.   (i) *The formula* (6) *in fact defines a map from the whole of* C_n *into* $\mathrm{Gr}^{\mathrm{ad}}$.

(ii) *Together these maps define a* bijection:

$$\bigcup_{n \geq 0} C_n \to \mathrm{Gr}^{\mathrm{ad}}.$$

Thus we have a kind of stratification of $\mathrm{Gr}^{\mathrm{ad}}$ in which the differences of successive strata are the hyper-Kähler varieties C_n. It is natural to ask: what is the geometrical meaning of the number n in this decomposition? The answer to this question is implicit in Sato's remark (see Refs. 3–6, 6, 7, 12, and 21) that the τ-function of a space W has an expansion

$$\tau_W(\mathbf{t}) = \sum_{\nu} \pi_{\nu} S_{\nu}(\mathbf{t}), \tag{7}$$

where the sum is over all partitions ν. Here S_{ν} is the Schur function of ν, and the π_{ν} are Plücker coordinates; the notation is that $p_i = -it_i$, where the p_i are the usual power sum variables. Now, the Grasmannian Gr_0 has a cell decomposition in which the n-dimensional cells are indexed by partitions of n; the open cell containing W is indexed by the unique nonzero π_{ν} of (maximal) weight n in the sum (6). It follows that for $W \in \mathrm{Gr}_0$, the weight n of ν is the degree of τ_W as a polynomial in t_1. Comparing with (4), we see that n is also the size of the matrices X and Z corresponding to W. In general the situation is similar: recall that $\mathrm{Gr}^{\mathrm{ad}}$ is a union of products of various $\mathrm{Gr}_\lambda(\lambda \in \mathbb{C})$, and each Gr_λ is isomorphic to Gr_0; hence $\mathrm{Gr}^{\mathrm{ad}}$ acquires a natural cell structure.

Theorem 3. *The image of the map* $C_n \to \mathrm{Gr}^{\mathrm{ad}}$ *in Theorem 2 is the union of all the open cells of dimension* n *in* $\mathrm{Gr}^{\mathrm{ad}}$.

The special spaces described in Example 1 also have a simple description in terms of cells: they are the points of $\mathrm{Gr}^{\mathrm{ad}}$ obtained by taking products of the (unique) 1-dimensional cells in the various Gr_λ. This may appear paradoxical, since the products of 1-cells form only a tiny part of $\mathrm{Gr}^{\mathrm{ad}}$, and C_n^d is a dense open subset of C_n. The resolution of the paradox is that if we topologize $\mathrm{Gr}^{\mathrm{ad}}$ as a cell complex, then the maps $C_n \to \mathrm{Gr}^{\mathrm{ad}}$ are not continuous: in fact they seem to be the opposite of continuous, in the sense that the smaller a part of C_n we look at, the larger the part of $\mathrm{Gr}^{\mathrm{ad}}$ that corresponds to it. Nevertheless, $\mathrm{Gr}^{\mathrm{ad}}$ does have another topology, in which these maps *are* continuous: it can be defined by regarding $\mathrm{Gr}^{\mathrm{ad}}$ as a subspace of some larger Grassmannian, say, the Hilbert space Grassmannian of Ref. 21. It seems to me an unusual feature of the Theorems 2 and 3 that it is hard to get any sense out of them without thinking simultaneously of these two topologies on $\mathrm{Gr}^{\mathrm{ad}}$: we need the cell-complex topology to give the simple characterization of the stratification in Theorem 3, and the other one to see that the differences of the strata form the beautiful varieties C_n.

4   Applications

4.1

The main application of Theorem 2, and the original motivation for the
work presented here, is to construct commutative algebras of *bispectral* linear differential operators. On the spaces C_n we have an obvious involution
b defined by

$$b(X, Z; v, w) = (Z^T, X^T; w^T, v^T).$$

The corresponding involution b on $\mathrm{Gr}^{\mathrm{ad}}$ looks very mysterious if one does
not know (5) (see Ref. 24). Setting $t_1 = x$, $t_2 = t_3 = \cdots = 0$ in (6), we
obtain the *stationary Baker function*:

$$\psi_W(x, z) = \{1 - w(X + xI)^{-1}(Z + zI)^{-1}v\}e^{xz}; \qquad (8)$$

from this formula it is clear that b acts on the Baker function by interchaging the roles of the variables x and z. It is well known (see Refs. 3–6, 6, 7, 12,
and 21) that ψ_W serves as a family of joint eigenfunctions for a commutative algebra of differential operators in the variable x (with multipliers that
are functions of z); the symmetry that we have just described shows that
equally ψ_W serves as a family of joint eigenfunctions for a commutative
algebra of differential operators in the variable z (with multipliers that are
functions of x). This is the phenomenon of *bispectrality* (see Ref. 9).

4.2

An unexpected application is to the theory of Schur functions; this is obtained by using the formula (4) in the case when X and Z are both nilpotent
(the corresponding spaces W are the centers of the cells in Gr_0). In this
case all but one of the Plücker coordinates in (7) is zero, so that τ_W is a
(scalar multiple of) a Schur function; thus (4) gives us a formula expressing
the Schur function of any partition of n as an $n \times n$ determinant involving the power sum variables. The formula obtained is well known in the
case when the Schur function is one of the elementary symmetric functions
(see Ref. 16, Chapter 1, Section 2, Example 8). but in general it appears
to be new.

5   The Spin Generalization

The results in this note are presumably the basic case of some fairly general
phenomenon. Perhaps the simplest generalization would be to the Sutherland system (cf. Ref. 13); instead, let us consider briefly what is likely to
happen when we modify the construction of Section 2 by letting v and w

be $n \times r$ and $r \times n$ matrices (respectively) for any $r \geq 1$. This leads to the "spin generalization" of the Calogero–Moser system first discussed (in exactly this way) by Gibbons and Hermesen [11]. The corresponding completed phase spaces $C_n(r)$ are again hyper-Kähler: changing the complex structure again gives us a known space, the moduli space of framed (that is, trivialized over the line at infinity) rank r torsion free sheaves over $\mathbb{CP}^2$ with $c_2 = n$ (cf. Refs. 19, 20, and 25); we recall (see Ref. 8) that the open subset of locally free sheaves is the space of framed $SU(r)$ instantons. The commutative group of Calogero–Moser flows that we had for $r = 1$ gets replaced by an action on $C_n(r)$ of the loop group $L^+ \, GL(r, \mathbb{C})$ of (say) entire functions of z with values in $GL(r, \mathbb{C})$. On the other side, the scalar KP hierarchy has to be replaced by its $r \times r$ matrix ("multicomponent") analog; the stationary Baker function is still given by the formula (8), and the relevant adelic Grassmannian $\mathrm{Gr}^{\mathrm{ad}}(r)$ is constructed just as before, but working with row vectors of length r of rational functions. The most interesting thing about this is that, just as the vector space $\mathbb{C}^\infty$ is isomorphic to $(\mathbb{C}^r)^\infty$, so $\mathrm{Gr}^{\mathrm{ad}}(r)$ is (probably) homeomorphic to the space $\mathrm{Gr}^{\mathrm{ad}}$ that we had in the scalar case. Thus the same Grassmannian $\mathrm{Gr}^{\mathrm{ad}}$ serves as a "generating space" (cf. Ref. 20) for any one of the sequence of spaces $\{M(n, r) : n \geq 0\}$, irrespective of what r is.

6 REFERENCES

1. H. Airault, H. P. McKean, and J. Moser, *Rational and elliptic solutions of the Korteweg–de Vries equation and a related many-body problem*, Commun. Pure Appl. Math. **30** (1977), No. 1, 95–148.

2. D. V. Chudnovsky and G. V. Chudnovsky, *Pole expansions of nonlinear partial differential equations*, Nuovo. Cimento **40B** (1977), No. 2, 339–350.

3. E. Date, M. Jimbo, M. Kashiwara, and T. Miwa, *Transformation groups for soliton equations. III. Operator approach to the Kadomtsev–Petviashvili equation*, J. Phys. Soc. Japan **50** (1981), No. 11, 3806–3812.

4. _____, *Transformation groups for soliton equations. VI. KP hierarchies of orthogonal and symplectic type*, J. Phys. Soc. Japan **50** (1981), No. 11, 3813–3818.

5. _____, *Transformation groups for soliton equations. IV. A new hierarchy of soliton equations of KP-type*, Physica **4D** (1981/1982), No. 3, 343–365.

6. _____, *Transformation groups for soliton equations. V. Quasiperiodic solutions of the orthogonal KP equation*, Publ. Res. Inst. Math. Sci. **18** (1982), No. 3, 1111–1119.

7. E. Date, M. Kashiwara, and T. Miwa, *Transformation groups for soli-ton equations. II. Vertex operators and τ functions*, Proc. Japan Acad. Ser. A Math. Sci. **57** (1981), No. 8, 387–392.

8. S. K. Donaldson, *Instantons and geometric invariant theory*, Commun. Math. Phys. **93** (1984), No. 4, 453–460.

9. J. J. Duistermaat and F. Grünbaum, *Differential equations in the spectral parameter*, Commun. Math. Phys. **103** (1986), No. 2, 177–240.

10. G. Ellingsrud and S. A. Strømme, *On the homology of the Hilbert scheme of points in the plane*, Invent. Math. **87** (1987), No. 2, 343–352.

11. J. Gibbons and T. Hermsen, *A generalisation of the Calogero–Moser system*, Physica **11D** (1984), No. 3, 337–348.

12. M. Kashiwara and T. Miwa, *Transformation groups for soliton equations. I. The τ function of the Kadomtsev–Petviashvili equation*, Proc. Japan Acad. Ser. A Math. Sci. **57** (1981), No. 7, 342–347.

13. D. Kazhdan, B. Kostant, and S. Sternberg, *Hamiltonian group actions and dynamical systems of Calogero type*, Commun. Pure Appl. Math. **31** (1978), No. 4, 481–507.

14. I. M. Krichever, *Rational solutions of the Kadomtsev–Petviashvili equation and integrable systems of N particles on a line*, Funct. Anal. Appl. **12** (1978), No. 1, 59–61.

15. _____, *On the rational solutions of the Zaharov–Shabat equations and completely integrable systems of N particles on the line*, J. Soviet Math. **21** (1983), 335–345.

16. I. G. Macdonald, *Symmetric Functions and Hall Polynomials*, 2nd ed., Oxford Math. Monogr., Oxford Univ. Press, New York, 1995.

17. J. Moser, *Three integrable Hamiltonian systems connected to isospectral deformations*, Adv. Math. **16** (1975), 197–220.

18. _____, *Various aspects of integrable Hamiltonian systems*, Dynamical Systems (Bressanone, 1978), Progr. Math., Vol. 8, Birkhäuser, Boston, MA, 1980, pp. 233–289.

19. H. Nakajima, *Heisenberg algebra and Hilbert schemes of points on projective surfaces*, Ann. of Math. **145** (1997), No. 2, 379–388.

20. _____, *Lectures on Hilbert schemes of points on surfaces*, 1999.

21. G. Segal and G. Wilson, *Loop groups and equations of KdV type*, Inst. Hautes Études Sci. Publ. Math. (1985), No. 61, 5–65.

22. T. Shiota, *Calogero–Moser hierarchy and Kp hierarchy*, J. Math. Phys. **35** (1994), No. 11, 5844–5849.

23. A. Treibich and J.-L. Verdier, *Variétés de kritchever des solitons elliptiques*, Proceedings of the Indo-French Conference on Geometry (Bombay, 1989) (S. Ramanan and A. Beauville, eds.), Hindustan Book Agency, Dehli, 1993, pp. 187–232.

24. G. Wilson, *Bispectral commutative ordinary differential operators*, J. Reine Angew. Math. **442** (1993), 177–204.

25. _____, *Collisions of Calogero–Moser particles and an adelic Grassmannian*, Invent. Math. **133** (1998), No. 1, 1–41.

36

Oscillator $9j$-Symbols, Multidimensional Factorization Method, and Multivariable Krawtchouk Polynomials

Alexei Zhedanov

ABSTRACT It is shown that $9j$-symbols of the oscillator algebra are expressed in terms of two-variable Krawtchouk polynomials. On the basis of multidimensional factorization method we study properties of N-variable Krawtchouk polynomials and show that they are common eigenstates of N commuting difference operators.

1 Introduction

$3nj$-symbols of Lie algebras are very useful tool in numerous theoretical problems. The importance of the study of such objects for pure mathematics and mathematical physics can be illustrated by finding of the Askey–Wilson polynomials [1], which are believed to be the "most general" orthogonal polynomials having nice properties. These polynomials were discovered, in particular, on the basis of known properties of $6j$-symbols for the $su(2)$ algebra (about connection between these objects see, e.g., Refs. 13, 15, and 21). Note that more simple $3j$-symbols (or Clebsch–Gordan coefficients) of su(2) algebra are expressed in terms of the Hahn polynomials [13].

S. Suslov showed [17, 18] that more complicated $9j$-symbols of su(2) algebra can be expressed in terms of some orthogonal polynomials in *two* discrete variables. However, explicit expression for such the polynomials is yet unknown.

On the other hand, some classes of orthogonal multivariable polynomials (and generally nonpolynomial multivariable functions) are intensively studied nowadays with respect to quantum integrable systems and combinatorial problems (see, e.g., Refs. 2, 10, and 11).

We show that $9j$-symbols of the oscillator algebra can be expressed in terms of Krawtchouk polynomials in two discrete arguments.

It is well known that both $3j$- and $6j$-symbols of the oscillator algebra are expressed in terms of the ordinary Krawtchouk polynomials in one discrete variable [8, 20]. So our result can be considered as some nontrivial extension of this result to the polynomials in two variables.

We also generalize 2-variable Krawtchouk polynomials to the multivariable case and show how the main properties of multivariable Krawtchouk polynomials can be found using some nontrivial generalization of the factorization method.

For details see Refs. 22 and 23.

2   Oscillator Algebra, Its Addition Rule, and $9j$-Symbols

The oscillator algebra is described by the commutation relations (we adopt notations that are slightly different from standard ones [12]):

$$[A_-, A_+] = a, \quad [A_0, A_\pm] = \pm A_\pm, \tag{1}$$

where A_0, $A_\pm$ are three generators, a is some positive constant. The Casimir operator of the oscillator algebra has the expression

$$Q = aA_0 - A_+ A_-. \tag{2}$$

Unitary irreducible representations of the algebra (1) are described by the basis $|n; a; \rho\rangle$ for which the action of the generators is described by

$$\begin{aligned}
A_0|n; a; \rho\rangle &= (n + \rho)|n; a; \rho\rangle \\
A_-|n; a; \rho\rangle &= \sqrt{an}|n - 1; a; \rho\rangle \\
A_+|n; a; \rho\rangle &= \sqrt{a(n + 1)}|n + 1; a; \rho\rangle \quad n = 0, 1, \ldots
\end{aligned} \tag{3}$$

where the representation parameter ρ is defined by the value of the Casimir operator on the given representation: $Q|n; a; \rho\rangle = \rho a|n; a; \rho\rangle$. The operators A_- and A_+ are Hermitian conjugated, whereas the operator A_0 is Hermitian given the representation (3). Note that the ordinary bose operators are a special case of the representations of the algebra (1) with $a = 1$, $\rho = 0$.

Now consider addition of two independent oscillator algebras $A_0^{(i)}$, $A_\pm^{(i)}$, a_i and $A_0^{(k)}$, $A_\pm^{(k)}$, a_k, [12] where a_i, a_k are two different arbitrary positive parameters, and all the operators with superscript i commute with the operators with superscript k. The addition rule has the form

$$A_p^{(ik)} = A_p^{(i)} + A_p^{(k)}, \quad a_{ik} = a_i + a_k, \tag{4}$$

where $p = 0, \pm$ and (super)subscript ik indicates the new oscillator algebra that is obtained by adding of two algebras.

The Casimir operator of the resulting algebra has the expression

$$Q_{ik} = a_i A_0^{(k)} + a_k A_0^{(i)} - A_+^{(i)} A_-^{(k)} - A_+^{(k)} A_-^{(i)} + Q_i + Q_k, \qquad (5)$$

where Q_i and Q_k are the Casimir operators of the adding algebras.

One can introduce the coupled basis $|n_{ik}; a_{ik}, \rho_{ik}\rangle$ for the resulting algebra where notations are the same as in (3). From the adding rules (4) we have the obvious relation

$$n_{ik} + \rho_{ik} = n_i + n_k + \rho_i + \rho_k, \qquad (6)$$

where the coupled representation parameter ρ_{ik} takes the values

$$\rho_{ik} = \rho_i + \rho_k + p_{12}, \quad p_{12} = 0, 1, 2 \dots. \qquad (7)$$

From (7) it is clear that the representation of the resulting algebra in coupled basis is uniquely defined by the fixing of 6 parameters ρ_i, ρ_k, a_i, a_k, n_{ik}, p_{12}.

Consider addition of 4 different oscillator algebras. There are two different schemes: $((1 \oplus 2) \oplus (3 \oplus 4))$ and $((1 \oplus 3) \oplus (2 \oplus 4))$. This leads to decomposition

$$|*; p_{13}, p_{24}, p_{13,24}\rangle = \sum_{p_{12}, p_{34}, p_{12,34}} F_{p_{12}, p_{34}, p_{12,34}}^{p_{13}, p_{24}, p_{13,24}} |*; p_{12}, p_{34}, p_{12,34}\rangle, \qquad (8)$$

where $*$ denotes that the decomposition is invariant with respect to changing of the representation numbers n_i. Note that the sum on the left-hand side of (8) is really reduced to a twofold sum, because of restrictions:

$$p_{12} + p_{34} + p_{12,34} = p_{13} + p_{24} + p_{13,24} = N = \rho_{1234} - \rho_1 - \rho_2 - \rho_3 - \rho_4. \qquad (9)$$

In what follows we redenote indices for brevity: $p_{13} = p$, $p_{24} = q$, $p_{13,24} = r$, $p_{12} = k$, $p_{34} = l$, $p_{12,34} = m$. Then the decomposition (8) is rewritten as

$$|*; p, q, r\rangle = \sum_{klm} F_{klm}^{pqr} |*; k, l, m\rangle, \qquad (10)$$

with the restrictions

$$p + q + r = k + l + m = N. \qquad (11)$$

Using method proposed in Ref. 6, one can obtain the identity [22]

$$\Psi^{pr}(u, v; N) = (1 - v - \alpha u)^p (1 + \beta v - \gamma u)^q (1 + u + \delta v)^r$$
$$= \sum_{klm} \Phi_{klm}^{pqr} u^k v^m, \qquad (12)$$

where $\alpha = a_2/a_1$, $\beta = a_4/a_2$, $\gamma = a_4/a_3$, $\delta = (a_1a_4 - a_2a_3)/a_2a_5$, $a_5 = a_1 + a_2 + a_3 + a_4$ and

$$\Phi_{klm}^{pqr} = \widetilde{F}_{klm}^{pqr} \left(\frac{a_2a_4a_{13}}{a_1a_3a_{24}}\right)^k \left(\frac{a_4}{a_{24}}\right)^m \left(-\frac{a_{24}}{a_4}\right)^p \left(\frac{a_{24}}{a_2}\right)^q \left(\frac{a_{12}a_{34}a_{24}}{a_2a_4a_5}\right)^r \qquad (13)$$

are modified $9j$-symbols. Note that the parameters α, β, γ, δ are not independent because of the relation $\delta = \beta(\gamma - \alpha)/(\gamma + \gamma\alpha + \alpha\beta + \gamma\alpha\beta)$.

The formula (12) yields the generation function $\Psi^{pr}(u, v; N)$ for the $9j$-symbols of oscillator algebra. This formula can be exploited in order to get many useful relations for $9j$-symbols.

In particular, from (12) it follows that $9j$-symbols are expressed as

$$\Phi_{klm}^{pqr} = Q_{pr}(k, m; N)\Phi_{klm}^{0N0}, \qquad (14)$$

where $Q_{pr}(k, m; N)$ are Krawtchouk orthogonal polynomials in two discrete variables k, m. The "vacuum amplitude" Φ_{klm}^{0N0} has the simple expression

$$\Phi_{klm}^{0N0} = \frac{N!}{k!l!m!}(-\gamma)^k(\beta)^m. \qquad (15)$$

The polynomials $Q_{pr}(k, m; N)$ possess some useful recurrence relations that can be easily derived from (12). For example

$$L^+(N)Q_{pr}(k, m; N - 1) = NQ_{p+1,r}(k, m; N), \qquad (16)$$

where $L^+(N)$ is the difference operator acting on the space of functions in two variables k, m by the formula

$$L^+(N) = \frac{\alpha k}{\gamma}T_k^- - \frac{m}{\beta}T_m^- + N - k - m, \qquad (17)$$

where we introduced elementary shift operators defined as $T_k^{\pm}Q(k, m) = Q(k \pm 1, m)$, $T_m^{\pm}Q(k, m) = Q(k, m \pm 1)$.

Analogously

$$M^+(N)Q_{pr}(k, m; N - 1) = NQ_{p,r+1}(k, m; N), \qquad (18)$$

where

$$M^+(N) = -\frac{k}{\gamma}T_k^- + \frac{\delta m}{\beta}T_m^- + N - k - m. \qquad (19)$$

Another pair of recurrence relations is

$$L^-Q_{pr}(k, m; N) = \frac{p\varepsilon}{N}Q_{p-1,r}(k, m; N - 1) \qquad (20)$$

and

$$M^-Q_{pr}(k, m; N) = \frac{r\varepsilon}{N}Q_{p,r-1}(k, m; N - 1), \qquad (21)$$

where the operators L^- and M^- are defined by

$$L^- = \gamma(\beta - \delta)T_k^+ - \beta(1 + \gamma)T_m^+ + \beta + \gamma\delta \tag{22}$$

$$M^- = -\gamma(1 + \beta)T_k^+ + \beta(\gamma - \alpha)T_m^+ + \alpha\beta + \gamma \tag{23}$$

From the relations (16), (18), (20), and (21) we can construct successively all the functions $Q_{pr}(k, m; N)$ starting from the trivial vacuum function $Q_{00}(k, m; N) = 1$. It is easily seen that then we indeed arrive at the polynomials in two discrete variables k, m having the degree $p + r$.

From orthogonality relation for 9j-symbols one can get orthogonality relation for the Krawtchouk polynomials $Q_{pr}(k, m; N)$ with respect to trinomial distribution [22]. Note that special class of two-variable Krawtchouk polynomials was constructed by G. Prizva [14]. (See also Ref 3 where some special class of the two-variable Hahn polynomials is constructed, the polynomials found in Ref. 14 can be obtained then by a limiting procedure).

The polynomials $Q_{pr}(k, m; N)$ possess some remarkable properties, in particular there is the Rodriguez formula for them [22] resembling analogous results for another multivariable polynomials connected with integrable systems (see, e.g., Ref. 11). Moreover, there are two commuting difference operators ("Hamiltonians") $H_1(N)$ and $H_2(N)$ such that the polynomials are common eigenfunctions of them [22].

These properties follow from some two-dimensional modification of the factorization method. It appears that this method allows generalization to N variables. In the next section we consider essential properties of this method.

3   General Scheme of Multidimensional Factorization Method

Consider a set of the operators $L_k(t)$ and $R_k(t)$. These operators depend on some (discrete) parameter t and are labeled with the index $k = 1, 2, \ldots, N$. By k-string we will call a subset of the operators $\{L_k(t), R_k(t)\}$ belonging to the same value of k. For the operators belonging to the k-string we postulate the same relations as for standard one-dimensional factorization scheme [7] (see also Refs. 12 and 16):

$$L_k(t)R_k(t) = R_k(t-1)L_k(t-1) + \varepsilon_k(t), \tag{24}$$

where

$$\varepsilon_k(t) = \nu_k(t) - \nu_k(t-1) \tag{25}$$

and $\nu_k(t)$ are some parameters.

Introduce N Hamiltonians,

$$H_k(t) = R_k(t)L_k(t), \quad k = 1, 2, \ldots, N. \tag{26}$$

Let $\psi_k(t)$ be an eigenstate of the Hamiltonian with an eigenvalue λ. Then the state $L_k(t)\psi(t)$ is eigenstate of the Hamiltonian $H_k(t-1)$ with eigenvalue $\lambda - \varepsilon_k(t)$. Analogously the state $R_k(t+1)\psi_k(t)$ is eigenstate of the Hamiltonian $H_k(t+1)$ with eigenvalue $\lambda + \varepsilon_k(t+1)$.

However, in order to construct a family of *commuting* Hamiltonians we should impose additional relations to the operators $R_k(t)$ and $L_k(t)$ belonging to different strings. We *postulate* the following relations:

$$L_k(t)R_k(t)L_j(t) - L_j(t)R_k(t)L_k(t) = \sigma_{kj}(t)L_j(t); \tag{27}$$
$$R_j(t)L_k(t)R_k(t) - R_k(t)L_k(t)R_j(t) = \sigma_{kj}(t)R_j(t), \tag{28}$$

where $\sigma_{kj}(t)$ are some real parameters. We assume that $j \neq k$ in (27) and (28). However, these relations are automatically valid for $j = k$ provided that $\sigma_{jj}(t) = 0$. In what follows we will call (24) as *in-string relations* (ISR) and (27), and (28) as *cross-string relations* (CSR).

From ISR and CSR we easily get the following:

Proposition 1. *All the Hamiltonians $H_k(t)$, $k = 1, 2, \ldots, N$ commute with one another (only for the same value of the time t).*

Another important feature of the factorization chain defined by ISR and CSR consists in *covariance properties* of the operators $L_k(t)$ and $R_k(t)$ with respect to a whole family of the Hamiltonians $H_j(t)$.

Proposition 2. *Let $\psi_j(t)$ be an eigenstate of the Hamiltonian $H_j(t)$ with an eigenvalue λ: $H_j(t)\psi_j(t) = \lambda\psi_j(t)$. Then*

$$H_j(t-1)L_k(t)\psi_j(t) = \big(\lambda + \sigma_{jk}(t) - \varepsilon_j(t)\big)L_k(t)\psi_j(t) \tag{29}$$
$$H_j(t+1)R_k(t+1)\psi(t) = \big(\lambda - \sigma_{jk}(t+1) + \varepsilon(t+1)\big)R_k(t+1)\psi_j(t) \tag{30}$$

for all possible pairs of k, j. In other words, the operator $L_k(t)$ transforms an eigenstate of any Hamiltonian $H_j(t)$ to an eigenstate of the Hamiltonian $H_j(t-1)$, whereas the operator $R_k(t+1)$ transforms it to the eigenstate of the Hamiltonian $H_j(t+1)$.

Thus starting from a known *common* eigenstate $\psi(t)$ of all the commuting Hamiltonians $H_j(t)$ (this eigenstate should be known for all possible values of the time t) we can construct arbitrary many common eigenstates for these Hamiltonians.

In particular, *assume* that there exists *a global vacuum state* $\psi_{\{0\}}(t)$ annihilated by all the operators $L_k(t)$:

$$L_k\psi_{\{0\}}(t) = 0, \quad k = 1, 2, \ldots, N. \tag{31}$$

Then obviously this state will be a common eigenstate for all the Hamiltonians with zero eigenvalues:

$$H_j(t)\psi_{\{0\}}(t) = 0. \tag{32}$$

Now acting repeatedly by the operators $R_k(t + n_k)$ to this global vacuum state we arrive at the another eigenstate:

$$\psi_{\{n\}}\left(t + \sum_{i=1}^{N} n_i\right)$$

$$= \prod_{k=1}^{N}\{R_k(t + n_k)R_k(t + n_k - 1)\cdots R_k(t + 1)\}\psi_{\{0\}}, \tag{33}$$

where we denote the set of nonnegative integers $\{n_1, n_2, \ldots, n_N\}$ by multi-index $\{n\}$. The state (33) is common eigenstate of the all the Hamiltonians $H_j(t + \sum_{i=1}^{N} n_i)$. In order to get the eigenstates of the "initial" Hamiltonians $H_j(t)$ we should merely shift the time $t \to t - \sum_{i=1}^{N} n_i$.

Of course, existence of the global vacuum state (31) is an additional assumption. Nevertheless, from theory of integrable many-particle systems it is known that such a vacuum state exists for a large class of problems [2]. We will not exploit further the multidimensional factorization chain for the generic situation but rather propose a simple algebraic ansatz providing a special solution for the ISR and CSR leading to multivariable Krawtchouk polynomials.

4   Algebraic Ansatz and a Special Solution of the Factorized Chain

In this section we propose an algebraic ansatz leading to a special solution of the multidimensional factorization chain.

Assume that the operators L_k don't depend on t, whereas the operator $R_k(t)$ depends linearly on t: $R_k(t) = R_k + t$, where R_k are some operators not depending on t. The operators L_k, R_k are assumed to satisfy the commutation relations

$$[L_i, R_k] = -L_i + \delta_{ik}, \tag{34}$$
$$[R_i, R_k] = R_i - R_k, \tag{35}$$
$$[L_i, L_k] = 0; \tag{36}$$

that is we assume that all the operators L_k commute with one another. Then, substituting these operators into ISR (24) and CSR (27), (28) we find that these relations are fulfilled with the parameters $\varepsilon_k(t) = \sigma_{jk}(t) = 1$

for all values of j, k (excepting the condition $\sigma_{kk} = 0$). Thus we get a special solution of the multidimensional factorization chain in terms of the operators satisfying commutation relations (34)–(36).

Assume the existence of a global vacuum state $\psi_{\{0\}}$ annihilated by all the operators L_k (31). Then from (33) we obtain the common eigenstate $\psi_{\{n\}}$ () of the Hamiltonians $H_j(\tilde{t}) = R_j L_j + \tilde{t} L_j$ where $\tilde{t} = t + \sum_{i=1}^{N} n_i$. Corresponding eigenvalues are $\{\lambda\} = \{n_1, n_2, \ldots, n_N\}$; that is, all the Hamiltonians possess equidistant spectra (of course, we consider here only a part of all possible spectra for the operators $H_j(t)$ connected with vacuum state).

For further applications it is useful to find a concrete realization of the operators L_k, R_k in terms of well-known operators. We present here one such realization in terms of the ordinary bose operators a_k, a_k^+, $k = 1, 2, \ldots, N$ satisfying the standard commutation relations:

$$[a_i, a_j^+] = \delta_{ij}, \quad [a_i, a_j] = [a_i^+, a_j^+] = 0. \tag{37}$$

The realization can be written as

$$L_k = \sum_{j=1}^{N} \beta_{kj} a_j; \tag{38}$$

$$R_k = \sum_{j=1}^{N} a_j^+ (\gamma_{jk} - a_j), \tag{39}$$

where the coefficients β_{ik} and γ_{ik} are chosen in such a manner that corresponding $N \times N$ matrices B and G are reciprocal to one another

$$BG = 1. \tag{40}$$

(as usual, the ik entries of the matrices B and G are β_{ik} and γ_{ik}).

Hence in order to construct a realization of a multidimension factorization chain it is sufficient to choose *arbitrary* nondegenerate $N \times N$ matrix B and write down the relations (38) and (39). Thus this realization have N^2 free parameters (i.e., the number of entries of the matrix B).

The global vacuum state in this case exists and coincides with the direct product of all oscillator vacuums $a_j |0\rangle = 0$, $j = 1, 2, \ldots, N$.

The algebraic ansatz (34)–(36) is direct multidimensional generalization of the hidden symmetry algebra for oscillator $9j$-symbols (see Ref. 22 corresponding to $N = 2$ case).

Note that P. Feinsilver [4] exploited algebraic relations similar to (34)–(36) to construct some families of orthogonal polynomials in several variables.

5 Difference Operators and Multivarible Polynomials

In this section we adopt the realization of the bose operators in terms of difference operators

$$a_k = \Delta_k = T_k^+ - 1;$$
$$a_k^+ = x_k T_k^-, \tag{41}$$

where the shift operators are defined on the space of functions of N variables by means of standard formulas:

$$T_k^\pm f(x_1, x_2, \ldots, x_k, \ldots, x_N) = f(x_1, x_2, \ldots, x_k \pm 1, \ldots, x_N).$$

Substituting (41) into (38) and (39) we get the realization of the operators $L_k(t)$, $R_k(t)$ in terms of difference operators acting on a space of functions of N discrete variables x_i. Note that the operators L_k in this realization are linear combinations of difference derivation operators Δ_k. This property is crucial in our approach, for it immediately leads to some class of multivariable polynomials.

Indeed, assume that the function $Q_M(x)$ is a polynomial in variables x_i of degree M (as usual, by degree of the multivariable polynomial we mean the maximal value of the sum of orders $n_1 + n_2 + \cdots + n_N$ among all elementary monomials $x_1^{n_1} x_2^{n_2} \cdots x_N^{n_N}$). Then any operator Δ_k transforms $Q_M(x)$ to another polynomial but with degree $M - 1$. Hence all the operators L_k also decrease the degree of a polynomial by 1. It is easily seen that conversely, any operator $R_k(t)$ transforms $Q_M(x)$ to another polynomial with degree $M + 1$. Hence the space of multivariable polynomials is invariant under the action of the operators $R_k(t)$, L_k.

The global vacuum state in our realization does exists and can be chosen as constant

$$Q_{\{0\}} = 1, \quad L_k Q_{\{0\}} = 0. \tag{42}$$

Acting to this vacuum by the operators $R_k(t)$ we arrive at the system of multivariable polynomials $Q_{\{n\}}(x; B; t)$ depending on N^2 parameters of the (arbitrary) matrix B and on discrete time t.

Choosing appropriate normalization constants we arrive at the difference-difference recurrence relations for these polynomials:

$$R_k(t) Q_{\{n\}}(x; t-1) = t Q_{\{n_k+1\}}(x; t); \tag{43}$$

$$L_k Q_{\{n\}}(x; t) = \left(\frac{n_k}{t}\right) Q_{\{n_k-1\}}(x; t-1), \tag{44}$$

where the operators $R_k(t)$, L_k are defined as

$$L_k = \sum_{i=1}^{N} \beta_{ki}\Delta_i; \tag{45}$$

$$R_k(t) = t + \sum_{i=1}^{N} x_i\left((\gamma_{ik}+1)T_i^- - 1\right) \tag{46}$$

and for brevity we adopt the notation $\{n_k \pm 1\} = \{n_1, n_2, \ldots, n_k \pm 1, \ldots, n_N\}$ for multiindex. We also omitted dependence on B in (43), (44).

In fact, the relations (43) and (44) *uniquely* determine all the polynomials $Q_{\{n\}}(x; t)$ starting from the vacuum state.

From the Proposition 1 we get the following:

Proposition 3. *The polynomials $Q_{\{n\}}(x; t)$ are simultaneous eigenfunctions of N commuting difference operators*

$$R_k(t)L_k Q_{\{n\}}(x; t) = n_k Q_{\{n\}}(x; t), \quad k = 1, 2, \ldots, N. \tag{47}$$

The degree of the polynomial is $\sum_{i=1}^{N} n_i$.

Note also a property

$$Q_{\{n\}}(0; t) = 1, \tag{48}$$

which is easily derived from (43). Here by $x = 0$ we mean that $x_k = 0$ for all values of k.

From (43) we arrive at Rodriguez formula for the polynomials $Q_{\{n\}}(x; t)$:

$$Q_{\{n\}}\left(x; t + \sum_{i=1}^{N} n_i\right) = \left\{\prod_{k=1}^{N} R_k(t+n_k)R_k(t+n_k-1)\cdots R_k(t+1)\right\}(1). \tag{49}$$

Note that the ordering of the operators R_k in (49) is essential because they are not commuting with one another.

In what follows we will assume that t is a *positive integer* such that $t > \sum_{i=1}^{N} n_i$. We will call the corresponding polynomials multivariable Krawtchouk polynomials. The reason is that for the case $N = 1$ these polynomials indeed coincide with the ordinary Krawtchouk polynomials in one discrete variable (see, e.g., Ref. 9). Special cases of these polynomials were studied earlier by M. Tratnik [19] and P. Feinsilver [5].

Introduce the functions $Y_1(z), Y_2(z), \ldots, Y_{N+1}(z)$ depending on N auxiliary variables $z = \{z_1, z_2, \ldots, z_N\}$:

$$Y_k(z) = 1 + \sum_{i=1}^{N}(1 + \gamma_{ki})z_i, \quad k = 1, 2, \ldots, N \tag{50}$$

$$Y_{N+1}(z) = 1 + z_1 + z_2 + \cdots + z_N,$$

where the coefficients γ_{ki} are the same as in (39).

Then the generating function for the polynomials $Q_n(x; B; t)$ is [23]

$$F(z, x; t) = \prod_{i=1}^{N+1} Y_i^{x_i}(z; t)$$

$$= \sum_{\{n\}} Q_n(x; B; t) \frac{t! z_1^{n_1} z_2^{n_2} \cdots z_N^{n_N}}{n_1! n_2! \ldots n_{N+1}!},\qquad (51)$$

where $x_{N+1} = t - \sum_{i=1}^N x_i$, $n_{N+1} = t - \sum_{i=1}^N n_i$, the arguments x_i are assumed to be nonnegative integers and summation in (51) is over all possible sets of $\{n_1, n_2, \ldots, n_N\}$ such that $0 \le n_{N+1} \le t$.

From the generating function (51) one can obtain many other important relations for the polynomials $Q_n(x; B; t)$: difference-difference, recurrence relations, duality property, etc. [23].

Another important property of the multivariable Krawtchouk polynomials is their biorthogonality.

Proposition 4. *The multivariable Krawtchouk polynomials and so-called adjacent polynomials [23] $\bar{Q}_n(x; B; t)$ satisfy the biorthogonality property*

$$\sum_{\{x\}} w(x) Q_n(x; B; t) \bar{Q}_m(x; B; t) = h_n \delta_{nm},\qquad (52)$$

where the weight function $w(x; B; t)$ coincides with multinomial distribution

$$w(x; t) = \kappa(t) \frac{t!}{x_{N+1}(t)!} \prod_{i=1}^{N} \frac{\mu_i^{x_i}}{x_i!},\qquad (53)$$

($\kappa(t)$ being a normalization constant) and the sum in (52) is over all the sets x_i, $i = 1, \ldots, N$ such that $0 \le x_{N+1} \le t$, h_n is an appropriate normalization constant, and $\delta_{nm} = \prod_{i=1}^{N} \delta_{n_i, m_i}$.

For explicit expression of the weight parameters μ_k and normalization constants $h_n(t; B)$ see Ref. 23.

Note that for some restrictions upon the matrix B one can obtain the pure orthogonality relation for the multivariate Krawtchouk polynomials [23]

6 Conclusion

In conclusion, it is naturally to propose the following:

Conjecture 1. *Generic $3nj$-symbols of the oscillator algebra are expressed in terms of multivariable Krawtchouk polynomials.*

Note that $3j$- and $6j$-symbols are known to be expressed in terms of the ordinary (one-variable) Krawtchouk polynomials [8, 20], whereas we showed that $9j$-symbols are expressed in terms of 2-variable Krawtchouk polynomials.

Acknowledgments: The author is grateful to V. Spiridonov, S. Suslov, and L. Vinet for discussion and to C. Dunkl and P. Feinsilver for stimulating communications.

7   REFERENCES

1. R. Askey and J. A. Wilson, *Some Basic Hypergeometric Orthogonal Polynomials that Generalize Jacobi Polynomials*, Mem. Amer. Math. Soc., Vol. 54, Amer. Math. Soc., Providence, RI, 1985.

2. J. F. van Diejen, *Commuting difference operators with polynomial eigenfunctions*, Compositio Math. **95** (1995), No. 2, 183–233.

3. C. F. Dunkl, *A difference equation and Hahn polynomials in two variables*, Pacific J. Math. **92** (1981), No. 1, 57–71.

4. P. Feinsilver, *Moment systems and orthogonal polynomials in several variables*, J. Math. Anal. Appl. **85** (1982), No. 2, 385–405.

5. P. Feinsilver and R. Schott, *Krawtchouk polynomials and finite probability theory*, Probability Measures on Groups. X (Oberwolfach, 1990) (H. Heyer, ed.), Plenum, New York, 1991, pp. 129–135.

6. Ya. I. Granovskiĭ and A. Zhedanov, *New construction of 3nj-symbols*, J. Phys. A **26** (1993), No. 17, 4339–4344.

7. L. Infeld and T. E. Hull, *The factorization method*, Rev. Mod. Phys. **23** (1951), 21–68.

8. J. Van der Jeugt, *Coupling coefficients for Lie algebra representations and addition formulas for special functions*, J. Math. Phys. **38** (1997), 2728–2740.

9. R. Koekoek and R. F. Swarttouw, *The Askey scheme of hypergeometric orthogonal polynomials and its q-analogue*, Tech. Report 94-05, Faculty of Technical Mathematics and Informatics, Delft University of Technology, 1994.

10. V. B. Kuznetsov and E. K. Sklyanin, *Separation of variables in the A_2 type Jack polynomials*, Sūrikaisekikenkyūsho Kōkyūroku **919** (1995), 27–43 (Japanese).

11. L. Lapointe and L. Vinet, *A Rodrigues formula for the Jack polynomials and the Macdonald–Stanley conjecture*, Internat. Math. Res. Notices (1995), No. 9, 419–424.

12. W. Miller, Jr., *Lie Theory and Special Functions*, Math. Sci. Engrg., Vol. 43, Academic Press, New York, 1968.

13. A. F. Nikiforov, Suslov, and V. B. Uvarov, *Classical orthogonal polynomials of a discrete variable*, Springer Ser. Comput. Phys., Springer, Berlin, 1991.

14. G. I. Prizva, *Some classes of orthogonal polynomials of certain discrete variables*, Dopovīdī Akad. Nauk Ukraïn. RSR Ser. A (1983), No. 8, 15–19 (Ukrainian).

15. Ya. A. Smorodinskii and S. K. Suslov, *6j symbols and orthogonal polynomials*, Sov. J. Nucl. Phys. **36** (1982), No. 4, 623–625.

16. V. Spiridonov, L. Vinet, and A. Zhedanov, *Difference Schrödinger operators with linear and exponential discrete spectra*, Lett. Math. Phys. **29** (1993), No. 1, 63–73.

17. S. K. Suslov, *The 9j-symbols as orthogonal polynomials in two discrete variables*, Sov. J. Nucl. Phys. **38** (1983), No. 4, 662–663.

18. _____, *On the theory of 9j-symbols*, Theoret. and Math. Phys. **88** (1991), No. 1, 720–724.

19. M. V. Tratnik, *Multivariable Meixner, Krawtchouk, and Meixner–Pollaczek polynomials*, J. Math. Phys. **30** (1989), No. 12, 2740–2749.

20. N. J. Vilenkin and A. U. Klimyk, *Representation of Lie Groups and Special Functions. Recent Advances*, Math. Appl., Vol. 316, Kluwer Acad. Publ., Dordrecht, 1995.

21. J. A. Wilson, *Some hypergeometric orthogonal polynomials*, SIAM J. Math. Anal. **11** (1980), No. 4, 690–701.

22. A. Zhedanov, *9j-symbols of the oscillator algebra and Krawtchouk polynomials in two variables*, J. Phys. A **30** (1997), No. 23, 8337–8353.

23. _____, *Multidimensional factorization method and multivariable Krawtchouk polynomials*, preprint, 1997.

9780387989686